# Logic, Methodology
# and
# Philosophy of Science

# Logic, Methodology
# and
# Philosophy of Science

edited by

Petr Hájek

Luis Valdés-Villanueva

Dag Westerståhl

ISBN 1-904987-21-4
King's College Publications
Scientific Director: Dov Gabbay
Managing Director: Jane Spurr
Department of Computer Science
Strand, London WC2R 2LS, UK
kcp@dcs.kcl.ac.uk

http://www.dcs.kcl.ac.uk/kcl-publications/

Cover design by Richard Fraser, www.avalonarts.co.uk
Printed by Lightning Source, Milton Keynes, UK

---

# CONTENTS

## Section B: General Philosophy of Science

## Section C: Philosophical Issues of Particular Sciences

# PREFACE

This volume contains papers based on invited lectures from the 12th International Congress of Logic, Methodology and Philosophy of Science, in Oviedo, Spain, August 2003. The congress was held under the auspices of the International Union of History and Philosophy of Science, Division of Logic, Methodology and Philosophy of Science (UHPS/DLMPS), at the invitation of Universidad de Oviedo in collaboration with the Sociedad de Lógica, Metodologia y Filosofia de la Ciencia en España. As always, the Congress combines invited lectures by leading scholars in the fields of logic, methodology and philosophy of science with a large number of papers contributed by researchers in these fields from all over the world. This book contains most of the invited talks, thus providing an overview of state of the art research in the area.

The invited lectures were distributed into fourteen sections, under the main headings of Logic (three sections), General Philosophy of Science (two sections), Philosophical Issues of Particular Sciences (six sections), and Ethical, Social and Historical Perspectives on Philosophy of Science (three sections). Furthermore, there were four plenary lectures and five special symposia, as well as a number of affiliated meetings. A complete list of the invited talks – in plenary sessions or in the sections and the special symposia – as well as of the individuals and committees involved in the organization of the Congress, appears below.

The papers in this volume have been reviewed by the editors. The editors wish to express their gratitude to Dov Gabbay, who offered to publish the book with King's College Publications at a time when we had been let down by the originally contracted publisher, and to Jane Spurr, who efficiently handled the practical matters of publication. Furthermore, we are most grateful to Dasa Harmancova, who has been responsible for most of the type-setting of the papers.

Prague, Oviedo, and Göteborg, June 2005.

*Petr Hájek*
*Luis M. Valdés-Villanueva*
*Dag Westerståhl*

# PLENARY LECTURES, SECTIONS AND SPECIAL SYMPOSIA

## A: LOGIC

### Section A.1. Mathematical Logic (Proof Theory, Recursion Theory, Model Theory, Set Theory)

*Section Program Committee:*

Jan Krajíček (chair, Czech Republic)
Rod Downey (New Zealand)
Stevo Todorčević (Canada)
Boris Zilber (United Kingdom)

*Invited speakers:*

J. Bagaria, *Natural Axioms for Set Theory that Decide Cantor's Continuum Problem*
P. Cholak, *The Computably Enumerable Sets: Recent Results and Future Directions*
D. R. Hirschfeldt, *Measures of Effective Randomness*
W. Hodges, *Definability and Automorphism Groups*
A. Kanamori, *Zermelo and Set Theory*
U. Kohlenbach, *Proof Theoretic Applications to Functional Analysis*
A. Pillay, *Finite Morley Rank Sets Definable in Differentially Closed Fields*
C. A. Di Prisco, *Colourings of the Real Numbers*
T. Scanlon, *Geometric Stability Theory in Geometry, Arithmetic and Logic*
W. H. Woodin, *A Structural Equiconsistency for AD-$\Re$*

### Section A.2. Philosophical Logic (Non-classical Logics, Logic and Language, Foundations of Logic)

*Section Program Committee:*

Johan van Benthem (chair, The Netherlands)
Larry Moss (USA)
Daniele Mundici (Italy)
Hans Rott (Germany)

*Invited speakers:*

R. Cignoli, *Glivenko Like Theorems in Natural Expansions of BCK-logic with Negation*
M. van Lambalgen, *Evolution of Higher Cognitive Functions: The Case of Logic*
E. Orlowska, D. Vakarelov, *Lattice-based Modal Logics and Modal Algebras*

G. Restall, *Multiple Conclusions*

H. Wansing, *On the Negation of Action Types: Constructive Concurrent PDL*

## Section A.3. Logic and Computation (Knowledge Representation and AI, Verification, Semantics of Programs, Interactive Proofs, Computational Linguistics)

*Section Program Committee:*

Joerg Flum (chair, Germany)
Dexter Kozen (USA)
Per Martin-Löf (Sweden)
Mogens Nielsen (Denmark)

*Invited speakers:*

L. Beklemishev, Long Games in Modal Logics and Probability Algebras
T. Coquand, *Dynamical Method in Algebra*
J. Esparza, *A False History of True Concurrence*
M. Otto, *Bisimulation Invariance and Expressive Completeness*
I. Walukiewicz, *Fixpoint Hierarchies*

## B: GENERAL PHILOSOPHY OF SCIENCE

## Section B.1. Methodology (Explanation, Causality, and Laws; Models, Experiment, and Theory)

*Section Program Committee:*

Ron Giere (chair, USA)
Mary S. Morgan (United Kingdom)
Mauricio Suárez (United Kingdom)

*Invited speakers:*

R. Ankeny, *Cases as Explanations: Modelling in the Biomedical and Human Sciences*
D. M. Bailer-Jones, *Models, Theories and Phenomena*
M. Boumans, *Truth versus Precision*
C. Hoefer, *Humean Effective Strategies*
P. Teller, *Conceptions of Truth: a Lesson from Philosophy of Science*
J. Woodward, *Invariance and Causal Explanation*

## Section B.2. Induction, Probability and Statistics (Induction, Statistical Inference, Learning Theory, Decision Theory)

*Section Program Committee:*

Theo Kuipers (chair, The Netherlands)
Donald Gillies (United Kingdom)
Deborah Mayo (USA)

*Invited speakers:*

C. Glymour, *Aggregation, Conditional Independence and Causal Inference, Illustrated by Gene Regulation*
I. Niiniluoto, *Verisimilitude and PAC-learning*
P. Milne, *Conditional Probability, Conditional Events and Single-Case Properties*

## C: PHILOSOPHICAL ISSUES OF PARTICULAR SCIENCES

### Section C.1. Philosophy of Mathematics

*Section Program Committee:*

Stuart Shapiro (chair, USA)
Penelope Maddy (USA)
Mark Steiner (Israel)

*Invited speakers:*

J. Floyd, *Wittgenstein on the Gödel Incompleteness Theorem*
H. Gaifman, *On the Notion of Intended Model*
I. Jané, *The Iterative Conception of Sets from a Cantorian Perspective*

### Section C.2. Philosophy of the Physical Sciences

*Section Program Committee:*

Andreas Kamlah (chair, Germany)
John Earman (USA)
Roberto Torretti (Chile)

*Invited speakers:*

J. Mosterín, *Anthropic Explanations in Cosmology*
F. Steinle, *Experiment and Concept Formulation*
J. Uffink, *Rereading Ludwig Boltzmann*

### Section C.3. Philosophy of the Biological Sciences

*Section Program Committee:*

Jean Gayon (chair, France)
Osamu Kanamori (Japan)
Alex Rosenberg (USA)

*Invited speakers:*

A. Leplège, *Measurement of Quality in the Health Sciences*
M. Morange, *What is Life? (revisited)*

## Section C.4. Philosophy of Cognitive Science and Artificial Intelligence (including Computational Perspectives in Psychology)

*Section Program Committee:*

Richard Grandy (chair, USA)
Daniel Kayser (France)
Stella Vosniadou (Greece)

*Invited speakers:*

J.-G. Ganascia, *Induction with Machine Learning Revisited*
B. Smith, *Cognitive Science and Biomedical Informatics. From Gene Ontology to Universal Medical Language*
T. Stone, *Evaluating Recent Work on Mental Simulations*

## Section C.5. Philosophy of Linguistics

*Section Program Committee:*

Jeff Pelletier (chair, Canada)
Uwe Reyle (Germany)
Tom Wasow (USA)

*Invited speakers:*

E. Keenan, *Linguistic Invariants*
G. Pullum, *Contrasting Applications of Logic in Natural Language Syntax*

## Section C.6. Philosophy of the Social Sciences (including Non-Computational Psychology)

*Section Program Committee:*

Uskali Mäki (chair, The Netherlands)
Felix Ovejero (Spain)
Don Ross (South Africa)

*Invited speaker:*

P. Pettit, *Group Agency*

## D: ETHICAL, SOCIAL AND HISTORICAL PERSPECTIVES ON PHILOSOPHY OF SCIENCE

## Section D.1. History of Logic, Methodology, and Philosophy of Science (History of the Topics Covered by the Programme of the Congress)

*Section Program Committee:*

Jan Woleński (chair, Poland)
Valentin Bazhanov (Russia)
Volker Peckhaus (Germany)

**Section D.2. Ethics of Science and Technology (Ethical Problems of Scientific Research, Applied Science, and Technology)**

*Section Program Committee:*

Wlodek Rabinowicz (chair, Sweden)
John Broome (UK)
Isaac Levi (USA)

*Invited speaker:*

R.-Z. Qiu, *Vulnerability as a Principle of Ethics of Science and Technology*
**Section D.3. Philosophical Questions Raised by the History and Sociology of Science**

*Section Program Committee:*

James Robert Brown (chair, Canada)
Catherine Chevalley (France)
Oswaldo Pessoa (Brasil)

*Invited speaker:*

J. McAllister, *Symmetries and Asymmetries in Science Studies*

## PLENARY SPEAKERS

M. O. Rabin, *Proofs and Persuasions from Computer Science* (opening lecture)
E. Sober, *Intelligent Design is Untestable. What about Natural Selection?*
W. H. Woodin, *The Axioms for Set Theory: Then and Now*
M. Garrido, *The New Scientific Image of Man and Philosophy* (closing lecture)

## SPECIAL SYMPOSIA

**Symposium 1. Scientific (Evidence-Based) Medicine, 19th-20th Centuries** (Joint DHS-DLMPS Symposium)

*Chair: Anne Fagot-Largeault*

C. Debru, *Biological, Clinical and Therapeutic Evidence in the Classification of Chronic Lymphocytic Leukemias*
A. Fagot-Largeault, *Evidence-based Medicine: Its History and Philosophy*
E. Giroux, *Conceptions of the Normal and the Pathological*
A. Leplège, *Measuring Qualities: How to Quantify Health, Pain, Well-being, etc.*
Z. Szawarski, *What it Means to Be Cured* (about the Polish School of Medicine)

**Symposium 2. Philosophy, Methodology and History of Technology** (Symposium arranged by DHS)

*Chairs: Alexandre Herlea and Juan José Saldaña*

G. Bechmann, *Technology as a Medium - A Constructivist Concept of Technology*
P. Brouzeng, *Interaction Science Technique dans l'Historie: Une Nouvelle Donnede la Culture en Europe*

C. Debru, *Philosophical Aspects of Biotechnologies in Europe and the USA*

V. Gorokhov, *The Problem of the Rational Analysis and Description of Technological Activity*

A. Herlea, J. J. Saldaña, *Philosophy, Methodology and History of Technology – Introductory Remarks*

L. Sofonea, *Some Considerations Concerning the Philosophical Dimension of Homo Technicus-Technologicus*

**Symposium 3. The Foundations of Evolution** (Plenary Session)

*Chair: Elliott Sober*

N. Elderedge, *What Drives Evolution?*

K. Sigmund, *Public Goods and Free Riders*

E. Sober, *Intelligent Design is Unstable. What about Natural Selection?*

**Symposium 4. Philosophy and Methodology of Empirical Modeling: Causation, Validation and Discovery**

*Chair: Deborah Mayo*

J.-G. Ganascia, *Rational Reconstruction of Wrong Theories*

C. Glymour, *Search, Not Confirmation*

A. Spanos, *Structural Equation Modeling (SEM), Causal Inference and Statistical Adequacy*

P. L. Spirtes, *Methodology in the Social Sciences*

J. Woodward, *Causal Mechanisms in Cognitive Psychology*

**Symposium 5. The Unusual Effectiveness of Logic in Computer Science**

*Chair: Martin Otto*

A. Dawar, *Fixed-Point Logics and Computation*

J. Esparza, *Logic in Automatic Verification*

U. Kohlenbach, *From Foundations to Functional Programming: Functionals of Higher Type in Computer Science*


**Executive Committee of the International Union of History and Philosophy of Science, Division of Logic, Methodology and Philosophy of Science (2000 - 2003)**

Michael Rabin (President, Israel), Graham Priest (First Vice-President, Australia), Deborah Mayo (Second Vice-President, USA), Dag Westerståhl (Secretary General, Sweden), Ulf Schmerl (Treasurer, Germany), Wesley Salmon[†] (Past President, USA), Xavier Caicedo (Assessor, Colombia), Roberto Cignoli (Assessor, Argentina), Anne Fagot-Largeault (Assessor, France), Peter Lipton (Assessor, UK), Ewa Orlowska (Assessor, Poland), Alexander Razborov (Assessor, Russia), Laszlo Szabo (Assessor, Hungary), Soshichi Uchii (Assessor, Japan).

**General Program Committee**

Petr Hájek (chair, Czech Republic), Peter Clark (United Kingdom), Maria Luisa Dalla Chiara (Italy), Toshio Ishigaki (Japan), Moshe Vardi (USA), Vladimir Vasyukov (Russia), Luis M. Valdés-Villanueva (Spain), Dag Westerståhl (Sweden).

**Local Organizing Committee**

Luis M. Valdés-Villanueva (chair), Eva Álvarez-Martino, Cipriano Barrio-Alonso, Roger Bosch i Bastardes, Javier Echeverría-Ezponda (CSIC), Carmen González del Tejo, José Antonio López-Cerezo, Alfonso García-Suárez, Miguel Lorente, Eulalia Pérez-Sedeño (Sociedad de Lógica, Metodología y Filosofía de la Ciencia en España), Jorge Rodríguez- Marqueze, Julián Velarde-Lombraña, Lorena Villamil-García.

**Supporting Institutions and Private Sponsors:**

Ministerio de Ciencia y Tecnología de España, Consejería de Educación y Cultura del Gobierno del Principado de Asturias, Ayuntamiento de Oviedo, Universidad de Oviedo, Caja de Ahorros de Asturias, Banco Herrero, Fundación Docente de Mineros Asturianos, El Gaitero, Iberia, Líneas Aéreas de España.

# President's Address

Dear Congress Attendees and Guests,

It is with great pleasure and excitement that I address you at the opening of the 12th Congress of Logic, Methodology and Philosophy of Science in this beautiful Spanish city of Oviedo.

I must start on a sad note, bringing up the memory of Wesley C. Salmon, friend, teacher, philosopher of science extraordinaire, and past President of IUHPS and DLMPS. Wesley is the person responsible for my election as President of DLMPS and has generously guided me with advice in fulfilling my duties in this capacity. He participated as Past President of the IUHPS/DLMPS Executive Committee in the initial phases of the organization of this Congress. In April of 2001 Wesley was taken from us in a terrible automobile accident. We all grieve for him and cherish his memory.

It is a happy duty to thank the people who, through their splendid efforts and dedicated work, made the arrangement of this Congress possible. First, the city fathers of Oviedo who have generously donated the use of this wonderful Congress Hall and other Oviedo facilities. Also, the leaders of the university who have provided use of university lecture halls. Luis Valdés is the Chair of the Local Organizing Committee; without him and his colleagues nothing would be possible. Petr Hájek is Chair of the Program Committee. He, the other members of the General Program Committee, the Chairs and members of the Section Program Committees, and the organizers of the special symposia, are responsible for the program of outstanding lectures in this Congress.

We, and especially I, must all be grateful to the members of the out-going Executive Committee: Deborah Mayo, Graham Priest, the Secretary of DLMPS Dag Westerståhl, and the Treasurer Ulf Schmerl. Dag, in particular, bore the brunt of the day-to-day work of DLMPS and is our main contact with other scientific organizations. Ulf takes meticulous care of DLMPS finances. Our Executive Committee meetings were intellectually exciting and everybody has greatly contributed to the shaping and content of this Congress. Elliot Sober stepped in after Wesley's departure and greatly helped with advice.

Let me turn briefly to the scientific aspects of the Congress and LMPS in general.

We are all aware of the fact that science has become very specialized and very fragmented, but at the same time we see efforts and trends of bridging between subdisciplines of the same branch of science and between the major branches of science. Thus physicists know little about biology or the areas of mathematics which are not of direct use to them, and solid state physicists have only a broad knowledge of the current developments in particle physics. As said, this state of affairs is changing. Scientists increasingly realize that they have a lot to learn from and to contribute to other disciplines. Thus in my field, computer science and mathematics, computer scientists reach out to problems in biology, economics, computational science as applied to physics or chemistry, to name just a few examples.

Our community of philosophers and historians of science, as well as the practitioners of specific scientific disciplines who are interested in philosophy and methodology of science, have a particular role to play in this process of convergence in the sciences. There are significant methodological and philosophical issues arising in the context of multi-disciplinary research, and our colleagues will surely apply themselves to these questions. But, in addition, our Congress is an ideal venue for getting together people from diverse scientific disciplines who wish to reach out and promote multi-disciplinary research. I hope that this trend will find expression in LMPS Congresses to come.

Speaking of the present Congress, the Executive Committee in outlining its scope did attempt to broaden the range of subjects covered in the Congress. We introduced numerous departures from the template of topics included in the Sections and Symposia, which stood essentially unchanged in the past twenty years. I feel that we were successful in these efforts and expect this evolution of the content of LMPS Congresses to continue.

Finally we must all thank the invited speakers and the presenters of contributed papers, for the effort and creativity that they have invested in this occasion. I am sure their excellent lectures will make this a great Congress.

*Michael O. Rabin*
*Harvard University and Hebrew University*

# PLENARY LECTURES

# Intelligent Design is Untestable. What about Natural Selection?

Elliott Sober[1]

*University of Wisconsin, Madison, USA*
*esober@wisc.edu*

**Abstract.** The argument from design is best understood as a likelihood inference. Its Achilles heel is our lack of knowledge concerning the aims and abilities that the putative designer would have; in consequence, it is impossible to determine whether the observations are more probable under the design hypothesis than they are under the hypothesis of chance. Hypotheses about the role played by natural selection in the history of life also can be evaluated within a likelihood framework, and here too there are auxiliary assumptions that need to be in place if the likelihoods of selection and chance are to be compared. I describe some problems that arise in connection with the project of obtaining independent evidence concerning those auxiliary assumptions.

## 1. What else could it be?

Defenders of the design argument sometimes ask "what else could it be?" when they observe a complex adaptive feature. The question is rhetorical; the point of asking it is to assert that intelligent design is the only mechanism that could possibly bring about the adaptations we observe. Contemporary evolutionists sometimes ask the same question, but with a different rhetorical point. Whereas intelligent design seems to some to be the only game in town, natural selection seems to others to be the only possible scientific explanation of adaptive complexity.

I will argue that intelligent design theorists and evolutionists are both wrong when they argue in this way. Whenever a hypothesis confers a probability on the observations without deductively entailing them, evaluating how well supported the hypothesis is requires that one consider alternatives. *Testing* the hypothesis requires *testing it against* competitors. Developing this point leads to a recognition of the crucial mistake that undermines the design argument. The question then arises as to whether evolutionary hypotheses about the process of natural selection fall prey to the same error. Although I'll begin by emphasizing the parallelism between intelligent design and natural selection, I emphatically do not think that they are on a par. The relevant difference is that intelligent design, as a claim about the adaptive features of organisms, is, at least as developed so far, an untestable hypothesis. Hypotheses describing the role of natural selection, on the other hand, can be tested. But *how* they are to be tested is an interesting question, as we shall see.

5

## 2. Likelihood and intelligent design

As mentioned, "what else could it be?" is a rhetorical question, whose point is to assert that some favored mechanism (H) is the only one that could possibly produce what we observe (O). This line of reasoning has a familiar deductive pattern, namely *modus tollens*:

(MT)
$$\frac{\begin{array}{l}\text{If H is false, then O will not be true.}\\ \text{O is true.}\end{array}}{\text{H is true.}}$$

Despite the allure of this line of reasoning, many defenders of the design argument have recognized that it is misguided. One of my favorite versions of the argument is due to John Arbuthnot [1], who was clear about this point. Arbuthnot tabulated birth records in London over 82 years and noticed that in each year, slightly more sons than daughters were born. Realizing that boys die in greater numbers than girls, he saw that the slight bias in the sex ratio at birth gradually subsides until there are equal numbers of males and females at the age of marriage. Arbuthnot took this to be evidence of intelligent design; God, in his benevolence, wants each man to have a wife and each woman to have a husband. To draw this conclusion, Arbuthnot considered what he took to be the relevant competing hypothesis – that the sex ratio at birth is determined by a chance process. Arbuthnot had something very specific in mind when he spoke of chance; he meant that each birth has a probability of $1/2$ of being a boy and a probability of $1/2$ of being a girl. According to the chance hypothesis, the probability that more boys than girls will be born in a given year is a little less than $1/2$. The chance hypothesis therefore entails that the probability of there being more boys than girls in each of 82 years is less than $(1/2)^{82}$ ([21], pp. 225-226).

Arbuthnot did not use *modus tollens* to defend intelligent design. I prefer to represent his thinking as a *likelihood inference*[2]:

(L)
$$\frac{\begin{array}{l}\text{Data}\\ \text{Pr( Data | Intelligent Design) is very high.}\\ \text{Pr( Data | Chance)} < (1/2)^{82}.\\ \text{The Law of Likelihood.}\end{array}}{\text{The Data favor Intelligent Design over Chance.}}$$

*The Law of Likelihood* [7,4,14] says that the data "favor" (lend more support to) the hypothesis that confers on them the greater probability. Here and in what follows, I use the terms "likelihood" and "likely" in the technical sense introduced by R. A. Fisher [5]. The likelihood of a hypothesis is not the probability it has in the light of the evidence; rather, it is the probability that the evidence has, given the hypothesis. Don't confuse Pr(Data | H) with Pr(H | Data). Understood in this way, Arbuthnot's argument does not purport to show that sex ratio is *probably* due to intelligent design. To obtain that result, he'd need further assumptions concerning the prior probabilities of the two hypotheses.[3] I omit these in my reconstruction of the design argument because I don't see how they can be understood as objective quantities.

The likelihood version of the design argument is modest. As just noted, it declines to draw conclusions about the probabilities of hypotheses. But it is modest in a second respect – it does not claim to evaluate all possible hypotheses. Arbuthnot considered Design and Chance, but could not have addressed the question of how Darwinian theory might explain sex ratio (Sober, forthcoming). Thus, even if Arbuthnot is right that Design "beats" Chance, it remains open that some third hypothesis might trump Design. There is no way to survey all possible explanations; we can do no more than consider the hypotheses that are available. The idea that there is a form of argument that sweeps all possible explanations from the field, save one, is an illusion.[4]

I conclude that the first premise in the *modus tollens* version of the design argument is false. It is false that intelligent design is the only process that could possibly produce the adaptations we observe. Long before Darwin, chance was on the table as a possible candidate, and after 1859 the hypothesis of evolution by natural selection provided a third possibility. What is needed is a comparative principle that applies when the observations are logically consistent with each of the competing hypotheses. The Law of Likelihood seems eminently suited to the task at hand.

### 3. What is wrong with the design argument?

To explain what is wrong with the design argument as an explanation of the complex adaptive features that we observe in organisms, it is useful to consider an application of this style of reasoning that works just fine. Here I have in mind William Paley's [11] famous example of the watch found on the heath. Construed as a likelihood inference, Paley's argument aims to establish two claims – that the watch's characteristics would be highly probable if the watch were built by an intelligent designer and that they would be very improbable if the watch were the product of chance. The latter claim I concede. But why are we so sure that the watch would probably have the features we observe if it were built by an intelligent designer?

To clarify this question, let's examine Table 1, which illustrates a set of possibilities concerning the abilities and desires that the putative designer of the watch might have had. The cell entries represent which hypothesis – intelligent design or chance – confers the higher probability on the watch's being made of metal and glass.[5] Which hypothesis wins this likelihood competition depends on which row and column is correct.[6] The observation that the watch is made of metal and glass would be highly probable if the designer wanted to make a watch out of metal and glass and had the know-how to do so, but not otherwise. If we have no knowledge of what these goals and abilities would be, we will not be able to compare the likelihoods of the two hypotheses.

The question we now are considering did not stop Paley in his tracks, nor should it have done. It is not an unfathomable mystery what goals and abilities the putative designer would have if the designer is a *human* designer. When Paley

| Table 1 | | Desires: What does the putative designer want the watch to be made of? | |
|---|---|---|---|
| | | metal and glass | not metal and glass |
| Abilities: What materials does the putative designer know how to use? | metal and glass | Design | Chance |
| | not metal and glass | Chance | Chance |

imagined walking across the heath and finding a watch, he already knew that his fellow Englishmen are able to build artifacts out of metal and glass and are rather inclined to do so. This is why he was entitled to assert that the probability of the observations, given the hypothesis of intelligent design, is reasonably high.

The situation with respect to the eye that vertebrates have is radically different. If an intelligent designer made this object, what is the probability that it would have the various features we observe? The probability would be extremely low if the designer in question were an $18^{th}$ century Englishman. But we all know that Paley had in mind a very different kind of designer. The problem is that this designer's *radical otherness* put Paley in a corner from which he was unable to escape. He was in no position to say what *this* designer's goals and abilities and raw materials would be, and so he was unable to assess the likelihood of the design hypothesis in this case.

The problem that Paley faced in his discussion of the eye is depicted in Table 2. If the putative designer were able to make the eye that vertebrates have (a "camera eye") and wanted to do so, then Design would have a higher likelihood than Chance. But if the designer were unable to do this, or if he were able to do whatever he pleased but preferred giving vertebrates the compound eye found in many insects, Chance would beat Design. Paley had no independent information about which row and which column is true (nor even about which are more probable and which are less).

| Table 2 | | Desires: What kind of eye does the putative designer want vertebrates to have? | |
|---|---|---|---|
| | | a camera eye | a compound eye |
| Abilities: What kind of eye is the putative designer able to give to vertebrates? | a camera eye | Design | Chance |
| | only a compound eye | Chance | Chance |

Thus, Paley's analogy between the watch and the eye is deeply misleading. In the case of the watch, we have independent knowledge of the characteristics the watch's designer would have if the watch were, in fact, made by an intelligent designer. This is precisely what we lack in the case of the eye. It does no good simply to *invent* assumptions about raw materials and desires and abilities; what is needed is *independent evidence* about them. Paley emphasizes in *Natural Theology* that he intends the design argument to establish no more than the *existence* of an intelligent designer, and that it is a separate question what *characteristics* that designer actually has. His argument runs into trouble because these two issues are not as separate as Paley would have liked.[7]

The criticism I have just described of the design argument does not require us to consider Darwinian theory. We don't need an alternative explanation of the adaptive contrivances of organisms to see that the intelligent design hypothesis – at least as it was developed by Arbuthnot and Paley, and as it is put forward by present-day intelligent design theorists – is untestable.[8]

## 4. The parallel challenge for selectionist explanations

Just as *modus tollens* is the wrong form of argument for creationists to use, so too should evolutionists avoid claiming that natural selection is the only process that could possibly give rise to the adaptive features we observe. In this case as well, we need to compare the likelihood of the hypothesis of natural selection with the likelihoods of alternative explanations. One obvious alternative is the idea of chance, which in modern evolutionary theory takes the form of the hypothesis of random genetic drift. The drift hypothesis says (roughly) that the alternative traits present in a population have nearly identical fitnesses and that trait frequencies change by random walk. Here we should follow Arbuthnot in what he said about chance in connection with sex ratio – it is *very improbable* (though not *impossible*) that the vertebrate eye should have the features we observe, if it arose by random genetic drift. We now need to consider what the probability of the eye's features would be, if the eye were produced by natural selection. That turns out to depend on further assumptions. Of course, these further assumptions do not concern the raw materials, goals, and abilities that a putative designer might have. To make it easier to explain what these further assumptions are, I'm going to change examples – from the much beloved vertebrate eye to the fact that polar bears have fur that is, let us say, 10 centimeters long. All the problems I'll describe in connection with explaining polar bear fur length also arise in connection with explaining the vertebrate eye.

First, I need to clarify the two hypotheses I want to compare. I will assume that evolution takes place in a finite population. This means that there is an element of drift in the evolutionary process, regardless of what else is going on. The question is whether selection also played a role. So we have two hypotheses – *pure drift* (PD) and *selection plus drift* (SPD). Were the alternative traits identical in fitness or were there fitness differences among them (and hence natural selection)? I will understand the idea of drift in a way that is somewhat nonstandard. The usual formulation is in terms of random *genetic* drift; however, the problem I want to address concerns a *phenotype* – the evolution of fur that is 10 centimeters long. To decide how random genetic drift would influence the evolution of this phenotype, we'd have to know how genes influence phenotypes. I am going to bypass these genetic details by using a purely phenotypic notion of drift; under the PD hypothesis, a population's probability of increasing its average fur length by a small amount is the same as its probability of reducing fur length by that amount.[9] I'll similarly bypass the genetic details in formulating the SPD hypothesis; I'll assume that the SPD hypothesis identifies some phenotype (O) as the optimal phenotype and says that an organism's fitness decreases monotonically as it deviates from that optimal value. This means, for example, that if 12 centimeters is the optimal fur length, then 11 is fitter than 10, 13 is fitter than 14, etc.[10] Given this singly-peaked fitness function, the SPD hypothesis says that a population's probability of moving a little closer to O exceeds its probability of moving a little farther away. The SPD hypothesis says that O is an *attractor* in the lineage's evolution.[11] For evolution to occur, either by pure drift or by selection plus drift, there must be variation. I'll assume that mutation always provides a cloud of variation around the population's average trait value.

We now need to assess the likelihoods of the two hypotheses. Given that present day polar bears have fur that is 10 centimeters long, what is the probability of this observation under the two hypotheses? The answer depends on the fur length that the ancestors of present day polar bears possessed and also on the optimal fur length towards which natural selection, if it occurred, would be pushing the lineage. Some of the options are described in Table 3. The lineage leading to the present population might begin with fur that is 2 or 8 or 10 centimeters long. And the optimal fur length might be 2 or 8 or 10 or 12 or 18 centimeters.

| Table 3 | | Possible optimal fur lengths | | | | |
|---|---|---|---|---|---|---|
| | | 2 | 8 | 10 | 12 | 18 |
| | 2 | PD | O | SPD | U | U |
| Possible initial states | 8 | PD | PD | SPD | U | U |
| | 10 | PD | PD | SPD | U | U |

Suppose the population's present value of 10 centimeters also happens to be the optimal value; this is the situation represented in the third column of Table 3. In this case, the initial state of the lineage does not matter. Regardless of which row we consider, the SPD hypothesis has a higher likelihood than the PD hypothesis – polar bears have a higher probability of exhibiting a trait value of 10 if selection is pushing them in that direction than they will have if fur length is the result of pure drift. In contrast, suppose that 2 is the optimal fur length. If the lineage starts evolving with a trait value of 8, then selection will work *against* its increasing to a value of 10. Reaching a value of 10 will then be *less* probable under the SPD hypothesis than it will be if the PD hypothesis is true. This is why PD beats SPD in the first column.

The cells in Table 3 with O or U in them are harder to evaluate. If the population began with a fur length of 2 and the optimal value is 8, then the population must *overshoot* this optimum if it is to exhibit a final state of 10. On the other hand, if the population begins with 2 and has 12 as its optimum, the population must *undershoot* the optimum if it is to end up with a trait value of 10. To understand these two harder cases, as well as the two easier cases already described, we need to further investigate the implications of the two hypotheses.

**Figure 1**

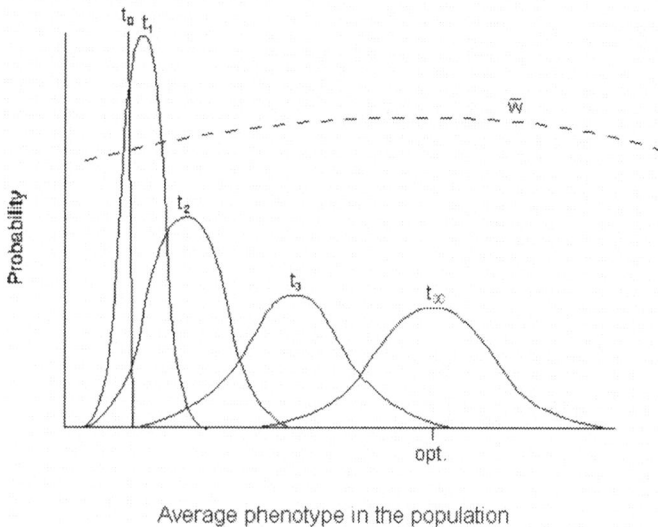

Average phenotype in the population

The dynamics of selection-plus-drift (SPD) are illustrated in Figure 1, adapted from Lande [8]. At the beginning of the process, at $t_0$, the average phenotype in the population has a sharp value. The state of the population at various later times is represented by different probability distributions. Notice that as the process unfolds, the mean value of the distribution moves in the direction of the optimum. The distribution also grows wider, reflecting the fact that the population's average phenotype becomes more uncertain as more time elapses. After infinite time (at $t_\infty$), the population is centered on the optimum. The speed at which the population moves towards this final distribution depends on the trait's heritability and on the strength of selection, which is represented in Figure 1 by the peakedness of the $\overline{w}$ curve. The width of the different distributions depends on the effective population size and on the strength of selection; the larger the product of these two, the narrower the bell curve. In summary, SPD can be described as *the shifting and squashing of a bell curve.*[12]

Figure 2 depicts the process of pure drift (PD); it involves just *the squashing of a bell curve.* Although uncertainty about the trait's future state increases with time, the mean value of the distribution remains unchanged. In the limit of infinite time (at $t_\infty$), the probability distribution is flat, indicating that all average phenotypes are equiprobable.[13]

We now are in a position to analyze when SPD will be more likely than PD. Figure 3a depicts the relevant distributions when there has been finite time since the lineage started evolving from its initial state (I). Notice that the PD distribution stays centered at I, whereas the SPD curve has moved in the direction of the putative optimum (O). Notice further that the PD curve has become more flattened than the SPD curve has; selection impedes spreading out. Figure 3b depicts the two distributions when there has been infinite time. The SPD curve is centered at the optimum while the PD curve is flat. Whether finite or infinite time has elapsed, the likelihood analysis is the same: *the SPD hypothesis is more likely than the PD hypothesis precisely when the population's actual value is "close" to the optimum.* Of course, what "close" means depends on how much time there has been between the lineage's initial state and the present, on the intensity of selection, on the trait's heritability, and on the effective population size. For example, if infinite time has elapsed (Figure 3b), the SPD curve will be more tightly centered on the optimum, the larger the population is. If 10 is the observed value of our polar bears, but 11 is the optimum, SPD will be more likely if the population is small, but the reverse will be true if the population is very large.

Figure 2

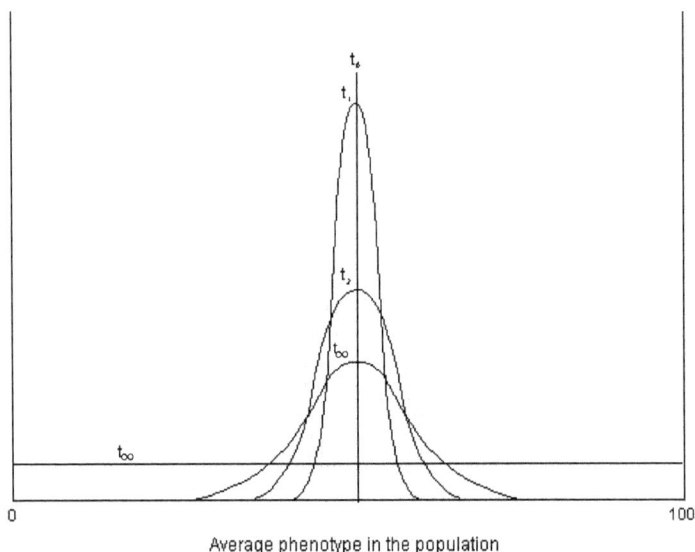

Average phenotype in the population

Let me summarize. If we are to test SPD against PD as possible explanations for why polar bears now have fur that is 10 centimeters long, we need to know what the optimal phenotype would be if the selection hypothesis were true. The four cases that must be considered are described in Table 3. In each, an arrow points from the population's initial state (I) to its present state (P); O is the optimum postulated by the SPD hypothesis. The first case (a) is the easiest; if the optimum (O) turns out to be 10, we're done – SPD has the higher likelihood. However, if the optimum differs from 10, even a little, we need more information. If we can

Figure 3

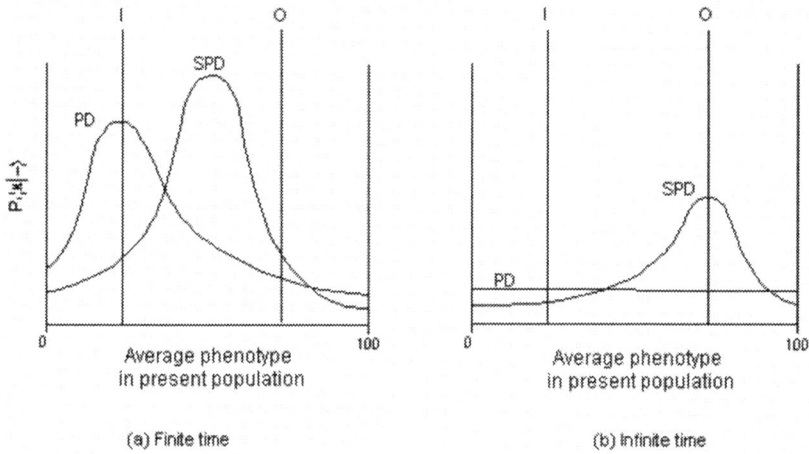

(a) Finite time  (b) Infinite time

discover what the lineage's initial state (I) was, and if this implies that (b) the population evolved *away* from the putative optimum, we're done – PD has the higher likelihood. But if our estimates of the values of I and O entail that there has been (c) overshooting or (d) undershooting, we need more information if we are to say which hypothesis is more likely. One surprise that emerges from this analysis is that undershooting is ambiguous – even if the population has evolved in the direction of the optimum, this, by itself, does not entail that SPD is more likely than PD.

| Table 3 | | Which hypothesis is more likely? |
|---|---|---|
| (a) present state coincides with the putative optimum | —\|————\|———<br>$I \rightarrow P = O$ | selection-plus-drift |
| (b) population evolves away from the putative optimum | —\|———\|——\|—<br>$P \leftarrow I \quad O$ | pure drift |
| (c) population overshoots the putative optimum | —\|———\|——\|—<br>$I \rightarrow O \rightarrow P$ | ? |
| (d) population undershoots the putative optimum | —\|———\|——\|—<br>$I \rightarrow P \quad O$ | ? |

Another interesting consequence of this analysis is that SPD is not automatically the favored hypothesis when the trait in question is a "complex adaptation." A complex adaptation that is not optimal may be sufficiently far from the optimum that the PD hypothesis has higher likelihood. Evolutionists who find it obvious that complex adaptations "must" have been produced by natural selection will not like this argument, but as far as I can see it is correct. As soon as a trait

– even a "complex" trait – departs from the optimum, even a little, further biological information is needed if we wish to discriminate between the SPD and PD hypotheses.

Earlier in this paper I chided intelligent design theorists for simply assuming that the observed traits of organisms *must* be what the putative intelligent designer intended and was able to achieve. The same epistemological point applies to the evolutionist. It does no good simply to assume that the observed fur length *must* be optimal because natural selection *must* have been the cause of the trait's evolution. This assertion is question-begging. What one needs is independent evidence about this and the other auxiliary assumptions that are required for the two hypotheses to generate testable predictions. I argued before that the assumptions that the design hypothesis needs are not independently supported. Is the evolutionist in a better situation than the creationist in this respect?

### 5. Independent evidence about the fitness function and the population's earlier trait value

If present day polar bears all have fur that is 10 centimeters long, how are we to discover what the fitness consequences would be of having fur that is longer or shorter? The most obvious way to address this question is to do an experiment. Let us dispatch a band of intrepid ecologists to the Arctic who will attach parkas to some polar bears, shave others, and leave others with their fur lengths unchanged. We then can monitor the survival and reproduction of these experimental subjects, and this will allow us to estimate the fitness values that attach to different fur lengths.

There is a second approach to the problem of identifying the fitness function, one that is less direct and more theoretical. Suppose there is an energetic cost associated with growing fur. We know that the heat loss an organism experiences depends on the ratio of its surface area and its volume. We also know that there is seasonal variation in temperature. Although it is bad to be too cold in winter, it also is bad to be too warm in summer. We also know something about the abundance of food. These and other considerations might allow us to construct a model that identifies what the optimal fur length is for organisms that have various other characteristics. Successful modeling of this type does not require the question-begging assumption that the bear's actual trait value is optimal or close to optimal.

Unfortunately, these two approaches face a problem. It is more obvious in connection with the experimental approach, but it attaches to both. The experiment, in the first instance, tells us about the fitness function that would be in place if there were variation in fur length among polar bears *now*. How is this relevant to our historical question concerning the processes that were at work as polar bears evolved? The same question attaches to the engineering approach, in that it uses assumptions about the *other* traits that polar bears have. For example, we probably will need information about the range of temperatures that exist in the bear's environment and about the bear's body mass and surface area. If we use data from current bears and their current environment, we need to consider whether these values provide good estimates of the values that were in place ancestrally.

    This leads to our second problem – how independent evidence about the lineage's ancestral fur length might be obtained. Of course, we can't jump in a time machine and go back in time to observe the characteristics of ancestors. Does this mean that the lineage's initial state is beyond the reach of evidence? We know that polar bears and other bears share common ancestors and we know this independently of our question about why polar bears now have fur that is 10 centimeters long. We can use other characteristics – for example, ones that have no adaptive significance for the organisms that have them – to infer the genealogical relationships that connect polar bears to other organisms; this allows us to specify a phylogenetic tree like the one depicted in Figure 4 in which polar bears and their relatives are tip species. We then can write down the fur lengths of polar bears and their near relatives on the tips of that tree. The observed character states of these tip species provide evidence about the character states of the ancestors, represented by interior nodes. What are the rules that govern this inference from present to past?

Figure 4

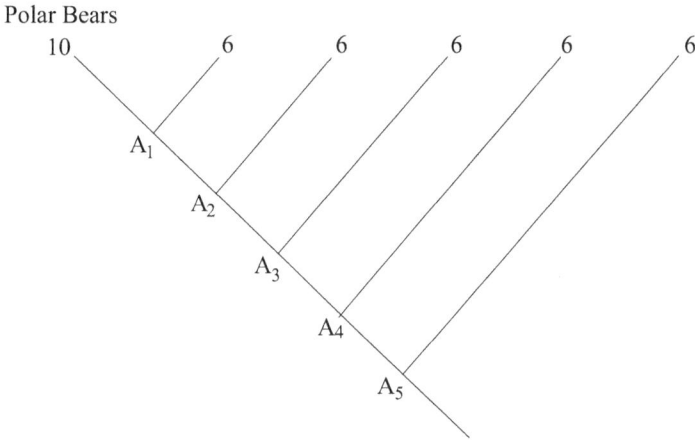

Figure 5

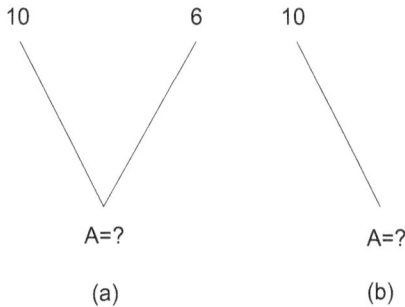

Before addressing that question, I want to describe why Figure 4 shows that our question about SPD versus PD needs to be spelled out in more detail. It is obvious that present day polar bears have multiple ancestors, and that different ancestors have different trait values. If these were all known, the problem of explaining why polar bears now have fur that is 10 centimeters long would decompose into a number of subproblems – why the fur length present at $A_5$ evolved to the length present at $A_4$, why $A_4$'s fur length evolved to the value found at $A_3$, etc. SPD may be a better answer than PD for some of these transitions, but the reverse might be true for others. Similarly, it is perfectly possible that SPD is better supported than PD as an answer to the question "why do polar bears have fur that is 10 centimeter long, given that their ancestor $A_i$ had a fur length of $f_1$" but that the reverse is true for the question "why do polar bears have fur that is 10 centimeter long, given that their ancestor $A_j$ had a fur length of $f_2$."

Now back to the problem of inferring the character states of ancestors. One standard method that biologists use is parsimony – we are to prefer the assignment of states to ancestors that minimizes the total amount of evolution that must have occurred to produce the trait values we observe in tip species. This is why assigning ancestor $A_1$ in Figure 4 a value of 8 and the other ancestors a value of 6 is said to have greater credibility than assigning them all a value of 10. But why should we use parsimony to draw this inference? Does the Law of Likelihood justify the Principle of Parsimony? If not, does the principle have some other justification? If not, does this mean that parsimony is an end in itself (perhaps being constitutive of the cast of mind we call "scientific") or should we rather conclude that a preference for parsimony is merely an unjustifiable prejudice? These are large questions, which I won't attempt to answer here. However, a few points may be useful. First, it turns out that if drift is the process at work in a phylogenetic tree, then the most parsimonious assignment of trait values to ancestors (where parsimony means minimizing the *squared* amount of change) is also the hypothesis of maximum likelihood [9]. On the other hand, if there is a directional selection process at work, parsimony and likelihood can fail to coincide [16]. This point can be grasped by considering the problem depicted in Figure 5a. Two descendants have trait values of 10 and 6; our task is to infer the character state of their most recent common ancestor A. Notice that if there is very strong directional selection for increased fur length towards an optimum of, say, 20 in both lineages, then the setting of the ancestor that maximizes the probability that the descendants will exhibit values of 10 and 6 will be something less than 6. The problem can be simplified even further, by considering just the descendant and ancestor depicted in Figure 5b. If the descendant has a trait value of 10, the most parsimonious assignment of character state to the ancestor is, of course, 10. But if the lineage has been undergoing strong selection for increasing its trait value, then the most likely assignment will be something less than 10. Imagine you want to swim across a river that has a very strong current. The way to maximize your probability of reaching a target on the other side is *not* to start directly across from it; rather, you should start a bit upstream.

It follows that parsimony does not provide evidence about ancestral character states that is *independent* of the hypotheses of chance and selection that we wish to test.[14] This problem concerning how ancestral fur length is to be inferred also is relevant to the question noted earlier about the fitness function – even if we can discover the fitness function that applies to polar bears *now*, why should we think that this function is the correct description of how selection would work *in ancestral populations*? Both problems have the same form – how are we to infer past from present without begging the question?[15]

## 6. Two ways out of the impasse

The analysis I gave of the polar bear fur problem suggests that nontrivial additional knowledge is needed if we are to test the SPD against the PD hypothesis. If this needed information is inaccessible, the evolutionary problem seems to fall prey to the same difficulty that I claim undermines the project of comparing the hypothesis of intelligent design and the hypothesis of chance.

I believe there are two ways out of this impasse. The first involves doing a *sensitivity analysis*. For example, even if we can't infer the character state of ancestors in a way that is independent of the hypotheses we wish to test, we still may be able to consider value ranges for the ancestral trait value and see how these affect the likelihood analysis. If the drift hypothesis entails that the best estimate of the ancestral character state is 6, while the selection hypothesis says that it is, say, 3, we can compare the likelihoods of SPD and PD by assuming that the ancestor had fur that was between 3 and 6 centimeters long (or that the range was even wider). Uncertainty about the heritability of fur length, the length of time the lineage has been evolving, etc. can be addressed in the same way. The task is then to determine in which regions of parameter space SPD has higher likelihood than PD and in which regions the reverse is true. It may turn out that the higher likelihood of SPD is *robust* over considerable variation in these auxiliary assumptions.

A second way to address the problem is widely used in biology; its rationale is discussed more fully in Sober and Orzack [20]. Suppose we know that the optimal fur length for bears living in colder climates is greater than the optimal fur length for bears living in warmer climates, even though we are unable to say what the optimal point value is for any organism in any environment. If we then observe that bears in colder climates tend to have longer fur than bears in warmer climates, this correlation counts as evidence in favor of the SPD hypothesis.[16] What is confirmed here is not the hypothesis that selection has given organisms optimal trait values, nor even that selection has provided them with fur lengths that are close to optimal; rather the favored hypothesis says that selection has caused trait values to evolve in the direction of their optima. The observed trend in trait values is more probable on the hypothesis of natural selection than it is on the hypothesis of chance. This is because the latter hypothesis predicts that fur length and ambient temperature should be independent. Notice that this solution to the problem involves changing the kind of data we seek to explain. Instead of trying to explain why polar bears have fur that is 10 centimeters long, we shift to the problem of explaining why bears in cold climates tend to have longer fur than bears in warmer climates. This permits us to finesse the problem of estimating the character state of ancestors, the optimal trait value, and other biological parameters.[17]

## 7. Conclusion

*Modus tollens* is a bad model for testing the design hypothesis. This is something that Arbuthnot, Paley, and other defenders of the design hypothesis realized, even if their modern-day epigones do not. A better model is likelihood; hypotheses confer different probabilities on the observations, and weight of evidence is assessed by comparing the degree to which different hypotheses probabilify the observations. This is where a Duhemian point becomes relevant.[18] The hypotheses we wish to test do not, by themselves, confer probabilities on the observations; they do so only when auxiliary assumptions are supplied. However, we can't merely invent auxiliary assumptions; rather, we need to find auxiliary assumptions that are independently attested. At this point the design argument runs into a wall. Paley reasoned well when he considered the watch found on the heath, but when he argued for intelligent design as an explanation of organic adaptation, he helped himself to assumptions that were not supported by evidence.

The evolutionary hypothesis of natural selection encounters the same logical challenge. A characteristic observed in a present day species might be explained by the hypothesis of natural selection, or by the hypothesis of drift, or by many other hypotheses that evolutionary theory allows us to construct. To decide whether selection makes the observations more probable than the chance hypothesis does, we need further information. If evolutionary biology can provide good estimates of relevant biological quantities, it avoids the fatal flaw that attaches to creationism. But if it cannot, evolutionary questions may become more tractable by shifting to a comparative framework. Instead of seeking to explain why a single species or group has some single trait value, we might set ourselves the task of explaining a pattern of variation. Fewer assumptions are required if the goal is to bring hypotheses into contact with comparative data.

## Notes

[1] My thanks to James Crow, Carter Denniston, Branden Fitelson, John Gillespie, Peter Godfrey-Smith, Alvin Goldman, Russell Lande, Richard Lewontin, Steve Orzack, Dmitri Petrov, Larry Shapiro, and Stephen Stich for useful discussion. I also am grateful to the National Science Foundation (Grant SES-9906997) for financial support.

[2] Most commentators on Arbuthnot think he was performing a significance test on the chance hypothesis, and rejecting it. For discussion, see Sober [19].

[3] It follows from Bayes' Theorem that $\Pr(\text{Design} \mid \text{Data}) > \Pr(\text{Chance} \mid \text{Data})$ if and only if $\Pr(\text{Data} \mid \text{Design}) \Pr(\text{Design}) > \Pr(\text{Data} \mid \text{Chance}) \Pr(\text{Chance})$.

[4] It is the goal of Dembski's [2] reconstruction of the design argument to show that all alternatives to design can be rejected, and the hypothesis of design left standing, without the design hypothesis' having to make any predictions at all. For criticisms, see Fitelson, Sober, and Stephens [6].

[5] Being made of metal and glass is just an example of the characteristics that the watch possesses. The same points would apply if we considered, instead,

the watch's ability to measure equal intervals of time, or the fact that it would probably not be able to do this if its internal assembly were changed at random.

6 I omit mention in this table of the *raw materials* the putative designer would have available; this constitutes a third dimension. Unfortunately, the piece of paper before you is flat.

7 For a more detailed treatment of the likelihood approach to the design argument, including replies to some objections, see Sober [17].

8 I do not claim that this must be a permanent feature of the hypothesis that organisms have their adaptive contrivances because of intelligent design. Perhaps the epistemic situation will change.

9 Except, of course, when the population has its minimum or maximum value. There is no way to have fur that is less than 0 centimeters long; I'll also assume that there is an upper bound on how long the fur can be (e.g., 100 centimeters).

10 It is not inevitable that a fitness function should be singly peaked. In addition, I'll help myself to the simplifying assumption that fitnesses are *frequency independent* – e.g., whether it is better for a bear to have fur that is 9 centimeters long or 8 does not depend on how common or rare these traits are in the population.

11 We might add to this the assumption that the intensity of selection is greater, the greater the population's distance from the optimum; this idea is often modeled as an Ornstein-Uhlenbeck ("rubber band") process.

12 I have conceptualized the SPD hypothesis as specifying an optimum that remains unchanged during the lineage's evolution. If selection were understood in terms of an optimum that itself evolves, the problem would be more complicated.

13 The case of infinite time makes it easy to see why an explicitly genetic model can generate predictions that radically differ from the purely phenotypic model considered here. Under the process of random *genetic* drift, each locus is homozygotic at equilibrium. In a one-locus two-allele model in which the population begins with each allele at 50%, there is a 0.5 probability that the population will be AA and a 0.5 probability that it will be aa. In a two-locus model, again with each allele at equal frequency at the start, each of the four configurations has a 0.25 probability – AABB, AAbb , aaBB, and aabb. Imagine that genotype determines phenotype (or that each genotype has associated with it a different average phenotypic value) and it becomes obvious that a genetic model can predict nonuniform phenotypic distributions at equilibrium. The model of selection-plus-drift is the same in this regard; there are genetic models that will alter the picture of how the average phenotype will evolve. See Turelli [22] for further discussion.

[14] This is the central problem with the protocol for testing adaptive hypotheses proposed by Ridley [13]; see Sober [16] for further discussion.

[15] The discovery of fossils is not a solution to this problem. Even if fur length could be inferred from a fossil find, it is important to remember that we can't assume that the fossils are ancestors of present day polar bears. They may simply be relatives. If so, the question remains the same – how is one to use these data to infer the character states of the most recent common ancestor that present day polar bears and this fossil share? The fact that the fossil is closer to the most recent common ancestor than is an organism that is alive today means that the fossil will provide stronger evidence. But the question of how unobserved cause is to be inferred from observed effect still must be faced.

[16] A proper test would have to control for the fact that the species in question are genealogically related, and so their trait values may fail to be independent; see Orzack and Sober [10] for discussion.

[17] If evolutionists get to "change the subject" (by seeking to explain cross-species correlations rather than the single trait value found in a single group), why can't intelligent design theorists do the same thing? They can, but the change doesn't get them out of the hot water described earlier. What is the probability that an intelligent designer would give bears in colder climates longer fur than bears in warmer climates? That still depends on the goals and abilities of the putative designer.

[18] I say that the point is Duhemian, rather than Duhem's, because Duhem [3] was thinking about deducing observational predictions, not probabilifying them; Quine [12] uses a deductivist formulation as well. Still, the same logical point applies, though it does not have the holistic epistemological consequences that Duhem and Quine claimed (Sober [15,18]).

## REFERENCES

1. J. Arbuthnot. An Argument for Divine Providence, taken from the Constant Regularity Observ'd in the Births of both Sexes. *Philosophical Transactions of the Royal Society of London*, 27:186-190, 1710.
2. W. Dembski. *The Design Inference*. New York: Cambridge University Press, 1998.
3. P. Duhem. *The Aim and Structure of Physical Theory*. Princeton: Princeton University Press, 1914.
4. A. Edwards. *Likelihood*. Cambridge: Cambridge University Press, 1972.
5. R. Fisher. *Statistical Methods for Research Workers*. Edinburgh: Oliver and Boyd, 1925.
6. B. Fitelson, C. Stephens and E. Sober. How Not to Detect Design – Critical Notice of W. Dembski's *The Design Inference*. *Philosophy of Science*, 66:472-488, 1999.

7. I. Hacking. *The Logic of Statistical Inference*. Cambridge: Cambridge University Press, 1965.
8. R. Lande. Natural Selection and Random Genetic Drift in Phenotypic Evolution. *Evolution*, 30:314-334, 1976.
9. W. Maddison. Squared-Change Parsimony Reconstructions of Ancestral States for Continuous-Valued Characters on a Phylogenetic Tree. *Systematic Zoology*, 40:304-314, 1991.
10. S. Orzack and E. Sober. Adaptation, Phylogenetic Inertia, and the Method of Controlled Comparisons. In: S. Orzack and E. Sober (eds.), *Adaptationism and Optimality*, New York: Cambridge University Press, 45-63, 2001.
11. W. Paley. *Natural Theology, or, Evidences of the Existence and Attributes of the Deity, Collected from the Appearances of Nature*. London: Rivington, 1802.
12. W. Quine. Two Dogmas of Empiricism. In: *From a Logical Point of View*. Cambridge: Harvard University Press, 20-46, 1953.
13. M. Ridley. *The Explanation of Organic Diversity*. Oxford: Oxford University Press, 1983.
14. R. Royall. *Statistical Evidence – a Likelihood Paradigm*. London: Chapman and Hall, 1997.
15. E. Sober. Quine's Two Dogmas. *Proceedings of the Aristotelean Society*, Supplementary Volume, 74:237-280, 2000.
16. E. Sober. Reconstructing Ancestral Character States – A Likelihood Perspective on Cladistic Parsimony. *The Monist*, 85:156-176, 2002.
17. E. Sober. The Design Argument. In: W. Mann (ed.), *The Blackwell Guide to Philosophy of Religion*, Oxford: Blackwell Publishers, 2004a.
18. E. Sober. Likelihood, Model Selection, and the Duhem-Quine Problem, *Journal of Philosophy*, 101:1-22, 2004.
19. E. Sober. Sex Ratio Theory, Ancient and Modern – The Debate Among, Arbuthnot, Bernouilli and DeMoivre, and the Evolutionary Ideas of Darwin, Düsing, Fisher, Williams, and Hamilton. J. Riskin (ed.), *The History and Philosophy of Artificial Life*, forthcoming.
20. E. Sober and S. Orzack. Common Ancestry and Natural Selection. *British Journal for the Philosophy of Science*, 54:423-437, 2003.
21. S. Stigler. *The History of Statistics*. Cambridge: Harvard University Press, 1986.
22. M. Turelli. Population Genetic Models for Polygenic Variation and Evolution. In: B. Weir, E. Eisen, M. Goodman, and G. Namkoong (eds.), *Proceedings of the Second International Conference on Quantitative Genetics*. Sunderland, MA: Sinauer, 601-618, 1988.

# The New Scientific Image of Man and Philosophy

Manuel Garrido

*Universidad Complutense de Madrid*
*P. O. Box 118, 28660, Boadilla del Monte, Madrid (Spain)*
*e-mail: mg@netma.com*

**Abstract.** The aim of this lecture is to make a survey of some of the main scientific and technological facts produced in the second half of $20^{th}$ century and to provide an overview of some of their philosophical implications. The author examines the two main products of the Technology of Information, the Computer and the Internet, and the Biotechnological Revolution and he suggests that, as a consequence of these facts, our traditional Modern Scientific Image of Man, modelled on the empiricist thought of Locke, has been replaced in this lapse of time by a new modern or Postmodern Scientific Image of Man

## Introduction: Russell's *Free Man Worship*

It was almost exactly one hundred years ago, on the third day of December 1903, when Bertrand Russell published his famous pamphlet *The Free Man Worship*[1] – a brief essay, with the format and style of a manifest – where he confronted with human values the main scientific facts of his time in order to sketch a personal code of behaviour, "an ideal of emancipation", which might be interesting for the mind and the heart of a free man.

At that time science was obviously, along with liberal democracy, one of the two basic facts of modernity. But both facts were already shadowed by some new discoveries and processes – the thermodynamic law of entropy, natural selection theory, revolutionary political trends – whose effect was to weaken the strong belief in the natural harmony between scientific progress and human happiness defended by the followers of the Enlightenment and the positivists. The Russellian manifest resulted to be an epochal expression of that ambivalent sentiment of its *Zeitgeist*. And this was perhaps one of the reasons for its legendary fame during the first three or four decades of the last century.

Russell felt himself confronted with "a strange mystery": on the one hand, "the world which Science presents for our belief" is "even more purposeless, more void of meaning" than that traditionally pictured by the Christian history of creation; on the other hand, "Man is yet free [...] and in this lies his superiority" to the material universe.

In the designing of his personal ideal of emancipation, he rejected, as a "submission to evil", "the worship of Force, to which Carlyle and Nietzsche and [...] Militarism have accustomed us", and even "the Stoic freedom", because "Christianity, in preaching the necessity of renunciation [...] has shown a wisdom exceeding that of the Promethean philosophy of rebellion."

But with "renunciation alone", insisted Russell, we cannot "build a temple for the worship of our own ideals." He appealed to the Platonic model of ideal truth and beauty to conclude that "in the contemplation of these things [i.e., truth and beauty] the vision of heaven will shape itself in our hearts, giving at once a touchstone to judge the world about us, and an inspiration by which to fashion to our needs whatever is not incapable of serving as a stone in the sacred temple. To abandon the struggle for private happiness, to expel all eagerness of temporary desire, to burn with passion for eternal things – this is emancipation, and this is the free man's worship."

Every reader of Russell's later books may appreciate that the personal proposals advanced by him in some essential paragraphs of this essay were excessively dependent on the mathematical Platonism of the first phase of his philosophical thought. And that explains why, a quarter of a century later, Russell himself wrote that "it [i.e. his manifest] depends upon a metaphysics which is more Platonic than that which I now believe in"[2]. Finally he decided to change the first title *The Free Man Worship* for the less pretentious *A Free Man Worship*.

As a landmark of the philosophical thought of the past century, the Russellian *Worship* certainly deserves to be remembered in its centenary. In the present lecture – dedicated to an historical survey of some of the main scientific and technological facts of the second half of that century and some of their philosophical implications – this *Worship* may also serve us both as a starting and a terminal point of reference and contrast with our approach.

## 1. The Fact and Concept of Technology

*The Explosion of Technology.*

One of the most impressive facts of our time is the current explosion of technology. Of course, the notion of *tekhne* as a fundamental dimension of human behaviour is older than Aristotle, and it is well known that the modern phenomenon of the rise of technology dates back to the first industrial revolution. But the fact remains that the current extraordinary development of technological inventions and processes – that we may characterize with the label "explosion of technology" – is a manifestation totally idiosyncratic of the past $20^{th}$ century, specially of its second half.

The atomic technology in the Fifties, the technology of space voyages with the anti-technological and ecological movements in the Sixties and Seventies, the technology of the personal computer and particle detectors in the Eighties, and the technology of Internet and genetic engineering in the Nineties are successively partial aspects of this global manifestation which has changed our world so profoundly.

*Technology and Philosophy of Science*

The mentioned explosion of technology forces us to state by default how limited have been the interests and the efforts of the philosophers of science of the second half of the $20^{th}$ century in paying attention and trying to explain such a phenomenon.

I have already recalled that Russell abandoned the Platonic ideal of the first phase of his thought. But it seems as if the weight of a certain Platonism, with its proverbial contempt for matter and the consequent and absolute hierarchic superiority of the function of *episteme* over the function of *tekhne*, had been constantly gravitating on the subsequent generations of philosophers of science. Peter Galison[3] has denounced that this anti-technological Platonism underlies several debates between the Neopositivists and their antagonist and successors, from Popper to Kuhn.

In the first years following the Second World War the Neopositivist view prevailed, which included as a central piece the belief in the ideal of pure immediate observational evidence substantially uncontaminated by every trait of conceptual theory. This evidence, that we consider today simply a myth, was then supposed to be able to be infallibly registered by the so-called "protocol sentences" and constituted for Carnap "the bedrock" of our scientific knowledge. As a reaction against such apotheosis of observational evidence a contrary exaltation of theory emerged, as defended not only by Popper, but also by his adversary Kuhn and his followers Lakatos and Feyerabend. While the basic epistemological unity was for the Neopositivists the protocol sentence, for Popper this unit was the theory. Moreover, as a result of a sort of dialectical development of the holistic trend of the new thought in philosophy of science against the atomism of the Neopositivist school, several conceptual constructions of holistic inspiration were successively proposed such as the Kuhnian paradigms and the Lakatosian and Popperian scientific and philosophical research programs, that seem to be a revival of the old philosophical *Weltanschauung* theories prior to the Second World War.

However in this long series of debates never, or very rarely, was it considered that the use of material instruments is no less characteristic of the empirical sciences than the use of logical arguments. As Galison has suggested[4], the material and instrumental factor in science could make a bigger contribution than logical argumentation to the current intercommunication between the Babelic aggregation of subcultures which constitute the actual development of science.

Speaking in general terms, we may state with Freeman Dyson that science, the *episteme*, has worked cooperatively in the course of history with two companions, philosophy and the crafts, wisdom and *tekhne*. In one or another sense, philosophy has been a source of concepts for science and the crafts a source of tools for it. Now, the philosophers of science of the past century have worked very much and very well in order to clarify the relations and the demarcation between science and philosophy, but little in order to clarify the relations and the demarcation between science and technology. The fact of the current explosion of technology and its penetration in the hard core of science forces us to admit, following Dyson, the relation of complementarity between them.

> [...] Two historians, Peter Galison and Thomas Kuhn, have explored in depth the process of scientific discovery in the modern age. Galison's great work, *Image and Logic,* was published in 1997. Kuhn's *The Structure of Scientific Revolutions* appeared thirty-five years earlier. Kuhn died before he had a chance to say what he thought of *Image and Logic.*
>
> Galison and Kuhn were both trained as physicists before they became historians. Both of them are primarily interested in the history of physics, and both have mastered the technical details of physics as well

as the scholarly craft of historiography. Yet their views of the history of science are totally different. Their two books have almost nothing in common. Galison's book contains hundreds of pictures of scientific apparatus; Kuhn's book contains only words. For Galison the process of scientific discovery is driven by new tools, for Kuhn by new concepts. Both pictures are true and neither is complete. The progress of science requires both new concepts and new tools.

[...] Some scientific revolutions arise from the invention of new tools for observing nature. Others arise from the discovery of new concepts for understanding nature.[5]

*The Philosophical Impact of Technology.*

The philosophers of the last century have been more diligent than the philosophers of science in taking into account the fact of technology. But their explanations of this fact are usually less than satisfactory. The most influential philosopher of technology in the 20th century, the Antiplatonist and Neonietzschean Martin Heidegger (1889-1976), has displayed a fatalistic and pessimist vision of this field in terms of a negative philosophical vision of technology which seems to be, essentially, a reproduction of the classical arguments of Marx, but applied by Heidegger to our current consumer way of life.

Heidegger's criticism of technology is by no means irrelevant to our present-day world, armed with nuclear weapons and controlled by vast technically-based organizations. But its level of abstraction is so high that its author cannot discriminate between electricity and nuclear bombs, between agricultural technologies and the holocaust[6].

In contrast with these generally negative attitudes of the great 20th century occidental philosophers regarding technology and given the special circumstance that Spain is the country in which the present Congress is taking place, I would like to dedicate some words to remember the last thoughts about the technique of the Spanish philosopher José Ortega y Gasset (1883-1955), thoughts which may be considered as an interesting optimistic vision of technology.

In an essay written in 1951 Ortega formulated the thesis of the technological condition of the human being but with the specific addition of his unlimited progress. The human condition, contends Ortega, is essentially a technological condition, and a condition, he insists, that may continue its development in an infinitely growing measure:

> One of the clearest laws of universal history is the fact that the technical movements of man have grown continuously in number and intensity, i.e., that the technical occupation of man, in this strict sense, has been developed with an indubitable progress; or, what is the same, that man, in a growing measure, is a technical being. And there is no concrete motive to believe that this process will not continue being so toward the infinite. So long as man lives, we must consider his technique as one of his constitutive and essentials traits, and we must proclaim the following thesis: man is technical.[7]

Ortega offers in this essay an interesting metaphor of technology as a "natural" prosthesis of man. He had previously pictured the human being as an "ill animal" ("ill" in the sense of deficient regarding instincts), and this picture, he said, may explain the emergence of technology as a necessary orthopaedics.

> This [i.e., the natural instinctive deficiency of the human animal] [...] shows us the triumph of technique, which aims to create a new world for us, because we do not match/fit in with the original world, where we have become ill. The new world of technique is, therefore, like "un enorme aparato ortopédico" [an enormous orthopaedic machine] that the technicians wish to create, and every technique has... the dramatic tendency and quality of being great and fabulous orthopaedics.[8]

## 2. From the Turing Man to the Network Paradigm

*The Computer and Internet.*

The role of technology as an inseparable companion in the main developments of natural sciences during the second half of the past century is notorious. In the case of information and computer sciences their technological connection seems to be even stronger and consequently it is more difficult to trace their demarcation line with technology. In fact, the two principal creatures of the revolution of this kind of sciences that emerged in this period of time have been just two technological products, the Computer, born in the late Thirties and plainly developed in the Sixties, and the Internet, conceived or preconceived in the Sixties and plainly developed in the Nineties. The implications of both products for the new scientific image of man and philosophy deserve our attention.

*The Turing Man: From the Computability Theory to the Computer Metaphor.*

If we look at the development of computer and information sciences during the Thirties and Forties, we may appreciate their strong connection – already mentioned – with technology. The pioneering Shannon work on circuit logic and the mathematical information theory, the cybernetics of Wiener or the design of classical computer architecture by von Neumann are eloquent examples. But what more particularly attracts our philosophical attention here is the contribution by Alan Turing "On computable numbers" (1936). On the one hand, this work represents a segment of pure theory; its limitation results constitute the "a priori zone" of computer sciences. On the other hand, one of the key notions introduced by Turing in his theory is the idea of a "technological" entity, the Turing *machine*, which has become one of those mathematical tools, alike Napier's logarithms or Abelian rings, which being as they are part of the common universal heritage of mathematical knowledge, are nevertheless labelled with the proper name of their respective authors [9]. The most significant example of this kind of artefacts, the "universal Turing machine", is the main demonstrative piece designed by Turing in order to achieve his limitation result on the *Entscheidungsproblem*. Everybody knows that fifteen years later this automatic ideal structure, his universal machine, was proposed by Turing – with the additional specification that such a device is structurally identical to a digital computer – as a model of our mind. This new proposal was formulated in a new essay, "Computing Machinery and Intelligence", originally published in 1950 in a journal, *Mind*, that is not specifically scientific nor technological, but mainly philosophical. And everybody also knows that such a proposal has inspired, at least in a general or indirect sense, the research programs of Artificial Intelligence, initiated in the mid-Fifties by the Dartmouth group, and of the vast hemisphere – to be developed later on – of cognitive sciences.

One of the main criticisms directed against the Turing model, a criticism which is correlatively extensible to the various research programs of similar inspiration, is that this model lacks the possibility of establishing a real contact with the external world. The underlying argument is that all the operative potential of a universal Turing machine is exclusively limited to the manipulation of elements of calculus, which, as a matter of indifference, can be minute stones (such as the *calculi* of the Romans) or the numerical or alphabetical symbols of our written language, but which do not provide nor can they provide real information about the world. It is usual to sum up this feature in a few words saying that the operations of the Turing machine are exclusively *syntactic*, i.e., unable to produce not even one bit of semantic content. The consequence is that such a device seems to remain prisoner of a solipsistic situation very similar to that pictured by Descartes for his subject of knowledge, conceived by him as a pure thinking thing or *res cogitans*.

This limitation is consistent with the original project of Turing, whose bet was the simulation of the activity of the human mind explicitly excluding every biological procedure. The deliberately intended result was the production of a sort of simulation of a mind "disembodied" from any biological body. The awareness of this condition generated in the Sixties a proliferation of programs oriented towards providing "semantic content" *via* the introduction of dictionaries and other kinds of material information in the memory of computers. And in the subsequent decades we have witnessed, on the one hand, the interest of AI and robotic researchers in programs which might simulate artificially our perception of physical reality and our controlled movements in space; and, on the other, the development of artificial frameworks which simulate our global or Gestaltic knowledge, the artificial reconstruction of our inductive procedures and even the invention of a "logic of circumscription" whose purpose is to contribute to making it possible to program a machine for solving problems in contexts where the change of circumstances and common sense play an essential role. In this recall, I do not consider it inopportune to dedicate some words of remembrance in the following footnote to the pioneering work on the automata theory by the Spanish engineer Leonardo Torres Quevedo[10].

*The Turing Man and the Traditionally Modern Scientific Image of Man.*

The Turing proposal of his universal machine as a model of our mind, a proposal that may be alternatively labelled as "the computer metaphor" or "the Turing man"[11], has introduced a substantial change in our modern scientific image of ourselves. The modern scientific image of man was shaped in its various general features and perspectives, not always mutually consistent, during the XVI and XVII centuries by the men which led the Great Scientific Revolution[12]. We may say, to simplify things, that until the past century the prevalent version of this image was provided by the empirical thought of Locke. In this version, the idea of a subject of knowledge is so structurally distant from the material objects of the world and from our technological products as in the thought of Aristotle. The Turing man offers a scientific picture of ourselves as thinking machines or computational beings that allows us to emphasize the relevance of our technological condition. Such an image was very far from being dreamed of by Russell and the men of his generation, and that makes a fundamental difference that separates them from other men born in the 20$^{\text{th}}$ century such as von Neumann, Gödel, Turing or Wiener. When the Neocartesian philosopher Popper proposed his metaphor of the programmer and his computer in order to explain the relation between "the Self and its brain", when the AI researcher Minsky spoke of the "society of mind" with the intention of explaining the structure of our mind as a recursively stratified society of computational atoms, or when von Neumann designed his self-reproducing automaton, their respective speculations and inventions can only be understood in the context of the seminal idea of the universal Turing machine or the Turing Man [13]. Russell died in 1970; but the two basic logical facts of the computability theory, introduced in the Thirties by the results of Gödel and Turing, were never seriously considered by the author of *Principia Mathematica*.

*The Network Paradigm: The Emergence of Internet.*
We have already spoken of Descartes. It is usual to say that from this French thinker onwards, modern philosophy has been so interested in the study of the subjectivity and the internal structure of thought, that it has forgotten the study of communication between the different subjects of knowledge. Only three hundred years later, in the Nineteenth century, did Hegel insist in the European continent and Peirce in the USA, that intercommunication between subjectivities represents an essential dimension of the human condition that concerns not only our social but also our scientific and theoretical activity.

It is a curious fact that the computer, whose history begins in the late Thirties and Forties of the past century, have required only thirty years to repeat the same journey. During the first three or four decades of their existence these devices have been conceived and constructed as machines which, apparently, simulated the introverted Cartesian solipsist. They were tools able to make prodigious computations, but unable to communicate between themselves. Only later on, during the Sixties, a few visionary researchers begin to attack the problem of achieving that these expansive thinking machines could be interconnected and therefore become able to open a flux of mutual information. This was the origin of the current computational net of nets called Internet, which may be briefly described as the marriage of the technology of the computer with the technology of telecommunication.

The story of this marriage consists of three chapters or sub-stories. The *first* one was the formation of Arpanet, the net of ARPA, the military agency created by the USA Congress at the end of the Fifties with the intention of finding a way out of the initial supremacy of the Russian Sputniks in the space race. It was ten years later, at the end of the Sixties, when ARPA, that had soon concentrated its efforts in the computational area, made a reality the project of a computational intercommunicative network. Its main artifices, the ARPA researchers Robert F. Taylor and Larry Roberts worked guided by the previous inspiration of several brilliant pioneers of that decade, mainly the visions of J. C. R. Licklider – whose legendary paper "Man-Computer Symbiosis"[14] forms part of the annals of the Internet story – and the revolutionary communication technique, simultaneously but independtly conceived by Donald Davis in U.K. and Paul Baran in the USA[15].

The military agency ARPA cooperated very closely with the main USA universities, displaying in each of them during the Seventies and Eighties a node of its powerful net. Later on there emerged, within and outside the USA and in a parallel way to Arpanet, a whole panoply of networks similar to it although with a different architecture. So arose the problem of their unification in a more complex device that would not be merely a new computer net but a net of nets. This is the main argument of the *second* sub-story: the transformation of Arpanet, the net of ARPA, into the present net of nets that is Internet.

The construction of an intercommunicative platform, a plattform able to overcome the differences of architecture and operative systems of the different computers in the different nets required, as the diplomatic tasks, to dispose of opportune protocols of common understanding. In their paper of 1974 "A Protocol for Packet Network Interconnection"[16], Robert Kahn and Vinton G. Cerf provided a *Transmission Control Protocol* (TCP), and with it the operative foundations of Internet, a name mentioned for the first time in their article.[17]

The transmutation of Arpanet in Internet was, obviously, a technological advance but also a factor of social And this is one of the keys of the *third* sub-story. The deep technological/social revolution initiated in the Seventies and developed in the Eighties with tools as popular as the UNIX operative system, the personal computer and the MODEM program, could not be alien to Internet. In his paper "The Cathedral and the Bazaar" Eric S. Raymond [18] encouraged for the art of software programming – and in the same way for Internet –, the bazaar model where everybody may test everything every time as opposed to the model of an untouchable "cathedral style" product that nobody may change.

In fact, the application which has most spectacularly incremented the world extension of Internet in the last decade, the World Wide Web is born in an institution, the CERN, much more similar to a cathedral than a bazaar. In the bosom of CERN and in order to solve the problem of retrieving as easily as possible the oceanic amount of reports, conferences and minutes of scientific meetings of the legions and legions of researchers which work in the CERN, the two brilliant minds of Tim Berner Lee and Robert Cailliau accomplished the ambitious project of a computational system able to automatically retrieve any kind of information of any number of scientific documents accumulated in an enormous set of big databases.[19] The system World Wide Web was officially presented in 1991 and soon unanimously accepted by the world community of physics.

But the bazaar factor claimed its part very soon. When the young researcher and hacker Marc Andreessen designed in 1993 in the University of Illinois his browser *Mosaic*, which incorporated the possibility of visualization and transmission of images, Tim Berners Lee considered the idea of such a sort of transmission through the net as something frivolous. But the sensational success of *Mosaic* in the informative traffic of Internet followed later on by Netscape – another creature of Andreessen – was decisive, although only for a short time. A bazaar is always open and exposed to the appetite of market forces, and the subsequent browser war has been clearly indicative of the power of these forces.

*The Sociological Impact of Internet.*

One of the more interesting, although perhaps not sufficiently discussed, socio-logical and philosophical theses of Marx is his contention about the determinant role of technology in the configuration of society. Internet's case illustrates this thesis in more than one aspect. One of them is the fact that in the last two decades several relevant social researchers, as Nasbit, Touraine, Giddens and very specially Castells, have returned to the path of the "grand theories" in sociology, seeing in the network a model partial or totally explicative of the actual forms of social configuration.

Between the several "megatrends", or great cosmopolitan lines of social development that shape our present technological society, Nasbit[20] considers the triple transition from the traditional modern technology to the high technology, from the industrial to the informational society, and from hierarchies to networking. In a similar context the Spanish social researcher Castells[21] speak of "the age of information" and of "the network society" to characterize the so called postmodern society, resultant of the emergency of the technology of information. And he sees in the concept of network, "the new information technology paradigm", the fundamental tool or category to understand the main actual economic, social, politic and cultural realities.

*The Dialectics Between Net and Self.*

The "logic" of the network is susceptible to be detected in the substitution of flexibility for hierarchy in the new forms of organisation of the new capitalism, with its new impressive methods able to reduce to nil the time required for the circulation of money and its new relative independence of spacial and geographical conditions; in the new global and cosmopolitans forms of concentration and distribution of political power, with the consequent power crisis of the old modern states; in the relevance of new symbolic systems of communication; and in the new culture of cyberspace and virtual reality.

But I would like here only to comment one of the more dramatic contrasts that we experience in the net. Dreyfus has said recently, speaking about Internet[22], that it produces in our selves the same feeling and effect of alienation that Kierkegaard denounced when he confronted the radical autonomy of his Self with the insubstantial universality of the Hegelian objective concepts of society and state. This remark seems to me out of the point. Every user of Internet have experienced by himself the tremendous difference between the zone or zones of the net mediatized and the not mediatized by officials powers, representing the last with enormous advantage over the ordinary media a really true public sphere as the last reduct of the ideal of free expression.

This contrast has been intelligently captured by Castells through his thesis of the increasinglily structuration of our society "around the bipolar opposition of the Net and the Self". The Net is source of the new spaces of marginalisation but also source of the new spaces of emancipation. In fact, much of the main successes of the more interesting actual emancipatory movements, as feminism or environmentalism, are due to their use of the World Wide Web:

> The ecological approach...emphasises the holistic character of all forms
> of matter and all information processing. Thus, the more we know,
> the more we sense the possibilities of our technology, and the more we
> realise the gigantic, dangerous gap between our enhanced productive
> capacities, and our primitive, unconscious and ultimately destructive
> social organisation. [...] Yet [...] this is to say that embryonic connec-
> tions between grassroots movements and symbol-oriented mobilisations
> on behalf of environmental justice bear the mark of alternative projects.
> These projects hint at superseding the exhausted social movements of
> industrial society, to resume, under historically appropriate forms, the
> old dialectics between domination and resistance, between *realpolitik*
> and utopia, between cynicism and hope.[23]

We may recapitulate in a few words, as a sort of first balance, everything until
now considered. With the explosion of technology, specially the computational
one, has emerged at the end of past millennium, between us and the physical
cosmos, a new brave and postmodern world that the men of Russell's generation
were very far to dream, a world that has changed with the idea of the Turing Man
our knowledge of our selves and that provides with Internet a formidable tool of
information, emancipation and even alienation that we don't know yet very well
how to control.

## 3. The Facts of Science

*Big Science versus Grand Theory.*
   The alliance between science, technology and politics has been, historically, a
recurrent juncture in war times. The effects of atomic and nuclear weapons with
the shadow of Hiroshima and Nagasaki at the end of World War II shapes, on the
one hand, the black profile of this juncture in its main visit to the 20th century. On
the other hand the plan of the nuclear bomb, the Manhattan Project, has evidenced
in physics since the forties, as his isomorphic brother, the Project Human Genoma
in biology since the Eighties and Nineties, the tremendous significance of the giant
budgets for the development of experimental sciences.
   During the second half of the past century we have assisted to the explosion of
technology in both natural sciences, physics and biology, and to their successive
acquisition of the status of "Big Science". But we have also assisted in both sciences, and in the same period of time, to their respective triumphs in the sphere
of the Grand Theory. The molecular biology revolution has allowed to explain
the genetic and hereditary mechanisms of living beings in molecular, i.e., *physical*
terms. (On its philosophical implications I shall make some comments in the next
section). And with regard to the physics, we find that the quantum field theory,
an admirable hybrid perhaps non entirely consistent of relativity theory and quantum mechanics, has shifted the attention of the research from the atomic to the
nuclear level and the particle physics; and, after a half century of brilliant results,
this theory has culminated in the seventies with the achievement of the Standard
Model of elementary particles, which give a spectacular account of the structure of
matter and a radically new and elegant map of the, until now, ultimate elements

of physical reality. The Standard Model describes correctly all processes where are interactive the electromagnetic, weak and nuclear forces. But such a model is not completely free of arbitrariness nor includes gravitatory interactions. Unfortunately, the attempts to overcome these limitations in a more ambitious theoretical system, as the String Theory and others, have not found up to the present enough experimental support. "For physicist", comments Weinberg, "the twentieth century seems to be ending sadly, but perhaps this is only the price we must pay for having already come so far"[24].

*The Dynamic Universe.*

When Russell published his essay *A Free Man Worship*, the two great scientific revolutions of the first half of 20th century, relativity theory and quantum mechanics, were to appear. During the following thirty years he worked seriously in the knowledge of both. Nevertheless, he did not show the same interest for the new Copernican revolution[25] (represented by the idea of an universe in expansion – a revolution implied by the Hubble's discoveries and the principles of general relativity – nor for the Big Bang theory.

But if quantum mechanics has been mainly the conceptual framework of the great achievements of the 20th century in particles physics, the general theory of relativity has been the conceptual framework of the new cosmological researches, that have offered to the men of this century an incredibly precise knowledge of the history and the structure of the cosmos which the following words may condense:

> We may conclude that during the 20th century our knowledge of the gravitation and of the structure of matter has made by first time deeply accessible to human reason the history and the structure of cosmos. It was 15.000 millions of years ago, when this story began with the Big Bang in the Planck era, dominated by quantum gravitational effects. After an inflationary period we arrive at a universe dominated by radiations, with the emergence at its end of the primordial nuclei (basically hydrogen and helium). Later, the radiation became disaccopled from the matter with the result that matter became itself transparent to radiation. The universe continues its process of expansion and cooling, and the gravitational effect began to be dominant, giving place to the formation of galaxies and stars, in whose interior several processes of thermonuclear fusion could give origin to the formation of new nuclei of helium and much other heavy nuclei, and to their later dissemination through the space as effect of the enormous explosions (supernovae), produced at the end of the life of many massive stars. These nuclei, with diverse material existent in the interstellar dust, give place, through new aggregations, to the formation of new generations of stars which this time could present planetary systems similar to our solar system, formed nearly 4.500 millions of years ago. Not much after, the physic-chemical conditions existent in our planet became suitable for the apparition of the first, eventually evolutionated, living beings., until the also eventual apparition of a species as the human, with the capacity to contemplate the cosmos and to reconstruct a part of its

history. If this improbable event has occurred only one or more times
in other place of the cosmos is matter of discussion.[26]

## 4. The Biotechnological Revolution and the New Scientific Image of Man

*The Enhancement Problem.*

In Russell's times the Darwinian revolution turned the attention of man toward
his past and true origins. In our times the biotechnological revolution turns our
eyes towards a new vision of man's present and towards an unedited vision of
his future. Not only may we now contemplate the landscape of currently genome
research, biomolecular medicine and genetic therapy, but the more impressive one
of an immediate future where the techniques of genetic manipulation might be
extended to our germ line; and perhaps not only with curative purposes but even
to confront the possibility of our enhancement.[27]

This possibility collides seriously with the until now traditionally modern scientific image of man. For this image, the ontological or natural qualities such as
intelligence or beauty, were matter of aleatory distribution of natural selection, and
the privileges of selfplasticity or acquisition of moral virtues of human beings was
exclusively a matter of ethics. Now we have serious reasons to believe that natural qualities may be enhanced by technological methods and that, although not
probably the hardcore of ethics, that twilight zone that the classics denominated
"intemperance" and today we label "weakness of will" might also be a matter of
genetic treatment and enhancement.

In Russell's most ambitious book on ethical affairs, *Marriage and Morals* (1929),
the traditional family and traditional sexual practices had no alternative, as in
Darwin's times, as the only way of specific propagation of the human kind. In our
times, when several animals have already been cloned and the sounds of the hour
of human cloning are already in the air, we begin to believe that sexual intercourse
may not be the only usual way of human propagation.

*Towards the Daedalus Paradigm.*

In fact, biology was never a preferred field of Russell's reflections. If, for example,
we compare his book *Marriage and Morals* with its contemporary fiction roman of
Aldous Huxley *Brave New World* (1932), we may conclude that this literary work,
where drug consumption and the techniques of artificial human incubation are operating, is more tuned in to our present-day scientific and philosophical problems
and preoccupations. But in this context I believe interesting to remark that the
inspirational source of Huxley's literary speculations was a prophetic and provocative essay published eighty years ago, in 1923, by the great biologist John Haldane
with the title *Daedalus or Science and the Future*. In this paper he forecasted:

> A time will however come (as I believe) when physiology will invade and
> destroy mathematical physics, as the latter have destroyed geometry.[28]

He focused that triumph on the discovery and consumption of new pharma-
cological products and artificial human incubation. And to plastically material-
ize his ideas, he opposed the two mythical figures of Prometheus and Daedalus.

Prometheus chosen as a symbol by Marx and also a symbol today for physicists, was the man who stole fire from the Gods and was condemned by them to eternal torment. Daedalus, first sculptor and then interested in biology, was, says Haldane, the first modern man. He invented the art of flying after planning and performing the first transgenical engineering experiment (the intercourse, through artificial methods, of the queen of Minos with a bull, resulting in the monster Minotaur). The *hybris* of Prometheus provokes our admiration, the *hybris* of Daedalus our repulsion:

> The chemical or physical inventor is always a Prometheus [...]. I fancy that the sentimental interest attached to Prometheus has unduly distracted our attention from the far more interesting figure of Daedalus. [...] There is no great invention, from fire to flying, which has not been hailed as an insult to some god. But if every physical and chemical invention is a blasphemy, every biological invention is a perversion. There is hardly one which, on first being brought to the notice of an observer from any nation which had not previously heard of their existence, would not appear to him as indecent and unnatural.[29]

### The Posthuman Problem.

The imagination of Haldane in his essay on Daedalus seems to anticipate in Einsteinian terms the revolution in moral principles that may be provoked today by the impressive results of the biomolecular revolution:

> The Condorcets, Benthams, and Marxs of the future will I think recognize [...] that perhaps in ethics as in physics, there are so to speak fourth and fifth dimensions that show themselves by effects which, like the perturbations of the planet Mercury, are hard to detect even in one generation, but yet perhaps in the course of ages are quite as important as the three dimensional phenomena.[30]

The present development of bioethics is directly proportional to the development of technology. Nevertheless the crucial questions implied here has been more acutely posed in the visionary lines of Haldane above reproduced. We need to think and to decide if the moral selfplasticity of man defended by traditional ethics must remain or not unchanged. The recent debate about the sense of the humanism between the provocative German thinker Sloterdijk[31] and the conservative Habermas is the best illustration of this problem. If the hard postmodernism of Turing connotes a Cartesian eco, the postmodernism of Sloterdijk has hard Neonietzschean resonances.

### Final Balance

If we compare our present worldview with the world viewed by Russell one hundred years ago, we may appreciate that, mainly as a consequence of the great scientific revolutions of the past century, the current landscape of the cosmos has become very different. We know today that not only the world of life but also the physical cosmos is a world in evolution and that the old celestial harmony of the Newtonian universe has been replaced by the most impressive changes and conflagrations in the theatre of the cosmos, including its catastrophic end.

But, on the other hand, we must also observe the invincible intrusion of the new brave world of technology, mainly in its two impressing/impressive manifestations of infosociety and technobiology, both totally alien to the mind of Russell and his contemporaries and both implying dramatic changes in our lives and our being whose consequences we are not yet able to suspect.

Paraphrasing ironically the memorable phrase of Marx: "Philosophers have only interpreted the world, in various ways; the point, however, is to change it", the Spanish social researcher Manuel Castells has remarked, and I personally agree, that the $20^{th}$ century has changed our world so much that what we desperately need now is, conversely, to interpret it.

I began this lecture by remembering Bertrand Russell. I would like to finish it doing the same regarding his friend and colleague Alfred North Whitehead. In 1929, the very same year in which his great work *Process and Reality* appeared, Whitehead published a little book entitled *The Function of Reason*[32], where he intended to incardinate, against Descartes, the reason in the cosmos, but vindicating nevertheless the role of its speculative dimension.

With this intention he pointed out "two main tendencies in the course of events", one "exemplified in the slow decay of physical nature", with the inevitable "degradation of energy", and the other exemplified "by the yearly renewal of nature in the spring, and by the upward course of biological evolution". In the cosmic frame of these two tendencies, of these two "contrasted aspects of history", renovation and degradation, there are, suggested Whitehead, the two traditional aspects of reason, the speculative or theoretical and the practical:

> The Greeks have bequeathed to us two figures, whose real or mythical lives conform to these two notions – Plato and Ulysses. The one shares Reason with the Gods, the other shares it with the foxes[33].

Practical reason, characteristic of Ulysses, is "Reason as seeking an immediate method of action", the "pragmatic agent" whose function consists in resolving the short-range problems of life. It embodies the use of methodical reason, the scientific procedure able to resolve with certainty, in a sort of routine way, a particular kind of problem, but not every kind of problem. To solve the long-term problems of global comprehension inherent to the drive for enhancement and self-improvement of life is the function concerning theoretical or speculative reason, which is "Reason as seeking a complete understanding".

In speculative reason we may find the source and the storage of new methods and models, and of the scientific and philosophical "conceptual schemas" responsible for the different totalizations of our experience. Its exercise might, says Whitehead, "save" the world from the anarchy resulting from the failure of the unilateral methods:

> In our experience, we find Reason and speculative imagination. This reign of Reason is vacillating, vague and dim. But it is there.

We have thus some knowledge, in a form specialized to the special attitudes of human beings, – we have some knowledge of that counter-tendency which converts the decay of an order into the birth of its successor[34].

Finally, I would like to add also this alternative opinion on science and technology written in 1947, near the end of his life, by Ludwig Wittgenstein, the most influent thinker of the second half of past century;

> Es könnte sein, dass die Wissenschaft und Industrie, und ihr Fortschritt, das Bleibendste der heutigen Welt ist. Dass jede Mutmassung eines Zusammenbruchs der Wissenschaft und Industrie einstweilen, und auf *lange* Zeit, ein blosser Traum sei, und dass Wissenschaft und Industrie nach und mit unendlichem Jammer die Welt einigen werden, ich meine, sie zu *einem* zusammenfassen werden, in welchem dann freilich alles eher als der Friede wohnen wird.
>
> Denn die Wissenschaft und die Industrie entscheiden doch die Kriege, oder so scheint es.[35]

## Notes

[1]This text had many editions. The vol. 12 of the *Collected Papers of Bertrand Russell*, specifically titled:*Contemplation and Action, 1902-14*, London, Allen and Unwin, 1985, contains the "Worship" in the pp. 62ff, with precedent commentaries of the editors of the vol., Richard A. Rempel, Andrew Brink and Margaret Moran.

[2] Russells words in 1929.

[3] *Image & Logic. A Material Culture of Microphysics*, University of Chicago Press,1997.

[4] *Id.*, p. 784ff.

[5] Freeman J. Dyson, *The Sun, the Genoma, and the Internet*, New York: New York Public Library, 1999, pp. 13-14.

[6] Cf Andrew Feenberg, *Transforming Technology. A Critical Theory Revisited*, Oxford: University Press, 2002.

[7] José Ortega y Gasset, "El mito del hombre allende la técnica", in: *Obras completas*, vol. IX, Madrid: Revista de Occidente, 1965, p. 618.

[8] *Id.*, p. 624. We may see here a discreet prelude of the Stock's metaman theory of today.

[9] On the machine's invention, cf. Andrew Hodges, *Alan Turing: The Enigma*, Walker & Co., 2000, pp. 92-110.

[10] *Excursus on Torres Quevedo*. Leonardo Torres Quevedo (1852-1936) is usually quoted in the books on history of technology as an internationally reputed constructor of dirigibles and ferries (a famous example of the latter is the still existent *Spanish Aerocar* in Niagara Falls). But he is less acknowledged for two facts: to have successfully substituted (in his time) the new electro-technical methods for the mechanical procedures of the old Babbage, and to have published in his sixties, in 1914 (more than twenty years before the emergence of the current computability

theory and the first computers, and more than forty years before the formulation of the Artificial Intelligence program) a visionary essay on automata, *Ensayos sobre Automática*, which deserves to be compared with the foundational essay of Turing on the computer metaphor appearing in 1950.

In the model of an *autómata con discernimiento* (automaton with discernment) proposed by Torres Quevedo in this essay there were no traces of Cartesian solipsism nor did such a model aim to be a mere abstract machine of logical and mathematical calculus. The need to warrant the communication of the automaton with the external world and its operations on it, i.e., the *nexus* as we say today of AI with robotics, was explicitly consigned by the Spanish engineer, who prescribed that the automaton might dispose not only of "senses" (thermometers, compasses, dynamometers, manometers ...), in order to obtain sensorial notice of the world, but also of "members" and "energy", in order to enable its movements and external actions. But the most important feature of Torres Quevedo's proposal is his demand, in a sense anticipatory of the current "logic of circumscription" of McCarthy, that the "autómata con discernimiento" must have a knowledge of the circumstances of its environment to be able to develop a behaviour of adaptation to the world.

"It is also necessary –and this is the main object of Automatics- for the automata to have *discernimiento*, that they can at each moment, *considering the impressions that they receive, and also, at times, those that they have received previously*, order the desired operation. *It is necessary that the automata imitate living beings, performing their acts, according to the impressions they receive and adapting their behaviour to the circumstances...*"

By way of example of the imaginative mind of Torres Quevedo I reproduce a passage of his *Essays*, where he discusses Descartes's opinion that an automaton could never maintain a reasonable dialogue. We may see here an intelligent anticipation of the nowadays famous "Chinese room argument" conceived by Searle in order to unmask and to invalidate the behaviourism underlying the test of Turing:

"Let us imagine a machine... in which, instead of three switches, there were, if necessary, thousands or millions; and that instead of three or four different positions, each switch had one position corresponding to each one of the signs of our writing (letters, coded figures, orthographic signs, etc.).

"It is perfectly easy to understand that it is possible, with the aid of these switches, to write any sentence, and even a more or less long discourse; that should depend on the number of switches that might be at our disposition"

"To each discourse there corresponds a position of the system, and, therefore an electromagnet."

"We may suppose that this device shoots a phonograph on which there is inscribed the answer to the question which has activated its movement, and so we will have an automaton able to discuss *de omni re scibile*."

"Certainly, the preliminary study of all possible questions, the redaction of the answer to each of them, and, finally, the construction of such a machine, would not be a very easy thing; but it would not be much more difficult than the construction of an ape or another animal, well enough imitated to induce the naturalist to classify it among the living species".

In an interview with the journal *Scientific American* in 1915 Torres Quevedo contended, as Turing did forty years later, that "at least in theory almost all the operations of a vast game of these could be performed by a machine, even those about which it is supposed that they need the intervention of a considerable intellectual capability".

[11] I take this denomination from the book of J. David Bolter *The Turing's Man*, London, Duckworth, 1984.

[12] Charles Taylor, *Sources of the Self*, Cambridge: University Press, 1985.

[13] In his paper of 1950 Turing considered an ingenious implication against formalisms power, derived of Gödel's theorem, as the main objection to his proposal. This a priori objection has been later developed by Lucas and Penrose. As Hume said commenting the ontological argument, tis objection is usually preferred by mathematical minds.

[14] J. C. R. Licklider, "Man-Computer Symbiosis", *IRE Transactions on Human Factor in Electronics*. Vol. 1, March 1960, pp. 4-11.

[15] The key of this technique consisted in the idea – which seemed to run against the classical principles of telephony – that the most efficient unities of circulation through the net channels are not the integral messages, but "packets" of segments of such messages which are able to be recomposed after their arrival at their point of destination. Baran conceived also the original project, equally decisive in the structure of Internet, of shaping a communication network with a deliberately non-hierarchical structure, whose force and capacity to survive were based on the ideas of decentralization and redundancy, in order to elude the risk (one of the Cold War obsessions) of collapse as a consequence of a selective attack.

[16] *IEEE Transactions in Communication*, 22, 5, pp. 637-48.

[17] A second and more important protocol elaborated six years later by both researchers and universally known as TCP/IP, added to a new TCP (in charge of the data management and the fragmentation of messages in packets) the so-called IP (*Internet Protocol*). This protocol localizes in the net the computer that is the point of destinatiion and sends the message to its direction. During the Eighties this protocol was incorporated into the operative system UNIX, then recently constructed and soon very popular. The great accessibility was for Internet a goal and a challenge but not, for military reasons, for Arpanet, that disappeared at the end of that decade.

[18] The text is in the net.

[19] Their system was a really perfect model, designed in a record time and essentially characterised by the "client-server" architecture, the introduction of new important protocols such as URL (*Uniform Resource Locator*), HTTP (*Hypertext Transport Protocol*) and HTML (*Hypertext Mark-up Language*), and the introduction of a browser program in each client computer of the net.

[20] John Naisbitt, *Megatrends: Ten New Directions Transforming our Lives*, New York: Warner Books, 1982.

[21] Manuel Castells, *The Information Age: Economy, Society and Culture*. Vol I: *The Rise of the Network Society*, Cambridge Ma., Oxford UK, Blackwell,1996; Vol II: *The Power of Identity*, Malden, Ma, Oxford, UK, Blackwell, 1997; Vol III: End of Millenium, Malden, Ma, Oxford, UK, Blackwell, 1998.

[22] Hubert L. Dreyfus, *On the Internet*. London and New York: Routledge, 2001.

[23] Vol II, p. 133.

[24] Steven Weinberg, *Facing Up*, Cambridge, Mass. – London, England: Harvard

University Press, 2001, p. 229.

[25] Antonio Dobado, "Física de partículas y cosmología en el Siglo XX (Hacia una historia integrada del universo), in: *El legado filosófico y científico del siglo XX,* eds. M. Garrido, L.M. Valdés & L. Arenas, Madrid: Cátedra (in printing).

[26] *Id. Ibid.*

[27] Gregory Stock, *Redesigning Humans*, Boston, New York: Houghton, Mifflin Company, 2003

[28] J.B.S. Haldane, "Daedalus or Science and the Future" (1923), in *Haldane's Daedalus Revisited*, ed. by Krishna R. Dronamraju, Oxford, New York, Tokio: Oxford University Press, 1995, p. 27.

[29] *Id.*, p. 36.

[30] *Id.*, p. 28.

[31] Peter Sloterdijk, Regeln für den Menschenpark, Franfurt am Main: Surhrkamp, 1999.

[32] Alfred North Whithehead, *The Function of Reason* (1929), Boston: Beacon Press, $4^{th}$ printing, 1966.

[33] *Id.* p. 10.

[34] *Id.* pp. 89-90.

[35] Ludwig Wittgenstein, *Vermischte Bemerkungen. Eine Auswahl aus dem Nachlass.* Hrsg von Georg Henrik von Wright unter Mitarbeit von Heikki Nyman. Neubearbeitumg des Textes durch Alois Pichler. Frankfurt a. M.: Suhrkamp, 1994, S. 123 ( Science and industry, and their progress, might turn out to be the most enduring thing in the modern world. Perhaps any speculation about a coming collapse of science and industry is, for the present and for a *long* time to come, nothing but a dream; perhaps science and industry, having caused infinite misery in the process, will unite the world – I mean condense it into a *single* unit, though in which peace is the last thing that will find a borne.

Because science and industry do decide wars, or so it seems. *Culture and Value*, translated by Peter Winch, Oxford: Blackwell, 1980, p. 63c)

# SECTION A:

# LOGIC

# Natural Axioms of Set Theory and the Continuum Problem

Joan Bagaria

*Institució Catalana de Recerca i Estudis Avançats (ICREA),*
*and Departament de Lògica, Història i Filosofia de la Ciència*
*Universitat de Barcelona. Baldiri Reixac, s/n. 08028 Barcelona*
*bagaria@ub.edu*

**Abstract.** As is well-known, Cantor's continuum problem, namely, what is the cardinality of $\mathbb{R}$? is independent of the usual ZFC axioms of Set Theory. K. Gödel ([12], [13]) suggested that new natural axioms should be found that would settle the problem and hinted at large-cardinal axioms as such. However, shortly after the invention of forcing, it was shown by Levy and Solovay [20] that the problem remains independent even if one adds to ZFC the usual large-cardinal axioms, like the existence of measurable cardinals, or even supercompact cardinals, provided, of course, that these axioms are consistent. While numerous axioms have been proposed that settle the problem–although not always in the same way–from the Axiom of Constructibility to strong combinatorial axioms like the Proper Forcing Axiom or Martin's Maximum, none of them so far has been recognized as a natural axiom and been accepted as an appropriate solution to the continuum problem. In this paper we discuss some heuristic principles, which might be regarded as *Meta-Axioms of Set Theory*, that provide a criterion for assessing the naturalness of the set-theoretic axioms. Under this criterion we then evaluate several kinds of axioms, with a special emphasis on a class of recently introduced set-theoretic principles for which we can reasonably argue that they constitute very natural axioms of Set Theory and which settle Cantor's continuum problem.

## 1. Introduction

> *There must be a first step in recognizing axioms, [...] a step which will make the axioms seem worth considering as axioms rather than merely as conjectures or speculations.*

W.N. Reinhardt ([27])

Cantor's continuum problem, namely, what is the cardinality of $\mathbb{R}$? has been the central problem in the development of Set Theory. Since Cantor's formulation in 1878 of the Continuum Hypothesis (CH), which states that every infinite subset of $\mathbb{R}$ is either countable or has the same cardinality as $\mathbb{R}$ ([8]), very dramatic and unexpected advances have been made by Set Theory towards the solution of the problem. As is well-known, neither CH nor its negation can be proved from the usual ZFC axioms of Set Theory, provided they are consistent. In Gödel's constructible universe CH holds, while Cohen's method of forcing allows to build models of ZFC in which the cardinality of $\mathbb{R}$ can be any cardinal, subject only to the necessary requirement that it have uncountable cofinality.

This situation, however, is far from satisfactory. Admittedly, some mathematicians, including Cohen himself (see [9]), have expressed the belief that no further, more satisfactory solution is attainable, and that one should be content with the independence results. But this is a rather uncommon position among mathematicians, and set theorists in particular, with respect to the continuum problem. Drawing on a realistic approach to Mathematics, the most common by far among mathematicians, one can argue that the only thing the results of Gödel and Cohen show is that the ZFC axioms, while sufficient for developing most of classical Mathematics, constitute too weak a formal system for settling Cantor's problem and they should, therefore, be supplemented with additional axioms. Indeed, Gödel himself formulated a program ([12],[13]) of finding new natural axioms which, added to the ZFC axioms, would settle the continuum problem, and he hinted that large cardinal axioms would do it. This has been known as *Gödel's program*. Unfortunately, however, it was soon noticed by Levy and Solovay [20] that the usual large cardinal axioms, like the existence of measurable cardinals, or even supercompact or huge cardinals, would not be enough. But this does not mean that Gödel's program is no longer defensible. Quite the contrary. It is still perfectly possible that new kinds of large-cardinal axioms, different from the ones that have been considered so far, could be relevant to the solution of the continuum problem. In fact, recent work by Woodin ([37]) shows that under large cardinals, any reasonable extension of the ZFC axioms that would settle all questions of the same complexity of CH, in a strong logic known as $\Omega$-logic, would refute CH (see [15] for a discussion of the relevance of Woodin's work on Gödel's program). But our purpose here is not to address the import of large-cardinal axioms to the continuum problem, at least not directly, but to introduce and discuss some heuristic principles, which might be regarded as *Meta-Axioms of Set Theory*, that provide a criterion for assessing the naturalness of the set-theoretic axioms. Under this criterion we then evaluate several kinds of axioms, including large cardinals, with a special emphasis on a class of set-theoretic principles that have been recently introduced, known as Bounded Forcing Axioms, for which one can reasonably argue that they constitute very natural axioms of Set Theory, and which settle Cantor's continuum problem.

## 2. Natural axioms of Set Theory

> *The central principle is the reflection principle, which presumably will be understood better as our experience increases.*

> K. Gödel ([35])

What should be counted as a *natural* axiom of Set Theory? Certainly any intuitively obvious fact about sets. Here we shall take for granted that the ZF axioms are of this sort. There is very little disagreement about this point. As for the Axiom of Choice, the reluctance regarding its full acceptance by some mathematicians is due more to some of its counter-intuitive consequences, rather than to its otherwise very natural character (see however [16]). It is a fact that no other universally (or almost-universally) accepted as intuitively obvious principles about sets have been proposed, perhaps with the only exception of the existence of small large cardinals, like the inaccessible cardinals.

If we accept that *being an intuitively obvious fact about sets* is a necessary requirement for a set-theoretic principle to be counted as an axiom, then no axioms other than the ZF (or ZFC) axioms, plus, perhaps, some small large-cardinal existence axioms should be accepted. So, if we were to look for additional axioms we should first try to sharpen our intuitions about sets until we were forced to accept some new principle as intuitively obvious, or at least intuitively reasonable. While this is *a priori* possible, and it would certainly be a remarkable achievement to discover such a new principle, there are at least two practical difficulties with this approach. First, it is well known that intuition may be easily confused with familiarity. For do we not end up finding reasonable whatever principle we have been using for a long time? Are we not eager to welcome as a new axiom any principle in which we have invested a considerable amount of time and effort, and for which we have developed, no doubt, a strong intuition? Second, in principle, incompatible intuitively reasonable principles could be found. For what prevents set-theoretic intuition to be developed in several irreconcilable ways? It may be replied that if this were the case, then all the better, for we would have several different set theories, all founded on intuition, albeit each on a different one. If this will be the case, then so be it. But we will see that, beyond intuition, there are other criteria which can be successfully used to find new axioms.

In his paper *What is Cantor's Continuum Problem?* ([12], [13]), Gödel considers two criteria for the acceptance of new axioms of Set Theory. One is that of *necessity or non-arbitrariness*. He uses this criterion to justify the existence of inaccessible cardinals. If we want to extend the operations of set formation beyond what is provable in ZFC, then we are forced to postulate the existence of an inaccessible cardinal (see our discussion of this point in section 4 below). Thus the existence of an inaccessible cardinal is a necessary, non-arbitrary assumption, for further extending the *set of* operation. Notice that the postulation of the existence of an inaccessible cardinal is analogous to the situation in which, starting from ZF-Infinity (i.e., Zermelo-Fraenkel Set Theory minus the Axiom of Infinity), we postulate the existence of an infinite set. Indeed, no matter how we extend the ZF-Infinity axioms by asserting the existence of new sets, we are forced to assert the existence of an infinite set, and so, in this sense, ZF is a necessary, non-arbitrary extension of ZF-Infinity. Once the existence of an inaccessible cardinal is accepted, then one is naturally led to the iteration of this principle, thus leading to hyperinaccessible cardinals, and beyond. But can larger cardinals be justified under the necessity criterion? In what sense, if any, are measurable cardinals necessary? We shall come back to this.

A second criterion used by Gödel in [12] for the acceptance as axioms of set-theoretic principles is *success*, that is, the *fruitfulness in their consequences*. This criterion is put forward as an alternative to *necessity* or *non-arbitrariness*. After over half a century of continued work on large cardinals, and especially since the discovery of the connections between large cardinals and determinacy in the eighties, it can be argued that the existence of large cardinals, at least up to Woodin cardinals, should be accepted as axioms of Set Theory, according to this criterion.

Indeed, Martin and Steel [25] showed that the Axiom of Projective Determinacy (PD), and in fact the axiom $AD^{L(\mathbb{R})}$, which asserts that all sets of reals definable fom ordinals and real numbers as parameters are determined, follows from axioms of large cardinals. Woodin showed that the existence of infinitely many Woodin Cardinals plus a measurable cardinal larger than all of them would suffice and, furthermore, that infinitely many Woodin cardinals are necessary to obtain PD (see [36] and [39]). As it became clear during the seventies through the spectacular advances made by Descriptive Set Theory under the assumption of PD, this principle appears to be the right one for developing the theory of projective sets of real numbers. Indeed, PD gives an essentially complete theory of the projective sets. Moreover, any known set-theoretic principle of at least the consistency strength of PD – for instance, the Proper Forcing Axiom – implies PD, which strongly suggests its necessity. The fruitfulness of large cardinal axioms is further exemplified by their numerous consequences in infinitary combinatorics (see [18]). It is now plainly clear that many desirable consequences, not only in Set Theory, but in all areas of Mathematics where set-theoretic methods are applied, follow from large-cardinal assumptions. Thus, strong large-cardinal principles have done very well under the fruitfulness criterion. But is this sufficient for accepting them as axioms of Set Theory? This may be so for the existence of infinitely many Woodin cardinals, since they have been shown to be both sufficient and necessary to obtain PD, thus yielding a rich and elegant theory for the projective sets of real numbers which extends the classical ZFC theorems of Descriptive Set Theory. For stronger large-cardinal principles, the situation is much less clear. The main problem in accepting large cardinal axioms is their consistency. After all, some large cardinal principles have been shown to be inconsistent and consequently rejected. Nevertheless, the so-called _inner model program_, which attempts to build canonical models for large cardinals, has developed very sophisticated methods for showing that, at least for large cardinals up to infinitely-many Woodin cardinals, one can construct canonical inner models with a well-developed fine structure, thereby building confidence in their consistency. So, in spite of some diverging opinions, we can fairly say that it is a widespread belief among set theorists that large-cardinal principles should be accepted as axioms of Set Theory provided there is a sufficiently well-developed inner model theory for them. This is already the case for infinitely many Woodin cardinals, but no such inner model theory has been yet developed for, e.g., supercompact cardinals.

But as has been pointed out before, large-cardinal axioms, in spite of their extraordinary success, are not sufficient for settling Cantor's Continuum Problem. So in the absence of any further intuitively obvious axioms, the question is whether there are any other kinds of axioms that are non-arbitrary and, if possible, that also satisfy the fruitfulness criterion.

Although the value of an axiom will ultimately be determined by its success, the criterion of _success_ can hardly be sufficient for accepting a new axiom. It should only be used to assess, _a posteriori_, the value of the axioms, which must be found according to other criteria.

H. Wang, in [34], and later in [35] section 8.7, quotes Gödel on his 1972 answer to the question of what should be the principles by which new axioms of Set

Theory should be introduced. According to Gödel there are five such principles: *Intuitive Range*, the *Closure Principle*, the *Reflection Principle*, *Extensionalization*, and *Uniformity*. The first, *Intuitive Range*, is the principle of intuitive set formation, which is embodied into the ZFC axioms. The *Closure Principle* can be subsumed into the principle of *Reflection*, which may be summarized as follows: The universe $V$ of all sets cannot be uniquely characterized, i.e., distinguished from all its initial segments, by any property expressible in any reasonable logic involving the membership relation. A weak form of this principle is the ZFC-provable reflection theorem of Montague and Levy (see [18]):

> *Any sentence in the first-order language of Set Theory that holds in $V$ holds also in some $V_\alpha$.*

Gödel's *Reflection* principle consists precisely of the extension of this theorem to higher-order logics, infinitary logics, etc.

The principle of *Extensionalization* asserts that $V$ satisfies an extensional form of the Axiom of Replacement and it is introduced in order to justify the existence of inaccessible cardinals. We will explain its role in the next section.

The principle of *Uniformity* asserts that the universe $V$ is uniform, in the sense that its structure is similar everywhere. In Gödel's words ([35], 8.7.5): *The same or analogous states of affairs reappear again and again (perhaps in more complicated versions).* He also says that this principle may also be called *the principle of proportionality of the universe*, according to which, analogues of the properties of small cardinals lead to large cardinals. Gödel claims that this principle makes plausible the introduction of measurable or strongly compact cardinals, insofar as those large-cardinal notions are obtained by generalizing to uncountable cardinals some properties of $\omega$.

Thus, following Gödel, in the search for new axioms beyond ZFC, we are to be guided by the criteria of *Necessity*, *Success*, *Reflection*, *Extensionalization*, and *Uniformity*, to which we should add that of *Consistency*, which Gödel certainly took for granted. The new axioms should be necessary in order to extend the operations of set formation beyond what is provable in ZFC, they should take the form of reflection principles, they should imply some kind of uniformity in the universe of all sets, and they should be both consistent and fruitful in their consequences.

In the next section we will discuss and attempt to further clarify these criteria so that they can be actually applied in the testing of – and the search for – new axioms. We will argue that all criteria reduce essentially to two: *Maximality* and *Fairness*. *Consistency* and *Success* play a complementary role, the first as a regulator and the second as a final test for value. All together, the criteria may be regarded as an attempt to define what *being a natural axiom of Set Theory* actually means. They may as well be viewed as a test for *necessity* or *non-arbitrariness*, since any set theoretic statement that satisfies the criteria will, in a precise sense, be forced upon us if we want to extend ZFC.

## 3. Meta-axioms of Set Theory

We are searching for additional axioms of Set Theory that extend ZFC, that is, for a sentence (or a recursive set of sentences) in the first-order language of

Set Theory. What are the criteria such a sentence should satisfy in order to be considered an axiom?

The first criterion is, of course, *Consistency*. We want the new axiom to be consistent with ZFC. Clearly, by Gödel's second incompleteness theorem, we can only hope for a proof of relative consistency. Namely, we should be able to prove that *if* ZFC is consistent, *then* so is ZFC plus the new axiom. There are many incompatible examples, e.g., $CON(ZFC)$ and $\neg CON(ZFC)$, the Axiom of Constructibility or its negation, the Continuum Hypothesis or its negation, Suslin's Hypothesis or its negation, etc. Thus, consistency cannot be the only criterion. Moreover, we should also entertain the possibility of accepting axioms whose consistency (modulo ZFC) cannot be proved in ZFC, simply because they can be shown to be, consistencywise, stronger than ZFC, but which nevertheless satisfy the other criteria.

Therefore, the criterion of *Consistency* can only play a regulatory role in the search and justification of new axioms. It puts a bound on the joint action of the other criteria. The mere fact that a set-theoretic principle can be shown to be consistent with ZFC does not make it automatically an axiom. But consistency with ZFC is certainly a necessary requirement. Moreover, if the new axiom is shown to be consistent modulo some large-cardinal assumption, then the consistency of such a large cardinal must follow from ZFC plus the new axiom, thus proving its necessity for the new axiom's consistency proof.

The second criterion is that of *Maximality*. Namely, the more sets the axiom asserts to exist, the better. Gödel already stated that: ...*Only a maximum property would seem to harmonize with the concept of set.*.(see [13]). The idea of maximizing has been defended by many people and it has been extensively discussed by P. Maddy (see [21] and [22]) in the context of her naturalistic philosophy of Set Theory. The maximality criterion has normally been used to provide a justification for the rejection of the Axiom of Constructibility, but here we intend to apply it systematically as a guiding criterion in the search for new axioms.

All large-cardinal axioms and all forcing axioms satisfy the *Maximality* criterion, in the weak sense that they all imply the existence of new sets. Thus, in such a generality this is clearly too vague a criterion, and therefore definitely useless. For if ZFC is consistent, then we can easily find statements that are consistent, modulo ZFC, and assert the existence of some new sets, but which are incompatible. Take, for instance, $CON(ZFC)$, which asserts the existence of a model of ZFC, and $\neg CON(ZFC)$, which asserts the existence of a (non-standard) proof of a contradiction from $ZFC$.

To attain a more concrete and useful form of the *Maximality* criterion it will be convenient to think about maximality in terms of models. Namely, suppose $V$ is the universe of all sets as given by ZFC, and think of $V$ as being properly contained in an *ideal* larger universe $W$ which also satisfies ZFC and contains, of course, some sets that do not belong to $V$ – and it may even contain $V$ itself as a set – and whose existence, therefore, cannot be proved in ZFC alone. Now the new axiom should imply that some of those sets existing in $W$ already exist in $V$, i.e., that some existential statements that hold in $W$ hold also in $V$. Since the sets in $V$ are already given we may as well allow for the existential statements to have parameters in $V$. Thus, *Maximality* leads to *Reflection* principles, namely, the existential statements (with parameters) that hold in the ideal extension $W$ *reflect* to $V$.

By repeated application of *Reflection*, something which the *Maximality* criterion forces us to do, the universe of all sets becomes more uniform. For instance, if some set $A$ is the solution of an existential sentence $\varphi(x)$ that holds in some ideal extension $W$ of $V$, then we may consider the sentence $(\varphi(x) \wedge \neg x = A)$, which contains $A$ as a parameter, and by applying *Reflection* again obtain another solution of $\varphi(x)$ different from $A$. Or if $\alpha$ is the rank of $A$, then by considering the sentence $(\varphi(x) \wedge rank(x) > \alpha)$ we obtain another solution of $\varphi(x)$ of higher rank, etc. Thus, *Reflection* leads to the existence of *many* solutions of any given existential statement, e.g., solutions of arbitrarily high rank. Gödel listed *Uniformity* as a separate principle. He understood it as a justification for the extrapolation to larger cardinals of some of the properties of small cardinals, like $\omega$. We do not consider this by itself as a sound criterion, since we do not see any need for arbitrary properties of, say, $\omega$ to hold for some larger cardinals. Some of its properties certainly do not hold for larger cardinals, like the property of being countable. So, some criterion should be given for choosing among all the distinct properties. In our remarks below regarding particular kinds of axioms we will see how a strong form of *Uniformity* does follow from the systematic application of the criterion of *Maximality*.

Notice that not all existential statements are maximizing principles in the same sense. Indeed, CH is an existential statement which asserts the existence of a function on $\omega_1$ that enumerates all the real numbers, but at the same time asserts the existence of *few* real numbers. So, does CH assert the existence of *more* sets or of *fewer* sets? On the other hand, not-CH is also an existential statement which asserts the existence of more than $\aleph_1$ many reals, while implying that, for instance, there are no diamond sequences. So, again it is unclear, *a priori*, whether not-CH is a *maximizing* or a *minimizing* principle. Which one of CH or its negation should we then accept according to the *Maximality* criterion? The difficulty of the question is best exemplified by the fact that it is easy to construct by forcing three models of ZFC, $M_1 \subseteq M_2 \subseteq M_3$, such that CH holds in both $M_1$ and $M_3$ and fails in $M_2$. The problem is that both CH and its negation are $\Sigma_2$ statements, and $\Sigma_2$ sentences, while asserting the existence of some sets, may in fact be limitative. The same applies to more complex existential sentences. The only unquestionably maximizing existential sentences are the $\Sigma_1$.

Another direct consequence of the *Maximality* criterion is Gödel's principle of *Extensionalization*. This can be stated as follows: We should require that $V$ satisfies all instances of the Replacement Axiom for functions with domain some set in $V$ and range contained in $V$ that are available in some ideal extension of $V$. To what extent is this a reasonable assumption? It is reasonable insofar as this is what we would like to have for $V$ itself. With $V$ the problem is that, besides the set-functions, there are no more such functions available other than those that are definable in $V$. But when more functions become available, even if they are ideal functions, there is no reason, a priori, why they should be excluded.

We may thus conclude that Gödel's principles of *Reflection, Extensionalization,* and *Uniformity* arise naturally from the systematic application of the criterion of *Maximality*.

We need a third criterion to help us sort out among all possible set existence statements that hold in some ideal extensions of $V$ those that will be taken as new axioms. Such a criterion may be called *Fairness*. We could also call it the *Equal Opportunity* criterion. It can be stated as:

*One should not discriminate against sentences of the same logical complexity.*

The rationale for this criterion is that in the absence of a clear intuition for the selection, among all the set-existence statements that hold in some ideal extension of the set-theoretic universe, of those that are true about sets, we have *a priori* no reason for accepting one or another. So, once we accept one, we must also accept all those that have the same logical complexity.

The logical complexity of a formula of the language of Set Theory is given by the Levy hierarchy, namely, the $\Sigma_n$ and $\Pi_n$ classes of formulas (see [17]).

If we are to allow parameters in our formulas, then we should also require that:

*One should not discriminate against sets of the same complexity.*

Now the complexity of a set may be defined in different ways, but the most natural measures of the complexity of a set are its rank and its hereditary cardinality.

Thus, a *fair* class of existential sentences will be one of the classes $\Sigma_n$ with parameters in some $V_\alpha$, $\alpha$ an ordinal, or some $H_\kappa$, $\kappa$ a cardinal. Classes of higher-order formulas, like the $\Sigma_n^m$, or formulas pertaining to some infinitary logic could also be considered. Moreover, the language could also be expanded by allowing new constants or predicates, etc.

Finally, there is the criterion of *Success*. As was remarked before, its main use is for evaluating the axioms that have been found by following the other criteria. A new axiom should not only be natural, but it should also be useful. Now, usefulness may be measured in different ways, but a useful new axiom must be able at least to decide some natural questions left undecided by ZFC. If, in addition, the new axiom provides a clearer picture of the set-theoretic universe, or sheds new light into obscure areas, or provides new simpler proofs of known results, then all the better.

In conclusion, once we agree on what kind of ideal extensions of $V$ we should be considering, by applying the three criteria above simultaneously (*Consistency, Maximality,* and *Fairness*), the crucial question becomes:

*Find a (largest possible) fair class $\Sigma$ of existential sentences such that the principle that asserts that all sentences in $\Sigma$ that hold in an ideal extension are true can be stated as a sentence (or a recursive set of sentences) in the first-order language of Set Theory and is consistent with ZFC.*

Once such a principle is found, we can reasonably argue that it constitutes a natural axiom of Set Theory. Its survival as a new axiom, in terms of being accepted and used by the set theorists, will then be largely determined by its success.

We shall now put to test our criteria in the case of large-cardinal axioms.

## 4. The naturalness of large-cardinal axioms

> *Whatever theory we have about what exists, it should be compatible with our understanding of our theory that the totality of existing things should be a set.*

> W.N. Reinhardt ([27])

Large cardinal axioms may be divided into two classes: the strong axioms of infinity, and the large cardinal axioms arising from elementary embeddings of $V$ into transitive proper classes, i.e., the measurable cardinals and above.

### 4.1. Strong axioms of infinity

The *strong axioms of infinity* originate when one considers ideal extensions of the universe $V$ of all sets, as given by ZFC, in which the transfinite sequence of all ordinals, and therefore the power set operation, is continued yet even further. In this ideal extension, the class $OR^V$ of all ordinals in $V$ would be an ordinal $\kappa$, and $V$ itself would be a set. We thus imagine $V$ to be actually some initial rank $V_\kappa$ of a larger universe so that $V_\kappa \models ZFC$.

We can introduce new axioms stating that sentences in a given fair class $\Sigma$ reflect to $V_\kappa$. These kinds of axiom, even though they satisfy our criteria, they may not have any large-cardinal strength and their consequences may be rather poor. For instance, the axiom that asserts that $V_\kappa$ satisfies ZFC and reflects all $\Sigma_n$ sentences, for some fixed $n$, follows from the existence of a stationary class of ordinals $\alpha$ such that $V_\alpha$ satisfies ZFC, a principle which has no large-cardinal strength and is consistent with the Axiom of Constructibility.

A crucial step forward in strength is obtained by requiring that $\kappa$ is a regular cardinal. Notice that if $V_\kappa$ is a model of ZFC, then $\langle V_\kappa, \in, \kappa \rangle \models$ "$\kappa$ is a regular cardinal". But $\kappa$ need not even be a cardinal in $V$. Requiring that $\kappa$ is a regular cardinal in $V$ amounts to requiring that $V_\kappa$ satisfies a bit of the second-order Replacement Axiom. Namely, Replacement for all functions with domain some ordinal less than $\kappa$ and values in $\kappa$, which need not be definable in $V_\kappa$. It turns out that since $V_\kappa \models ZFC$, satisfying this bit of second-order Replacement implies that $V_\kappa$ satisfies the full second-order Replacement Axiom. This form of *extensional Replacement* is exactly the content of Gödel's principle of *Extensionalization*, which we have already discussed in the previous section; we argued its naturalness under the *Maximality* criterion.

Now for $\kappa$ a regular cardinal, the following are equivalent:

1. $V_\kappa \models ZFC$

2. $V_\kappa \prec_{\Sigma_1} V$

   i.e., $V_\kappa$ reflects all $\Sigma_1$ sentences with parameters, which means that for every $a_1, ..., a_k \in V_\kappa$ and every $\Sigma_1$-formula $\varphi(x_1, ..., x_k)$,

   $$V_\kappa \models \varphi(a_1, ..., a_k) \quad \text{iff} \quad \varphi(a_1, ..., a_k).$$

A regular cardinal satisfying (1) or (2) above is *inaccessible*. Thus according to our criteria the existence of an inaccessible cardinal is a natural axiom of Set Theory.

If we want to continue, *yet one more step*, the iterative construction of $V$, we are forced to accept the existence of an inaccessible cardinal. The existence of an inaccessible cardinal is the first of the large cardinal axioms.

The existence of an inaccessible cardinal cannot be proved in ZFC, for if $\kappa$ is inaccessible, then $V_\kappa$ is a model of ZFC. Hence, the consistency of ZFC cannot imply the consistency of ZFC plus the existence of an inaccessible cardinal. The sentence that asserts the existence of an inaccessible cardinal $\kappa$, as every other large cardinal axiom, has greater consistency-strength than ZFC. Therefore, it cannot satisfy the criterion of *Consistency* in its basic form, but of course it trivially satisfies it modulo large cardinals. It does however satisfy the other two criteria of *Maximality* and *Fairness* for the class of $\Sigma_1$ formulas with parameters in $V_\kappa = H_\kappa$.

The next step is to consider the class of $\Sigma_2$ sentences, namely, suppose that $\kappa$ is inaccessible and

$$V_\kappa \prec_2 V$$

i.e., it reflects all $\Sigma_2$ sentences with parameters. Then $\kappa$ is an inaccessible cardinal, a limit of inaccessible cardinals, and much more.

More generally, for every $n$ one may consider the existence of a regular cardinal $\kappa$ such that

$$V_\kappa \prec_n V$$

Such a cardinal is called *$n$-reflecting*. The axioms that assert the existence of $n$-reflecting cardinals do satisfy the criteria of *Maximality* and *Fairness*. But if $n < m$, then $ZFC$ plus the existence of an $m$-reflecting cardinal implies the consistency of $ZFC$ plus there is a $n$-reflecting cardinal. Thus, those axioms are strictly increasing in consistency strength.

Notice that since for $n < m$, if $\kappa$ is an $m$-reflecting cardinal then it is also $n$-reflecting, asserting the existence of an $m$-reflecting cardinal makes the universe larger than just asserting the existence of an $n$-reflecting cardinal.

For each $n$, the sentence: *There exists a $n$-reflecting cardinal*, can be written as a first-order sentence. However, by Tarski's theorem on the undefinability of truth, there cannot be a definable $\kappa$ such that $V_\kappa$ reflects all sentences. Moreover, the sentence: *There exists a cardinal $\kappa$ that reflects all $\Sigma_n$ sentences, all $n$*, cannot even be written in the first-order language of Set Theory.

We conclude that the set of all sentences of the form: *There exists a $n$-reflecting cardinal*, $n$ an integer, forms a recursive set of natural axioms of Set Theory (modulo its consistency with ZFC). In fact, by the same arguments, and following the principle of *Maximality*, we are led to the acceptance as a natural recursive set of axioms the set of all sentences of the form: *There exists a proper class of $n$-reflecting cardinals*, $n$ an integer (modulo its consistency with ZFC).

A strengthening of the notion of inaccessibility is that of a Mahlo cardinal: $\kappa$ is a *Mahlo* cardinal if it is regular and the set of inaccessible cardinals below $\kappa$ is stationary, i.e., every closed and unbounded subset of $\kappa$ contains an inaccessible cardinal. Notice that since inaccessible cardinals are regular, we cannot hope to have a club of inaccessible cardinals below $\kappa$, but we may have the next best thing, namely, a stationary set of them. This is a natural assumption according to the principle of *Maximality*. The point is that, provably in ZFC, every sentence $\varphi$ that holds in $V$ reflects to a club class of $V_\alpha$. So, there should be an inaccessible cardinal $\kappa$ such that $V_\kappa$ satisfies $\varphi$. Once the existence of inaccessible cardinals is accepted, we should also accept that there are as many of them as possible, and this means a stationary class of them.

A Mahlo cardinal cardinal $\kappa$ is inaccessible, and in $V_\kappa$ there is a stationary class of $\Sigma_\omega$-reflecting cardinals, i.e., $\Sigma_n$-reflecting for every $n$. Notice that $\kappa$ is Mahlo iff $\kappa$ is regular, $V_\kappa \models ZFC$, and the set of regular $\lambda < \kappa$ such that $V_\lambda \models ZFC$ is stationary. Thus, once inaccessible cardinals and reflecting cardinals are accepted, Mahlo cardinals are the next natural step in the process of extending the reflection properties of the universe of all sets.

By allowing higher-order formulas one obtains the so-called *indescribable cardinals*, which form a hierarchy, according to the complexity and the order of the formulas reflected: $\kappa$ is $\Sigma_n^m$-*indescribable* ($\Pi_n^m$-*indescribable*) if for every $A \subseteq V_\kappa$ and every $\Sigma_n^m$-sentence ($\Pi_n^m$-sentence) $\varphi$, if $\langle V_\kappa, \in, A \rangle \models \varphi$, then there is $\lambda < \kappa$ such that $\langle V_\lambda, \in, A \cap V_\lambda \rangle \models \varphi$.

We have that $\kappa$ is $\Sigma_1^1$-indescribable iff it is inaccessible. A minimal strengthening of this property yields the $\Pi_1^1$-indescribable cardinals. $\Pi_1^1$-indescribable cardinals are also known as *weakly-compact cardinals*. Every weakly-compact cardinal $\kappa$ is Mahlo and the set of Mahlo cardinals below $\kappa$ is stationary.

Above all those cardinals are the *totally indescribable* cardinals. i.e., $\kappa$ is totally indescribable if for every $A \subseteq V_\kappa$ and every sentence, of any complexity and any order, that holds in $\langle V_\kappa, \in, A \rangle$ it already holds in some $\langle V_\lambda, \in, A \cap V_\lambda \rangle$, $\lambda < \kappa$.

Totally indescribable cardinals seem to be the end in the direction of extending the reflection properties of $V$ obtained by considering ideal extensions of the sequence of ordinals. We may have a stationary class of totally indescribable cardinals, but no stronger forms of reflection seem possible.

It can be shown that if the large cardinal axioms considered so far are consistent with ZFC, then they are also consistent with ZFC plus $V = L$. This is not surprising since those axioms arise without making any assumptions on the structure of $V$ beyond ZFC, and for all we know $V$ might just be $L$.

## 4.2. Large cardinal axioms

One obtains much stronger axioms by considering another kind of ideal extension of $V$. Even though $V$ contains *all* sets, we may think of $V$ as included in a larger transitive universe $M$ having the same ordinals as $V$ so that $M$ is fatter than $V$, in the sense that for every ordinal $\alpha$, $V_\alpha$ is included in $M_\alpha$, and for some $\alpha$ – hence also for all ordinals greater than $\alpha$ – the inclusion is proper. According to the *Fairness* criterion, we would like to say that every $\Sigma_1$ sentence, possibly with parameters in $V$, that holds in $M$, already holds in $V$. But this is not possible. No transitive proper class $V$ different from $M$ can be a $\Sigma_1$-elementary substructure of $M$. The reason is that if this were the case, then $M_\alpha = V_\alpha$, for all $\alpha$, contradicting the assumption that $M$ was fatter than $V$. The problem here is twofold. On one hand we assumed $M$ contains some sets that do not belong to $V$, while having the same ordinals. On the other hand we allowed arbitrary parameters in our $\Sigma_1$ sentences. But there is a more fundamental problem: in considering ideal extensions of $V$ which contain the same ordinals, we just do not know what are the ideal sets that exist in $M$ but not in $V$. In the case of the strong axioms of infinity, when we considered ideal extensions where the ordinals extended beyond all the ordinals of $V$, we knew what the new sets could be like, namely, the constructible sets built at the ideal ordinal stages. But in the present situation, where the ordinals of $V$ and $M$ are the same and $V$ is contained in $M$, we just do not have any clue as to what the ideal sets in $M$ might be. In other words, for all we know $V$, and therefore $M$, might just be $L$.

One possible way out of this difficulty is to take $M$ to be a subclass of $V$, so that there are really no new sets, but view $V$ as properly contained in $M$. This is possible if we think of $V$ as *embedded* into $M$. By transitively collapsing $M$ we may just assume that $M$ is transitive. So, suppose that $M$ is a transitive class and there exists an embedding $j : V \to M$ which is not the identity and is $\Sigma_1$-elementary, i.e., for every $\Sigma_1$ sentence $\varphi(x_1, ..., x_n)$, and every $a_1, ..., a_n$,

$$\varphi(a_1, ..., a_n) \quad \text{iff} \quad M \models \varphi(j(a_1), ..., j(a_n)).$$

Then there is a least cardinal such that $j(\kappa) \neq \kappa$, called the critical point of $j$. $\kappa$ is the first ordinal where $j''V$ and $M$ start to differ. Indeed, we have that $j \restriction V_\kappa$ is the identity. Such a cardinal is measurable, i.e., there exists a two-valued $\kappa$-complete measure $\mathcal{U}$ on $\kappa$, namely $\mathcal{U} = \{X \subseteq \kappa : \kappa \in j(X)\}$. In fact, the existence of a measurable cardinal is equivalent to the existence of a $\Sigma_1$-elementary embedding, different from the identity, of $V$ into a transitive class $M$. The class $M$ is the transitive collapse of the ultrapower $V^\kappa/\mathcal{U}$, and the embedding is given by $j(x) = \pi([c_x]_\mathcal{U})$, where $c_x : \kappa \to \{x\}$ is the constant function $x$ and $\pi$ is the Mostowski transitive collapsing function.

If $\kappa$ is a measurable cardinal, then it is the $\kappa$-th inaccessible cardinal. However, it need not even be $\Sigma_2$-reflecting.

As it turns out, if $j : V \to M$ is $\Sigma_1$-elementary, then it is fully elementary, i.e., for every formula $\varphi(x_1, ..., x_n)$ and every $a_1, ..., a_n$,

$$\varphi(a_1, ..., a_n) \quad \text{iff} \quad M \models \varphi(j(a_1), ..., j(a_n)).$$

Although the sentence *There exists an embedding from V into M* is not first-order expressible, we can assert the existence of an elementary embedding from $V$ into some class $M$ just by asserting the existence of a measurable cardinal $\kappa$, which is first-order expressible.

Thus, we conclude that the axiom that asserts the existence of a measurable cardinal satisfies the criteria of *Maximality* and *Fairness* and is, therefore, a natural axiom of Set Theory (modulo its consistency with ZFC).

$M$ cannot be $V$ itself, since by a famous result of Kunen (see [17]), one cannot have a non-trivial elementary embedding $j : V \to V$. $M$ cannot be $L$ either, since as it was observed by Scott (see [17]) otherwise we would have $V = L$ and, if $\kappa$ is the least measurable and $j$ the associated embedding, by elementarity, in $L$ $j(\kappa)$ would be the least measurable cardinal, thus contradicting the fact that $\kappa < j(\kappa)$. Thus, unlike in the case of $\Sigma_n$-reflecting cardinals, the existence of a measurable cardinal implies that $V \neq L$.

The larger $M$, the closer it is to $V$, the stronger is the axiom obtained. This is not surprising, since the richer $M$ is, the richer is any substructure elementarily embedded into it. The upper bound is when $M$ is $V$ itself, which leads to inconsistency, by Kunen's result. Some possible strengthenings are the following: first, we may require that $M$ contain arbitrarily large initial segments of $V$, namely,

*There is a cardinal $\kappa$ such that for every ordinal $\alpha$ there is an elementary embedding $j : V \to M$, $M$ transitive, with critical point $\kappa$ and with $V_\alpha \subseteq M$.*

Such a cardinal $\kappa$ is known as a *strong cardinal*. If $\kappa$ is strong, then it is the $\kappa$-th measurable cardinal. Unlike the case of measurable cardinals, the existence of a strong cardinal $\kappa$ cannot be formulated in terms of the existence of a certain measure on $\kappa$. However, a formulation in the first-order language of Set Theory is still possible, although somewhat more involved (see [18]). If there exists a strong cardinal, then $V \neq L(A)$, for every set $A$. In particular, $V \neq L(V_\alpha)$, for every $\alpha$. Thus the existence of a strong cardinal could never be obtained by just ideally extending the ordinal sequence. A further strengthening is given by the following:

*There is a cardinal $\kappa$ such that for every ordinal $\alpha$ there is an elementary embedding $j : V \to M$, $M$ transitive, with critical point $\kappa$ and with $^\alpha M \subseteq M$.*

Such a $\kappa$ is called a *supercompact cardinal*. If $\kappa$ is supercompact, then it is strong. Consistency-wise, the existence of a supercompact cardinal is much stronger than the existence of a strong cardinal. Many other variations and further strengthenings are possible (see [18]), yielding ever stronger axioms. Specially important for their essential role in Descriptive Set Theory are the Woodin cardinals, which are consistency-wise between strong and supercompact cardinals.

We already remarked that the upper limit of the axioms of this sort is given by Kunen's proof of the impossibility of having a non-trivial elementary embedding $j : V \to V$. But by fusing together the two kinds of ideal extensions of $V$ considered so far, namely, the extension of the ordinal sequence and the existence of elementary embeddings of $V$ into some transitive classes, we could ask for the existence of some non-trivial elementary embedding $j : V_\alpha \to V_\alpha$, for some $\alpha$. This turns out to be an extremely strong axiom, although so far no inconsistency has been derived from it. But this axiom does satisfy the two criteria of *Maximality* and *Fairness*, and so, modulo its consistency, is a natural axiom of Set Theory.

As with the axioms of strong infinity, in the case of axioms of large cardinals, once we are led to the acceptance of the existence of a certain large cardinal, by applying the principle of *Maximality* we are naturally led to the acceptance of a (stationary) proper class of them.

Let us stop here our discussion of the axioms of large cardinals, since the above examples are sufficient for our present purposes. We just wanted to illustrate the fact that the usual large cardinal axioms are nothing else but the natural axioms – *natural* meaning that they satisfy the criteria of *Maximality* and *Fairness* – one obtains by asserting the existence of those sets that would exist in ideal extensions of $V$ obtained by either expanding the ordinal sequence or by viewing $V$ as embedded in yet a larger universe having the same ordinals, but which is, in fact, a subclass of $V$. It has been repeatedly argued that the remarkable fact that large cardinal axioms, in spite of the initially different motivations for their introduction, have been shown to fall into a linearly ordered hierarchy, lends them naturalness and contributes to their justification as additional axioms of Set Theory. But this is a misleading perspective. There is nothing remarkable about the fact that the large cardinal axioms fall into a linear hierarchy, for this is an immediate consequence of their being equivalent to ever stronger reflection principles from ideal expansions of the universe into $V$. What are remarkable, in any case, are the results that characterize them as reflection principles, thus revealing their true nature.

Another possible solution to the difficulties of finding fair axioms arising from ideal extensions of $V$ which contain the same ordinals is provided by the method of *forcing*. Forcing is actually the only general method we know of which, starting with a model of ZFC, allows to build a larger new model of ZFC.

## 5. Suslin's Hypothesis and Forcing Axioms

> *Forcing is a method to make true statements about something of which we know nothing.*
>
>                                                                    K. Gödel ([35])

Arguably, the second most important problem for the development of Set Theory (the first being, of course, Cantor's continuum problem) has been Suslin's Hypothesis: *Every complete dense and without endpoints linear ordering with the countable chain condition is order-isomorphic to* $\mathbb{R}$. The proof of its failure in $L$ by Jensen led to his discovery of the $\diamondsuit$ principle and all the subsequent combinatorial principles in $L$, the development of fine structure theory, etc. On the other hand, the proof of its consistency by Solovay and Tennenbaum [30] gave birth to the theory of iterated forcing with all its developments and applications. The special relevance of Suslin's Hypothesis to our discussion lies in the fact that, as we shall see, it is in the proof of its consistency that we find the origin of the class of set-theoretic principles that we want to discuss.

The proof of the consistency of Suslin's Hypothesis using iterated forcing led to the isolation by D. Martin [24] of a set-theoretic principle which has been known as *Martin's Axiom* (MA). In spite of its name, at first glance the principle can be hardly recognized as an *axiom*. It states the following:

*For every partially-ordered set* $\mathbb{P}$ *with the countable chain condition, and for every family* $\mathcal{D}$ *of cardinality less than the cardinality of the continuum of dense open (in the order topology) subsets of* $\mathbb{P}$, *there is a filter* $\mathcal{F} \subseteq \mathbb{P}$ *that intersects all sets in* $\mathcal{D}$.

This axiom can also be seen as a generalization of the Baire Category Theorem, for it is equivalent to the following:

*In every compact Hausdorff ccc space, the intersection of fewer than the cardinality of the continuum dense open sets is dense.*

Since its formulation in 1970, MA has been widely used not only within Set Theory, but it has also been successfully applied to the solution of many problems in Combinatorics, General Topology, Measure Theory, Real Analysis, etc. (see [10]). However, in spite of its success as a technical tool, the prevalent opinion has been that it is by no means an axiom, in the same sense that the other ZFC axioms are, namely, an intuitively obvious fact about sets (see, for instance, [19]).

In the late seventies, and as an outgrowth of his study of Jensen's forcing which was used to prove the consistency of Suslin's Hypothesis with the generalized Continuum Hypothesis, Shelah introduced the notion of *Proper Forcing* (see [28]). *Properness* is a property of partially-ordered sets weaker than the countable chain condition (ccc). It is a rather natural notion that arises when one wants to perform forcing iterations with partial orderings that are not ccc without collapsing $\omega_1$.

Several weaker notions than the ccc had already been considered in the literature before Shelah's notion of properness, and the corresponding stronger forms of MA had been formulated and applied. Especially successful was Baumgartner's Axiom A, a property of partial orderings weaker than the ccc which encompassed many of the partial orderings used in forcing constructions involving the continuum. Since properness is an even weaker condition than the Axiom A property, Baumgartner naturally formulated the *Proper Forcing Axiom* (PFA), that is, MA for the class of proper posets with the necessary restriction that the family $\mathcal{D}$ of dense open subsets of the partial ordering $\mathbb{P}$ be of cardinality at most $\aleph_1$. Without this restriction the axiom would just be inconsistent with ZFC. Baumgartner also showed that PFA is consistent with ZFC, assuming the consistency of ZFC with the existence of a supercompact cardinal.

An even weaker notion than properness was introduced by Shelah in [28], namely, *semi-properness*, which is essentially the weakest property that a partial ordering must have in order to iterate it without collapsing $\omega_1$. The corresponding axiom, the *Semi-Proper Forcing Axiom* (SPFA), was subsequently formulated by Shelah and proved to be consistent modulo a supercompact cardinal. In a rather surprising result, however, Shelah [29] showed that SPFA was actually equivalent to the maximal possible extension of MA, introduced by Foreman, Magidor and Shelah in [23] and known as *Martin's Maximum* (MM). This is MA for the class of partial orderings that do not collapse stationary subsets of $\omega_1$ (and for $\mathcal{D}$ of cardinality at most $\aleph_1$, a necessary assumption as it was pointed out before). Many consequences of MM are proved in [23], the most remarkable for our purposes being that the size of the continuum is $\aleph_2$.

Thus, MM, the strongest consistent (modulo the existence of a supercompact cardinal) generalization of MA settles the continuum problem, and in a way that was already predicted by Gödel, namely that its size is $\aleph_2$. This result was later improved by Todorčević and Veličković by showing that PFA (actually MA for a class much smaller than the Axiom A partial orderings, a principle consistent modulo the existence of a weakly-compact cardinal, suffices) implies already that the continuum has size $\aleph_2$ (see [7]). The question therefore arises as to what extent these are natural axioms of Set Theory.

On the one hand, they are generalizations of ZFC-provable statements, for they generalize $MA_{\aleph_1}$ which is itself a generalization of the Baire Category Theorem. Further, they have been shown to be consistent modulo some large cardinal axioms. But generalizing some ZFC theorems should certainly not be taken as a sufficient condition for being considered as axioms, for the simple reason that ZFC theorems may be generalized in incompatible ways. To be counted as natural axioms we need to see that they satisfy the criteria of *Maximality* and *Fairness*.

## 5.1. Forcing axioms as principles of generic absoluteness

We have already remarked that forcing axioms were regarded, until recently, as ad hoc principles, very useful indeed as technical tools for proving the consistency of mathematical statements without having to use forcing directly, but by no means real axioms. However, some recent results show that, in fact, certain *bounded* forms of the forcing axioms are real axioms. The first indication of this is a result first proved by J. Stavi and J. Väänänen, which shows that Martin's Axiom is equivalent to the following statement:

*Every $\Sigma_1$ sentence with parameters in $H_{2^{\aleph_0}}$ that can be forced to hold by a ccc forcing notion, is true.*

Unfortunately, the result remained unpublished for many years, but it was later independently discovered and first published in [4]. The Stavi-Väänänen paper containing the result has now also been published ([31]).

This result shows that by considering ideal forcing extensions of the universe, $MA$ can be seen to satisfy the criteria of *Maximality* and *Fairness*.

As for stronger forcing axioms, S. Fuchino [11] gave the following surprising characterization of PFA in terms of potential embeddings:

*PFA is equivalent to the statement that for any two structures $\mathcal{A}$ and $\mathcal{B}$, with $\mathcal{A}$ of cardinality $\aleph_1$, if a proper forcing notion forces that there is an embedding of $\mathcal{A}$ into $\mathcal{B}$, then such an embedding exists.*

The same characterization holds for the axioms SPFA and MM, replacing *proper* by *semi-proper* or by *preserving stationary subsets of $\omega_1$*, respectively.

Given two structures $\mathcal{A}$ and $\mathcal{B}$, the sentence: *There exists an embedding of $\mathcal{A}$ into $\mathcal{B}$*, is $\Sigma_1$ in the parameters $\mathcal{A}$ and $\mathcal{B}$. Thus, PFA satisfies to some extent the criterion of *Maximality*, for it asserts the existence of certain sets, namely, embeddings between structures, that would exist in an ideal forcing extension of the universe by a proper poset. But it does not seem to satisfy the *Fairness* criterion, since the class of existential sentences that assert the existence of embeddings between structures appears to be too restrictive. Similar considerations apply to the axioms SPFA and MM.

## 5.2. Bounded Forcing Axioms

PFA can also be formulated as follows: *For every proper partial ordering $\mathbb{P}$ and every family $\mathcal{D}$ of size $\aleph_1$ of maximal antichains of $\mathbb{B} =_{df} r.o.(\mathbb{P}) \setminus \{0\}$, there is a filter $\mathcal{F} \subseteq \mathbb{B}$ that intersects every antichain in $\mathcal{D}$.*

M. Goldstern and S. Shelah [14] introduced the Bounded Proper Forcing Axiom (BPFA) which is like PFA, as formulated above, but with the additional requirement that the maximal antichains of $\mathcal{D}$ have size at most $\aleph_1$. Fuchino's argument shows that BPFA is actually equivalent to the statement that for any two structures $\mathcal{A}$ and $\mathcal{B}$ of size $\aleph_1$, if a proper forcing notion forces that there is an embedding of $\mathcal{A}$ into $\mathcal{B}$, then such an embedding exists. Notice that in this formulation we may assume that the structures $\mathcal{A}$ and $\mathcal{B}$ belong to $H_{\omega_2}$.

Unlike the case of structures of arbitrarily large size, the set of $\Sigma_1$-sentences that assert the existence of an embedding between structures of size $\aleph_1$ as parameters is not restrictive, for if any such sentence that can be forced is true, then the same applies to any other $\Sigma_1$ sentence with parameters in $H_{\omega_2}$. Thus we have the following characterization of BPFA ([5]):

*BPFA is equivalent to the statement that every $\Sigma_1$ sentence with parameters in $H_{\omega_2}$ that is forced by a proper forcing notion is true.*

More generally, given a class of forcing notions $\Gamma$, let the Bounded Forcing Axiom for the class $\Gamma$, written $BFA(\Gamma)$, be the following statement:

*Every $\Sigma_1$ sentence with parameters in $H_{\omega_2}$ that is forced by a forcing notion in $\Gamma$ is true.*

That is, for every $\mathbb{P} \in \Gamma$, if $\varphi$ is a $\Sigma_1$ sentence, possibly with parameters in $H_{\omega_2}$, that has $r.o.(\mathbb{P})$-Boolean value $\mathbf{1}$, then $\varphi$ holds.

Thus, MA for families of dense open sets of size $\aleph_1$ is just $BFA(\Gamma)$, where $\Gamma$ is the class of ccc posets. Also, we can formulate the bounded forms of SPFA and MM. Namely: The Bounded Semi-proper Forcing Axiom (BSPFA) and the Bounded Martin's Maximum (BMM) are the axioms $BFA(\Gamma)$, where $\Gamma$ is the class of semi-proper posets or the class of posets that preserve stationary subsets of $\omega_1$, respectively.

Goldstern and Shelah ([14]) showed that BPFA is consistent relative to the consistency of the existence of a $\Sigma_2$-reflecting cardinal, and that this is its exact consistency strength. The same applies to BSPFA. Further, Woodin proved the consistency of BMM ([37]) relative to the existence of large cardinals much weaker than a supercompact ($\omega + 1$-many Woodin cardinals suffices). As for consistency strength, R. Schindler has shown that BMM implies that for every set $X$ there is an inner model with a strong cardinal containing $X$. Thus, BMM is, consistency-wise, much stronger that SPFA and PFA. Schindler has also shown, modulo large cardinals, that BPFA does not imply BSPFA. Therefore, the axioms BPFA, BSPFA, and BMM form a strictly increasing chain in strength.

Of course, there are no real extensions of the universe of all sets, and therefore no real forcing extensions. But given a forcing notion $\mathbb{P}$, we can define the Boolean-valued model $V^{\mathbb{B}}$, where $\mathbb{B} = r.o.(\mathbb{P})$, and view $V$ as contained in $V^{\mathbb{B}}$ via the canonical embedding given by $x \longmapsto \check{x}$. Thus, if we want to maximize all $\Sigma_1$ sentences that hold in $V^{\mathbb{B}}$ or, equivalently, that would hold in any ideal extension of $V$ by $\mathbb{B}$, allowing both a fair class of parameters as large as possible and a class of forcing extensions as wide as possible, this is exactly what the Bounded Forcing Axioms do.

It is worth noting that it is a theorem of ZFC that all $\Sigma_1$ sentences that hold in some Boolean-valued model $V^{\mathbb{B}}$, allowing only sets in $H_{\omega_1}$ as parameters, are true. So, the Bounded Forcing Axioms are just natural generalizations of this fact to $H_{\omega_2}$. Moreover, this is the most we can hope for. We cannot have the same for $\Sigma_2$ formulas since, for instance, both CH and its negation are of this sort. Moreover, as we pointed out in the last section, $V$ cannot be a $\Sigma_1$-elementary substructure of $V^{\mathbb{B}}$ for any non-trivial $\mathbb{B}$. In fact, for many $\mathbb{B}$ we cannot even allow as parameters of the $\Sigma_1$ formulas all sets in $H_{\omega_3}$ (see [6] for a thorough discussion of the limitations of Bounded Forcing Axioms). Furthermore, if we want $\Gamma$ to be the class of *all* forcing notions, then we cannot even have $\omega_1$ as a parameter, since we can easily collapse $\omega_1$ to $\omega$, and saying that $\omega_1$ *is countable* is $\Sigma_1$ in the parameter $\omega_1$. Even $BFA(\Gamma)$ for the class of forcing notions that preserve $\omega_1$ is inconsistent with ZFC. For if $S$ is a stationary and co-stationary subset of $\omega_1$, then we can add a club $C \subseteq S$ by forcing and at the same time preserve $\omega_1$. But saying that $S$ *contains a club* is $\Sigma_1$ in the parameter $S$, and so the axiom would imply that such a club exists in the ground model, which is impossible.

So, a natural question is what is the maximal class $\Gamma$ for which $BFA(\Gamma)$ is consistent with ZFC. This class has been singled out by D. Asperó [1]: Let $\Gamma$ be the class of all posets $\mathbb{P}$ such that for every set X of cardinality $\aleph_1$ of stationary subsets of $\omega_1$ there is a condition $p \in \mathbb{P}$ such that $p$ forces that $S$ is stationary for every $S \in X$. This class coincides with the class of forcing notions that preserve stationary subsets of $\omega_1$ if and only if the ideal of the non-stationary subsets of $\omega_1$ is $\omega_1$-dense. The axiom $BFA(\Gamma)$ is maximal, i.e., if $\mathbb{P} \notin \Gamma$, then the Bounded Forcing Axiom for $\mathbb{P}$ fails. Asperó also shows that the axiom can be forced assuming the existence of a $\Sigma_2$-reflecting cardinal which is the limit of strongly compact cardinals.

We conclude that Bounded Forcing Axioms are *the* natural axioms of Set Theory arising from the application of the criteria of *Maximality* and *Fairness* to ideal forcing extensions of $V$. Bounded Forcing Axioms are axioms of generic absoluteness for $H_{\omega_2}$. Generally speaking, an axiom of generic absoluteness asserts that whatever statement can be forced is true, subject only to the requirement that it be consistent. Axioms of generic absoluteness for $H_{\omega_1}$, i.e., axioms that state that whatever statements with parameters in $H_{\omega_1}$ can be forced they are true, appear naturally in Descriptive Set Theory, and they are a consequence of large cardinals (see [6]). Thus, the Bounded Forcing Axioms constitute the next level, i.e., for $H_{\omega_2}$, of this kind of axioms. Since the continuum problem is decided in $H_{\omega_2}$, it is reasonable to expect that the Bounded Forcing Axioms will be the appropriate kind of axioms for solving the problem.

## 6. Bounded Forcing Axioms and the continuum problem

Many consequences, mostly combinatorial, of the axioms BPFA, BSPFA, and BMM are known (see [2] and [33]). But the relevance of Bounded Forcing Axioms to our present discussion is that, unlike the axioms of large cardinals, they do settle Cantor's continuum problem.

Woodin [37] showed that if there exists a measurable cardinal, then BMM implies that there is a well-ordering of the reals in length $\omega_2$ which is definable in $H_{\omega_2}$ with an $\omega_1$-sequence of stationary subsets of $\omega_1$ as a parameter, and hence the cardinality of the continuum is $\aleph_2$. D. Asperó and P. Welch [3] obtained the same result from a weaker large-cardinal hypothesis. Finally, Todorcevic [32] proved that BMM implies that there is a well-ordering of the reals in length $\omega_2$ which is definable in $H_{\omega_2}$ with a $\omega_1$-sequence of real numbers as a parameter, and so the cardinality of the continuum is $\aleph_2$.

Showing that BMM implies that the size of the continuum is $\aleph_2$ requires some method for coding reals by ordinals less than $\omega_2$. Two such methods were devised by Woodin – assuming the existence of a measurable cardinal – and Todorcevic, respectively. Very recently, Justin T. Moore [26] has discovered a new coding method which further improves on the aforementioned chain of results of Woodin,

Asperó-Welch, and Todorcevic, namely: BPFA implies that there is a well-ordering of the reals in length $\omega_2$ which is definable in $H_{\omega_2}$ with an $\omega_1$-sequence of countable ordinals as a parameter, and hence the cardinality of the continuum is $\aleph_2$.

Since, as we have already argued, Bounded Forcing Axioms are natural axioms of Set Theory, the results that show that they imply that the cardinality of the continuum is $\aleph_2$ constitute a natural solution to Cantor's continuum problem.

There still remains the question of the consistency of the Bounded Forcing Axioms with ZFC. We already observed that BPFA and BSPFA are consistent relative to the existence of a $\Sigma_2$-reflecting cardinal, a very weak large-cardinal hypothesis in the large-cardinal hierarchy. The consistency strength of BMM is not known, this being one of the most interesting open questions in the area. BMM may even imply PD, i.e., that every projective set of real numbers is determined, and so its consistency strength would be roughly at the level of infinitely-many Woodin cardinals. It is also an open question whether Asperó's maximal bounded forcing axiom is actually equivalent to BMM. Further open questions are the following: It would be interesting to know whether there is any Bounded Forcing Axiom, for a natural class of forcing notions, that implies that the cardinality of the continuum is $\aleph_2$ and whose consistency strength is just ZFC. It would also be of great interest to find, under some form of Bounded Forcing Axiom, a coding of reals by ordinals less than $\omega_2$ using a single real as parameter.

Bounded Forcing Axioms are at least as natural as the axioms of large cardinals. Both kinds of axioms satisfy the criteria of *Maximality* and *Fairness*. But Bounded Forcing Axioms are in a sense more natural than the axioms of large cardinals, for the ideal extensions on which they are based, namely, the ideal forcing extensions of the universe, are more intuitive than the ideal extensions obtained by viewing a transitive class $M$, which is already included in $V$, as an extension of $V$ via the trick of embedding $V$ into it.

All known large-cardinal axioms are compatible with Bounded Forcing Axioms. Thus it is reasonable to work with both kinds of axioms simultaneously. Woodin has isolated an axiom we may call *Woodin's Maximum* (WM), that brings together the power of large cardinals and the Bounded Forcing Axioms. WM has the astonishing property that it decides in $\Omega$-logic the whole theory of $H_{\omega_2}$ (see [38]). WM asserts the following:

1. There exists a proper class of Woodin cardinals, and

2. A strong form of BMM holds in every inner model M of ZFC that contains $H_{\omega_2}$ and thinks that there is a proper class of Woodin cardinals.

The strong form of BMM of (2) says: Every $\Sigma_1$ sentence (with parameters) in the language of the structure $\langle H_{\omega_2}, \in, NS_{\omega_1}, X \rangle$ – where $NS_{\omega_1}$ is the non-stationary ideal and $X$ is any set of reals in $L(\mathbb{R})$ – that holds in some (ideal) forcing extension of $V$ via a forcing notion that preserves stationary subsets of $\omega_1$ holds already in $V$.

Woodin [37] has shown that the consistency strength of WM is essentially that of the existence of a proper class of Woodin cardinals. Moreover, assuming the existence of a proper class of Woodin cardinals and an inaccessible limit of Woodin cardinals, he proved that WM is $\Omega$-consistent. So, if the $\Omega$-conjecture is true, then WM holds in some (ideal) forcing extension of the universe $V$. This would certainly contribute to making WM, according to our criteria, a natural axiom of Set Theory.

## REFERENCES

1.  Asperó, D. 2002. A Maximal Bounded Forcing Axiom. *J. Symbolic Logic* 67(1):130–142.
2.  Asperó, D. and J. Bagaria 2001. Bounded forcing axioms and the continuum. *Ann. Pure Appl. Logic* 109(3):179–203.
3.  Asperó, D. and P. Welch 2002. Bounded Martin's Maximum, weak Erdös cardinals, and $\psi_{AC}$. *J. Symbolic Logic* 67(3):1141–1152.
4.  Bagaria, J. 1997. A characterization of Martin's Axiom in terms of absoluteness. *J. Symbolic Logic* 62:366–372.
5.  Bagaria, J. 2000. Bounded forcing axioms as principles of generic absoluteness. *Arch. Math. Logic* 39(6):393–401.
6.  Bagaria, J. 2003. Axioms of Generic Absoluteness. *CRM Preprints* 563:1–25.
7.  Bekkali, M. 1991. *Topics in Set Theory* volume 1476 of *Lecture Notes in Mathematics*. : Springer-verlag.
8.  Cantor, G. 1878. Ein beitrag zur mannigfaltigkeitslehre. *J. f. Math.* 84:242–258.
9.  Cohen, P. J. 1971. Comments on the foundations of set theory. In *Axiomatic Set Theory*, ed. Scott, D. S., volume 13 of *Proceedings of Symposia in Pure Mathematics* 9–15. : Amer. Math. Soc.
10. Fremlin, D. 1984. *Consequences of Martin's Axiom* volume 84 of *Cambridge Tracts in Math.* : Cambridge Univ. Press.
11. Fuchino, S. 1992. On potential embeddings and versions of Martin's Axiom. *Notre Dame J. Formal Logic* 33:481–492.
12. Gödel, K. 1947. What is Cantor's Continuum Problem? *American Mathematical Monthly , USA* 54:515–525.
13. Gödel, K. 1983. What is Cantor's Continuum Problem? In *Philosophy of Mathematics. Selected Readings*, ed. Benacerraf, P. and H. Putnam, Cambridge University Press 470–485 : .
14. Goldstern, M. and S. Shelah 1995. The Bounded Proper Forcing Axiom. *J. Symbolic Logic* 60:58–73.
15. Hauser, K. 2000. Gödel's Program Revisited. Preprint.
16. Hauser, K. 2004. Is Choice Self-Evident? Preprint.
17. Jech, T. 2003. *Set Theory. The Third Millenium Edition, Revised and Expanded.* Springer Monographs in Mathematics. : Springer-Verlag.
18. Kanamori, A. 1994. *The Higher Infinite* volume 88 of *Perspectives in Mathematical Logic.* : Springer-Verlag.
19. Kunen, K. 1980. *Set Theory: An Introduction to Independence Proofs.* : North-Holland.
20. Levy, A. and R. M. Solovay 1967. Measurable cardinals and the Continuum Hypothesis. *Israel J. Math.* 5:234–248.
21. Maddy, P. 1996. Set-theoretic naturalism. *J. Symbolic Logic* 61:490–514.
22. Maddy, P. 1998. V=L and MAXIMIZE. In *Logic Colloquium 95* volume 11 of

*Lecture Notes Logic* 134–152. : Springer.

23. Magidor,, M. F. M. and S. Shelah 1988. Martin's Maximum, saturated ideals, and non-regular ultrafilters. Part I. *Annals of Mathematics* 127:1–47.

24. Martin, D. A. and R. Solovay 1970. Internal Cohen extensions. *Ann. Math. Logic* 2:143–178.

25. Martin, D. A. and J. R. Steel 1989. A proof of Projective Determinacy. *Journal of the American Math. Soc.* 2:71–125.

26. Moore, J. 2003. Set mapping reflection. *Preprint* 77:731–736.

27. Reinhardt, W. N. 1974. Remarks on reflection principles, large cardinals, and elementary embeddings. *Proc. of Symp. in Pure Math.* 13, Part II:189–205.

28. Shelah, S. 1982. *Proper Forcing* volume 940 of *Lecture Notes in Math.* : Springer-Verlag.

29. Shelah, S. 1987. Semiproper Forcing Axiom implies Martin Maximum but not $PFA^+$. *J. Symbolic Logic* 52:360–367.

30. Solovay, R. and S. Tennenbaum 1971. Iterated Cohen extensions and Souslin's problem. *Ann. Math.* 94:201–245.

31. Stavi, J. and J. Väänänen 2002. Reflection Principles for the Continuum. In *Logic and Algebra*, ed. Zhang, Y., volume 302 of *Contemporary Math.* 59–84. : AMS.

32. Todorčević, S. 2002. Generic absoluteness and the continuum. *Math. Res. Lett.* 9(4):465–471.

33. Todorčević, S. 2002. Localized reflection and fragments of PFA. 58:135–148.

34. Wang, H. 1974. *From Mathematics to Philosophy.* : Routledge and Kegan Paul.

35. Wang, H. 1996. *A logical journey: From Gödel to Philosophy.* : MIT Press. Cambridge, Mass.

36. Woodin, W. H. 1988. Supercompact cardinals, sets of reals, and weakly homogeneous trees. *Proc. Nat. Acad. Sci.* 85:6587–6591.

37. Woodin, W. H. 1999. *The Axiom of Determinacy, Forcing Axioms and the Nonstationary Ideal* volume 1 of *de Gruyter Series in Logic and Its Applications*. : Walter de Gruyter.

38. Woodin, W. H. 2001. The Continuum Hypothesis. Parts I and II. *Notices of the Amer. Math. Soc.* 48(6 and 7):567–576 and 681–690.

39. Woodin, W. H., A. R. D. Mathias, and K. Hauser. *The Axiom of Determinacy.* To appear.

# Veblen Hierarchy in the Context of Provability Algebras

Lev D. Beklemishev[*]

*Steklov Mathematical Institute, Moscow and Utrecht University*
*Lev.Beklemishev@phil.uu.nl*

**Abstract.** We study an extension of Japaridze's polymodal logic **GLP** with transfinitely many modalities and develop a provability-algebraic ordinal notation system up to the ordinal $\Gamma_0$.

In the papers [1,2] a new algebraic approach to the traditional proof-theoretic ordinal analysis was presented based on the concept of *graded provability algebra*. The graded provability algebra of a formal theory $T$ is its Lindenbaum boolean algebra equipped with additional unary operators $\langle n \rangle$ mapping a sentence $\varphi$ to the sentence $n\text{-}\mathsf{Con}(\varphi)$ expressing that $T + \varphi$ is $n$-consistent. The $n$-consistency operators, together with their dual $n$-provability operators, satisfy a particular modal logic **GLP** described by G. Japaridze (see [3]). In this framework, an ordinal notation system up to the ordinal $\epsilon_0$ naturally emerges from the closed fragment of **GLP**. This allows for a transparent proof-theoretic analysis of Peano arithmetic PA, including a characterization of its class of provably total computable functions and a consistency proof à la Gentzen. More generally, it yields a characterization of provable $\Pi_n$-sentences of PA, for any $n \geq 1$, in terms of iterated reflection principles (Schmerl's theorem) and leads to an interesting combinatorial independent principle (see [2]).

It appears to be a natural project to extend this approach to theories stronger than PA. The next stage are theories of the strength of predicative analysis whose proof-theoretic ordinal is the Feferman–Schütte ordinal $\Gamma_0$. In this paper we make the first step in this direction and present a construction of *autonomous expansions* of provability algebras that leads to an ordinal notation system up to $\Gamma_0$. We postpone the proof-theoretic analysis of concrete theories to a later paper. Here we stay within the context of modal logic and ordinal notation systems. As a side result we obtain a normal form theorem for the closed fragment of an extension of Japaridze's logic **GLP**. A similar result for **GLP** itself is due to Ignatiev [5].

## 1. GLP and its arithmetical interpretation

We first introduce and study a variant of Japaridze's polymodal logic **GLP** in a language with transfinitely many modalities.

Let $\mathcal{L}_\Lambda$ be the language of propositional polymodal logic with modalities $[x]$ labelled by ordinals $x$ from a subclass $\Lambda \subseteq \mathsf{On}$. As usual, $\langle x \rangle \varphi$ abbreviates $\neg[x]\neg\varphi$.

[*]Supported by the Russian Foundation for Basic Research.

The system $\mathbf{GLP}_\Lambda$ is an analog of Japaridze's logic $\mathbf{GLP}$ in this language:

**Axioms:**   (i) Boolean tautologies;

      (ii) $[x](\varphi \to \psi) \to ([x]\varphi \to [x]\psi)$;

     (iii) $[x]([x]\varphi \to \varphi) \to [x]\varphi$;

     (iv) $[x]\varphi \to [y][x]\varphi$, for $x \le y$.

      (v) $\langle x\rangle\varphi \to [y]\langle x\rangle\varphi$, for $x < y$.

    (vi) $[x]\varphi \to [y]\varphi$, for $x \le y$;

**Rules:** modus ponens, $\varphi \vdash [x]\varphi$.

The original Japaridze's system $\mathbf{GLP}$ is just $\mathbf{GLP}_\omega$. The logic given by Axioms (i)–(v) and the same inference rules was isolated by Ignatiev [5] and is denoted $\mathbf{GLP}_\Lambda^-$ in this paper. If $\Lambda = \mathrm{On}$ we omit the subscript $\Lambda$, so $\mathbf{GLP}$ and $\mathbf{GLP}^-$ actually mean $\mathbf{GLP}_{\mathrm{On}}$ and $\mathbf{GLP}_{\mathrm{On}}^-$. It is easy to see that in the presence of Axiom (vi), Axiom (iv) becomes redundant.

Given a sufficiently strong arithmetical theory $T$, a provability interpretation for $\mathbf{GLP}_\omega$ (see [2]) is given by reading the modal formula $[n]\varphi$ as the sentence "$\varphi$ is provable from $T$ together with all true $\Pi_n$-sentences."

For ordinals $x > \omega$ the logic $\mathbf{GLP}_x$ can be interpreted in formal theories in which the hyperarithmetical hierarchy up to level $x$ can be defined. The intuitive meaning of the formula $[y]\varphi$ is then "$\varphi$ is provable from $T$ together with all true hyperarithmetical sentences of level $y$."

We do not want to give precise definitions of the appropriate theories in this paper. However, we mention that the construction of hyperarithmetical hierarchy is explicit in various systems of *ramified analysis* (see [4,8,7]) and second order theories like $(\Pi_1^0\text{-CA})_{<x}$ and ATR.

## 2. Normal forms for words

We study the closed (or letterless) fragment of $\mathbf{GLP}$. The systems $\mathbf{GLP}_\Lambda$ are very similar to $\mathbf{GLP}_\omega$. A normal form theorem for closed formulas of $\mathbf{GLP}_\omega$ was obtained in [5]. Since in every formula only finitely many modalities can occur, essentially the same theorem holds in any $\mathbf{GLP}_\Lambda$. However, to make the paper self-contained, we present a short proof of this theorem here along the lines of [2]. Our treatment also simplifies the one in [5] for the case $\mathbf{GLP}_\omega$.

Let $S$ denote the class of all words in the alphabet On, including the empty word $\epsilon$. $S_x$ will denote the class of all words in the alphabet

$$\Lambda = \{y \in \mathrm{On} : x \le y\}.$$

We shall reserve Greek letters $\alpha, \beta, \gamma, \ldots$ for words, and Latin letters $x, y, z \ldots$ for ordinals. To each element $\alpha = x_1 x_2 \ldots x_k$ of $S$ we associate its *modal interpretation*, that is, the closed modal formula

$$\langle x_1\rangle\langle x_2\rangle \cdots \langle x_k\rangle\top, \tag{1}$$

We do not distinguish between the word $\alpha$ and formula (1). We also identify $\epsilon$ with $\top$.

Below we use $\vdash$ to denote provability in **GLP**. We write $\alpha \sim \beta$ if $\vdash \alpha \leftrightarrow \beta$. $\alpha = \beta$ means graphical identity.

For each $x$ there is an ordering $<_x$ on $S$ defined by

$$\alpha <_x \beta \iff \vdash \beta \to \langle x \rangle \alpha.$$

It is immediately seen that $<_x$ is transitive. One can show in two different ways that it is irreflexive. In view of Axiom (iii), $\alpha <_x \alpha$ iff $\vdash \alpha \to \langle x \rangle \alpha$ iff $\vdash \neg \alpha$. One way to show that this is impossible is to appeal to the provability interpretation of $\alpha$ and to see that $\alpha$ must be a true sentence. Another way would be to produce a Kripke model for **GLP** at some world of which $\alpha$ is true.

We shall later see that $<_x$ is well-founded and our task will be to determine its ordinal for logics of the form $\mathbf{GLP}_y$, for $y > x$.

We notice an obvious 'shifting' property of these orderings. Let $x \uparrow \alpha$ be the result of replacing in $\alpha$ every ordinal $y$ by $x + y$. Clearly, all axioms of **GLP** are stable under this mapping, so

$$\alpha <_0 \beta \quad \Rightarrow \quad x \uparrow \alpha <_x x \uparrow \beta. \tag{2}$$

The converse implication also holds, and in fact $x \uparrow \cdot$ is an isomorphism of $(S_0, <_0)$ onto $(S_x, <_x)$ (see Corollary 7 below). The inverse mapping will be denoted $x \downarrow \cdot$.

*Width* $w(\alpha)$ of a word $\alpha \in S$ is the number of different letters occurring in it. We shall often define functions on words by induction on their width. The length of $\alpha$ is denoted $|\alpha|$ and $\min(\alpha)$ denotes the smallest ordinal occurring in $\alpha$.

Some of the elements of $S$ are pairwise equivalent, so we first define a subclass $NF \subset S$ of *normal forms*.

- $\epsilon$ and any word of width 1 belongs to $NF$.

- Assume $w(\alpha) > 1$ and let $x = \min(\alpha)$. Then graphically $\alpha = \alpha_1 x \cdots x \alpha_k$, where all $\alpha_i$ do not contain $x$ and hence $w(\alpha_i) < w(\alpha)$ for $1 \leq i \leq k$. Then $\alpha \in NF$ iff all $\alpha_i \in NF$ and, for all $1 \leq i < k$, $\alpha_{i+1} \not<_{x+1} \alpha_i$. (Note that $\alpha_i \in S_{x+1}$.)

**Lemma 1**  *(i) If $x < y$, then $\mathbf{GLP}^- \vdash (\langle y \rangle \varphi \wedge \langle x \rangle \psi) \leftrightarrow \langle y \rangle (\varphi \wedge \langle x \rangle \psi)$;*

*(ii) If $x < y$, then $\mathbf{GLP}^- \vdash (\langle y \rangle \varphi \wedge [x] \psi) \leftrightarrow \langle y \rangle (\varphi \wedge [x] \psi)$;*

*(iii) If $\alpha \in S_{x+1}$, then $\mathbf{GLP}^- \vdash \alpha \wedge x\beta \leftrightarrow \alpha x\beta$.*

*(iv) If $\alpha_1, \alpha_2 \in S_{x+1}$ and $\alpha_1 \sim \alpha_2$, then $\alpha_1 x\beta \sim \alpha_2 x\beta$.*

**Proof.** Statements (i) and (ii) essentially follow from Axioms (v) and (iv), respectively. Statement (iii) follows by repeated application of (i). Statement (iv) follows from (iii). $\boxtimes$

**Lemma 2** *Let $\alpha = \alpha_1 x \alpha_2 x \cdots x \alpha_k$, where all $\alpha_i \in S_{x+1}$. If $\alpha_1 >_{x+1} \alpha_2$, then*

$$\alpha \sim \alpha_1 x \alpha_3 x \cdots x \alpha_k.$$

**Proof.** Let $\beta = \alpha_3 x \cdots x \alpha_k$. Since $\alpha_1, \alpha_2 \in S_{x+1}$, using Lemma 1 (iii) we obtain

$$\mathbf{GLP}^- \vdash \alpha = \alpha_1 x \alpha_2 x \beta \;\; \leftrightarrow \;\; \alpha_1 \wedge x \alpha_2 x \beta$$
$$\rightarrow \;\; \alpha_1 \wedge x \beta, \quad \text{by Axiom (iv)}$$
$$\rightarrow \;\; \alpha_1 x \beta.$$

On the other hand, if $\vdash \alpha_1 \rightarrow \langle x + 1 \rangle \alpha_2$, then

$$\vdash \alpha_1 x \beta \;\; \leftrightarrow \;\; \alpha_1 \wedge x \beta$$
$$\rightarrow \;\; \langle x + 1 \rangle \alpha_2 \wedge x \beta$$
$$\rightarrow \;\; \langle x + 1 \rangle (\alpha_2 \wedge x \beta)$$
$$\rightarrow \;\; \langle x + 1 \rangle \alpha_2 x \beta$$
$$\rightarrow \;\; x \alpha_2 x \beta, \quad \text{by Axiom (vi).}$$

Hence,

$$\vdash \alpha_1 x \beta \;\; \leftrightarrow \;\; \alpha_1 \wedge x \beta$$
$$\leftrightarrow \;\; \alpha_1 \wedge x \alpha_2 x \beta$$
$$\leftrightarrow \;\; \alpha_1 x \alpha_2 x \beta.$$

Therefore, $\alpha_1 x \beta \sim \alpha_1 x \alpha_2 x \beta$. $\boxtimes$

**Proposition 3** *Every word $\alpha \in S$ can be brought into an equivalent normal form, that is, there is an $\alpha' \in NF$ such that $\alpha' \sim \alpha$.*

**Proof.** The word $\alpha'$ can be constructed by induction on the length of $\alpha$. Write $\alpha$ in the form $\alpha_1 x \alpha_2 x \cdots x \alpha_k$ with all $\alpha_i \in S_{x+1}$ and $x = \min(\alpha)$. By the induction hypothesis we may assume the word $\alpha_2 x \cdots x \alpha_k$ to be in a normal form. Secondly, using Lemma 1 (iv) we may also assume that $\alpha_1 \in NF$. Hence, if $\alpha_1 \not{\succ}_{x+1} \alpha_2$, $\alpha$ is already in a normal form. Otherwise, apply Lemma 2 and bring the word $\alpha_1 x \alpha_3 x \cdots x \alpha_k$ to a normal form using the induction hypothesis. $\boxtimes$

**Proposition 4** *Any two normal forms $\alpha, \beta \in S_x$ are $<_x$-comparable, that is,*

$$\alpha <_x \beta \;\; \text{or} \;\; \beta <_x \alpha \;\; \text{or} \;\; \beta = \alpha. \tag{$*$}$$

**Proof.** Without loss of generality, we may assume that $x$ occurs in $\alpha$ or $\beta$, hence $x = \min(\alpha\beta)$. (If $x < \min(\alpha\beta)$ the claim only becomes weaker.) In view of (2), we may also assume that $x = 0$ (otherwise, consider $x{\downarrow}\alpha$ and $x{\downarrow}\beta$).

We reason by induction on $w(\alpha\beta)$. For unary words the claim is obvious, so we consider the case that $w(\alpha\beta) > 1$.

As before, $\alpha$ and $\beta$ can be written in the form

$$\alpha = \alpha_k 0 \alpha_{k-1} 0 \cdots 0 \alpha_1, \qquad \beta = \beta_m 0 \beta_{m-1} 0 \cdots 0 \beta_1,$$

where all $\alpha_i$ and $\beta_j$ do not contain 0. By the induction hypothesis we obtain

$$\alpha_1 <_1 \beta_1 \;\; \text{or} \;\; \beta_1 <_1 \alpha_1 \;\; \text{or} \;\; \beta_1 = \alpha_1.$$

**Claim.**

If $\alpha_1 <_1 \beta_1$, then $\alpha <_0 \beta$. Symmetrically, if $\alpha_1 >_1 \beta_1$, then $\alpha >_0 \beta$.

We only prove the first part. Let $\overline{\alpha}_i = \alpha_i 0 \cdots 0\alpha_1$. We prove by induction on $i$ that $\overline{\alpha}_i <_1 \beta_1$, for all $i \leq k$. It is obvious that $\beta_1 \leq_0 \beta$ and $<_1$ is stronger than $<_0$, so the Claim will follow.

Notice that $\vdash \alpha_{i+1}0\overline{\alpha}_i \leftrightarrow (\alpha_{i+1} \wedge 0\overline{\alpha}_i)$. On the other hand, $\beta_1 >_1 \alpha_{i+1}$ by the transitivity of $<_1$ and because $\alpha \in NF$. By the induction hypothesis we have $\beta_1 >_1 \overline{\alpha}_i$, hence

$$
\begin{aligned}
\vdash \beta_1 &\rightarrow \langle 1\rangle \alpha_{i+1} \wedge \langle 0\rangle \overline{\alpha}_i \\
&\rightarrow \langle 1\rangle (\alpha_{i+1} \wedge 0\overline{\alpha}_i) \\
&\rightarrow \langle 1\rangle \alpha_{i+1}0\overline{\alpha}_i,
\end{aligned}
$$

which proves the induction step.

Continuing the proof of Proposition 4 from the Claim we can conclude that the disjunction $(*)$ can only be false if $\alpha_1 = \beta_1$. In this case we have to compare $\alpha_2$ and $\beta_2$ using the induction hypothesis again. Assume w.l.o.g. that $\alpha_2 <_1 \beta_2$. Then we have $\alpha_2 0\alpha_1 <_1 \beta_2 0\beta_1$, because

$$
\vdash \beta_2 \wedge 0\alpha_1 \rightarrow \langle 1\rangle (\alpha_2 \wedge 0\alpha_1).
$$

Following the proof of the Claim we then obtain $\overline{\alpha}_i <_1 \beta_2 0\beta_1 \leq_0 \beta$, for all $i > 1$. It follows that in this case $\alpha <_0 \beta$. Using the symmetry, the only remaining case is that both $\alpha_1 = \beta_1$ and $\alpha_2 = \beta_2$, and the reasoning can be continued. If $\alpha \neq \beta$, at the end we come to the situation when one of the two words, say $\alpha$, is a proper end segment of the other. Then obviously $\beta >_0 \alpha$. $\boxtimes$

From the proof of Proposition 4 one can extract the following recursive comparison algorithm. Given $\alpha, \beta \in S$ consider three cases:

1. $x = \min(\alpha) < \min(\beta)$. Then write $\alpha$ in the form $\alpha = \alpha_2 x \alpha_1$ with $\alpha_1 \in S_{x+1}$ and $<_{x+1}$-compare $\alpha_1$ with $\beta$.

2. $x = \min(\beta) < \min(\alpha)$. This is symmetrical.

3. $x = \min(\beta) = \min(\alpha)$. Write $\alpha = \alpha_2 x \alpha_1$ and $\beta = \beta_2 x \beta_1$ with $\alpha_1, \beta_1 \in S_{x+1}$. $<_{x+1}$-compare $\alpha_1$ with $\beta_1$. If $\alpha_1 = \beta_1$, $<_x$-compare $\alpha_2$ with $\beta_2$.

This can be considered as a definition by recursion on $|\alpha| + |\beta|$, because in all cases $|\alpha_i| < |\alpha|$ and $|\beta_i| < |\beta|$, for $i = 1, 2$.

We also infer the following corollaries.

**Corollary 5** *The normal form of a word is graphically unique.*

**Corollary 6** *For any $\alpha, \beta \in S_x$, either $\vdash x\alpha \rightarrow x\beta$ or $\vdash x\beta \rightarrow x\alpha$ and this can be effectively decided.*

**Corollary 7** *$(S, <_0)$ is isomorphic to $(S_x, <_x)$:*

$$
\alpha <_0 \beta \iff x \uparrow \alpha <_x x \uparrow \beta.
$$

**Proof.** If $\alpha \not<_0 \beta$, then $\alpha \sim \beta$ or $\beta <_0 \alpha$. Then we have $x \uparrow \alpha \sim x \uparrow \beta$ or $x \uparrow \beta <_x x \uparrow \alpha$. In both cases $x \uparrow \alpha <_x x \uparrow \beta$ would contradict the irreflexivity of $<_x$. $\boxtimes$

It also follows from Proposition 4 that on $S_x$ the orderings $<_x$ and $<_0$ coincide.

**Corollary 8** *For all $\alpha, \beta \in S_x$, $\alpha <_x \beta$ iff $\alpha <_0 \beta$.*

**Proof.** If $\alpha <_x \beta$ clearly $\alpha <_0 \beta$. Conversely, assume $\alpha, \beta \in S_x$, $\alpha <_0 \beta$ and $\alpha \not<_x \beta$. Then by Proposition 4, $\beta <_x \alpha$ or $\alpha \sim \beta$. So, in the first case we have $\alpha <_0 \beta <_0 \alpha$. In the second case, obviously $\alpha <_0 \beta \sim \alpha$, both cases contradicting the irreflexivity of $<_0$. $\boxtimes$

**Lemma 9** *For all $\alpha, \beta \in S_x$ there is an (effectively constructible) $\gamma \in S_x$ such that $\vdash \gamma \leftrightarrow (\alpha \wedge \beta)$.*

**Proof.** We reason by induction on the width of $\alpha\beta$. Without loss of generality assume that $x = \min(\alpha\beta)$. We can write $\alpha$ and $\beta$ in the form $\alpha = \alpha_1 x \alpha'$ and $\beta = \beta_1 x \beta'$ with $\alpha_1, \beta_1 \in S_{x+1}$. We then have

$$\alpha \wedge \beta \sim \alpha_1 \wedge x\alpha' \wedge \beta_1 \wedge x\beta'.$$

From Corollary 6 we know that either $\vdash x\alpha' \to x\beta'$ or $\vdash x\beta' \to x\alpha'$. Assume $x\alpha'$ is stronger. By the induction hypothesis we can find a $\gamma_1 \in S_{x+1}$ such that $\gamma_1 \sim \alpha_1 \wedge \beta_1$. Therefore

$$
\begin{aligned}
\alpha \wedge \beta \quad &\sim \quad \alpha_1 \wedge \beta_1 \wedge x\alpha' \\
&\sim \quad \gamma_1 \wedge x\alpha' \\
&\sim \quad \gamma_1 x\alpha',
\end{aligned}
$$

which has the required form. $\boxtimes$

## 3. Normal forms for arbitrary closed formulas of GLP

Here we prove that an arbitrary closed formula of **GLP** is equivalent to a boolean combination of words and give a decision procedure for the closed fragment of **GLP**. However, for the construction of the ordinal notation system this result is not needed, so the reader only interested in ordinal notations can skip this section.

**Lemma 10** *Suppose $\alpha, \alpha_1, \ldots, \alpha_k \in S_x$. Then there is $\beta \in S_x \cup \{\bot\}$ such that*

$$\beta \sim \langle x \rangle (\alpha \wedge \bigwedge_i \neg\alpha_i).$$

**Proof.** We consider two cases.
    CASE 1. For some $i$, $\vdash \alpha \to \alpha_i$. Then $(\alpha \wedge \bigwedge_i \neg\alpha_i) \sim \bot$ and hence $\langle x \rangle (\alpha \wedge \bigwedge_i \neg\alpha_i) \sim \bot$.
    CASE 2. For all $i$, $\nvdash \alpha \to \alpha_i$. Consider any $\alpha_i$. By Lemma 1, $\alpha \wedge \alpha_i$ is equivalent to a word $\gamma \in S_x$. By Proposition 4, one of the three cases holds: $\gamma \sim \alpha$, $\vdash \gamma \to \langle x \rangle \alpha$, or $\vdash \alpha \to \langle x \rangle \gamma$.

By our assumption the first case is impossible. The third case is impossible, because $\vdash \gamma \rightarrow \alpha$, so one would get $\vdash \alpha \rightarrow \langle x \rangle \alpha$, contradicting the irreflexivity of $<_x$.

So, we conclude that the second case must hold, that is,

$$\vdash \alpha \wedge \alpha_i \rightarrow \langle x \rangle \alpha.$$

This holds for all $i$, so

$$\vdash \alpha \wedge \bigvee_i \alpha_i \rightarrow \langle x \rangle \alpha.$$

Hence, we obtain

$$
\begin{aligned}
\vdash [x](\alpha \rightarrow \bigvee_i \alpha_i) \quad &\leftrightarrow \quad [x](\alpha \rightarrow (\alpha \wedge \bigvee_i \alpha_i)) \\
&\rightarrow \quad [x](\alpha \rightarrow \langle x \rangle \alpha) \\
&\rightarrow \quad [x]\neg\alpha \\
&\rightarrow \quad [x](\alpha \rightarrow \bigvee_i \alpha_i).
\end{aligned}
$$

So, $\langle x \rangle(\alpha \wedge \bigwedge_i \neg \alpha_i) \sim \langle x \rangle \alpha$. $\boxtimes$

**Lemma 11** *Let $\varphi(\alpha_1, \ldots, \alpha_k)$ be a boolean combination of words $\alpha_1, \ldots, \alpha_k$. Then $\langle x \rangle \varphi$ is equivalent to a boolean combination of words.*

**Proof.** Every nonempty $\alpha_i$ can be represented in the form $\alpha_i = \alpha_i' y_i \alpha_i''$, where $\alpha_i' \in S_x$ and $y_i < x$. We have $\alpha_i \sim (\alpha_i' \wedge y_i \alpha_i'')$. This means that we can w.l.o.g. assume that, for every $i$, either $\alpha_i \in S_x$ or $\alpha_i$ begins with a letter $y_i < x$.

Write $\varphi$ in disjunctive normal form, that is, $\varphi \sim \bigvee_l \varphi_l$, where $\varphi_l = \bigwedge_i \pm\alpha_i$. We have $\langle x \rangle \varphi \sim \bigvee_l \langle x \rangle \varphi_l$, so it is sufficient to bring $\langle x \rangle \varphi_l$ to the required form.

Lemma 1 enables us to use the following identities, for $y < x$:

$$\langle x \rangle (\psi \wedge \langle y \rangle \theta) \sim (\langle x \rangle \psi \wedge \langle y \rangle \theta),$$
$$\langle x \rangle (\psi \wedge \neg \langle y \rangle \theta) \sim (\langle x \rangle \psi \wedge \neg \langle y \rangle \theta).$$

It follows that

$$\vdash \langle x \rangle \bigwedge_i \pm\alpha_i \leftrightarrow (\bigwedge_{i \notin I} \pm\alpha_i \wedge \langle x \rangle \bigwedge_{j \in I} \pm\alpha_j),$$

where $I$ consists of those $i$ for which $\alpha_i \in S_x$ (hence, all the other $\alpha_i$ begin with some $y_i < x$).

Now we are in a position to apply Lemma 10, which yields that $\langle x \rangle \bigwedge_{i \in I} \pm\alpha_i$ is equivalent to a word (or falsity). $\boxtimes$

**Corollary 12** *Every closed formula of **GLP** is equivalent to a boolean combination of words.*

**Corollary 13** *Let $\Lambda$ be a recursive well-ordering. There is an effective decision algorithm for the closed fragment of **GLP**$_\Lambda$.*

**Proof.** We have to decide if $\vdash \varphi(\alpha_1, \ldots, \alpha_n)$, for a given boolean combination $\varphi$ of words $\alpha_1, \ldots, \alpha_n$. Write $\varphi$ in *conjunctive* normal form. The provability of $\varphi$ is equivalent to the provability of every conjunct, so it is sufficient to test the provability of the formulas of the form $\bigvee_i \pm\alpha_i$. Since the set of words modulo **GLP** is closed under conjunction, we can simplify this to a formula of the form

$$\alpha \to \bigvee_j \alpha_j,$$

where $\alpha$ accumulates all negatively occurring words.

We claim: $\vdash \alpha \to \bigvee_j \alpha_j$ iff $\vdash \alpha \to \alpha_j$, for some $j$. This is similar to the proof of Lemma 10, where the implication from right to left is obvious.

Suppose $\nvdash \alpha \to \alpha_j$, and let $\beta_j$ be a word equivalent to $\alpha \wedge \alpha_j$. By Proposition 4 we obtain $\vdash \beta_j \to \langle 0 \rangle\alpha$. This holds for all $j$, so

$$\vdash \alpha \wedge \bigvee_j \alpha_j \to \langle 0 \rangle\alpha.$$

Hence, if $\vdash \alpha \to \bigvee_j \alpha_j$, then $\vdash \alpha \to \langle 0 \rangle\alpha$, which is impossible.

Finally, we can test if $\vdash \alpha \to \alpha_j$ by bringing the words $\alpha$ and $\alpha \wedge \alpha_j$ to a normal form and testing if they coincide. ⊠

As another corollary we obtain the following statement.

**Corollary 14** *Let $\varphi, \psi$ be any closed formulas of* **GLP**. *Then* $\vdash \langle 0 \rangle\varphi \to \langle 0 \rangle\psi$ *or* $\vdash \langle 0 \rangle\psi \to \langle 0 \rangle\varphi$.

**Proof.** By Lemma 10 both $\langle 0 \rangle\varphi$ and $\langle 0 \rangle\psi$ are equivalent to some words or falsity. The case of falsity or truth immediately validates one of the implications. We may also assume that both words begin with 0, because a formula of the form $\langle 0 \rangle\varphi$ does not imply any other (nonempty) word. Of any two words beginning with 0 one of the two implies the other, by Corollary 6. ⊠

## 4. Well-foundedness of the orderings $<_x$ on words

Here we shall give a simple but not very constructive proof that the ordering $<_0$ of the set *NF* of normal forms is well-founded. In the following sections we shall compute its order type for the set of words in the logic **GLP**$_x$, for $x \in$ On. The formulas obtained will then allow for a more constructive well-foundedness proof. This makes the present proof superfluous, except that it allows for a clear derivation of these formulas which otherwise would appear rather unmotivated.

**Theorem 1** *The ordering $<_0$ on NF is well-founded.*

**Proof.** We use the idea of a minimal sequence coming from the proof of Kruskal theorem [6].

Suppose there is an infinite decreasing chain $\gamma_1 >_0 \gamma_2 >_0 \cdots$ of words in *NF*. We can assume that this chain is minimal in the following sense. $\gamma_1$ has minimal length of all the words such that there is an infinite decreasing chain starting from $\gamma_1$. $\gamma_2$ has minimal length of all the words such that there is an infinite decreasing chain starting from $\gamma_1, \gamma_2$. Etcetera.

We can also assume that 0 occurs in some $\gamma_k$. Otherwise, if $x$ is the minimal letter occurring in some of the words $\gamma_i$, consider the decreasing sequence $\delta_i := x \downarrow \gamma_i$, for $i < \omega$.

Let $\gamma_k = \alpha_k 0 \beta_k$ with $\beta_k \in S_1$. We can also write every $\gamma_i$, for $i > k$, in the form $\gamma_i = \alpha_i 0 \beta_i$, assuming that $\beta_i \in S_1$ and the part $\alpha_i 0$ can be empty. By the recursive definition of the ordering we obviously have: $\beta_k \geq_0 \beta_{k+1} \geq_0 \dots$.

For the sequence $\beta_k \geq_0 \beta_{k+1} \geq_0 \dots$ there are two possibilities.

1. There is an infinite strictly decreasing subsequence $\beta_k >_0 \beta_{k_1} >_0 \beta_{k_2} >_0 \dots$. Then the chain $\gamma_1 >_0 \dots >_0 \gamma_{k-1} >_0 \beta_k >_0 \beta_{k_1} >_0 \beta_{k_2} >_0 \dots$ is strictly decreasing. This contradicts the minimality of the length of $\gamma_k$.

2. The sequence $\beta_i$ stabilizes at some stage $s$, that is, $\beta_s = \beta_{s+1} = \dots$. Then by the recursive definition of $<_0$ the chain $\gamma_1 >_0 \dots >_0 \gamma_{s-1} >_0 \alpha_s >_0 \alpha_{s+1} >_0 \alpha_{s+2} >_0 \dots$ is strictly decreasing. This contradicts the minimality of the length of $\gamma_s$.

Both possibilities lead to a contradiction. ⊠

As an immediate corollary we also conclude that the orderings $(S_x \cap NF, <_x)$ for any $x$ are well-founded.

## 5. The Veblen hierarchy

Recall the standard definition of the Veblen hierarchy (see [8]). Given a class $X \subseteq \mathrm{On}$ let $\mathrm{en}_X$ denote its enumerating function. Let $X'$ denote the class of fixed points of $\mathrm{en}_X$, that is, $X' = \{x \in \mathrm{On} : \mathrm{en}_X(x) = x\}$. Define by transfinite induction on $x$ the so-called critical classes:

$$\mathrm{Cr}_0 = \{\omega^{1+y} : y \in \mathrm{On}\}$$
$$\mathrm{Cr}_{x+1} = (\mathrm{Cr}_x)'$$
$$\mathrm{Cr}_x = \bigcap_{y<x} \mathrm{Cr}_y, \quad \text{if } x \text{ is a limit ordinal.}$$

Let $\varphi_x$ be the enumerating function of $\mathrm{Cr}_x$. In particular, $\varphi_0(y) = \omega^{1+y}$ and $\varphi_1$ enumerates the fixed points of $\varphi_0$, that is, $\varphi_1(y) = \epsilon_y$.

Our definition of $\mathrm{Cr}_0$ and $\varphi_0$ deviates slightly from the standard one, because we start counting with $\omega$, not with 1. However, this does not change the definitions of $\mathrm{Cr}_x$ for $x > 0$.

It is easy to verify that for all $x$ the classes $\mathrm{Cr}_x$ are closed and unbounded, and that the functions $\varphi_x$ are continuous.

The least ordinal $x$ such that $x \in \mathrm{Cr}_x$ is the Feferman-Schütte ordinal $\Gamma_0$. It can also be characterized as the limit of the sequence $\varphi_0(0), \varphi_{\varphi_0(0)}(0), \dots$, in other words, as the first ordinal closed under the operation $x \mapsto \varphi_x(0)$.

## 6. Representation of the critical classes

In this section we shall derive formulas for the order types of segments of $<_0$. The same formulas can then be used to establish the well-foundedness of $<_0$.

Let $o_x(\alpha)$ denote the order type of $\alpha$ in the ordering $(S_x, <_x)$. $o(\alpha)$ is short for $o_0(\alpha)$. Obviously, we have $o_x(\alpha) = o(x \downarrow \alpha)$. For a set $X \subseteq S_x$ we also let $o_x(X) = \{o_x(\alpha) : \alpha \in X\}$.

**Lemma 15**    *(i) $o(0^n) = n$, for any $n$.*

   *(ii) If $\alpha = \alpha_1 0 \ldots 0 \alpha_n$, where all $\alpha_i \in S_1$ and not all of them empty, then $o(\alpha) = \omega^{o_1(\alpha_n)} + \cdots + \omega^{o_1(\alpha_1)}$.*

**Proof.** We only prove (ii). Firstly, bringing $\alpha$ to the normal form does not change either $o(\alpha)$ or the value of the expression on the right hand side, so we may assume $\alpha$ to be in the unique normal form, in particular, $\alpha_1 \leq_1 \alpha_2 \leq_1 \ldots \leq_1 \alpha_n$.

Secondly, the ordering $(S \cap NF, <_0)$ is isomorphic to the lexicographic ordering of such sequences $(\alpha_n, \ldots, \alpha_1)$ of elements of $(S_1 \cap NF, <_1)$. Therefore, the mapping $f(\alpha) = \omega^{o_1(\alpha_n)} + \cdots + \omega^{o_1(\alpha_1)}$ is order-preserving:

$$\alpha <_0 \beta \Rightarrow f(\alpha) < f(\beta).$$

It is also obvious that $f$ is onto, hence an isomorphism between $(S_0 \cap NF, <_0)$ and On. (Here we use the fact that $o_1$ is an isomorphism from $(S_1 \cap NF, <_1)$ to On.) $\boxtimes$

   Define $S_x^+ = S_x \setminus \{\epsilon\}$.

**Lemma 16** *If $f$ enumerates $o(S_x^+)$ and $g$ enumerates $o(S_y^+)$, then $o(S_{x+y}^+)$ is enumerated by $f \circ g$.*

**Proof.** We have: $S_{x+y}^+ \subseteq S_x^+ \subseteq S_0^+$. Look at the logic $\mathbf{GLP}_\Lambda$ with $\Lambda = \{y : y \geq x\}$. As noted above, this system is isomorphic to $\mathbf{GLP}$ by the function mapping every modality $\langle y \rangle$ to $\langle x+y \rangle$. Now, $g(z)$ is the $z$-th element of $o(S_y^+)$, hence of $o_x(S_{x+y}^+)$. Similarly, $f(u)$ is the $u$-th element of $o_x(S_x^+)$. Hence, $f(g(z))$ is the $z$-th element of $o(S_{x+y}^+)$. $\boxtimes$

   Now we use these lemmas to prove the following theorem.

**Theorem 2** $o(S_{\omega^x}^+) = \mathrm{Cr}_x$.

**Proof.** Transfinite induction on $x$. The basis of the induction, $o(S_1^+) = \mathrm{Cr}_0 = \{\omega^{1+y} : y \in \mathrm{On}\}$, follows from Lemma 15.

   Let us now prove that $o(S_{\omega^{x+1}}^+) = \mathrm{Cr}_{x+1}$. By the induction hypothesis we have

$$o(S_{\omega^x}^+) = \mathrm{Cr}_x = \{\varphi_x(y) : y \in \mathrm{On}\}.$$

Lemma 16 yields that $o(S_{\omega^x + \omega^x}^+)$ is enumerated by $\varphi_x \circ \varphi_x$, that is, by $y \mapsto \varphi_x(\varphi_x(y))$. Similarly,

$$o(S_{\omega^x \cdot n}^+) = \{\varphi_x^{(n)}(y) : y \in \mathrm{On}\}.$$

However, the set

$$C := \bigcap_{n > 0} \{\varphi_x^{(n)}(y) : y \in \mathrm{On}\} = \mathrm{Cr}_{x+1}.$$

Indeed, every $z \in \mathrm{Cr}_{x+1}$ is a fixed point of $\varphi_x$, hence $\varphi_x^{(n)}(z) = z$, for all $n$, that is, $z \in C$. In the converse direction, if $z \in C$, consider the sequence $z_n$ such that $\varphi_x^{(n)}(z_n) = z$. We have $z_0 \geq z_1 \geq z_2 \geq \cdots$, because $\varphi_x$ is monotone. Hence, there must exist an $n$ such that $z_n = z_{n+1}$. Then

$$\varphi_x^{(n+1)}(z_n) = z = \varphi_x^{(n)}(z_n),$$

that is $\varphi_x(z) = z$, q.e.d.

Finally, if $x$ is a limit ordinal, $w^x = \sup\{w^y : y < x\}$. Accordingly, we have

$$S^+_{w^x} = \bigcap_{y<x} S^+_{w^y} = \bigcap_{y<x} \mathrm{Cr}_y = \mathrm{Cr}_x.$$

This completes the proof. ⊠

## 7. Calculating ordinals

From the previous result we can derive a formula for the function $o(\alpha)$. If $x \le z$, let $-x + z$ denote the unique ordinal $y$ such that $x + y = z$.

**Lemma 17** *Assume $\alpha \ne \epsilon$ and $x = w^{x_1} + \cdots + w^{x_k}$ in Cantor normal form, $x > 0$.*
*Then*

$$o(x \uparrow \alpha) = \varphi_{x_1}(\ldots(\varphi_{x_k}(-1 + o(\alpha)))\ldots).$$

**Proof.** By the previous lemma and Theorem 2, $\varphi_{x_1}(\ldots(\varphi_{x_k}(y))\ldots)$ is the $y$-th element of $o(S^+_x) = o(\{x \uparrow \alpha : \alpha \in S^+_0\})$. However, the enumeration of $o(S^+_0)$ starts with 0 and hence the place of $\alpha$ in the enumeration is $-1 + o(\alpha)$. ⊠

**Example 18** If $\alpha \ne \epsilon$, then

1. $o(1 \uparrow \alpha) = \varphi_0(-1 + o(\alpha)) = w^{1+(-1+o(\alpha))} = w^{o(\alpha)}$.

2. $o(2 \uparrow \alpha) = \varphi_0(\varphi_0(-1 + o(\alpha))) = w^{w^{o(\alpha)}}$.

3. $o(w \uparrow \alpha) = \varphi_1(-1 + o(\alpha))$.

4. $o(ww) = o(w \uparrow 00) = \varphi_1(-1 + 2) = \epsilon_1$.

5. $o(w + w) = \varphi_1(\varphi_1(-1 + o(0))) = \varphi_1(\varphi_1(0)) = \epsilon_{\epsilon_0}$.

A combination of Lemma 15 and Lemma 17 provides a recursive calculation procedure for the ordinal of any $\alpha \in S$ by recursion on the width. If 0 occurs in $\alpha$ use Lemma 15. Otherwise, find $x = \min(\alpha)$, find $o(x \downarrow \alpha)$ using Lemma 15 and then apply Lemma 17 to compute $o(\alpha)$. We illustrate this by two examples.

**Example 19**

$$
\begin{aligned}
o(w2w) &= \varphi_0(\varphi_0(-1 + o(w0w))) \\
&= w^{w^{o(w0w)}} \\
&= w^{w^{w^{-1+o(w)}+w^{-1+o(w)}}} \\
&= w^{w^{\epsilon_0+\epsilon_0}}
\end{aligned}
$$

**Example 20**

$$
\begin{aligned}
o((w + 1)(w + w)) &= \varphi_1(\varphi_0(-1 + o(0w))) \\
&= \varphi_1(w^{\epsilon_0+1}) \\
&= \epsilon_{w^{\epsilon_0+1}} = \epsilon_{\epsilon_0 \cdot w}
\end{aligned}
$$

Notice that a more constructive way to prove the well-foundedness of $<_0$ would be to show that the mapping $o$ defined by the above recursive procedure is order-preserving.

## 8. An ordinal notation system up to $\Gamma_0$

The provability algebraic view suggests the following notion of *autonomous expansion* of provability algebras. The structures considered will be free 0-generated algebras, in other words, Lindenbaum algebras of the closed fragments of the logics $\mathbf{GLP}_\Lambda$. The construction will be an expansion in the sense that the languages of the algebras will grow. It will also be an extension in the sense that more elements will be added to the structure at each step.

We start with the ordinal 0 and consider the free 0-generated provability algebra with the only modality $\langle 0 \rangle$. The ordering $<_0$ on the set of words provides an ordinal notation system[2] for all ordinals $< \omega$.

Next we consider the free 0-generated provability algebra with modalities labelled by ordinals up to $\omega$ (or rather, by their notations given by the previous algebra). This is, essentially, the graded provability algebra for $\mathbf{GLP}_\omega$ and it provides an ordinal notation system for $\epsilon_0$.

We further consider the algebra with modalities labelled by ordinals up to $\epsilon_0$, etc. At each step the previously constructed ordinal notations $x$ are used to define new operators $\langle x \rangle$ which allow to freely generate more elements of the algebra and, thus, more ordinal notations.

In this way we obtain an ordinal notation system up to the least ordinal closed under the following operation $F$: given an ordinal $x$ consider the free 0-generated provability algebra with modalities labelled by ordinals up to $x$ and compute the order type $F(x)$ of the ordering $(NF, <_0)$. We call the least ordinal closed under $F$ the first *modally inaccessible* ordinal.

From the previous section we obtain that the first modally inaccessible ordinal is $\Gamma_0$.

**Corollary 21** *The least ordinal $x$ such that $F(x) = x$ is $\Gamma_0$.*

**Proof.** If $x < \Gamma_0$, then so is $\omega^x$. The free 0-generated algebra with modalities labelled by ordinals $< \omega^x$ provides an ordinal notation system for all ordinals up to $\min(o(S^+_{\omega^x}))$, that is, has order type $\varphi_x(0) < \Gamma_0$.

In the other direction, it is easy to see that any ordinal $x < \Gamma_0$ cannot be closed under $F$, because $x \neq \varphi_x(0)$. $\boxtimes$

The process of autonomous expansion naturally leads to a system of ordinal notation up to $\Gamma_0$. One can describe this system as follows.

*Ordinal notations* are essentially the balanced bracket expressions. More formally, the set of notations $\Gamma$ is defined using the primitive symbols ( and ) by the following rules:

1. $\epsilon \in \Gamma$ (empty expression).

2. If $\alpha, \beta \in \Gamma$, then $(\alpha)\beta \in \Gamma$.

Intuitively, the brackets are just the usual modal operator brackets $\langle \cdot \rangle$, where one allows the modalities to be labelled by modal formulas themselves (obtained

---

[2]In this trivial case ordinal notations are just sequences of zeros.

at some previous stage). We omit the symbols $\top$ everywhere, so only the brackets remain.

Formally, we define a translation $* : \Gamma \to S$ in such a way that an element $\alpha \in \Gamma$ will denote the ordinal $o(\alpha^*)$. We set

$$\epsilon^* = \epsilon, \quad ((\alpha)\beta)^* = o(\alpha^*)\beta^*.$$

For $\alpha \in \Gamma$ we also define $o^*(\alpha) := o(\alpha^*)$.

**Example 22**   1. $()^* = o(\epsilon) = 0$, hence $o^*(()) = 1$.

2. $()()$ translates to $00$ and denotes $2$.

3. $(())^* = o(()^*) = o(0) = 1$, so $o^*((())) = o(1) = \omega$.

4. $((()())()()(()))^* = 2001$, so $((()())()()(()))$ denotes $o(2001) = o(2) = \omega^\omega$.

5. $(((())))^* = o((())^*) = o(1) = \omega$, hence $((()))$ denotes $o(\omega) = \epsilon_0$.

One can also view the elements of $\Gamma$ as ordered trees. $\epsilon$ is a single node tree and $(\alpha)\beta$ is obtained by planting the tree $\alpha$ immediately above the root and to the left of the rest of the tree $\beta$. The bracket expression corresponding to a tree can be obtained by a leftmost path search walk through the tree and writing ( when going up and ) when going down, respectively. In this way, e.g., $\epsilon_0$ is represented by a linear tree of height 3.

Some notations from $\Gamma$ denote the same ordinal, but there is a unique normal form theorem. Recursively, one defines $(\alpha)\beta$ to be in normal form, if so are $\alpha$, $\beta$ and the word $o(\alpha^*)\beta^* \in NF$ in the sense of Section 2. Let $NF(\Gamma)$ denote the set of all normal forms in $\Gamma$.

**Lemma 23** *If* $\alpha, \beta \in \Gamma$ *are in normal form and* $o^*(\alpha) = o^*(\beta)$, *then* $\alpha = \beta$.

**Proof.** Induction on the depth of nesting of brackets in $\alpha$ and $\beta$. By definition of the normal form, both $\alpha^*, \beta^* \in NF$. Besides, we have $o(\alpha^*) = o(\beta^*)$, so $\alpha^* = \beta^*$. Write $\alpha$ in the form $(\alpha_1) \cdots (\alpha_k)$. Graphical equality $\alpha^* = \beta^*$ yields that $\beta$ has the form $(\beta_1) \cdots (\beta_k)$ with $o(\alpha_i^*) = o(\beta_i^*)$, for all $i$. Hence, by the induction hypothesis $\alpha_i = \beta_i$, for all $i$, that is, $\alpha = \beta$. $\boxtimes$

For $\alpha, \beta \in \Gamma$ define

$$\alpha <_0 \beta \iff \alpha^* <_0 \beta^*.$$

From the above lemma we obtain the following corollary.

**Proposition 24** *The ordering* $(NF(\Gamma), <_0)$ *is a well-ordering of order type* $\Gamma_0$.

**Proof.** We have to prove that $o^* : NF(\Gamma) \to \Gamma_0$ is an isomorphism. Lemma 23 implies that this mapping is injective. The mapping is order-preserving by the definition of $<_0$ and since the function $o : NF \to On$ is an embedding:

$$\alpha <_0 \beta \iff \alpha^* <_0 \beta^* \iff o(\alpha^*) < o(\beta^*).$$

The mapping $o^*$ is surjective by Corollary 21. $\boxtimes$

## REFERENCES

1. L.D. Beklemishev. Provability algebras and proof-theoretic ordinals, I. Logic Group Preprint Series 208, University of Utrecht, March 2001. `http://preprints.phil.uu.nl/lgps/`.

2. L.D. Beklemishev. The Worm principle. Logic Group Preprint Series 219, University of Utrecht, March 2003. `http://preprints.phil.uu.nl/lgps/`.

3. G. Boolos. *The Logic of Provability.* Cambridge University Press, Cambridge, 1993.

4. S. Feferman. Systems of predicative analysis. *The Journal of Symbolic Logic*, 29:1–30, 1964.

5. K.N. Ignatiev. On strong provability predicates and the associated modal logics. *The Journal of Symbolic Logic*, 58:249–290, 1993.

6. C.St.J.A. Nash-Williams. On well-quasi-ordering finite trees. *Proc. Cambridge Phil. Soc.*, 59:833–835, 1963.

7. U.R. Schmerl. A proof-theoretical fine structure in systems of ramified analysis. *Archive for Mathematical Logic*, 22:167–186, 1982.

8. K. Schütte. *Proof Theory.* Springer-Verlag, Berlin, Heidelberg, New York, 1977.

# A Completeness Proof for Geometrical Logic

Thierry Coquand

*Computer Science, Chalmers University, SE-412 96 Göteborg, Sweden*
*www.cs.chalmers.se/~coquand*

**Abstract.** Given a geometrical theory, we give a site model, defined as a forcing relation, which is complete for this theory. This model is what is called also the *generic* model of a geometric theory [4]. What is interesting is that this model can be defined without references to logic and the forcing conditions are simply finite sets of atomic formulae, contrary to the model construction in [4,18]. This model is inspired by [8].

## 1. Geometrical Theory and Geometrical Logic

A *geometric* or *dynamical* theory is a set of geometric formulae. A *geometric formula* is a first-order formula, without parameters, of the form

$$\phi_0 \rightarrow (\exists \overrightarrow{v_1})\phi_1 \vee \ldots \vee (\exists \overrightarrow{v_k})\phi_k$$

where the formulae $\phi_i$ are finite conjunctions of atomic formulae. There may be free variables present in the formulae $\phi_i$ and they are, as usual, implicitly universally quantified. We don't assume any range restriction, so there may be free variables appearing in the conclusion not appearing in the hypothesis $\phi_0$. A special case is when $k = 0$ in which case the formula becomes $\phi_0 \rightarrow \perp$ and expresses the negation of $\phi_0$. Another special case is when $k = 1$ and $\overrightarrow{v_1}$ is empty, in which case the formula is of the form $\phi_0 \rightarrow \phi_1$ and can be seen as a conjunction of Horn clauses. Finally the conjunction $\phi_0$ itself may be empty.

We let **V** be the set of all variables $x, y, z, \ldots$ and **P** be the set of all parameters $a_0, a_1, \ldots$ Atomic formulae are of the form $R(u_0, \ldots, u_{n-1})$, where $u_0, \ldots, u_{n-1}$ are terms built from variables, parameters and function symbols and $R$ a predicate symbol of arity $n$. A *sentence* is a closed first-order formula (not necessarily geometric). A *fact* is an atomic sentence, i.e. a closed atomic formula.

In the following we fix a geometric theory $T$. We now describe the notion of *dynamical proof* with respect to this theory [8]. We look at the formulae of the theory $T$ as a collection of *rules*. The purpose of a dynamical proof is to establish the correctness of a fact with reference to some given set of facts $X$ and the dynamical rules belonging to $T$ starting from a given set of facts. A dynamical proof shows when a given fact $F$ is a consequence of the given set of facts $X$. Formally, a dynamical proof is a rooted tree. At the root of the tree is the set of facts $X$ we start with. Each node consists of a set of facts, representing a state of information. The sets increase monotonically along the way from the root to the leaves. The successors of a node are determined by the dynamical rules that add new information to the set of already available atomic formulas. The different immediate successors of a node correspond to case distinctions. Every leaf of a

dynamical proof contains either a contradiction or the fact under investigation
$F$. If all leaves contain a contradiction then the given set of atomic formulas is
contradictory.

Here is an example of a dynamical proof. The geometrical theory is

1. $P(x) \wedge U(x) \rightarrow Q(x) \vee \exists y. R(x, y)$

2. $P(x) \wedge Q(x) \rightarrow \perp$

3. $P(x) \wedge R(x, y) \rightarrow S(x)$

4. $P(x) \wedge T(x) \rightarrow U(x)$

5. $U(x) \wedge S(x) \rightarrow V(x) \vee Q(x)$

and the following tree is a derivation of $V(a_0)$ from $P(a_0)$ and $T(a_0)$

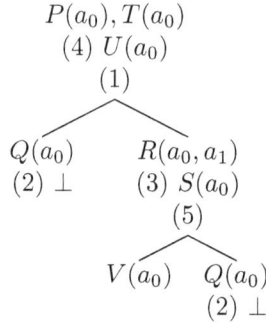

$$
\begin{array}{c}
P(a_0), T(a_0) \\
(4)\ U(a_0) \\
(1)
\end{array}
$$

$$
\begin{array}{cc}
Q(a_0) & R(a_0, a_1) \\
(2)\ \perp & (3)\ S(a_0) \\
& (5)
\end{array}
$$

$$
\begin{array}{cc}
V(a_0) & Q(a_0) \\
& (2)\ \perp
\end{array}
$$

The main goal of this note is to show that this method of proof is complete
w.r.t. intuitionistic derivation. The method of proof is interesting since it involves
a model construction. We are going to build a model of $T$, which is intuitively a
"generic" model. It is defined by a forcing relation, $X \Vdash \phi$, where $X$ is a state of
information, i.e. a finite set of facts and where $\phi$ is now an *arbitrary* first-order
sentence. The definition will be such that, if $\phi$ is a fact, $X \Vdash \phi$ means exactly that
there is a dynamical proof of $\phi$ from $X$.

## 2. A Complete Site Model

### 2.1. The forcing relation

The *conditions* will be pairs $X = (I; L)$ where $I$ is a finite set of parameters
and $L$ is a finite set of facts, with only parameters in $I$. We let $D(X)$ be the set
of parameters $I$, and $T(X)$ the set of closed terms built from the parameters in
$I$ and $C(X) = L$ the set of facts in $X$. We write $X \subseteq Y$ iff $D(X) \subseteq D(Y)$ and
$C(X) \subseteq C(Y)$.

We define inductively the relation $X \lhd U$ which expresses that a finite set of
conditions $U$ *covers* a condition $X$. The intuition behind this definition is the
following: think of $X$ as the initial facts in a dynamical proof, and let $X_0, \dots, X_{n-1}$
be the set of all branches of this tree, identifying a branch with the finite set of facts
appearing in it. Then we have that $X_0, \dots, X_{n-1}$ cover $X$. The precise definition

is that $X \lhd \{X\}$ and that if $X = (I; L)$ and we have a closed instance of an axiom of $T$

$$\phi_0 \to (\exists \overrightarrow{v_1})\phi_1 \vee \ldots \vee (\exists \overrightarrow{v_k})\phi_k$$

with all parameters in $I$ such that all conjuncts of $\phi_0$ are in $L$ and for all $i$

$$(I, \overrightarrow{m_i}; L, \phi_i(\overrightarrow{v_i} = \overrightarrow{m_i})) \lhd U_i$$

where $\overrightarrow{m_i}$ are new parameters not in $I$, then we have $X \lhd \cup U_i$. It may be that $k = 0$ in which case we have $X \lhd \emptyset$. Notice that if $X \lhd U$ and $Y \in U$ then $X \subseteq Y$.

The conditions should be thought of as finite presentations of a "potential" model of the theory $T$. A condition $(I; L)$ specifies indeed a finite set of generators $I$ and a finite set of atomic relations $L$. Given a condition $X$, a covering $X \lhd U$ can be thought of as a possible finite exploration of a model satisfying $X$. The branching reflects the fact that there may be non-canonical choices in building the model from the finite information $X$.

If $\phi$ is a fact, to say that there is a dynamical proof of $\phi$ from $X$ means exactly that there exists a covering $X \lhd X_1, \ldots, X_n$ with $\phi \in C(X_i)$ for all $i$.

We define a *map* $f:(I; L) \to (J; M)$ between two conditions to be a one-to-one map $f:I \to J$ (renaming) such that $\phi f \in M$ if $\phi \in L$. If $\phi$ is a formula with only parameters in $I$, we write $\phi f$ the formula obtained by replacing in $\phi$ the parameter $a$ by the parameter $f(a)$.

If $\phi$ is a first-order sentence with only parameters in $D(X)$, we define $X \Vdash \phi$ by induction on $\phi$.

If $\phi$ is atomic, $X \Vdash \phi$ if $X \lhd X_0, \ldots, X_{n-1}$ and $\phi \in C(X_i)$ for all $i < n$

If $\phi$ is $\phi_1 \to \phi_2$ we have $X \Vdash \phi$ if for any $f:X \to Y$ we have $Y \Vdash \phi_2 f$ whenever $Y \Vdash \phi_1 f$

If $\phi$ is $\phi_1 \wedge \phi_2$ we have $X \Vdash \phi$ if $X \Vdash \phi_1$ and $X \Vdash \phi_2$

If $\phi$ is $\phi_1 \vee \phi_2$ we have $X \Vdash \phi$ if $X \lhd U$ and for all $Y \in U$ we have $Y \Vdash \phi_1$ or $Y \Vdash \phi_2$

If $\phi$ is $(\forall x)\psi$ we have $X \Vdash \phi$ if for any $f:X \to Y$ and $a \in T(Y)$ we have $Y \Vdash \psi f(x = a)$

If $\phi$ is $(\exists x)\psi$ we have $X \Vdash \phi$ if we have $X \lhd X_0, \ldots, X_{n-1}$ and $a_i \in T(X_i)$ such that $X_i \Vdash \psi(x = a_i)$ for all $i < n$

If $\phi$ is $\bot$ we have $X \Vdash \phi$ if $X \lhd \emptyset$

The clause for $X \Vdash (\exists x)\phi$ reflects the fact that we may have to reason by cases to build a witness for an existential statement.

## 2.2. Correctness

The main result is the following.

**Theorem 2.1.** *If $\vdash_T \phi$ then for any $\rho:\mathbf{P} \to T(X)$ we have $X \Vdash \phi\rho$. More generally, if $\phi_1, \ldots, \phi_n \vdash_T \phi$ and we have $X \Vdash \phi_1\rho, \ldots, X \Vdash \phi_n\rho$ then $X \Vdash \phi\rho$.*

**Lemma 2.2.** *If $X \lhd U$ and $f:X \to Y$ then there exists $V$ such that $Y \lhd V$ and for all $Y' \in V$ there exists $X' \in U$ with $g:L \to M$ which extends $f$.*

*Proof.* We prove this by induction on the construction of $X \lhd U$.

If $U = \{X\}$ we can take $V = \{Y\}$.

If $X = (I; L)$ and there exists a closed instance of an axiom of $T$

$$\phi_0 \to (\exists \overrightarrow{v_1})\phi_1 \vee \ldots \vee (\exists \overrightarrow{v_k})\phi_k$$

such that all conjuncts of $\phi_0$ are in $L$ and for all $i$

$$X_i = (I, \overrightarrow{m_i}; L, \phi_i(\overrightarrow{v_i} = \overrightarrow{m_i})) \lhd U_i$$

we extend the renaming $f$ by choosing $\overrightarrow{b_i}$ not in $D(Y)$ and by taking $g_i(\overrightarrow{m_i}) = \overrightarrow{b_i}$. We then define

$$Y_i = (D(Y), \overrightarrow{b_i}; L, \phi_i(\rho, \overrightarrow{v_i} = \overrightarrow{b_i}))$$

so that $g_i : X_i \to Y_i$. By induction hypothesis, we can find $V_i$ such that $Y_i \lhd V_i$ and for all $Y' \in V_i$ there exists $X' \in U_i$ with $h : X' \to Y'$ which extends $g_i$. We then take $V = \cup V_i$. $\square$

**Lemma 2.3.** *If $X \lhd X_0, \ldots, X_{n-1}$ and $X_i \lhd V_i$ for all $i < n$ then $X \lhd \cup_{i<n} V_i$.*

*Proof.* This is direct by induction on the proof of $X \lhd X_0, \ldots, X_{n-1}$. $\square$

**Lemma 2.4.** *If $X \Vdash \phi$ and $f : X \to Y$ then $Y \Vdash \phi f$.*

*Proof.* We prove this by induction on $\phi$.

If $\phi$ is atomic we have $U$ such that $X \lhd U$ and $\phi \in C(Z)$ for all $Z \in U$. By lemma 2.2 we can find $V$ such that $Y \lhd V$ and for all $T \in V$ there is $Z \in U$ and $g : Z \to T$ which extends $f$. Since $\phi \in C(Z)$ we have $\phi g \in C(T)$ and since $g$ extends $f$ and $\rho$ takes its values in $T(X)$ we have $\phi g = \phi f$. It follows that we have $\phi f \in C(T)$ for all $T \in V$ and hence $Y \Vdash \phi f$.

If $\phi$ is $\phi_1 \to \phi_2$ and $g : Y \to Z$ we have $Z \Vdash \phi_2 f g$ whenever $Z \Vdash \phi_1 f g$ and hence $Y \Vdash \phi f$. (Notice that we don't use any induction hypothesis in this case.)

If $\phi$ is $\phi_1 \wedge \phi_2$ we have, by induction hypothesis, $Y \Vdash \phi_1 f$ and $Y \Vdash \phi_2 f$ and so $Y \Vdash \phi f$.

If $\phi$ is $\phi_1 \vee \phi_2$ we have $U$ such that $X \lhd U$ and for all $Z \in U$ we have $Z \Vdash \phi_1$ or $Z \Vdash \phi_2$. By lemma 2.2 we can find $V$ such that $Y \lhd V$ and for all $T \in V$ there is $Z \in U$ and $g : Z \to T$ which extends $f$. By induction hypothesis, we have then $T \Vdash \phi_1 g$ or $T \Vdash \phi_2 g$ since $g$ extends $f$ and $\rho$ takes its values in $T(X)$, and hence $\phi_i g = \phi_i f$. Thus for all $T \in V$ we have $T \Vdash \phi_1 f$ or $T \Vdash \phi_2 f$ and hence $Y \Vdash \phi f$.

If $\phi$ is $(\forall x)\psi$ we have for any $g : Y \to Z$ and $a \in T(Z)$ that $Z \Vdash \psi f g(x = a)$. Hence, $Y \Vdash \phi f$. (Notice that we don't use any induction hypothesis in this case.)

If $\phi$ is $(\exists x)\psi$ we have $U$ such that $X \lhd U$ and for all $Z \in U$ we have $Z \Vdash \psi(x = a)$ for some $a \in T(Z)$. By lemma 2.2 we can find $V$ such that $Y \lhd V$ and for all $T \in V$ there is $Z \in U$ and $g : Z \to T$ which extends $f$. We have $Z \Vdash \psi(x = a)$ for some $a \in T(Z)$ and so, by induction hypothesis, $T \Vdash \psi g(x = g(a))$. Since $g$ extends $f$ and $\rho$ takes its values in $T(X)$ we have $\psi g = \psi f$. Thus we get $T \Vdash \psi f(x = g(a))$. This shows that $Y \Vdash \phi f$.

If $\phi$ is $\bot$ we have $X \lhd \emptyset$. By lemma 2.2 we have also $Y \lhd \emptyset$ and so $Y \Vdash \phi f$. $\square$

**Lemma 2.5.** *If $X \lhd X_0, \ldots, X_{n-1}$ and $X_i \Vdash \phi$ for all $i < n$ then $X \Vdash \phi$.*

*Proof.* We prove this by induction on $\phi$.

If $\phi$ is atomic we have for all $i < n$ a set $V_i$ such that $X_i \lhd V_i$ and $\phi \in C(Y)$ for all $Y \in V_i$. By lemma 2.3 we have $X \lhd \cup_{i<n} V_i$ and hence $X \Vdash \phi$.

If $\phi$ is $\phi_1 \to \phi_2$ and $f : X \to Y$ is such that $Y \Vdash \phi_1 f$ we have by lemma 2.2 $V$ such that $Y \lhd V$ and for all $Z \in V$ there exists $i < n$ and $g : X_i \to Z$ extending $f$. By lemma 2.4 we have $Z \Vdash \phi_1 f$. Since $g$ extends $f$ we have $\phi_1 f = \phi_1 g$ and so $Z \Vdash \phi_1 g$. Since $X_i \Vdash \phi$ we get that $Z \Vdash \phi_2 g$ and hence $Z \Vdash \phi_2 f$. Since this holds for all $Z \in V$ we get by induction hypothesis $Y \Vdash \phi_2 f$. Hence $X \Vdash \phi$.

If $\phi$ is $\phi_1 \wedge \phi_2$ we have by induction hypothesis $X \Vdash \phi_1$ and $X \Vdash \phi_2$ and so $X \Vdash \phi$.

If $\phi$ is $\phi_1 \vee \phi_2$ we have for all $i < n$ a set $V_i$ such that $X_i \lhd V_i$ and $Y \Vdash \phi_1$ or $Y \Vdash \phi_2$ for all $Y \in V_i$. By lemma 2.3 we have $X \lhd \cup_{i<n} V_i$ and hence $X \Vdash \phi$.

If $\phi$ is $(\forall x)\psi$ and $f : X \to Y$ and $a \in T(Y)$ we have by lemma 2.2 $V$ such that $Y \lhd V$ and for all $Z \in V$ there exists $i < n$ and $g : X_i \to Z$ extending $f$. Then $Z \Vdash \psi g(x = a)$. Since $g$ extends $f$ we have $\psi f = \psi g$ and so $Z \Vdash \psi f(x = a)$. Since this holds for all $Z \in V$ we get by induction hypothesis $Y \Vdash \psi f(x = a)$. Hence $X \Vdash \phi$.

If $\phi$ is $(\exists x)\psi$ we have for all $i < n$ a set $V_i$ such that $X_i \lhd V_i$ and for all $Y \in V_i$ we have $m \in T(Y)$ such that $Y \Vdash \psi(x = m)$. By lemma 2.3 we have $X \lhd \cup_{i<n} V_i$ and hence $X \Vdash \phi$.

If $\phi$ is $\perp$ we have $X_i \lhd \emptyset$ for all $i < n$ and by lemma 2.3 we get $X \lhd \emptyset$ and so $X \Vdash \phi$. $\qquad\square$

We can now prove the main theorem. If $\Gamma$ is a finite set of sentences and $\phi$ is a sentence we define inductively $\Gamma \vdash \phi$ by the clauses (this is a convenient formulation of the usual intuitionistic natural deduction for first-order logic)

1. $\Gamma \vdash \phi$ if $\phi \in \Gamma$

2. $\Gamma \vdash \phi_1 \to \phi_2$ if $\Gamma, \phi_1 \vdash \phi_2$

3. $\Gamma \vdash \phi_1 \wedge \phi_2$ if $\Gamma \vdash \phi_1$ and $\Gamma \vdash \phi_2$

4. $\Gamma \vdash \phi_1 \vee \phi_2$ if $\Gamma \vdash \phi_1$ or $\Gamma \vdash \phi_2$

5. $\Gamma \vdash (\forall x)\psi$ if $\Gamma \vdash \psi(x = a)$ for some fresh parameter $a$

6. $\Gamma \vdash (\exists x)\psi$ if $\Gamma \vdash \psi(x = t)$ for some term $t$

7. $\Gamma \vdash \phi_2$ if $\Gamma \vdash \phi_1 \to \phi_2$ and $\Gamma \vdash \phi_1$

8. $\Gamma \vdash \phi_i$ if $\Gamma \vdash \phi_1 \wedge \phi_2$

9. $\Gamma \vdash \phi$ if $\Gamma \vdash \phi_1 \vee \phi_2$ and $\Gamma, \phi_i \vdash \phi$

10. $\Gamma \vdash \phi$ if $\Gamma \vdash (\exists x)\psi$ and $\Gamma, \psi(x = a) \vdash \phi$ for some fresh $a$

11. $\Gamma \vdash \phi(x = t)$ if $\Gamma \vdash (\forall x)\phi$

12. $\Gamma \vdash \phi$ if $\Gamma \vdash \perp$

If $\Gamma = \{\phi_1, \ldots, \phi_n\}$ we let $X \Vdash \Gamma$ mean that $X \Vdash \phi_i$ for all $i = 1, \ldots, n$.

We now prove that if $X \Vdash \Gamma\rho$ then $X \Vdash \phi\rho$ by induction on the proof of $\Gamma \vdash \phi$.

Suppose $\phi \in \Gamma$ then $X \Vdash \Gamma\rho$ implies directly $X \Vdash \phi\rho$.

If $\phi$ is $\phi_1 \rightarrow \phi_2$ and $\Gamma, \phi_1 \vdash \phi_2$. Assume $X \Vdash \Gamma\rho$ and $f{:}X \rightarrow Y$ and $Y \Vdash \phi_1\rho f$. By lemma 2.4 we have $Y \Vdash \Gamma\rho f$. Hence $Y \Vdash (\Gamma, \phi_1)\rho f$. Hence by induction $Y \Vdash \phi_2\rho f$. This shows $X \Vdash \phi\rho$.

If $\phi$ is $\phi_1 \wedge \phi_2$ and $\Gamma \vdash \phi_1$, $\Gamma \vdash \phi_2$ and $X \Vdash \Gamma\rho$, by induction, we have $X \Vdash \phi_1\rho$ and $X \Vdash \phi_2\rho$ and hence $X \Vdash \phi\rho$.

If $\phi$ is $\phi_1 \vee \phi_2$ and $\Gamma \vdash \phi_1$ or $\Gamma \vdash \phi_2$ and $X \Vdash \Gamma\rho$, by induction, we have $X \Vdash \phi_1\rho$ or $X \Vdash \phi_2\rho$ and hence, since $X \lhd \{X\}$, we have $X \Vdash \phi\rho$.

If $\phi$ is $(\forall x)\psi$ and $\Gamma \vdash \psi(x = a)$ for some fresh $a$ and $f{:}X \rightarrow Y$ and $m \in T(Y)$ then, by lemma 2.4, we have $Y \Vdash \Gamma\rho f$. We define $\nu{:}\mathbf{P} \rightarrow T(Y)$ by taking $\nu(u) = f(\rho(u))$ if $u \neq a$ and $\nu(a) = m$. We have by induction $Y \Vdash \psi(x = a)\nu$ which is $Y \Vdash \psi\rho f(x = m)$. This shows $X \Vdash \phi\rho$.

If $\phi$ is $(\exists x)\psi$ and $\Gamma \vdash \psi(x = t)$ for some term $t$ and $X \Vdash \Gamma\rho$ we have, by induction, $X \Vdash \psi(x = t)\rho$ which is $X \Vdash \psi\rho(x = t\rho)$. Since $X \lhd \{X\}$, we get $X \Vdash \phi\rho$.

If $\Gamma \vdash \psi \rightarrow \phi$ and $\Gamma \vdash \psi$ and $X \Vdash \Gamma\rho$ then, by induction, we have $X \Vdash (\psi \rightarrow \phi)\rho$ and $X \Vdash \psi\rho$ which implies $X \Vdash \phi\rho$.

If $\Gamma \vdash \phi_1 \wedge \phi_2$ and $X \Vdash \Gamma\rho$ then, by induction, we have $X \Vdash (\phi_1 \wedge \phi_2)\rho$ and hence $X \Vdash \phi_i\rho$ for $i = 1, 2$.

If $\Gamma \vdash \phi_1 \vee \phi_2$ and $\Gamma, \phi_i \vdash \phi$ for $i = 1, 2$ and $X \Vdash \Gamma\rho$ then, by induction we have $X \Vdash (\phi_1 \vee \phi_2)\rho$. Hence we have $X \lhd X_0, \ldots, X_{n-1}$, with $X_i \Vdash \phi_1\rho$ or $X_i \Vdash \phi_2\rho$ for each $i < n$. Since $X \subseteq X_i$ we have by lemma 2.4 that $X_i \Vdash \Gamma\rho$. Also $X_i \Vdash (\Gamma, \phi_1)\rho$ or $X_i \Vdash (\Gamma, \phi_2)\rho$. By induction, this implies $X_i \Vdash \phi\rho$. By lemma 2.5, we get $X \Vdash \phi\rho$.

Suppose $\Gamma \vdash (\exists x)\psi$ and $\Gamma, \psi(x = a) \vdash \phi$ with $a$ fresh and $X \Vdash \Gamma\rho$. By induction we have $X \Vdash ((\exists x)\psi)\rho$. Hence we have $X \lhd X_0, \ldots, X_{n-1}$ and $m_i \in T(X_i)$ with $X_i \Vdash \psi\rho(x = m_i)$. Since $X \subseteq X_i$ we have by lemma 2.4 that $X_i \Vdash \Gamma\rho$. If we define $\nu_i{:}\mathbf{P} \rightarrow T(Y)$ by $\nu_i(u) = \rho(u)$ if $u \neq a$ and $\nu_i(a) = m_i$ we have $\Gamma\nu_i = \Gamma\rho$ and so $X_i \Vdash \Gamma\nu_i$ and $X_i \Vdash \psi(x = a)\nu_i$ since $\psi(x = a)\nu_i = \psi\rho(x = m_i)$. Hence by induction $X_i \Vdash \phi\nu_i$. Hence for all $i$ we have $X_i \Vdash \phi\rho$ since $\phi\rho = \phi\nu_i$. It follows that we have $X \Vdash \phi\rho$ by lemma 2.5.

If $\Gamma \vdash (\forall x)\phi$ and $X \Vdash \Gamma\rho$ then, by induction, we have $X \Vdash (\forall x)\phi\rho$. This implies $X \Vdash \phi\rho(x = t\rho)$ which is $X \Vdash \phi(x = t)\rho$.

If $\Gamma \vdash \bot$ and $X \Vdash \Gamma\rho$ then, by induction, $X \Vdash \bot \rho$ and hence $X \lhd \emptyset$. By lemma 2.5 this implies $X \Vdash \phi\rho$.

This concludes the proof of the main theorem.

## 2.3. Simplification

**Lemma 2.6.** *If we have that $Y \Vdash \phi_1$ implies $Y \Vdash \phi_2$ for all $Y \supseteq X$ then $X \Vdash \phi_1 \rightarrow \phi_2$. If $Y \Vdash \phi(x = a)$ for all $Y \supseteq X$ and $a \in T(Y)$ then $X \Vdash (\forall x)\phi$.*

*Proof.* We treat only the case of implication, since the case of universal quantification has a similar justification. Assume that $Y \Vdash \phi_1$ implies $Y \Vdash \phi_2$ for all $Y \supseteq X$ and that $f{:}X \rightarrow Y$ is such that $Y \Vdash \phi_1 f$. There exists then $Y_0 \supseteq X$ with a *bijective* map $f_0 : Y_0 \rightarrow Y$ extending $f$. By lemma 2.4, we have $Y_0 \Vdash \phi_1 f f_0^{-1}$. Since $f_0$ extends $f$ this implies $Y_0 \Vdash \phi_1$. Hence by hypothesis, we have also $Y_0 \Vdash \phi_2$. By lemma 2.4, this implies $Y \Vdash \phi_2 f_0$ and since $f_0$ extends $f$ we have also $Y \Vdash \phi_2 f$ as desired. □

This shows that in the definition of the forcing relation, for the clauses for implication and universal quantification, we can limit ourselves to renamings that are inclusions. Hence the definition of forcing can be stated without references to renaming. A similar remark is made in [2].

## 2.4. Propositional case

The definition of forcing simplifies: the conditions are now finite sets of atomic propositions

1. $X \Vdash \phi$ if $X \lhd U$ and $\phi \in C(Y)$ for all $Y \in U$

2. $X \Vdash \phi_1 \to \phi_2$ if for any $Y \supseteq X$ we have $Y \Vdash \phi_2$ whenever $Y \Vdash \phi_1$

3. $X \Vdash \phi_1 \wedge \phi_2$ if $X \Vdash \phi_1$ and $X \Vdash \phi_2$

4. $X \Vdash \phi_1 \vee \phi_2$ if $X \lhd U$ and for all $Y \in U$ we have $Y \Vdash \phi_1$ or $Y \Vdash \phi_2$

5. $X \Vdash \perp$ if $X \lhd \emptyset$

and our definition of $X \Vdash F$ becomes similar to the one of hyper-resolution [19]. The method of trees in this case can be traced back to Lewis Carroll [1].

## 2.5. Related work

Our definition of the syntactical site is similar to the one presented in [18,4]. However one main difference is that our notion of morphism in this site is simply *renaming* and hence does not refer to the theory $T$, as in these references. This is important for instance if we want to use our model to show the *consistency* of the theory $T$. In [17] there is another construction, attributed to Coste, closer to our definition, which is also given in [8,12]. There morphisms are algebra morphisms, and the objects are finitely presented structures of a suitable subtheory of the theory $T$. It is not emphasized however there that this gives a purely syntactical, and constructive, completeness proof of the notion of dynamical proof for geometrical formulae[1]. A related construction is presented in [2], attributed to Buchholz, which applies to any theory (not necessarily geometrical). In the references [5] and [7], we present a completeness proof for topological models using a *generalised* inductive definition. It is remarkable that the present completeness proof uses only ordinary inductive definitions.

Our completeness theorem can be compared to theorem 1.1 of [8], which is a cut-elimination theorem. Both results can be seen as algorithms to transform a usual proof into a dynamical proof. It would be interesting to compare these two algorithms on simple examples.

The notion of dynamical proof is quite close to the tableau method [22]. Since it is possible to write any first-order formula in a geometrical way, essentially by naming each subformula and its negation, our completeness result actually shows also the completeness of the tableau method. In [5], we present an example showing the possible interest of the notion of dynamical proof for automatic deduction. Similar ideas, with an implementation in Prolog, appeared already in [16].

---

[1]For instance, in [8] a similar construction is presented as a non-constructive model construction.

## 3. Examples

### 3.1. Infinite model

The following theory

$\neg(x < x)$
$x < y \wedge y < z \rightarrow x < z$
$(\exists y)[x < y]$

is consistent but has no finite model. In this case, finite presentations define finite posets and we can build directly a forcing model by taking finite posets as conditions. A direct extension of a poset $X$ is obtained by choosing $x \in X$ and adding a new element $y$ to $X$ with the only constraint that $y > x$. We write $X \lhd Y$ if we get $Y$ from $X$ by successive direct extensions. The forcing relation becomes

$X \Vdash \phi$ if $\phi$ holds in $X$
$X \Vdash \phi_1 \rightarrow \phi_2$ if for any map $f{:}X \rightarrow Y$ we have $Y \Vdash \phi_2 f$ whenever $Y \Vdash \phi_1 f$
$X \Vdash \phi_1 \wedge \phi_2$ if $X \Vdash \phi_1$ and $X \Vdash \phi_2$
$X \Vdash \phi_1 \vee \phi_2$ if $X \lhd Y$ and we have $Y \Vdash \phi_1$ or $Y \Vdash \phi_2$
$X \Vdash (\forall x)\phi$ if for any map $f{:}X \rightarrow Y$ and $a \in Y$ we have $Y \Vdash \phi f(x = a)$
$X \Vdash (\exists x)\phi$ if $X \lhd Y$ and we have $Y \Vdash \phi(x = a)$ for some $a \in Y$

Furthermore, though the theory has no finite models, the consistency is established by considering only finitely presented, and hence in this case finite, structures. This seems connected to similar remarks in [20].

### 3.2. Theory of fields

The theory of fields has terms built from $1, 0, +, -, \times$ and only one predicate symbols $Z(t)$, which stands for $t = 0$. We can then write $t_1 = t_2$ for $Z(t_1 - t_2)$ and can consider the terms modulo the usual equations for rings. We have the three axioms for rings

$Z(0)$
$Z(a) \wedge Z(b) \rightarrow Z(a + b)$
$Z(a) \rightarrow Z\,(ab)$

In order to get the theory of fields we add the axioms $\neg Z(1)$ and

$$Z(x) \vee (\exists y)Z(xy - 1)$$

The conditions can be thought of as finite presentations of rings. We can then simplify the site model by taking as conditions finitely presented rings and as morphisms finitely presented extensions (adding finitely many new parameters and new equations). Starting from a ring $A$ with an element $a \in A$ the basic covering corresponding to the axiom of field is then obtained by taking the two extensions $A \rightarrow A/<a>$ and $A \rightarrow A[x]/<ax - 1>$.

Another possible geometric axiom that we can add to the theory of ring is

$$(\exists y)(xy = 1) \vee (\exists y)((1 - x)y = 1)$$

which expresses that the ring is a *local* ring. In this case the basic coverings are obtained by the two extensions $A \rightarrow A[x]/<ax-1>$ and $A \rightarrow A[x]/<(1-a)x-1>$.

Here is a remark, due to Kock [15], which shows an interesting consequence of the main theorem 2.1. The following non geometrical formula is forced in this theory

$$\Vdash \neg(\wedge x_i = 0) \to \vee_i(\exists y)(x_i y = 1) \qquad (*)$$

Indeed, we have $A \Vdash \neg(\wedge x_i = 0)$ iff $1 \in <x_0, \ldots, x_{n-1}>$ in $A$.[2] It is also clear that if $1 \in <x_0, \ldots, x_{n-1}>$ in $A$ then we have $\vee_i(\exists y)(x_i y = 1)$ if $A$ is a local ring. It follows then from the main theorem that if a geometrical formula can be proved with $(*)$ then it can be proved without.

### 3.3. Consistency versus Quantifier Elimination

As seen in the two examples above, one interest of the method is to allow the construction of models, and hence to analyse the consistency of a theory. This may be interesting even if the theory admits quantifier elimination, because the consistency proof may be simpler than the proof of quantifier elimination. We believe that Herbrand had something similar in mind when he alluded to a proof of quantifier elimination for proving the consistency of the theory of real closed fields and then added that his model construction provides a simpler consistency argument [10].

We shall treat the example of the theory of algebraically closed fields. The argument is reminiscent of the one used by Skolem [21] in his analysis of the theory of the projective plane. In both cases, the crucial step is to show that the introduction of "auxiliary elements" allowed by existential axiom does not prove new facts about the old elements.

**Lemma 3.1.** *If $a, b \in A$ then $b$ is nilpotent in $A[x]/<ax - 1>$ iff $ab$ is nilpotent in $A$.*

*Proof.* If $b^m$ is 0 in $A[x]/<ax - 1>$ it is 0 in $A[1/a]$. This implies that for some $n$ we have $a^n b^m = 0$ in $A$ and hence $ab$ is nilpotent. $\square$

**Corollary 3.2.** *If $a, b \in A$ and $b$ is nilpotent in $A[x]/<ax - 1>$ and in $A/<a>$ then $b$ is nilpotent in $A$.*

**Lemma 3.3.** *If $p$ is a monic non-constant polynomial in $A[x]$ and $a \in A$ then $a$ is nilpotent in $A$ if, and only if, it is nilpotent in $A[x]/<p>$.*

*Proof.* Since $p$ is monic and non-constant, an equality $a^n = pq$ for $q \in A[x]$ implies $q = 0$ and hence $a^n = 0$ in $A$. $\square$

The geometric theory of algebraically closed fields is obtained from the theory of rings by adding to the axiom of fields the axiom schema, for $n \geq 1$

$$(\exists x) Z(x^n + a_{n-1}x^{n-1} + \ldots + a_0)$$

The forcing conditions are arbitrary finitely presented rings, and we add as a basic covering the extension $A \to A[x]/<p>$ for each monic and non-constant $p \in A[x]$.

---

[2]For this it is enough to consider the finitely presented extension $A \to A/<x_0, \ldots, x_{n-1}>$ which corresponds to adding the facts $Z(x_0), \ldots, Z(x_{n-1})$. This shows that to have $A \Vdash \neg(\wedge x_i = 0)$ implies that the ring $A/<x_0, \ldots, x_{n-1}>$ is trivial.

**Theorem 3.4.** *If $a \in A$ then $A \Vdash Z(a)$ iff $a$ is nilpotent.*

*Proof.* Direct from corollary 3.2 and lemma 3.3.                                    □

It follows from our main theorem that, for any ring $A$ and any formula intuitionistically provable from the (geometric) theory of algebraically closed fields and the positive diagram of $A$ we have $A \Vdash \phi$.

In particular, suppose that $Z(a)$ is derivable from the geometric theory of algebraically closed fields and the positive diagram of $A$. Then we have $A \Vdash Z(a)$ and hence $a$ is nilpotent. Thus, if $Z(1)$ is derivable $A$ should be a trivial ring. This shows the consistency of the theory of algebraically closed fields. Furthermore this consistency proof can be interpreted as building effectively a *non-standard* model of the theory.

In the reference [13] is sketched an argument for the consistency of this theory which proves also quantifier elimination. A complete argument is presented in [8]. In the context of our paper, the result of quantifier elimination can be interpreted as follows: the consistency of a branch in a dynamical proof in the theory of algebraically closed fields is *decidable*.

### Acknowledgement

I would like to thank Wolfgang Ahrendt for his comments on a preliminary version of this paper.

### REFERENCES

1.  F. Abeles. Lewis Carroll's method of trees: its origins in Studies in logic. *Modern Logic* 1 (1990), no. 1, 25–35.
2.  J. Avigad. Algebraic proofs of cut elimination. *J. Log. Algebr. Program.* 49 (2001), no. 1-2, 15–30.
3.  J. Avigad, J. Helzner. Transfer principles in nonstandard intuitionistic arithmetic. *Archive for Mathematical Logic*, 41:581-602, 2002.
4.  J. L. Bell. *Toposes and local set theories. An introduction.* Oxford Logic Guides, 14. Oxford University Press, New York, 1988.
5.  M. Bezem, Th. Coquand. Newman's lemma — a case study in proof automation and geometric logic. *Bull. Eur. Assoc. Theor. Comput. Sci.* EATCS No. 79 (2003).
6.  A. Blass. Topoi and computation, *Bulletin of the EATCS* **36**, October 1988, pp. 57–65.
7.  Th. Coquand, S. Sadocco, G. Sambin and J.Smith. Formal topologies on the set of first-order formulae. *J. Symbolic Logic* 65 (2000), no. 3, 1183–1192.
8.  M. Coste, H. Lombardi, and M-F. Roy. Dynamical methods in algebra: effective Nullstellensätze, *Annals of Pure and Applied Logic* **111**(3):203–256, 2001.
9.  A. Grzegorczyk. A philosophically plausible formal interpretation of intuitionistic logic. *Indag. Math.* 26 1964 596–601.
10.  J. Herbrand. *Ecrits Logiques.* Presses Universitaires de France, Paris 1968.

11. P. Johnstone. *Stone Spaces*, Cambridge Studies in Advanced Mathematics **3**, Cambridge University Press, 1982.

12. P. T. Johnstone. Rings, fields, and spectra. *J. Algebra* 49 (1977), no. 1, 238–260.

13. A. Joyal. Le théorème de Chevalley-Tarski. Cahiers de Topologie et Géometrie Differentielle, (1975).

14. S. C. Kleene. *Mathematical logic,* John Wiley & Sons, Inc., New York-London-Sydney, 1967.

15. A. Kock. Universal projective geometry via topos theory. *J. Pure Appl. Algebra* 9 (1976/77), no. 1, 1–24.

16. R. Manthey and F. Bry. SATCHMO: a theorem prover implemented in Prolog. *Proc. of 9th Conf. on Automated Deduction* , LNAI 310, 1988.

17. M. Makkai and G. Reyes. *First order categorical logic.* Model-theoretical methods in the theory of topoi and related categories. Lecture Notes in Mathematics, Vol. 611.

18. E. Palmgren. Constructive sheaf semantics. *Math. Logic Quart.* 43 (1997), no. 3, 321–327.

19. J. A. Robinson. Automatic deduction with hyper-resolution. *Internat. J. Comput. Math.* 1 1965 227–234.

20. A. Robinson. Formalism 64. in *Logic, Methodology and Philos. Sci.* pp. 228–246, 1965, North-Holland, Amsterdam

21. T. Skolem. Logisch-kombinatorische Untersuchungen uber die Erfullbarkeit oder Beweisbarkeit mathematischer Satze nebst einem Theoreme uber dichte Mengen. Skrifter utgit av Videnskappsellkapet i Kristiania, 4:4–36. Reprinted in *Selected Works in Logic,* edited by J.E. Fenstad, Universitetsforlaget, Oslo.

22. R. M. Smullyan. *First-order logic.* Corrected reprint of the 1968 original. Dover Publications, Inc., New York, 1995.

23. G. Wraith. Intuitionistic algebra: some recent developments in topos theory. in *Proceedings of the International Congress of Mathematicians* (Helsinki, 1978), pp. 331–337, Acad. Sci. Fennica, Helsinki, 1980.

# The Computably Enumerable Sets: Recent Results and Future Directions

Peter A. Cholak*

*University of Notre Dame, USA*

**Abstract.** We survey some of the recent results on the structure of the computably enumerable (c.e.) sets under inclusion. Our main interest is on collections of c.e. sets which are closed under automorphic images, such as the orbit of a c.e. set, and their (Turing) degree theoretic and dynamic properties. We take an algebraic viewpoint rather than the traditional dynamic viewpoint.

## 1. Introduction

One of the core themes in mathematical logic is the interplay between *definability* and *computability*[2]. We are going to focus on this interplay within the structure of the computably enumerable (c.e.) sets under inclusion. This structure is denoted $\mathcal{E}$. $\mathcal{E}$ is a natural structure as, for example, one might think of a c.e. set as the solution set to a diophantine equation or the theorems generated from a finite set of axioms. As auxiliary structures we will consider true arithmetic, $(\mathbb{N}, +, \times, 0, 1)$, and the c.e. Turing degrees.

Our primary interest is on collections of c.e. sets which are closed under automorphic images, such as the orbit of a c.e. set. When are these collections definable in $\mathcal{E}$? by an elementary formula? The idea that one can learn about a structure by studying the properties invariant under automorphisms goes back to 1872 and Klein's Erlangen Program.

We can also ask computability theoretic questions. How are these collections related degree theoretically? How are they related in terms of the jump? What is their complexity in the arithmetic and hyperarithmetic hierarchy as an index set? We consider these questions as a computability theoretic subtheme of our work.

Another subtheme to our work will be that of dynamic properties. A c.e. set is enumerated at some rate. How does this enumeration compare with the other enumerations? with the standard enumeration of all c.e. sets? As we will see this issue is interrelated with the above questions.

Our notation will follow Soare [22] unless otherwise noted. More or less this is a survey paper. But we make no claims that it is a complete picture of all the recent work in this area. For that one would have to find a paper or book which

---

*Most of the current work discussed is joint with Leo Harrington. My thanks to the organizers and program committees of the 12th International Congress of Logic, Methodology and Philosophy of Science. Cholak's research was partially supported NSF Grants DMS 96-34565, 99-88716, and 02-45167.*

[2]As suggested by Soare [23] we use "computability theory" rather than "recursion theory", "computable" rather than "recursive" and "computably enumerable" rather than "recursively enumerable".

combines parts of Cholak [4], Soare [24], Cholak and Harrington [7] plus this paper. Unless otherwise noted the theorems and their proofs appear in one of Cholak and Harrington [10], Cholak and Harrington [9] or a paper with Harrington too young to have a working title.

## 2. The Structures $\mathcal{E}$ and $\mathcal{E}^*$

$W_e$ is the domain of the $e$th Turing machine. Hence $\mathcal{E} = (\{W_e : e \in \omega\}, \subseteq)$. Since $0, 1, \cup,$ and $\cap$ are definable from $\subseteq$ in $\mathcal{E}$, one might consider $\mathcal{E}$ as a lattice. $\mathcal{F}$ is the filter of finite sets (within $\mathcal{E}$). $\subseteq^*$ is inclusion modulo finite difference. $\mathcal{E}^*$ is $\mathcal{E}$ modulo $\mathcal{F}$. $X^*$ is the equivalence class of $X$ in $\mathcal{E}^*$.

We will consider the two structures $\mathcal{E}$ and $\mathcal{E}^*$ as interchangeable. This can be justified. A nontrivial orbit in $\mathcal{E}$ gives rise to an nontrivial orbit in $\mathcal{E}^*$. The following theorem shows the converse is true and hence when considered as collections of c.e. sets these structures have the same orbits.

**Theorem 1** (Soare [21]). *If $\Phi \in Aut(\mathcal{E}^*)$, $\Phi(A^*) = \widehat{A}^*$ and $A$ is infinite and coinfinite then there is a $\Psi \in Aut(\mathcal{E})$ such that $\Psi(A) = \widehat{A}$.*

Recall $X$ is computable iff $X$ and $\overline{X}$ are both c.e. sets. A set $X$ is computable iff $X$ and $\overline{X}$ are c.e. iff there is c.e. set $Y$ such that $X \cup Y = \omega$ and $X \cap Y = \emptyset$. So being computable or not is definable in $\mathcal{E}$.

Being finite is definable in $\mathcal{E}$. A set $X$ is finite iff every subset of $X$ is computable. Therefore a definable collection of c.e. sets in $\mathcal{E}^*$ is a definable collection of c.e. sets in $\mathcal{E}$. If we have a definable collection of c.e. sets in $\mathcal{E}$, we can closed that collection under finite difference and still have a definable collection in $\mathcal{E}$ which is also definable in $\mathcal{E}^*$. So when considered as collections of c.e. sets closed under finite difference $\mathcal{E}$ and $\mathcal{E}^*$ have the same definable collections.

## 3. The Creative Sets

The simplest automorphism of $\mathcal{E}$ is a computable permutation of $\omega$, a 1-reduction. The creative sets are 1-complete and hence the collection of creative sets is closed under computable permutations. Is this collection also closed under automorphic images? Surprisingly, in the early 80's, Harrington showed the answer was yes; the property of being creative is definable in $\mathcal{E}$.

**Theorem 2** (Harrington, see Soare [22] Theorem XV.1.1). *A computably enumerable set $A$ is creative iff $(\exists C \supset A)(\forall B \subseteq C)(\exists R)[R$ is computable and $R \cap C$ is not computable and $R \cap A = R \cap B]$.*

**Corollary 3.** *The creative sets form an orbit.*

Based on these results one might ask "is it enough just to consider just computable permutations when dealing with collections of c.e. sets which are closed under automorphic images? In the early 80's, in unpublished work, Harrington showed the answer is no.

**Theorem 4** (Harrington, Unpublished). *The creative sets are the only orbit of $\mathcal{E}$ which remains an orbit when we restrict the allowable automorphisms to computable permutations.*

## 4. Automorphisms of $\mathcal{E}^*$

Hence we need to further explore the automorphisms of $\mathcal{E}^*$. First we should note that there are many automorphisms of $\mathcal{E}^*$.

**Theorem 5** (Lachlan, see Soare [22] Theorem XV.2.2). *There are $2^{\aleph_0}$ automorphisms of $\mathcal{E}^*$.*

However this proof uses a cohesiveness argument which does not provide "useful" automorphisms. So we must find other ways of constructing automorphisms.

Given the sucess of the last section, one might ask can we continue using permutations as automorpisms?

**Theorem 6** (Soare [21]). *Every automorphism $\Phi$ of $\mathcal{E}^*$ is induced by a permutation $p$ of $\omega$, i.e., $\Phi(W_e) =^* p(W_e)$.*

But the converse is not true. There is a permutation which takes the even numbers to $K$, the halting set. Hence this permutation takes a computable set to a non-computable set and cannot induce an automorphism of $\mathcal{E}^*$. So it is better to think of $\Phi(W_e) =^* W_{g(e)}$, for some function $g$. We say $\Phi$ is $\Delta_n^0$ iff $g$ is $\Delta_n^0$.

To make life notationally easier lets consider two copies of $\mathcal{E}$. One living in $\omega$, the standard copy, and living in a hatted copy of $\omega$, $\widehat{\omega}$. In the hatted copy, everything wears a hat.

So to build an automorphism of $\mathcal{E}^*$ it is enough to build $g$ and $h$ such that $\Phi(W_e) =^* \widehat{W}_{g(e)}$ and $\Phi^{-1}(\widehat{W}_e) = W_{h(e)}$. Lets try to construct $g$ and $h$ via a back and forth argument similar to how we might show all countable dense linear orders are isomorphic.

We start with $W_0$ and find a c.e. image, $\widehat{W}_{g(0)}$. We must ensure $W_0$ is finite iff $\widehat{W}_{g(0)}$ is finite, $W_0$ is cofinite iff $\widehat{W}_{g(0)}$ is cofinite, and $W_0$ is infinite iff $\widehat{W}_{g(0)}$ is infinite. Now given $\widehat{W}_0$ we will find a c.e. preimage, $W_{h(0)}$. But now we must find $W_{h(0)}$ such that $\widehat{W}_{g(0)} \cap \widehat{W}_0$ is infinite (finite or cofinite) iff $W_0 \cap W_{h(0)}$ is infinite (finite or cofinite) and similarly for $\widehat{W}_{g(0)} - \widehat{W}_0$ and $W_0 - W_{h(0)}$, $\widehat{W}_0 - \widehat{W}_{g(0)}$ and $W_{h(0)} - W_0$, and $\widehat{\omega} - (\widehat{W}_{g(0)} \cup \widehat{W}_0)$ and $\omega - (W_0 \cup W_{h(0)})$.

We can try and continue but things will get out of hand quickly. We need to know if a difference of c.e. (d.c.e.) sets is infinite (finite or cofinite). In general, this is a $\Pi_3^0$-complete question. As we add more sets, the number of d.c.e. sets we must consider grows exponentially. Furthermore we must somehow find whether these sets are infinite in some fashion which allows us to build all *computably enumerable* images and preimages.

The solution is to modify our back and forth argument and do it using a tree construction. The tree provides a good framework to organize all information needed. In computability theory, trees are normally $\Pi_2^0$ branching. That is, the true path is computable in $\mathbf{0}''$. In addition, we are allowed finite injury along the true path. This corresponds to asking $\Sigma_3^0$ questions.

A d.c.e. set is of the form $X - Y$ where $X$ and $Y$ are c.e. Computably enumerable sets come with an enumeration. We can use these enumerations to stagewise approximate $X - Y$. So $(X - Y)_s = X_s - Y_s$. Let $U = \cup_s(X_s - Y_s)$. The set $U$ is c.e. and $X - Y \subseteq U$. It is $\Pi_2^0$ to see if $U$ is infinite. If $U$ is infinite we say that

$X - Y$ is *well-visited*. If $X - Y$ is well-visited then infinitely many balls look like they will be in $X - Y$. Just because $X - Y$ is well-visited does not mean $X - Y$ is infinite. The set $X - Y$ is *well-resided* if infinitely many balls in $X_s - Y_s$ remain in $X_t - Y_t$ for all later stages $t$. We can assume $X - Y$ is infinite unless $X - Y$ is *not well-resided*. The set $X - Y$ is not well-resided if for almost all $x$ and $s$ there is a stage $t$ such that if $x \in (X_s - Y_s)$ then $x \notin (X_t - Y_t)$. So whether a d.c.e. set is not well-resided is a $\Sigma_3^0$ question. Hence it is possible to use a tree to determine if a d.c.e. set is well-visited and/or not well-resided and from this information, using a tree, construct an automorphism.

As a result most previous constructions of automorphisms are tree constructions. At this point, but for the following observations, we are not going to discuss these constructions. We refer the reader to the papers mentioned below. We will bring up the issue of automorphisms later but in a different tone.

The art of the state in automorphism constructions has changed over the years but they all, in some form, use the idea of well-visited and not well-resided. As a result they all are very dynamic. They depend on dynamic properties rather than order theoretic properties of $\mathcal{E}$. Sometimes these theorems are labeled as *extension theorems*, as they extend an isomorphism between substructures of $\mathcal{E}^*$ to an automorphism of $\mathcal{E}^*$. Generally an extension uses *entry states*. The issues of extension theorems and entry sets will come up again in more detail.

The first automorphism construction is due to Soare [21]. A similar construction can be found in Soare [22], Section XV.4-6. Soare's original argument was not a tree construction. In the 80s, in unpublished work, Harrington placed the construction on a tree and proved a number of results, such as Theorems 2 and 4. The first published version of the tree method or, as it is also called, the $\Delta_3^0$ method, for constructing automorphisms of $\mathcal{E}^*$ appeared in Cholak [2] and later in Cholak [3]. The current published state of the art appears in Harrington and Soare [15]. We will discuss some yet unpublished papers on automorphisms in later sections.

## 5. Some Previous Results

Before turning to our recent work we want to use some older results to highlight ours themes. The classical example is that of the maximal sets:

### 5.1. Maximal Sets

**Definition 7.** $M$ is *maximal* if for all c.e. sets $W$ either $W \subseteq^* M$ or $M \cup W =^* \omega$.

**Theorem 8** (Soare [21]). *If $M$ and $\widehat{M}$ are maximal sets then there is a $\Delta_3^0$ automorphism $\Phi$ of $\mathcal{E}$ such that $\Phi(M) = \widehat{M}$. Hence the maximal sets form an orbit.*

Note: in this case we say $M$ is *automorphic* to $\widehat{M}$ and write $M \approx_{\Delta_3^0} \widehat{M}$.

**Theorem 9** (Martin [19]). *A computably enumerable degree $\mathbf{h}$ is high iff there is a maximal set $M$ such that $M \in \mathbf{h}$.*

The proof of the above theorem captures dynamic properties of the high degrees. Hence we have a collection of results about the maximal sets which involve definability, automorphisms and orbits, degree theoretic issues and some dynamic properties.

## 5.2. Post's Program

Post's Problem and the corresponding program has a great impact on the development of computability theory.

**Post's Problem.** *Find an incomplete, noncomputable computably enumerable degree.*

Post's Problem was solved by Friedberg [13] and Mučnik [20]. These solutions introduced the finite injury method and the use of priority and changed the face of computability theory. But these solutions did not solve the problem in the way that Post intended.

**Post's Program.** *Find a "thinness" property of a computably enumerable set which guarantees incompleteness.*

Various solutions have been posed. The first was Marčenkov [18] which ensures that a computably enumerable set is low$_2$. There have been other suggested solutions. For example, one by Ambos-Spies and Nies [1] which ensures that a set is cappable.

However it is unclear what Post meant by a "thinness property". It seems (at least to this author) that only Marčenkov [18] was truly offered as a solution to Post's Program given this ambiguity and has the most potential to satisfy Post.

One way to interpret Post's Program is find a definable property in $\mathcal{E}$ which ensures incompleteness. The following theorem implies that all the previous posed solutions are not solutions in this sense. None of these solutions are definable in $\mathcal{E}$.

**Theorem 10** (Cholak [3], Harrington and Soare [15]). *Every computably enumerable set is automorphic to a high set.*

However Harrington and Soare were able to come up with a solution to Post's Program when considered this way.

**Definition 11.**

$$(\exists C)_{A \subseteq_m C}(\forall B \subseteq C)(\exists D \subseteq C)(\forall S)_{S \sqsubset C}[[B \cap (S - A) = D \cap (S - A)] \Rightarrow$$
$$(\exists T)[\overline{C} \subset T \wedge A \cap (S \cap T) = B \cap (S \cap T)]], \qquad Q(A)$$

where $S \sqsubset T$ iff $(\exists \check{S})[S \cap \check{S} = \emptyset \wedge S \cup \check{S} = T]$ and $A \subseteq_m C$ iff $C - A$ is not finite and $(\forall W)[\overline{C} \subseteq W \Rightarrow \overline{A} \subseteq W]$.

**Theorem 12** (Harrington and Soare [14]). *$Q(A)$ implies $A$ is incomplete and $Q(A)$ holds for some $A$.*

## 5.3. Automorphic to a Complete Set

Post's Program and the possible solutions prompted the question which computably enumerable sets are automorphic to a complete set (in the same orbit as a complete set). For example, all maximal and creative sets are automorphic to a complete set. The best known answer involves dynamic properties. In order to avoid some complex notation, we present a predecessor to the best known result.

**Definition 13.** A set $A$ is *promptly simple* iff there is a computable function $p$, and for all computably enumerable sets $W$, if $W$ is infinite then $(\exists x)(\exists s)[x \in W_{\text{at } s} \cap A_{p(s)}]$.

**Theorem 14** (Cholak et al. [6] (presumed by Harrington and Soare [15])). *All promptly simple sets (almost prompt) are automorphic to a complete set.*

**Question 15.** *Which computably enumerable sets are automorphic to a complete set?*

**Conjecture 16.** *The index set of computably enumerable sets automorphic to a complete set is either $\Sigma_1^1$-complete (or reasonably low in the arithmetical hierarchy, say $\Delta_{10}^0$).*

In the latter half of this paper we will present some evidence supporting this conjecture. We direct the reader to Section 10.

### 5.4. Degree Theoretic Control

As we have seen we do not have a handle on controlling the degree of $\widehat{A}$. Harrington (see Harrington and Soare [15]) showed we can avoid a lower cone. Can we avoid an upper cone?

**Question 17.** *Let $A$ and $\mathbf{d}$ be incomplete. Does there exist an $\widehat{A}$ such that $A \approx \widehat{A}$ and $\mathbf{d} \not\leq_T \widehat{A}$?*

Like the last question, this question can be shaped into an index set question.

### 5.5. Controlling the Double Jump

Whereas we do not have control of the Turing degree of $\widehat{A}$ we do have more control over its double jump. This work leads to a large number of invariant degree classes. We will skip this area in interest of time. We direct the reader to Cholak and Harrington [7] and Cholak and Harrington [8] for more information.

## 6. Automorphisms, Again

Theorems previously mentioned exhibited an automorphic vs. elementary definable difference dichotomy. For example, thirty years ago it was thought every orbit contained a complete set but we now know there is a definable property which implies incompleteness. Our goal is to make this dichotomy more explicit. For future results we need to know more about how to construct automorphisms. We will move towards an algebraic extension theorem.

**Definition 18.** $\mathcal{L}^*(A)$ is $\{W \cup A : W \text{ a c.e. set}\}$ under $\subseteq$ modulo the ideal of finite sets ($\mathcal{F}$). (The outside of a set.)

**Definition 19.** $\mathcal{E}^*(A)$ is $\{W \cap A : W \text{ a c.e. set}\}$ under $\subseteq$ modulo $\mathcal{F}$. (The inside of a set and dual of Definition 18.)

If $A \approx \widehat{A}$ then $\mathcal{E}^*(A) \cong \mathcal{E}^*(\widehat{A})$ & $\mathcal{L}^*(A) \cong \mathcal{L}^*(\widehat{A})$. But what about the other direction? It turns out that it is always the case that $\mathcal{E}^*(A) \cong \mathcal{E}^*$; use a computable one-to-one function whose range is $A$ to induce the desired isomorphism.

So the question becomes what conditions are necessary to ensure that $\mathcal{L}^*(A) \cong \mathcal{L}^*(\widehat{A})$ implies $A \approx \widehat{A}$. By Cholak [3] when $\mathcal{L}^*(A)$ is infinite, extra conditions are needed. We will discuss the case when $\mathcal{L}^*(A)$ is finite shortly. When these conditions hold we say that we have extended the isomorphism between $\mathcal{L}^*(A)$ and $\mathcal{L}^*(\widehat{A})$ to an automorphism of $\mathcal{E}^*$.

## 6.1. The Structure $\mathcal{S}_{\mathcal{R}}(A)$

The following is necessary for our algebraic extension theorem.

**Definition 20.** $\mathcal{S}(A) = \{B : \exists C (B \sqcup C = A)\}$. $\mathcal{R}(A) = \{R : R \subseteq A \text{ and } R \text{ is computable}\}$. Let $\mathcal{S}_{\mathcal{R}}(A)$ be the quotient structure $\mathcal{S}(A)$ modulo $\mathcal{R}(A)$.

$\mathcal{S}(A)$ is the splits of $A$ and $\mathcal{S}(A)$ forms a Boolean algebra. $\mathcal{R}(A)$ is the computable subsets of $A$ and is an ideal of $\mathcal{S}(A)$.

**Theorem 21** (Nies, see Cholak and Harrington [10]). $\mathcal{S}_{\mathcal{R}}(A)$ *is a $\Sigma_3^0$ atomless Boolean algebra.*

We are interested in extendible substructures, $\mathcal{B}$, of $\mathcal{S}_{\mathcal{R}}(A)$.

**Definition 22.** A subalgebra, $\mathcal{B}$, of $\mathcal{S}_{\mathcal{R}}(A)$ is an *extendible* subalgebra of $\mathcal{S}_{\mathcal{R}}(A)$ if there is a $\Delta_3^0$ set $B$, an effective listing of splits of $A$, $\{S_i\}_{i<\omega}$ and $\{\check{S}_i\}_{i<\omega}$ such that for all $i$, $S_i \sqcup \check{S}_i = A$ and $\{S_i\}_{i \in B}$ generates $\mathcal{B}$.

We are interested in a particular extendible subalgebra. Given a computably enumerable set $W$, $A$ breaks into two disjoint computably enumerable pieces, $W \searrow A$ (the integers which enter $W$ first and then $A$) and $A\backslash W$ (the integers which enter $A$ first and then maybe enter $W$). The sets of the form $W \searrow A$ are called *entry sets*. The integers which are in some state $\nu$ as they enter $A$ form a computably enumerable set $W$. In this case $W = W \searrow A$. $\nu$ is an entry state iff $W \searrow A$ is infinite. Now $W \searrow A \sqcup A\backslash W = A$ and hence the entry sets form splits of $A$.

**Definition 23.** $\mathcal{E}_A$ is the Boolean algebra generated by the entry sets.

**Lemma 24.** $\mathcal{E}_A$ *is an* extendible *subalgebra of $\mathcal{S}_{\mathcal{R}}(A)$.*

## 7. The Algebraic Extension Theorems

The statement of the following theorem is an algebraic version of an extension theorem. While the statement is algebraic, the proof of this theorem is not algebraic. This is the one place (in this paper) we have to get our hands dirty and use the automorphism method mentioned earlier. The theorem is equivalent to Soare's New Extension Theorem (see Soare [24]) and the extension theorem stated in Soare [22].

**Definition 25.** Let $\mathcal{B}$ be an extendible algebra witnessed by $B$, $\{S_i\}_{i<\omega}$ and $\{\check{S}_i\}_{i<\omega}$. Similarly for $\widehat{\mathcal{B}}$. $\mathcal{B}$ and $\widehat{\mathcal{B}}$ are *extendibly $\Delta_3^0$ isomorphic* via a $\Delta_3^0$ function $\Theta$ if the map $S_i$ goes to $\widehat{S}_{\Theta(i)}$, for $i \in B$, and $\widehat{S}_i$ goes to $S_{\Theta^{-1}(i)}$, for $i \in \widehat{B}$, induces an isomorphism between $\mathcal{B}$ and $\widehat{\mathcal{B}}$.

**Theorem 26.** *Let $\mathcal{B} \subseteq \mathcal{S}_{\mathcal{R}}(A)$ and $\widehat{\mathcal{B}} \subseteq \mathcal{S}_{\mathcal{R}}(\widehat{A})$ be two extendible Boolean algebras which are extendibly $\Delta_3^0$ isomorphic via $\Theta$ (and both $A$ and $\widehat{A}$ are not finite). Then there is a $\Phi$ such that $\Phi$ is a $\Delta_3^0$ isomorphism between $\mathcal{E}^*(A)$ and $\mathcal{E}^*(\widehat{A})$ and $\Phi$ extends $\Theta$ ($\Phi$ and $\Theta$ agree modulo $\mathcal{R}(A)$ on $\widehat{\mathcal{B}}$).*

This gives us an isomorphism $\Phi$ between $\mathcal{E}^*(A)$ and $\mathcal{E}^*(\widehat{A})$ extending a given $\Theta$ between $\mathcal{B}$ and $\widehat{\mathcal{B}}$. (We will come back to $\Theta$ later.) How do we extend this into an automorphism $\Lambda$ such that $\Lambda(A) = \widehat{A}$?

Lets assume that $\mathcal{L}^*(A)$ and $\mathcal{L}^*(\widehat{A})$ are isomorphic via $\Psi$. (Otherwise $A$ and $\widehat{A}$ cannot be in the same orbit.) Now $W = (W - A) \sqcup (W \cap A)$. What about letting $\Lambda(W) = (\Psi(W \cup A) - \widehat{A}) \sqcup \Phi(W \cap A)$. $\Lambda$ is clearly order preserving.

But why is $\Lambda(W)$ and $\Lambda^{-1}(W)$ computably enumerable? For this we need the notion of supports.

## 7.1. Supports

**Definition 27.** $S \in \mathcal{S}(A)$ *supports* $X$ iff $S \subseteq X$ and $(X - A) \sqcup S$ is a computably enumerable set.

**Lemma 28.** $W \smallsetminus A$ *supports* $W$.

*Proof.* $W = (W - A) \sqcup (W \smallsetminus A) \sqcup (A \smallsetminus W)$ and $(W - A) \sqcup (W \smallsetminus A) = W \backslash A$ is a computably enumerable set.                                                              □

**Definition 29.** An extendible subalgebra $\mathcal{B}$ *supports* $\mathcal{L} \subseteq \mathcal{L}^*(A)$ if for all $W \in \mathcal{L}$ there is an $S \in \mathcal{B}$ such that $S$ supports $W$.

**Lemma 30.** $\mathcal{E}_A$ *supports* $\mathcal{L}^*(A)$.

**Definition 31.** Assume that $\mathcal{L}^*(A)$ and $\mathcal{L}^*(\widehat{A})$ are isomorphic via $\Psi$, $\mathcal{B}$ and $\widehat{\mathcal{B}}$ are isomorphic via $\Theta$, $\mathcal{B}$ supports $\mathcal{L}$, and $\widehat{\mathcal{B}}$ supports $\widehat{\mathcal{L}}$. Then the isomorphisms $\Psi$ and $\Theta$ *preserve* the supports of $\mathcal{L}$ and $\widehat{\mathcal{L}}$ if $W \in \mathcal{L}$, $B \in \mathcal{B}$, and $B$ supports $W$ then $(\Psi(W \cup A) - \widehat{A}) \sqcup \Theta(B)$ is a computably enumerable set and if $\widehat{W} \in \widehat{\mathcal{L}}$, $\widehat{B} \in \widehat{\mathcal{B}}$, and $\widehat{B}$ supports $\widehat{W}$ then $(\Psi^{-1}(\widehat{W} \cup \widehat{A}) - A) \sqcup \Theta^{-1}(\widehat{B})$ is a computably enumerable set. Notice we do not worry about inclusion for the image (the first clause of the definition of support). For shorthand if $\mathcal{L} = \mathcal{L}^*(A)$ and $\widehat{\mathcal{L}} = \mathcal{L}^*(\widehat{A})$ we just say isomorphisms $\Psi$ and $\Theta$ *preserve supports*.

## 7.2. A Stronger Extension Theorem
**Theorem 32.** *Assume that*

1. *$\mathcal{L}^*(A)$ and $\mathcal{L}^*(\widehat{A})$ are isomorphic via $\Psi$;*

2. *$\mathcal{B}$ and $\widehat{\mathcal{B}}$ are extendible algebras which are extendibly $\Delta_3^0$ isomorphic via $\Theta$;*

3. *$\mathcal{B}$ supports $\mathcal{L}^*(A)$;*

4. *$\widehat{\mathcal{B}}$ supports $\mathcal{L}^*(\widehat{A})$;*

5. *$\Psi$ and $\Theta$ preserve supports.*

*Then $\Lambda(W) = (\Psi(W \cup A) - \widehat{A}) \sqcup \Phi(W \cap A)$ is an automorphism of $\mathcal{E}$ such that $\Lambda(A) = \widehat{A}$, $\Lambda \upharpoonright \mathcal{L}^*(A) = \Psi$, and $\Lambda \upharpoonright \mathcal{E}^*(A)$ is $\Delta_3^0$. ($\Phi$ is from Theorem 26.)*

This theorem follows from Theorem 26 by purely algebraic means.

## 8. Using the Algebraic Extension Theorems

### 8.1. Splits, again

Assume that $A$ and $\widehat{A}$ are automorphic via $\Psi$. Hence $\mathcal{L}^*(A)$ and $\mathcal{L}^*(\widehat{A})$ are isomorphic via $\Psi$. So the structures $\mathcal{S}_{\mathcal{R}}(A)$ and $\mathcal{S}_{\mathcal{R}}(\widehat{A})$ are isomorphic via isomorphism induced by $\Psi$. What is surpising is this isomorphism is $\Delta_3^0$

**Theorem 33.** $\mathcal{S}_{\mathcal{R}}(A)$ and $\mathcal{S}_{\mathcal{R}}(\widehat{A})$ are $\Delta_3^0$-isomorphic structures via an isomorphism $\Theta$ induced by $\Psi$.

Note $\mathcal{S}_{\mathcal{R}}(A)$ and $\mathcal{S}_{\mathcal{R}}(\widehat{A})$ are $\Sigma_3^0$ atomless Boolean algebras so we know they are $\Delta_4^0$ isomorphic. So this theorem is an improvement of one quantifier or jump. The proof heavily uses coding, is difficult, and very technical.

### 8.2. Automorphisms to Automorphisms

Lets assume that $A$ and $\widehat{A}$ are in the same orbit and see what the above theorems tell us.

**Theorem 34.** *If $A$ and $\widehat{A}$ are automorphic via $\Psi$ then*

1. $\mathcal{L}^*(A)$ *and* $\mathcal{L}^*(\widehat{A})$ *are isomorphic via* $\Psi$;

2. *there is an extendible* $\mathcal{B}$ *supporting* $\mathcal{L}^*(A)$

3. *there is an extendible* $\widehat{\mathcal{B}}$ *supporting* $\mathcal{L}^*(\widehat{A})$;

4. $\mathcal{B}$ *and* $\widehat{\mathcal{B}}$ *are extendibly* $\Delta_3^0$ *isomorphic via* $\Theta$;

5. *the isomorphisms* $\Psi$ *and* $\Theta$ *preserve supports.*

We are given $\Theta$ via Theorem 33. It turns out that $\Theta(\mathcal{E}_A)$ is an extendible algebra which is extendible $\Delta_3^0$ isomorphic to $\mathcal{E}_A$ via a $\Theta'$ induced by $\Theta$ and similarly for $\Theta^{-1}(\mathcal{E}_A)$. It is not hard to show that the join of extendible algebras is extendible. So $\mathcal{B} = \mathcal{E}_A \oplus \Theta^{-1}(\mathcal{E}_{\widehat{A}})$ and $\widehat{\mathcal{B}} = \Theta(\mathcal{E}_A) \oplus \mathcal{E}_{\widehat{A}}$ are extendible. It is straightforward to show they extendibly isomorphic via an $\Theta'$ induced via $\Theta$. We will just drop $\Theta'$ in favor of $\Theta$.

Applying Theorem 32 to the above theorem results in the following:

**Theorem 35.** *If $A$ and $\widehat{A}$ are automorphic via $\Psi$ then they are automorphic via $\Lambda$ where $\Lambda \restriction \mathcal{L}^*(A) = \Psi$ and $\Lambda \restriction \mathcal{E}^*(A)$ is $\Delta_3^0$.*

In all the previously known techniques for constructing automorphisms $\Psi$ it is *always* the case that $\Psi \restriction \mathcal{E}^*(A)$ is $\Delta_3^0$. One wondered if this was a limitation in our techniques or the automorphisms themselves.

### 8.3. Preserving the Computable Subsets

Our current goal is to show that there is an isomorphism $\Lambda$ between $\mathcal{E}^*(A)$ and $\mathcal{E}^*(\widehat{A})$ which preserve the computable sets. Hence $\Lambda(R)$ is computable iff $R$ is computable. This is more subtle than it appears. For example, if we will find a computable permutation $p$ taking $A$ to $\widehat{A}$ then defining $\Lambda(W) = p(W)$ does the desired job. But $p$ only exists if $A$ and $\widehat{A}$ have the same 1-degree.

Recall $\mathcal{R}(A)$ is the computable subset of $A$. Let $\overline{\mathcal{R}}(A) = \{\overline{R} : R \in \mathcal{R}(A)\}$.

**Theorem 36.** *Assume that*

1. $\mathcal{B}$ *and* $\widehat{\mathcal{B}}$ *are extendible algebras which are extendibly* $\Delta_3^0$ *isomorphic via* $\Theta$;

2. *for all* $R \in \mathcal{R}(A)$ *there is an* $S$ *such that* $S \in \mathcal{B} \cap \mathcal{R}(A)$ *and* $R \subseteq S$;

3. *For all* $\widehat{R} \in \mathcal{R}(\widehat{A})$ *there is an* $\widehat{S}$ *such that* $\widehat{S} \in \widehat{\mathcal{B}} \cap \mathcal{R}(\widehat{A})$ *and* $\widehat{R} \subseteq \widehat{S}$.

4. $\Theta$ *preserves the computable subsets.*

*Then there is a* $\Lambda$ *such that* $\Lambda$ *is a* $\Delta_3^0$ *isomorphism between* $\mathcal{E}^*(A) \cup \overline{\mathcal{R}}(A)$ *and* $\mathcal{E}^*(\widehat{A}) \cup \overline{\mathcal{R}}(\widehat{A})$. $\Lambda$ *preserves the computable subsets.*

Item 2 says that the computable subset of $\mathcal{B}$ supports the complements of the computable subsets of $A$. Item 3 is the dual of Item 2. Like Theorem 32, this theorem follows from Theorem 26 by purely algebraic means. The hypotheses of Theorem 36 are easy to meet.

**Lemma 37.** *Let* $A$ *and* $\widehat{A}$ *be two noncomputable computably enumerable sets. Then there are extendible Boolean algebras* $\mathcal{B}$ *and* $\widehat{\mathcal{B}}$ *and a* $\Delta_3^0$ $\Theta$ *such that*

1. $\mathcal{B}$ *and* $\widehat{\mathcal{B}}$ *are extendibly isomorphic via* $\Theta$;

2. *for all* $R \in \mathcal{R}(A)$ *there is an* $S$ *such that* $S \in \mathcal{B} \cap \mathcal{R}(A)$ *and* $R \subseteq S$;

3. *for all* $\widehat{R} \in \mathcal{R}(\widehat{A})$ *there is an* $\widehat{S}$ *such that* $\widehat{S} \in \widehat{\mathcal{B}} \cap \mathcal{R}(\widehat{A})$ *and* $\widehat{R} \subseteq \widehat{S}$;

4. $\Theta$ *preserves the computable subsets.*

**Corollary 38.** *If* $A$ *and* $\widehat{A}$ *are two noncomputable computably enumerable sets then there is a* $\Lambda$ *such that* $\Lambda$ *is a* $\Delta_3^0$ *isomorphism between* $\mathcal{E}^*(A) \cup \overline{\mathcal{R}}(A)$ *and* $\mathcal{E}^*(\widehat{A}) \cup \overline{\mathcal{R}}(\widehat{A})$. $\Lambda$ *preserves the computable subsets.*

Corollary 38 is not a new result. It was known to Herrmann but unpublished. It follows from Theorem 26 which is equivalent to almost all the previous extension theorems and Lemma 37 which is equivalent to Soare's Order-Preserving Enumeration Theorem (Soare [21] see XV.5.1 of [22]).

### 8.4. The Maximal Sets, again

**Theorem 39.** *Assume that* $M$ *and* $\widehat{M}$ *are maximal. Then an isomorphism* $\Lambda$ *between* $\mathcal{E}^*(M) \cup \overline{\mathcal{R}}(M)$ *and* $\mathcal{E}^*(\widehat{M}) \cup \overline{\mathcal{R}}(\widehat{M})$ *(as given by Corollary 38) induces an automorphism* $\Psi$ *taking* $M$ *to* $\widehat{M}$.

*Proof.* If $W \subseteq^* M$ let $\Psi(W) = \Lambda(W)$. If $\overline{M} \subseteq W$ then $M \cup W = \omega$ and there is a computable set $R \subset M$ such that $\overline{R} \subseteq W$ which implies $W = \overline{R} \sqcup (W \cap R)$. In this case, let $\Psi(W) = \Lambda(\overline{R}) \sqcup \Lambda(W \cap R)$. Determining which case applies can be done computable in $\mathbf{0}''$. □

**Claim 40.** *All of the known definable orbits have a similar algebraic proof using Corollary 38. However, in many cases, dealing with the sets outside or disjoint from* $A$ *takes more than an oracle for* $\mathbf{0}''$.

One particular orbit we should mention is that of the hemimaximal sets, nontrivial splits of maximal sets. Downey and Stob [12] first showed that the hemimaximal sets formed an orbit by using the standard methods known at that time. But Herrmann claimed to have found an algebraic proof. Herrmann never published this proof. The first algebraic proof of this result can be found in Cholak et al. [5]. Herrmann's work and the work in [5] can be seen as a forerunner to the current work.

## 9. New Orbits

First we need to generalize the structure $\mathcal{L}^*(A)$.

**Definition 41** (The sets disjoint from $A$). Let $\mathcal{D}(A) = \{B : \exists W (B \subseteq A \cup W$ and $W \cap A =^* \emptyset)\}$. Let $\mathcal{E}_{\mathcal{D}(A)}$ be $\mathcal{E}$ modulo $\mathcal{D}(A)$. $A$ is $\mathcal{D}$-*hhsimple* iff $\mathcal{E}_{\mathcal{D}(A)}$ is a Boolean algebra. $A$ is $\mathcal{D}$-*maximal* iff $\mathcal{E}_{\mathcal{D}(A)}$ is the two element Boolean algebra.

**Lemma 42.** *If $A$ is simple then $\mathcal{E}_{\mathcal{D}(A)} \cong_{\Delta_3^0} \mathcal{L}^*(A)$.*

The lemma follows since if $A$ is simple the only computably enumerable subsets of it's complement are finite. Except for the creative sets, up to this year all known orbits were orbits of $\mathcal{D}$-hhsimple sets. For example, the hemimaximal sets are $\mathcal{D}$-maximal.

### 9.1. A "Definable" $\Delta_5^0$ Orbit which is Not a $\Delta_3^0$ Orbit

**Definition 43.** $F$ is $A$-special if $F = A$ or $F - A$ is computably enumerable and for all computably enumerable $V$ if $V \cap A = \emptyset$ then $V - F$ is computably enumerable.

If $F$ is $A$-special then $F \in \mathcal{D}(A)$.

**Definition 44.** A computably enumerable set $A$ is *pseudo hemi $\mathcal{D}$-maximal*: iff

1. $\forall W$ if $(W \cap A = \emptyset)$ then $\exists F$ such that $F$ is $A$-special and $W \subseteq F$.

2. $\forall F$ if $F$ is $A$-special then $\exists F_1$ such that $F_1$ is $A$-special, $F \cap F_1 = \emptyset$, and $F_1$ is not computable.

3. $\forall W \exists F$ such that $F$ is $A$-special and either $W \subseteq^* F \cup A$ or $W \cup F \cup A =^* \omega$.

Using Item 2, we can find a disjoint sequence of $A$-special sets, $\{F_i\}$, such that $A = F_0$ and for all $e$ either $W_e \subseteq \bigsqcup_{i \le e} F_i$ or $W_e \cup \bigsqcup_{i \le e} F_i =^* \omega$. It turns out that the $A$-special sets are also pseudo hemi $\mathcal{D}$-maximal. Hence the set $\bigsqcup_{i \in \omega} F_i$ behaves like a maximal set (it is not c.e.) and the $F_i$s behave like splits of $\bigsqcup_{i \in \omega} F_i$. Item 3 implies that $A$ is $\mathcal{D}$-maximal. This explains why these sets are called pseudo hemi $\mathcal{D}$-maximal. Clearly being pseudo hemi $\mathcal{D}$-maximal is definable.

**Theorem 45.** *The pseudo hemi $\mathcal{D}$-maximal sets form a definable $\Delta_5^0$ orbit. Furthermore there are two pseudo hemi $\mathcal{D}$-maximal sets which are not in the same $\Delta_3^0$-orbit.*

Let $A$ and $\widehat{A}$ be two different pseudo hemi $\mathcal{D}$-maximal sets. We can use Corollary 38 to build the desired automorphism $\Psi$ like we did for the maximal sets.

Apply Corollary 38 to $F_i$ and the corresponding $\widehat{F}_i$ to get $\Lambda_i$. Now if $W_e \subseteq \bigsqcup_{i \le e} F_i$ then let $\Psi(W_e) = \bigsqcup \Lambda_i(W_e \cap F_i)$. Otherwise $W_e \cup \bigsqcup_{i \le e} F_i =^* \omega$. In which case there is a computable subset $R$ of $\bigsqcup_{i \le e} F_i$ such that $W_e \cup R =^* \omega$. So $W_e = \overline{R} \cup \bigsqcup_{i \le e} (W_e \cap R \cap F_i)$. Now $R \cap F_i$ is computable. So $\Psi(R) = \bigsqcup_{i \le e} (\Lambda_i(R \cap F_i))$ is computable. Let $W_e = \overline{\Psi(R)} \cup \bigsqcup_{i \le e} (\Lambda(W_e \cap R \cap F_i))$.

The reason the automorphism is $\Delta_5^0$ is that we need $0^{(4)}$ to find the sequence $\{F_i\}$. To show that it is not an orbit under $\Delta_3^0$ we build two pseudo hemi $\mathcal{D}$-maximal sets, $A$ and $\widehat{A}$, one such that there is $\Delta_3^0$ sequence of disjoint $A$-special sets like above and one such that there is no $\Delta_3^0$ sequence of disjoint $A$-special sets. While these sets are in the same orbit they cannot be in the same $\Delta_3^0$ orbit.

## 9.2. Some Restrictions

Note the complexity in the above orbit arises because of the sequence of $A$-special sets. One might wonder if this is necessary. Using Theorem 33 we can show:

**Theorem 46.** *If $A$ is $\mathcal{D}$-hhsimple and $A$ and $\widehat{A}$ are in the same orbit then $\mathcal{E}_{\mathcal{D}(A)}$* $\cong_{\Delta_3^0} \mathcal{E}_{\mathcal{D}(\widehat{A})}$.

Hence the complexity in the orbit must come from how $A$ interacts with the sets in $\mathcal{D}(A)$ at least in the case that $A$ is $\mathcal{D}$-hhsimple. Using a different coding we can improve this to all sets $A$.

**Theorem 47.** *If $A$ and $\widehat{A}$ are automorphic then $\mathcal{E}_{\mathcal{D}(A)}$ and $\mathcal{E}_{\mathcal{D}(\widehat{A})}$ are $\Delta_6^0$-isomorphic.*

These theorems plus some work of Maass [17] allow us to classify the complexity of the orbits of several different sets.

**Theorem 48.** *If $A$ is $\mathcal{D}$-hhsimple and simple (i.e., hhsimple) then $A \approx \widehat{A}$ iff $\mathcal{L}^*(A) \cong_{\Delta_3^0} \mathcal{L}^*(\widehat{A})$. (The "only if" is by Maass [17]).*

**Theorem 49.** *If $A$ is simple then $A \approx \widehat{A}$ iff $A \approx_{\Delta_6^0} \widehat{A}$.*

**Theorem 50.** *If $A$ and $\widehat{A}$ are both promptly simple then $A \approx \widehat{A}$ iff $A \approx_{\Delta_3^0} \widehat{A}$.*

## 9.3. The $r$-maximal sets

Lets reflect on what our results say about the orbit of $r$-maximal sets.

**Theorem 51.** *Let $A$ and $\widehat{A}$ be $r$-maximal. Then $\Lambda(A) = \widehat{A}$ iff*

1. *$\mathcal{L}^*(A)$ and $\mathcal{L}^*(\widehat{A})$ are isomorphic via a $\Phi$ which is $\Delta_6^0$.*

2. *There is a $\Theta$, extendible $\mathcal{B}$ and $\widehat{\mathcal{B}}$ such that $\mathcal{B}$ supports $\mathcal{L}^*(A)$, $\widehat{\mathcal{B}}$ supports $\mathcal{L}^*(\widehat{A})$, $\mathcal{B}$ and $\widehat{\mathcal{B}}$ are extendibly $\Delta_3^0$ isomorphic via $\Theta$, $\Theta$ is $\Delta_3^0$, and $\Phi$ and $\Theta$ preserve supports.*

Every item but the last one in Item 2 are always true; see Theorem 32 and the first few sentences after Theorem 34. Hence the $\Phi$ is the issue. One approach to algebraically building $\Phi$ was more or less suggested in Cholak and Nies [11].

## 10. Slaman-Woodin Conjecture

In 1990, Slaman and Woodin make the following conjecture.

**The Slaman-Woodin Conjecture.** *The set $\{\langle i,j \rangle : W_i \approx W_j\}$ is $\Sigma_1^1$-complete.*

Clearly this set is in $\Sigma_1^1$. It is a natural question to ask whether this set is complete for this class. Slaman and Woodin were led to this conjecture by a series of results.

**Theorem 52** (Remmel/Folklore). *Let $\mathcal{B}_i$ be a listing of computable Boolean algebras. The set $\{\langle i,j \rangle : \mathcal{B}_i \cong \mathcal{B}_j\}$ is $\Sigma_1^1$-complete.*

**Theorem 53** (Lachlan [16]). *There is a computably enumerable set $H_i$ such that $\mathcal{L}^*(H_i) \cong \mathcal{B}_i$.*

**Corollary 54** (Slaman and Woodin). *The set $\{\langle i,j \rangle : \mathcal{L}^*(H_i) \cong \mathcal{L}^*(H_j)\}$ is $\Sigma_1^1$-complete.*

This led to the Slaman-Woodin Conjecture. The idea was to replace "$\mathcal{L}^*(H_i) \cong \mathcal{L}^*(H_j)$" with "$H_i \approx H_j$". By Theorem 48, this is impossible.

In 1995, the conjecture was shown to be true. However the proof is very difficult and most likely will never be properly written up.

**Theorem 55** (Cholak, Downey, and Harrington). *The set $\{\langle i,j \rangle : W_i \approx W_j\}$ is $\Sigma_1^1$-complete.*

There is, however, a slightly stronger result with some very interesting corollaries.

**Theorem 56.** *There is a (computably enumerable) set $A$ such that the index set $\{i : W_i \approx A\}$ is $\Sigma_1^1$-complete.*

## Corollary 57.

1. *Not all orbits of $\mathcal{E}$ are elementarily definable.*

2. *$\mathcal{E}$ is not a prime model.*

3. *There is no arithmetic description of all orbits of $\mathcal{E}$.*

4. *Scott rank of $\mathcal{E}$ is $\omega_1^{CK} + 1$.*

5. *(corollary of the proof) For all $\alpha \geq 10$, there is a properly $\Delta_\alpha^0$ orbit.*

To prove this theorem we encode into the orbit of $A$ the isomorphisms of a computably branching tree. Given a (certain) $\Delta_3^0$ list of parameters, $\mathcal{B}_i \subset \mathcal{S}_\mathcal{R}(A)$, we can define a definable invariant, $(\mathcal{N}(A), \prec)$. If $A \approx \widehat{A}$ then $(\mathcal{N}(A), \prec) \cong (\mathcal{N}(\widehat{A}), \prec)$.

Modulo some unmentioned (here) definable properties, if, for all $i$, there is an (certain) extendible $\widehat{\mathcal{B}}_i$ and a $\Delta_3^0$ isomorphism between $\mathcal{B}_i$ and $\widehat{\mathcal{B}}_i$ and $(\mathcal{N}(A), \prec) \cong (\mathcal{N}(\widehat{A}), \prec)$ then $A \approx \widehat{A}$. The proof is similar to the proof that pseudo hemi $\mathcal{D}$-maximal sets form an orbit.

Now given a computably branching tree $(T, \prec)$ we can construct an $A$ (and the $\mathcal{B}_i$) such that $(\mathcal{N}(A), \prec) \cong (T, \prec)$. Depending on $T$, $\{\widehat{T} : \widehat{T} \cong T\}$ ranges from properly $\Delta_\alpha^0$ to $\Sigma_1^1$-complete and so does the index set of the orbit of $A$.

## 10.1. Last Question

$(\mathcal{N}(A), \prec)$ is a definable (with the parameters $\mathcal{B}_i$s) invariant which determines the orbit of $A$. But it only works for some $A$, those with the correct $\mathcal{B}_i$s.

**Question 58.** *Can we define the invariant $(\mathcal{N}(A), \prec)$ (or something similar) without using any parameters?*

A positive answer would allow us to describe all orbits in terms of a tree.

## References

[1] Klaus Ambos-Spies and André Nies. Cappable recursively enumerable degrees and Post's program. *Arch. Math. Logic*, 32(1):51–56, 1992. ISSN 0933-5846.

[2] Peter Cholak. *Automorphisms of the lattice of recursively enumerable sets.* PhD thesis, University of Wisconsin, 1991.

[3] Peter Cholak. Automorphisms of the lattice of recursively enumerable sets. *Mem. Amer. Math. Soc.*, 113(541):viii+151, 1995. ISSN 0065-9266.

[4] Peter Cholak. The global structure of computably enumerable sets. In *Computability theory and its applications (Boulder, CO, 1999)*, volume 257 of *Contemp. Math.*, pages 61–72, Providence, RI, 2000. Amer. Math. Soc.

[5] Peter Cholak, Rod Downey, and Eberhard Herrmann. Some orbits for $\mathcal{E}$. *Ann. Pure Appl. Logic*, 107(1-3):193–226, 2001. ISSN 0168-0072.

[6] Peter Cholak, Rod Downey, and Michael Stob. Automorphisms of the lattice of recursively enumerable sets: promptly simple sets. *Trans. Amer. Math. Soc.*, 332(2):555–570, 1992. ISSN 0002-9947.

[7] Peter Cholak and Leo Harrington. Definable encodings in the computably enumerable sets. *Bull. Symbolic Logic*, 6(2):185–196, 2000. ISSN 1079-8986.

[8] Peter Cholak and Leo Harrington. On the definability of the double jump in the computably enumerable sets. *J. Math. Log.*, 2(2):261–296, 2002. ISSN 0219-0613.

[9] Peter Cholak and Leo Harrington. Extension theorems, orbits, and automorphisms of the computably enumerable sets. To appear., 2003.

[10] Peter Cholak and Leo Harrington. Isomorphisms of splits of computably enumerable sets. *J. of Symbolic Logic*, 68(3):1044–1064, 2003.

[11] Peter Cholak and André Nies. Atomless $r$-maximal sets. *Israel J. Math.*, 113: 305–322, 1999. ISSN 0021-2172.

[12] Rod Downey and Michael Stob. Automorphisms of the lattice of recursively enumerable sets: orbits. *Adv. Math.*, 92(2):237–265, 1992. ISSN 0001-8708.

[13] Richard Friedberg. A criterion for completeness of degrees of unsolvability. *J. Symb. Logic*, 22:159–160, 1957.

[14] Leo Harrington and Robert I. Soare. Post's program and incomplete recursively enumerable sets. *Proc. Nat. Acad. Sci. U.S.A.*, 88(22):10242–10246, 1991. ISSN 0027-8424.

[15] Leo Harrington and Robert I. Soare. The $\Delta_3^0$-automorphism method and noninvariant classes of degrees. *J. Amer. Math. Soc.*, 9(3):617–666, 1996. ISSN 0894-0347.

[16] Alistair Lachlan. On the lattice of recursively enumerable sets. *Trans. Amer. Math. Soc.*, 130:1–37, 1968.

[17] Wolfgang Maass. On the orbits of hyperhypersimple sets. *J. Symbolic Logic*, 49(1):51–62, 1984. ISSN 0022-4812.

[18] Sergey Marčenkov. A certain class of incomplete sets. *Mat. Zametki*, 20(4): 473–478, 1976.

[19] Donald Martin. Completeness, the recursion theorem, and effectively simple sets. *Proc. Amer. Math. Soc.*, 17:838–842, 1966.

[20] A. A. Mučnik. On the unsolvability of the problem of reducibility in the theory of algorithms. *Dokl. Akad. Nauk SSSR (N.S.)*, 108:194–197, 1956.

[21] Robert I. Soare. Automorphisms of the lattice of recursively enumerable sets. I. Maximal sets. *Ann. of Math. (2)*, 100:80–120, 1974.

[22] Robert I. Soare. *Recursively enumerable sets and degrees*. Perspectives in Mathematical Logic. Springer-Verlag, Berlin, 1987. ISBN 3-540-15299-7. A study of computable functions and computably generated sets.

[23] Robert I. Soare. Computability and recursion. *Bull. Symbolic Logic*, 2(3): 284–321, 1996. ISSN 1079-8986.

[24] Robert I. Soare. Extensions, automorphisms, and definability. In *Computability theory and its applications (Boulder, CO, 1999)*, volume 257 of *Contemp. Math.*, pages 279–307. Amer. Math. Soc., Providence, RI, 2000.

# Definability and Automorphism Groups

Wilfrid Hodges

*School of Mathematical Sciences, Queen Mary, University of London*
*Mile End Road, London E1 4NS*
*w.hodges@qmul.ac.uk*

**Abstract.** Suppose $B$ is a structure with a relativised reduct $A$. We consider set-theoretic definitions that find, for each structure $A'$ isomorphic to $A$, a structure $B'$ isomorphic to $B$ such that $A'$ is the relativised reduct of $B'$. There is a group homomorphism $\nu$ from the automorphism group of $B$ to that of $A$. We state some theorems relating properties of $\nu$ with properties of the definitions. One very plausible relationship remains unproved. But we can prove enough of it to show that there is no set-theoretic definition which, provably in ZFC, finds for each field a unique algebraic closure of that field.

The recent publication of Ibn Sina's 'Book of definitions' [9] is a reminder that logic has traditionally had two arms, deduction and definition. Up to the middle of the twentieth century there were leading logicians who regarded the general theory of definition as a central topic; they included Leibniz, Frege, Leśniewski, Tarski and Beth among others.

During the second half of the twentieth century, definition didn't go away. But while the logical calculi of Gentzen and others formed a focus for both teaching and research in proof theory, there was no corresponding general formalism of definition. Instead, separate disciplines developed definition in a range of contexts. One should include at least:

(a) **Descriptive set theory**. Defining relations, functions etc. on $\mathbb{R}$.

(b) **Model theory**. Defining classes of structures, and relations within structures.

(c) **Formal specification**. Defining the behaviour of systems.

As yet there is no grand unification on offer for these kinds of definition; maybe there never will be. But we can prove some results that cut across these areas. In this paper I shall state and prove some facts about structural definability in the universe of sets. I start with easy cases and move to cases which are still problematic, generalising from the easy cases when possible.

My warm thanks to the organisers of the Congress in Logic, Methodology and Philosophy of Science at Oviedo in August 2003. The results of §6 below were worked out during a visit to the Mathematical Logic Year at the Mittag-Leffler Institute in 2000/1; much thanks to the organisers of the Year. The area that we study was first opened up by Haim Gaifman [4] in 1974, and it was an honour to have him in the audience at Oviedo.

## 1. Kinds of definition

The word 'structure' has two uses in model theory. First, one uses it as a count noun, as in

An abelian group can be regarded as a structure.

Second, one uses it as a noncount noun, for example in

An isomorphism is a bijection that preserves structure.

We shall need both senses. But some locutions, for example 'the structure', could be taken in either sense, and it would be better to avoid this ambiguity. I shall use *constituent structure* to mean structure in the second sense.

Every structure $A$ in the first sense has a signature, i.e. the family $\sigma$ of nonlogical symbols $S$ that have interpretations $S_A$ in $A$; we say that $A$ is a $\sigma$-structure. For each symbol $S$ in the family, $\sigma$ carries the information whether $S$ is an individual constant, a function symbol or a relation symbol, and in the second and third cases, what the arity of $S$ is. We assume that structures are constructed set-theoretically in such a way that they determine their signatures. For some of the examples below it is convenient to allow many-sorted signatures, but I leave details of this to the reader.

If $A$ is a $\tau$-structure and $\sigma$ is a subset of $\tau$, we write $(A|\sigma)$ for the reduct of $A$ to $\sigma$; this is the $\sigma$-structure got from $A$ by ignoring the symbols not in $\sigma$. The structures $A$ and $(A|\sigma)$ have the same elements. Suppose $\theta(x)$ is an atomic formula in the signature $\tau$; then we write $(A|_\theta\sigma)$ for the substructure of $(A|\sigma)$ whose domain is the set of elements that satisfy $\theta$ in $A$. We say then that $(A|_\theta\sigma)$ is a *relativised reduct* of $A$, and (less happily) that $A$ is a *relativised expansion* of $(A|_\theta\sigma)$. Reducts are the special case of relativised reducts where $\theta(x)$ is $x = x$.

Henceforth $V$ is the universe of well-founded sets. Let $\phi(x, y, z)$ be a formula in the first-order language of sets. Let $c$ be any set. Then $\phi$ and $c$ provide us with two kinds of definition, which for simplicity we can call *set definition* and *class definition* respectively.

*Set definition:* The *domain* of $\phi(x, y, c)$ is the class of all sets $a$ such that

$$V \models \exists_{=1} x \phi(x, a, c).$$

For each $a$ in the domain, let $b_a$ be the unique set such that $V \models \phi(b_a, a, c)$. Then $\phi(x, y, c)$ defines the operation taking each $a$ in the domain to $b_a$.

*Class definition:* $\phi(x, y, c)$ defines the operation taking each set $a$ to the class of sets $b$ such that $V \models \phi(b, a, c)$. We say that $a$ is in the *domain* of $\phi(x, y, c)$ if this class is nonempty.

The set $c$ in $\phi(x, y, c)$ is called the *parameter* of the definition (in both cases). We shall sometimes consider definitions where there is more than one parameter, but the account above covers these since $c$ can be a sequence of sets.

Sometimes the variable $y$ doesn't occur free in $\phi$, so that $\phi(x, c)$ defines a set or a class outright. Lacking any other sensible term I shall say that the definition is of *type I* in such cases. When $y$ occurs free in $\phi$, the definition is of *type II*.

We can illustrate these notions from the three areas (a)–(c) mentioned above.

Given the structure $\mathbb{R}$ of reals, we can define a borel set $X$ of reals by a suitable formula $\psi(x, r)$ with parameter $r$; in general $\psi$ will be infinitary and $r$ will be an infinite sequence of reals. One way to handle this is to introduce a new 1-ary relation symbol $P$; let $\sigma$ be the signature of $\mathbb{R}$ and let $\sigma^+$ be $\sigma$ with $P$ added. Then to define $X$ is to define a $\sigma^+$-structure $B$ that is an expansion of $\mathbb{R}$ with $P_B = X$. Set-theoretically the definition is a set definition of type I, $\phi(x, \langle \psi, r \rangle)$, which says

$$(x \text{ is a } \sigma^+\text{-structure}) \wedge ((x|\sigma) = \mathbb{R}) \wedge (\forall s \in \mathbb{R})(x \models P(s) \leftrightarrow \mathbb{R} \models \psi(s, r)).$$

There is no need to list $\sigma^+$, $\sigma$ and $\mathbb{R}$ as parameters since they are definable in $V$ without parameters.

Definitions of this kind are frequent in descriptive set theory. In model theory we often meet the same situation but with a structure $A$ in place of $\mathbb{R}$. If the structure is fixed, it can go as a parameter. But another common situation is that we define expansions for all models $A$ of some fixed theory $T$. In this case the definition defines an operation taking each model $A$ of $T$ to an expansion $A^+$, and a better form for the definition is a type II set definition $\phi(x, y, \langle \sigma, \sigma^+, T, \psi \rangle)$ saying

$$(x \text{ is a } \sigma^+\text{-structure}) \wedge ((x|\sigma) = y) \wedge (y \models_\sigma T) \wedge$$
$$(\forall s \in \text{dom}(y))(x \models_{\sigma^+} P(s) \leftrightarrow y \models_\sigma \psi(s)).$$

where now $\sigma$ is the signature appropriate for $T$, and $\models_{\sigma^+}$, $\models_\sigma$ are the satisfaction relations for the two signatures.

Another kind of definition in model theory is the definition of the class of all models of a theory. This can only be a class definition. One normally meets it in a form that works simultaneously for all languages of a certain logic, for example first-order. The definition is a class definition $\phi(x, \langle z_1, z_2 \rangle)$ of type I:

$$(z_1 \text{ is a signature}) \wedge$$
$$(z_2 \text{ is a theory in the first-order language of signature } z_1) \wedge$$
$$(x \text{ is a } z_1\text{-structure} \wedge x \models_{z_1} z_2).$$

The situation with formal specification is a little more complicated and I sketch only one or two of the possibilities. Initial model specifications define, for an equational theory or more generally a strict universal Horn theory $T$, the unique initial model of $T$. From a category theory point of view the initial model is defined only up to isomorphism, suggesting we might have to use a class definition of type I. But in fact there is always a canonical choice, namely the structure generated by the closed terms of $T$, and this allows a set definition of type I. (This is a foretaste of issues that we discuss below.) Sometimes workers in the area refer to 'parametrised' initial model specifications; these are of type II, and often the defined structure is a relativised expansion of the structure it is defined from.

Set-theoretic specifications are not as amenable as one might assume; some-times they require a different universe of sets. But one can single out within Z, for example, a class of specifications that are naturally expressed as set definitions of type II. This yields translations into initial model specifications (Hodges [5], Kirchner and Mosses [10]).

## 2. Structural definitions

We make some assumptions for the rest of this paper. Unless otherwise stated, the formula $\phi(x, y, c)$ is a type II definition with parameter $c$. We assume signatures $\sigma$ and $\sigma^+$ are given (for example as part of the parameter). The signature $\sigma^+$ consists of $\sigma$ with new relation symbols added. We shall assume:

If $V \models \phi(B, A, c)$, then $B$ is a $\sigma^+$-structure and $A$ is a relativised reduct    (1)
$(B|_\theta \sigma)$ of $B$; the relativising formula $\theta$ depends only on $\phi$ and $c$, not on $A$ or $B$. If also $\phi(B', A', c)$ and $A$ is isomorphic to $A'$, then $B$ is isomorphic to $B'$.

For purposes of this paper we shall say that the definition $\phi(x, y, c)$ is a *structural* definition if it satisfies (1). When $\phi(B, A, c)$ implies that $A$ is a reduct of $B$, we shall say that the definition is of *reduct type*.

We write $\mathrm{Aut}(A)$ for the group of all automorphisms of the structure $A$.

Suppose for example that $\phi(x, y, c)$ is a structural definition. If $V \models \phi(B, A, c)$, then $A$ is a relativised reduct of $B$, and it follows that every automorphism $\beta$ of $B$ restricts to an automorphism $\nu(\beta)$ of $A$. The map $\nu : \mathrm{Aut}(B) \to \mathrm{Aut}(A)$ is a group homomorphism. In the case where $A$ is a reduct of $B$, $\mathrm{Aut}(B)$ is a subgroup of $\mathrm{Aut}(A)$ and $\nu$ is the inclusion map. We use this notation $\nu$ throughout the remainder of this paper.

In ordinary mathematical practice the algebraic closure $\bar{K}$ of a field $K$ is determined only up to isomorphism over $K$, and so the definition of the algebraic closure operator has to be a class definition. For Michael Makkai this is more than just a matter of practice; his 'principle of isomorphism' requires us to extend the same treatment even to functors [11]:

> when singling out an object with a certain property, we should be con-tent with determining the object up to isomorphism only. ... The idea behind the notion of *anafunctor* ... is that the same principle should extend to values of functors: their object-values are to be determined up to isomorphism only.

This is an interesting view, but I hope it doesn't forbid us to ask whether or when the structures in question can be pinned down more precisely.

Thus, given a class definition $\phi(x, y, c)$, we can ask whether there is a set definition $\phi'(x, y, c')$ that uniformises $\phi(x, y, c)$, in the sense that for every structure $A$ in the domain of $\phi(x, y, c)$ there is a unique $B$ such that $V \models \phi'(B, A, c')$, and for this $B$, $V \models \phi(B, A, c)$. A definable global choice function gives us such a uniformisation immediately.

One important slogan about set definitions is that the larger the domain, the nicer the defined operation. We shall cash in this slogan in several ways. The following *surjectivity property* is one kind of niceness:

If $V \models \phi(B, A, c)$, $V \models \phi(B', A', c)$ and $i : A \to A'$ is an isomorphism, $\quad$ (2)
then there is an isomorphism $j : B \to B'$ such that $j$ extends $i$.

Given (1), one can show that (2) holds if and only if $\nu : \text{Aut}(B) \to \text{Aut}(A)$ is surjective. Parametrised initial model specifications usually have the surjectivity property. Algebraic closure of fields has it too, as do most well-known closure operations on structures.

We write ZFCU for Zermelo-Fraenkel set theory ZFC with a set of urelements (i.e. objects that are not sets and have no members). If $u$ is a set of urelements, we can construct a universe $V(u)$ of sets which is a model of ZFCU, with $u$ as its set of urelements. The construction is as follows, by induction on the ordinals:

$$
\begin{aligned}
V_0(u) &= u \\
V_{i+1}(u) &= V_i(u) \cup \mathcal{P}(V_i(u)) \\
V_\delta(u) &= \bigcup_{i < \delta} V_i(u) \quad (\delta \text{ limit}) \\
V(u) &= \bigcup_{i \text{ an ordinal}} V_i(u).
\end{aligned}
$$

Each set in $V$ appears also in $V(u)$; we write $\check{} : V \to V(u)$ for the inclusion map.

Each permutation $\alpha$ of the set $u$ induces an automorphism $\alpha^\star$ of $(V(u), \in)$, by induction on rank:

$$\alpha^\star(a) = \{\alpha^\star(b) : b \in a\}.$$

This applies in particular when $u$ contains the domain of a structure $A$. For example each relation $R_A$ of $A$ is a member of $V(u)$. If $\alpha$ is a permutation of $u$ that fixes $\text{dom}(A)$ setwise, then $\alpha^\star(R_A)$ will be a relation on $\text{dom}(A)$, and it will be equal to $R_A$ if the restriction of $\alpha$ to $\text{dom}(A)$ is an automorphism of $A$.

On the other hand $V$ is rigid, and hence $\alpha^\star(\check{a}) = \check{a}$ for each set $a$ in $V$. We shall assume that expressions $\psi$ from languages are in $V$, so that $\alpha^\star(\psi) = \psi$.

When $u$ is a set in $V$, we can imitate this construction within $V$, regarding the elements of $u$ as urelements. But each set in $u$ will reappear at some point in the construction, and this will spoil its disguise as an urelement. A suitable coding of sets will get around this problem, though in general the embedding $\check{} : V \to V(u)$ will no longer be the identity. When $\alpha$ is an automorphism of a structure $A$, we can imagine $\text{dom}(A)$ taken as a set of urelements in this sense, so that $\alpha^\star$ acts on sets 'above $\text{dom}(A)$', but fixes linguistic expressions. Thus if $\beta$ is a permutation of $\text{dom}(A)$ such that $\beta^\star(A) = A$, then for each relation symbol $R$, $R_{\beta^\star(A)} = R_A$, so $\beta$ is an automorphism of $A$.

## 3. Set definitions of reduct type

The following theorem is classical. It describes structural set definitions of reduct type whose domain is the class of all reducts to $\sigma$ of models of a complete first-order theory $T$ of signature $\sigma^+$, where $\sigma^+$ consists of $\sigma$ together with new relation symbols. Clause (a) is the surjectivity property.

**Theorem 1** *Let $T$, $\sigma$, $\sigma^+$ be as above. The following are equivalent:*

(a) *If B is any model of T, then $\nu : \mathrm{Aut}(B) \to \mathrm{Aut}(B|\sigma)$ is surjective.*

(b) *For each relation symbol R in $\sigma^+$ but not in $\sigma$, T has a consequence of the form*

$$\forall \bar{x} \; (R\bar{x} \leftrightarrow \psi(\bar{x}))$$

*where $\psi(\bar{x})$ is a first-order formula of signature $\sigma$.*

(c) *If B, B' are models of T with $(B|\sigma) = (B'|\sigma)$, then $B = B'$.*

For the **proof**, see for example Hodges [6] pp. 515f. The equivalence of (b) and (c) is due to Evert Beth, and their equivalence to (a) was noted by Lars Svenonius. Theorem 1 depends very strongly on the fact that the domain includes all models of a complete first-order theory. The results below make no use of this kind of assumption. In fact we shall eliminate it by assuming that *all structures in the domain of the definition are isomorphic.*

The chief interest of the next theorem is the implication (c) $\Rightarrow$ (a), which tells us that every set definition of reduct type that is structural in enough models of ZFCU has the surjectivity property. I labour some details of the proof because we shall need to consider generalisations of them later.

**Theorem 2** *Let A be a $\sigma$-structure and B an expansion of A formed by adding relations to A. Then the following are equivalent:*

(a) *The homomorphism $\nu : \mathrm{Aut}(B) \to \mathrm{Aut}(A)$ is surjective.*

(b) *Each relation $R_B$ of B is definable in A by a formula $\psi(\bar{x})$ of $L_{\infty,\infty}(\sigma)$.*

(c) *There are a formula $\phi(x,y,z)$ of set theory and a set c in V such that if $V(u)$ is a model of ZFCU satisfying "There are at least $|\mathrm{dom}(\check{B})|$ urelements", then $\phi(x,y,\check{c})$ is a structural set definition in $V(u)$, the domain of $\phi(x,y,\check{c})$ is the class of structures isomorphic to $\check{A}$ in $V(u)$, and $V(u) \models \phi(\check{B}, \check{A}, \check{c})$.*

(d) *See below.*

**Proof.** (a) $\Rightarrow$ (b): Assume (a). Suppose for example that $\sigma^+$ consists of $\sigma$ together with a binary relation symbol $R$, so that $\psi$ will have the form $\psi(x,y)$. Write $\Delta(c_a : a \in A)$ for the diagram of $A$, i.e. the set of atomic or negated atomic sentences true in $A$ when we add new constants $c_a$ to name the elements $a$ of $A$. Then $\psi(x,y)$ can be

$$(\exists x_a)_{a \in A} \left( \forall x \bigvee_{a \in A}(x = x_a) \wedge \right.$$
$$\left. \bigwedge \Delta(x_a : a \in A) \wedge \bigvee_{Rab \text{ holds}} (x = x_a \wedge y = x_b) \right).$$

(b) $\Rightarrow$ (a): Suppose $\psi(x)$ defines $R_B$ in $A$. Then for every automorphism $\alpha$ of $A$, $\psi$ defines $\alpha^\star(R_B)$, so $\alpha^\star(R_B) = R_B$.

(a) $\Rightarrow$ (c): Assuming (a), let $c$ be $\langle B, \sigma \rangle$ and let $\phi(x,y,\langle z_1, z_2 \rangle)$ say

There is a signature $\sigma^+$ extending $z_2$, such that $z_1$ and $x$ are isomorphic $\sigma^+$-structures, $y$ is $(x|z_2)$, and every isomorphism from $(z_1|z_2)$ to $y$ is an isomorphism from $z_1$ to $x$.

(a) guarantees that the structure $x$ defined by this formula is independent of the choice of isomorphism from $(x|\sigma)$ to $y$.

(c) $\Rightarrow$ (a): Assume (c). Consider a model $V(u)$ of $ZFCU$, and let $A'$ be a structure isomorphic to $\check{A}$, with $\mathrm{dom}(A) \subseteq u$. By assumption there are a structure $B'$ and a map $i$ such that

$$V(u) \models \phi(B', A', \check{c}) \wedge (i : \check{B} \to B' \text{ is an isomorphism}). \tag{3}$$

Each automorphism $\alpha$ of $\check{A}$ induces a permutation $\alpha'$ of $u$ by the formula

$$\alpha'(x) = \begin{cases} i\alpha(a) & \text{if } x = ia, \ a \in \check{A}, \\ x & \text{otherwise.} \end{cases} \tag{4}$$

We write $\alpha^\star$ for $(\alpha')^\star$. If $b$ is an element of $\check{B}$, $\alpha$ is an automorphism of $\check{A}$ and $V(u) \models (ib = d)$, then

$$V(u) \models \alpha^\star(i)(b) = \alpha^\star(d) = (\alpha^\star \circ i)(b),$$

and hence

$$\alpha^\star(i) = \alpha^\star \circ i. \tag{5}$$

From (4) and (5) we infer, for each $a \in \check{A}$,

$$(\alpha^\star i)(a) = \alpha^\star(ia) = \alpha'(ia) = i(\alpha a). \tag{6}$$

Applying $\alpha^\star$ to (3) we get:

$$V(u) \models \quad \phi(\alpha^\star(B'), \alpha^\star(A'), \alpha^\star(\check{c})) \wedge \\ (\alpha^\star(i) : \alpha^\star(\check{B}) \to \alpha^\star(B') \text{ is an isomorphism}). \tag{7}$$

Now $\alpha^\star$ leaves all members of $\check{V}$ (the image of $\check{\ }$) fixed. Also by (3) and (4), $\alpha^\star(A') = A'$. So we infer

$$V(u) \models \phi(\alpha^\star(B'), A', \check{c}) \wedge (\alpha^\star(i) : \check{B} \to \alpha^\star(B') \text{ is an isomorphism}).$$

By (3), the first conjunct above and the assumption on $\phi$, $\alpha^\star(B') = B'$, and so by the second conjunct

$$V(u) \models \alpha^\star(i) : \check{B} \to B' \text{ is an isomorphism}.$$

Reasoning in $V(u)$, $i^{-1} \circ \alpha^\star(i)$ is an automorphism of $\check{B}$; by (6),

$$(i^{-1} \circ \alpha^\star(i))(a) = i^{-1}i(\alpha a) = \alpha(a) \text{ for all } a \in A.$$

Hence $i^{-1} \circ \alpha^\star(i)$ is an automorphism of $\check{B}$ extending $\alpha$. Pulling back this information to $V$ via $\check{\ }$, we have (a). $\qquad \square$

The missing clause (d) is an equivalent to (c) for ZFC in place of ZFCU. The proof of (a) $\Rightarrow$ (c) transfers word for word to ZFC. But for (d) $\Rightarrow$ (a) we can't simply replace $V(u)$ by $V$. For example if the universe $V$ carries a global well-ordering $<$, then we can provide a structural definition as in (c) by defining $B'$, for each $A'$ isomorphic to $A$, to be the $<$-first $\sigma^+$-structure isomorphic to $B$ for which $(B'|\sigma) = A'$. This structural definition is available regardless of whether (a) is true or false.

So any candidate for the missing clause (d) must neutralise all such well-orderings. One naturally looks at boolean extensions of $V$. When $A$ and $B$ are countable (or more strictly, when they have countable transitive closures), we can choose a generic ultrafilter and construct a countable transitive model around $A$ and $B$, at the price of assuming there are countable models of ZFC.

**Theorem 2** *(continued)*

(d1) *There are a formula $\phi(x,y,z)$ of set theory and a set $c$ such that for every complete boolean algebra $\mathbb{B}$ in $V$,*

$\|\phi(x,y,\check{c})$ *is a structural definition with domain the class of structures isomorphic to $\check{A}$, and $\phi(\check{B},\check{A},\check{c})\|_{\mathbb{B}} = 1$.*

*Assuming ZFC has countable transitive models, if $B$ above is countable then (d1) is also equivalent to (d2):*

(d2) *There are a formula $\phi(x,y,z)$ of set theory and a set $c$ such that for every countable transitive model $M$ of ZFC with $B$ in $M$, $\phi(x,y,\check{c})$ is a structural definition in $M$ with domain the class of structures isomorphic in $M$ to $A$, and $M \models \phi(B,A,c)$.*

**Proof.** (a) $\Rightarrow$ (d1): The same definition as in (c) works.

(d1) $\Rightarrow$ (a): Assume (d1). By analogy with the argument from (c), we want $\mathrm{Aut}(A)$ and $\mathrm{Aut}(B)$ to have the same values in the boolean universe as they have in $V$. So we put $\lambda = 2^{|\mathrm{dom}\,B|}$ and we choose a $\lambda$-closed notion of forcing. Also we want to construct with boolean value 1 an isomorphic copy $A'$ of $A$, so that automorphisms of $A'$ lift to automorphisms of the boolean universe. There is a standard technology for this. We let $\mathbb{P}$ be the set of all partial maps $p :$ $\lambda^+ \times \lambda^+ \times \lambda^+ \to 2$ with domain of cardinality $\leqslant \lambda$, partially ordered by reverse inclusion; we write $\mathbb{B}$ for the regular open algebra of $\mathbb{P}$.

Now we reason in the boolean universe $V^{\mathbb{B}}$. With boolean value 1, there is a generic family $G$ of $\lambda^+$ independently generic sets, each of them a set of $\lambda^+$ independently generic subsets of $\lambda^+$. If $a, b$ are sets of subsets of $\lambda^+$, we write $a \equiv b$ if the symmetric difference of $a$ and $b$ has cardinality $\leqslant \lambda$. We write $u$ for the set of equivalence classes of elements of $G$. From this point on, the proof marches in step with the proof of (c) $\Rightarrow$ (a). One difference is that while every permutation $\alpha'$ of $u$ does lift to an automorphism of the boolean universe, the lifting is by no means unique or canonical. But in the present argument we only need to know that it exists.

(d1) $\Leftrightarrow$ (d2) is a standard forcing argument.                                                      $\square$

## 4. Relativised reducts

Henceforth we drop the assumption that the relativising formula is $x = x$. Now a set definition of $B$ has to account for the new elements of $B$ as well as the new constituent structure.

**Theorem 3** *Let $A$ be a $\sigma$-structure which is a relativised reduct of a structure $B$. Then each of (c), (d) implies (a).*

*(a) The homomorphism $\nu : \mathrm{Aut}(B) \to \mathrm{Aut}(A)$ is surjective.*

*(c) As Theorem 2(c).*

*(d) As Theorem 2(d1,d2)*

**Proof.** (c) $\Rightarrow$ (a): Assume (c). There are in $V(u)$ a structure $B'$ and a map $i$ such that $i$ is an isomorphism from $\check{A}$ to $A' = (B'|_\theta\sigma)$, $\mathrm{dom}(A') \subseteq u$ and $i$ extends to an isomorphism $j : \check{B} \to B'$. We have

$$V(u) \models \quad \phi(B', A', \check{c}) \wedge (j : \check{B} \to B' \text{ is an isomorphism} \\ \text{extending the isomorphism } i : \check{A} \to A').$$

The rest of the argument is virtually identical with that in Theorem 2. For example if $b$ is an element of $\check{B}$, $\alpha$ is an automorphism of $\check{A}$ and $V(u) \models (jb = d)$, then

$$\alpha^\star(j) = \alpha^\star \circ j. \tag{8}$$

The same argument as before shows that $j^{-1} \circ \alpha^\star(j)$ is an automorphism of $\check{B}$ extending $\alpha$.

(d) $\Rightarrow$ (a) is due to Harvey Friedman [3]. $\qquad\qquad\square$

Theorem 3 is less satisfactory than Theorem 2 in three ways. First, there is no counterpart of Theorem 2(b). Second, we don't prove (c) from (a). Third, we don't prove (d) from (a). It turns out that we can remedy the first two defects quickly by strengthening (a). The third defect remains problematic, but we can narrow the gap.

Begin by noticing that the proof of (c) $\Rightarrow$ (a) in Theorem 2 proved more than (a). In the notation of that proof, define a map $s : \mathrm{Aut}(\check{A}) \to \mathrm{Aut}(\check{B})$ by

$$s\alpha = j^{-1} \circ \alpha^\star \circ j = j^{-1} \circ (\alpha^\star j)$$

(cf. (8)). Then $s\alpha$ extends $\alpha$. We claim that $s$ is a group homomorphism. Thus:

$$\begin{aligned}
(s\alpha)(s\beta)) &= j^{-1}(\alpha^\star)jj^{-1}(\beta^\star)j \\
&= j^{-1}(\alpha^\star\beta^\star)j \\
&= j^{-1}((\alpha\beta)^\star)j \\
&= s(\alpha\beta).
\end{aligned} \tag{9}$$

The equation $\alpha^\star\beta^\star = (\alpha\beta)^\star$ follows directly from the definition of the map $\alpha \mapsto \alpha^\star$.

A group homomorphism $f : G \to H$ is said to be *split surjective* if there is a homomorphism $g : H \to G$ such that $fg = 1_G$; the homomorphism $g$ is called a *splitting* of $f$. In Theorem 4 below we repair Theorem 3 by adding the word 'split' in clause (a). Note that the identity map is always split surjective, so that the addition of 'split' to (a) in Theorem 2 makes no difference there.

Before we state the theorem, we need to prepare clause (b). We say that a signature $\tau^\theta$ is a *$\theta$-expansion* of $\tau$ if $\tau^\theta$ consists of $\tau$ together with some function symbols (possibly infinitary) where the new function symbols are interpreted as partial functions defined only on some sequences of elements satisfying $\theta$. We call these new symbols the *coordinatising symbols*. An atomic formula containing an undefined function is counted as false; so $F(\bar{a}) = F(\bar{a})$ expresses that $\bar{a}$ is in the domain of $F$. By a $\theta$-*atomic* formula we mean the result of replacing one or more variables in some atomic formula of signature $\tau$ by terms $F(\bar{y})$ where the $F$ are coordinatising symbols.

**Theorem 4** *Let $A$ be a $\sigma$-structure which is a relativised reduct of a structure $B$. Then (a)–(c) are equivalent, and (a) entails (d).*

(a) *The homomorphism $\nu : \mathrm{Aut}(B) \to \mathrm{Aut}(A)$ is split surjective.*

(b) *There is a $\theta$-expansion $\sigma^{+\theta}$ of $\sigma^+$ and for each $\theta$-atomic formula $\xi(\bar{x})$ there is a formula $\psi_\xi(\bar{x})$ of $L_{\infty\infty}(\sigma)$, such that $B$ can be expanded to a model $B^\theta$ of all the sentences*

$$(\forall \bar{x} \text{ satisfying } \theta)(\xi(\bar{x}) \leftrightarrow \psi_\xi(\bar{x}))$$

*such that $\mathrm{dom}(A)$ is a set of generators for $B^\theta$.*

(c) *As Theorem 2(c).*

(d) *As Theorem 2(d1,d2)*

**Proof**. We already have (c) $\Rightarrow$ (a) by the previous theorem and (9).

(b) $\Rightarrow$ (c), (d): The sentences in (b) provide an explicit description, for any isomorphism $i : A \to A'$, of a structure $B'$ that is isomorphic to $B$ by an isomorphism extending $i$. Split surjectivity guarantees that $B'$ depends only on $A$ and not on the choice of isomorphism $i$.

(a) $\Rightarrow$ (b): Assume (a) and let $s$ be a splitting of $\nu$. We use $\bar{x}$ for arrays $(x_a : a \in A)$ indexed by the elements of $A$; $\bar{a}$ is the array where every element indexes itself. For each element $b$ of $B$ that is not in $A$, introduce a coordinatising symbol $F_b(\bar{x})$, and interpret

$$(F_b)_{B^\theta}(\bar{c}) = \begin{cases} (s\alpha)b & \text{if for some } \alpha \in \mathrm{Aut}(A), \bar{c} = \alpha\bar{a}, \\ \text{undefined} & \text{if no such } \alpha. \end{cases}$$

Let $\Delta(x_a)_{a\in A}$ be the diagram of $A$. To illustrate the definitions $\psi$, suppose $\xi(\bar{y}, \bar{z}, w)$ is the $\theta$-atomic formula $R(F_b(\bar{y}), F_c(\bar{z}), w)$. Then $\psi_\xi(\bar{y}, \bar{z}, w)$ is the formula

$$\exists \bar{u} \left( \bigwedge \Delta(\bar{u}) \wedge \text{ "}\bar{u} \text{ lists all the elements satisfying } \theta\text{" } \wedge \right.$$

$$\left. \bigvee_{\beta,\gamma\in\mathrm{Aut}(A), d\in A, B^\theta \models R(s\beta b, s\gamma c, d)} \left(w = u_d \wedge \bigwedge_{a\in A}(y_a = u_{\beta a} \wedge z_a = u_{\gamma a})\right) \right).$$

We check that this works. In one direction, suppose $B^\theta \models \xi(\bar{b}, \bar{c}, d)$ where $\bar{b}, \bar{c}, d$ lie in $A$. Then $F_b(\bar{b})$ is defined in $B^\theta$, and so there is a (necessarily unique) $\beta \in \mathrm{Aut}(A)$ such that $\bar{b} = \beta\bar{a}$, and hence $F_b(\bar{b})$ is the element $(s\beta)b$. Likewise $F_c(\bar{c})$ is $s\gamma c$ for a corresponding $\gamma \in \mathrm{Aut}(A)$. Thus $B^\theta \models R(s\beta b, s\gamma c, d)$. From this information, taking $\bar{a}$ for $\bar{u}$, we have $B \models \psi_\xi(\bar{b}, \bar{c}, d)$.

Conversely suppose $B \models \psi_\xi(\bar{b}, \bar{c}, d)$, where again $\bar{b}, \bar{c}, d$ lie in $A$. Let $\bar{e}$ be a sequence for $\bar{u}$ in $\psi_\xi$. Then there are a $\alpha \in \mathrm{Aut}(A)$ such that $\bar{e} = \alpha\bar{a}$, and $\beta, \gamma \in \mathrm{Aut}(A)$ and $a'$ in $A$ such that

$$B \models R(s\beta b, s\gamma c, a') \wedge d = e_{\alpha a'} \wedge \bigwedge_{a \in A} (b_a = e_{\beta a} \wedge c_a = e_{\gamma a}). \tag{10}$$

Since $s$ is a splitting, acting on the first conjunct by $s\alpha$ gives

$$B \models R(s(\alpha\beta)(b), s(\alpha\gamma)(c), \alpha a').$$

Hence

$$B^\theta \models R(F_b(\alpha\beta\bar{a}), F_c(\alpha\gamma\bar{a}), \alpha a').$$

Decoding (10) then yields $B^\theta \models \xi(\bar{b}, \bar{c}, d)$. $\qquad\square$

A natural question is whether there is any theorem about first-order theories that stands to Theorem 4 as Theorem 1 stood to Theorem 2. There are partial positive answers, and they are interesting. See above all the survey of Evans, Macpherson and Ivanov [2] and the papers cited there.

## 5. Weak splittings

Theorem 4 ought to extend to ZFC. But so far this result eludes us. Transferring the argument from ZFCU, the problem is that in the boolean universe the liftings $\alpha^\star$ are not unique, and there is no way of ensuring that $\alpha^\star\beta^\star = (\alpha\beta)^\star$. Shelah suggests looking at large families of liftings, and using the $\Delta$-system lemma to find large coherent subfamilies. This approach gives useful information, but less than (d) $\Rightarrow$ (a).

Let $\nu : G \to H$ be a surjective group homomorphism. By a *weak splitting* of $\nu$ we mean a map $s : H \to G$ such that

(a) $\nu s$ is the identity on $H$;

(b) there is a commutative subgroup $G_0$ of $G$ such that if $f_1, \ldots, f_k$ are elements of $H$ for which $f_1^{\varepsilon_1} \ldots f_k^{\varepsilon_k} = 1$ (where $\varepsilon_i$ is each either 1 or $-1$), then $s(f_1)^{\varepsilon_1} \ldots s(f_k)^{\varepsilon_k} \in G_0$.

If $G_0 = \{1\}$ in this definition, we have the definition of a splitting of $\nu$. Thus every splitting is a weak splitting. We say that $\nu$ is *weakly split* if it has a weak splitting. The following theorem will appear in Hodges and Shelah [8]; the proof is due to Shelah.

**Theorem 5** *(d) below implies (a):*

(a) *The homomorphism* $\nu : \mathrm{Aut}(B) \to \mathrm{Aut}(A)$ *is a weakly split surjection.*

(d) *As Theorem 2(d1,d2).*

## 6. Applications

**Example 1.** Let $G$ be the multiplicative group of $3 \times 3$ upper unitriangular matrices over the ring $\mathbb{Z}/(8\mathbb{Z})$. Let $H$ be the corresponding group over $\mathbb{Z}/(2\mathbb{Z})$, and let $\nu : G \to H$ be the canonical surjection. We show that $\nu$ doesn't weakly split.

Suppose for contradiction that $s$ is a weak splitting of $\nu$. Let $g_1, g_2$ be the two matrices

$$g_1 = \begin{pmatrix} 1 & 1 & 0 \\ 0 & 1 & 0 \\ 0 & 0 & 1 \end{pmatrix}, \quad g_2 = \begin{pmatrix} 1 & 0 & 0 \\ 0 & 1 & 1 \\ 0 & 0 & 1 \end{pmatrix}$$

in $G$, and write $f_1 = \nu(g_1)$, $f_2 = \nu(g_2)$. Now $f_1^2 = f_2^2 = 1$ in $H$, so the weak splitting property tells us that $s(f_1)^2$ and $s(f_2)^2$ commute in $G$. But it is easily checked (using the fact that all entries of $s(f_i) - f_i$ are divisible by 2) that $s(f_1)^2$ and $s(f_2)^2$ don't commute.

**Example 2.** Let $m$ and $n$ be positive integers with $n \geqslant 3$, and let $p$ be a prime with $p^m > 3$. Let $G$ (resp. $H$) be the multiplicative group of invertible $n \times n$ matrices over the ring $\mathbb{Z}/(p^{3m}\mathbb{Z})$ (resp. $\mathbb{Z}/(p^m\mathbb{Z})$), and let $\nu : G \to H$ be the canonical surjection. We shall show that $\nu$ doesn't weakly split.

We write $I$ for the identity element in $G$ and in $H$. The kernel of $\nu$ is the group of matrices of the form $I + p^m f$ where $f$ is in $G$. For any $i, j$ with $1 \leqslant i < j \leqslant n$ let $\delta_{ij}$ be the $n \times n$ matrix which has 1 in the $ij$-th place and 0 elsewhere; then $I + \delta_{ij}$ is an element of $G$ and $\nu(I + \delta_{ij})$ has order $p^m$. The liftings of $\nu(I + \delta_{ij})$ to $G$ are the matrices of the form $I + \delta_{ij} + p^m f$ with $f$ in $G$. Now we repeat a calculation from Evans, Hodges and Hodkinson [1] Prop. 3.7. The element $(I + \delta_{ij} + p^m f)^{p^m}$ is

$$I + \binom{p^m}{1}(\delta_{ij} + p^m f) + \binom{p^m}{2}(\delta_{ij} + p^m f)^2 + \binom{p^m}{3}(\delta_{ij} + p^m f)^3 + \dots$$

Since $\delta_{ij}\delta_{ij} = 0$, $p^{3m}x = 0$ in $\mathbb{Z}/(p^{3m}\mathbb{Z})$ and $p^m > 3$, this multiplies out to

$$I + p^m\delta_{ij} + p^{2m}f + \frac{p^{2m}(p^m - 1)}{2}(\delta_{ij}f + f\delta_{ij}) + \frac{p^{2m}(p^m - 1)(p^m - 2)}{6}\delta_{ij}f\delta_{ij}.$$

Now take

$$g_1 = I + \delta_{12}, \quad g_2 = I + \delta_{23}$$

in $H$, and let $s$ be a weak splitting of $\nu$. Then

$$s(g_1) = I + \delta_{12} + p^m f_1, \quad s(g_2) = I + \delta_{23} + p^m f_2$$

for some $f_1, f_2$ in $G$. Since $s$ is a weak splitting,

$$s(g_1)^{p^m} s(g_2)^{p^m} = s(g_2)^{p^m} s(g_1)^{p^m}.$$

But our calculations show at once that

$$s(g_1)^{p^m} s(g_2)^{p^m} - s(g_2)^{p^m} s(g_1)^{p^m} = p^{2m} \delta_{13} \neq 0.$$

This contradiction proves that $\nu$ doesn't weakly split.

**Example 3**. Let $G$ and $H$ be as in Example 1. Since $n \times n$ upper triangular matrix groups are nilpotent of class $n - 1$, $G$ is a finite soluble group. So by Shafarevich [12] there is a Galois extension $K$ of the field $\mathbb{Q}$ of rationals such that $G$ is the Galois group of $K/\mathbb{Q}$. Let $k$ be the fixed field of the kernel $G_0$ of $\nu : G \to H$. Then $H$ is the Galois group of the extension $k/\mathbb{Q}$.

**Corollary 6** *There is no formula $\psi(x, y)$ of set theory such that, provably from ZFC, if $K$ is a field then there is a unique set $b$ such that $\psi(b, K)$, and this set $b$ is an algebraic closure of $K$.*

**Proof**. Suppose to the contrary that there is such a formula $\psi$. Then for each field $k'$ isomorphic to $k$ as in Example 3, $\psi$ finds an algebraic closure $\bar{k}'$ of $k'$. Since $K/\mathbb{Q}$ is a Galois extension of finite degree, there is a unique subfield $K'$ of $\bar{k}'$ isomorphic to $K$, and a formula of set theory (without parameters) finds $K'$ from $\bar{k}'$. Putting all this together, we have a structural set definition $\phi(x, y)$ with domain the class of fields isomorphic to $k$, that defines $K$ from $k$ in all models of ZFC. By Theorem 5 it follows that $\nu : \text{Aut}(K) \to \text{Aut}(k)$ weakly splits. Since $\mathbb{Q}$ is rigid, $\nu$ here is the same as the $\nu$ of Examples 1 and 3, and we saw in Example 1 that $\nu$ doesn't weakly split; contradiction. $\square$

**Example 4**. Let $G$ and $H$ be as in Example 2. Let $B$ (resp. $A$) be the direct sum of $n$ copies of the abelian group $\mathbb{Z}/(p^{3m}\mathbb{Z})$ (resp. $\mathbb{Z}/(p^m\mathbb{Z})$), and identify $A$ with $p^{2m}B$. Let the relation symbol $P$ pick out $A$ within $B$. Then $G$ (resp. $H$) is the automorphism group of $B$ (resp. $A$), and $\nu : G \to H$ is the map induced by restriction.

**Corollary 7** *There is no formula $\psi(x, y)$ of set theory such that, provably from ZFC, if $A$ is an abelian group then there is a unique set $b$ such that $\psi(b, A)$, and this set $b$ is a divisible hull of $A$.*

The **proof** is analogous to that of Corollary 6, using Examples 2 and 4 in place of 1 and 3. $\square$

Neither of these corollaries is in the least bit surprising. The surprise was that they were so difficult to prove.

**REFERENCES**

1. David M. Evans, Wilfrid Hodges and I. M. Hodkinson, 'Automorphisms of bounded abelian groups', *Forum Mathematicum* 3 (1991) 523–541.
2. David M. Evans, Dugald Macpherson and Alexandre A. Ivanov, 'Finite covers', in *Model Theory of Groups and Automorphism Groups*, ed. David M. Evans, Cambridge University Press 1997.

3.  Harvey Friedman, 'On the naturalness of definable operations', *Houston Journal of Mathematics* 5 (1979) 325–330.
4.  Haim Gaifman, 'Operations on relational structures, functors and classes. I', in *Proceedings of the Tarski Symposium*, ed. Leon Henkin et al., American Mathematical Society, Providence RI 1974, pp. 21–39.
5.  Wilfrid Hodges, 'The meaning of specifications II: Set-theoretic specification', in *Semantics of Programming Languages and Model Theory*, ed. Manfred Droste and Yuri Gurevich, Gordon and Breach, Yverdon 1993, pp. 43–68.
6.  Wilfrid Hodges, *Model Theory*, Cambridge University Press, Cambridge 1993.
7.  Wilfrid Hodges and Saharon Shelah, 'Naturality and definability, I', *Journal of London Mathematical Society* 33 (1986) 1–12.
8.  Wilfrid Hodges and Saharon Shelah, 'Naturality and definability, II' (in preparation).
9.  Kiki Kennedy-Day, *Books of Definition in Islamic Philosophy*, RoutledgeCurzon, London 2003.
10. Hélène Kirchner and Peter D. Mosses, 'Algebraic specifications, higher-order types and set-theoretic models', *Journal of Logic and Computation* 11 (2001) 453–481.
11. Michael Makkai, 'Avoiding the axiom of choice in general category theory', *Journal of Pure and Applied Algebra* 108 (1996) 109–173.
12. I. R. Shafarevich, 'On the construction of fields with a given Galois group of order $\ell^{a}$' (Russian), *Izv. Akad. Nauk SSSR Ser. Mat.* 18 (1954) 261–296.

# Evolutionary Considerations on Logical Reasoning

Michiel van Lambalgen[*]

*Department of Philosophy and Cognitive Science Center Amsterdam, University of Amsterdam*
*M.vanLambalgen@uva.nl*

**Abstract.** A famous series of experiments in evolutionary psychology, due to Leda Cosmides [4] attempts to show that (1) human cannot have evolved a capacity for general logical reasoning, and (2) humans are able to do correct logical reasoning only in cases which were once evolutionary advantageous; in particular she focusses on social contracts. We show that these experiments are deeply flawed, because they rest on a misunderstanding of logic. In the second half of the paper, another account of a possible evolutionary origin of logical reasoning is sketched, which is based on the intimate connection between neural networks and a particular form of non-monotonic reasoning, namely logic programming with negation as failure.

## 1. Introduction: psychology of reasoning and logic

Psychology and logic have a fraught relationship. Following Frege, who wrote

> The logicians ... are too much caught up in psychology ... Logic is in no way a part of psychology. The Pythagorean theorem expresses the same thought for all men, while each person has its own representations, feelings and resolutions that are different from those of every other person. Thoughts are not psychic structures, and thinking is not an inner producing and forming, but an apprehension of thoughts which are already objectively given[2].

logicians have for the most part declined to look into the role that logic plays in cognitive processes.

On the other hand, psychologists, noticing that human reasoning often does not conform to the norms of classical logic, have declared formal logic to be irrelevant for the study of actual human reasoning. This rift between logic and psychology has had detrimental effects on cognitive science. Here I want to look at one particular instance, the evolutionary explanation for the origin of logical reasoning proposed by Cosmides [4]. This explanation has been held up as a model of the power of evolutionary theorizing, and has high public profile; see for example Steven Pinker's *How the mind works* [13]. Nevertheless the proposed explanation is deeply flawed. One could point to errors in the biological reasoning, but for our present purposes it is more informative to point out the severe misrepresentation

---

[*]Based on an invited lecture at the 12th International Congress of Logic, Methodology and Philosophy of Science, Oviedo, August 2003. The research reported here is joint work with Keith Stenning (Edinburgh). The author is grateful to the Netherlands Organization for Scientific Research (NWO) for support under grant 360-80-000.
[2]G. Frege, letter to Husserl; see Vol.VI, p. 113 of [8]

of logic in evolutionary psychology . Accordingly, in the first part of the paper we present theoretical and experimental results which show that the evolutionary psychologists' interpretation of their experiments is untenable. By no means do we wish to imply that evolutionary considerations do not apply to logic, and so in the second part of the paper a different approach to the evolution of our reasoning capacity is outlined. Very briefly, whereas evolutionary psychology tries to explain logical reasoning as an adaptation, we would bet our money on logical reasoning being an exaptation.

The remainder of this section is devoted to a very brief synopsis of the psychological background. The psychology of reasoning is concerned with the experimental study of reasoning patterns also studied by logicians, and in the literature we may find experimental research on, for example,

- reasoning with syllogisms in adults

- reasoning with propositional connectives in adults

- acquisition of connectives and quantifiers in children

- reasoning in subjects with various psychiatric or cognitive impairments

- brain correlates of reasoning

- reasoning in 'primitive' societies

For example, an adult subject may be presented with the premises

> If Julie has an essay, she studies late in the library.
> Julie does not study late in the library.

and is then asked: what, if anything, follows? In this case (*modus tollens*) it may happen that half of the subjects reply that nothing can be concluded. In contrast, the analogous experiment for *modus ponens*, with minor premiss

> Julie has an essay.

typically yields success scores of around 95%. The psychologist is then interested in explaining the difference in performance, and believes that differences such as this actually provide a window on the cognitive processes underlying logical reasoning. What is distinctive about the psychology of reasoning is that it views its task as uncovering *the mechanism* of logical reasoning: what goes on in the brain (and where) when it makes an inference? The field has fragmented into different schools, each identified by what it takes to be the mechanism underlying reasoning.

### Mental logic

This school maintains that logical reasoning is the application of formal rules, more or less like natural deduction. Here is an example (from [15]): the theory tries to explain why humans tend to have difficulty with *modus tollens*, by assuming that this is not a primitive rule, unlike *modus ponens*; *modus tollens* has to be derived each time it is used, therefore it leads to longer processing time.

## Mental models

The founding father of the 'mental models' school is Phil Johnson-Laird; applications of 'mental models' to logic can be found in Johnson-Laird and Byrne [9]. The main claim of this school is that reasoners do not apply content-independent formal rules (such as for example *modus ponens*), but construct models for sentences and read off conclusions from these, which are then subject to a process of validation by looking for alternative models. Errors in reasoning are typically explained by assuming that subjects read off a conclusion from the initial model which is not true in all models of the premises. The 'mental models' school arose as a reaction against 'mental logic' because it was felt that formal, content-less rules would be unable to explain the so-called 'content-effects' in reasoning.

## Darwinian algorithms

Evolutionary psychology has also tried to shed its light on logic, beginning with the famous (or notorious) paper 'The logic of social exchange: Has natural selection shaped how humans reason? Studies with the Wason selection task' by Leda Cosmides [4]. Here the main claim is that there is no role for (formal, domain–independent) logic in cognition; whenever we appear to reason logically, this is because we have evolved strategies ('Darwinian algorithms') to solve a problem in a particular domain (such as social contracts).

There are more 'schools' than have been mentioned here. But for now the most important point is that talk about *the mechanism* is apt to be highly confusing, because it does not take into account that any cognitive phenomenon can be studied at various levels. The classical discussion of this issue is David Marr's [11], where he points out that cognitive science should distinguish at least the following three levels of inquiry

1. identification of the information processing task as an input–output function

2. specification of an algorithm which computes that function

3. neural implementation of the algorithm specified

These distinctions between levels are of course familiar from computer science, as is the observation that there is no reason to stick to three levels only, since a program written in one language (say Prolog) may be implemented in another language (say C), and so on all the way down to an assembly language. Furthermore, it is tempting to think that the neural implementation provides some kind of rock bottom, the most fundamental level of inquiry; but of course the neural implementation uses a *model* of what actual neurons do, and not the real things themselves.

The upshot of this discussion is that it makes no sense to ask for 'the' mechanism underlying reasoning without first specifying a level at which intends to study this question. As a consequence, a superficially sensible distinction such as that between mental logic and mental models may turn out to be empirically meaningless after all, at least in the form it is usually stated. The fact that the argument patterns studied by psychologists typically come from logics with a completeness theorem should already give one pause: at the input–output level manipulations with rules and manipulations with models cannot be distinguished. One can look at subtler measures such as error rates or reaction times; e.g. mental modelers like to point to a correlation between the difficulty of a syllogistic figure (as measured by error rate), and the number of models that the premisses of the figure allow (see Stenning [17] for a discussion of this type of argumentation).

Since the seventies, the psychology of reasoning has turned its back on the insights of logic, mainly under the influence of experimental results purportedly showing that subjects' behaviour in reasoning tasks is completely unrelated to logical form. The most celebrated of these results is Wason's four card task [23]. This task is concerned with reasoning about a conditional. Subjects are presented with the following form:

> Below is depicted a set of four cards, of which you can see only the exposed face but not the hidden back. On each card, there is a number on one of its sides and a letter on the other.

> Also below there is a rule which applies only to the four cards. Your task is to decide which if any of these four cards you *must* turn in order to decide if the rule is true. Don't turn unnecessary cards. Tick the cards you want to turn.

> **Rule:** *If there is a vowel on one side, then there is an even number on the other side.*

**Cards:**

| A | K | 4 | 7 |
|---|---|---|---|

The results are striking. If we represent the rule in propositional logic[3] as an implication $p \rightarrow q$, the observed pattern of results is typically given by the following table

- 0–5% $p, \neg q$

- 45% $p, q$

- 35% $p$

- 7% $p, q, \neg q$      .

- rest miscellaneous

Wason claimed that the logically correct answer in this case should be $p, \neg q$, an answer given by a tiny minority, and he therefore considered the vast majority to be irrational. Here is an excerpt from his own description of the experiment

> Our basic paradigm has the enormous advantage of being artificial and novel; in these studies we are not interested in everyday thought, but in the kind of thinking which occurs when there is minimal meaning in the things around

---

[3]We will later see many reasons why the standard analysis of this experiment is wrong. Let us note already at this stage that the presence of the combined deictic/anaphoric expression 'one side – other side' calls for a formalization in predicate not propositional logic.

us. On a much smaller scale, what do our students' remarks remind us of in real life? They are like saying 'Of course, the earth is flat', 'Of course, we are descended from Adam and Eve', 'Of course, space has nothing to do with time'. The old ways of seeing things now look like absurd prejudices, but our highly intelligent student volunteers display analogous miniature prejudices when their premature conclusions are challenged by the facts. As Kuhn has shown, old paradigms do not die in the face of a few counterexamples. In the same way, our volunteers do not often accommodate their thought to new observations, even those governed by logical necessity, in a deceptive problem situation. They will frequently deny the facts, or contradict themselves, rather than shift their frame of reference. [. . . ] [T]he present interpretation, in terms of the development of dogma and its resistance to truth, reveals the interest and excitement generated by research in this area [25].

The list of students' cognitive sins is itself interesting: one might think that the relation between space and time (still a subject of deep physical and philosophical enquiry) is not quite in the same category as the shape of the earth (a settled issue). In fact, we would claim that logic is more analogous to the former than to the latter.

Paradoxically, what really damaged the role of logic in the psychology of reasoning was the observation, by Wason and Johnson-Laird [24], that in some cases reasoning with a rule which has clear semantic content elicits a high percentage of $p, \neg q$ answers in the four card task, typically 75% or higher. An example would be the rule

If you want to drink alcohol on these premisses, you have to be over 18

with cards laid out as follows

| whisky | juice | 19 | 12 |
|--------|-------|----|----|

This result was interpreted as showing that logical form is not a determinant of human reasoning: here we have two conditionals, obviously of the same logical form, and the same reasoning task; in one case performance is disastrous, in the other case it is as it should be, given that *modus tollens* is felt to be much harder than *modus ponens*. Surely this shows that logical form, hence logic, plays no role in cognition? This point was picked up in Cosmides' celebrated paper [4], where, in combination with her own results, it was used to argue that this pattern of results actually provides a clue to the evolutionary origin of the human reasoning capacity, in so far as it exists.

## 2. Evolutionary psychology and reasoning: Cosmides' 1989 study of the Wason selection task

Much can be said about the ideological roots of evolutionary psychology, but for this we refer the reader to the relevant literature (see for example Malik's fairly dispassionate study [10], and the references cited therein). Suffice it to say here that evolutionary psychology considers itself to be an improved version of sociobiology, because it uses the methods of experimental psychology, and may thus be immune to reproaches of coming up with 'just so stories'. But the ultimate aim is the same as that of sociobiology: to show that the mind is *in toto* a product of natural selection. One might think: 'Indeed, how can it be otherwise?', but the bite is in a particular interpretation of what 'product' here means.

Broadly speaking, one may distinguish two types of evolutionary process: *adaptation* and *exaptation*. The textbook example of an adaptation is that of the melanic moth, which arose by natural selection on the blackened trees of industrial revolution Manchester, outcompeting its lighter relatives. The textbook example of an exaptation is the use of feathered wings for flight: originally feathers were selected for because they led to superior thermal insulation, and then some bird serendipitously discovered its potential for airborne movement. Only then a process of secondary adaptation set in, favouring birds whose wings were better equipped for flight.

Evolutionary psychology claims that complex functions such as those found in cognition can only be explained as adaptations, responses to *specific* environmental pressures.

> [Content-specific mechanisms] will be far more efficient than general purpose mechanisms ... [content-independent systems] could not evolve, could not manage their own reproduction, and would be grossly inefficient and easily out-competed if they did [5].

This puts logic in an awkward position: reasoning, like learning and memory, seems to be domain-general. Just as we seem to be able to store information about any topic, we are apparently able to reason about any topic. In fact, the very definition of validity appears to emphasize the domain-independence of logic: 'whatever you substitute for the non-logical terms, if the premises are true, then so is the conclusion'.

The previous paragraph abounds with 'seems' and 'appears' because we believe that evolutionary psychology completely misrepresents the nature of logic. However, for the moment we will continue their line of reasoning, which is that

(1) logic is content-independent by definition,

(2) content-independent mechanisms cannot have evolved, as opposed to content-dependent mechanisms, so that therefore

(3) logic must be shown to be content-dependent after all, and moreover

(4) content-dependent because it arose from a very specific environmental pressure.

The pressure that Cosmides [4] focusses on is the need to police social contracts. Society can maintain cohesion only if its members are mutually bound by social contracts, and, furthermore, if each member is capable of unmasking those who cheat on a contract. That is, humankind will have evolved *cheater detectors*, a genetically determined module whose dedicated function is to unmask cheaters. Cosmides and, in her wake evolutionary psychology, claims that cheater detection is the evolutionary root of logical reasoning. Moreover, the traces of this evolutionary root can still be found in today's logical reasoning: logical reasoning is successful if and only if it takes the form of cheater detection. Due to space limitations it is impossible to give a full discussion of all of Cosmides' experiments, but we hope the following gives the reader a flavour of her paradigm.

Cosmides' main tool is a variant of the four card task involving a social contract in a fictional setting. The fictional setting is introduced to eliminate the hypothesis that successful logical reasoning is due to familiarity with the content of the rule.

> You are an anthropologist studying the Kaluame, a Polynesian people who live in small, warring bands on Maku Island in the Pacific. You are interested in how Kaluame "big men" – chieftains –wield power.
>
> Big Kiku is a Kaluame big man who is known for his ruthlessness. As a sign of loyalty, he makes his own subjects put a tattoo on their face. Members of other Kaluame bands never have facial tattoos. Big Kiku has made so many enemies in other Kaluame bands, that being caught in another village with a facial tattoo is, quite literally, the kiss of death.
>
> Four men from different bands stumble in Big Kikus village, starving and desperate. They have been kicked out of their respective villages for various misdeeds, and have come to Big Kiku because they need food badly. Big Kiku offers each of them the following deal:
>
> > 'If you get a tattoo on your face, then I'll give you cassava root.'
>
> You learn that Big Kiku hates some of these men for betraying him to his enemies. You suspect he will cheat and betray some of them. Thus, this is the perfect opportunity for you to see first hand how Big Kiku wields his power.
>
> The cards below have information about the fates of the four men. Each card represents one man. One side of a card tells whether or not the man went through with the facial tattoo that evening and the other side of the card tells whether or not Big Kiku gave that man cassava root the next day.

| tattoo | no tattoo | cassava | no cassava |
|--------|-----------|---------|------------|

In this experiment, performed on Stanford undergraduates, 75% chooses the 'tattoo' and 'no cassava' cards, i.e. the $p, \neg q$ answer. Cosmides concludes from this that familiarity hypothesis is refuted, and the activation of the cheater detection module is responsible for the observed results. She tries to corroborate this explanation in two ways. In one experimental condition, she claims that cheater detection and logic actually lead to different predictions. Suppose the rule in the above experimental condition is changed to

> 'If I give you cassava root, you must get a tattoo on your face'

Cosmides claims that the logical form of the rule dictates the choice of the 'cassava' and 'no tattoo' cards, whereas again 75% of subjects choose the 'tattoo' and 'no cassava' cards, as in the previous condition. Thus, logic would actually be a bad guide to reasoning.

She then entertains the possibility that there is after all a 'logic of social contracts', a kind of deontic logic, which facilitates reasoning in the above condition. The experimental condition to refute this suggestion is the 'altruism' experiment, in which subjects are asked to investigate whether said Big Kiku has behaved altruistically toward the four men. The condition starts out as before, but continues

> ... You learn that Big Kiku hates some of these men for betraying him to his enemies. You suspect he will cheat and betray some of them. However, you

have also heard that Big Kiku sometimes, quite unexpectedly, shows great
*generosity* towards others – that he is sometimes quite *altruistic*. Thus, this
is the perfect opportunity for you to see first hand how Big Kiku wields his
power. ... Did Big Kiku behave *altruistically* towards any of these four men?
Indicate only those card(s) you definitely need to turn over to see if Big Kiku
has behaved *altruistically* to any of these four men.

| tattoo | no tattoo | cassava | no cassava |

Cosmides claims that in this case the correct answer is the set 'no tattoo' and
'cassava', but, although no figures are given, she writes that few subjects choose
this answer. This would show that humans have no 'altruism detection module',
and no 'logic of social contracts', but only the cheater detection module.

## 3. What's wrong with this?

One may criticize Cosmides' experiments because her nullhypotheses (i.e. what
logic predicts) are wrong. For example, in the case of the 'switched social contract',
logic and cheater detection only give different predictions if the rule is formulated as
a (one-way) conditional. But obviously a contract is a symmetric affair, so that the
true logical form is more like a biconditional[4], and the answer sets are determined
by the perspective enforced by the instructions. This points to a deeper problem:
the tendency, endemic in the psychology of reasoning, to equate logical form with
surface form. Similarly, in the altruism experiment it is by no means clear what the
'correct' answer should be. One may argue: 'A true altruist gives without asking
anything in return. Therefore the cards as laid out already show that Big Kiku has
not behaved altruistically', i.e. *no* card has to be chosen. One may also argue that
*all* cards must be chosen; for Cosmides' suggested answer allows that Big Kiku
still cheats while being altruistic. This then already gives three possible answer
sets, and we haven't even taken into account that the *any* in the instruction is a
negative polarity item, bringing in complications of its own.

Here we want to focus on the deeper problem indicated above: the role of logical
form. As we have seen, a typical argument in the psychology of reasoning goes
like this: the rule in Wason's abstract task and in a task with 'content' such as
Cosmides' have the same logical form; success scores in these cases are very different, *ergo* logical form is not a determinant of success. But one may well question
whether the logical forms are indeed the same. Logical form is not determined by
looking at the surface form, noticing an 'if ... then', and concluding: 'Ah yes, this
must be the material implication'. Assigning logical form to a natural language
expression at least involves

1. choosing a formal language

2. choosing a formal expression to match the natural language expression

3. choosing a semantics for the language

4. choosing a definition of valid argument

---

[4]Or rather what corresponds to this in deontic logic.

For each parameter there are many different possibilities, and a subject in an experiment must engage in a considerable amount of reasoning to determine what, in the given context, the most appropriate choices are. In the case at hand, the difference between an abstract Wason-type rule, and a rule like Cosmides' shows up in the semantics. Basically, the difference is this: the abstract rule, which is descriptive, can be true or false, whereas in the other cases we are concerned with deontic rules, norms which can only be violated, but not shown to be false. This simple distinction can actually explain a number of the differences in performance that have been observed, the idea being that descriptive conditionals occasion much more processing difficulties than deontic conditionals. We will now present some examples from an experimental program pursued jointly with Keith Stenning, in which subjects are engaged in a dialogue while working through the abstract four card task. These dialogues provide a huge amount of information on the difficulties subjects experience while trying to make sense of, indeed imposing logical form on, the instructions. Here is a list of the problems we could identify; there may be more.

- what is truth?

- what is falsity?

- pragmatics: the authority of the source of the rule

- rules and exceptions

- reasoning and planning

- interaction between interpretation and reasoning

- truth of the rule vs. 'truth' of a case

- cards as viewed as a sample from a larger domain

- obtaining evidence for the rule versus evaluation of the cards

- subjects' understanding of propositional connectives generally

For a full discussion of these difficulties, with illustrative excerpts from dialogues, we refer the reader to Stenning and van Lambalgen [18]. Here we concentrate on two of these: the interaction between reasoning and planning, and the interaction between interpretation and reasoning.

### 3.1. Planning one's card choices

In daily life reasoning and reactive planning are intertwined. When planning a trip, one does not engage in backwards chaining from the goal to the present state in order to come up with a fully worked out plan, but one builds in possibilities for observation, and allows one's plan to reactively depend on the outcome of the observation. The peculiarity of Wason's task is that such reactive planning is not allowed. One must choose the cards *a priori*, without the possibility to turn the cards and see what is on the other side. Some subjects find this impossible to do:

*Subject* 10.

*S.* OK so if there is a vowel on this side then there is an even number, so I can turn A to find out whether there is an even number on the other side or I can turn the 4 to see if there is a vowel on the other side.

*E.* So would you turn over the other cards? Do you need to turn over the other cards?

*S.* I think it just depends on what you find on the other side of the card. No I wouldn't turn them....

*E.* So you are inclined to turn this over [the A] because you wanted to check?

*S.* Yes, to see if there is an even number.

*E.* And you want to turn this over [the 4]?

*S.* Yes, to check if there is a vowel, but if I found an odd number [on the back of the A], then I don't need to turn this [the 4].... Well, I'm confused again because I don't know what's on the back, I don't know if this one ...

*E.* What about the 7?

*S.* Yes the 7 could have a vowel, then that would prove the whole thing wrong. So that's what I mean, do you turn one at a time or do you ...?

If one modifies the instruction in the Wason task so that it reads

> Your task is to decide which of these four cards you *must* turn (if any) in order to decide if the rule is true. *Assume that you have to decide whether to turn each card before you get any information from any of the turns you choose to make.*

$p, \neg q$ scores increase to 25%. The moral is that some subjects find the task in the original form impossible, because they bring to it their common sense understanding of the relation between reasoning and planning, and nothing in the instructions tells them not to. But note that in the case of deontic rules, this difficulty cannot arise. Whether one card violates the norm is independent of whether another card violates the norm, so there is no planning involved.

### 3.2. The unbearable lightness of interpretations

A common assumption in psychological experiments on reasoning is that one can distinguish a stage in which the instructions are semantically interpreted, followed by a stage in which the subject reasons with the material thus interpreted. We will now see that this assumption is false. Some subjects suit the semantic interpretation to the reasoning task, presumably to ease the processing load. We investigated this phenomenon by manipulating the interpretation of the deictic/anaphoric expression 'one side ... other side'. The rule

> If there is a vowel on one side, then there is an even number on the other side.

can be decomposed into

> (1) If there is a vowel *on the visible face*, then there is an even number *on the invisible back* [first anaphora condition]
> (2) If there is a vowel *on the invisible back*, then there is an even number *on the visible face* [second anaphora condition]

The second condition may cause subjects great trouble, and may lead them to change their interpretation of the conditional. Here is an example.

> Subject 16 [has correctly chosen A in first anaphora condition.]
>
> *E.* The next one says that if there is a vowel on the back of the card, so that's the bit you can't see, then there is an even number on the face of the card, so that's the bit you can see; so that again is slightly different, the reverse, so what would you do?
>
> *S.* Again I'd turn the 4 so that would be proof but not ultimate proof but some proof ...
>
> *E.* With a similar reasoning as before?
>
> *S.* Yes, I'm pretty sure what you are after ... I think it is a bit more complicated this time, with the vowel on the back of the card and the even number, that suggests that *if and only if* there is an even number there can be a vowel, I think I'd turn others just to see if there was a vowel, so I think I'd turn the 7 as well.

This is of course very curious: the subject appears not to be hampered by a compositional notion of meaning in which 'if ... then' has a fixed meaning. Instead, she believes that the meaning may depend upon other material.

Below is an even more striking instance of this phenomenon.

> Subject 23 [Standard Wason task]
>
> *S.* Then for this card [4/K] the statement is not true[5].
>
> *E.* Could you give a reason why it is not?
>
> *S.* Well, I guess this also assumes that the statement is reversible, and if it becomes the reverse, then instead of saying if there is an A on one side, there is a 4 on the other side, it's like saying if there was a 4 on one side, then there is an A on the other ...
>
> *E.* Now we'll discuss the issue of symmetry, you said you took this to be symmetrical.
>
> *S.* Well, actually it's effectively symmetrical because you've got this either exposed or hidden clause, for each part of the statement. So it's basically symmetrical.
>
> *E.* But there are two levels of symmetry involved here. One level is the symmetry between visible face and invisible back, and the other aspect of symmetry is involved with the direction of the statement 'if ... then'.
>
> *S.* Right, o.k. so I guess in terms of the 'if ... then' it is not symmetrical ... In that case you do not need that one [4], you just need A.

The subject thus infers the symmetric, biconditional nature of the 'if ... then' from the symmetry of 'one side ... other side'. We then correct him, by pointing out that these two notions of symmetry are actually distinct. Because of this, he comes to realize that the conditional is asymmetric, but there is a strong suggestion that he then switches to an asymmetric interpretation of 'one side ... other side' as well,

---

[5]The notation [4/K] means: upon turning the real card with a 4 on the face, the subject found a K.

by choosing an answer which is appropriate only for the asymmetric case where 'one side' means 'visible face', and 'other side' means 'invisible back'. We next led him through a different experiment, the two-rule task, in which this suggestion was strongly reinforced:

> Below is depicted a set of four cards, of which you can see only the exposed face but not the hidden back. On each card, there is a number (either 3 or 8) on one of its sides and a letter (either U or I) on the other. Also below there appear two rules. One rule is true of all the cards, the other isn't. Your task is to decide which cards (if any) you *must* turn in order to decide which rule holds. Don't turn unnecessary cards. Tick the cards you want to turn.
>
> **Rule 1:** *If there is a vowel on one side, then there is an even number on the other side.*
>
> **Rule 2:** *If there is a consonant on one side, then there is an even number on the other side.*
> **Cards:**

| U | I | 3 | 8 |
|---|---|---|---|

[Same subject in two-rule experiment; while attempting the task he makes some notes which indicate that he is still aware of the symmetry of the cards]

*S.* For U, if there is an 8 on the other side, then rule one is true, and you'd assume that rule two is false. And with I, if you have an 8, then rule one is false and rule two is true.

[The subject has turned the U and I cards, which both carry 8 on the back, and proceeds to turn the 3 and 8 cards.]

*S.* Now the 3, it's a U and it's irrelevant because there is no reverse of the rules. And the 8, it's an I and again it's irrelevant because there is no reverse of the rules. ... Well, my conclusion is that the framework is wrong. I suppose rules one and two really hold for the cards.

*E.* We are definitely convinced only one rule is true ...

*S.* Well ... say you again apply the rules, yes you could apply the rules again in a second stab for these cards [3 and 8] here.

*E.* What do you mean by 'in a second stab'?

*S.* Well I was kind of assuming before you could only look at the cards once based on what side was currently shown to you. ... This one here [8] in the previous stab was irrelevant, because it would be equivalent to the reverse side when applied to this rule, I guess now we can actually turn it over and find the 8 leads to I, and you can go to this card again [3], now we turn it over and we apply this rule again and the U does not lead to an 8 here. So if you can repeat turns rule two is true for all the cards.

*E.* You first thought this card [3] irrelevant.

*S.* Well it's irrelevant if you can give only one turn of the card.

The subject has now become set in his ways: the suspicion that he interprets 'one ... other side' asymmetrically is confirmed and he goes through all kinds of mental gyrations to reconcile his understanding of the task with the experimenter's. We thus see a reciprocal influence of the interpretation of the conditional and of the anaphora. Hence the interpretation is far from stable. But note that, again, this particular problem cannot arise for the deontic examples, because the expression 'one side ... other side' is lacking there.

The upshot of this is that the differences in logical form between descriptive and deontic conditionals imply differences in processing loads, favouring deontic conditionals. Therefore the observed results do not argue against a human capacity for reasoning based on logical form. At most they show that processing difficulties may conspire to make correct performance in the descriptive task hard to attain. This points to another flaw in the proposed evolutionary explanation: it treats logical reasoning in isolation from other cognitive functions, in particular memory systems. Here we see a drawback of the 'massive modularity' advocated by evolutionary psychology: it has no room for the interaction between modules. But perhaps it is precisely this interaction which gives a clue to the evolutionary origin of logical reasoning?

## 4. Can planning tell us something about the origin of logic?

In some ways classical logic is the nemesis of working memory. Consider what is involved in checking semantically whether an argument of the form $\varphi_1, \varphi_2 / \psi$ is classically valid. One has to construct a model $\mathcal{M}$, then check whether $\mathcal{M} \models \varphi_1, \varphi_2$; if not, discard $\mathcal{M}$; otherwise, proceed to check whether $\mathcal{M} \models \psi$, and repeat until all models have been checked. This procedure puts heavy demands on working memory, because the models which have to be constructed are generally not saliently different, so are hard to tell apart; by the same token, it is not easy to check whether one has looked at all relevant models. The fact that classical logic does not fit harmoniously with the operation of working memory suggests that when speculating about the evolutionary origin of logic, we should not take classical logic as a starting point. Classical logic may indeed be an acquired trick, because it requires overcoming the tyranny of working memory. There may however be other logics which are very much easier on working memory, for instance because the number of models to be considered is much lower, or because these models exhibit salient differences.

In this section we look at the logic inherent in planning, and we claim that this logic is a fitting subject for evolutionary enquiry, both because it poses fewer demands on working memory, and because there exists experimental evidence showing that this logic is naturally applied in reasoning tasks. Before we let out the secret and explain to the reader what the logic inherent in planning is, let us dwell a little on the evolutionary importance of planning.

By definition, planning consists in the construction of a *sequence* of actions which will achieve a given goal, taking into account properties of the world and the agent, and also events that might occur in the world. Both humans and non-human primates engage in planning. It has even been attested in monkeys. In recent experiments with squirrel monkeys by McGonigle, Chalmers and Dickinson [12], a monkey has to touch all shapes appearing randomly on a computer screen. The shapes come in different colours, and the interesting fact is that, after extensive training, the monkey comes up with the plan of touching all shapes of a particular colour, and doing this for each colour. This example clearly shows the hierarchical nature of planning: a goal is to be achieved by means of actions which are themselves composed of actions. It is precisely the hierarchical, 'recursive' nature of planning which has led some researchers to surmise that planning has been co-opted by the language faculty, especially syntax (Greenfield [7]; Steedman [16]). It is consistent with this that Broca's area is immediately adjacent to areas for motor planning, although this argument loses some force in view of modern evidence that Broca's area is not the sole locus for syntax. There is also a route from planning to language that goes via semantics. There is a live possibility that a distinguishing feature of human language vis à vis ape language is the ability to engage in discourse. Chimpanzees can produce single sentences, which when read charitably show some signs of syntax. But stringing sentences together into a discourse, with all the anaphoric and temporal relations that this entails, seems to be beyond the linguistic capabilities of apes. One can make a good case, however, that constructing a temporal ordering of events out of a discourse involves an appeal to the planning faculty (see van Lambalgen and Hamm [20]).

The preceding considerations lend some plausibility to the suggestion that planning, a function we share with the nonhuman primates, has been co-opted for a higher cognitive function, viz. language. We will now consider the possibility that planning has been important in the evolution of logical thinking. We first look at a body of data on reasoning which, when interpreted properly, can be seen to show planning at work.

## 5. The 'suppression effect' (Byrne [3], Dieussaert et al. [6])– standardly conceived

Suppose one presents a subject with the following innocuous premisses:

(1)    a. *If she has an essay to write she will study late in the library.*

      b. *She has an essay to write.*

In this case roughly 95% of subjects draw the conclusion 'She will study late in the library' (one wonders what the remaining 5% are thinking). Next suppose one adds the premiss

(2) *If the library is open, she will study late in the library.*

and one asks again: what follows? In this case, only 60% concludes 'She will study late in the library'.

However, if instead of the above, the premiss

(3) *If she has a textbook to read, she will study late in the library*

is added, then the percentage of 'She will study late in the library'–conclusions is again 95%.

These observations are due to Ruth Byrne [3], and they were used by her to argue against a rule-based account of logical reasoning such as found in, e.g., Rips [14]. For if valid arguments can be suppressed, then surely logical inference cannot be a matter of blindly applying rules; and furthermore the fact that suppression depends on the content of the added premiss is taken to be an argument against the role of logical form in reasoning. We believe that this type of argumentation is wildly off the mark, but for the moment we will not comment on it, preferring to continue with the presentation of Byrne's tantalizing data.

Byrne investigated not only *modus ponens* (MP), but also *modus tollens* (MT), and the 'fallacies' *affirmation of the consequent* (AC), and *denial of the antecedent* (DA), with respect to both types of added premisses, 5 and 5.

### AC, premiss 5

If she has an essay to write she will study late in the library.
If the library stays open then she will study late in the library.
She will study late in the library.

What follows? Here 50% concludes 'She has an essay to write', comparable to the two-premise case 5.

### DA, premiss 5

If she has an essay to write she will study late in the library.
If the library stays open then she will study late in the library.
She doesn't have an essay to write.

50% responds 'She will not study late in the library', again comparable to the two-premise case 5.

### MT, premiss 5

If she has an essay to write she will study late in the library.
If the library stays open then she will study late in the library.
She will not study late in the library.

44% concludes 'She does not have an essay to write', compared to 70% in the two-premise case 5 – a clear case of suppression.

We now move on to consider the second type of premiss.

### AC, premiss 5

If she has an essay to write she will study late in the library.
If she has some textbooks to read, she will study late in the library.
She stays late in the library.

Now 16% responds 'She has an essay to write', compared to 55% in 5; hence also fallacies can be suppressed.

**DA, premiss 5**

If she has an essay to write she will study late in the library.
If she has some textbooks to read, she will study late in the library.
She does not have an essay to write.

22% concludes 'She will not study late in the library', compared to 50% in 5 –
again a clear case of the suppression of a fallacy.

**MT, premiss 5**

If she has an essay to write she will study late in the library.
If she has some textbooks to read, she will study late in the library.
She will not study late in the library.

70% concludes 'She does not have an essay to write', the same percentage as in the
two-premise case 5.

## 6. The 'suppression effect' as an instance of planning

We will now indicate how the observed non-classical answers can be analyzed as
applications of planning. Here we concentrate on *modus ponens*; for a full treatment
we refer the reader to Stenning and van Lambalgen [19].

### 6.1. What is planning?

We defined planning as setting a goal and devising a *sequence* of actions that will
achieve that goal, taking into account events in, and properties of the world and the
agent. In this definition, 'will achieve' cannot mean: '*provably* achieves', because of
the notorious frame problem: it is impossible to take into account all eventualities
whose occurrence might be relevant to the success of the plan. Therefore the
question arises: what makes a good plan? A reasonable suggestion is: the plan
works to the best of one's present knowledge. Viewed in terms of models, this
means that the plan achieves the goal in a 'minimal model' of reality, where, very
roughly speaking, every proposition is false which you have no reason to assume to
be true. In particular, in the minimal model no events occur which are not forced
to occur by the data. This makes planning a form of non-monotonic reasoning: the
fact that

'goal $G$ can be achieved in circumstances $C$'

does not imply

'goal $G$ can be achieved in circumstances $C + D$'

The book van Lambalgen and Hamm [20] formalizes the computations performed by
the planning faculty by means of logic programming with negation as failure. The
purpose of [20] is to show that the semantics of tense and aspect in natural language
can be explained on the assumption that temporal notions are encoded in such a
way as to subserve planning. For our present purposes we may abstract from the
temporal component of planning, and concentrate on the inference engine required
for planning. Non-monotonic logics abound, of course, but logic programming is

singled out by being both expressive and computationally efficient. Below we shall see that it also has an appealing implementation in neural nets.

## 6.2. Suppression as an application of planning logic

The first step in showing that the suppression effect may be seen as an instance of planning, is to provide a decent formalization of the natural language conditional. As has been remarked above, the literature on the psychology of reasoning tends to assume that the conditional should be formalized as a material implication; from this point of view the suppression effect is indeed paradoxical. An interpretation which is much more in line with the natural language understanding of the conditional, is one which allows the conditional to have exceptions. We may then interpret the natural language expression 'If $A$, then $B$' as a logic programming clause

$$A \wedge \neg ab \ \to \ B,$$

where $ab$ is a designated proposition letter whose intended meaning is 'something abnormal is the case'. An important difference between the conditional just defined and the material implication is that the former cannot be false, and hence cannot meaningfully be iterated; this is because we take the conditional to be part of a logic program, hence as something given. One might at first think that this militates against the proposed formalization of the conditional: surely one occasionally wants to prove a conditional false, for example in the four card task? Yes, to be sure; but in the case of *cooperative* communication one takes what one's interlocutor says as given, and one is mostly concerned with integrating a (usually highly compressed) discourse into a meaningful whole. The claim is here both that subjects interpret a task such as the suppression task in a cooperative manner, and that the logic of such cooperative situations is not classical logic. The latter is appropriate in the context of *adversarial* communication, where one challenges one's interlocutor to justify what he says – thus entertaining the possibility that what he says is false.

We assume that proposition letters of the type $ab$ are governed by the *closed world assumption*: if there is no positive evidence for $ab$, conclude $\neg ab$. We need not assume this principle applies across the board, i.e. for all proposition letters, but as an assumption governing human reasoning about abnormalities it seems plausible.

Now look at the basic argument, 5. The conditional premise is formalized as $A \wedge \neg ab \ \to \ B$, and in the absence of further information we may set $ab$ equal to $\bot$. This means that $A \wedge \neg ab \ \to \ B$ reduces to $A \ \to \ B$ , and $B$ follows from the given minor premise $A$.

Let us now integrate the third premise, starting with 5. We thus have the premise set

> *If she has an essay, she studies late in the library.*
> *She has an essay.*
> *If the library is open, she studies late in the library.*

The formal representation of the set of three premises is given by the *four* clauses

1. $A \wedge \neg ab \;\rightarrow\; B$

2. $A$

3. $C \wedge \neg ab' \;\rightarrow\; B$

4. $\neg C \rightarrow ab$

The fourth clause represents a side-effect of integrating the third premiss, namely establishing a relation between the disabling condition of the first premiss and the lexical material in the antecedent of the third premiss. We might also add a clause $\neg B \rightarrow ab'$, but this does not affect the outcome of the analysis.

Now consider the following computation. Closed world reasoning applied to the fourth clause gives

$$\neg C \;\leftrightarrow\; ab,$$

which is equivalent to

$$C \;\leftrightarrow\; \neg ab.$$

After substitution in the first premise we get

$$A \wedge C \;\rightarrow\; B,$$

from which $B$ does not follow if given $A$ only.

The second kind of premiss, of type 5, leads to the premiss set

> *If she has an essay, she studies late in the library.*
> *She has an essay.*
> *If she has a textbook to read, she studies late in the library.*

The formal representation of the set of three premisses is now given by the *three* clauses

1. $A \wedge \neg ab \;\rightarrow\; B$

2. $A$

3. $C \wedge \neg ab' \;\rightarrow\; B$

That is, integration of the third premiss does not lead to the addition of information on $ab$ or $ab'$. When we now start computing, we see that $ab$ and $ab'$ are both set to $\bot$, and this reduces the premiss set to $A$, $A \vee C \;\rightarrow\; B$, from which $B$ follows.

Space limitations forbid us to treat the other inference types, for which see [19]. Before we proceed to discuss the relevance of these results for cognition, let us consider what they mean for Byrne's argument in [3]. Because the conditionals contain a parameter of the form $ab$, the logical form of the set of premisses is only determined after those parameters have been set. Logical form cannot be read off from the premisses directly, but has to be determined by means of a reasoning process. Therefore Byrne's interpretation of these results, that logical reasoning cannot be a matter of applying formal rules, is not warranted: it is just that formal rules do not apply to the surface form, but to the logical form that is the result of interpretation.

### 6.3. Planning logic and working memory

We will now show that the above computations can actually be performed very fast in suitable neural networks. The observation that there is a strong connection between logic programming and neural nets is not new (see d'Avila Garcez, Broda and Gabbay [1]), but what is new here is a very straightforward modeling of closed world reasoning (negation as failure) by means of coupled neural nets. This exploits the soundness and completeness of negation as failure with respect to Kleene's three-valued logic.

Intuitively the idea is this. The nonmonotonic consequence relation that forms the background of the above computations is defined by means of the notion of completion of a program.

**Definition 1.** *A program is a finite set of conditionals of the form $A_1 \wedge \ldots \wedge A_n \wedge \neg ab \rightarrow B$, together with the clauses $\bot \rightarrow ab$ for all proposition letters of the form ab occurring in the conditionals.*

**Definition 2.** *The completion of a program $P$ is given by the following procedure:*

1. *take all clauses $\varphi_i \rightarrow q$ whose head is $q$ and form the expression $\bigvee_i \varphi_i \rightarrow q$*

2. *replace the $\rightarrow$'s by $\leftrightarrow$'s*

3. *this gives the completion of $P$, which will be denoted by $comp(P)$.*

*If $P$ is a logic program, define the nonmonotonic consequence relation $\approx$ by*

$$P \approx \varphi \text{ iff } comp(P) \models \varphi.$$

In terms of $\approx$, the formal representation of the suppression of *modus ponens*[6] can be given by (we omit the superfluous $\bot$'s)

(4)    a. $p,\ p \wedge \neg ab \rightarrow q \approx q$

   b. $p,\ p \wedge \neg ab \rightarrow q,\ r \wedge \neg ab' \rightarrow q,\ \neg r \rightarrow ab \not\approx q.$

The models of interest are thus models of the completion of a program $P$. It is well-known that these models can be obtained as fixed points of a suitable (Kleene) three-valued consequence operator $T_P$ associated to $P$.

**Definition 3.** *Let $P$ be a program in the sense of definition 1. Given a three-valued model $\mathcal{M}$, $T_P(\mathcal{M})$ is the model determined by*

1. *$T_P(\mathcal{M})(q) = 1$ iff there is a clause $\varphi \rightarrow q$ such that $\mathcal{M} \models \varphi$*

2. *$T_P(\mathcal{M})(q) = 0$ iff there is a clause $\varphi \rightarrow q$ in $P$ and for all such clauses, $\mathcal{M} \models \neg \varphi$*

---

[6] $\approx$ captures the observed 'forward' inferences, MP and DA. For 'backward' inferences (MT and AC) a different logic programing technique must be used, so-called integrity constraints. Interestingly, whereas forward inferences correspond neurally to feed forward computations, the backward inferences correspond to the backpropagation algorithm, which changes weights in a network. Details can be found in [19].

The preceding definition (together with the definition of program) ensures that unrestricted negation as failure applies only to propositions of the form $ab$; other proposition letters about which there is no information may remain undecided.

**Lemma 1.** *Let $P$ be a program.*

    a. *$\mathcal{M}$ is a model of the comp$(P)$ iff it is a fixed point of $T_P$.*

    b. *The least fixed point of $T_P$ is reached in finitely many steps ($n + 1$ if the program consists of $n$ clauses).*

What is of importance here is that the relevant models can also be viewed as stable states of a neural network, obtained by a feed forward computation mimicking the action of the consequence operator. Here are the pertinent definitions.[7]

**Definition 4.** *A computational unit, or unit for short, is a function with the following input-output behaviour*

    1. *inputs are delivered to the unit via links, which have weights $w_j \in \mathbb{R}$*

    2. *the inputs can be both excitatory or inhibitory; let $x_1 \dots x_n \in \mathbb{R}$ be excitatory, and $y_1 \dots y_m \in \mathbb{R}$ inhibitory*

    3. *if one of the $y_i$ fires, i.e. $y_i \neq 0$, the unit is shut off, and outputs 0*

    4. *otherwise, the quantity $\sum_{i=1}^{i=n} x_i w_i$ is computed; if this quantity is greater than or equal to a threshold $\theta$, the unit outputs 1, if not it outputs 0*

    5. *we assume that this computation takes one time-step.*

**Definition 5.** *A spreading activation network is a directed graph on a set of units, whose (directed) edges are called links.*
*A (feed forward) neural network is a spreading activation network with two distinguished sets of units, $I$ (input) and $O$ (output), with the added condition that there is no path from a unit in $O$ to one in $I$.*

Represent the three truth values $\{u, 0, 1\}$ in Kleene's logic as pairs $(0, 0) = u$, $(0, 1) = 0$ and $(1, 0) = 1$, ordered lexicographically via $0 < 1$. We shall refer to the first component in the pair as the $+$ (or 'true') component, and to the right component as the $-$ (or 'false') component. Interpret a 1 neurally as 'activation', and 0 as 'no activation'. A three-valued binary AND can then be represented as a pair of units as in figure 1.

What we see here is two coupled neural nets, labeled $+$ (above the separating sheet) and $-$ (below the sheet). Each proposition letter is represented by a pair of units, one in the $+$ net, and one in the $-$ net. Each such pair will be called a *node*. The thick vertical lines indicate inhibitory connections between units in the $+$ and $-$ nets; the horizontal arrows represent excitatory connections. The threshold of the AND$+$ unit is 2, and that of the AND$-$ unit is 1.

---

[7]For expository purposes we consider only very simple neurons, whose thresholds are numbers, instead of functions such as the sigmoid. A more realistic version can be found in [1].

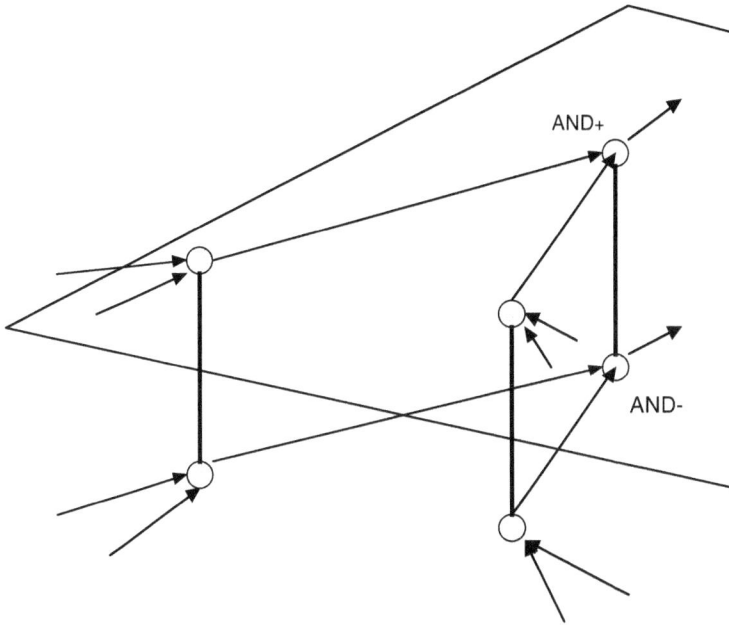

Figure 1. Three-valued AND

As an example, suppose the two truth values $(1,0)$ and $(0,0)$ are fed into the unit. The sum of the plus components is 1, hence AND+ does not fire. The sum of the $-$ components is 0, so AND$-$ likewise does not fire. The output is therefore $(0,0)$, as it should be. There is an inhibitory link between the $+$ and $-$ units belonging to the same proposition letter (or logic gate) because we do not want the truth value $(1,1)$, i.e. both units firing simultaneously.

We next have to associate a neural net to a logic program. We mean 'associate' in two senses here. In the first sense, we have to show formally that a logic programming computation starting from a program $P$ can be performed by a suitable neural net, depending on $P$. The second sense of 'associate' is that while processing a set of conditionals, formalized as a logic program $P$, a temporary neural network (associated with $P$ in the first sense) is set up in working memory, which then does the computation. 'Temporary' here means that the links involved exist only briefly, as opposed to the long-term links of declarative memory. Below we shall discuss a possible mechanism for the creation of such networks, due to Bienenstock and von der Malsburg; but first we indicate the formal structure of the requisite networks.

Consider a program consisting of a single clause of the form $p \land q \rightarrow r$. The associated net is obtained by representing the proposition letters as pairs of units, joining $p, q$ by means of the three-valued AND, and linking the output of AND to $r$. The input nodes are $p, q$ and $r$ is the output node. In the simple feed forward computations studied here, all links can be taken to have weight 1. The $\rightarrow$ in the clause is thus coded as a *link*, not as a pair of units. This will be of some importance later.

Here is a more elaborate example, a net corresponding to the suppression of *modus ponens* as discussed above. For the sake of readability, we give only the + net; the diagram should be extended with a − net as in the diagram for AND above.

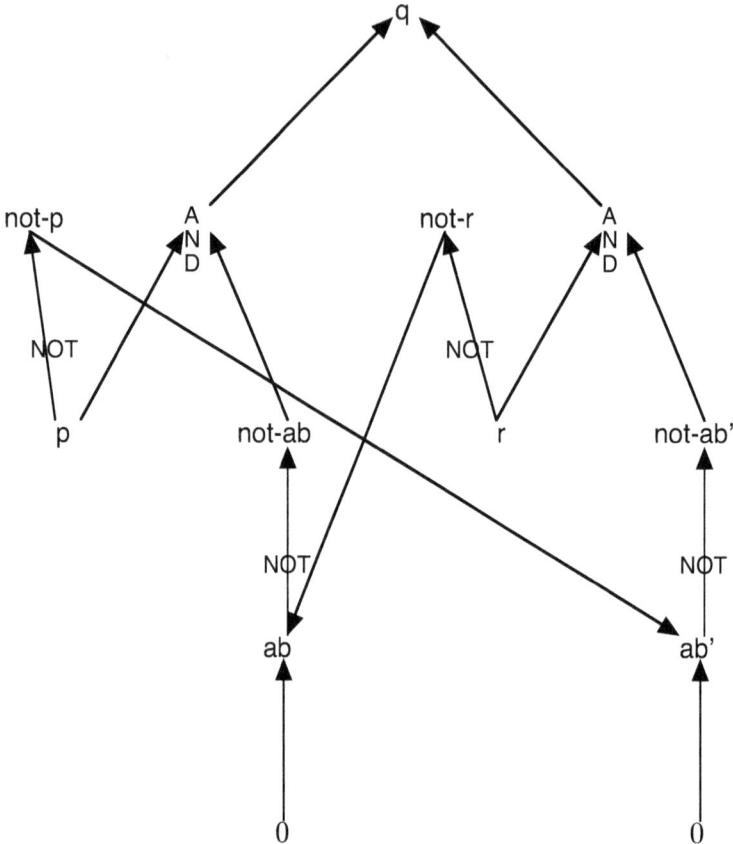

Figure 2. Network for the suppression of MP

In this picture, the links of the form $0 \to ab$ represent the + part of the link from ⊥ to the pair of units corresponding to $ab$. A NOT written across a link indicates that the link passes through a node which reverses (1,0) and (0,1), and leaves (0,0) in place. AND indicates a three-valued conjunction as depicted above. The output node $q$ implicitly contains an OR gate: its + threshold is 1, its − equals the number of incoming links. The abnormality nodes likewise contain an implicit OR.

We now trace the course of the computation of a stable state of the network, showing that $q$ is not true in the minimal model of the program. Initially all nodes have activation (0,0). Then the input $p$ is fed into the network, i.e. (1,0) is fed into the $p$ node. This causes the $ab'$ node to update its signal from (0,0) to (0,1), so that $\neg ab'$ changes its signal to (1,0). But no further updates occur and a stable state has been reached, in which $q$ outputs (0,0). If we view this state of activation as a (three-valued) model, we see that $p$ is true, and all other proposition letters are undecided. Not surprisingly, this model is also the least fixed point of the three-valued consequence operator associated to the program.

It is actually the least fixed point that is of paramount importance in these considerations, for the relation $\approx$ is completely determined by what happens there. Larger fixed points differ in that some values (0,0) in the least fixed point have been changed to (0,1) or (1,0) in the larger fixed point; but by the monotonicity property (with respect to truth values) of Kleene's logic this has no effect on the output unit pairs, in the sense that an output value (1,0) cannot be changed into (0,1) (or conversely). Therefore the relation $P \approx \varphi$, for $\varphi$ containing only $\neg, \wedge, \vee$, is determined by the minimal model of the completion of $P$. We thus see that the relation $\approx$ is determined by a *single* model, which moreover is computable deterministically by means of a simple neural network. In this sense $\approx$ is in principle easier on working memory than classical $\models$. The picture is complicated slightly by backward inferences like MT, which require an additional computation to change weights in the network, but the basic principle is the same.

We will now indicate briefly how such networks may be set up in working memory, following the highly suggestive treatment in a series of papers by Bienenstock and von der Malsburg [2,22,21]. They observed that, apart from the 'permanent' connection strengths between nodes created during storage in declarative memory, one also needs variable connection strengths, which vary on the psychological time scale of large fractions of a second. The strength of these so-called dynamical links increases when the nodes which a link connects have the same state of activation; networks of this type are therefore described by a modified Hopfield equation. Applied to the suppression task, we get something like the following. Declarative memory, usually modeled by some kind of spreading activation network, contains a node representing the concept 'library', with links to nodes representing concepts like 'open' , 'study', 'essay' and 'book'. These links have positive weights. Upon being presented with the conditional 'if she has an essay, she will study late in the library', these links become temporarily reinforced, and the system of nodes and links thereby becomes part of working memory, forming a network like the ones studied above. Working memory then computes the stable state of the network, and the state of the output node is passed on to the language production system. Modification of the connection strengths is an automatic (albeit in part probabilistic) process, and therefore the whole process, from reading the premises to producing an answer, proceeds automatically.

### 6.4. Logic and evolution

Let us retrace our steps. From a logical point of view, what we have shown is that the observed behaviour in the suppression task can be explained on the assumption that subjects apply a logic suitable for planning to the reasoning problems at hand. In principle, 'apply' can mean different things here: subjects may construct a derivation of the required conclusion using some such rule as resolution together with negation as failure, or they may construct the minimal model corresponding to the premises, and read off what is true there. These two different data structures lead to the same input-output relation, and hence cannot be distinguished starting from behavioural data as obtained in the suppression task. Byrne's claim to have refuted the view that logical reasoning is a matter of applying rules, is therefore correct only insofar as the rules applied in this case are not those of classical logic. But the rules applied may well be the ones appropriate to planning, i.e. backward chaining from a goal and negation as failure[8].

There is an interesting procedural difference between 'rules' and 'models' in this case. Backward chaining is in principle an indeterministic process, since a given goal may unify with the head of several clauses. This form of goal-oriented thinking seems to occur mostly consciously, with explicit selection of clauses to be unified with the goal, and memorization of those paths in the tree that have already been explored[9]. On the other hand, the process of constructing a model and reading off what is true there is completely automatic, and can be easily mimicked by a neural network.

An exaptationist account of the origin of logical reasoning might then runs as follows. Planning is a capability shared by humans and nonhuman primates, even monkeys. If the above picture of the operation of working memory is correct, it requires the animal to represent goals and actions as nodes in declarative memory, and causal influences as links between those nodes. Humans have language in which to formulate goals and actions, but language also accesses the representations of goals and actions in declarative memory. Therefore one could suppose that the process subserving planning in animals also allows humans to draw quick conclusions, and to modify these conclusions if the need arises. Since the process is automatic, it need not be accessible to consciousness. That is, if for the moment we abstract from what we know about humans, it might have been the case that logical inference is more like a reflex, a form of low-level processing. In such a case, it would be impossible to argue about a putative conclusion. But here it may be of some importance that the conclusions arrived at by automatic processing can also be derived more laboriously, by a conscious process of backward chaining. In this way one could become conscious of the main assumption underlying this form of reasoning, that of a closed world, and thus there would be room for exploring alternatives. Such a reconsideration would of course be prompted when two speakers, whose closed worlds happened to differ, are engaged in a dialogue.

## REFERENCES

1. A. d' Avila Garcez, K. B. Broda, and D. Gabbay. *Neural-symbolic learning systems: foundations and applications.* Springer, London, 2002.
2. E. Bienenstock and C. von der Malsburg. A neural network for invariant pattern recognition. *Europhysics Letters*, 4(1):121–126, 1987.

---

[8]Note that inference rules such as MP and MT, extensively investigated in the psychology of reasoning, become derived (and defeasible) in this setup. The contrast that the 'mental rules' school [15] draws between MP and MT (the former primitive, the latter derived, hence less easily accessible) resurfaces here as a difference in complexity between the underlying computations: feed forward in the case of MP, backpropagation in the case of MT.

[9]See [20] for a discussion of the relation between planning and our conscious sense of time, and the effect this has on the linguistic encoding of time.

3. R. M. J. Byrne. Suppressing valid inferences with conditionals. *Cognition*, 31:61–83, 1989.
4. L. Cosmides. The logic of social exchange: Has natural selection shaped how humans reason? studies with the Wason selection task. *Cognition*, 31:187–276, 1989.
5. L. Cosmides and J. Tooby. Cognitive adaptations for social exchange. In *The adapted mind: evolutionary psychology and the generation of culture*. Oxford University Press, 1992.
6. K. Dieussaert, W. Schaeken, W. Schroyen, and G. d'Ydewalle. Strategies during complex conditional inferences. *Thinking and reasoning*, 6(2):125–161, 2000.
7. P. M. Greenfield. Language, tools and the brain: the ontogeny and phylogeny of hierarchically organized sequential behavior. *Behavioral and brain sciences*, 14:531–595, 1991.
8. E. Husserl. *Briefwechsel, Volumes I–X*. Kluwer, 1994.
9. P. N. Johnson-Laird and R. M. Byrne. *Deduction*. Lawrence Erlbaum Associates, Hove, Sussex., 1991.
10. K. Malik. *Man, beast and zombie*. Rutgers University Press, New Brunswick, 2002.
11. D. Marr. *Vision: A Computational investigation into the human representation and processing of visual information*. W.H. Freeman, San Fransisco, 1982.
12. B. McGonigle, M. Chalmers, and A. Dickinson. Concurrent disjoint and reciprocal classification by *cebus apella* in serial ordering tasks: evidence for hierarchical organization. *Animal Cognition*, In press.
13. S. Pinker. *How the mind works*. Norton, 1997.
14. L. J. Rips. Cognitive processes in propositional reasoning. *Psychological Review*, 90:38–71, 1983.
15. L. J. Rips. *The psychology of proof*. The M.I.T. Press, Cambridge, MA, 1994.
16. M. Steedman. Plans, affordances and combinatory grammar. *Linguistics and Philosophy*, 26, 2003.
17. K. Stenning. *Seeing reason. Image and language in learning to think*. Oxford University Press, Oxford, 2002.
18. K. Stenning and M. van Lambalgen. A little logic goes a long way: basing experiment on semantic theory in the cognitive science of conditional reasoning. To appear in *Cognitive Science*, 2004.
19. K. Stenning and M. van Lambalgen. A working memory model of relations between interpretation and reasoning. 2003. Submitted to *Cognitive Science*.
20. M. van Lambalgen and F. Hamm. *The proper treatment of events*. To appear with Blackwell Publishing, Oxford and Boston, 2004. Until publication, manuscript available at http://staff.science.uva.nl/~michiell.
21. C. von der Malsburg. Pattern recognition by labeled graph matching. *Neural networks*, 1:141–148, 1988.
22. C. von der Malsburg and E. Bienenstock. A neural network for the retrieval of superimposed connection patterns. *Europhysics Letters*, 3(11):1243–1249, 1987.
23. P. C. Wason. Reasoning about a rule. *Quarterly Journal of Experimental Psychology*, 20:273–281, 1968.

24. P. C. Wason and P. N. Johnson-Laird. *Psychology of Reasoning: Structure and Content.* Harvard University Press, Boston, 1972.
25. P. C. Wason. Problem solving. In R.L. Gregory, editor, *The Oxford Companion to the Mind*, pages 641–644. Oxford University Press, Oxford, 1987.

# Lattice-Based Modal Algebras and Modal Logics

Ewa Orłowska [a] and Dimiter Vakarelov [b]

[a] *National Institute of Telecommunications, Warsaw, Poland*
*orlowska@itl.waw.pl*

[b] *Department of Mathematics, Sofia University, Sofia, Bulgaria*
*dvak@fmi.uni-sofia.bg*

**Abstract.** We study not-necessarily distributive lattices with modal operators of possibility, necessity, sufficiency (or equivalently negative necessity), and dual sufficiency (negative possibility), and the corresponding logics. We present representation theorems, relational semantics, and complete axiomatisation.

## 1. INTRODUCTION

The motivation and inspiration for this work comes from the three sources. First, we follow the line of research on Boolean algebras with additional operators; second, we get inspiration from the methods of reasoning with incomplete information; and third, a background for our technical developments is provided by representation theory for lattices and lattice-based logics.

Algebraic treatment of binary relations initiated by Tarski (Tarski 1941) is based on the assumption that algebras of relations are Boolean algebras with additional operations and constants specific to binary relations such as relative product, converse, identity, etc. This approach inspired a study of Boolean algebras with arbitrary additive and normal operators (Jónsson and Tarski 1952), which are the predecessors of modal algebras with possibility operators. Developments in modal correspondence theory have shown that possibility operators are not always sufficient for expressing relational properties. For that reason new classes of operators (e.g. sufficiency and dual sufficiency operators also referred to as negative necessity and negative possibility operators, respectively) have been proposed. The corresponding logics (Humberstone 1983, Goranko 1990) and algebras (Orłowska 1995, Düntsch and Orłowska 2001, 2002, Demri and Orłowska 2002) are again based on Boolean algebras.

In this paper we propose to weaken the basic structure of these algebras and logics by replacing the underlying Boolean algebra by a not necessarily distributive lattice. Distributive lattices with negative modalities have been developed in Vakarelov (1976, 1989), and those with the usual positive modalities in Vakarelov (1980). Distributive lattices with operators have also been studied in Dunn (1990), Gehrke and Jónsson (1994), Sofronie-Stokkermans (2000). Then a hierarchy of modal-like algebras and logics can be developed such that its bottom level consists of lattice-based structures, and any higher level is an axiomatic and/or signature extension of its predecessors. A step towards these studies is Düntsch, Orłowska, and Radzikowska (2003).

We introduce and investigate lattices with the four modalities: possibility $\Diamond$, necessity $\Box$, sufficiency (negative necessity) $\boxminus$, and dual sufficiency (negative possibility) $\Diamondblack$. In the classical modal algebras and logics based on Boolean algebras the operations of possibility and necessity are duals of each other due to the presence of the complement. In case of lattices the situation is different. Since there is no complement, possibility and necessity are independent operators. Moreover, possibility and necessity must be split into "positive" and "negative" operators, because otherwise we do not have any means for expressing negative modalities such as "possibly false" and "necessarily false" in terms of the ordinary positive modalities "possibly true" and "necessarily true". We study these modalities in an algebraic and logical framework. On the algebraic side, the focus is on a representation theory. We extend Urquhart representation theorem for not necessarily distributive lattices (Urquhart 1978) following the method presented in Allwein and Dunn (1993). On the logical side, we present Gentzen-style proof systems and Kripke-style semantics.

On the application side, it can be observed that a lattice structure is a minimal ingredient of a great variety of information structures dealt with in theories of incompleteness and uncertainty such as multiple valued logics (e.g. Hajek 1998) and probabilistic theories (e.g. van Lambalgen 2001), which are based on the concept of degree of truth. On the other hand, the theories of incompleteness focused on the concept of approximation often employ modal-like operators treated as formal counterparts to approximations (see e.g., Demri and Orłowska 2002). Putting a lattice structure and a modal structure together would lead to hybrid theories in which both graded information and approximate information could be modelled. Some extensions of such algebras meaningful for representation of incomplete information would then be, for example, various kinds of residuated lattices (see e.g., Höhle 1996, Orłowska and Radzikowska 2001).

## 2. DOUBLY ORDERED SETS AND STABLE SETS

In this section we list some definitions and facts which are repeatedly used throughout the paper. Most of them are from Urquhart (1978). The proofs of the lemmas are straightforward, some of them can be found in the Urquhart paper. Throughout the paper we often use the same symbol for denoting an algebra or a relational system and their corresponding universes.

By a *doubly ordered set* we mean a relational system $(X, \leq_1, \leq_2)$, where X is a non-empty set, $\leq_1$ and $\leq_2$ are quasi orders on X such that for all x,y$\in$X, if x$\leq_1$y and x$\leq_2$y, then x=y.

Given a doubly ordered set, the mappings l: $2^X \to 2^X$ and r: $2^X \to 2^X$ are defined as follows. For any A$\subseteq$X:

l(A) = {x$\in$X: for every y$\in$X, if x $\leq_1$y, then y$\notin$A},
r(A) = {x$\in$X: for every y$\in$X, if x $\leq_2$y, then y$\notin$A}.

A set A$\subseteq$X is said to be *l-stable* (resp. *r-stable*) whenever lr(A) = A (resp. rl(A) = A). A is $\leq_i$-*increasing* set, i = 1,2, whenever for all x, y$\in$X, if x$\in$A and x$\leq_i$y, then y$\in$A.

Lemma 2.1
For every doubly ordered set $(X, \leq_1, \leq_2)$ and for all A, B $\subseteq$ X the following

conditions are satisfied:

   I. $l(A)$ is a $\leq_1$-increasing set,

  II. $r(A)$ is a $\leq_2$-increasing set,

 III. $l(A)$, $r(A) \subseteq$ -A.

**Lemma 2.2**
The family of $\leq_i$-increasing sets, i=1,2, forms a distributive lattice, where join and meet are union and intersection of sets.

Observe that mapping l treated as an operation of the lattice of $\leq_1$-increasing sets is an intuitionistic negation, and so is r treated as an operation of the lattice of $\leq_2$-increasing sets.

**Lemma 2.3 (Urquhart)**
For every doubly ordered set $(X, \leq_1, \leq_2)$ the mappings l and r form a Galois connection between the lattice of $\leq_1$-increasing subsets of X and the lattice of $\leq_2$-increasing subsets of X.

This lemma amounts to saying that $A \subseteq l(B)$ iff $B \subseteq r(A)$ for all A and B such that A is a $\leq_1$-increasing set and B is a $\leq_2$-increasing set.

**Lemma 2.4**
For every doubly ordered set $(X, \leq_1, \leq_2)$ and for all A, B $\subseteq$ X the following conditions are satisfied:

   I. If A is a $\leq_2$-increasing set, then $l(A)$ is an l-stable set,

  II. If A is a $\leq_1$-increasing set, then $r(A)$ is an r-stable set,

 III. $lrlr(A) = lr(A)$,

 IV. $rlrl(A) = rl(A)$,

  V. If A and B are l-stable sets, then $r(A) \cap r(B)$ is an r-stable set,

 VI. If A is l-stable, then $r(A)$ is r-stable and if A is r-stable, then $l(A)$ is l-stable.

Given a doubly ordered set $(X, \leq_1, \leq_2)$, by $L(X)$ we denote the family of l-stable subsets of X. We define the following operations and constants on $L(X)$. For all A, $B \in L(X)$:

$$A \wedge^c B = A \cap B,$$
$$A \vee^c B = l(r(A) \cap r(B)),$$
$$0^c = \emptyset, \ 1^c = X.$$

**Lemma 2.5**
For every doubly ordered set $(X, \leq_1, \leq_2)$ and for all A,B$\in$L(X), A$\wedge^c$B, A$\vee^c$B, X, and $\emptyset$ are l-stable sets.

The system $(L(X), \vee^c, \wedge^c, 1^c, 0^c)$ with the operations and constants defined above is referred to as the *complex algebra of a doubly ordered set* $(X, \leq_1, \leq_2)$.

**Lemma 2.6**
For every doubly ordered set $(X, \leq_1, \leq_2)$, its complex algebra $(L(X), \vee^c, \wedge^c, 1^c, 0^c)$ is a lattice.

Complex algebras are not necessarily distributive.

## 3. URQUHART REPRESENTATION OF LATTICES

In this section we give a brief survey of the Urquhart results leading to the representation theorem for not necessarily distributive lattices. We use a modal logic terminology in formulating the representability result. Since in this paper we are interested in the interplay between logics and algebras, we do not assume any topological structure in the frames. As a consequence, we use a weaker formulation of the representation theorems than in the original Urquhart result, due to the lack of the compactness assumption.

Let $(W, \vee, \wedge, 1, 0)$ be a bounded lattice. By a *filter-ideal pair* of W we mean a pair $(x_1, x_2)$ such that $x_1$ is a filter of W, $x_2$ is an ideal of W and $x_1 \cap x_2 = \emptyset$. We define an ordering $\leq$ on the family of filter-ideal pairs of lattice W:

$(x_1, x_2) \leq (y_1, y_2)$ iff $x_1 \subseteq y_1$ and $x_2 \subseteq y_2$.

A filter-ideal pair $(x_1, x_2)$ is said to be maximal whenever it is maximal with respect to the partial order $\leq$. Let X(W) be the family of the maximal filter-ideal pairs of a lattice $(W, \vee, \wedge, 1, 0)$. We define relations $\subseteq_1$ and $\subseteq_2$ on X(W) as follows. Let $x = (x_1, x_2)$ and $y = (y_1, y_2)$ be elements of X(W), then:

$x \subseteq_1 y$ iff $x_1 \subseteq y_1$,

$x \subseteq_2 y$ iff $x_2 \subseteq y_2$.

The system $(X(W), \subseteq_1, \subseteq_2)$ is referred to as a *canonical frame of a lattice* $(W, \vee, \wedge, 1, 0)$.

Lemma 3.1
For every lattice $(W, \vee, \wedge, 1, 0)$, its canonical frame $(X(W), \subseteq_1, \subseteq_2)$ is a doubly ordered set.

The condition saying that for all $x, y \in X(W)$, if $x \subseteq_1 y$ and $x \subseteq_2 y$, then $x = y$, follows from the maximality of the pairs x and y.

Observe that if a lattice $(W, \vee, \wedge, 1, 0)$ is distributive and $(x_1, x_2) \in X(W)$, then $x_1$ and $x_2$ are a prime filter and a prime ideal of W, respectively, $x_1 \cap x_2 = \emptyset$, and $x_1 \cup x_2 = W$. It follows that $\subseteq_2 = \subseteq_1^{-1}$.

Consider the complex algebra L(X(W)) of the doubly ordered set $(X(W), \subseteq_1, \subseteq_2)$ as defined in section 2. Define the mapping h: $W \to 2^{X(W)}$ as follows. For every $a \in W$:

$h(a) = \{x = (x_1, x_2) \in X(W): a \in x_1\}$.

Theorem 3.2 (Urquhart)
For every lattice $(W, \vee, \wedge, 1, 0)$ and for every $a \in W$ the following assertions hold:
  I. $rh(a) = \{x = (x_1, x_2) \in X(W): a \in x_2\}$,
 II. $h(a)$ is an l-stable set,
III. h is a lattice embedding.

As a corollary we obtain the following weak form of the Urquhart representation theorem.

Theorem 3.3 (Representation theorem)
Every bounded lattice is isomorphic to a subalgebra of the complex algebra of its canonical frame.

## 4. LATTICE-BASED PROPOSITIONAL LOGIC LAT

In this section we present a basic propositional logic LAT whose algebraic semantics is determined by the class of lattices. As usual with the logics which do not have tautologies, a deduction system for the logic LAT is fomulated as a simple sequent-style system.

The formulas of the language of logic LAT are built in a usual way from propositional variables of a countable infinite set VAR and from propositional constants T (true) and F (false) with the propositional connectives $\vee$ and $\wedge$ of disjunction and conjunction, respectively. Let FOR be the set of formulas of LAT. By abusing a notation, throughout the paper we denote the lattice operations and the logical connectives with the same symbols.

*Algebraic semantics* of the language of logic LAT is determined by the class of lattices. Let $(W, \vee, \wedge, 1, 0)$ be a lattice and let $\leq$ be its natural ordering. By a valuation in W we mean a function v: $VAR \cup \{T, F\} \rightarrow W$ such that $v(T) = 1$ and $v(F) = 0$. We extend valuation v to all the formulas in the usual way:

$v(A \vee B) = v(A) \vee v(B)$, $v(A \wedge B) = v(A) \wedge v(B)$.

By a *sequent* we mean an expression of the form $A \vdash B$, where A and B are formulas of LAT. A sequent $A \vdash B$ is satisfied by a valuation v whenever $v(A) \leq v(B)$; $A \vdash B$ is true in a lattice W whenever for every valuation v in W, $v(A) \leq v(B)$.

A deductive system for LAT consists of the following axioms and rules:

(Ax1)  $A \vdash A$
(Ax2)  $F \vdash A$  (Ax3)  $A \vdash T$
(Ax4)  $A \wedge B \vdash A$  (Ax5)  $A \wedge B \vdash B$
(Ax6)  $A \vdash A \vee B$  (Ax7)  $B \vdash A \vee B$

$$(R1) \quad \frac{A \vdash B, \ B \vdash C}{A \vdash C} \qquad (R2) \quad \frac{C \vdash A, \ C \vdash B}{C \vdash A \wedge B} \qquad (R3) \quad \frac{A \vdash C, \ B \vdash C}{A \vee B \vdash C}$$

A *derivation* of a sequent $A \vdash B$ in LAT is a finite sequence S of sequents such that each of them is either an axiom or is obtained from some earlier sequents of S using a rule, and the last sequent of S is $A \vdash B$. A sequent $A \vdash B$ is *provable* in LAT whenever there is a derivation of $A \vdash B$ in LAT.

Theorem 4.1 (Soundness)
For all formulas A and B, if $A \vdash B$ is provable in LAT, then it is true in all lattices.

Proof: As usual, the proof consits in showing that the axioms are true in all lattices and the rules preserve truth.

We define a binary relation $\approx$ in set FOR:
$A \approx B$ iff $A \vdash B$ and $B \vdash A$ are provable in LAT.

Lemma 4.2
I. $\approx$ is an equivalence relation,
II. $\approx$ is a congruence with respect to $\vee$ and $\wedge$.

Proof: By way of example we prove (II) for $\vee$. Assume that (i) $A_1 \approx B_1$ and (ii) $A_2 \approx B_2$. By (i) sequent $A_1 \vdash B_1$ is provable and from (Ax6) sequent $B_1 \vdash B_1 \vee B_2$

is provable. Applying rule (R1) we get (iii) $A_1 \vdash B_1 \vee B_2$ is provable. Similarly, from (ii) and provability of $B_2 \vdash B_1 \vee B_2$ we obtain (iv) $A_2 \vdash B_1 \vee B_2$ is provable. Then applying rule (R3) to (iii) and (iv) we have $A_1 \vee A_2 \vdash B_1 \vee B_2$ is provable. In a similar way we obtain provability of $B_1 \vee B_2 \vdash A_1 \vee A_2$, which completes the proof.

We define the Lindenbaum algebra (FOR/$\approx$, $\vee$, $\wedge$, $|T|$, $|F|$) of LAT with the operations defined as follows:

$|A| \vee |B| = |A \vee B|$, $|A| \wedge |B| = |A \wedge B|$, $|T| = \{A: A \approx T\}$, $|F| = \{A: A \approx F\}$.

**Lemma 4.3**

I. The Lindenbaum algebra (FOR/$\approx$, $\vee$, $\wedge$, $|T|$, $|F|$) of LAT is a lattice with the unit element $|T|$ and the zero element $|F|$,

II. The natural ordering of this lattice is set inclusion, i.e., $|A| \vee |B| = |B|$ iff $|A| \subseteq |B|$.

The proof is by an easy verification.

We define a canonical valuation $v^c$: VAR $\cup$ $\{T,F\} \rightarrow$ FOR/$\approx$ as $v^c(p) = |p|$ for every $p \in$ VAR, $v^c(T) = |T|$, and $v^c(F) = |F|$. By induction on the structure of a formula one can prove that for every formula A, $v^c(A) = |A|$.

**Lemma 4.4**

For all formulas A and B the following conditions are equivalent:

I. $A \vdash B$ is provable in LAT,

II. $|A| \subseteq |B|$.

Proof: (I) $\rightarrow$ (II) By Theorem 4.1 and Lemma 4.3(I), $A \vdash B$ must be true in the Lindenbaum algebra, so by Lemma 4.3 (II) we have $|A| \subseteq |B|$.

(II) $\rightarrow$ (I) The proof is by induction on the structure of formulas A and B. By way of example we consider the case when $A = A_1 \vee A_2$. By the assumption for every formula C, if $A_1 \vee A_2 \vdash C$ and $C \vdash A_1 \vee A_2$ are provable, then $C \vdash B$ and $B \vdash C$ are provable. Putting $C = A_1 \vee A_2$ we get that $A_1 \vee A_2 \vdash B$ is provable. In the remaining cases the proofs are similar.

**Theorem 4.5 (Completeness)**

For all formulas A and B of logic LAT, if a sequent $A \vdash B$ is true in all lattices, then it is provable in LAT.

Proof: Suppose that $A \vdash B$ is not provable. By Lemma 4.4, not $|A| \subseteq |B|$. Hence, not $v^c(A) \subseteq v^c(B)$, which means that $A \vdash B$ is not true in the Lindenbaum algebra, a contradiction with the assumption.

A different Gentzen-style system for lattice-based logics can be found in Takano (2002). The systems for quantum logics also include the rules relevant for lattices (see e.g. Battilotti and Fagian 2003).

## 5. KRIPKE-STYLE SEMANTICS FOR LOGIC LAT

In this section we present a Kripke-style semantics for logic LAT. A Kripke model for LAT is a system $M = (X, \leq_1, \leq_2, m)$ such that $(X, \leq_1, \leq_2)$ is a doubly ordered set and m: $VAR \rightarrow 2^X$ is a meaning function such that m(p) is an l-stable subset of X. The satisfiability of formulas by states in a model is defined as follows:

$M, x \models p$ iff $x \in m(p)$,

$M, x \models T$, not $M, x \models F$,

$M, x \models A \wedge B$ iff $M, x \models A$ and $M, x \models B$,

$M, x \models A \vee B$ iff for every $y \in X$, if $x \leq_1 y$, then there is $z \in X$ such that $y \leq_2 z$ and $M, z \models A$ or there is $t \in X$ such that $y \leq_2 t$ and $M, t \models B$.

We extend meaning function m to all formulas of LAT:

$m(A) = \{x \in X: M, x \models A\}$.

Lemma 5.1

For all formulas A and B of logic LAT the following assertions hold:

  I. $m(A \wedge B) = m(A) \cap m(B)$,

  II. $m(A \vee B) = l(r(m(A)) \cap r(m(B)))$,

  III. $m(T) = X, m(F) = \emptyset$.

The proof is by an easy verification.

We say that a sequent $A \vdash B$ is true in a model $M = (X, \leq_1, \leq_2, m)$ whenever $m(A) \subseteq m(B)$.

Lemma 5.2

  I. For every formula A of LAT, m(A) is an l-stable set,

  II. For every model $M = (X, \leq_1, \leq_2, m)$ for LAT the family $\{m(A): A$ is a formula of LAT$\}$ is a lattice of l-stable subsets of X.

Proof: Condition (I) follows from Lemma 2.5, and condition (II) from Lemma 2.6.

Lemma 5.3

For all formulas A and B of LAT the following conditions are equivalent:

  I. A sequent $A \vdash B$ is true in all models for LAT,

  II. For every doubly ordered set $(X, \leq_1, \leq_2)$ the sequent $A \vdash B$ is true in the complex algebra of X.

Proof: The theorem follows from the fact that due to Lemmas 5.1 and 5.2, a meaning function m of any Kripke model $(X, \leq_1, \leq_2, m)$ can be considered as a valuation in the complex algebra of $(X, \leq_1, \leq_2)$.

The following theorem states an equivalence of the algebraic and Kripke-style semantics for LAT.

Theorem 5.4

For all formulas A and B of LAT the following conditions are equivalent:

I. A sequent A ⊢ B is true in all lattices,

II. A ⊢ B is true in all Kripke models for LAT.

Proof: (I) → (II) It follows from (I) that, in particular, A ⊢ B is true in the complex algebras of doubly ordered sets. By Lemma 5.3, condition (II) follows.

(II) → (I) Suppose that for some lattice W and a valuation v in W we have not v(A)≤v(B). Consider the canonical frame X(W) of lattice W. Let h be an embedding of W in L(X(W)) guaranteed by Theorem 3.3. It follows that not h(v(A)) ⊆ h(v(B)). Consider a Kripke model M based on the canonical frame X(W) such that m(p) = h(v(p)) for every propositional variable p. It can be easily shown that for every formula A, m(A) = h(v(A)). Thus A ⊢ B is not true in M, a contradiction.

As a corollary of Theorems 4.1, 4.5, and 5.4 we obtain the soundness and completeness theorem for LAT with respect to the Kripke semantics.

Theorem 5.5

For all formulas A and B of LAT the following conditions are equivalent:

I. A sequent A ⊢ B is provable in LAT,

II. A ⊢ B is true in all Kripke models for LAT.

The Kripke semantics presented in this section is equivalent to the three-valued Kripke semantics developed in Allwein and Dunn (1993).

## 6. LATTICE-BASED POSSIBILITY (MODAL) ALGEBRAS

In this section we begin introducing the lattice-based algebras with additional unary operators. As stated in section 1, lacking a complement we need to consider four independent basic modalities. One of them is a possibility operator dealt with in this section.

By an $L-$ *possibility algebra* we mean an algebra P = (W, ∨, ∧, 1, 0, ◊), where (W, ∨, ∧, 1, 0) is a bounded lattice and ◊ is a unary operation on W satisfying for all a,b∈W:

(P1)   ◊(a∨b) = ◊a ∨◊b            additive

(P2)   ◊0 = 0                          normal

Observe, that any possibility operator is isotone. The operators ⟨R⟩ and ⟨−R⟩ of the Boolean modal logic, where R is a binary relation and − is a Boolean complement, are examples of such possibility operators.

For every A⊆W we define:

◊A = {a∈W: ◊a∈A},

□$^{-1}$A = {a∈W: there is b∈A such that ◊b ≤ a},

where ≤ is the natural ordering of the lattice W.

By a filter (resp. ideal) of an L-possibility algebra (W, ∨, ∧, 1, 0, ◊) we mean a filter (resp. ideal) of the underlying lattice (W, ∨, ∧, 1, 0).

Lemma 6.1

For every L-possibility algebra P = (W, ∨, ∧, 1, 0, ◊) and for all A,B ⊆ W the following conditions are satisfied:

   I. If A is an ideal of P, then so is $\Diamond A$,

  II. If $A \subseteq B$, then $\Diamond A \subseteq \Diamond B$,

 III. If A is a filter of P, then so is $\Box^{-1}A$,

 IV. $\Box^{-1}A \subseteq B$ implies $A \subseteq \Diamond B$; If B is a filter of P, then $A \subseteq \Diamond B$ implies $\Box^{-1}A \subseteq B$.

Proof: By way of example we prove (III) and (IV).

Proof of (III): Let a, b$\in \Box^{-1}A$. By definition, there is c$\in$A such that $\Diamond c \leq a$, and there is d$\in$A such that $\Diamond d \leq b$. Since A is a filter, we have (i) $c \wedge d \in A$. Since $c \wedge d \leq c$, we also have $\Diamond(c \wedge d) \leq \Diamond d$. Hence, $\Diamond(c \wedge d) \leq a$ and $\Diamond(c \wedge d) \leq b$. It follows that (ii) $\Diamond(c \wedge d) \leq a \wedge b$. From (i) and (ii) we get $a \wedge b \in \Box^{-1}A$. Now assume that $a \leq b$ and a$\in \Box^{-1}A$. It follows that there is c$\in$A such that $\Diamond c \leq a$. Hence $\Diamond c \leq b$, and therefore b$\in \Box^{-1}A$.

Proof of (IV): We have to show that for every a$\in$W, if a$\in$A then $\Diamond a \in B$. By the assumption, if $\Diamond a \in \Box^{-1}A$ then $\Diamond a \in B$. By definition, $\Diamond a \in \Box^{-1}A$ iff (i) there is c$\in$A such that $\Diamond c \leq \Diamond a$. So if a$\in$A, then taking c=a condition (i) is satisfied, which completes the proof of the first part of (IV). To prove the second part, let a$\in \Box^{-1}A$. It follows that there is b$\in$A such that $\Diamond b \leq a$. By the assumption $\Diamond b \in B$. Since B is a filter, we have a$\in$B.

The proofs of the remaining conditions are similar.

On the logical side, we introduce the notion of possibility frame. A *possibility frame* is a relational system $(X, \leq_1, \leq_2, R_\Diamond, S_\Diamond)$ such that $(X, \leq_1, \leq_2)$ is a doubly ordered set and $R_\Diamond$ and $S_\Diamond$ are binary relations on X satisfying the following conditions for all x,y,x',y'$\in$ X:

  (Mono $R_\Diamond$) If $x \leq_1 x'$, $xR_\Diamond y$, and $y' \leq_1 y$, then $x'R_\Diamond y'$,

  (Mono $S_\Diamond$) If $x' \leq_2 x$, $xS_\Diamond y$, and $y \leq_2 y'$, then $x'S_\Diamond y'$,

  (SC $R_\Diamond S_\Diamond$) If $xR_\Diamond y$, then there is $y' \in X$ such that $y \leq_1 y'$ and $xS_\Diamond y'$,

  (SC $S_\Diamond R_\Diamond$) If $xS_\Diamond y$, then there is $x' \in X$ such that $x \leq_2 x'$ and $x'R_\Diamond y$.

The conditions (Mono $S_\Diamond$) and (Mono $R_\Diamond$) are referred to as *possibility monotonicity conditions*. The conditions (SC $R_\Diamond S_\Diamond$) and (SC $S_\Diamond R_\Diamond$) are referred to as *possibility stability conditions*. They provide relationships between relations $S_\Diamond$ and $R_\Diamond$.

We define unary operators $[S_\Diamond]$ and $\langle R_\Diamond \rangle$ on $2^X$ as follows. For every $A \subseteq X$:

$[S_\Diamond]A = \{x \in X:$ for every $y \in X$, if $xS_\Diamond y$ then $y \in A\}$,

$\langle R_\Diamond \rangle A = \{x \in X:$ there is $y \in X$ such that $xR_\Diamond y$ and $y \in A\}$.

**Lemma 6.2**

For every possibility frame $(X, \leq_1, \leq_2, R_\Diamond, S_\Diamond)$ and for every $A \subseteq X$ the following conditions are satisfied:

  I. $[S_\Diamond]A$ is a $\leq_2$-increasing set,

 II. $\langle R_\Diamond \rangle A$ is a $\leq_1$-increasing set.

Proof: (I) Assume that (i) $x \in [S_\Diamond]A$, (ii) $x \leq_2 y$, and suppose that $y \notin [S_\Diamond]A$. It follows that there is z such that (iii) $yS_\Diamond z$ and $z \notin A$. From (i) we obtain (iv) for every y, if $xS_\Diamond y$, then $y \in A$. From (ii), (iii), (Mono $S_\Diamond$), and reflexivity of $\leq_2$ we get $xS_\Diamond z$. From (iv) $z \in A$, a contradiction.

The proof of (II) is similar.

Lemma 6.3

For every possibility frame $(X, \leq_1, \leq_2, R_\Diamond, S_\Diamond)$ and for every $A \subseteq X$, if A is r-stable then:

    I. $[S_\Diamond]A = r\langle R_\Diamond \rangle l(A)$,
    II. $[S_\Diamond]A$ is r-stable.

Proof:

(I) ($\subseteq$) Let $x \in X$ and assume that (i) $x \in [S_\Diamond]A$. Suppose that (ii) $x \notin r\langle R_\Diamond \rangle l(A)$. (ii) means that there is $y \in X$ such that (iii) $x \leq_2 y$ and (iv) $y \in \langle R_\Diamond \rangle l(A)$. From (iv) there is $z \in X$ such that (v) $yR_\Diamond z$ and (vi) $z \in l(A)$. From (v) and (SC $R_\Diamond S_\Diamond$) there is $z' \in X$ such that (vii) $z \leq_1 z'$ and (viii) $yS_\Diamond z'$. From (vii) and (vi) we have (ix) $z' \notin A$. From (iii), (viii), and (Mono $S_\Diamond$) we obtain (x) $xS_\Diamond z'$. From (i) and (x) we get $z' \in A$, which contradicts (ix).

($\supseteq$) Now assume that (i) $x \in r\langle R_\Diamond \rangle l(A)$ and suppose that (ii) $x \notin [S_\Diamond]A$. By (ii) there is $y \in X$ such that (iii) $xS_\Diamond y$ and (iv) $y \notin A$. Since A is r-stable, we have A $= rl(A)$, and from (iv) there is $y' \in X$ such that (v) $y \leq_2 y'$ and (vi) $y' \in l(A)$. From (iii), (v), and (Mono $S_\Diamond$) we obtain (vii) $xS_\Diamond y'$. Furthermore, by (vii) and (SC $S_\Diamond R_\Diamond$) there is $x' \in X$ such that (viii) $x \leq_2 x'$ and (ix) $x'R_\Diamond y'$. (ix) and (vi) yield (x) $x' \in \langle R_\Diamond \rangle l(A)$. From (x) and (viii) we get $x \notin r\langle R_\Diamond \rangle l(A)$, which contradicts (i).

(II) By Lemma 6.2(II) $\langle R_\Diamond \rangle l(A)$ is a $\leq_1$-increasing set. By Lemma 2.4(II) $r\langle R_\Diamond \rangle l(A)$ is r-stable. Using the equality from (I) we conclude that (II) holds.

Given a possibility frame, we define its complex algebra as follows. A *complex algebra of a possibility frame* $(X, \leq_1, \leq_2, R_\Diamond, S_\Diamond)$ is an algebra $(L(X), \vee^c, \wedge^c, 1^c, 0^c, \Diamond^c)$, where $(L(X), \vee^c, \wedge^c, 1^c, 0^c)$ is the complex algebra of the doubly ordered set $(X, \leq_1, \leq_2)$ as defined in section 2, and $\Diamond^c$ is a unary operator defined as:
$$\Diamond^c A = l[S_\Diamond]r(A).$$

Lemma 6.4

For every possibility frame $(X, \leq_1, \leq_2, R_\Diamond, S_\Diamond)$ and for every $A \subseteq X$, if A is l-stable then:

    I. $\Diamond^c A$ is l-stable,
    II. $\Diamond^c A = lr\langle R_\Diamond \rangle A$.

Proof: (I) follows from Lemma 6.2(I) and Lemma 2.4(I). (II) follows immediately from Lemma 6.3(I).

Theorem 6.5

For every possibility frame $(X, \leq_1, \leq_2, R_\Diamond, S_\Diamond)$, its complex algebra $(L(X), \vee^c, \wedge^c, 1^c, 0^c, \Diamond^c)$ is an L-possibility algebra.

Proof: We show that $\Diamond^c(A \vee^c B) = \Diamond^c A \vee^c \Diamond^c B$ and $\Diamond^c 0^c = 0^c$.

$$\lozenge^c(A \lor^c B) = l[S_\lozenge]r(A \lor B) \qquad \text{definition of } \lozenge^c$$
$$= lr\langle R_\lozenge \rangle lr(A \lor^c B) \qquad \text{Lemma 6.3(I)}$$
$$= lr\langle R_\lozenge \rangle lr(l(r(A) \cap r(B))) \qquad \text{definition of } \lor^c$$
$$= lr\langle R_\lozenge \rangle l(r(A) \cap r(B)) \qquad \text{Lemma 2.4 (V)}$$
$$= l[S_\lozenge](r(A) \cap r(B)) \qquad \text{Lemma 6.3(I)}$$
$$= l([S_\lozenge]r(A) \cap [S_\lozenge]r(B)) \qquad \text{distributivity of } [S_\lozenge] \text{ over } \cap$$

Since A and B are l-stable sets, by Lemma 2.4(VI) r(A) is r-stable. Then by Lemma 6.3.(I) $[S_\lozenge]r(A)$ is r-stable. Hence, $[S_\lozenge]r(A) = rl[S_\lozenge]r(A)$. By Lemma 6.3(I) and the definition of $\lozenge^c$, $rl[S_\lozenge]r(A) = r\lozenge^c A$. The similar reasoning shows that $rl[S_\lozenge]r(B) = r(\lozenge^c B)$. Hence, $\lozenge^c(A \lor^c B) = l(r\lozenge^c A \cap r\lozenge^c B) = \lozenge^c A \lor^c \lozenge^c B$.

We also have $\lozenge^c 0^c = l[S_\lozenge]r(\emptyset) = l[S_\lozenge]W = l(W) = \emptyset$.

Now we define canonical frames of L-possibility algebras. Let $P = (W, \lor, \land, 1, 0, \lozenge)$ be an L-possibility algebra. By a filter-ideal pair of P we mean a filter-ideal pair of the underlying lattice $(W, \lor, \land, 1, 0)$. Let X(P) be the family of all the maximal filter-ideal pairs of P and let $(X(P), \subseteq_1, \subseteq_2)$ be the canonical frame of the lattice reduct of algebra P defined as in section 3.

We define binary relations $S_\lozenge^c$ and $R_\lozenge^c$ on X(P) as follows. Let $x = (x_1, x_2)$ and $y = (y_1, y_2)$ be elements of X(P), then:
$$xS_\lozenge^c y \text{ iff } \lozenge x_2 \subseteq y_2,$$
$$xR_\lozenge^c y \text{ iff } y_1 \subseteq \lozenge x_1.$$

**Lemma 6.6**
For every L-possibility algebra $P = (W, \lor, \land, 1, 0, \lozenge)$, for every $x \in X(P)$, and for every $a \in W$ the following conditions are equivalent:

I. $\lozenge a \in x_2$,

II. For every $y \in X(P)$, if $xS_\lozenge^c y$ then $a \in y_2$.

Proof:

($\rightarrow$) This part follows directly from the definition of $S_\lozenge^c$.

($\leftarrow$) Let $P = (W, \lor, \land, 1, 0, \lozenge)$ be an L-possibility algebra. Suppose that (i) $\lozenge a \notin x_2$. We show that there is $y \in X(P)$ such that $xS_\lozenge^c y$ and $a \notin y_2$. Let $[a) = \{b \in W: a \leq b\}$, where $\leq$ is the natural ordering of the lattice W. The set $[a)$ is a filter of W generated by a. By (i) we have $a \notin \lozenge x_2$ which implies $[a) \cap \lozenge x_2 = \emptyset$. It follows that $([a), \lozenge x_2)$ is a filter-ideal pair. Let $y = (y_1, y_2)$ be its extension to a maximal filter-ideal pair. Then $[a) \subseteq y_1$, and hence $a \in y_1$. Consequently, $a \notin y_2$. Since $\lozenge x_2 \subseteq y_2$, we have $xS_\lozenge^c y$, which completes the proof.

The system $(X(P), \subseteq_1, \subseteq_2, R_\lozenge^c, S_\lozenge^c)$ is called the *canonical frame of an L-possibility algebra* P.

**Lemma 6.7**
For every L-possibility algebra $P = (W, \lor, \land, 1, 0, \lozenge)$, the relations of its canonical frame satisfy the possibility monotonicity conditions.

Proof: By way of example we prove the monotonicity condition for relation $S_\lozenge^c$. We have to show that for all $x, y, x', y' \in X(P)$, if $x' \subseteq_2 x$, $xS_\lozenge^c y$, and $y \subseteq_2 y'$, then x'$S^c_\Diamond$y'. Assume that (i) $x_1 \subseteq$ x'$_1$, (ii) $y_1 \subseteq \Diamond x_1$, and (iii) y'$_1 \subseteq y_1$. From (i) by Lemma 6.1(II) we get (iv) $\Diamond x_1 \subseteq \Diamond$x'$_1$. From (ii) and (iv) we obtain (v) $y_1 \subseteq \Diamond$x'$_1$. From (iii) and (v) we have y'$_1 \subseteq \Diamond$x'$_1$, which completes the proof.

The proof of the monotonicity condition for $R^c_\Diamond$ is similar.

### Lemma 6.8

For every L-possibility algebra P = (W, $\vee$, $\wedge$, 1, 0, $\Diamond$), the relations of its canonical frame satisfy the possibility stability conditions.

Proof: By way of example we prove condition (SC $S_\Diamond R_\Diamond$). We have to show that for all x,y $\in$ X(P), if x$S^c_\Diamond$y, then there is x'$\in$ X(P) such that x $\subseteq_2$ x' and x'$R^c_\Diamond$y. Assume that $\Diamond x_2 \subseteq y_2$. We have to show that there is an x'$\in$X(P) such that $x_2 \subseteq$x'$_2$ and $y_1 \subseteq \Diamond$x'$_1$. We show that $\Box^{-1}y_1 \cap x_2 = \emptyset$. For suppose conversely, then for some a$\in$W the conditions (i) a$\in \Box^{-1}y_1$ and (ii) a$\in x_2$ are satisfied. From (i) there is b$\in y_1$ such that $\Diamond$b$\leq$a. By (ii) we get $\Diamond$b$\in x_2$, and hence b$\in \Diamond x_2$. By the assumption we obtain b$\in y_2$. It follows that $y_1 \cap y_2 \neq \emptyset$, a contradiction. By Lemma 6.1(III), $\Box^{-1}y_1$ is a filter, so ($\Box^{-1}y_1,x_2$) is a filter-ideal pair. We extend it to a maximal filter-ideal pair, say x'=(x'$_1$,x'$_2$). Then we have $\Box^{-1}y_1 \subseteq$x'$_1$ and $x_2 \subseteq$x'$_2$. From Lemma 6.1(IV) we obtain $y_1 \subseteq \Diamond$x'$_1$, which completes the proof.

As a corollary we have the following theorem:

### Theorem 6.9

For every L-possibility algebra, its canonical frame is a possibility frame.

Now we extend the Urquhart representation theorem for lattices to L-possibility algebras.

Let P = (W, $\vee$, $\wedge$, 1, 0, $\Diamond$) be an L-possibility algebra and let h: W $\rightarrow 2^{X(P)}$ be the mapping defined as in section 3: h(a) = {x$\in$X(P): a$\in x_1$}.

### Lemma 6.10

For every L-possibility algebra P = (W, $\vee$, $\wedge$, 1, 0, $\Diamond$) and for every a$\in$W, h($\Diamond$a) = $\Diamond^c$h(a).

Proof: Let x$\in$X(P). We have x$\in \Diamond^c$h(a) iff x$\in$l[$S^c_\Diamond$]rh(a) iff for every y$\in$X(P), if x$\subseteq_1$y, then y$\notin$[$S^c_\Diamond$]rh(a) iff for every y, if x$\subseteq_1$y, then there is z$\in$X(P) such that y$S^c_\Diamond$z and (i) z$\notin$rh(a). By Theorem 3.2(I), (i) is equivalent to a$\notin z_2$. So we will prove that x$\in$h($\Diamond$a) iff for every y, if x$\subseteq_1$y, then there is z$\in$X(P) such that y$S^c_\Diamond$z and a$\notin z_2$.

($\rightarrow$) Assume that $\Diamond$a$\in x_1$ and suppose that $x_1 \subseteq y_1$. Then $\Diamond$a$\in y_1$, and hence $\Diamond$a$\notin y_2$. By Lemma 6.6 there is z$\in$X(P) such that y$S^c_\Diamond$z and a$\notin z_2$.

($\leftarrow$) Assume that (i) $\Diamond$a$\notin x_1$. We show that there is y$\in$X(P) such that $x_1 \subseteq y_1$ and for every z$\in$X(P), if y$S^c_\Diamond$z, then a$\in z_2$. Let ($\Diamond$a] be an ideal of P generated by $\Diamond$a. From (i) we get ($\Diamond$a] $\cap x_1 = \emptyset$, and hence ($x_1$,($\Diamond$a]) is a filter-ideal pair of P. Let y=($y_1,y_2$) be its extension to a maximal filter-ideal pair. Then $x_1 \subseteq y_1$ and ($\Diamond$a]$\subseteq y_2$, so $\Diamond$a$\in y_2$. By Lemma 6.6 the required condition follows.

**Theorem 6.11** (Representation theorem)
Every L-possibility algebra is isomorphic to a subalgebra of the complex algebra of its canonical frame.
Proof: The theorem follows from Theorem 3.3 and Lemma 6.10.

## 7. LATTICE-BASED POSSIBILITY LOGIC LATP

The language of possibility logic LATP is obtained from the language of logic LAT by addition of a unary propositional connective $\Diamond$ of possibility and by a suitable extension of the notion of a formula and a sequent. The deductive system for LATP consists of the axioms and rules of the system for LAT and, moreover, the following axioms and a rule are added:

(Ax8$\Diamond$)    $\Diamond F \vdash F$,
(Ax9$\Diamond$)    $\Diamond(A \lor B) \vdash \Diamond A \lor \Diamond B$.

(R4$\Diamond$)    $$\frac{A \vdash B}{\Diamond A \vdash \Diamond B}$$

**Theorem 7.1** (Soundness and Completeness)
For all formulas A and B of logic LATP the following conditions are equivalent:
  I. A sequent $A \vdash B$ is provable in LATP,
  II. $A \vdash B$ is true in all possibility algebras.

The proof can be easily obtained by appropriately extending the respective proofs for logic LAT.

Kripke-style semantics for logic LATP is defined in terms of possibility frames. By a model for the language of logic LATP we mean a system $M = (X, \leq_1, \leq_2, R_\Diamond, S_\Diamond, m)$ such that $(X, \leq_1, \leq_2, R_\Diamond, S_\Diamond)$ is a possibility frame and m: VAR$\rightarrow 2^X$ is a meaning function such that m(p) is an l-stable subset of X. The satisfiability of formulas by states in a model is defined as in section 5 and, moreover, for the formulas built with the possibility operation we define:
  $M, x \models \Diamond A$ iff for every $y \in X$, if $x \leq_1 y$ then there is $z \in X$ such that $y S_\Diamond z$ and there is $t \in X$ such that $z \leq_2 t$ and $M, t \models A$.

**Lemma 7.2**
For every model $(X, \leq_1, \leq_2, R_\Diamond, S_\Diamond, m)$ for LATP the following conditions are satisfied:
  I. $m(\Diamond A) = l[S_\Diamond]r(m(A))$,
  II. The family {m(A): A is a formula of LATP} is an L-possibility algebra of l-stable subsets of X.

Proof: (I) follows directly from the corresponding definitions. (II) follows from Lemmas 5.2 and 6.4(I).

The following theorems similar to Lemma 5.3 and Theorems 5.4 and 5.5 hold for LATP.

**Lemma 7.3**
For all formulas A and B of LATP the following conditions are equivalent:

I. A sequent $A \vdash B$ is true in all Kripke models for LATP,

II. For every possibility frame $(X, \leq_1, \leq_2, R_\Diamond, S_\Diamond)$ the sequent $A \vdash B$ is true in all complex algebras of X.

Kripke semantics for LATP is equivalent to algebraic semantics as the following theorem says.

Theorem 7.4

For all formulas A and B of LATP the following conditions are equivalent:

I. A sequent $A \vdash B$ is true in all L-possibility algebras,

II. $A \vdash B$ is true in all Kripke models for LATP.

Theorem 7.5 (Soundness and Completeness)

For all formulas A and B of LATP the following conditions are equivalent:

I. A sequent $A \vdash B$ is provable in LATP,

II. $A \vdash B$ is true in all Kripke models for LATP.

The proofs of the above theorems can be obtained in the similar way as the proofs of the corresponding theorems for LAT.

## 8. LATTICE-BASED NECESSITY ALGEBRAS AND LOGICS

In this section we present without proofs all the notions and constructions leading to a representation theorem for lattice-based necessity algebras, and we present an extension of logic LAT to a lattice-based necessity logic.

By an *L-necessity algebra* we mean an algebra $N = (W, \vee, \wedge, 1, 0, \Box)$, where $(W, \vee, \wedge, 1, 0)$ is a bounded lattice and $\Box$ is a unary operation on W satisfying for all $a, b \in W$:

(N1)  $\Box(a \wedge b) = \Box a \wedge \Box b$            multiplicative

(N2)  $\Box 1 = 1$                                        dual normal

Operators [R] and [-R] of the Boolean modal logic are examples of such necessity operators.

For every $A \subseteq W$ we define:

  $\Box A = \{a \in W: \Box a \in A\}$,

  $\Diamond^{-1}A = \{a \in W: \text{there is } b \in A \text{ such that } a \leq \Box b\}$,

where $\leq$ is the natural ordering of the lattice W.

A *necessity frame* is a relational system $(X, \leq_1, \leq_2, R_\Box, S_\Box)$ such that $(X, \leq_1, \leq_2)$ is a doubly ordered set, and $R_\Box$ and $S_\Box$ are binary relations on X satisfying the following conditions for all $x, y, x', y' \in X$:

(Mono $R_\Box$) If $x' \leq_1 x$, $xR_\Box y$, and $y \leq_1 y'$, then $x'R_\Box y'$,

(Mono $S_\Box$) If $x \leq_2 x'$, $xS_\Box y$, and $y' \leq_2 y$, then $x'S_\Box y'$,

(SC $R_\Box S_\Box$) If $xR_\Box y$ then there is $x' \in X$ such that $x \leq_1 x'$ and $x'S_\Box y$,

(SC $S_\Box R_\Box$) If $xS_\Box y$ then there is $y' \in X$ such that $y \leq_2 y'$ and $xR_\Box y'$.

The operator $[R_\Box]$ of classical necessity determined by relation $R_\Box$ and the operator $\langle S_\Box \rangle$ of classical possibility determined by $S_\Box$, both acting on subsets of

X, are defined as usual. Then a *complex algebra of a necessity frame* $(X, \leq_1, \leq_2,$ $R_\square, S_\square)$ is an algebra $(L(X), \vee^c, \wedge^c, 1^c, 0^c, \square^c)$, where $(L(X), \vee^c, \wedge^c, 1^c, 0^c)$ is the complex algebra of the doubly ordered set $(X, \leq_1, \leq_2)$, and $\square^c$ is a unary operator defined as follows:

$\square^c A = [R_\square]A.$

It can be shown that for any l-stable set $A$, $\square^c A = l\langle S_\square \rangle r(A)$, and the complex algebras of necessity frames are L-necessity algebras.

A *canonical frame of the L-necessity algebra* $N = (W, \vee, \wedge, 1, 0, \square)$ is the system $(X(N), \subseteq_1, \subseteq_2, R_\square^c, S_\square^c)$ such that $(X(N), \subseteq_1, \subseteq_2)$ is a canonical frame of the lattice reduct of $N$ and the relations $R_\square^c$ and $S_\square^c$ on $X(N)$ are defined as follows:

$xR_\square^c y$ iff $\square x_1 \subseteq y_1,$

$xS_\square^c y$ iff $y_2 \subseteq \square x_2.$

A canonical frame of an L-necessity algebra is a necessity frame. A representation theorem analogous to Theorem 6.11 holds for L-necessity algebras.

A lattice-based necessity logic LATN is a formal system whose language is obtained from the language of logic LAT by adding a unary propositional connective $\square$ of necessity and by suitably extending the notion of formula and sequent. The deductive system for LATN consists of the axioms and rules of the system for LAT and, moreover, the following axiom and a rule are added:

(Ax8□)    $T \vdash \square T,$

(Ax9□)    $\square A \wedge \square B \vdash \square(A \wedge B)$

(R4□)    $\dfrac{A \vdash B}{\square A \vdash \square B}$

Kripke semantics for LATN is determined by the class of necessity frames. By a model for the language of logic LATN we mean a system $M = (X, \leq_1, \leq_2, R_\square,$ $S_\square, m)$ such that $(X, \leq_1, \leq_2, R_\square, S_\square)$ is a necessity frame and $m: VAR \to 2^X$ is a meaning function such that $m(p)$ is an l-stable subset of X. The satisfiability of formulas by states in a model is defined as in section 5 and, moreover, for the formulas built with the necessity operation we define:

$M, x \models \square A$ iff for every $y \in X$, if $xR_\square y$ then $M, y \models A.$

The theorem on the equivalence of algebraic and Kripke semantics and the completeness theorem analogous to Theorems 7.4 and 7.5, respectively, hold for logic LATN.

## 9. LATTICE-BASED SUFFICIENCY (NEGATIVE NECESSITY) ALGEBRAS AND LOGICS

By an *L-sufficiency (negative necessity) algebra* we mean an algebra $S = (W, \vee,$ $\wedge, 1, 0, \boxminus)$, where $(W, \vee, \wedge, 1, 0)$ is a bounded lattice and $\boxminus$ is a unary operation on W satisfying for all $a, b \in W$:

(S1)    $\boxminus(a \vee b) = \boxminus a \wedge \boxminus b$      coadditive

(S2)    $\boxminus 0 = 1$      conormal

Observe that any sufficiency operator is antitone. The sufficiency operators $[\![R]\!]$ and the operators $[R]\neg$ obtained as a composition of the classical necessity operator $[R]$ with the classical negation are examples of such operators. We recall that, given a Kripke frame $(X, R)$, where $R$ is a binary relation on $X$ and $A \subseteq X$, $[R]A = \{x \in X:$ for all $y$, if $y \in A$ then $(x,y) \in R\}$.

These examples motivate that this class of algebras is named sufficiency or negative necessity algebras.

For every $A \subseteq W$ we define:

$\square\, A = \{a \in W: \square\, a \in A\}$,

$\diamond^{-1}A = \{a \in W:$ there is $b \in A$ such that $a \leq \square\, b\}$,

where $\leq$ is the natural ordering of the lattice $W$.

As previously, by a filter (resp. ideal) of an L-sufficiency algebra $(W, \vee, \wedge, 1, 0, \square\,)$ we mean a filter (resp. ideal) of the underlying lattice $(W, \vee, \wedge, 1, 0)$.

Lemma 9.1

For every L-sufficiency algebra $S = (W, \vee, \wedge, 1, 0, \square\,)$ and for all $A, B \subseteq W$ the following conditions are satisfied:

   I. If $A$ is a filter of $S$, then $\square\, A$ is an ideal of $S$,

  II. If $A \subseteq B$, then $\square\, A \subseteq \square\, B$,

 III. If $A$ is a filter of $S$, then $\diamond^{-1}A$ is an ideal of $S$,

 IV. $\diamond^{-1}A \subseteq B$ implies $A \subseteq \square\, B$; If $B$ is an ideal of $S$, then $A \subseteq \square\, B$ implies $\diamond^{-1}A \subseteq B$.

The proof is by an easy verification using the corresponding definitions.

Now we define sufficiency frames. A *sufficiency frame* is a system $(X, \leq_1, \leq_2, R_\square, S_\square)$ such that $(X, \leq_1, \leq_2)$ is a doubly ordered set and the relations $R_\square$ and $S_\square$ satisfy the following conditions:

(Mono $R_\square$)    If $x' \leq_1 x$, $xR_\square y$, and $y \leq_2 y'$, then $x'\ R_\square y'$,

(Mono $S_\square$)    If $x \leq_2 x'$, $xS_\square y$, and $y' \leq_1 y$, then $x'\ S_\square y'$,

(SC $R_\square S_\square$)   If $xR_\square y$ then there is $x' \in X$ such that $x \leq_1 x'$ and $x'S_\square y$,

(SC $S_\square R_\square$)   If $xS_\square y$ then there is $y' \in X$ such that $y \leq_1 y'$ and $xR_\square y'$.

In analogy to the respective conditions for the relations in necessity frames, the conditions (Mono $R_\square$) and (Mono $S_\square$) are referred to as *sufficiency monotonicity conditions*, and conditions (SC $R_\square S_\square$) and (SC $S_\square R_\square$) as *sufficiency stability conditions*.

We define unary operators $[R_\square]$ and $\langle S_\square \rangle$ on $2^X$ as follows. For every $A \subseteq X$:

$[R_\square]A = \{x \in X:$ for every $y \in X$ if $xR_\square y$ then $y \in A\}$,

$\langle S_\square \rangle A = \{x \in X:$ there is $y \in X$ such that $xS_\square y$ and $y \in A\}$.

Lemma 9.2

For every sufficiency frame $(X, \leq_1, \leq_2, R_\square, S_\square)$ and for every $A \subseteq X$ the following conditions are satisfied:

  I. $[R_\square]A$ is a $\leq_1$-increasing set,

 II. $\langle S_\square \rangle A$ is a $\leq_2$-increasing set.

Proof: (I) Assume that (i) $x \in [R_\square]A$ and (ii) $x \leq_1 y$ and suppose that (iii) $y \notin [R_\square]A$. (iii) means that there is $z$ such that (iv) $yR_\square z$ and (v) $z \notin A$. From (ii), (iv) and (Mono $R_\square$) we get (vi) $xR_\square z$. Using (i) and (vi) we conclude that $z \in A$, a contradiction with (v).

The proof of (II) is similar.

A *complex algebra of a sufficiency frame* $(X, \leq_1, \leq_2, R_\square, S_\square)$ is an algebra $(L(X), \vee^c, \wedge^c, 1^c, 0^c, \square^c)$ such that $(L(X), \vee^c, \wedge^c, 1^c, 0^c)$ is the complex algebra of the doubly ordered set $(X, \leq_1, \leq_2)$ and $\square^c$ is a unary operator defined as follows:
$$\square^c A = [R_\square] r(A).$$

Observe that this definition supports the intuition of negative necessity. Indeed, the mapping r plays the role of an intuitionistic-like negation interpreted in Kripke-style in terms of the relation $\leq_2$. Due to the forthcoming representation theorem 9.10, $\square$ A can be read 'necessarily, A is false'.

In the following lemma we present an equivalent formulation of the operation $\square^c$ which uses the relation $S_\square$.

### Lemma 9.3
For every sufficiency frame $(X, \leq_1, \leq_2, R_\square, S_\square)$ and for every l-stable set $A \subseteq X$ the following conditions are satisfied:

   I. $\square^c A = l \langle S_\square \rangle (A)$,

   II. $\square^c A$ is an l-stable set.

Proof: (I) ($\subseteq$) Assume that (i) $x \in [R_\square] r(A)$ and suppose that (ii) $x \notin l \langle S_\square \rangle (A)$. (ii) means that there is $y \in X$ such that (iii) $x \leq_1 y$ and (iv) $y \in \langle S_\square \rangle (A)$. From (iv) there is $z \in X$ such that (v) $y S_\square z$ and (vi) $z \in A$. From (v) and the stability condition (SC $S_\square R_\square$) there is $z' \in X$ such that (vii) $z \leq_1 z'$ and (viii) $y R_\square z'$. Since by Lemma 2.1(I) A is $\leq_1$-increasing, it follows from (vi) and (vii) that $z' \in A$. From (i), (iii), and Lemma 9.2(I) we have (ix) $y \in [R_\square] r(A)$. From (ix) and (viii) we obtain (x) $z' \in r(A)$. It follows from Lemma 2.1(III) that $z' \notin A$, a contradiction.

($\supseteq$) Assume that (i) $x \in l \langle S_\square \rangle A$ and suppose that (ii) $x \notin [R_\square] r(A)$. (ii) means that there is $y \in X$ such that (iii) $x R_\square y$ and (iv) $y \notin r(A)$. (iv) means that there is $z \in X$ such that (v) $y \leq_2 z$ and (vi) $z \in A$. From (iii), (vi), and the monotonicity condition (Mono $R_\square$) we get (vii) $x R_\square z$. From the stability condition (SC $R_\square S_\square$) there is x' such that (viii) $x \leq_1 x'$ and (ix) $x' S_\square z$. From (i), (viii), and the definition of the mapping l we obtain (x) $x' \notin \langle S_\square \rangle A$. On the other hand, from (vi), (ix), and the definition of $\langle S_\square \rangle$ we get $x' \in \langle S_\square \rangle A$ which contradicts (x).

(II) By Lemma 9.2(II) $\langle S_\square \rangle A$ is a $\leq_2$-increasing set. It follows from Lemma 2.4(I) that $l \langle S_\square \rangle A$ is an l-stable set, so by (I) $\square^c A$ is an l-stable set.

### Theorem 9.4
For every sufficiency frame $(X, \leq_1, \leq_2, R_\square, S_\square)$, its complex algebra $(L(X), \vee^c, \wedge^c, 1^c, 0^c, \square^c)$ is an L-sufficiency algebra.

Proof: The proof is by showing that $\square$ satisfies the axioms (S1) and (S2). Let A, B $\in L(X)$. We have $\square^c (A \vee^c B) = [R_\square] r(A \vee^c B) = [R_\square] rl(r(A) \cap r(B)) = [R_\square](r(A) \cap r(B))$ (since by Lemma 2.4 (V) $r(A) \cap r(B)$ is r-stable) $= [R_\square] r(A) \cap [R_\square] r(B) = \square^c A \wedge^c \square^c B$. Similarly, $\square^c \emptyset = l \langle S_\square \rangle \emptyset = l(\emptyset) = X$.

Let S $= (W, \vee, \wedge, 1, 0, \square)$ be an L-sufficiency algebra. By a filter-ideal pair of S we mean a filter-ideal pair of the underlying lattice $(W, \vee, \wedge, 1, 0)$. Let X(S) be the family of all the maximal filter-ideal pairs of S and let $(X(S), \subseteq_1, \subseteq_2)$ be a canonical frame of the lattice reduct of S as defined in section 3.

We define binary relations $R_\square^c$ and $S_\square^c$ on X(S) as follows. Let $x = (x_1, x_2)$ and $y = (y_1, y_2)$ be elements of X(S), then:

$xR_\square^c y$ iff $\square x_1 \subseteq y_2$,

$xS_\square^c y$ iff $y_1 \subseteq \square x_2$.

By the *canonical frame of an L-sufficiency algebra* S we mean the system $(X(S), \subseteq_1, \subseteq_2, R_\square^c, S_\square^c)$.

### Lemma 9.5

For every L-sufficiency algebra $S = (W, \vee, \wedge, 1, 0, \square)$, for every $x \in X(S)$, and for every $a \in W$ the following conditions are equivalent:

I. $\square a \in x_1$,

II. For every $y \in X(S)$, if $xR_\square^c y$ then $a \in y_2$.

Proof: $(\rightarrow)$ Assume that $\square a \in x_1$ and $xR_\square^c y$. From the corresponding definitions we get $a \in \square x_1$ and $\square x_1 \subseteq y_2$. Hence, $a \in y_2$.

$(\leftarrow)$ Now assume that $\square a \notin x_1$ which means (i) $a \notin \square x_1$. Let $[a) = \{b \in W: a \le b\}$ be the filter of S generated by a, where $\le$ is the natural ordering of the lattice W. By Lemma 9.1(I) $\square x_1$ is an ideal. It follows from (i) that $\square x_1 \cap (a] = \emptyset$. Hence $([a), \square x_1)$ is a filter-ideal pair. We extend it to a maximal filter-ideal pair, say $y = (y_1, y_2)$. So we have (ii) $[a) \subseteq y_1$ and (iii) $\square x_1 \subseteq y_2$. From (ii) we have (iv) $a \in y_1$. Since $y_1 \cap y_2 = \emptyset$, from (iv) we obtain $a \notin y_2$. Moreover, (iii) means that $x R_\square^c y$. It follows that (II) does not hold, which completes the proof.

### Lemma 9.6

For every sufficiency algebra $S = (W, \vee, \wedge, 1, 0, \square)$, the relations of its canonical frame satisfy the sufficiency monotonicity conditions.

Proof: By way of example we prove the monotonicity condition for relation $R_\square^c$. We have to show that for all $x, y, x', y' \in X(S)$, if $x' \subseteq_1 x$, $xR_\square^c y$, and $y \subseteq_2 y'$, then $x'R_\square^c y'$. Assume that (i) $x'_1 \subseteq x_1$, (ii) $\square x_1 \subseteq y_2$, and (iii) $y_2 \subseteq y'_2$. From (i) and Lemma 9.1(II) we obtain (iv) $\square x'_1 \subseteq \square x_1$. From (iv), (ii), and (iii) we get $\square x'_1 \subseteq y'_2$.

The proof of the monotonicity condition for $S_\square^c$ is similar.

### Lemma 9.7

For every L-sufficiency algebra $S = (W, \vee, \wedge, 1, 0, \square)$, the relations of its canonical frame satisfy the sufficiency stability conditions.

Proof: We show $(SC\ R_\square S_\square)$. We have to prove that for all $x, y \in X(S)$, if $xR_\square^c y$ then there is $x' \in X(S)$ such that $x \subseteq_1 x'$ and $x'S_\square^c y$. Assume that $\square x_1 \subseteq y_2$. We will show that $x_1 \cap \diamondsuit^{-1} y_1 = \emptyset$. For suppose conversely, then there is $a \in W$ such that (i) $a \in x_1$ and (ii) $a \in \Diamond^{-1} y_1$. By (ii) there is $b \in W$ such that (iii) $b \in y_1$ and (iv) $a \leq \Box b$. Since $x_1$ is a filter, from (i) and (iv) we obtain (v) $\Box b \in x_1$, and hence (vi) $b \in \Box x_1$. From (vi) and the assumption we get $b \in y_2$ which is in contradiction with (iii). So $(x_1, \Diamond^{-1} y_1)$ is a filter-ideal pair. Let $x' = (x'_1, x'_2)$ be its extension to a maximal filter-ideal pair, that is $x_1 \subseteq x'_1$ and $\Diamond^{-1} y_1 \subseteq x'_2$ which means that $y_1 \subseteq \Box x'_2$ by Lemma 9.1(IV). We conclude that $x' S_\Box y$.

Now we prove (SC $S_\Box R_\Box$), that is for all $x,y \in X(S)$, if $x S^c_\Box y$ then there is $y' \in X(S)$ such that $y \subseteq_1 y'$ and $x R^c_\Box y'$. Assume that $y_1 \subseteq \Box x_2$. We show that $\Box x_1 \cap y_1 = \emptyset$. For suppose conversely, then there is $a \in W$ such that (i) $a \in \Box x_1$ and (ii) $a \in y_1$. From (i) we get (iii) $\Box a \in x_1$. The assumption and (ii) yield (iv) $\Box a \in x_2$. From (iii) and (iv) we have $x_1 \cap x_2 \neq \emptyset$, a contradiction. Since by Lemma 9.1(I) $\Box x_1$ is an ideal, $(y_1, \Box x_1)$ is a filter-ideal pair of S. Let $y' = (y'_1, y'_2)$ be its extension to a maximal filter-ideal pair. It follows that $\Box x_1 \subseteq y'_2$, and hence we have $x R^c_\Box y'$. Similarly, since $y_1 \subseteq y'_1$, we get $y \subseteq_1 y'$, which completes the proof.

We conclude with the following theorem.

**Theorem 9.8**
For every L-sufficiency algebra, its canonical frame is a sufficiency frame.

Let $S = (W, \vee, \wedge, 1, 0, \Box)$ be an L-sufficiency algebra and let h be the mapping defined as in section 3: $h(a) = \{x \in X(S) : a \in x_1\}$.

**Lemma 9.9**
For every L-sufficiency algebra $S = (W, \vee, \wedge, 1, 0, \Box)$ and for every $a \in W$, $h(\Box a) = \Box^c h(a)$.

Proof: The theorem follows from the equivalence of the following statements: $x \in h(\Box a)$ iff $\Box a \in x_1$ iff for every $y \in X(S)$, $x R^c_\Box y$ implies $a \in y_2$ (by Lemma 9.5) iff for every $y \in X(S)$, $x R^c_\Box y$ implies (i) for every $z \in X(S)$, if $y_2 \subseteq z_2$ then $a \in z_2$. Since $(z_1, z_2)$ is a maximal filter-ideal pair, $a \in z_2$ is equivalent to $a \notin z_1$. We conclude that (i) is equivalent to $y \in r(h(a))$, and hence $x \in \Box^c h(a)$.

We conclude that the following representation theorem holds for L-sufficiency algebras.

**Theorem 9.10 (Representation theorem)**
Every L-sufficiency algebra is isomorphic to a subalgebra of the complex algebra of its canonical frame.

Proof: The theorem follows from Theorem 3.3 and Lemma 9.9.

A lattice-based sufficiency logic LATS has the following specific axioms and a rule:

(Ax8$\Box$)    $T \vdash \Box F$,
(Ax9$\Box$)    $\Box A \wedge \Box B \vdash \Box (A \vee B)$,

(R4$\Box$)    $\dfrac{A \vdash B}{\Box B \vdash \Box A}$

Algebraic semantics of logic LATS is determined by the class of L-sufficiency algebras and the Kripke semantics is determined by the class of sufficiency frames. The equivalence of algebraic and Kripke semantics and the completeness theorem can be proved as the analogous theorems in section 7.

## 10. LATTICE-BASED DUAL SUFFICIENCY (NEGATIVE POSSIBILITY) ALGEBRAS AND LOGICS

In this section we present without proofs all the notions and constructions leading to a representation theorem for lattice-based dual sufficiency algebras. We also present an axiomatic system of lattice-based dual sufficiency logic.

By an *L-dual sufficiency (negative possibility) algebra* we mean an algebra DS $= (W, \vee, \wedge, 1, 0, \Diamond)$, where $(W, \vee, \wedge, 1, 0)$ is a bounded lattice and $\Diamond$ is a unary operation on W satisfying for all a,b∈W:
(DS1)   $\Diamond(a \wedge b) = \Diamond a \vee \Diamond b$          comultiplicative
(DS2)   $\Diamond 1 = 0$                          codual normal
The classical dual sufficiency operators $\langle R \rangle$ and the operators $\langle R \rangle \neg$ obtained by composing the classical possibility operator with the classical negation are examples of such operators. We recall that, given a Kripke frame (X, R), where R is a binary relation on X and A⊆X, $\langle R \rangle A = \{x \in X$: there is $y \notin A$ such that $(x,y) \notin R\}$. These examples motivate the names of this class of algebras.

For every A⊆W we define:
   $\Diamond A = \{a \in W: \Diamond a \in A\}$,
   $\square^{-1}A = \{a \in W$: there is $b \in A$ such that $\Diamond b \le a\}$,
   where $\le$ is the natural ordering of the lattice W.

A *dual sufficiency frame* is a system $(X, \le_1, \le_2, R_\Diamond, S_\Diamond)$ such that $(X, \le_1, \le_2)$ is a doubly ordered set and the relations $R_\Diamond$ and $S_\Diamond$ satisfy the following conditions:
   (Mono $R_\Diamond$)     If $x \le_1 x'$, x $R_\Diamond$ y, and $y' \le_2 y$, then $x' R_\Diamond y'$,
   (Mono $S_\Diamond$)     If $x' \le_2 x$, x $S_\Diamond$ y, and $y \le_1 y'$, then $x' S_\Diamond y'$,
   (SC $R_\Diamond$ $S_\Diamond$)    If x $R_\Diamond$ y then there is $y' \in X$ such that $y \le_2 y'$ and x $S_\Diamond y'$,
   (SC $S_\Diamond$ $R_\Diamond$)    If $xS_\Diamond$ y then there is $x' \in X$ such that $x \le_2 x'$ and $x' R_\Diamond y$.
The operator $[S_\Diamond]$ of classical necessity determined by relation $S_\Diamond$ and the operator $\langle R_\Diamond \rangle$ of classical possibility determined by relation $R_\Diamond$ are defined as usual.

A *complex algebra of a dual sufficiency frame* $(X, \le_1, \le_2, R_\Diamond, S_\Diamond)$ is an algebra $(L(X), \vee^c, \wedge^c, 1^c, 0^c, \Diamond^c)$, where $(L(X), \vee^c, \wedge^c, 1^c, 0^c)$ is the complex algebra of the doubly ordered set $(X, \le_1, \le_2)$ and $\Diamond^c$ is a unary operator defined as follows:
   $\Diamond^c A = l[S_\Diamond]A$.
Since the mapping l is an intutitionistic-like negation interpreted in terms of the relation $\le_1$, the above definition confirms the intuition of negative possibility.

For every dual sufficiency frame $(X, \leq_1, \leq_2, R_\lozenge, S_\lozenge)$ and for every l-stable set $A \subseteq X$ we have $[S_\lozenge]A = r\langle R_\lozenge \rangle r(A)$. The complex algebras of dual sufficiency frames are L-dual sufficiency algebras.

By a *canonical frame of the dual sufficiency algebra DS* we mean a system $(X(DS), \subseteq_1, \subseteq_2, R^c_\lozenge, S^c_\lozenge)$ such that $(X(DS), \subseteq_1, \subseteq_2)$ is a canonical frame of the lattice reduct of DS and binary relations $R^c_\lozenge$ and $S^c_\lozenge$ on X(DS) are defined as follows:

$xR^c_\lozenge y$ iff $y_2 \subseteq \lozenge x_1$,

$xS^c_\lozenge y$ iff $\lozenge x_2 \subseteq y_1$.

For every L-dual sufficiency algebra, its canonical frame is a dual sufficiency frame. A representation theorem analogous to Theorem 6.11 holds for L-dual sufficiency algebras.

A lattice-based dual sufficiency logic LATDS has the following specific axioms and a specific rule:

(Ax8$\lozenge$)      $\lozenge T \vdash F$,

(Ax9$\lozenge$)      $\lozenge(A \wedge B) \vdash \lozenge A \vee \lozenge B$,

(R4$\lozenge$)      $\dfrac{A \vdash B}{\lozenge B \vdash \lozenge A}$

Algebraic semantics of logic LATDS is determined by the class of L-dual sufficiency algebras and the Kripke semantics is determined by the class of dual sufficiency frames. As with the logics developed in previous sections, the equivalence of algebraic and Kripke semantics and the completeness theorem hold.

## 11. A HIERARCHY OF LATTICE-BASED MODAL ALGEBRAS AND LOGICS

Observe that if $(W, \vee, \wedge, 1, 0)$ is a Boolean lattice, then every L-possibility algebra $P = (W, \vee, \wedge, 1, 0, \lozenge)$ is a MOA (modal) algebra, every L-necessity algebra $N = (W, \vee, \wedge, 1, 0, \square)$ is a DMOA (dual modal) algebra, every L-sufficiency algebra is a SUA (sufficiency) algebra, and every L-dual sufficiency algebra is a DSUA (dual sufficiency) algebra according to the terminology of Düntsch and Orłowska (2001) and Demri and Orłowska (2002). In MOA and DMOA specific (non-Boolean) operators are duals of each other and so are the operators of SUA and DSUA. Consequently, the results about the class MOA easily translate into the results about the class DMOA. The same concerns the classes SUA and DSUA.

With the lattice based algebras considered in this paper the situation is not that symetric. Therefore we should consider either four independent classes of algebras, as defined in sections 6, 8, 9, and 10, or the classes of algebras of the form $(W, \vee, \wedge, 1, 0, \lozenge, \square)$ and $(W, \vee, \wedge, 1, 0, \lozenge, \boxdot)$ such that $(W, \vee, \wedge, 1, 0, \lozenge)$ is an L-possibility algebra, $(W, \vee, \wedge, 1, 0, \square)$ is an L-necessity algebra, $(W, \vee, \wedge, 1, 0, \lozenge)$ is an L-dual sufficiency algebra, and $(W, \vee, \wedge, 1, 0, \boxdot)$ is an L-sufficiency algebra. Then the lattice-based mixed algebras would be of the form $(W, \vee, \wedge, 1, 0, \lozenge, \square, \lozenge, \boxdot)$. The other classes of lattice-based algebras can be obtained from those algebras by adding axioms reflecting properties of the operators added to a lattice and/or properties of the underlying lattice. The mixed algebras based on Boolean algebras are discussed in Düntsch and Orłowska (2001, 2004).

On the logical side, in analogy to the weakest modal logic K, we define a lattice-based modal logic LATK as a join of LATP and LATN, and the weakest lattice-based logic of sufficiency would be a join of LATS and LATDS. Consequently, the language of LATK is obtained from the language of LAT by extending it with the connectives $\Diamond$ and $\Box$, and the deduction system of LATK consists of the axioms and rules of LAT together with the respective axioms and rules specific for the two modal connectives $\Diamond$ and $\Box$, as presented in sections 7 and 8. LATK is a weakest lattice-based modal logic and it is a basis for several other logics.

For example, a lattice-based counterpart to the logic T is a logic LATT obtained from LATK by adding the following axioms:

(AxT$\Diamond$)     $A \vdash \Diamond A$,

(AxT$\Box$)     $\Box A \vdash A$.

Similarly, a lattice-based counterpart to the logic S4 is a logic LATS4 obtained from LATT by adding the axioms:

(AxS4$\Diamond$)     $\Diamond\Diamond A \vdash A$,

(AxS4$\Box$)     $\Box A \vdash \Box\Box A$.

It is clear that many other classes of logics, analogous to the classical Boolean modal logics can be defined in this manner. All of them belong to the lower levels of a hierarchy of modal logics which starts with lattice-based structures and ends with Boolean logics. The problems of Kripke semantics and modal definability for the logics from such a hierarchy requires further studies.

## 12. CONCLUSION

In this paper we introduced four classes of (not necessarily distributive) lattices with additional operators: modal operators of possibility and necessity, and negative modalities of sufficiency and dual sufficiency. We also presented the logics associated with these classes of algebras. We proved representation theorems for the classes of algebras in question, and we developed Kripke semantics and complete axiom systems for the logics. The representation theorems explicitly reflect the modal aspect of the operators and motivate in a formal way the intuition of positive and negative modalities.

We suggested a hierarchy of lattice-based modal algebras and logics. A systematic study of the correspondence theory and decidability of these logics requires further work.

## ACKNOWLEDGMENTS

The cooperation for this work was partially supported by EU COST Action 274 "Theory and Applications of Relational Structures as Knowledge Instruments" (TARSKI) and NATO Collaborative Linkage Grant PST CLG 977641.

The second author was partially supported by the RILA 12 project sponsored by the Bulgarian Ministry of Science and Education.

## REFERENCES

1.  Allwein, G. and Dunn, M. (1993) Kripke models for linear logic. *Journal of Symbolic Logic 58*, No 2, 514-545.
2.  Battilotti, G. and Faggian, C. (2003) *Quantum logic and cube logics.* Manuscript.
3.  Demri, S. and Orłowska, E. (2002) *Incomplete Information: Structure, Inference, Complexity.* EATCS Monographs in Theoretical Computer Science, Springer, 2002.
4.  Dunn, M. (1990) Gaggle theory: an abstraction of Galois connections and residuation, with applications to negation, implication, and various logical operations. Lecture Notes in Artificial Intelligence 478, Springer, 31-51.
5.  Düntsch, I. and Orłowska, E. (2001) Beyond modalities: sufficiency and mixed algebras. In: E. Orłowska and A. Szalas (eds) *Relational Methods for Computer Science Applications*, Physica Verlag, Heidelberg, 263-285.
6.  Düntsch, I. and Orłowska, E. (2004) Boolean algebras arising from information systems. *Proceedings of the Tarski Centenary Conference*, Banach Centre, Warsaw, June 2001. Annals of Pure and Applied Logic 127, No 1-3, 77-98.
7.  Düntsch, I., Orłowska, E., and Radzikowska, A. (2003) Lattice-based relation algebras and their representability. In: H. de Swart, E. Orłowska, G. Schmidt, and M. Roubens (eds) *Theory and Applications of Relational Structures as Knowledge Instruments*. Lecture Notes in Computer Science 2929, Springer, 231-255.
8.  Gehrke, M. and Jónsson, B. (1994) Bounded distributive lattices with operators. *Mathematica Japonica* 40, No 2, 207-215.
9.  Goranko, V. (1990) Modal definability in enriched languages. *Notre Dame Journal of Formal Logic* 31, 81-105.
10. Hájek, P. (1998) *Metamathematics of Fuzzy Logic.* Trends in Logic vol. 4, Kluwer, Dordrecht.
11. Humberstone, L. (1983) Inaccessible worlds. *Notre Dame Journal of Formal Logic* 24, 346-352.
12. Jónsson, B. and Tarski, A. (1952) Boolean algebras with operators I. *American Journal of Mathematics* 73, 891-939.
13. Orłowska, E. (1995) Information algebras. Lecture Notes in Computer Science 639, *Proceedings of AMAST'95*, Montreal, Canada, 1995, 50-65.
14. Orłowska, E. and Radzikowska, A.(2001) Double residuated lattices and their applications. Lecture Notes in Computer Science 2561, Springer, 171-189.
15. Sofronie-Stokkermans, V. (2000) Duality and canonical extensions of bounded distributive lattices with operators, and applications to the semantics of non-classical logics. Studia Logica 64, Part I 93-132, Part II 151-172.
16. Takano, M. (2002) Strong completeness of lattice-valued logic. *Archive of Mathematical Logic* 41, No 5, 497-505.
17. Tarski, A. (1941) On the calculus of relations. *Journal of Symbolic Logic* 6, 73-89.
18. Urquhart, A. (1978) A topological representation theory for lattices. *Algebra Universalis* 8, 45-58.

19. Vakarelov, D. (1976) Theory of negation in certain logical systems. Algebraic and semantic approach. Ph.D dissertation, University of Warsaw, Department of Mathematics.
20. Vakarelov, D. (1980) Simple examples of incomplete logics. *Comptes rendus de l'Academie Bulgarian des Sciences* 33, No 5, 587-589.
21. Vakarelov, D. (1989) Consistency, completeness and negation. In: G. Priest, R. Routley, and J. Norman (eds) *Paraconsistent Logic. Essays on the Inconsistent.* Analytica Philosophia Verlag, München-Hamden-Wien, 328-363.
22. Van Lambalgen, M. (2001) Conditional quantification, or poor man's probability. *Journal of Logic and Computation* 11, No 2, 295-335.

# Mathematics versus Metamathematics in Ramsey Theory of the Real Numbers

Carlos Augusto Di Prisco*

*Instituto Venezolano de Investigaciones Científicas, Caracas, Venezuela*
cdiprisc@ivic.ve

**Abstract.** The study of partition properties of the set of real numbers in several of its different presentations has been a very active field of research with interesting and sometimes surprising results. These partition properties are of the following form: if the set of real numbers (or a related space) is partitioned into a finite collection of pieces, there is a "large" collection of reals contained in one of the pieces. Different notions of largeness have been considered, and they give rise to properties of varied combinatorial character. The interplay between metamathematical questions and combinatorial problems has been present throughout the development of the theory. Most of these properties are, in their full generality, inconsistent with the axiom of choice, but versions of them where only partitions into simple pieces are considered, for example, Borel pieces, can be proved to be true. Nevertheless, the unrestricted versions are consistent with weak forms of the axiom of choice. We present here an overview of results about some of these partition properties, some old, some recent, and we mention several open problems.

## 1. Introduction

The theory of partitions as an area of combinatorial set theory originated in the 1930's with F. P Ramsey's famous theorem [34]. Some earlier results of Schur and Van der Waerden have some of the same combinatorial flavor, but it was Ramsey's theorem which attracted wide interest in partitions (see [15]). In the 1950's Erdös and Rado gave shape to the theory and extended Ramsey's result in several different directions with their partition calculus [12].

Ramsey's Theorem asserts that given positive integers $n$ and $k$, for every partition of $\mathbb{N}^{[n]} = \{a \subseteq \mathbb{N} : |a| = n\}$ into $k$ many pieces, there is an infinite $H \subseteq \mathbb{N}$ such that $H^{[n]}$, the collection of all of its $n$-element subsets, is contained in one piece. Such a set $H$ is said to be homogeneous for the partition. A finitary version of Ramsey's theorem is stated as follows, given positive integers $n, k, m$, there is a positive integer $N$ such that for every set $A$ of $N$ elements and every partition of $A^{[n]} = \{a \subseteq A : |a| = n\}$ into $k$ many pieces, there is a subset $H \subseteq A$ with $|H| = m$, for which $H^{[n]}$ is contained in one of the pieces (i.e. $H$ is homogeneous for the partition).

Ramsey proved his theorem as a technical tool to answer a question about decidability of first order logic [34]. This technical result turned out to be extremely interesting in itself, many different applications have been found, and the multiple

---

*This article was written while the author was a Visiting Researcher of Institució Catalana de Recerca i Estudis Avançats (ICREA) at the Centre de Recerca Mathemàtica (CRM), Barcelona.

171

ways it has been generalized or adapted to other contexts constitute a rich theory with applications in other areas of mathematics. The books [15,31] provide a good sample of the degree of development of Ramsey Theory. Several other developments in this area have also started from metamathematical considerations. In these other cases as well, the solution of a metamathematical problem led to the development of concepts and theories of a purely combinatorial character. On the other hand, some natural questions about partitions have required a metamathematical analysis, for example to establish consistency results.

The obvious infinite dimensional generalization of Ramsey's theorem to partitions of the set $\mathbb{N}^{[\infty]}$ of infinite subsets of $\mathbb{N}$, instead of the collection of subsets of a specified finite size, is false. Nevertheless, partitions of $\mathbb{N}^{[\infty]}$ into a finite number of "topologically simple" pieces always admit an infinite homogeneous set, i.e. an infinite set $H \subseteq \mathbb{N}$ such that $H^{[\infty]} = \{A \subseteq H : A \text{ infinite}\}$ is contained in one of the pieces.

Since we can identify subsets of $\mathbb{N}$ with their characteristic functions, the set $\mathbb{N}^{[\infty]}$ corresponds to a subset of the Cantor space $2^{\mathbb{N}}$, which, with the inherited topology, is homeomorphic to the set of irrational numbers, and also to the Baire space, the set $\mathbb{N}^{\infty}$ of all infinite sequences of natural numbers endowed with the product topology. In this sense, we are dealing with partitions of the set of real numbers, considering these numbers represented by infinite subsets of $\mathbb{N}$, or infinite sequences of natural numbers.

Partition properties are frequently stated in terms of colorings. A $k$-coloring of a set $S$ is simply a function $c : S \to K$ where $K$ is a set of size $k$. Clearly, every $k$-coloring of $S$ determines a partition of $S$ into $k$ pieces. For example, Ramsey's Theorem states that for every $n, k \in \mathbb{N}$ and every $k$-coloring of $\mathbb{N}^{[n]}$, there is $H \in \mathbb{N}^{[\infty]}$ such that $H^{[n]}$ is monochromatic.

We will consider partitions (or colorings) of $\mathbb{N}^{[\infty]}$ the set of infinite subsets of $\mathbb{N}$, and partitions of infinite products of various structures, finite or infinite, and the existence of different types of monochromatic sets for them. Different types of monochromatic sets usually give rise to corresponding partition properties of different strengths. We will be interested in the interrelationship between them, looking both into metamathematical and purely combinatorial aspects.

We mention some open questions, some quite old, like the questions about the necessity of inaccessible cardinals for the consistency of the Ramsey property $\omega \to (\omega)^{\omega}$, or if this partition property follows from the axiom of determinacy $AD$; and some more recent questions, perhaps easier to answer.

The notation used is standard. $\mathbb{N}$ is the set of natural numbers, which is identified with $\omega$, the first infinite ordinal. $\mathbb{N}^{\infty}$ is the set of infinite sequences of natural numbers; the topological space obtained giving it the product topology (obtained from $\mathbb{N}$ with the discrete topology) is called the Baire space. $\mathbb{N}^{<\infty} = \bigcup_{n=0}^{\infty} \mathbb{N}^n$ is the set of finite sequences of natural numbers. Given a set $A$ and $n \in \mathbb{N}$, $A^{[n]}$ is the collection of those subsets of $A$ which have exactly $n$ elements. The collection of finite subsets of $\mathbb{N}$ is denoted by $\mathbb{N}^{[<\infty]}$. For every infinite $A \subseteq \mathbb{N}$, we use $A^{[\infty]}$ to denote the collection of infinite subsets of $A$, accordingly, $\mathbb{N}^{[\infty]}$ is the set of infinite subsets of $\mathbb{N}$. If $A \in \mathbb{N}^{[\infty]}$ and $a \in \mathbb{N}^{[<\infty]}$, then $A/a = \{n \in A : \max a < n\}$.

We will use the letters $n, m, k, l, \ldots$ to denote natural numbers, and $A, B, C, H$, $X, Y, \ldots$ to denote infinite sets of natural numbers. The letters $a, b, c, \ldots$ will be used to denote finite sets of natural numbers, and $s, t, r, \ldots$ to denote finite sequences.

The author would like to thank Stevo Todorcevic for suggestions and comments which helped to improve this article.

## 2. The Ramsey property

The partition symbol

$$\omega \to (\omega)^\omega,$$

stands for the statement

"For every coloring $c : \mathbb{N}^{[\infty]} \to \{0, 1\}$ there is an infinite $H \subseteq \mathbb{N}$ such that $H^{[\infty]}$ is monochromatic".

If we restrict ourselves to partitions

$$c : \mathbb{N}^{[\infty]} \to \{0, 1\}$$

measurable with respect to a certain $\sigma$-field $\mathcal{C}$ of subsets of $\mathbb{N}^{[\infty]}$, we use the notation

$$\omega \to_\mathcal{C} (\omega)^\omega,$$

for the corresponding property.

As mentioned in the introduction, using the axiom of choice a partition of $\mathbb{N}^{[\infty]}$ can be given for which there is no infinite homogeneous set. Consider, for example, the equivalence relation defined on $\mathbb{N}^{[\infty]}$ by $A \sim B$ if and only if $A \Delta B$, the symmetric difference of $A$ and $B$, is finite. Pick one element from each equivalence class, and define a partition of $\mathbb{N}^{[\infty]} = \mathcal{A} \cup \mathcal{B}$ as follows. Put $A$ in $\mathcal{A}$ if and only if it differs from the chosen representative of its class in a (finite) set of even cardinality. Clearly, no set of the form $H^{[\infty]} = \{Y \subseteq H : Y \text{ infinite}\}$ is included in one of the pieces, as for every $Y \in \mathbb{N}^{[\infty]}$, $Y$ and $Y \setminus \{\min Y\}$ lie in different pieces [2].

The non constructive character of counterexamples like this one, or the one given by Erdös and Rado in [11], suggests asking if such a counterexample can be given without using the axiom of choice. This question was in fact posed in a seminar on Ramsey theory conducted by D. Scott at Stanford University in 1967. Mathias recalls[3] that shortly after, several people, including P. Cohen, A. Ehrenfeucht and F. Galvin, had shown that for every open subset $\mathcal{O} \subseteq \mathbb{N}^{[\infty]}$ there is an infinite set $A \in \mathbb{N}^{[\infty]}$, such that $A^{[\infty]} \subseteq \mathcal{O}$ or $A^{[\infty]} \cap \mathcal{O} = \emptyset$, and thus a partition into an open set and its complement cannot be a counterexample.

Nash-Williams [30] proved a result that implies that if $\mathbb{N}^{[\infty]}$ is partitioned into a finite number of sets which are simultaneously open and closed, there is $H \in \mathbb{N}^{[\infty]}$ such that $H^{[\infty]}$ is contained in one of the pieces. In [13], Galvin extends Nash-Williams' result to partitions into an open set and its complement, a step towards the proof of the following theorem of [14] from which follows that infinite homogeneous sets exist for partitions of $\mathbb{N}^{[\infty]}$ into a finite number of Borel subsets.

---

[2]This example is due to A. R. D. Mathias. In his thesis it appears in a slightly different presentation which requires only the use of the axiom of choice for pairs

[3]In a communication presented at the CRM, Barcelona, January of 2004.

**Theorem 1.** *(Galvin and Prikry [14]) For every Borel-measurable 2-coloring of* $\mathbb{N}^{[\infty]}$ *there is a set* $H \in \mathbb{N}^{[\infty]}$ *such that* $H^{[\infty]}$ *is monochromatic. In symbols,*

$$\omega \rightarrow_{Borel} (\omega)^{\omega}.$$

Note that this theorem can be stated as a property of Borel subsets of $\mathbb{N}^{[\infty]}$, namely, for every Borel set $\mathcal{A} \subseteq \mathbb{N}^{[\infty]}$ there is an infinite set $H \subseteq \mathbb{N}$ such that $H^{[\infty]} \subseteq \mathcal{A}$ or $H^{[\infty]} \cap \mathcal{A} = \emptyset$.

Consider the sets of the form

$$[a, B] = \{X : a \subseteq X \subseteq a \cup B\},$$

with $a \in \mathbb{N}^{[<\infty]}, B \in (\mathbb{N}/a)^{[\infty]}$.

**Definition 1.** *A set* $\mathcal{X} \subseteq \mathbb{N}^{[\infty]}$ *is Ramsey if for every* $[a, A]$ *there is* $B \in A^{[\infty]}$ *such that*

$$[a, B] \subseteq \mathcal{X} \text{ or } [a, B] \cap \mathcal{X} = \emptyset.$$

Galvin and Prikry actually showed in [14] that every Borel subset of $\mathbb{N}^{[\infty]}$ is Ramsey (they used "completely Ramsey" to name this property), from where Theorem 1 follows applying the definition of Ramsey to $[\emptyset, \mathbb{N}]$.

Silver ([36]) extended this result proving that analytic sets are Ramsey. Recall that a subset of $\mathbb{N}^{[\infty]}$ is analytic if it is the image of $\mathbb{N}^{[\infty]}$ by a continuous function from $\mathbb{N}^{[\infty]}$ to itself. Silver's proof uses metamathematical methods; Ellentuck ([10]) gave a topological proof of Silver's result using the topology on $\mathbb{N}^{[\infty]}$ generated by the basic sets $[a, A]$. This topology, frequently called Ellentuck's Topology, refines the product topology, and it characterizes the Ramsey sets as follows.

**Theorem 2.** *(Ellentuck)[10] A subset of* $\mathbb{N}^{[\infty]}$ *is Ramsey if and only if it has the Baire property (with respect to Ellentuck's topology).*

*The* $\sigma$*-field of Ramsey sets is closed under Souslin's operations, and therefore contains the analytic sets.*

The Ramsey property can be viewed as another regularity property of subsets of $\mathbb{N}^{[\infty]}$, as Lebesgue measurability, the Baire property, or the perfect subset property. The existence of non-measurable sets of reals, of sets without the property of Baire, and of uncountable sets which do not contain perfect subsets are consequences of the axiom of choice, just as the existence of non-Ramsey sets. Solovay [37] constructed a model of set theory where every set of real numbers is Lebesgue measurable, has the property of Baire and, if uncountable, contains a perfect subset. Obviously, the axiom of choice does not hold in this model, but only a weak version of this axiom called "Axiom of Dependent Choices" (DC). The construction of the model relies on the assumption of the existence of an inaccessible cardinal (see [20,21]).

Mathias [28] showed that in Solovay's model all subsets of $\mathbb{N}^{[\infty]}$ are Ramsey. Therefore, the partition relation $\omega \rightarrow (\omega)^{\omega}$ is consistent in the following precise sense.

**Theorem 3.** *(Mathias)[28]*

$$ZF + DC + \omega \to (\omega)^\omega$$

*is consistent provided*

$$ZFC + \text{``There exists an inaccessible cardinal''}$$

*is consistent.*

**Question 1.** *One of the problems of the theory, which has remained open for several decades, is whether the hypothesis about inaccessible cardinals is necessary in Theorem 3.*

Shelah has shown that this hypothesis is necessary for the consistency of "all sets of reals are Lebesgue measurable", but not for the consistency of "all sets of real numbers have the property of Baire". It was known earlier that the hypothesis is necessary for the consistency of "every uncountable set of real numbers contains a perfect subset".

An argument of [27] shows that the existence of a non-principal ultrafilter on $\mathbb{N}$, which is a consequence of the axiom of choice, also provides a counterexample for the partition relation $\omega \to (\omega)^\omega$:

Let $\mathcal{U}$ be a non-principal ultrafilter on $\mathbb{N}$. Every set $X \in \mathbb{N}^{[\infty]}$ determines an infinite collection of consecutive intervals of $\mathbb{N}$ as follows: if $X = \{x_0, x_1, \dots\}$ is the increasing enumeration of $X$, let for every $n \in \mathbb{N}$, $I_n = [x_n, x_{n+1}) = \{k : x_n \leq k < x_{n+1}\}$. Define a partition $\mathbb{N}^{[\infty]} = \mathcal{A} \cup \mathcal{B}$ putting $X \in \mathcal{A}$ if and only if $\bigcup_{n \in \mathbb{N}} [x_{2n}, x_{2n+1}) \in \mathcal{U}$. For no $H \in \mathbb{N}^{[\infty]}$ the set $H^{[\infty]}$ is homogeneous, because $X \in \mathcal{A}$ if and only if $X \setminus \{\min X\} \in \mathcal{B}$. This is so because if $Y$ is obtained removing from $X$ its first element, then $y_n = x_{n+1}$, so $\bigcup_i [y_{2i}, y_{2i+1})$ differs from the complement of $\bigcup_i [x_{2i}, x_{2i+1})$ only on a finite set, so exactly one of those two unions belong to the ultrafilter $\mathcal{U}$. The fact that the existence of non-principal ultrafilters on $\mathbb{N}$ implies the negation of $\omega \to (\omega)^\omega$ will be used below.

## 3. Perfect set properties

As we have seen, the Ramsey property concerns the existence of monochromatic sets of the form $H^{[\infty]}$, with $H \in \mathbb{N}^{[\infty]}$, for finite partitions of $\mathbb{N}^{[\infty]}$. We will now consider a weaker property, stated in terms of the existence of monochromatic perfect sets.

The symbol

$$\omega \to (\text{perfect})^\omega$$

stands for the statement

"for every coloring $c : \mathbb{N}^{[\infty]} \to \{0, 1\}$ there is a perfect monochromatic subset of $\mathbb{N}^{[\infty]}$".

It is well known that $\mathbb{N}^{[\infty]}$ can be partitioned in two pieces neither of which contain a perfect set [1]. Nevertheless, every Borel-measurable coloring of $c : \mathbb{N}^{[\infty]} \to \{0, 1\}$ there is a monochromatic perfect set. This is expressed by the partition symbol

$$\omega \to_{Borel} (\text{perfect})^\omega.$$

Notice that $\omega \to (\omega)^\omega$ implies $\omega \to (\text{perfect})^\omega$, since sets of the form $H^{[\infty]}$ are perfect. We saw in the previous section that the existence of a non principal ultrafilter on $\mathbb{N}$ gives a counterexample to $\omega \to (\omega)^\omega$. On the other hand, $\omega \to (\text{perfect})^\omega$ is consistent with $ZF + DC +$ "there is an ultrafilter on $\mathbb{N}$ (see [3]), and therefore it is strictly weaker than $\omega \to (\omega)^\omega$.

**Question 2.** *Is there a choice principle equivalent to the negation of $\omega \to (\text{perfect})^\omega$ ?*

Halpern and Läuchli proved in [17] a deep and powerful partition property about perfect trees. We need some definitions before stating the theorem.

A tree is a partially ordered set $(T, \prec)$ such that for every $u \in T$, the set $\{v : v \prec u\}$ is well ordered by $\prec$. The order type of $\{v : v \prec u\}$ is called the height of $u$ in the tree. We consider trees in which every element has finite height. The height of the tree is the supremum of the heights of its elements. For every $n \in \mathbb{N}$, $T(n)$ is the collection of elements of $T$ of height $n$, or the $n$-th level of the tree. A tree is perfect if for every $u \in T$, there are $v, w \in T$ such that $u \prec v$, $u \prec w$, with $v$ and $w$ incomparable with respect to the order $\prec$. If $A \subseteq \mathbb{N}$, $T \upharpoonright A$ is the collection $\{u \in T : \text{height}(u) \in A\}$.

If $d \in \mathbb{N}$ and $T_i$ is a tree for each $i < d$, then $\otimes_{i<d} T_i$ is the tree

$$\{(t_0, \ldots, t_{d-1}) \in \prod_{i<d} T_i : \text{height}(t_0) = \cdots = \text{height}(t_{d-1})\}$$

with the ordering $(t_0, \ldots, t_{d-1}) \prec (t'_0, \ldots, t'_{d-1})$ if for every $i < d$ $t_i \prec_i t'_i$ (where $\prec_i$ is the ordering of the tree $T_i$).

**Theorem 4.** *([17]) Let $d \in \mathbb{N}$, and for every $i < d$ let $T_i$ be a perfect tree of height $\omega$. For every*

$$c : \otimes_{i<d} T_i \to \{0, 1\},$$

*there is $A \in \mathbb{N}^{[\infty]}$, and a perfect subtree $R_i \subseteq T_i$ for every $i < d$ such that $c$ is constant on $\otimes_{i<d}(R_i \upharpoonright A)$.*

This result was proved to solve a question regarding the Axiom of Choice and one of its consequences, the Boolean prime ideal theorem, which says that there is a prime ideal in every Boolean algebra. The Halpern-Laüchli theorem was obtained in order to construct a model of set theory where the Boolean prime ideal theorem holds but not the axiom of choice (see [18]). This is, then, another example of a deep combinatorial principle obtained to answer a metamathematical question.

Laver in [22] extended the Halpern-Laüchli theorem to infinite products of perfect trees.

## 4. Monochromatic sublattices of $\mathcal{P}(\mathbb{N})$

We consider now a partition property defined in terms of a different type of homogeneity for partitions of $\mathbb{N}^{[\infty]}$. Instead of requiring for every coloring $c : \mathbb{N}^{[\infty]} \to \{0, 1\}$ the existence of an infinite set $B$ with all of its infinite subsets of the same color, we only require that all subsets of $B$ containing a fixed subset $A$ have the same color.

The partition symbol

$$\omega \to ((\omega))^\omega$$

means that for every coloring $c : \mathbb{N}^{[\infty]} \to \{0,1\}$ there is a pair $(A, B)$ with $A \subseteq B \in \mathbb{N}^{[\infty]}$ such that

$$[A, B] = \{X : A \subseteq X \subseteq B\}$$

is monochromatic (we put no requirements on the size of $A$). This type of homogeneous sets were studied in [6]. Partition properties of certain classes of lattices has been considered previously from various points of view, see, for example, [33].

Since every such sublattice $[A, B]$ is a perfect subset of $\mathbb{N}^{[\infty]}$, it follows immediately from the definitions that

$$\omega \to (\omega)^\omega \text{ implies } \omega \to ((\omega))^\omega \text{ implies } \omega \to (\text{perfect})^\omega.$$

Using the same argument presented at the end of Section 2 for the property $\omega \to (\omega)^\omega$, it can be shown that a non-principal ultrafilter on $\mathbb{N}$ provides a counterexample for the property $\omega \to ((\omega))^\omega$, and so, the second implication is strict. Although the exact relation between $\omega \to (\omega)^\omega$ and $\omega \to ((\omega))^\omega$ is still unknown, most likely the first implication is also strict, since the consistency of $(ZF + DC + \omega \to ((\omega))^\omega)$ follows just from the consistency of $ZFC$, with no hypothesis involving inaccessible cardinals. This is so since $\omega \to_{Baire} ((\omega))^\omega$ holds (see [6]), and the consistency of "every subset of $\mathbb{N}^\infty$ has the Baire property" has been established by Shelah assuming just the consistency of $ZFC$ [35].

J. Brendle, L. Halbeisen and B. Lowe have studied the property $\omega \to ((\omega))^\omega$ restricted to $\Sigma_2^1$ sets, $\Delta_2^1$ sets, and projective sets in general (see, for example, [2], and [16], where it is shown that $\omega \to_{\Sigma_2^1} ((\omega))^\omega$ does not imply $\omega \to_{\Sigma_2^1} (\omega)^\omega$ ).

**Question 3.** *What is the exact relationship between the properties $\omega \to (\omega)^\omega$ and $\omega \to ((\omega))^\omega$ ?*

## 5. Polarized partitions

Dealing with partitions of the space $\mathbb{N}^{[\infty]}$, we have considered several types of homogeneous sets. A different form is obtained as follows. Given a finite coloring of $\mathbb{N}^{[\infty]}$, we require the existence of a sequence of finite sets $\{H_i\}_{i=0}^\infty$ of certain required cardinalities such that for every $i$, $\max(H_i) < \min(H_{i+1})$, and such that the collection $\{X \in \mathbb{N}^{[\infty]} : \forall i \, |X \cap H_i| = 1\}$ is monochromatic. A convenient way to treat this is to consider partitions of the Baire space $\mathbb{N}^\infty = \mathbb{N} \times \mathbb{N} \times \ldots$, the set of infinite sequences of natural numbers with the product topology obtained when $\mathbb{N}$ is considered as a discrete space.

Given

$$c : \mathbb{N}^\infty \to \{1, 2\},$$

we want a monochromatic product $\prod_{i=0}^\infty H_i$, with $H_i \subseteq \mathbb{N}$ of some specified size for every $i$. ([4]).

We use the partition symbol

$$\begin{pmatrix} \omega \\ \omega \\ \vdots \end{pmatrix} \longrightarrow \begin{pmatrix} m_0 \\ m_1 \\ \vdots \end{pmatrix}$$

to express that for every coloring $c : \mathbb{N}^\infty \to 2$, there is a monochromatic product $\prod_{i=0}^\infty H_i$ with $H_i \subseteq \mathbb{N}$ and $|H_i| = m_i$ for every $i$.

Notice that here we do not require the sequence $\{H_i\}_{i=0}^\infty$ to be increasing, but as shown in [4] this is not essential.

It should be clear that we cannot require $H_0$ and $H_1$ to be both infinite: let $c : \mathbb{N}^\infty \to \{0,1\}$ be defined by $c(x) = 0$ if and only if $x(0) < x(1)$. Clearly, a product $\prod_{i=0}^\infty H_i$ with $H_i \subseteq \mathbb{N}$ and with $H_0$ and $H_1$ both infinite cannot be monochromatic for $c$. In the same fashion a counterexample can be given if we require infinite sets in any two specified (fixed) coordinates. If no coordinates are specified, it is possible to get for every $c : \mathbb{N}^\infty \to \{0,1\}$ a monochromatic product $\prod_{i=0}^\infty H_i$ with at least two infinite factors. The position of the infinite factors depending on the partition. This follows from results of G. Moran and D. Strauss [29]. They prove that for every $k \in \mathbb{N}$ and every partition of $\mathbb{N}^\infty$ into two pieces, there is a monochromatic product $\prod_{i=0}^\infty H_i$ with $k$ factors that are infinite (in fact, each of them is the whole set $\mathbb{N}$) and the rest of the factors are singletons. It is not known if we can also require the finite factors to be non-trivial, having at least two elements each.

**Question 4.** *Is the following statement consistent? (We mean consistent with ZF, provided that ZF is consistent). For every partition $c : \mathbb{N}^\infty \to \{1,2\}$, there is a homogeneous product $\prod_{i=0}^\infty H_i$ such that for every $i$, $|H_i| > 1$ and there are $m, n \in \mathbb{N}$, $m \neq n$ with $|H_m| = |H_n| = \aleph_0$. (see [6]).*

Henle proved in [19] that $\omega \to (\omega)^\omega$ implies that every such partition admits a homogeneous product $\prod_{i=0}^\infty H_i$ such that for every $i$, $|H_i| > 1$ and $H_0$ is infinite.

Identifying infinite subsets of $\mathbb{N}$ with their increasing enumeration, it is easy to verify that $\omega \to (\omega)^\omega$ implies $\begin{pmatrix} \omega \\ \omega \\ \vdots \end{pmatrix} \longrightarrow \begin{pmatrix} m_0 \\ m_1 \\ \vdots \end{pmatrix}$ for every sequence $\{m_i\}_{i=0}^\infty$ of positive integers.

The exact relationship between the partition properties $\omega \to (\omega)^\omega$ and $\begin{pmatrix} \omega \\ \omega \\ \vdots \end{pmatrix} \longrightarrow \begin{pmatrix} m_0 \\ m_1 \\ \vdots \end{pmatrix}$ (for a given sequence $\{m_i\}_{i=0}^\infty$ of positive integers) turned out to be an interesting problem. The two partition properties are not equivalent since polarized partition property is consistent with the existence of non principal ultrafilters on $\mathbb{N}$ (see section 9). To prove this, it was necessary to develop a theory of partitions of products of finite sets and their parametrized versions. This theory, of intrinsic combinatorial interest, is another example of a combinatorial development motivated by a metamathematical question. The next three sections are devoted to present some aspects of this theory (see [7,8,38]).

## 6. Products of finite sets

In the previous section we considered partitions of $\mathbb{N}^\infty$. Here we will deal with partitions of subspaces of $\mathbb{N}^\infty$ of the form $\prod_{i=0}^\infty n_i$, where $\{n_i\}_{i=0}^\infty$ is a sequence of positive integers and, as usual, we identify each $n$ with the set $\{0,1,\ldots,n-1\}$. In other words, we will consider partitions of the set of infinite sequences of positive integers bounded by some sequence $\{n_i\}_{i=0}^\infty$.

Given a sequence $\{n_i\}_{i=0}^\infty$, invoking the axiom of choice we can define a coloring $c : \prod_{i=0}^\infty n_i \to \{0,1\}$ such that for no sequence $\{H_i\}_{i=0}^\infty$ with $H_i \subseteq n_i$ and $|H_i| = 2$ for every $i$, the product $\prod_{i=0}^\infty H_i$ is monochromatic. Restricting the class of colorings to be considered, some positive results are obtained. For example, there is a sequence $\{n_i\}_{i=0}^\infty$ such that every Borel coloring $c : \prod_{i=0}^\infty n_i \to \{0,1\}$ admits a monochromatic product of pairs ([24,7]). An interesting connection appeared here between these partition relations restricted to definable classes of colorings and the Grzegorczyk hierarchy of primitive recursive functions (see Section 8 below).

In more general terms, for every sequence $\{m_i\}_{i=0}^\infty$ of positive integers, there is a sequence $\{n_i\}_{i=0}^\infty$ of positive integers such that for every Souslin-measurable coloring $c : \prod_{i=0}^\infty n_i \to \{0,1\}$ there is a monochromatic product $\prod_{i=0}^\infty H_i$ where for every $i$, $H_i \subseteq n_i$ has size $m_i$ (Corollary 1).

Partitions of infinite products of finite sets were considered in [4,23], with particular interest in the question of the consistency of the existence some sequence $\{n_i\}_{i=0}^\infty$ such that for every partition of $\prod_{i=0}^\infty n_i$ into two pieces there is a monochromatic product of pairs. In Section 9 we come back to this question, and see how Solovay's model provides a positive answer.

In this section we describe, for each sequence $\{m_i\}_{i=0}^\infty$, a class of products of finite sets and a class of colorings which admit homogeneous products whose factors have sizes determined by $\{m_i\}_{i=0}^\infty$. First we need some definitions. Given a sequence $\vec{n} = \{n_i\}_{i=0}^\infty$ of natural numbers, a sequence $\vec{H} = \{H_i\}_{i=0}^\infty$ of finite sets of natural numbers is said to be of type $\vec{n}$, or a $\vec{n}$-sequence, if for every $i$, $|H_i| = n_i$. We also say in this case that the product $\prod_{i=0}^\infty H_i$ is an $\vec{n}$-product.

Whenever $\{H_i\}_{i=0}^\infty$ and $\{J_i\}_{i=0}^\infty$ are sequences of finite sets of natural numbers we say that $\prod_{i=0}^\infty J_i$ is a sub-product of $\prod_{i=0}^\infty H_i$ if $J_i \subseteq H_i$ for every $i$. We write $\vec{J} \leq_k \vec{H}$ if $J_i \subseteq H_i$ for every $i \in \mathbb{N}$ and $J_i = H_i$ for $i < k$.

Given sequences of positive integers $\vec{m} = \{m_i\}_{i=0}^\infty$ and $\vec{n} = \{n_i\}_{i=0}^\infty$,

$$(\vec{n}) \to (\vec{m})$$

expresses that for every partition $c : \prod_{i=0}^\infty H_i \to \{0,1\}$ such that $|H_i| = n_i$ for all $i$, there is a sequence $\{J_i\}_{i=0}^\infty$ with $J_i \subseteq H_i$, and $|J_i| = m_i$ for all $i$, such that $c$ is constant on $\prod_{i=0}^\infty J_i$. More concisely, for every 2-coloring of a product of type $\vec{n}$, there is a monochromatic sub-product of type $\vec{m}$. As before, if we restrict ourselves to colorings measurable with respect to a certain $\sigma$-field $\mathcal{C}$ of subsets of $\mathbb{N}^{[\infty]}$, we use the notation

$$(\vec{n}) \to_{\mathcal{C}} (\vec{m})$$

for the corresponding property.

Notice that $(\vec{n}) \to (\vec{m})$ does not follow from $\omega \to (\omega)^\omega$, as

$$\begin{pmatrix} \omega \\ \omega \\ \vdots \end{pmatrix} \to \begin{pmatrix} m_0 \\ m_1 \\ \vdots \end{pmatrix}$$

does (for sequences $\{m_i\}_{i=0}^\infty$ and $\{n_i\}_{i=0}^\infty$ of positive integers).

For partitions of finite products we use the notation

$$\begin{pmatrix} n_0 \\ n_1 \\ \vdots \\ n_k \end{pmatrix} \to \begin{pmatrix} m_0 \\ m_1 \\ \vdots \\ m_k \end{pmatrix}$$

which means that for every partition $c : \prod_{i=0}^k H_i \to \{0,1\}$ such that $|H_i| = n_i$ for all $i \le k$, there is a sequence $\{J_i\}_{i=0}^k$ with $J_i \subseteq H_i$, and $|J_i| = m_i$ for all $i \le k$, such that $c$ is constant on $\prod_{i=0}^k J_i$.

**Definition 2.** *Let $S : \mathbb{N}^{<\infty} \to \mathbb{N}$ be defined as follows*

$$S(m_0) = 2m_0 - 1$$

$$S(m_0, ..., m_{i+1}) = 2(m_{i+1} - 1) \left[ \prod_{k=0}^i \left( \frac{m_k}{S(m_0, ..., m_k)} \right) \right] + 1.$$

The function $S$ has the following property which can be verified by induction.

**Lemma 1.** *Let $(m_i) \in \mathbb{N}^\infty$ and let $n_i = S(m_0, ..., m_i)$ for all $i$. Then for every $k$*

$$\begin{pmatrix} n_0 \\ n_1 \\ \vdots \\ n_k \end{pmatrix} \to \begin{pmatrix} m_0 \\ m_1 \\ \vdots \\ m_k \end{pmatrix}.$$

It follows that we can get monochromatic sub-products of type $\vec{m}$ for every continuous 2-coloring of $\prod_{i=0}^\infty S(m_o, \ldots, m_i)$ (note that a continuous coloring gives a partition into clopen pieces).

The existence of monochromatic $\vec{m}$-sub-products for semicontinuous 2-colorings, i.e. partitions into a closed set and its complement, also follows from the Lemma (see [7]). To extend this result to a wider class of colorings, for example to Borel-measurable colorings, iterates of the function $S$ are defined in order to carry out certain diagonalization arguments.

The iterates of the function $S$ are defined recursively as follows.

$$S^{(0)}(m_0, ..., m_i) = S(m_0, ..., m_i),$$
$$S^{(p+1)}(m_0, ..., m_i) =$$
$$S(S^{(p)}(m_0), S^{(p)}(m_0, m_1), ..., S^{(p)}(m_0, ..., m_i)).$$

Using the function $S$ and its iterates, for every sequence $\{m_i\}_{i=0}^{\infty}$ of positive integers a family $\mathcal{H}(\vec{m})$ of $\vec{m}$-sequences of finite sets with the following properties can be defined :

1. For every closed $\mathcal{X} \subseteq \mathbb{N}^{\infty}$, every $\vec{H} \in \mathcal{H}(\vec{m})$ and every $k \in \mathbb{N}$, there is $\vec{J} \in \mathcal{H}(\vec{m})$ such that $\vec{J} \leq_k \vec{H}$ and $\mathcal{X} \cap \prod_{i=0}^{\infty} J_i$ is clopen (in $\prod_{i=0}^{\infty} J_i$).

2. Given a sequence

$$\vec{H}^0 \leq_{l_0} \vec{H}^1 \leq_{l_1} \cdots \leq_{l_{j-1}} \vec{H}^j \leq_{l_j} \cdots$$

of elements of $\mathcal{H}(\vec{m})$, there is $\vec{H} \in \mathcal{H}(\vec{m})$ such that for every $j$, $\vec{H} \leq_{l_j} \vec{H}^j$.

Now, for each sequence $\vec{m}$, we can define a corresponding $\sigma$-field of subsets of $\mathbb{N}^{\infty}$: $\mathcal{C}(\vec{m})$ is the collection of all $\mathcal{X} \subseteq \mathbb{N}^{\infty}$ such that for every $\vec{H} \in \mathcal{H}(\vec{m})$ and for every $n \in \mathbb{N}$ there is $\vec{J} \in \mathcal{H}(\vec{m})$ such that $\vec{J} \leq_n \vec{H}$ and $\mathcal{X} \cap \prod_i J_i$ is clopen in $\prod_i J_i$.

**Theorem 5.** *([9]) For every sequence $\vec{m}$, $\mathcal{C}(\vec{m})$ is a $\sigma$-field which contains the closed sets and is closed under Souslin's operation.*

The argument given in [9] to prove that $\mathcal{C}(\vec{m})$ is closed under Souslin's operation is metamathematical. It uses the decomposition of analytic sets into Borel sets and a forcing notion preserving $\aleph_1$. Once analytic sets are shown to be in $\mathcal{C}(\vec{m})$, it follows that $\mathcal{C}(\vec{m})$ is closed under Souslin's operation. S. Todorcevic has recently found a combinatorial proof of this fact ([39]).

By the comments in the paragraphs following Lemma 1, we have the following.

**Corollary 1.** *For every sequence $\vec{m} = \{m_i\}_{i=0}^{\infty}$, there is a sequence $\vec{n} = \{n_i\}_{i=0}^{\infty}$ such that*

$$(\vec{n}) \rightarrow_{\mathcal{C}(\vec{m})} (\vec{m}).$$

*In particular, for any Souslin-measurable*

$$c : \prod_{i=0}^{\infty} n_i \rightarrow \{0,1\}$$

*there exist $H_i \subseteq N_i$ with $|H_i| = m_i$ for all $i$, such that $c$ is constant on $\prod_{i=0}^{\infty} H_i$.*

## 7. Parametrized partitions of products of finite sets

The following lemma from [8] is a more uniform version of Lemma 1. Besides being interesting in its own right, it is useful to parametrize the partition property $\omega \rightarrow (\omega)^{\omega}$ with partition properties of the form $\vec{n} \rightarrow (\vec{m})$. The proof of the lemma given in [8] uses the hypothesis "all $\Sigma_2^1$-sets are Ramsey", which for some applications, like the one given in Section 9, can then be eliminated.

**Lemma 2.** *There is $R : \mathbb{N}^{<\infty} \rightarrow \mathbb{N}$ such that for every infinite sequence $\vec{m} = \{m_i\}_{i=0}^{\infty}$ of positive integers and every coloring*

$$c : \bigcup_k \prod_{i<k} R(m_0, \ldots, m_i) \rightarrow \{0,1\},$$

*there exist $H_i \subseteq R(m_0, \ldots, m_i)$, $|H_i| = m_i$ for all $i$, and an infinite $A \subseteq \mathbb{N}$, such that $c$ is constant on $\bigcup_{k \in A} \prod_{i<k} H_i$.*

We cannot hope to get even more uniformity, since it is easy to define a partition for which there is no $\vec{m}$-sequence $\vec{H}$ such that for every $k$, the sub-product $\prod_{i=0}^{k} H_i$ monochromatic.

The iterates of the function $R$ given by Lemma 2 are defined as those for $S$ in the previous section, and for every sequence $\{m_i\}_{i=0}^{\infty}$ of positive integers, a family $\mathcal{H}_R(\vec{m})$ of sequences of finite sets can be defined diagonalizing through the function $R$ and its iterates applied to $\vec{m}$. Using this collection $\mathcal{H}_R(\vec{m})$, a field $\mathcal{PC}(\vec{m})$ of subsets of $\mathbb{N}^{\infty} \times \mathbb{N}^{[\infty]}$ is defined as follows: a subset $\mathcal{X} \subseteq \mathbb{N}^{\infty} \times \mathbb{N}^{[\infty]}$ is in $\mathcal{PC}(\vec{m})$ if for every sequence $\vec{H} \in \mathcal{H}_R(\vec{m})$, every $n$, every $A \in \mathbb{N}^{[\infty]}$, every $a \in \mathbb{N}^{[<\infty]}$, there exist a sequence $\vec{J} \in \mathcal{H}_R(\vec{m})$ such that $\vec{J} \leq_n \vec{H}$, $k \geq n$, and $B \subseteq A$ such that $\forall s \in \prod_{i<k} J_i$

$$[s, \vec{J}] \times [a, B] \subseteq \mathcal{X} \text{ or } [s, \vec{J}] \times [a, B] \cap \mathcal{X} = \emptyset.$$

Here, $[s, \vec{J}]$ is the collection of elements of $\prod_{i=0}^{\infty} J_i$ which extend $s$.

For partitions of the product space $\mathbb{N}^{\infty} \times \mathbb{N}^{[\infty]}$, $\mathcal{PC}(\vec{m})$ plays a rôle analogous to that of the field $\mathcal{C}(\vec{m})$ of the previous section. The properties satisfied by the families $\mathcal{H}_R(\vec{m})$ are then used to prove the following.

**Theorem 6.** *([9]) For every sequence $\vec{m}$, $\mathcal{PC}(\vec{m})$ is a $\sigma$-field of subsets of $\mathbb{N}^{\infty} \times \mathbb{N}^{[\infty]}$ containing the open sets and closed under Souslin's operation.*

The proof of this theorem is much harder than the proof of Theorem 5. To show that open subsets of $\mathbb{N}^{\infty} \times \mathbb{N}^{[\infty]}$ are in $\mathcal{PC}(\vec{m})$, the notion of "barrier" of Nash-Williams is used, and a combinatorial forcing the developed along the lines of [14].

As for the case of the field $\mathcal{C}(\vec{m})$, the argument of [9] to show that every $\mathcal{PC}(\vec{m})$ is closed under Souslin's operation uses metamathematical tools. There is a recent combinatorial proof due to S. Todorcevic.

**Corollary 2.** *Every analytic subset of $\mathbb{N}^{\infty} \times \mathbb{N}^{[\infty]}$ is in $\mathcal{PC}(\vec{m})$.*

It can be verified ([8] 5.2) that for every sequence $\{m_i\}_{i=0}^{\infty}$ there is a sequence $\{n_i\}_{i=0}^{\infty}$ such that given a set $\mathcal{X} \subseteq \prod_{i=0}^{\infty} n_i \times \mathbb{N}^{[\infty]}$ in $\mathcal{PC}(\vec{m})$, there is an $(m_i)$-sequence $\{H_i\}_{i=0}^{\infty}$ and an infinite set $H$ such that

$$\prod_{i=0}^{\infty} H_i \times H^{[\infty]} \subseteq \mathcal{X} \text{ or } \prod_{i=0}^{\infty} H_i \times H^{[\infty]} \cap \mathcal{X} = \emptyset.$$

Thus, we obtain the following.

**Corollary 3.** *For every sequence $\{m_i\}_{i=0}^{\infty}$, there is a sequence $\{n_i\}_{i=0}^{\infty}$ such that for every analytic $\mathcal{X} \subseteq \prod_{i=0}^{\infty} n_i \times \mathbb{N}^{[\infty]}$, there is an $(m_i)$-sequence $\{H_i\}_{i=0}^{\infty}$ and an infinite set $H$ such that*

$$\prod_{i=0}^{\infty} H_i \times H^{[\infty]} \subseteq \mathcal{X} \text{ or } \prod_{i=0}^{\infty} H_i \times H^{[\infty]} \cap \mathcal{X} = \emptyset.$$

**Question 5.** *Is the hypothesis "all $\Sigma_2^1$-sets are Ramsey" necessary for Lemma 2?*

## 8. Rates of growth

It is shown in [7] that for the constant sequence $m_i = 2$ for all $i$, the sequence $n_i = 2^{2^{2^{i+1}}}$ satisfies

$$(\vec{n}) \rightarrow_{Clopen} (\vec{m}).$$

More generally,

**Theorem 7.** *([38]) For every primitive recursive sequence $\{m_i\}_{i=0}^{\infty}$, there is a primitive recursive sequence $\{n_i\}_{i=0}^{\infty}$ such that*

$$(\vec{n}) \rightarrow_{Borel} (\vec{m}).$$

**Question 6.** *Is there a primitive recursive sequence $\{n_i\}_{i=0}^{\infty}$ such that for every Souslin measurable coloring*

$$c : (\prod_{i=0}^{\infty} n_i) \times \mathbb{N}^{[\infty]} \rightarrow \{0,1\}$$

*there exist $H_i \subseteq n_i$, $|H_i| = 2$ for all $i$, and an infinite $H \subseteq \mathbb{N}$, such that the product $(\prod_{i=0}^{\infty} H_i) \times H^{[\infty]}$ is monochromatic?*

## 9. Some consistency results

In this section we indicate how to obtain a model where the partition property

$$\begin{pmatrix} \omega \\ \omega \\ \vdots \end{pmatrix} \rightarrow \begin{pmatrix} m_0 \\ m_1 \\ \vdots \end{pmatrix}$$

holds for every sequence $\{m_i\}_{i=0}^{\infty}$, together with the existence of a non-principal ultrafilter on $\mathbb{N}$. We assume familiarity with Solovay's model where all sets of real numbers are Lebesgue measurable [37]. The reader can consult [20] for a presentation of this model and its main properties.

We say that a model $M$ is a Solovay model over a ground model $V$ if $M = L(\mathbb{R})$, the class of sets constructible from $\mathbb{R}$, where $\mathbb{R}$ is the set of reals in a generic extension of $V$ obtained using the Levy order to collapse an inaccessible cardinal of $V$ to $(\omega_1)^M$.

The following theorem establishes that in a Solovay model, the fields of sets $\mathcal{C}(\vec{m})$ and $\mathcal{PC}(\vec{m})$, defined in the previous sections. include all subsets of the corresponding spaces. Since in a Solovay model all subsets of $\mathbb{N}^{[\infty]}$ are Ramsey, Lemma 2 holds there.

**Theorem 8.** *([9]) In every Solovay model, for every sequence $\{m_i\}_{i=0}^{\infty}$ of positive integers, every subset of $\mathbb{N}^{\infty}$ is in $\mathcal{C}(\vec{m})$, and every subset of $\mathbb{N}^{\infty} \times \mathbb{N}^{[\infty]}$ is in $\mathcal{PC}(\vec{m})$.*

The main ingredients of the proof are several well known properties of Solovay models, in particular the fact that in a Solovay model every subset of $\mathbb{N}^{\infty}$ or of $\mathbb{N}^{\infty} \times \mathbb{N}^{[\infty]}$ is the union of $\aleph_1$ analytic sets.

From the fact about $\mathcal{C}(\vec{m})$ follows that in a Solovay model, for every sequence $\{m_i\}_{i=0}^{\infty}$, there is a sequence $\{n_i\}_{i=0}^{\infty}$ such that $(\vec{n}) \rightarrow (\vec{m})$. Using $\mathcal{PC}(\vec{m})$ we obtain the parametrized version.

**Corollary 4.** *In every Solovay model, for every sequence* $\{m_i\}_{i=0}^{\infty}$ *there exists a sequence* $\{n_i\}_{i=0}^{\infty}$ *such that for every coloring*

$$c : \prod_{i=0}^{\infty} n_i \times \mathbb{N}^{[\infty]} \to \{0,1\}$$

*there exist* $H_i \subseteq n_i$ , $|H_i| = m_i$ *for every* $i$, *and an infinite* $H \subseteq \mathbb{N}$ *such that* $(\prod_{i=0}^{\infty} H_i) \times H^{[\infty]}$ *is monochromatic.*

Notice that this implies that in every Solovay model, for every sequence $\{m_i\}_{i=0}^{\infty}$ and for every coloring

$$c : \mathbb{N}^{\infty} \times \mathbb{N}^{[\infty]} \to \{0,1\}$$

there exist $H_i \subseteq \mathbb{N}$ , $|H_i| = m_i$ for every $i$, and an infinite $H \subseteq \mathbb{N}$ such that $(\prod_{i=0}^{\infty} H_i) \times H^{[\infty]}$ is monochromatic.

The consistency of this parametrized partition property can be used to show that the property $\omega \to (\omega)^{\omega}$ is not implied by the polarized partition property

$$\begin{pmatrix} \omega \\ \omega \\ \vdots \end{pmatrix} \to \begin{pmatrix} m_0 \\ m_1 \\ \vdots \end{pmatrix} .$$

This is achieved starting from a Solovay model $L(\mathbb{R})$, and adding a generic selective ultrafilter $U$ to obtain the generic extension $L(\mathbb{R})[U]$. The parametrized partition property in $L(\mathbb{R})$ is then used to show that this generic extension satisfies the polarized partition property

$$\begin{pmatrix} \omega \\ \omega \\ \vdots \end{pmatrix} \to \begin{pmatrix} m_0 \\ m_1 \\ \vdots \end{pmatrix} .$$

Since in the presence of non principal ultrafilters on $\mathbb{N}$ there are non Ramsey subsets of $\mathbb{N}^{[\infty]}$, the property $\omega \to (\omega)^{\omega}$ does not hold in the generic extension $L(\mathbb{R})[U]$. Therefore, the property

$$\begin{pmatrix} \omega \\ \omega \\ \vdots \end{pmatrix} \to \begin{pmatrix} m_0 \\ m_1 \\ \vdots \end{pmatrix}$$

does not imply $\omega \to (\omega)^{\omega}$.

We end with a question related to the relationship between the polarized partition property

$$\begin{pmatrix} \omega \\ \omega \\ \vdots \end{pmatrix} \to \begin{pmatrix} m_0 \\ m_1 \\ \vdots \end{pmatrix}$$

and the property $\omega \to (\text{perfect})^{\omega}$ of Section 3.

**Question 7.** *It is clear that $\omega \to (\text{perfect})^\omega$ follows from*

$$\begin{pmatrix} \omega \\ \omega \\ \vdots \end{pmatrix} \to \begin{pmatrix} 2 \\ 2 \\ \vdots \end{pmatrix},$$

*since every infinite product of pairs is perfect. Are these two partition properties equivalent? Both are consistent with the existence of ultrafilters on $\mathbb{N}$, but perhaps there is a stronger consequence of the Axiom of Choice which can be used to distinguish between them.*

## 10. Determinacy and partitions

Given a set $\mathcal{X} \subseteq \mathbb{N}^\infty$, consider the game $G_\mathcal{X}$ played by two players, I and II, who alternate playing natural numbers. I plays $n_0$, then II plays $n_1$, I plays $n_2$, etc., forming an infinite sequence $x = \langle n_0, n_1, n_2, \ldots \rangle$. Player I wins the game if $x \in \mathcal{X}$. A winning strategy for player I is a function $\sigma : \mathbb{N}^{<\infty} \to \mathbb{N}$ such that any run of the game $G_\mathcal{X}$ in which I's moves are determined using $\sigma$ produces an element of $\mathcal{X}$. Analogously, we can define winning strategy for II. A set $X \subseteq \mathbb{N}^\infty$ is determined if one of the players has a winning strategy for the game $G_\mathcal{X}$. The Axiom of Determinacy is the statement "all subsets of $\mathbb{N}^\infty$ are determined". The axiom of choice implies that there are non-determined sets, in fact, if neither $\mathcal{X}$ nor $\mathbb{N}^\infty \setminus \mathcal{X}$ contain a perfect set, then $\mathcal{X}$ is not determined. Nevertheless, all Borel sets are determined [25].

There are some unresolved questions regarding the connections between the partition properties we have considered here and the Axiom of Determinacy ($AD$).

$AD$ implies that every uncountable subset of $\mathbb{N}^\infty$ contains a perfect set and that every subset of $\mathbb{N}^\infty$ has the Baire property. Therefore, $AD$ implies $\omega \to (\text{perfect})^\omega$ and $\omega \to ((\omega))^\omega$. It is unknown if $\omega \to (\omega)^\omega$ follows from $AD$. Prikry [32] showed that $\omega \to (\omega)^\omega$ is a consequence of $AD_\mathbb{R}$, determinacy of subsets of $\mathbb{R}^\infty$. The property $\omega \to (\omega)^\omega$ is also a consequence of $AD$ and $V = L(\mathbb{R})$ [26].

**Question 8.** *Is $\omega \to (\omega)^\omega$ a consequence of AD? Is the partition property $\begin{pmatrix} \omega \\ \omega \\ \vdots \end{pmatrix} \to$*

$\begin{pmatrix} 2 \\ 2 \\ \vdots \end{pmatrix}$ *a consequence of AD?*

## REFERENCES

1. Bernstein, F., Zur Theorie der Trigonometrischen Reiche. *Berichte über die Verhandlungen der Königlich Sächsischen Gesellschaft der Wissenschaften zu Leipzig Mathematisch-Physiche Klasse* 60 (1908) 325-338.
2. Brendle, J., L. Halbeisen and B. Löwe, Silver measurability and its relations to other regularity properties, (preprint).

3. Di Prisco, C. A., Partition properties and perfect sets. *Notas de Lógica Matemática*, Universidad Nacional del Sur, Bahía Blanca, Argentina, 38 (1993)119-127.
4. Di Prisco, C. A. and J. Henle, Partitions of products. *Journal of Symbolic Logic* 58 (1993), 860-871.
5. Di Prisco, C. A. and J. Henle, Partitions of the reals and choice. In: Models, algebras and proofs (X. Caicedo and C. H. Montenegro, Eds.) Marcel Dekker, Inc. 1999, 13-23.
6. Di Prisco, C. A. and J. Henle, Doughnuts, floating ordinals, square brackets, and ultraflitters. *Journal of Symbolic Logic* 65 (2000) 461-473.
7. Di Prisco, C. A., J. Llopis and S. Todorcevic, Borel partitions of products of finite sets and the Ackermann function. *Journal of Combinatorial Theory, Series A*, 93 (2001) 333-349.
8. Di Prisco, C. A., J. Llopis y S. Todorcevic, Parametrized partitions of products of finite sets. *Combinatorica* (to appear).
9. Di Prisco, C.A. and S. Todorcevic, Souslin partitions of products of finite sets. *Advances in Mathematics*, 176 (2003) 145-173.
10. Ellentuck, E., A new proof that analytic sets are Ramsey. *Journal of Symbolic Logic* 39 (1974), 256-290.
11. Erdös, P. and R. Rado, Combinatorial theorems on classification of subsets of a given set. *Proceedings of the London mathematical Society* 3 (1952) 417-439.
12. Erdös, P. and R. Rado, A partition calculus in set theory, *Bulletin of the Amer. Math. Soc.* 62 (1956), 427-489. (Abstract 68T-368).
13. Galvin, F., A generalization of Ramsey's Theorem. *Notices of the American Mathematical Society* 15 (1968) 548.
14. Galvin, F. and K. Prikry, Borel sets and Ramsey's theorem, *Journal of Symbolic Logic* 38 (1973), 193-198.
15. Graham, R. L., B. L. Rotschild and J. H. Spencer, Ramsey Theory. John Wiley and sons, 1990.
16. Halbeisen, L., Making doughnuts of Cohen reals. *Mathematical Logic Quarterly* 49 (2003) 173–178.
17. Halpern, J.D. and H. Läuchli, A partition theorem, *Transactions of the American Mathematical Society* 124 (1966)360-367.
18. Halpern, J.D. and A. Levy, The Boolean prime ideal theorem does not imply the axiom of choice. In "Axiomatic Set Theory". Proc. Symp. Pure Mathematics 13 I (D. Scott, Ed.) pp 83-134. American Mathematical Society, 1971.
19. Henle, J., The consistency of one fixed omega, *Journal of Symbolic Logic* 60 (1995), 172-177.
20. Jech, T., Set Theory. Springer Verlag, 1997.
21. Kanamori, A., The higher infinite, Springer Verlag, 1997.
22. Laver, R., Products of infinitely many perfect trees, *Journal of the London Mathematical Society* (2), 29(1984) 385-396.
23. Llopis, J., A note in polarized partitions, *Notas de Lógica Matemática*, Universidad Nacional del Sur, Bahía Blanca, Argentina, 39 (1994), 89-94.
24. Llopis, J. and S. Todorcevic, Borel partitions on products of finite sets, *Acta Científica Venezolana* 47 (1996), 85-88.
25. D. A. Martin, Borel Determinacy. *Annals of Mathematics* 102 (1975) 2363-371.

26. Martin, D. A. and J. R. Steel, The extent of scales in $L(\mathbb{R})$, in Cabal Seminar 79-81 (A. S. Kechris, D. A. Martin and Y. Moschovakis, Eds.). Lecture Notes in Mathematics 1019, 1983

27. Mathias, A. R. D., A remark on rare filters, in: Infinite and finite sets (A. Hajnal. R. Rado and Vera Sós, Eds.). Colloquia Mathematica Societatis Janos Bolyai 10, , North Holland, 1975.

28. Mathias, A.R.D., Happy Families. *Annals of Pure and Applied Logic* 12 (1977) 59-111.

29. G. Moran, G. and D. Strauss, Countable partitions of product spaces, *Mathematika* 27 (1980) 213-224.

30. Nash-Williams, C.St. J. A., On well quasi-ordering transfinite sequences. *Proceedings of the Cambridge Philosophical Society* 61 (1965) 33-39.

31. Nešetřil, J. and V. Rödl, Eds., Mathematics of Ramsey Theory. Springer Verlag, 1990.

32. Prikry, K.L., Determinateness and partitions. *Proceedings of the American Mathematical Society* 54 (1976)303-306.

33. Prömel, H. J. and B. Voigt, Recent results in partition (Ramsey) theory for finite lattices. Dicrete Mathematics 35 (1981) 185-198.

34. Ramsey, F. P., On a problem of formal logic. *Proc. of the London Mathematical Society, Ser. 2* 30, Part 4 (1928) 338–384.

35. Shelah, S., Can you take Solovay's inaccessible away? *Israel Journal of mathematics* 48 (1984) 1-47.

36. Silver, J., Every analytic set is Ramsey, *Journal of Symbolic Logic* 35 (1970), 60-64.

37. Solovay, R., A model of set theory where every set of reals is Lebesgue measurable. *Annals of Mathematics* 92 (1970) 1-56.

38. Todorcevic, S., A new quantitative analysis of some basic principles in the theory of functions of a real variable. *Bull. de l'Acad. Serbe des Sci.* vol. CXXII, No, 26 (2001), 133-144.

39. Todorcevic, S., Introduction to Ramsey Spaces (in preparation).

.

# Multiple Conclusions

Greg Restall*

*Philosophy Department, The University of Melbourne*
*restall@unimelb.edu.au*

**Abstract.** I argue for the following four theses. (1) Denial is not to be analysed as the assertion of a negation. (2) Given the concepts of assertion and denial, we have the resources to analyse logical consequence as relating arguments with *multiple* premises and *multiple* conclusions. Gentzen, Gerhard's multiple conclusion calculus can be understood in a straightforward, motivated, non-question-begging way. (3) If a broadly anti-realist or inferentialist justification of a logical system works, it works just as well for *classical* logic as it does for *intuitionistic* logic. The special case for an anti-realist justification of intuitionistic logic over and above a justification of classical logic relies on an unjustified assumption about the shape of proofs. Finally, (4) this picture of logical consequence provides a relatively neutral shared vocabulary which can help us und erstand and adjudicate debates between proponents of classical and non-classical logics.

* * *

Our topic is the notion of logical consequence: the link between premises and conclusions, the glue that holds together deductively valid argument. How can we understand this relation between premises and conclusions? It seems that any account begs questions. Painting with very broad brushtrokes, we can sketch the landscape of disagreement like this: "Realists" prefer an analysis of logical consequence in terms of the preservation of *truth* [29]. "Anti-realists" take this to be unhelpful and offer alternative analyses. Some, like Dummett, look to preservation of *warrant to assert* [9,36]. Others, like Brandom [5], take inference as primitive, and analyse other notions in terms of it. There is plenty of disagreement on the "realist" side of the fence too. It is one thing to argue that logical consequence involves preservation of truth. It is another to explain how far truth must be preserved. Is the preservation essentially *modal* (in all circumstances [25]) or *analytic* (vouchsafed by the meanings of the terms involved) or *formal* (guaranteed by the logical structure of the premises and conclusions [28,29]), or do we need a combination of these factors [12]? If there is to be some kind of privileged logical vocabulary, what is the principle of demarcation for that vocabulary [32]?

---

*Many thanks to Allen Hazen, Graham Priest and Barry Taylor for fruitful discussions while I was preparing a this paper. Thanks also to audiences at La Trobe University, the University of Melbourne, the 2003 Australasian Association for Logic Conference in Adelaide, and the 12th International Congress for Logic, Methodology and Philosophy of Science in Oviedo—including Diderik Batens, Thierry Coquand, Jen Davoren, Philip Ebert, Joke Meheus, David Miller, Peter Milne, Peter Schroeder–Heister, John Slaney and Tim Oakley—for comments on presentations of this material, and to JC Beall, Richard Home, Ben Boyd, Jeremy St. John, Luke Howson and Charlie Donahue for comments on drafts of the paper. ¶ This research is supported by the Australian Research Council, through grant DP0343388.

Even then, if we manage to find agreement on the significance and ground of logical consequence and the scope of logical vocabulary, there is scope for further disagreement. There are different accounts of the valid arguments, even those couched in the simple propositional connectives of conjunction, disjunction, the conditional and negation. Do we admit the law of the excluded middle as a truth of logic, or not [8,10,15,37]? Is it legitimate to infer anything you like from a contradiction, or is this argument form invalid [1,2,27,37]? Can one distribute conjunctions over disjunctions, or do quantum-mechanical experiments provide a counterexample to this inference [4,13]? In the midst of all of this disagreement, is there *any* hope for finding a shared vocabulary between parties of these disagreements? In this paper I attempt to unearth some common ground where many have thought there is none, and to show how that from this common ground we can clarify some of these debates.

* * *

First, I will argue that the speech-act of *denial* is best not analysed in terms of *assertion* and *negation* but rather, that denial is, in some sense, prior to negation. I will provide three different arguments for this position. The first involves the case of an agent with a limited logical vocabulary. The second argument, closely related to the first, involves the case of the proponent of a non-classical logic. The third will rely on general principles about the way logical consequence rationally constrains assertion and denial.

ARGUMENT ONE: Parents of small children are aware that the ability to *refuse*, *deny* and *reject* arrives very early in life. Considering whether or not something is the case – whether to accept that something is the case or to reject it – at least *appears* to be an ability children acquire quite readily. At face value, it seems that the ability to assert and to deny, to say *yes* or *no* to simple questions, arrives earlier than any ability the child has to form sentences featuring negation as an operator. It is one thing to consider whether or not $A$ is the case, and it is another to take the *negation* $\sim A$ as a further item for consideration and reflection, to be combined with others, or to be supposed, questioned, addressed or refuted in its own right. The case of early development lends credence to the claim that the ability to deny can occur prior to the ability to form negations. If this is the case, the denial of $A$, in the mouth of a child, is perhaps best not analysed as the assertion of $\sim A$.

So, we might say that denial may be *acquisitionally prior* to negation. One can acquire the ability to deny before the ability to form negations.

ARGUMENT TWO: Consider a related case. Sometimes we are confronted with theories which propose non-standard accounts of negation, and sometimes we are confronted with people who endorse such theories. These will give us cases of people who appear to reject $A$ without accepting $\sim A$, or who appear to accept $\sim A$ without rejecting $A$. If things are as they appear in these cases, then we have further reason to reject the analysis of rejection as the acceptance of a negation. I will consider just two cases.

*Supervaluationism*: The supervaluationist [10,17,37] account of truth-value gaps enjoins us to allow for claims which are not determinately true, and not determinately false. These claims are those which are true on some valuations and false on others. In the case of the supervaluational account of *vagueness*, borderline cases of vague terms are a good example. If Fred is a borderline case of baldness, then on some valuations "Fred is bald" is true, and on others, "Fred is bald" is false. So, "Fred is bald" is not true under the *super*valuation, and it is to be rejected. However, "Fred is not bald" is similarly true on some valuations and false on others. So, "Fred is not bald" is not true under the supervaluation, and it, too, is to be rejected. Truth value gaps provide examples where denial and the assertion of a negation come apart. The supervaluationist rejects $A$ without accepting $\sim A$. When questioned, she will deny $A$, and she will *also* deny $\sim A$. She will not accept $\sim A$. The supervaluationist seems to be a counterexample to the analysis of denial as the assertion of a negation.

*Dialetheism*: The dialetheist provides is the dual case [19,20,22,23,27]. A dialetheist allows for truth-value *gluts* instead of truth-value *gaps*. Dialetheists, on occasion, take it to be appropriate to assert both $A$ and $\sim A$. A popular example is provided by the semantic paradoxes. Graham Priest's analysis of the liar paradox, for example, enjoins us to accept both the liar sentence and its negation, and to reject neither. In this case, it seems, the dialetheist accepts a negation $\sim A$ *without* rejecting $A$, the proposition negated. When questioned, he will assert $A$, and he will *also* assert $\sim A$. He will not reject $\sim A$. The dialetheist, too, seems to be a counterexample to the analysis of denial as the assertion of a negation.

In each case, we seem to have reason to take denial to be something other than the assertion of a negation, at least in the mouths of the supervaluationist and the dialetheist. This argument is not conclusive: the proponent of the analysis may well say that the supervaluationist and the dialetheist are confused about negation, and that their denials really *do* have the content of a negation, despite their protestations to the contrary. Although this is a possible response, there is no doubt that it does violence to the positions of both the supervaluationist and the dialetheist. We would do better to see if there is an understanding of the connections between assertion, denial, acceptance, rejection and negation which allows us to take these positions at something approaching face value. This example shows that denial may be *conceptually separated* from the assertion of a negation.

ARGUMENT THREE: The third argument is more extensive than the other two. We will consider the relationship between logical consequence and assertion and denial. It is common ground that logical consequence, whatever it amounts to, has some kind of grip on assertion and denial, acceptance and rejection. It makes sense for us to analyse and to criticise or laud our own beliefs, or the beliefs of others, using canons of deductive consequence. But not only our beliefs fall under logic's gaze. So also our hypotheses, suppositions, stories and flights of fancy may also be evaluated using logical norms. We measure all such things for coherence or consistency. We look for consequences, for what leads *on* from what we have considered, and we look for premises, for what might lead *to* what we consider now. Logical notions are nothing if they have no applicability to regulate the cognitive states of agents like us, and the content of such states.

Consider, then, how logic might apply to the case of a cognitive agent, and consider the case of a simple deductively valid argument, with one premise $A$ and one conclusion $B$. (We represent the validity thus: '$A \vdash B$.') What *grip* could this inference have on an agent? When could an agent fall *foul* of this inference, and when could an agent *comply* with it?

If an agent accepts $A$, then it is tempting to say that the agent also *ought* accept $B$, because $B$ follows from $A$. But this is too strong a requirement to take seriously. Let's consider why not.

(1) The requirement as I have naïvely expressed it is ludicrous if read as it stands. Consider the circumstance in which an agent might accept $A$ for no good reason. But the argument from $A$ to $A$ is valid, and the mere fact that the agent *happens* to accept $A$ gives the agent no *reason* to accept $A$. So, the requirement that you ought to accept the consequences of your beliefs is altogether too strong as it stands, as we shall see.

This error in the requirement is corrected with a straightforward scope distinction. Instead of saying that if $A$ entails $B$ and if you accept $A$ then you ought to accept $B$, we should perhaps say that if $A$ entails $B$ then it ought to be the case that if you accept $A$ you accept $B$. But this, too, is altogether too strong, as the following considerations show.

(2) There are consequences of which we are unaware. As a result, logical consequence on its own provides no obligation to believe. Here is an example: I accept all of the axioms of Peano arithmetic (PA). I do not believe all of the consequences of those axioms. Goldbach's conjecture (GC) could well be a consequence of those axioms, but I am not aware of this if it is the case, and I do not accept GC. If GC is a consequence of PA, then there is a sense in which I have not lived up to some kind of standard if I fail to accept it. My beliefs are not as comprehensive as they could be. If I believed GC, then in some important sense I would not make any more mistakes than I have already made, because GC is a consequence of my prior beliefs. However, it is by no means clear that comprehensiveness of this kind is desirable.

(3) In fact, comprehensiveness is *undesirable* for limited agents like us. If the inference from $A$ to $A \vee B$ is valid, and if our beliefs are always to be closed under logical consequence, then for any belief we must have infinitely many more. But consider a very long disjunction, in which *one* of the disjuncts we already accept. In what sense is it desirable that we accept this? The belief may be too complex to even *consider*, let alone, to believe or accept or assert.

Notice that it is not a sufficient repair to demand that we merely accept the *immediate* logical consequences of our beliefs. It may well be true that logical consequence in general may be analysed in terms of chains of immediate inferences we all accept when they are presented to us. The problems we have seen hold for immediate consequence. The inference from the axioms of PA to Goldbach's conjecture might be decomposable into steps of immediate inferences. This would not make Goldbach's conjecture any more rationally obligatory, if we are unaware of that proof. If the inference from $A$ to $A \vee B$ is an immediate inference, then logical closure licenses an infinite collection of (irrelevant) beliefs.[2]

---

[2]This point is not new, Gilbert Harman, for example, argues for it in *Change in View* [14].

(4) Furthermore, logical consequence is sometimes impossible to check. If I must accept the consequences of my beliefs, then I must accept all tautologies. If logical consequence is as complex as consequence in classical first-order logic, then the demand for closure under logical consequence can easily be *uncomputable*. For very many sets of statements, there is no algorithm to determine whether or not a given statement is a logical consequence of that set. Closure under logical consequence cannot be underwritten by algorithm, so demanding it goes beyond what we could rightly expect for an agent whose capacities are computationally bounded.

So, these arguments show that logical *closure* is too strict a standard to demand, and failure to live up to it is no failure at all. Logical consequence must have some other grip on agents like us. But what could this grip be? Consider again the case of the valid argument from $A$ to $B$, and suppose, as we did before, that an agent accepts $A$. What can we say about the agent's attitude to $B$? The one thing we can say about the agent's current attitude is that if she *rejects* $B$, she has made a mistake.

If an agent's cognitive state, in part, is measured in terms of those things she accepts and those she rejects, then valid arguments constrain those combinations of acceptance and rejection. As we have seen, a one-premise, one-conclusion argument from $A$ to $B$ constrains acceptance/rejection by ruling out accepting $A$ and rejecting $B$. This explanation of the grip of valid argument has the advantage of symmetry. A valid argument from $A$ to $B$ does not, except by force of habit, have to be read as *establishing* the conclusion. If the conclusion is unbelievable, then it could just as well be read as *undermining* the premise. Reading the argument as constraining a pattern of acceptance and rejection gives this symmetry its rightful place.

It follows from this reflection that if there are reasoning and representing agents who do not have the concept of negation, and if it is still appropriate for us to analyse their reasoning using a notion of logical consequence, then we ought to take those agents as possessing the ability to *deny* without having the ability to *negate*. This seems plausible. As an agent accepts and rejects, it is filtering out information and ruling out possibilities. If the agent accepts $A$ and $B$ and also *rejects* the conjunction $A \wedge B$, then it has made a mistake, and this mistake can be explained without resorting to taking the agent to having a competence with manipulating *negations* as well as *conjunctions*.

What more can I say about the relationship between accepting and rejecting and the cognate speech-acts of assertion and denial? I leave some of the details to the next section, but here is some of what this picture involves. To accept $A$ is to (in part) close off the possibility of rejecting $A$. To accept $A$ and then to go on to *reject* $A$ will result in a *revision* of your commitments, and not a mere *addition* to them. Similarly, to reject $A$ is to (in part) close off the possibility of accepting $A$. To reject $A$ and then to go on to *accept* $A$ will result in a revision of your commitments, and not a mere addition to them.

I will close this section responding to the Fregean argument against the position I have just taken. Frege took it that denial is best analysed as the assertion of a negation because it seems that rejecting this analysis results in unnecessary proliferation of rules of inference.[3] I will use Dummett's example from his discussion

---

[3] Allen Hazen informs me that Meinong's *assumptions* play the same role as Frege's *contents* [18].

of Frege's point [7, pp. 316–317]. Consider the argument from the premises 'If he is not a philosopher, he won't understand the question' and 'He is not a philosopher' to the conclusion 'He won't understand the question.' In this argument, the instance of 'he is not a philosopher' in the antecedent of the *conditional* premise is clearly a *negation*. A denial does not embed inside conditionals in this manner. However, it seems that the other premise, and the conclusion, may be treated as *denials* and not assertions of negations. If this is the case, then we must explain the connection between these denials and the *negation* found in the conditional premise. It seems better, and simpler, to treat the premises and the conclusion as assertions, for then the argument has the form of *modus ponens*, as it manifestly appears to be. Does this not pose a problem for any view which takes denial to be prior to negation?

There are a number of responses to this problem already available in the literature. Price's "Why 'Not'?" [21] proposes two-factor analysis of negation which allows an utterance of "he is not a philosopher" to be *both* an assertion of a negation and a denial. This would certainly dull the objection but it would not entirely defeat the nagging worry that any analysis of negation which utilises denial is committed to there being rather more arguments presented in Dummett's example than the simple *modus ponens* which appears on the surface.

Instead of a two-factor response, I propose an alternative picture of the situation. Arguments and argument forms do not, at the first instance, connect assertions or denials. Argument forms connect the content of these assertions and denials: propositions. The argument form of *modus ponens* connects two propositions as premises ($A$ and $A \supset B$) and one conclusion ($B$). Those contents may be accepted or rejected (and asserted or denied), or we agnostic (or silent) about them. As we have seen, the validity of *modus ponens* tells us that the assertion of the premises $A$ and $A \supset B$ together with the denial of the conclusion $B$ is, in some sense to be explained, a bad thing. But one can utilise the argument of *modus ponens* without asserting the premises, or while denying the conclusion.

What of Dummett's argument? It is an instance of *modus ponens*, pure and simple. The argument involves premises and a conclusion, and these include negations. Frege's point is a sharp one when wielded against the view that takes all outermost sentential negations to express denials (as in the view Frege targeted, of the orthodox Aristotelian logic of his day), but it has no effect on views which agree with his reading of the structure of the argument. On this view, there is but one argument there, but nonetheless, the argument in and of itself does not tell us whether to assert the premises (and thereby to rule out rejecting the conclusion) or to deny the conclusion (and thereby rule out accepting both premises).

In taking this view of the structure of arguments, it should be clear that I also distance myself from the superficially similar approach of Smiley's "Rejection" [35]. Smiley proposes an account of logical consequence where the unit of argument is not the proposition but the *judgement*: a proposition signed with a marker for acceptance or rejection. While there is a formal correspondence between this account of proof and the picture I prefer, Smiley's system seems to fall foul of the considerations entertained earlier in this section. Take an argument from a premise I accept to an impossibly complex conclusion which is a consequence of this argument. The system as it stands commends that if I accept the premise

I ought to accept the conclusion. We have already seen that requiring this is altogether too strong. This is another reason to take arguments as connecting contents and not their assertions or denials.

*   *   *

In this section I will explain how this perspective on agents motivates the structural rules of the classical multiple premise, multiple conclusion sequent calculus of Gentzen. But before we get to the formal details of how one might understand the particular logical connectives, we need to spend a little more time considering the behaviour of assertion and denial, and the corresponding states of acceptance and rejection.

In what follows, we will use the notion of a STATE. Given a particular language – which may be rich, containing many different notions, including logical constants, but which may also be completely devoid of any logical constants at all – a STATE expressed in that language is a pair of sets of statements expressed in that language. We will use the notation '$[X : Y]$' to represent states, where $X$ and $Y$ are sets of statements. A state might be used to represent the *outlook* of an agent which we take to *accept* each statement in $X$ and *reject* each statement in $Y$. We might also use a state to represent the *context* in some dialogue or discourse at which each statement $X$ is *asserted* and each statement in $Y$ is *denied*.

We will avail ourselves of the usual notational shorthand of proof theory, by taking $[A : B]$ to be the state consisting of the singleton set $\{A\}$ accepted and the singleton set $\{B\}$ rejected. Similarly, if $[X : Y]$ is some state, we will take $[X, A : B, Y]$ to be the state which adds the statement $A$ to the left set $X$ and adds the statement $B$ to the right set $Y$. Furthermore, we will simply use *nothing* to denote the empty set of statements, so $[X : \ ]$ is a state in which nothing is denied (or rejected) and $[\ : Y]$ is a state in which nothing is asserted (or accepted). It follows that $[\ : \ ]$ is the minimal state which accepts nothing and rejects nothing.

With the notion of a state at hand we may begin to consider how we might *evaluate* states. Even with this thin notion of state as the focus of our discussion, we can lay down some criteria for evaluating states. Not all states are on a par, for some states are self-defeating. In particular, if a state contains a statement in both the left set and the right set, then this state undermines itself. If the state represents the cognitive architecture of an agent, then this agent both *accepts* and *rejects* some statement. If the state represents the state of play in some dialogue or discourse, then some statement has both been *asserted* and *denied*. The state is undermined.

We must take care in expressing this feature of states, if we are to keep the discussion relatively neutral. This requirement is not the same as the requirement of *consistency* or *non-contradiction* rejected by the dialetheist. The dialetheist recommends that we accept both a statement $A$ and its negation $\sim A$, not that we simultaneously accept and reject $A$. Nothing in this requirement need be seen as inimical to the friend of contradictions. Priest's own account of the relationship between assertion and denial indicates that a denial expresses a refusal to accept, not the acceptance of a negation [24,26].

Similarly, nothing in this requirement need be seen as inimical to the anti-realist, or to the quasi-realist who might prefer that we explain our primitive notions without appealing to a prior notion of *truth*. We do not explain the consistency requirement in terms of the impossibility of $A$ being both *true* and *false* at the same time. While we might wish to explain the coherence or incoherence of a state in terms of truth, this is by no means required at this early stage of the discussion.

The fact that a state where the left- and right-sets overlap is self-defeating is the first of a number of observations about how states can undercut themselves. Instead of continuing to call these states self-defeating, we will call them *incoherent* because we will also talk about states which are not self-defeating, and seems more pleasing to call these states *coherent* than to call them *non-self-defeating*. We will also use a suggestive notation for calling states incoherent. If $[X : Y]$ is incoherent, we will write '$X \vdash Y$.'[4]

None of this discussion should suggest that given a particular language there is only one notion of coherence or one notion of logical consequence. There may be different criteria for measuring the coherence of combinations of assertions and denials [3]. In the considerations that follow, we are examining the features of *any* notion of coherence. Here are features one might plausibly take to be constitutive of a relation of coherence. We start with the consistency requirement we have already discussed.

CONSISTENCY: The state $[A : A]$ is incoherent. In other words, $A \vdash A$.

The next requirement trades on the features of collections. If there is an incoherence in the state $[X : Y]$ then that incoherence remains no matter what we *add* to the left- and right-sets. The only way to transform the incoherent $[X : Y]$ into a coherent state is to remove something from $X$ or something from $Y$.

SUBSTATE: If $[X : Y]$ is coherent, and if $X' \subseteq X$ and $Y' \subseteq Y$, then $[X' : Y']$ is also coherent. In other words (and contrapositively), if $X \vdash Y$, $X \subseteq X'$ and $Y \subseteq Y'$, then $X' \vdash Y'$.

Those familiar with substructural logics [30] will be aware that the substate requirement is equivalent, on this reading, with the structural rule of *weakening*: If $X \vdash Y$ then $X, A \vdash Y$. A form of this rule is rejected in standard relevant logics such as R, on grounds of relevance. If we can infer from $X$ to $Y$ we need not use $A$ in an inference from $X, A$ to $Y$. The conflict here is merely apparent. Accepting *our* form of weakening does not mean accepting *all* forms of weakening. Nothing said here counts against the existence of a form of premise combination for which weakening is unacceptable.[5]

The next requirement is potentially more controversial. If we have a coherent state $[X : Y]$ then either its extension to assert $A$ or its extension to deny $A$ is coherent.

EXTENSIBILITY: If $[X : Y]$ is coherent, then so is one of $[X, A : Y]$ and $[X : A, Y]$. In other words (and contrapositively) if $X \vdash A, Y$ and $X, A \vdash Y$ then $X \vdash Y$.

This may appear controversial because it appears to endorse a form of the law of the excluded middle. It tells us that if $A$ is *undeniable* in the context of the

---

[4]Note that once one reads this turnstile as a form of *consequence* from $X$ to $Y$, one must read $X$ and $Y$ differently—it is the *conjunction* of all $X$ which entails the *disjunction* of all $Y$.

[5]The importance of allowing different forms of premise combination is clearly explained in Slaney's "A General Logic" [34].

state $[X : Y]$ then it is coherent to assert $A$, provided that was $[X : Y]$ is already coherent. However, this does not rule out truth-value gaps and it does not implicitly endorse the law of the excluded middle.[6] On the contrary, this requirement follows from the intuitive picture of the connection between assertion and denial. To deny $A$ is to place it out of further consideration. To go on and to accept $A$ is to change one's mind. Dually, to accept $A$ is to place its *denial* out of further consideration. To go on to deny $A$ is to change one's mind. If one *cannot* coherently assert $A$, in the context of a coherent state $[X : Y]$, then it must at least be *coherent* to place it out of further consideration, for the inference relation itself has already, in effect, done so. Any move to accept $A$ must take a step back by withdrawing some of the background state $[X : Y]$.

EXTENSIBILITY underwrites the transitivity of entailment. If $A \vdash B$ and $B \vdash C$, then by the SUBSTATE condition, we have $A \vdash B, C$ and $A, B \vdash C$. By EXTENSIBILITY, then, it follows that $A \vdash C$.

One might consider yet another structural feature for coherence.

LOCALITY: If $[X : Y]$ is incoherent, then there are finite $X' \subseteq X$ and $Y' \subseteq Y$ such that $[X' : Y']$ is incoherent.

According to LOCALITY, incoherence never requires an infinite body of assertions and denials. Just as EXTENSIBILITY is the coherence version of the *cut* rule, CONSISTENCY is *identity* and SUBSTATE is *thinning*, the rule of LOCALITY corresponds to *compactness*. Although locality is an important feature of a logicality, it will not play any role in the discussion that follows.

We have just motivated all of the structural rules of a standard multiple conclusion consequence relation as rules for the constraint of assertion and denial (or accepting and rejecting). (None of this, of course, counts against logics with *different* collections of structural rules [30]. The only consequence for these logics is that premise or conclusion combination is not to be read as joint assertion or joint denial.) Before going on to consider the significance of this for the choice of a logical system, and for the evaluation of different rules for each connective, we would do well to linger a while to see what can be expressed in this vocabulary.

If $X \vdash Y$ then it is incoherent to assert all of $X$ and deny all of $Y$. This has a number of special cases worth spelling out:

– If $A \vdash$ then it is incoherent to assert $A$.

– If $A, B \vdash$ then it is incoherent to assert both $A$ and $B$.

– If $\vdash B$ then it is incoherent to deny $B$.

– If $\vdash A, B$ then it is incoherent to deny both $A$ and $B$.

---

[6]As we will see later, all of this may be used to present the proof-theory of intuitionistic logic. The position is compatible with an anti-realist account of intuitionistic logic. The reading of this account of assertion and denial is a subtle one, for the intuitionist. Our sense of denial is not as strong as the intuitionist's assertion of a negation, but not as weak as the intuitionist's mere failure to assert. The requirement is that to deny, in our sense, is to *refuse* to accept. A statement is rejected if any move to accept it would be a change of mind, and not merely a supplementation with new information.

– If $A \vdash B$ then it is incoherent to assert $A$ and deny $B$.

Notice that the multiple-premise, multiple-conclusion structure enables us to represent both notions of *inference* ($A \vdash B$), *incompatibility* or *contrariety* ($A, B \vdash$) and *sub-contrariety* ($\vdash A, B$). On this picture there is no need for separate fundamental abilities to infer and to register incompatibility. These are all species of the larger phenomenon of regulating patterns of acceptings and rejectings.

Now notice that '$A \vdash$' does not commit us to rejecting $A$. It just rules out (on pain of incoherence) accepting $A$. Similarly, $\vdash B$ does not commit us to accepting $B$. It just rules out (on pain of incoherence) rejecting $B$. Consider, then, the sense in which '$A \vdash B$' tells us that $B$ follows *from* $A$. If all it does is rule out the case in which we assert $A$ and deny $B$, there seems to be little room for *consequence*.

Appearances are deceptive, in this case. If $A \vdash B$ and we accept $A$, then given the choice between accepting or rejecting $B$ (and keeping our attitude to $A$ fixed) we must accept $B$ if we are to maintain coherence. If we reject $B$, then we fall into incoherence. The feature undergirding the *consequence* behind '$A \vdash B$' is the consideration of $B$. Once $B$ is up for consideration, we can consider what our present commitments bring to bear. If it is the case that $A \vdash B$ and we already accept $A$ then rejecting $B$ is out of the question, unless we revise our opinion of $A$. As it is coherent to either accept $B$ or to reject it (provided that our current state is coherent) then we may accept $B$ at no further cost to coherence. EXTENSIBILITY tells us that any incoherence in $[X, A, B : Y]$ is already present in $[X, A : Y]$, provided that $A \vdash B$. Adding new consequences of already accepted items maintains coherence, no matter what the background assumptions might be. Similarly, if $A \vdash B$ and we have rejected $B$, then accepting $A$ is not an option (if I am to continue to reject $B$) but rejecting $A$ comes at no cost to coherence, no matter what the background assumptions might be. Provided that we may work with the notions of *assertion* and *denial*, we may express a relation of logical consequence relating multiple premises and multiple conclusions.

$$* \quad * \quad *$$

Not everyone is happy with multiple conclusion presentations of logical consequence. Here is a representative critical passage, from Tennant's *The Taming of the True* [36].

> ... the classical logician has to treat of sequents of the form $X : Y$ where the succedent $Y$ may in general contain more than one sentence. In general, this smuggles in non-constructivity through the back door. For provable sequents are supposed to represent acceptable arguments. In normal practice, arguments take one from premises to a single conclusion. There is no acceptable interpretation of the 'validity' of a sequent $X : Q_1, \ldots, Q_n$ in terms of preservation of warrant to assert when $X$ contains only sentences involving no disjunctions. If one is told that $X : Q_1, \ldots, Q_n$ is 'valid' in the extended sense for multiple-conclusion arguments, the intuitionist can demand to know precisely which disjunct $Q_i$, then, proves to be derivable from $X$. No answer to such a question can be provided in general with the multiple-conclusion sequent calculus of the classical logician. It behooves us, then, to stay with a natural deduction system, and to present it in sequent form only if we observe

the requirement that sequents should not have multiple conclusions. [36, page 320]

There are two criticisms of multiple conclusion consequence in this passage. The first, implicit, criticism concerns 'normal practice.' According to Tennant, in normal practice an argument has multiple premises and a single conclusion, and sequents are to be used to represent the structure of such arguments. A sequent $X : A$ represents the periphery of an argument, with premises $X$ at the leaves of a tree, and $A$ at its conclusion, the root.

This point about the structure of everyday arguments and proofs is not straightforward. Everyday arguments are most often not *explicitly* presented in tree form, but linearly. Just as we might find upward branching implicit in a tree (with multiple premises and single conclusion) we *might* also find downward branching present in the structure of linear proofs. The standard classical multiple conclusion proof of the intuitionistically invalid sequent $\forall x(Fx \lor Gx) \vdash \forall xFx \lor \exists xGx$ is a case in point.

Suppose everyone is either *happy* or *tired*. Choose a person. It follows that this person is either happy or tired. There are two cases. Case (i) this person is happy. Case (ii) this person is tired, and as a result someone is tired. As a result, either this person is happy or *someone* (namely that person) is tired. But the person we chose was arbitrary, so either *everyone* is happy or someone is tired.

Case-based reasoning, like this, can be represented in a multiple-conclusion sequent calculus. Here is a straightforward multiple-conclusion sequent proof of the target sequent.

$$
\frac{\dfrac{\dfrac{\dfrac{\dfrac{\forall x(Fx \lor Gx) \vdash \forall x(Fx \lor Gx)}{\forall x(Fx \lor Gx) \vdash Fa \lor Ga}}{\forall x(Fx \lor Gx) \vdash Fa, Ga}}{\forall x(Fx \lor Gx) \vdash Fa, \exists xGx}}{\forall x(Fx \lor Gx) \vdash \forall xFx, \exists xGx}}{\forall x(Fx \lor Gx) \vdash \forall xFx \lor \exists xGx}
$$

The sequent $\forall x(Fx \lor Gx) \vdash Fa, \exists xGx$ in this proof represents the stage in the English-language proof where we have two cases active, one concluding in $Fa$ and the other, in $\exists xGx$. This demonstration appears to keep two conclusions ($Fa$ and $Ga$) active at the one time, and it makes available a proof of the constructively invalid distribution principle. However, Tennant's complaint that this is "smuggles in non-constructivity through the back door" is at the very least too swift. You *could* complain that this proof is somehow non-constructive, but that point does not count uniquely against the proof structure we have chosen. Constructivity can equally be restored by restricting the application of the universal quantifier introduction rule to sequents with only one formula on the right. (This will be discussed further, below.) The structure of proof itself is does not dictate what rules one employs for connectives using this structure. It certainly makes non-constructive proof *available*, if the vocabulary allows it, but it does not *mandate* it.

So, the argument from the structure of everyday argument is not conclusive. Shoesmith and Smiley's classic *Multiple-Conclusion Logic* contains much more discussion of this issue [33], and I refer the reader there for more details. It certainly *appears* that we can use multiple conclusion reasoning to represent certain structures in everyday proof, but this point is not conclusive.

Tennant's second argument is more important for our purposes. He claims that we cannot explain the validity of multiple conclusion sequents in an anti-realistically acceptable fashion.[7] Tennant is right that the criterion of a preservation of warrant to assert will not do to explain the validity of a multiple conclusion sequent. But as we have seen, this is not the only option for reading such sequents. Preservation of warrant to assert is no more acceptable in reading multiple conclusion sequents than is converse preservation of warrant to deny for multiple premise sequents. Given a valid sequent $A, B : C$, and given that we have warrant to deny $C$, the reasoner (intuitionist or not) can demand to know precisely which conjunct $A$ or $B$, then, proves to be refutable from $C$. The reasoner can demand this as much as she wishes, but no answer will be forthcoming. The valid sequent $A, B : C$ does not wear on its face which premise $A$ or $B$ ought be denied, and neither does the valid sequent $A : B, C$ wear on its face which conclusion ought be asserted.

So, the presence of multiple conclusions, on their own, does not make the reading of sequents any less acceptable to the anti-realist. The explanation of the significance of these sequents is given purely in terms of the norms governing assertion and denial, and nothing we have seen so far entails that these norms must be explained in the terms of truth, reference or correspondence. So, the picture of inferential relations as governing a norm of combinations of assertions and denials makes available a reading of sequents which does not lean upon truth or reference in the first instance.

Indeed, the reading of coherence of states that we have given thus far does not lean upon any specific notion of *warrant*. However, it could if we wished to use warrant as a guide to coherence. It is be open to us to define the coherence for states in the following way. Coherent states are those underwritten by a possible warrant. So, on this picture, $[X : Y]$ is coherent if and only if it is possible for there to be a warrant to assert all of $X$ and deny all of $Y$. This will validate CONSISTENCY and SUBSTATE trivially. The only wrinkle is the verification that assertion, denial and warrant are connected in such a way as to validate EXTENSIBILITY. For this, we need to show that if there is some warrant to assert $X$ and deny $Y$ then there is some warrant to either assert $X, A$ and deny $Y$ or to assert $X$ and deny $A, Y$. If there is no possible warrant to assert $X, A$ and deny $Y$ then this looks suspiciously like there is good reason for anyone committed to accepting $X$ and denying $Y$ to deny $A$.[8] Instead of continuing to develop this point here, we will now proceed

---

[7]This objection is not restricted Tennant's writing. Dummett has similar critical comments in the *Logical Basis of Metaphysics* [9, page 187].

[8]Here I demonstrate the disjunction $A \lor B$ by assuming $\sim A$ and deriving $B$. Constructivists might quibble with this argument form, but this alone is no reason to reject this instance of that form. All valid arguments are instances of invalid argument forms. If this argument is really

to sketch what this might say for the choice of one's principles governing logical connectives.

* * *

We have defended the priority of denial in an analysis of the inferential properties of negation. Given this move, the obvious next step is to notice that we can *define* the behaviour of negation like this: An assertion of $\sim A$ has the same significance as the denial of $A$, and a denial of $\sim A$ has the same significance as the assertion of $A$. End of story [21,35]. This is not our approach, for we have seen to many examples of inferential practices where this identification is rejected. Supervaluationists and some intuitionists are happy to deny both $A$ and $\sim A$. Dialetheists at the very least appear to assert both $A$ and $\sim A$. Do we have a means of *explaining* the divergences in practice in such a way as to clarify what is at stake in the disagreement between these parties?

The classical rules for negation take the following form.

$$\frac{X \vdash A, Y}{X, \sim A \vdash Y} \; (\sim\text{L}) \qquad \frac{X, A \vdash Y}{X \vdash \sim A, Y} \; (\sim\text{R})$$

These rules tell us that in the $[X : Y]$, asserting $\sim A$ has the same effect as a denial of $A$ and denying $\sim A$ has the same effect as the assertion of $A$. Clearly these rules are not acceptable to all parties. The intuitionist and supervaluationist both reject $(\sim\text{R})$ in the case where $Y$ is non-empty, because it licenses the following derivation:

$$\frac{A \vdash A}{\vdash A, \sim A} \; (\sim\text{R})$$

For the intuitionist and supervaluationist, sometimes it is appropriate to deny both $A$ and $\sim A$, so they take themselves to have a counterexample to $(\sim\text{R})$ as it stands. The dialetheist, similarly, rejects the inference $(\sim\text{L})$ because it licenses the following inference

$$\frac{A \vdash A}{A, \sim A \vdash} \; (\sim\text{L})$$

But for the dialetheist, sometimes it is appropriate to assert both $A$ and $\sim A$, so $(\sim\text{L})$ cannot be accepted in its full generality.

If we restrict the $(\sim\text{R})$ rule to the case of an empty consequent $Y$, (and do the same for the $(\supset\text{R})$ rule, which we will not consider here) but leave other rules as they are, we have a system for intuitionstic logic. This system has a feature. If we add *another* negation connective (we will write it like this: '$-$') satisfying the classical $(-\text{L})$ and $(-\text{R})$ rules, then the two negations collapse: $\sim$ inherits its classical brother's features.

$$\frac{\dfrac{A \vdash A}{A, \sim A \vdash} \; (\sim\text{L})}{\sim A \vdash -A} \; (-\text{R}) \qquad \frac{\dfrac{A \vdash A}{A, -A \vdash} \; (-\text{L})}{-A \vdash \sim A} \; (\sim\text{R})$$

---

invalid, it has counterexample. I leave it to constructivists to provide a plausible counterexample to the inference. Suppose there is a possible warrant for $[X : Y]$. Exactly *how* can the claim that either there is a possible warrant for $[X, A : Y]$ or there is a possible warrant for $[X : A, Y]$ fail?

So, the *restricted* intuitionistic rules maintain their distinctive features only when other connectives are barred from entry into the system.

Dummett notices this feature in another context. One can consider a restriction to the disjunction and conjunction rules, as follows:

$$\frac{A \vdash Y \quad B \vdash Y}{A \vee B \vdash Y} \; (\vee L') \qquad \frac{X \vdash A \quad Y \vdash B}{X \vdash A \wedge B} \; (\wedge R')$$

In the absence of a left context $X$ in $(\vee L')$ and a right context $Y$ in $(\wedge R')$, the distribution of conjunction over disjunction $(A \wedge (B \vee C) \vdash (A \wedge B) \vee (A \wedge C))$ cannot be proved. Yet if you add a *new* disjunction with the traditional $(\vee L)$ rule, the two disjunctions collapse and distribution can be proved [9, page 290].[9] Dummett takes this to be a failing for the rules for non-distributing conjunction and disjunction. The rules do not serve to *fix* the interpretation of conjunction or disjunction, because given the addition of new rules, the behaviour of the old connectives change. If this were to be a valid criticism in the case of non-distributive disjunction and conjunction, it would be a criticism in the case of intuitionistic negation too, for we have seen the cases to be completely parallel.

Dummett has a possible response to this issue in the case of intuitionistic negation: he can attempt to argue that the new rules added (for Boolean negation, '$-$') are illegitimate. He must argue that the manipulation of multiple formulas on the right violates some kind of constraint on the structure of proof. However, if this succeeds in the case of intuitionistic logic, it leaves the way open for a *parallel* case in for non-distributive conjunction and disjunction. Perhaps there are *other* constraints on the structure proof, other than the multiple-premise single-conclusion constraint assumed by the intuitionist. We have known all along that the behaviour of connectives is supremely sensitive to the structural rules available for the manipulation of sequents [30]. This is another case where this sensitivity arises.

If no convincing case can be found for the restriction of formulas in contexts in the statements of rules, then it is tempting to conclude that we should accept the completely unrestricted classical inference principles, because these rules are perfectly general, simple and can be conservatively added to any non-logical vocabulary. The discussion of conservative extension here is quite complicated, because we have already seen examples of where it fails. If we have weak logical principles already in our vocabulary, the addition of strong logical principles can result in a non-conservative extension. However, there is a sense that if we have a purely non-logical vocabulary, the addition of classical inference principles can be totally conservative over the old language, as can be shown by a traditional cut-elimination or normalisation proof. It would be very comforting to think, then, that logical vocabulary can be always conservatively added over some discourse and that it can, in Brandom's suggestive phrase, be purely *expressive* of the inferential commitments already endorsed in the base vocabulary without adding any new commitments in the old vocabulary [5,6].

---

[9]The case is identical if we lift the restriction on the *conjunction right* rule too, but Dummett does not consider this case, presumably because he does not take there to be any restriction in the rule $\wedge R'$.

Unfortunately, the matter is not so simple. There seem to be properly incompatible extensions to a single base language and consequence relation. The dialetheist's motivating examples provide us with a pertinent case. Consider adding to one's vocabulary a predicate $\in$ and variable-binding term forming operator $\{\ :\ \}$ satisfying the following rules, which are a form of Frege's BASIC LAW (V) for class membership.

$$\frac{X, \phi(a) \vdash Y}{X, a \in \{x : \phi(x)\} \vdash Y} \, (\in L) \qquad \frac{X \vdash \phi(a), Y}{X \vdash a \in \{x : \phi(x)\}, Y} \, (\in R)$$

It is trivial to show that this addition to the language is coherent if the language contains only predicates and names and no other logical vocabulary.

Given *this* collection of rules, it is impossible to add Boolean negation and preserve the transitivity of inference and consistency, as the Russell paradox shows. Let $r$ be the term $\{x : \sim(x \in x)\}$: We have the following two proofs.

$$\frac{\dfrac{r \in r \vdash r \in r}{r \in r \vdash \sim(r \in r)} \, (\in R)}{\vdash \sim(r \in r)} \, (\sim R) \qquad \frac{\dfrac{r \in r \vdash r \in r}{\sim(r \in r) \vdash r \in r} \, (\in L)}{\sim(r \in r) \vdash} \, (\sim L)$$

Given the transitivity of entailment, we have the empty sequent ' $\vdash$ ', and by weakening, triviality results. It follows that conservative extension criteria, on their own, do not suffice to choose one logical system over another. One's starting point matters.[10]

* * *

Here are some concluding points.

¶ Everyone has an opportunity to account for the meaning of their logical vocabulary in terms the way it constrains assertion and denial. This justification is independent of issues of realism or anti-realism. If you like, you can explain these constraints on assertion and denial in terms of what is true or what can possibly be true. If you like, you can explain these constraints in terms of what is warranted or possibly warranted. Or you could give some other account. Or you can be agnostic. Whatever approach you choose for the explication of deductively valid inference, common ground between positions is found in the way that logical consequence constrains assertion and denial.

¶ The multiple-conclusion calculus, long thought to be formally useful as an account of the preservation of *truth* and a formal setting for classical consequence [11,31] can also be understood and appropriated by those who take the right semantic theory to be expressed in the anti-realist vocabulary of norms of assertion and denial rather than the realist talk of truth, correspondence and reference.

¶ Non-classical logicians, such as dialetheists, supervaluationsts and intuitionists cannot be charged with incoherence. There seems to be enough shared vocabulary to understand and evaluate their positions. These different theories are different

---

[10]This is not to say, of course, that there are no *other* reasons to favour one approach to another. We might reject the $\in$ rules because they seem to be existentially committing, for example.

proposals for the logic of negation, and thereby, for the way that negation constrains assertion and denial.

¶ It is often thought that intuitionistic logic fares especially well when it comes to proof-theoretic justification of logical consequence. The justification here is seriously incomplete. The intuitionist has to do more work to explain the priority of assertion over denial and the resulting restriction of connective rules on the right. This restriction might be plausible and defensible, but if that kind of explanation is possible for the intuitionist, it might also be possible for others.

¶ The debates over the paradoxes of self-reference can be seen disagreement over what vocabulary is more important to save, and what kind of inferential machinery is most important. *No-one* can have it all. Triviality results from the combination of $(\in L)$, $(\in R)$, $(\sim L)$ and $(\sim R)$. Something must go.

¶ We have not answered all of the questions, but we have at least managed to see how different people are playing on the same field, with something like the same rules.

## REFERENCES

1. Alan Ross Anderson and Nuel D. Belnap. *Entailment: The Logic of Relevance and Necessity*, volume 1. Princeton University Press, Princeton, 1975.
2. Alan Ross Anderson, Nuel D. Belnap, and J. Michael Dunn. *Entailment: The Logic of Relevance and Necessity*, volume 2. Princeton University Press, Princeton, 1992.
3. JC Beall and Greg Restall. "Logical Pluralism". *Australasian Journal of Philosophy*, 78:475–493, 2000.
4. E. Beltrametti and G. Cassinelli. *The Logic of Quantum Mechanics*. van Nostrand, 1981.
5. Robert B. Brandom. *Making It Explicit*. Harvard University Press, 1994.
6. Robert B. Brandom. *Articulating Reasons: an introduction to inferentialism*. Harvard University Press, 2000.
7. Michaelt Dummett. *Frege: Philosophy of Language*. Duckworth, 1973.
8. Michael Dummett. *Elements of Intuitionism*. Oxford University Press, Oxford, 1977.
9. Michael Dummett. *The Logical Basis of Metaphysics*. Harvard University Press, 1991.
10. Kit Fine. "Vagueness, Truth and Logic". *Synthese*, 30:265–300, 1975. Reprinted in *Vagueness: A Reader* [16].
11. Ian Hacking. "What is Logic?". *The Journal of Philosophy*, 76:285–319, 1979.
12. William H. Hanson. "The Concept of Logical Consequence". *The Philosophical Review*, 106:365–409, 1997.
13. Gary Hardegree and P. J. Frazer. "Charting the Labyrinth of Quantum Logics: A Progress Report". In E. Beltrametti and Bas C. van Fraassen, editors, *Current Issues in Quantum Logic*, pages 53–76. Plenum Press, 1981.
14. Gilbert Harman. *Change In View: Principles of Reasoning*. Bradford Books. MIT Press, 1986.

15. Arend Heyting. *Intuitionism: An Introduction*. North Holland, Amsterdam, 1956.
16. Rosanna Keefe and Peter Smith. *Vagueness: A Reader*. Bradford Books. MIT Press, 1997.
17. Vann McGee. *Truth, Vagueness and Paradox*. Hackett Publishing Company, Indianapolis, 1991.
18. Alexius Meinong. *On Assumptions*. University of California Press, Berkeley and Los Angeles, California, 1983. Translated and edited by James Heanue.
19. Terence Parsons. "Assertion, Denial, and the Liar Paradox". *Journal of Philosophical Logic*, 13:137–152, 1984.
20. Terence Parsons. "True Contradictions". *Canadian Journal of Philosophy*, 20:335–354, 1990.
21. Huw Price. "Why 'Not'?". *Mind*, 99:222–238, 1990.
22. Graham Priest. "The Logic of Paradox". *Journal of Philosophical Logic*, 8:219–241, 1979.
23. Graham Priest. "Inconsistencies in Motion". *American Philosophical Quarterly*, 22:339–345, 1985.
24. Graham Priest. "Can Contradictions be True?, II: Yes". *Proceedings of the Aristotelian Society*, Supplementary Volume 67:35–54, 1993.
25. Graham Priest. "Validity". In Achillé C. Varzi, editor, *The Nature of Logic*, volume 4 of *European Review of Philosophy*, pages 183–205. CSLI Publications, Stanford, 1999.
26. Graham Priest. "What Not? A Defence of Dialetheic Theory of Negation". In Dov Gabbay and Heinrich Wansing, editors, *What is Negation?*, volume 13 of *Applied Logic Series*, pages 101–120. Kluwer Academic Publishers, 1999.
27. Graham Priest, Richard Sylvan, and Jean Norman, editors. *Paraconsistent Logic: Essays on the Inconsistent*. Philosophia Verlag, 1989.
28. Willard van Orman Quine. "Carnap and Logical Truth". In *The Ways of Paradox and Other Essays*, pages 100–134. Random House, New York, 1966. (Published initially in 1954).
29. Willard van Orman Quine. *Philosophy of Logic*. Prentice-Hall, Englewood Cliffs, NJ, 1970.
30. Greg Restall. *An Introduction to Substructural Logics*. Routledge, 2000.
31. Dana Scott. "On Engendering an Illusion of Understanding". *Journal of Philosophy*, 68:787–807, 1971.
32. Gila Sher. *The Bounds of Logic*. MIT Press, 1991.
33. D. J. Shoesmith and T. J. Smiley. *Multiple Conclusion Logic*. Cambridge University Press, Cambridge, 1978.
34. John K. Slaney. "A General Logic". *Australasian Journal of Philosophy*, 68:74–88, 1990.
35. T. J. Smiley. "Rejection". *Analysis*, 56:1–9, 1996.
36. Neil Tennant. *The Taming of the True*. Clarendon Press, Oxford, 1997.
37. Achillé Varzi. *An Essay in Universal Semantics*, volume 1 of *Topoi Library*. Kluwer Academic Publishers, Dordrecht, Boston and London, 1999.

# On the Negation of Action Types: Constructive Concurrent PDL

Heinrich Wansing

*Dresden University of Technology, Institute of Philosophy, 01062 Dresden, Germany*
*Heinrich.Wansing@mailbox.tu-dresden.de*

**Abstract.** The negation of generic actions and its use in reasoning about actions and obligation is discussed. Complementation is rejected as a negation operation on action types. Instead of complementation, a constructive negation operation is suggested, which is interpreted as *refraining*. Concurrent propositional dynamic logic, CPDL, is extended by this new operation. The resulting dynamic logic, constructive concurrent PDL, CCPDL, is axiomatized and shown to be decidable and complete with respect to standard models.

## 1. Introduction

There seems to exist semantic opposition between expressions for action types. In the lexicon we have pairs like 'to dress' versus 'to undress' and 'to robe' versus 'to disrobe', 'to compress' versus 'to decompress' etc. We also have in English pairs like 'to smoke' versus 'to refrain from smoking' and 'to steal' versus 'to prevent from stealing'. It thus seems to make sense to talk about (forms of) negation of generic actions. Clearly, the negation of an action type should be a one-place action type forming operation. If action types are conceived of as binary relations understood as sets of input-output pairs representing performances of action types, then the unary operations of complement and converse seem to be natural candidates for negation operations on generic actions. In this paper I shall consider the complement of binary relations and shall argue

1. that complement and restricted complement fail to be action type forming, and

2. that the negation of generic atomic actions is *primitive* and not defined by any particular relational operation.

As to the converse of binary relations, let me just remark that not all generic actions are like dressing, robing, or compressing. It is not only that the strings 'unkill' or 'unsmoke' fail to be English words, ordinary agents certainly, as a matter of fact, cannot perform unkillings or unsmokings. Hence, the converse is only a partial function, its domain is not the class of *all generic actions*. Moreover, it is the complement that has been used in certain interesting applications of dynamic logic to reasoning about action and change, action and norms, and in other application areas. It is complement formation to which I shall turn now.

## 2. The problem of expressing alternative action

There are state transitions that fail to be instantiations of action types. The radioactive decay of an atomic nucleus, for example, seems to be a happening in which no decision making human or artificial agent is involved. If variants of dynamic logic are used as representation languages and means for reasoning about *actions* and not about change in general, action types play the role of programs, and instead of operations on programs, operations on action types are studied. In the literature, the notion of an action complement has played an important role in the application of dynamic propositional logics to actions. In particular, complement and versions of a restricted complement have been suggested as suitable for expressing the concept of an alternative action. De Giacomo and Lenzerini [7], for instance, emphasize that their dynamic propositional logic $\mathcal{DIFR}$ allows:

> specifying the "non-execution" of atomic actions ... formulated by interpreting it as "the execution of some action other than a given one". ... [I]t is essentially this feature that allows us to provide a compact representation of the frame axioms

that is, axioms expressing what is not affected by actions. Non-execution of an atomic action type is thus understood as the execution of an alternative generic action, and this idea is used to obtain a monotonic solution of the frame problem (see also [6] and references therein).

The complement with respect to the universal relation on a given state space is, as it seems, a prominent candidate for modeling the notion of alternative action. Let $U$ be the universal relation $S \times S$ on a state space $S$. The idea is that if $\alpha$ denotes a generic action, the formula

$$A \supset [U \setminus \alpha]A$$

may be read as "performing an action other than $\alpha$ in a state at which $A$ is true, cannot result in a state at which $A$ is not true". In other words, a transition from a state at which $A$ is true to a state where $A$ is not true requires performing the action type $\alpha$. Instead of using the complement $U \setminus \alpha$, de Giacomo and Lenzerini define a restricted complement **any** $\setminus \rho$, where **any** $\subseteq U$ is a designated atomic action thought of as the most general atomic action, and $\rho$ is an atomic action. De Giacomo and Lenzerini use the symbol '¬' to denote both Boolean negation and action negation. A frame axiom then has the form $A \supset [\neg\rho]A$.

**Example 1** We here modify an example from [7]: Lifting both sides of a table. A child sits on top of a rectangular table (with two sides, a left and a right one). If just one side of the table is lifted, the child will jump on the floor. However, if both sides of the table are simultaneously lifted, the child will remain sitting.[1] We use the following primitive propositions:

- *child–on–table, down–left–side, down–right–side*

---

[1] In De Giacomo's and Lenzerini's example, there is a vase on the table. The vase will slide down and fall on the floor if only one side of the table is lifted and not both. The example has been changed, because sliding down, though changing the situation, does not seem to be an action.

and the following primitive generic actions:

- *child–jumps–down, lift–left, lift–right*

We have the following preconditions that must be satisfied in order to execute the actions under consideration:

- $\langle$lift–left$\rangle\top \equiv$ down–left–side
- $\langle$lift–right$\rangle\top \equiv$ down–right–side
- $\langle$child–jumps–down$\rangle\top \equiv$ (child–on–table $\wedge$ ((down–left–side $\wedge$ ¬down–right–side) $\vee$ (down–right–side $\wedge$ ¬down–left–side)))

The effect of performing the three primitive actions can now be described as follows:

- [lift–left] ¬ down–left–side
- [lift–right] ¬ down–right–side
- [child–jumps–down] ¬ child–on–table

The crucial point is the formulation of frame axioms; it is here where the negation of action types enters the picture:

- child–on–table $\supset$ [¬ child–jumps–down] child–on–table
- down–right–side $\supset$ [¬ lift–right] down–right–side
- down–left–side $\supset$ [¬ lift–left] down–left–side

**Example 2** In dynamic deontic logic , initiated by Meyer [11], the deontic operators are applied not to formulas, but to action type expressions. In particular, obligation is defined in terms of prohibition and action negation:

$$O\alpha \equiv F\neg\alpha \quad \text{(It is obligatory to do } \alpha \text{ iff it is forbidden to do } not\text{–}\alpha.)$$

Also Meyer defines action negation as a form of complement, though not unrestricted complement.

The appealing availability of complement formation with respect to the universal relation notwithstanding, some authors have called into question the viability of this approach in principle. Krister Segerberg [14, p. 377], for example, remarks that (notation slightly adjusted):

> [b]y contrast with intersection, the question concerning complement is intricate and involves much extratheoretical considerations: Do we humans really think in terms of complements? Does the analysis of human languages suggest that we do? ... [i]s it not the case that the choice between two actions $\alpha$ and $U \setminus \alpha$ is often a choice between $\alpha$ and some action $\beta$ that is a proper subset of $U \setminus \alpha$? Before these questions have been answered, this author feels a certain unease about the unrestricted acceptance of closure under complement.

This quotation is a useful starting point for discussing the negation of action types. We may extract four questions:

Q1 Is the notion of complement appropriate insofar as humans in fact employ this concept when reasoning about actions?

Q2  Is the closure of action types under complement reflected in human languages?

Q3  Is it appropriate to form the complement with respect to the *universal* relation?

Q4  Is complementation at all action type forming?

Recently, Jan Broersen [3], [4] has taken Segerberg's reservations as a basis for arguing in favour of a restricted action complement as a means of expressing the notion of alternative action. Unlike de Giacomo and Lenzerini, Broersen does not consider the complement of atomic actions with respect to a most general atomic action **any** but the complement with respect to the following set of input-output pairs:

$$U' := \{\langle u, s\rangle \mid \langle u, s\rangle \text{ belongs to some action type } \beta\}$$

The interpretation of the complement thus depends on the available atomic actions and the operations on action types and their properties. This overcomes a problem with the unrestricted complement $\neg$. Broersen points out that $A \supset [\neg a]A$ does not only express that actions other than $a$ cannot lead from a state where $A$ holds to a state where $A$ is not true, but also expresses that if $A$ is true, *all states* at which $A$ is not true, can be reached by $a$. In this sense, $[\neg a]A$ is a 'window operator'.

Although I believe that the idea of relativized action complement is interesting, I share Segerberg's general hesitation with respect to closure under complement, because I think that relativized action complement also fails to be action type forming. As we shall see, it seems that the answer to question Q2 is negative. The negative answer to question Q2 may be taken to indicate that the answer to question Q1 is negative, too. But then it is inappropriate to use the unrestricted or the restricted complement for representing reasoning about negative generic *actions*. Hence, there are reasons to believe that the answer to questions Q1–Q4 is: *No*.

To see that complementation fails to be action type forming, it may be helpful to consider a distinction discussed by von Wright [17] that has more recently been highlighted in *stit* theory, the modal logic of *concrete* actions interpreted in models of branching time (or space-time), see [1] and the references therein. There is a difference between *not acting* on the one hand and *refraining from seeing to it that something is the case* on the other hand. Consider an action type agent $a$ is, as a matter of fact, unable to execute, say, jumping two meters high. Imagine $a$ walking down the street and consider the following descriptions of this scene:

>(a)     $a$ does not jump two meters high.

>(b) ?   $a$ performs a non-two-meters-high-jumping.

>(c)     $a$ refrains from jumping two meters high.

Whereas statement (a) is a true description of the imagined situation, (c) is false in the imagined situation. In order to refrain from seeing to it that she jumps two meters high, $a$ must be *able* to jump so high, but we assumed that $a$ is not capable of doing this. Hence, in general not seeing to it that $Q$ is different from refraining from seeing to it that $Q$. Moreover, whereas statement (a) clearly fails to be an agentive sentence, (c) is an action report. Agent *a* refrains from jumping two meters high only if *a* has, consciously or not, decided not to jump so high. Statements (b) and (c) differ in information content. Statement (c) provides a description of which action *a* performs, whereas (b) leaves this open. Statement (b) is not a description of any particular *action*, and, moreover, given that its grammaticality is dubious, I conclude that there is reason to doubt that (b) is an action report at all. Therefore, I doubt that the complement is action type forming. According to Belnap, Perloff and Xu, the reification of action types like the non-two-meters-high-jumping enriches the ontology "without having any sense that we know what we are talking about" [1, p. 42]. The reasoning is independent of whether we conceive of non-two-meters-high-jumping as the complement or the restricted complement of jumping two meters high.

If an agent *a* prevents an agent *b* from seeing to it that *Q*, this may be thought of as *a* seeing to it that *b* does not see to it that *Q*. Refraining then becomes a special case of preventing: if an agent refrains from seeing to it that *Q*, the agent prevents *herself* from seeing to it that *Q*. There thus appear to be two distinct categories of deed types: actions and preventions, see [15].

**Examples 1 and 2 (continued).** Let $\alpha$ be an action type. Under the interpretation of the negation $-\alpha$ as the refraining from $\alpha$, the crucial formulas from Examples 1 and 2 receive the following reading:

- child–on–table $\supset$ [$-$ child–jump–down] child–on–table
  If the child is on the table, then it remains on the table after every refraining from jumping down.
- down–right–side $\supset$ [$-$ lift–right] down–right–side
  If the right side of the table is down, it remains down after every refraining from lifting the right side.
- down–left–side $\supset$ [$-$ lift–left] down–left–side
  If the left side of the table is down, it remains down after every refraining from lifting the left side.

- $O\alpha \equiv F - \alpha$
  It is obligatory to do $\alpha$ iff it is forbidden to refrain from (to prevent oneself from) doing $\alpha$.

I take it that these readings are natural and express what is required in the applications. The first formula, for example, says that if the child is on the table, then only by not refraining from jumping from the table, the child will not be on the table.[2] One might object that "not refraining from jumping from the table" fails to describe an action type as much as "non-jumping from the table" does. Whereas refraining from refraining to see to it that *Q* implies seeing to it that *Q*, not refraining from seeing to it that *Q* does not imply seeing to it that *Q*, so that the first formula fails to express that the child can leave the table only by jumping down. The child cannot jump two meters high and hence it cannot refrain from jumping two meters high. Therefore, the child does not refrain from jumping two

---

[2]Note that in our example, we do not consider primitive generic actions such as remove–child–from–table.

meters high, but this does not mean that it jumps two meters high. However, if we assume that the child *is able* to jump down from the table, then, intuitively, not refraining from jumping down implies jumping down. This intuition is formally reinforced by the *dstit* operator, the deliberative *stit* operator, see [1], [10, p. 26]: $\neg[a\,dstit:\neg[a\,dstit:Q]]$ ($a$ does not refrain from seeing to it that $Q$) and $\Diamond[a\,dstit:Q]$ ($a$ is able to see to it that $Q$) together entail $[a\,dstit:Q]$ ($a$ sees to it that $Q$). Therefore the formula from our example is appropriate, at least if we think of types of concrete actions such that action reports are interpreted using the dstit operator, which means that the moment of choice or action and the moment of evaluation are the same.

If performing $\alpha$ is obligatory, it may still be not forbidden to do something else than $\alpha$, say, in parallel. The fourth formula expresses that performing $\alpha$ is obligatory if and only if it is forbidden to refrain from executing $\alpha$. This seems to be exactly what is intended by defining obligation in terms of prohibition.

In this paper I shall

1. treat generic actions and refrainings on a par, so that the negation of atomic actions is not defined by any specific relational operation, and

2. treat the negation of action types in a constructive manner, so that the negation of compound action types is defined separately for each kind of compound action type.

The result is an extension of concurrent PDL, referred to as constructive concurrent PDL, CCPDL.[3]

It is well-known that the validity problem for PDL with unrestricted complement is undecidable, see [9, p. 269]. As we shall see, constructive concurrent PDL has the finite model property and hence *is* decidable. We thus obtain a decidable extension of PDL including action type negation.[4]

## 3. CCPDL, constructive concurrent PDL

The presentation of CCPDL may and will for reasons of both convenience and comparison closely follow Goldblatt's treatment of concurrent PDL, CPDL, and the decidability and completeness proof in [8]. The language of CCPDL extends the language of CPDL by including a unary operation of action type negation. We shall use prefix notation for unary operations on action types. The language is defined as follows:

---

[3]Whereas CCPDL uses a constructive action algebra, in the Constructive Concurrent Dynamic Logic of Wijesekera and Nerode [16] the underlying classical logic is replaced by intuitionistic logic.

[4]Another kind of negation of generic actions is van Benthem's dynamic negation, [2]. Dynamic negation is a test for the impossibility of performing a generic action: $\sim_d \alpha \equiv_{def} ?[\alpha]\bot$. Returning to our example, we would obtain the axiom: child–on–table $\supset [\sim_d$ child–jump–down] child–on–table. Thus, if the child is on the table, then after every successful test for the impossibility of jumping down, it remains on the table. This is not a frame axiom. Hence, although $\sim_d \alpha$ *is* an action type, $\sim_d$ appears not to be suitable for part of the intended applications of the concept of negative action.

atomic formulas:                          $p \in \Phi$
atomic action type terms:      $\pi \in \Pi$
formulas:                                  $A \in Form(\Phi, \Pi)$
action type terms:               $\alpha \in Act(\Phi, \Pi)$

$$A \quad ::= \quad p \mid \bot \mid (A_1 \supset A_2) \mid \langle \alpha \rangle A \mid [\alpha] A$$

$$\alpha \quad ::= \quad \pi \mid -\alpha \mid *\alpha \mid (\alpha_1 ; \alpha_2) \mid (\alpha_1 \cup \alpha_2) \mid (\alpha_1 \cap \alpha_2) \mid ?A$$

The classical connectives $\top$, $\wedge$, $\vee$, and $\equiv$ are defined as usual. Although $-\pi$ is a compound expression, nothing specific will be said about the relation between $\pi$ and $-\pi$. 'Action-literals' $\pi$ and $-\pi$ are the basic building blocks of complex action type terms. The intended readings of 'dynamic' formulas $\langle \alpha \rangle A$, $[\alpha] A$ and the action type operations are:

| | |
|---|---|
| $\langle \alpha \rangle A$ | after some execution of $\alpha$, $A$ is the case |
| $[\alpha] A$ | after every execution of $\alpha$, $A$ is the case |
| $-\alpha$ | the refraining from $\alpha$ |
| $*\alpha$ | the iterated execution of $\alpha$ for some finite number ($\geq 0$) of times |
| $(\alpha_1 ; \alpha_2)$ | the composition of $\alpha_1$ and $\alpha_2$ |
| $(\alpha_1 \cup \alpha_2)$ | the indeterministic choice between $\alpha_1$ and $\alpha_2$ |
| $(\alpha_1 \cap \alpha_2)$ | the concurrent combination of $\alpha_1$ and $\alpha_2$ |
| $?A$ | the test for $A$ |

A CCPDL-model is a CPDL-model, namely a structure

$$\mathcal{M} = (S, \{R_\alpha : \alpha \in Act(\Phi, \Pi)\}, v),$$

where $S$ is a nonempty set of states, and for each action type term $\alpha$, $R_\alpha$ is a *reachability relation* on $S$, i.e., $R_\alpha \subseteq S \times \mathcal{P}(S)$, and $v : \Phi \longrightarrow 2^S$ is a valuation function. The truth of a formula $A$ at a state $s \in S$ in $\mathcal{M}$ ($\mathcal{M}, s \models A$) is defined as follows:

$\mathcal{M}, s \not\models \bot$
$\mathcal{M}, s \models (A \supset B)$    iff    $\mathcal{M}, s \models A$ implies $\mathcal{M}, s \models B$
$\mathcal{M}, s \models \langle \alpha \rangle A$    iff    there exists $T \subseteq S$ such that $sR_\alpha T$ and $T \subseteq A^{\mathcal{M}}$
$\mathcal{M}, s \models [\alpha] A$    iff    for all $T \subseteq S$, $sR_\alpha T$ implies $T \subseteq A^{\mathcal{M}}$

where $A^{\mathcal{M}} = \{u \in S : \mathcal{M}, u \models A\}$. We say that $A$ is true in $\mathcal{M}$ iff $A^{\mathcal{M}} = S$. Validity is defined as truth in all models. Note that in this semantics $[\alpha]$ and $\langle \alpha \rangle$ are not interdefinable using negation.[5]

What is meant by composition, combination and by iteration of action types understood as reachability relations is made precise by the following definitions. Let $R$ and $Q$ be reachability relations. The composition $R \cdot Q \subseteq S \times \mathcal{P}(S)$ is defined by

$s(R \cdot Q)T$    iff    there exists $U \subseteq S$ such that $sRU$, and a set of subsets
                $\{T_u : u \in U\}$ of $T$ with $uQT_u$ for all
                $u \in U$, such that $T = \bigcup\{T_u : u \in U\}$.

---

[5] In Peleg's [13] original version of concurrent dynamic logic, $[\alpha] A$ is defined as $\neg \langle \alpha \rangle \neg A$. This gives $\mathcal{M}, s \models [\alpha] A$ iff for all $T \subseteq S$, $sR_\alpha T$ implies $T \cap A^{\mathcal{M}} \neq \emptyset$.

For concurrent combination $R \otimes Q$ we have:

$$R \otimes Q = \{(s, T \cup W) : sRT \text{ and } sQW\}.$$

The iteration $R^{(*)}$ is defined as follows. Let $Id = \{(s, \{s\}) : s \in S\}$, and define an infinite sequence of reachability relations $R^{(n)}$ by the following induction:

$$
\begin{array}{rcl}
R^{(0)} & = & Id \\
R^{(n+1)} & = & Id \cup R \cdot R^{(n)} \\
R^{(*)} & = & \bigcup\{R^{(n)} : n \in \omega\}
\end{array}
$$

Note that $n \leq m$ implies $R^{(n)} \subseteq R^{(m)}$.

A reachability relation $R$ can be reduced to a binary relation $\overline{R}$ in a sense, see [8, p. 120 f.], by defining:

$$s\overline{R}t \quad \text{iff} \quad t \in \bigcup\{T : sRT\}$$

Then for any CCPDL-model $\mathcal{M}$,

$$\mathcal{M}, s \models [\alpha]A \quad \text{iff} \quad (s\overline{R_\alpha}t \text{ implies } \mathcal{M}, t \models A)$$

The defined relations satisfy a number of properties:[6]

**Theorem 1** ([8, Theorem 10.14]) *Let $R^*$ denote the reflexive transitive closure of $R$, let $\circ$ denote the composition of binary relations and let id denote the identity relation.*

*(1)* $\overline{\bigcup_{i \in I} R_i} = \bigcup_{i \in I} \overline{R_i}$

*(2)* $R \subseteq Q$ *implies* $\overline{R} \subseteq \overline{Q}$

*(3)* $\overline{R \cdot Q} \subseteq \overline{R} \circ \overline{Q}$

*(4)* $Id \subseteq Q$ *implies* $\overline{R \cdot Q} = \overline{R} \circ \overline{Q}$

*(5)* $\overline{R^{(n+1)}} = id \cup \overline{R} \circ \overline{R^{(n)}}$

*(6)* $\overline{R^{(n)}} = \overline{R}^0 \cup \ldots \cup \overline{R}^n$

*(7)* $\overline{R^{(*)}} = \overline{R}^*$

It follows that $\mathcal{M}, s \models [*\alpha]A$ iff $(s\overline{R_\alpha}^* t$ implies $\mathcal{M}, t \models A)$.

A CCPDL-model $\mathcal{M} = (S, \{R_\alpha : \alpha \in Act(\Phi, \Pi)\}, v)$, is a *standard* model if it satisfies the following conditions:

---

[6] For $n \geq 0$, the binary relations $R^n$ are defined as follows: $sR^0t$ iff $s = t$, and $sR^{n+1}t$ iff $\exists u \, (sR^n u \text{ and } uRt)$.

$$
\begin{aligned}
R_{(\alpha \cup \beta)} &= R_\alpha \cup R_\beta \\
R_{(\alpha \cap \beta)} &= R_\alpha \otimes R_\beta \\
R_{(\alpha ; \beta)} &= R_\alpha \cdot R_\beta \\
R_{*\alpha} &= R_\alpha^{(*)} \\
R_{?A} &= \{(s, \{s\}) ; \mathcal{M}, s \models A\} \\
R_{-(\alpha \cup \beta)} &= R_{-\alpha} \otimes R_{-\beta} \\
R_{-(\alpha \cap \beta)} &= R_{-\alpha} \cup R_{-\beta} \\
R_{-(\alpha ; \beta)} &= R_{-\alpha} \cdot R_{-\beta} \\
R_{-*\alpha} &= R_{*-\alpha} \\
R_{-?A} &= \{(s, \{s\}) : \mathcal{M}, s \not\models A\} = R_{?\neg A} \\
R_{--\alpha} &= R_\alpha
\end{aligned}
$$

The condition $R_{*\alpha} = R_\alpha^{(*)}$ is carefully explained in [8, p. 119 f.]. The last six conditions are new. The first two of them hardly need any justification. To refrain from executing nondeterministically either $\alpha$ or $\beta$ amounts to refraining from $\alpha$ and refraining from $\beta$ in parallel. To refrain from the parallel execution of $\alpha$ and $\beta$ amounts to refraining from $\alpha$ or refraining from $\beta$ nondeterministically. In order to refrain from executing the composition of $\alpha$ and $\beta$, it may not be enough just to refrain from $\alpha$ or to refrain from $\beta$. Whatever an agent does to prevent himself from performing $\alpha$ or $\beta$ may not be such that it also prevents himself from executing the composition of $\alpha$ and $\beta$. If this has been realized, also the fourth condition is clear. The fifth condition is perhaps less clear. What does it mean to prevent oneself from testing $A$? If to test means to continue if $A$ is true and to fail otherwise, this is prevented by continuing if $A$ is not true and failing otherwise, hence condition five. The sixth condition reflects the ref-ref thesis from *stit* theory: refraining from refraining amounts to doing.

Note that we postulate no connection between an atomic action type $\pi$ and its negation $-\pi$. What can be considered the refraining from an atomic action may depend on a particular application context. Also in this sense our treatment of action negation is constructive. However, further elaboration is, perhaps, required. Refraining from an execution of $\pi$ is an action, so, intuitively, executing $-\pi$ amounts to executing some generic action other than $\pi$. Which action type this is, i.e., what counts as refraining from $\pi$, is not a question to be settled by the underlying logic.[7] For example, you may refrain from $\pi$, lifting the right side of the table, by lifting the left side. But this does not mean that you and I cannot execute both action types, lifting the right side and lifting the left side, in parallel. Whatever may be an adequate description of $-\pi$, there is no reason to assume that you and I together cannot perform $\pi \cap -\pi$, the parallel execution of $\pi$ and the refraining form $\pi$ (provided you are able, say, to perform $\pi$ and I am able to execute $-\pi$).[8] In general the interaction between refraining as a negative generic action and concurrent combination is not such that the validity of $\langle \alpha \rangle \top \wedge \langle -\alpha \rangle \top \supset \langle \alpha \cap -\alpha \rangle \top$ should be problematic, see axiom $D - Comb$ in Table 1.

---

[7] Consequently, we admit models in which the sense of "other than" is entirely non-extensional: $\pi$ and $-\pi$ may be interpreted by the same reachability relation.

[8] Although we should not overstrain the connections with stit theory, note that whereas $[a \; dstit : Q] \wedge \neg [a \; dstit : \neg [a \; dstit : Q]]$ is unsatisfiable, $[a \; dstit : Q] \wedge [b \; dstit : \neg [b \; dstit : Q]]$, where $a \neq b$, is satisfiable.

CCPDL is defined as the smallest subset of $Form(\Phi, \Pi)$ that comprises all classical tautologies, is closed under uniform substitution, modus ponens, and the necessitation rule for $[\alpha]$, for every $\alpha \in Act(\Phi, \Pi)$, and contains the axiom schemata listed in Table 1. Provability in CCPDL, $\vdash$, is defined in the usual way. From this axiomatization of CCPDL it is evident that CPDL is properly contained in CCPDL.

Soundness with respect to standard models can be shown by induction on proofs in CCPDL.

**Theorem 2** *If $\vdash A$, then $A$ is true in all standard CCPDL-models.*

As in the case of CPDL and PDL, to prove completeness, a canonical model is defined. The canonical model fails to be a standard model . However, there exist finite filtrations of the canonical model that are standard. Hence CCPDL turns out to be complete with respect to the class of finite standard models. The states of the canonical model consist of the CCPDL-maximal sets of formulas. A set $\Delta \subseteq Form(\Phi, \Pi)$ is CCPDL-maximal iff it is CCPDL-consistent and for every $A \in Form(\Phi, \Pi)$, $A \in \Delta$ or $\neg A \in \Delta$. Let $S^m = \{\Delta : \Delta$ is CCPDL-maximal$\}$. For every formula $A$, define

$$\|A\| = \{s \in S^m : A \in s\}$$

For every action type term $\alpha$ and every $s \in S^m$, define

$$s_\alpha = \{A : [\alpha]A \in s\}, \quad \|s_\alpha\| = \{t \in S^m : s_\alpha \subseteq t\}.$$

**Theorem 3** *([8, Theorem 10.18])*

*(1) $\vdash A$ iff $\|A\| = S^m$*

*(2) $\vdash A \supset B$ iff $\|A\| \subseteq \|B\|$*

*(3) $\|A \vee B\| = \|A\| \cup \|B\|$*

*(4) $\|A \wedge B\| = \|A\| \cap \|B\|$*

*(5) $\|s_\alpha\| \subseteq \|A\|$ implies $[\alpha]A \in s$*

*(6) If $\|s_\alpha\| \cap \|B\| \subseteq \|A\|$ and $\langle\alpha\rangle B \in s$, then $\langle\alpha\rangle A \in s$*

*(7) If $s, u \in S^m$ and $s_\alpha \subseteq u$, then $\|u_\beta\| \subseteq \|s_{\alpha;\beta}\|$*

*(8) $\|s_{\alpha \cup \beta}\| = \|s_\alpha\| \cup \|s_\beta\|$*

*(9) If $\langle\alpha\rangle\top, \langle\beta\rangle\top \in s$, then $\|s_{\alpha \cap \beta}\| = \|s_\alpha\| \cup \|s_\beta\|$*

The canonical model for CCPDL is the CCPDL-model

$$\mathcal{M}^m = (S^m, \{R_\alpha : \alpha \in Act(\Phi, \Pi)\}, v^m),$$

where

$$
\begin{aligned}
sR_\alpha T \quad &\text{iff} \quad \text{there exists } B \text{ such that } \langle\alpha\rangle B \in s \text{ and } T = \|s_\alpha\| \cap \|B\| \\
v^m(p) \quad &= \quad \|p\|
\end{aligned}
$$

**Theorem 4** *[8, Theorem 10.19]) Let $s \in S^m$ and $T \subseteq S^m$.*

*(1) $\langle\alpha\rangle A \in s$ iff there exist $T$ such that $sR_\alpha T$ and $T \subseteq \|A\|$*

| | |
|---|---|
| $B - K$ | $[\alpha](A \supset B) \supset ([\alpha]A \supset [\alpha]B)$ |
| $B - Comp$ | $[\alpha; \beta]A \equiv [\alpha][\beta]A$ |
| $B - Alt$ | $[\alpha \cup \beta]A \equiv [\alpha]A \wedge [\beta]A$ |
| $B - Comb$ | $[\alpha \cap \beta]A \equiv (\langle \alpha \rangle \top \supset [\beta]A) \wedge (\langle \beta \rangle \top \supset [\alpha]A)$ |
| $B - Mix$ | $[*\alpha]A \supset (A \wedge [\alpha][*\alpha]A)$ |
| $B - Ind$ | $[*\alpha](A \supset [\alpha]A) \supset (A \supset [*\alpha]A)$ |
| $B - Test$ | $[?A]B \equiv (A \supset B)$ |
| $B - negComp$ | $[-(\alpha; \beta)]A \equiv [-\alpha][-\beta]A$ |
| $B - negAlt$ | $[-(\alpha \cup \beta)]A \equiv (\langle -\alpha \rangle \top \supset [-\beta]A) \wedge (\langle -\beta \rangle \top \supset [-\alpha]A)$ |
| $B - negComb$ | $[-(\alpha \cap \beta)]A \equiv [-\alpha]A \wedge [-\beta]A$ |
| $B - negMix$ | $[- * \alpha]A \supset (A \wedge [-\alpha][- * \alpha]A)$ |
| $B - negInd$ | $[- * \alpha](A \supset [-\alpha]A) \supset (A \supset [- * \alpha]A)$ |
| $B - negTest$ | $[-?A]B \equiv (\neg A \supset B)$ |
| $B - negNeg$ | $[- - \alpha]A \equiv [\alpha]A$ |
| | |
| $D - K$ | $[\alpha](A \supset B) \supset (\langle \alpha \rangle A \supset \langle \alpha \rangle B)$ |
| $D - Comp$ | $\langle \alpha; \beta \rangle A \equiv \langle \alpha \rangle \langle \beta \rangle A$ |
| $D - Alt$ | $\langle \alpha \cup \beta \rangle A \equiv (\langle \alpha \rangle A \vee \langle \beta \rangle A)$ |
| $D - Comb$ | $\langle \alpha \cap \beta \rangle A \equiv (\langle \alpha \rangle A \wedge \langle \beta \rangle A)$ |
| $D - Mix$ | $(A \vee \langle \alpha \rangle \langle *\alpha \rangle A) \supset \langle *\alpha \rangle A$ |
| $D - Ind$ | $[*\alpha](\langle \alpha \rangle A \supset A) \supset (\langle *\alpha \rangle A \supset A)$ |
| $D - Test$ | $\langle ?A \rangle B \equiv (A \wedge B)$ |
| $D - negComp$ | $\langle -(\alpha; \beta) \rangle A \equiv \langle -\alpha \rangle \langle -\beta \rangle A$ |
| $D - negAlt$ | $\langle -(\alpha \cup \beta) \rangle A \equiv (\langle -\alpha \rangle A \wedge \langle -\beta \rangle A)$ |
| $D - negComb$ | $\langle -(\alpha \cap \beta) \rangle A \equiv (\langle -\alpha \rangle A \vee \langle -\beta \rangle A)$ |
| $D - negMix$ | $(A \vee \langle -\alpha \rangle \langle - * \alpha \rangle A) \supset \langle - * \alpha \rangle A$ |
| $D - negInd$ | $[- * \alpha](\langle -\alpha \rangle A \supset A) \supset (\langle - * \alpha \rangle A \supset A)$ |
| $D - negTest$ | $\langle -?A \rangle B \equiv (\neg A \wedge B)$ |
| $D - negNeg$ | $\langle - - \alpha \rangle \equiv \langle \alpha \rangle$ |
| | |
| $B - D$ | $[\alpha]\bot \vee \langle \alpha \rangle \top$ |

Table 1
Axioms of CCPDL

(2) $\langle\alpha\rangle\top \in s$ implies $sR_\alpha\|S_\alpha\|$

(3) $s\overline{R_\alpha}t$ iff $s_\alpha \subseteq t$

(4) $[\alpha]A \in s$ iff $sR_\alpha T$ implies $T \subseteq \|A\|$

(Note that the proof of 3. appeals to Axiom $B - D$.) It follows that if there is a $t$ such that $s\overline{R_\alpha}t$, then $\langle\alpha\rangle\top \in s$.

**Lemma 1** Let $A \in Form(\Phi, \Pi)\}$. Then $\mathcal{M}^m, s \models A$ iff $A \in s$.

*Proof.* By induction on $A$, using Theorem 4 (1) and (4). ∎

**Corollary 1** $\{\mathcal{M}^m\} \models A$ iff $\vdash A$.

**Lemma 2** In $\mathcal{M}^m$ the following holds:

(1) $sR_{\alpha;\beta}T$ implies $s(R_\alpha \cdot R_\beta)W$ for some $W \subseteq T$

(2) $sR_{\alpha\cup\beta}T$ implies $s(R_\alpha \cup R_\beta)W$ for some $W \subseteq T$

(3) $R_{\alpha\cap\beta} \subseteq R_\alpha \otimes R_\beta$

(4) Tests are standard: $sR_{?A}T$ iff $T = \{s\}$ and $s \in A^{\mathcal{M}^m}$

(5) Negated tests are standard: $sR_{-?A}T$ iff $T = \{s\}$ and $s \in \neg A^{\mathcal{M}^m}$

(6) Negated negations are standard: $sR_{--\alpha}T$ iff $sR_\alpha T$

(7) $sR_{-(\alpha;\beta)}T$ implies $s(R_{-\alpha} \cdot R_{-\beta})W$ for some $W \subseteq T$

(8) $R_{-(\alpha\cup\beta)} \subseteq R_{-\alpha} \otimes R_{-\beta}$

(9) $sR_{-(\alpha\cap\beta)}T$ implies $s(R_{-\alpha} \cup R_{-\beta})W$ for some $W \subseteq T$.

*Proof.* For (1)–(4), see the proof of [8, Theorem 10.22]. The proof of (5) is similar to the proof of (4). For (6) note that $\|s_\alpha\| = \|s_{--\alpha}\|$. For (7)–(9), the proofs follow patterns of the proofs of (1)–(3). For example, (8): Suppose $sR_{-(\alpha\cup\beta)}T$. Then $T = \|s_{-(\alpha\cup\beta)}\| \cap \|A\|$ for some $A$ such that $\langle-(\alpha\cup\beta)\rangle A \in s$. By $D-negAlt$, $\langle-\alpha\rangle A, \langle-\beta\rangle A \in s$. Therefore, $sR_{-\alpha}(\|s_{-\alpha}\|\cap\|A\|)$ and $sR_{-\beta}(\|s_{-\beta}\|\cap\|A\|)$. Hence $s(R_{-\alpha} \otimes R_{-\beta})U$, where

$$U = (\|s_{-\alpha}\| \cap \|A\|) \cup (\|s_{-\beta}\| \cap \|A\|) = (\|s_{-\alpha}\| \cup \|s_{-\beta}\|) \cap \|A\|.$$

Since $\langle-\alpha\rangle A \in s$ and $\vdash \langle-\alpha\rangle A \supset \langle-\alpha\rangle\top$, we have $\langle-\alpha\rangle\top \in s$. Analogously, $\langle-\beta\rangle\top \in s$. It follows by Theorem 3 (9) that $U = T$. ∎

The following definition gives a representation of action composition by the composition $\circ$ of binary relations. Let $\mathcal{M} = (S, \{R_\alpha : \alpha \in Act(\Phi, \Pi)\}, v)$, be a CCPDL model. Define a family of relations $\{R_\alpha^+ : \alpha \in Act(\Phi, \Pi)\}$ on $S$ inductively as follows:

$$
\begin{aligned}
R_\pi^+ &= \overline{R_\pi} & R_{-\pi}^+ &= \overline{R_{-\pi}} \\
R_{?A}^+ &= \overline{R_{?A}} & R_{-?A}^+ &= \overline{R_{-?A}} \\
R_{*\alpha}^+ &= (R_\alpha^+)^* & R_{-*\alpha}^+ &= R_{*-\alpha}^+ \\
R_{\alpha;\beta}^+ &= R_\alpha^+ \circ R_\beta^+ & R_{-(\alpha;\beta)}^+ &= R_{-\alpha;-\beta}^+ \\
R_{\alpha\cup\beta}^+ &= R_\alpha^+ \cup R_\beta^+ & R_{-(\alpha\cup\beta)}^+ &= R_{-\alpha\cap-\beta}^+ \\
& & R_{--\alpha}^+ &= R_\alpha^+ \\
& & R_{-(\alpha\cap\beta)}^+ &= R_{-\alpha\cup-\beta}^+
\end{aligned}
$$

$sR_{\alpha\cap\beta}^+ t$ iff for some $T$, either

(i) $sR_\alpha^+ t$ and $sR_\beta T$, or

(ii) $sR_\alpha T$ and $sR_\beta^+ t$

**Theorem 5** *In a CCPDL model that is standard except possibly for tests and negated tests, $\overline{R_\alpha} \subseteq R_\alpha^+$.*

*Proof.* By induction on the degree $d(\alpha)$ of $\alpha$, where $d(\alpha)$ is inductively defined by:

$$
\begin{aligned}
d(\pi) &= d(-\pi) = d(?A) = d(-?A) = 1\\
d(*\alpha) &= d(\alpha) + 1\\
d(--\alpha) &= d(\alpha) + 1\\
d(-*\alpha) &= d(-\alpha) + 2\\
d(\alpha \sharp \beta) &= max(d(\alpha), d(\beta)) + 1, \; \sharp \in \{;, \cup, \cap\}\\
d(-(\alpha \sharp \beta)) &= max(d(-\alpha), d(-\beta)) + 2, \; \sharp \in \{;, \cup, \cap\}
\end{aligned}
$$

The case $d(\alpha) = 1$ is obvious. For $d(\alpha) > 1$ and $\alpha \neq -\beta$, see [8, p. 128]. For $d(\alpha) > 1$ and $\alpha = -\beta$, just use the induction hypothesis. For example:

$$
\begin{aligned}
\overline{R_{-(\alpha \cup \beta)}} &= \overline{R_{-\alpha \cap -\beta}} \quad \text{by standardness}\\
&\subseteq R_{-\alpha \cap -\beta}^+ \quad \text{induction hypothesis}\\
&= R_{-(\alpha \cup \beta)}^+ \quad \blacksquare
\end{aligned}
$$

Theorem 5 is used to prove the following theorem, which is applied in the proof of the Filtration Lemma.

**Theorem 6** *Let $\mathcal{M}$ be a model that is standard except possibly for tests and negated tests. For every $\alpha \in Act(\Phi, \Pi)$ and $A \in Form(\Phi, \Pi)$,*

$$
\mathcal{M}, s \models [\alpha]A \quad \text{iff} \quad sR_\alpha^+ t \; \text{implies} \; \mathcal{M}, t \models A.
$$

*Proof.* Analogous to the proof of Theorem 10.24 in [8] using now also the definition of $R_{-\alpha}^+$. The direction from left to right is shown by induction on the degree of $\alpha$. We here consider only negated iterations. Suppose $\mathcal{M}, s \models [-*\alpha]A$. We show that for any $n$,

$$
s(R_{-\alpha}^+)^n t \; \text{implies} \; \mathcal{M}, t \models [-*\alpha]A.
$$

If $n = 0$, then $s = t$ and the claim holds by assumption. Suppose the claim holds for $n$ and assume that $s(R_{-\alpha}^+)^{n+1} t$. Then there exists $u$ with $s(R_{-\alpha}^+)^n u$ and $uR_{-\alpha}^+ t$. By the induction hypothesis on $n$, $\mathcal{M}, u \models [-*\alpha]A$. By standardness for negated iteration, $\mathcal{M}$ verifies $B - negMix$ and thus $\mathcal{M}, u \models [-\alpha][-*\alpha]A$. By the induction hypothesis on $-\alpha$, $\mathcal{M}, t \models [-*\alpha]A$, so that the displayed claim is proved. If now $sR_{-*\alpha}^+ t$, then $s(R_{-\alpha}^+)^* t$ and thus $s(R_{-\alpha}^+)^n t$ for some $n$. Hence $\mathcal{M}, t \models [-*\alpha]A$. Verification of $B - negMix$ implies $\mathcal{M}, t \models A$.        $\blacksquare$

A set of formulas $\Delta$ satisfies the extended Fischer-Ladner conditions iff

$\Delta$ is closed under subformulas

$[?A]B \in \Delta \Rightarrow A \in \Delta$

$[\alpha; \beta]B \in \Delta \Rightarrow [\alpha][\beta]B \in \Delta$

$[\alpha \cup \beta]B \in \Delta \Rightarrow [\alpha]B, [\beta]B \in \Delta$

$[\alpha \cap \beta]B \in \Delta \Rightarrow [\alpha]B, [\beta]B, \langle\alpha\rangle\top, \langle\beta\rangle\top \in \Delta$

$[*\alpha]B \in \Delta \Rightarrow [\alpha][*\alpha]B \in \Delta$

$[-?A]B \in \Delta \Rightarrow \neg A \in \Delta$

$[-(\alpha;\beta)]B \in \Delta \Rightarrow [-\alpha][-\beta]B \in \Delta$

$[-(\alpha \cup \beta)]B \in \Delta \Rightarrow [-\alpha]B, [-\beta]B, \langle -\alpha \rangle \top, \langle -\beta \rangle \top \in \Delta$

$[-(\alpha \cap \beta)]B \in \Delta \Rightarrow [-\alpha]B, [-\beta]B \in \Delta$

$[- * \alpha]B \in \Delta \Rightarrow [-\alpha][- * \alpha]B \in \Delta$

$[- - \alpha]B \in \Delta \Rightarrow [\alpha]B \in \Delta$

$\langle ?A \rangle B \in \Delta \Rightarrow A \in \Delta$

$\langle \alpha;\beta \rangle B \in \Delta \Rightarrow \langle \alpha \rangle \langle \beta \rangle B \in \Delta$

$\langle \alpha \cup \beta \rangle B \in \Delta \Rightarrow \langle \alpha \rangle B, \langle \beta \rangle B \in \Delta$

$\langle \alpha \cap \beta \rangle B \in \Delta \Rightarrow \langle \alpha \rangle B, \langle \beta \rangle B \in \Delta$

$\langle * \alpha \rangle B \in \Delta \Rightarrow \langle \alpha \rangle \langle * \alpha \rangle B \in \Delta$

$\langle -?A \rangle B \in \Delta \Rightarrow \neg A \in \Delta$

$\langle -(\alpha;\beta) \rangle B \in \Delta \Rightarrow \langle -\alpha \rangle \langle -\beta \rangle B \in \Delta$

$\langle -(\alpha \cup \beta) \rangle B \in \Delta \Rightarrow \langle -\alpha \rangle B, \langle -\beta \rangle B \in \Delta$

$\langle -(\alpha \cap \beta) \rangle B \in \Delta \Rightarrow \langle -\alpha \rangle B, \langle -\beta \rangle B \in \Delta$

$\langle - * \alpha \rangle B \in \Delta \Rightarrow \langle -\alpha \rangle \langle - * \alpha \rangle B \in \Delta$

$\langle - - \alpha \rangle B \in \Delta \Rightarrow \langle \alpha \rangle B \in \Delta$

**Lemma 3** *For any $A \in Act(\Phi, \Pi)$, there exists a finite set $\Delta$ such that $\Delta$ satisfies the extended Fischer-Ladner conditions and $A \in \Delta$.*

*Proof.* Similar to the proof of the corresponding lemma for PDL [5]. A relation $\prec$ between formulas is defined such that if $[\alpha]A \prec [\beta]B$ or $\langle \alpha \rangle A \prec \langle \beta \rangle B$ , then $\beta \notin \Pi \cup \{-\pi : \pi \in \Pi\}$ and $\alpha$ contains less symbols than $\beta$. For example

> if $B = [-(\alpha;\beta)]C$, then $[-\alpha][-\beta]C \prec B$, $[-\beta]C \prec B$
> if $B = [-?C]D$, then $A \prec B$ for every subformula $A$ of $\neg C$
> if $B = [-(\alpha \cup \beta)]C$, then $[-\alpha]C \prec B$, $[-\beta]C \prec B$, $\top \prec B$
> if $B = [- * \alpha]C$, then $[-\alpha][- * \alpha]C \prec B$

We may then put $\Delta =$ the smallest set of formulas that contains $A$ and is closed under subformulas of $A$ and $\prec$. ∎

Let $\Delta$ be a finite set satisfying the extended Fischer-Ladner closure conditions. For $s, t \in S^m$ define

$s \sim_\Delta t$ iff $s \cap \Delta = t \cap \Delta$,

[i.e., $s \sim_\Delta t$ iff $A \in \Delta$ implies $(\mathcal{M}^m, s \models A$ iff $\mathcal{M}^m, t \models A)$]

$|\, s \,| = \{t \in S^m : s \sim_\Delta t\}$,

$S_\Delta = \{|\, s \,| : s \in S^m\}$.

For $T \subseteq S^m$ and $X \subseteq S_\Delta$ define

$|\, T \,| = \{|\, s \,| : s \in T\}$, $\quad S_X = \{s \in S^m : |\, s \,| \in X\}$.

The definition of filtrations and the proof of the Filtration Lemma are exactly as in Goldblatt's proof for CPDL [8, p. 130 f.].

**Definition 1** *(Filtrations of the canonical model) Consider*

$$\mathcal{M} = (S_\Delta, \{\rho_\alpha : \alpha \in Act_\Delta\}, V_\Delta),$$

*where $Act_\Delta$ = the smallest set of generic actions that includes all atomic and negated atomic actions and all tests and negated tests occurring in $\Delta$, and that is closed under $-$, $*$, $;$, $\cap$ and $\cup$. Moreover, $V_\Delta : \Phi \cap \Delta \longrightarrow 2^{S_\Delta}$ is defined by*

$$\mid s \mid \in V_\Delta(p) \ \text{iff} \ s \in v^m(p) \ (\text{iff} \ p \in s).$$

*The reachability relation $\rho_\alpha$ is called a $\Delta$-filtration of $R_\alpha$ from $\mathcal{M}^m$ iff*

(B1) $s\overline{R_\alpha}t$ *implies* $\mid s \mid \rho_\alpha^+ \mid t \mid$

(B2) $\mid s \mid \overline{\rho_\alpha} \mid t \mid$ *implies* $\{B : [\alpha]B \in s \cap \Delta\} \subseteq t$

(D1) $sR_\alpha T$ *implies* $\mid s \mid \rho_\alpha X$ *for some* $X \subseteq \mid T \mid$

(D2) *if* $\mid s \mid \rho_\alpha X$ *and* $S_X \subseteq \|B\|$, *then* $\langle \alpha \rangle B \in \Delta$ *implies* $\langle \alpha \rangle B \in s$

*The relation $\rho_\alpha$ is said to be strong if $sR_\alpha T$ implies $\mid s \mid \rho_\alpha \mid T \mid$. Finally, $\mathcal{M}$ is said to be a $\Delta$-filtration of the canonical model $\mathcal{M}^m$ iff for every $\alpha \in Act_\Delta$, $\rho_\alpha$ is a $\Delta$-filtration of $R_\alpha$.*

Any strong relation satisfies not only (D1) but also (B1) if the model $\mathcal{M}$ is standard except possibly for tests and negated tests.

**Lemma 4** *(Filtration Lemma) If $\mathcal{M}$ is a $\Delta$-filtration of $\mathcal{M}^m$ and standard except possibly for tests and negated tests, then for any $A \in \Delta$ and $s \in S^m$,*

$$s \in A^{\mathcal{M}^m} \ \text{iff} \ \mid s \mid \in A^{\mathcal{M}}.$$

Note that since $\Delta$ is finite, $S_\Delta$ is finite, too. Moreover, it can be shown that $\Delta$-filtrations of $R_\alpha$ from $\mathcal{M}^m$ exist, see [8, Theorem 10.28]. What we need to show is that there exists a filtration of $\mathcal{M}^m$ that is a *standard* model.

**Definition 2** *Let $\Delta$ be a finite set satisfying the extended Fischer-Ladner conditions. Define*

$$\mathcal{M}_\Delta = (S_\Delta, \{\rho_\alpha : \alpha \in Act_\Delta\}, V_\Delta)$$

*by requiring that (i) $\rho_\pi$ is any $\Delta$-filtration of $R_\pi$ (ii) $\rho_{-\pi}$ is any $\Delta$-filtration of $R_{-\pi}$, (iii)*

$$\rho_{?A} = \{(\mid s \mid, \{\mid s \mid\}) : s \in A^{\mathcal{M}^m}\} \ \text{and} \ \rho_{-?A} = \{(\mid s \mid, \{\mid s \mid\}) : s \in \neg A^{\mathcal{M}^m}\},$$

*and (iv) otherwise $\rho_\alpha$ is defined by the conditions specifying standard models. Clearly, $\mathcal{M}_\Delta$ is standard except possibly for tests and negated tests.*

Note that the Fischer-Ladner closure conditions for tests and negated tests guarantee that the relations $\rho_{?A}$ and $\rho_{-?A}$ are well-defined.

**Theorem 7** $\mathcal{M}_\Delta$ *is a $\Delta$-filtration of $\mathcal{M}^m$.*

*Proof.* See the Appendix. ∎

**Corollary 2** $\mathcal{M}_\Delta$ *is a standard model.*

*Proof.* By the previous theorem and the Filtration Lemma,

$$\rho_{?A} = \{(\mid s \mid, \{\mid s \mid\}) : s \in A^{\mathcal{M}_\Delta}\} \text{ and } \rho_{-?A} = \{(\mid s \mid, \{\mid s \mid\}) : s \in \neg A^{\mathcal{M}_\Delta}\},$$

so that $\mathcal{M}_\Delta$ is standard for tests and negated tests.                ∎

By completely familiar arguments we obtain the desired result:

**Theorem 8** *The logic CCPDL is (i) decidable and (ii) sound and complete with respect to the class of all standard models.*

## 4. Conclusion

In this paper, modalities from a *constructive* action algebra have been added to *classical* propositional logic. Decidability of the resulting system CCPDL and its completeness with respect to standard models have been shown. I take it that the action algebra of CCPDL is well-motivated: whereas neither complement nor restricted complement are suitable for representing a notion of negative generic *action*, constructive negation in the sense of refraining is. Therefore, CCPDL appears to be a promising system of dynamic logic. It suggests itself as part of a system of description logic for reasoning about actions and obligation. To the extent to which non-classical, constructive description logics may play a role in knowledge representation, see [12], it should be of interest to graft the constructive action algebra of CCPDL also on systems of constructive description logic.

## 5. Appendix

To prove Theorem 7, it must be shown that $\rho_\alpha$ is a $\Delta$-filtration of $R_\alpha$, for every $\alpha \in Act_\Delta$ with $\alpha \notin \Pi \cup \{-\pi : \pi \in \Pi\}$. The proof is by induction on the degree $d(\alpha)$ of $\alpha$. For the cases in which $\alpha$ is not negated, see the proof of Theorem 10.29 in [8]. The proof for negated $\alpha$ follows analogous patterns, which we here partly repeat.

Negated tests. Suppose $-?B \in Act_\Delta$. If $sR_{-?B}T$, by Lemma 2 (5), $T = \{s\}$ and $\mathcal{M}^m, s \models \neg B$. Therefore $\mid T \mid = \{\mid s \mid\}$ and $\mid s \mid \rho_{-?B} \mid T \mid$ by definition of $\rho_{-?B}$. In other words, $\rho_{-?B}$ is strong and hence (B1) and (D1) are satisfied. (B2): Let $\mid s \mid \overline{\rho_{-?B}} \mid t \mid$, so that $\mid s \mid = \mid t \mid$ and $\neg B \in s$. If $[-?B]D \in s \cap \Delta$, then by $B - negTest$, $D \in s$. Since $s \sim_\Delta t$, we have $D \in t$. (D2): Use $D - negTest$.

Negated composition. Suppose that $-(\alpha;\beta) \in Act_\Delta$ and that $\rho_{-\alpha}$ and $\rho_{-\beta}$ are $\Delta$-filtrations of $R_{-\alpha}$ and $R_{-\beta}$. (B1): For every $s \in S^m$, and every $\rho_\alpha^+$, there exists a formula $A_s$ such that

$$A_s \in t \text{ iff } \mid s \mid \rho_\alpha^+ \mid t \mid,$$

see [8, Theorem 9.7]. Therefore, to show that $s\overline{R_\alpha}t$ implies $\mid s \mid \rho_\alpha^+ \mid t \mid$, it is enough to show that $[\alpha]A_s \in s$, because $s\overline{R_\alpha}t$ iff $s_\alpha \subseteq t$, see Theorem 4 (3). To show that $[-(\alpha;\beta)]A_s \in s$, it is enough to show that $[-\alpha][-\beta]A_s \in s$, by $B - negComp$. Now, if $s\overline{R_{-\alpha}u R_{-\beta}}t$, then $\mid s \mid \rho_{-\alpha}^+ \mid u \mid \rho_{-\beta}^+ \mid t \mid$, by (B1) for $-\alpha$ and $-\beta$. Hence $\mid s \mid \rho_{-(\alpha;\beta)}^+ \mid t \mid$ by definition of $\rho_{-(\alpha;\beta)}^+$. Thus, $A_s \in t$. Therefore $[-\alpha][-\beta]A_s \in s$ and, by $B - negComp$, $[-(\alpha;\beta)]A_s \in s$. (B2): Suppose $\mid s \mid \overline{\rho_{-(\alpha;\beta)}} \mid t \mid$. Then by standardness $\mid s \mid \overline{\rho_{-\alpha} \cdot \rho_{-\beta}} \mid t \mid$ and by Theorem 1 (3), $\mid s \mid \overline{\rho_{-\alpha} \circ \rho_{-\beta}} \mid t \mid$. Hence there exists $u$ with $\mid s \mid \overline{\rho_{-\alpha}} u \mid$ and $\mid u \mid \overline{\rho_{-\beta}} \mid t \mid$. If now $[-(\alpha;\beta)]B \in s \cap \Delta$, then by extended Fischer-Ladner closure and

$B - negComp$, also $[-\alpha][-\beta]B \in s \cap \Delta$. By (B2) for $-\alpha$ and $-\beta$, we have $[-\beta]B \in u$ and therefore $B \in t$. (D1): Let $sR_{-(\alpha;\beta)}T$. Then by Lemma 2 (7) and definition of composition, there is a $U \subseteq S^m$ such that $sR_{-\alpha}U$ and for every $u \in U$, there exists $T_u \subseteq T$ with $uR_{-\beta}T_u$. By (D1) for $-\alpha$, there exists $X \subseteq S_\Delta$ such that $| s | \rho_{-\alpha}X$ with $X \subseteq | U |$. If $x \in X$, then $x = | u |$ for some $u \in U$. By (D1) for $-\beta$, there exists $Y_x \subseteq S_\Delta$ such that $x\rho_{-\beta}Y_x \subseteq | T_u | \subseteq | T |$. Putting

$$Z = \bigcup \{Y_x : x \in X\}$$

we obtain $| s | \rho_{-\alpha} \cdot \rho_{-\beta}Z$ and therefore $| s | \rho_{-\alpha;-\beta}Z \subseteq | T |$. (D2): Suppose $S_X \subseteq \|B\|$, $\langle -(\alpha;\beta)\rangle B \in \Delta$ and $| s | \rho_{-(\alpha;\beta)}X$. By standardness for negated compositions, $| s | \rho_{-\alpha} \cdot \rho_{-\beta}X$. In other words, there exists $Y \subseteq S_\Delta$ such that $| s | \rho_{-\alpha}Y$, $X = \bigcup \{X_y : y \in Y\}$, and $y\rho_{-\beta}X_y$ for all $y \in Y$. If $t \in S_Y$, then $| t | \in Y$ and $S_{X_{|t|}} \subseteq S_x \subseteq \|B\|$. Since $| t | \rho_{-\beta}X_{|t|}$ and, by extended Fischer-Ladner closure, $\langle -\beta\rangle B \in \Delta$, by (D2) for $-\beta$, we obtain $\langle -\beta\rangle B \in t$. Therefore $S_Y \subseteq \|\langle -\beta\rangle B\|$. Since, by extended Fischer-Ladner closure $\langle -\alpha\rangle\langle -\beta\rangle \in \Delta$, and $| s | \rho_{-\alpha}Y$, (D2) for $-\alpha$ gives $\langle -\alpha\rangle\langle -\beta\rangle B \in s$. By $D - negComp$, finally, $\langle -\alpha;-\beta\rangle B \in s$.

Negated choice. (B1): Let $A_s$ be such that

$$A_s \in t \text{ iff } | s | \rho^+_{-(\alpha \cup \beta)}| t |$$

We show that

$$(\ddagger) \quad (\langle -\alpha\rangle\top \supset [-\beta]A_s), (\langle -\beta\rangle\top \supset [-\alpha]A_s) \in s,$$

from which by $B - negAlt$, the desired $[-(\alpha \cup \beta)]A_s \in s$ follows. Suppose $\langle -\alpha\rangle\top \in s$. Then for some $T$, $sR_{-\alpha}T$. By (D1) for $-\alpha$, $| s | \rho_{-\alpha}X$ for some $X$. If now $s\overline{R}_{-\beta}t$, by (B1) for $-\beta$, $| s | \rho^+_{-\beta}| t |$. Thus, by definition of $\rho^+_{-(\alpha \cup \beta)}$, $| s | \rho^+_{-(\alpha \cup \beta)}| t |$. Therefore $A_s \in t$ and hence $[-\beta]A_s \in s$. The proof that $(\langle -\beta\rangle\top \supset [-\alpha]A) \in s$ is similar. (B2): Suppose $| s |$ $\overline{\rho_{-(\alpha \cup \beta)}} | t |$, i.e., $| s | \overline{\rho_{-\alpha \cap -\beta}} | t |$, i.e., $| s | \overline{\rho_{-\alpha} \otimes \rho_{-\beta}} | t |$. Then there are $X, Y$ such that $| s | \rho_{-\alpha}X$, $| s | \rho_{-\beta}Y$, and $| t | \in X$ or $| t | \in Y$. Let $[-(\alpha \cup \beta)]B \in s \cap \Delta$. By extended Fischer -Ladner closure, $\langle -\alpha\rangle\top$, $\langle -\beta\rangle\top \in \Delta$. Because $S_X$, $S_Y \subseteq \|\top\|$, by (D2) for $-\alpha$ and $-\beta$ we obtain $\langle -\alpha\rangle\top$, $\langle -\beta\rangle\top \in s$. Axiom $B - negAlt$ gives $[-\alpha]B$, $[-\beta]B \in s$. If $| t | \in X$, then $| s | \overline{\rho_{-\alpha}} | t |$ and hence $B \in t$ by (B2) for $-\alpha$. Similarly, $| t | \in Y$ implies $B \in t$. (D1): If $sR_{-(\alpha \cup \beta)}T$, by Lemma 2 (7), there exist $W_1, W_2$ such that $sR_{-\alpha}W_1$, $sR_{-\beta}W_2$, and $T = W_1 \cup W_2$. Because of (D1) for $-\alpha$ and $-\beta$, there exist $X_1, X_2$ such that $| s | \rho_{-\alpha}X_1 \subseteq | W_1 |$ and $| s | \rho_{-\beta}X_2 \subseteq | W_2 |$. Thus, by standardness for negated choices,

$$| s | \rho_{-(\alpha \cup \beta)}(X_1 \cup X_2) \subseteq | W_1 | \cup | W_2 | \subseteq | T |.$$

(D2): Suppose $| s | \rho_{-(\alpha \cup \beta)}X$, $S_X \subseteq \|B\|$, and $\langle -(\alpha \cup \beta)\rangle B \in \Delta$. By standardness for negated choices, there exist $Y, Z$ such that $| s | \rho_{-\alpha}Y$, $| s | \rho_{-\beta}Z$, and $X = Y \cup Z$. By extended Fischer-Ladner closure, $\langle -\alpha\rangle B$, $\langle -\beta\rangle B \in \Delta$. Since, moreover, $S_Y$, $S_Z \subseteq S_X \subseteq \|B\|$, by (D2) for $-\alpha$ and $-\beta$, $\langle -\alpha\rangle B$, $\langle -\beta\rangle B \in s$. By axiom $D - negAlt$, $\langle -(\alpha \cup \beta)\rangle B \in s$.

Negated combination. Fairly straightforward.

Negated iteration. (B1): Let $A_s$ be such that $A_s \in t$ iff $| s | \rho^+_{-*\alpha}| t |$. Then $A_s \in s$ because $| s | (\rho^+_{-\alpha})^0 | s |$. It suffices to show that $[- * \alpha]A_s \in s$. To show the latter, it is enough to prove $\vdash A_s \supset [-\alpha]A_s$, since by necessitation for $[- * \alpha]$, $\vdash [- * \alpha](A_s \supset [-\alpha]A_s)$, and then by $B - negInd$, $(A_s \supset [- * \alpha]A_s) \in s$. Now, suppose $A_s \in t \in S^m$. By definition of $\rho^+_{-*\alpha}$, $| s | (\rho^+_{-\alpha})^* | t |$, and therefore, $| s | (\rho^+_{-\alpha})^n | t |$ for some $n \geq 0$. If $t\overline{R}_{-\alpha}u$, then by (B1) for $-\alpha$, $| t | \rho^+_{-\alpha}| u |$, and so $| s | (\rho^+_{-\alpha})^{n+1}| u |$. Therefore, $| s | \rho^+_{-*\alpha}| u |$ and hence $A_s \in u$. Therefore $[-\alpha]A_s \in t$. (B2): By definition of $\rho^+_{-*\alpha}$ and Theorem 1 (7), it suffices to show that $| s |(\overline{\rho^{+}_{-\alpha}})^{*}| t |$ implies $\{B : [- * \alpha]B \in s \cap \Delta\} \subseteq t$. To this end, it is shown that for every $n \geq 0$,

$$(\dagger\dagger) \quad | s |(\overline{\rho^{+}_{-\alpha}})^{n}| t | \text{ implies } \{[- * \alpha]B : [- * \alpha]B \in s \cap \Delta\} \subseteq t,$$

because then $| s |(\overline{\rho^{+}_{-\alpha}})^{*}| t |$ gives $| s |(\overline{\rho^{+}_{-\alpha}})^{n}| t |$ for some $n$, and $[- * \alpha]B \in s \cap \Delta$ implies $[- * \alpha]B \in t$. Using $B - negMix$, we obtain $B \in t$. $(\dagger\dagger)$ is proved by induction on $n$. If $n = 0$, $s \cap \Delta = t \cap \Delta$, and the claim is obvious. Suppose that the claim holds for $n$, and assume that $| s |(\overline{\rho^{+}_{-\alpha}})^{n+1}| t |$. Then for some $u$, $| s |(\overline{\rho^{+}_{-\alpha}})^{n}| u |$ and $| u |(\overline{\rho^{+}_{-\alpha}})| t |$. By the induction hypothesis for $n$, $[- * \alpha]B \in s \cap \Delta$ implies $[- * \alpha]B \in u$. By $B - negMix$ and extended Fischer-Ladner closure, $[-\alpha][- * \alpha]B \in u \cap \Delta$. But then, by (B2) for $-\alpha$, $[- * \alpha]B \in t$. (D1): If $T \subseteq S^{m}$, then let $A_{T}$ be such that for every $s \in S^{m}$,

$$A_{T} \in s \text{ iff } | s |\rho_{-*\alpha}X \text{ for some } X \subseteq | T |.$$

It is enough to show

$$(\natural) \quad T \subseteq \|A_{T}\| \quad \text{and} \quad (\natural\natural) \quad \vdash \langle -\alpha\rangle A_{T} \supset A_{T}.$$

If we then have $sR_{-*\alpha}T$, by $(\natural)$ and Theorem 4 (1), $\langle - * \alpha\rangle A_{T} \in s$. Necessitation for $- * \alpha$, $(\natural\natural)$, and $D - negInd$ yield $\vdash \langle - * \alpha\rangle A_{T} \supset A_{T}$. Therefore $A_{T} \in s$ and, by our assumption, $| s |\rho_{-*\alpha}X$ for some $X \subseteq | T |$. $(\natural)$: If $t \in T$, then $\{| t |\} \subseteq | T |$. Since $Id \subseteq \rho^{(*)}_{-\alpha} = \rho_{-*\alpha}$, $| t | \rho_{-*\alpha}\{| t |\}$. For $X = \{| t |\}$, we obtain $A_{T} \in t$ and thus $t \in \|A_{T}\|$. To verify $(\natural\natural)$, suppose that $s \in S^{m}$ and $\langle -\alpha\rangle A_{T} \in s$. Then, by Theorem 4 (1), $sR_{-\alpha}U$ for some $U \subseteq \|A_{T}\|$. (D1) for $-\alpha$ yields $| s |\rho_{-\alpha}X$ for some $X \subseteq | U |$. Since $S_{\Delta}$ is finite, there exists $k \in \omega$ such that $X = \{| u_{0} |, \ldots, | u_{k-1} |\}$, for some $u_{0}, \ldots, u_{k-1} \in U$. Since $U \subseteq \|A_{T}\|$, $A_{T} \in u_{i}$, for every $u_{i}$ with $0 \leq i < k$. Therefore, $| u_{i} |\rho_{-*\alpha}Y_{i}$ for some $Y_{i} \subseteq | T |$. By standardness for negated iterations, $| u_{i} |\rho^{n_{i}}_{-\alpha}Y_{i}$ for some $n_{i}$. Denote by $n$ the maximum of $n_{0}, \ldots, n_{k-1}$. Since $l \leq m$ implies $R^{(l)} \subseteq R^{(m)}$, for every $i < k$, $| u_{i} |\rho^{n}_{-\alpha}Y_{i}$. For $Y = \bigcup\{Y_{i} : 0 \leq i < k\}$, $| s |(\rho_{-\alpha} \cdot \rho^{(n)}_{-\alpha})Y$. Hence $| s |\rho^{n+1}_{-\alpha}Y$ and hence $| s |\rho^{(*)}_{-\alpha}Y$ and $A_{T} \in s$. (D2): Suppose $| s |\rho_{-*\alpha}X$. Since then $| s |\rho^{(n)}_{-\alpha}X$ for some $n$, it is enough to show that for all $n$, and every $s \in S^{m}$,

$$(\ddagger\ddagger) \quad \text{if } | s |\rho^{(n)}_{-\alpha}X \text{ and } S_{X} \subseteq \|B\|, \text{ then } \langle - * \alpha\rangle B \in \Delta \text{ implies } \langle - * \alpha\rangle B \in s.$$

If $n = 0$, then $| s |\rho^{(0)}_{-\alpha}X$ implies $X = \{| s |\}$, and thus $s \in S_{X}$. Since $S_{X} \subseteq \|B\|$, we have $B \in s$ and then $\langle - * \alpha\rangle B \in s$, by axiom $D - negMix$. Suppose now that $(\ddagger\ddagger)$ holds for $n$, and assume that $| s |\rho^{(n+1)}_{-\alpha}X$, $S_{X} \subseteq \|B\|$, and $\langle - * \alpha\rangle B \in \Delta$. If $| s |\rho^{(0)}_{-\alpha}X$, the wanted result follows as before. If $n \neq 0$, then $| s |(\rho_{-\alpha} \cdot \rho^{(n)}_{-\alpha})X$. In that case, there is a $Y$ such that $| s |\rho_{-\alpha}Y$ and $X = \bigcup\{X_{y} : y \in Y\}$ with $y\rho^{(n)}_{-\alpha}Y$ for every $y \in Y$. Since $| t |\rho^{(n)}_{-\alpha}X_{|t|}$ and $S_{X_{|t|}} \subseteq S_{X} \subseteq \|B\|$, the induction hypothesis on $n$ yields $\langle - * \alpha\rangle B \in t$. Therefore, $S_{Y} \subseteq \|\langle - * \alpha\rangle B\|$. By extended Fischer-Ladner closure, $\langle -\alpha\rangle\langle - * \alpha\rangle B \in \Delta$, and since $| s |\rho_{-\alpha}Y$, (D2) for $-\alpha$ gives $\langle -\alpha\rangle\langle - * \alpha\rangle B \in s$. By $D - negMix$, it follows that $\langle - * \alpha\rangle B \in s$.

Negated negation. (B1): Let $s\overline{R_{--\alpha}}t$. Since $\mathcal{M}^{m}$ is standard for negated negations, $s\overline{R_{\alpha}}t$, and hence, by (B1) for $\alpha$ and definition of $\rho^{+}_{--\alpha}$, $| s |\rho^{+}_{--\alpha}| t |$. (B2): If $| s | \overline{\rho_{--\alpha}}| t |$, then $| s | \overline{\rho_{\alpha}} | t |$. By $B - negNeg$, $[- - \alpha]B \in s$ iff $[\alpha]B \in s$. (B2) for $\alpha$ gives $\{B : [- - \alpha]B \in s \cap \Delta\} \subseteq t$. (D1): Suppose $sR_{--\alpha}T$. By standardness of $\mathcal{M}^{m}$ for negated negations, $sR_{\alpha}T$. By (D1) for $\alpha$, $| s |\rho_{\alpha}X$ for some $X \subseteq | T |$. By definition of $\mathcal{M}_{\Delta}$, $| s |\rho_{--\alpha}X$ for some $X \subseteq | T |$. (D2): Suppose $| s |\rho_{--\alpha}X$, $S_{X} \subseteq \|B\|$, and $\langle - - \alpha\rangle B \in \Delta$. Then $| s |\rho_{\alpha}X$ and, by extended Fischer-Ladner closure, $\langle\alpha\rangle B \in \Delta$. By (D2) for $\alpha$, $\langle\alpha\rangle B \in s$ and hence, by $D - negNeg$, $\langle - - \alpha\rangle B \in s$. ∎

**REFERENCES**

1. N. Belnap, M. Perloff, and M. Xu, *Facing the Future*, Oxford University Press, Oxford, 2001.
2. J. van Benthem, Minimal Deontic Logics, `Bulletin of the Section of Logic* 8 (1979), 36–42.
3. J. Broersen, Relativized Action Complement for Dynamic Logics, in: P. Balbianai et al. (eds.), *Advances in Modal Logic, Vol. 4*, King's College Publications, London, 2003, 51–69.
4. J. Broersen, *Modal Action Logics for Reasoning about Reactive Systems*, PhD thesis, Free University of Amsterdam, 2003.
5. M. Fischer and R. Ladner, Propositional Dynamic Logic of Regular Programs, *J. of Computer and System Sceinces* 18 (1979), 194–211.
6. N. Foo, D. Zhang, Y. Zhang, S. Chopra and B. Vo, Encoding Solutions of the Frame Problem in Dynamic Logic, in: T. Eiter, W. Faber, and M. Truszczynski (eds.), *Logic Programming and Nonmonotonic Reasoning (LPNMR'01)*, LNAI 2173, Springer-Verlag, Berlin, 2001, 240–253.
7. G. de Giacomo and M. Lenzerini, PDL-based Framework for Reasoning about Action, in: *Proc. the 4th Congress of the Italian Association for Artificial Intelligence (AI\*IA'95)*, Lecture Notes in AI 992, Springer-Verlag, Berlin, 1995, 103–114.
8. R. Goldblatt, *Logics of Time and Computation*, CSLI Lecture Notes No. 7, Stanford, second edition, 1992.
9. D. Harel, D. Kozen, and J. Tiuryn, *Dynamic Logic*, MIT Press, Cambridge/Massachusetts, 2000.
10. J. Horty, *Agency and Deontic Logic*, Oxford University Press, Oxford, 2001.
11. J.-J. Meyer, A Different Approach to Deontic Logic, *Notre Dame J. of Formal Logic* 29 (1988), 109–136.
12. S.P. Odintsov and H. Wansing, Inconsistency-tolerant Description Logic: Motivation and Basic Systems, in: V.F. Hendricks and J. Malinowski (eds.), *Trends in Logic: 50 Years of Studia Logica*, Kluwer Academic Publishers, Dordrecht, 2003, 301–335.
13. D. Peleg, Concurrent Dynamic Logic, *J. Assoc. Comput. Mach.* 34 (1987), 450–479.
14. K. Segerberg, Outline of a Logic of Action. In: F. Wolter et al. (eds.), *Advances in Modal Logic. Vol. 3*, World Scientific Publishers, Singapore/London, 2002, 365–387.
15. H. Wansing, Actions and Preventions, in: J. Faye, U. Scheffler, and M. Urchs (eds.), *Logic and Causal Reasoning*, Akademie Verlag, Berlin, 1994, 131–140.
16. D. Wijesekera and A. Nerode, Tableaux for Constructive Concurrent Dynamic Logic, 2001, to appear in: *Theoretical Computer Science*.
17. G.-H. von Wright, *Norm and Action: A Logical Inquiry*, Routledge and Kegan Paul, London, 1963.

# Section B:

# General Philosophy of Science

# Case-Based Reasoning in the Biomedical and Human Sciences: Lessons from Model Organisms[1]

Rachel A. Ankeny

*Director and Senior Lecturer, Unit for History and Philosophy of Science, University of Sydney, Carslaw F07, Sydney 2006 NSW AUSTRALIA*
*rankeny@science.usyd.edu.au*

**Abstract.** This paper examines the epistemological role of case-based reasoning in the biological sciences, focusing on model organisms as a central example. I argue that many of the same goals served by using case studies in medicine are implicit in the design of and principles underlying the early stages of model organism research, particularly in developmental and molecular biology. Focusing on cases and models as components of scientific reasoning points us to the importance of coming to a fuller understanding of medicine (and other fields where case-based reasoning is ubiquitous) in order to have a more precise and richer account of scientific reasoning in biology and the human sciences.

## 1. Introduction

Human sciences, such as medicine and psychology, have an extensive history of use of the case study as an object through which knowledge and explanations are created. The case is a way to capture clinical and empirical data and communicate findings to other practitioners and researchers, among other purposes. Although this type of reasoning has been examined in a variety of disciplinary contexts, little work has been done to connect it to the parallel use of this form of reasoning within the biological sciences, despite increased attention in recent years to 'model organisms' which I have argued elsewhere are a relatively recent concept and depend on reasoning via cases (e.g., [2]).[2]

This paper examines the epistemological role of case-based reasoning in the biological sciences, focusing on model organisms as a central example. I argue that many of the same goals served by using case studies to communicate information in medicine can be seen as implicit in the design of and principles underlying the early stages of model organism research, particularly in developmental and molecular biology. Not surprisingly, this research also is subject to many of the same epistemic limitations that have come to be recognized in other contexts where case-based reasoning is prevalent. Focusing on cases and models as components of scientific reasoning and explanation also helps to reveal the connections between the biomedical and human sciences and other fields of science. It points us to the importance of coming to a fuller understanding of medicine (and other fields where case-based reasoning is ubiquitous) in order to have a more precise and richer account of scientific reasoning in biology and the human sciences, rather than isolating medicine as a messy field with distinct concerns, fundamentally different than the pure (or more precisely, less applied) sciences.

The paper first reviews the literature on the case as used in medicine, and then explores previous discussions of the use of cases in science by philosophers of science, especially as related to the philosophy of medicine. Based on my previous historical and philosophical research on model organisms (see, e.g., [1,3], and forthcoming), I outline the stages of research necessary to make and work with a model organism, and argue that this process clearly reflects the use of case-based reasoning. In conclusion, the history and philosophy of the biomedical sciences is reflectively examined, and especially the currently burgeoning field of model organism studies, to show how our focus on cases is both essential and problematic, and how increased attention to case-based reasoning within science could profitably shift philosophical discussions about the biomedical sciences.

Some cautionary notes on what will *not* be explored in this paper. First, the arguments presented should not be interpreted as implying that *all* or even *most* reasoning in the human and biomedical sciences occurs via cases, only as providing support for the claim that case-based reasoning plays a central role in the pursuit and communication of knowledge in these fields. Second, my focus is on science as practiced, drawing on literature from the history of these sciences. Hence the claims made are descriptive and epistemological, not normative or prescriptive, although some of the potential limitations as well as advantages inherent in case-based reasoning have implications for methodological principles that could serve as guides for fruitful scientific practice.

## 2. The Role of the Case in Medicine

Medicine has a notoriously complex and conflicted relationship with regard to its use of cases and their epistemic status. On the one hand, case studies and reports, and at the extreme, so-called 'syndrome' letters or pedagogical anecdotes (see [24,25]) remain essential ways of providing information about particular clinical phenomena, usually as observed in a single or a few individuals under uncontrolled circumstances [34]. They allow practitioners to recognize similar patterns as new patients present themselves, and to expand their background knowledge beyond their experiences of the typical or the usual in the clinic. On the other hand, single cases are seen as problematic in as much as they are deviations even from the norm of what is abnormal, as it were. They are exceptions rather than rules, and heighten the practitioners' awareness that their field is in fact a 'science of particulars' (to use the term coined by [19]), or even as often claimed, an art rather than a science.

This tension is associated with an underlying epistemic issue in medicine. Its practice is focused on individual human beings and their health, and the particularities inherent in instances of illness and its treatment or care. Where science is characterized (or perhaps caricatured) as based in analysis, controlled experimentation, and so on, medicine clearly must rely primarily on the clinical and the empirical as the knowledge basis from which explanations are developed. Due to ethical and practical limitations, oftentimes the types of experimental trials that are traditionally claimed as a hallmark of any 'scientific' field are not available to it, or only available through extrapolation from experimentation with non-human subjects. Although there are of course theories and explanations (understood in the more traditional sense) that underlie the practice of medicine, applying theories to individuals who present in the clinic is not something that is easily taught. It is a sort of practical wisdom (*phronesis*), that comes from recognition and practice with regard to the particulars, and it is arguable that in this domain there is not knowledge (*episteme*) in its purest Aristotelian sense (see [16], 4), but rather knowledge of a weaker sort, that which is constantly evolving, incomplete, and uncertain (e.g., see [25], 241).

Although scientific knowledge may also be useful in addressing practical questions, when the process involved is situation-dependent and associated with a particular case, general knowledge is too abstract and not specific enough (see [26], 229–230). Explanations tend toward explanation sketches with limited domains of applicability. Nonetheless, there is a rational method at play in the practice of medicine, and differential diagnosis and other medical reasoning processes certainly have the status of scientific methods for their practitioners. As has been noted in different ways by commentators on medical reasoning, the underlying epistemology of medical diagnosis and treatment is less like that of the 'hard' sciences and more similar to that of the social sciences, including history, in that hypotheses (if they are ever explicitly formulated) typically come from observations ([25], 244; see also [7]); observations then fit together into larger patterns. A similar trend has been described in the history of the use of statistics and epidemiological patterns in the practice of medicine and the social sciences: in the early 1800s, there was recognition that seemingly isolated facts could be grouped together so that they revealed patterns (e.g., [12]). Thus the starting point for additional investigations of such patterns, and eventually hypothesis testing followed by the development of explanations, is a rich, descriptive case.

It is helpful to provide a brief review of the general form of case-based reasoning as used in medicine and elsewhere to ground discussion of the comparison to the biomedical sciences (on case-based reasoning generally, see [27]). The basic method is that an index case is constructed, in more or less detail depending on the goals of the particular situation, and then retrieved (or perhaps retrieved in a general sense as an approximation of an index case, and its relevant details then articulated so that the case can be utilized in the current context). Then the new instance or phenomenon of interest is examined in relation to the index case. The result is a feedback loop between processes of justification of the fit between the two, as well as adaptation of the index case as appropriate, particularly via assessment of similarity and identity relations. Physicians learn to take in a wide range of clinical data presented by a patient, differing radically in terms of kind, relevance, and reliability ([7]; see also [15]), and then synthesize them with background knowledge of index cases in order to reach the necessary conclusions regarding actions such as diagnosis or treatment.

Therefore in medicine when a case study is published, it usually highlights a new disorder that a practitioner has been unable to map onto existing disease categories. The details provided in the initial description of this base case highlight what is thought to be essential to understanding this case as distinct and also for identifying other cases of the same disorder. Oftentimes no hypotheses or explanations are provided about the mechanisms of disease causation. In the background is the index or paradigm case, which is that of the human being who is 'normal' with respect to the abnormal features noted in the index case. For instance, consider the disease cystic fibrosis (for a review of this case study in the context of the reduction of diseases in molecular medicine, see [4]). Initially what was noticed was a cluster of disease symptoms not found in normal persons: difficulties breathing caused by pulmonary fluid accumulation; pancreatic malfunction; and so on. Eventually it was found that the disease was genetically associated, caused by the inheritance of two mutated copies of a particular allele.

Over time, as evidence develops, what serves as the index case or paradigm for this disorder may in fact be altered; essential features may prove to be incidental or unique to an individual. So in the case of cystic fibrosis, it was known that many men with the disease also were infertile (due to congenital absence of the vas deferens). However once the allele associated with cystic fibrosis was identified, infertility clinics found that many men without any or only very mild pulmonary disease in fact had mutations associated with cystic fibrosis. Consequently, there was the need to revise the 'spectrum' of disease associated with cystic fibrosis, to include what was called 'genital cystic fibrosis.'

Finally, what we consider to be the normal index case may also be altered over time, as we discover the range of variants or errors in what we had assumed to be the shared or common attributes. What is essential to this form of reasoning is the feedback loop that exists between the index case (the descriptive model of the normal) and the case of interest (the abnormal condition). Newly gathered evidence can change what is considered to be the index case or whether something should be considered a unique case at all. As the causal links are articulated between the index case and the case of interest, explanations and theories can begin to be generated. In section 4, I return to this form of case-based reasoning to examine model organism work, but first in section 3, examine previous arguments that focus on cases as an epistemic stage in theory/explanation generation in science.

## 3. Cases in the History and Philosophy of Science

Suggestions about recognizing the case as an exemplar form of reasoning the philosophy of science date back to discussions in the 1970s about the relevance for philosophy of medicine for the philosophy of science (among others, [36]; [37]), but these arguments were not pursued in any detail (a notable exception is Schaffner's work, e.g. [31,32]). Of course the importance of the case has been recognized in different domains within HPS, for instance implicitly in discussions of statistical thinking and probabilistic methods by Ian Hacking [20,21] among others (e.g., see also [28,13,18]) and explicitly by John Forrester [16] in his work on the human sciences, particularly psychoanalysis. They have recognized ways in which our understanding of the relevant level of investigation has transformed in the $20^{th}$ century as well as the tensions that are implicit in various scientific developments in this period. For instance, we know that species exist in some sense as general or type categories, but in reality they are groupings of individuals that vary even within the species and that are in many sense contingent outcomes of the evolutionary process. So we quickly encounter questions about what we can know about: can we have knowledge of that which is not general or universal? Even if we can in some sense know about individual instances, and indeed medicine and psychology among other fields must consider and treat individuals, this knowledge is not the sort that the philosophy of science normally takes as part of its domain.

On the other hand, there has been growing awareness of the role of idealization in science (e.g., [11]) and the realization of the need to consider how science comes to recognize and characterize phenomena, as distinct from the raw material that constitute the input data (e.g., [9,39]; although I find this distinction sometimes confusing due to differing uses of the terms by practicing scientists, it is useful for my purposes here). This sort of transformation from data to phenomena is parallel to the process that takes place in the biomedical sciences; individuals provide raw data that through collation and comparison with additional data begin to form what we might call phenomena, or more simply, a body of knowledge about which theories and explanations can be formulated. Medical practice needs to respond primarily to individuals, and philosophical examinations of these practices may indeed be different in kind from the most traditional philosophy of science. However, such philosophy is not different from many other philosophical and historical examinations of the processes of *doing* science, with similar concerns about interplay between science and values, tensions introduced by applications of science, and so on.

So too is much of medical teaching based on case studies, from the clinical case as used in the early $20^{th}$ century in medical schools and journals, to more contemporaneous 'problem-based learning' techniques. Some have argued (notably [35]) that in fact in a world of pluralism with regard to values that play an essential role in medical practice, we must resort to casuistic or case-based reasoning. Similarly, as I argue in the next section, we have discovered that we must focus on cases when analyzing particular practices within the biomedical sciences, in large part because of the various values and complexities that enter into any biological science viewed as a practice, but particularly work with model organisms.

More generally, what are the main characteristics of a science or a scientific practice which reveal that case-based reasoning is in play? Theories of case-based reasoning from artificial intelligence, law, and elsewhere (e.g., [30,22,27,40]) point to the intuition that the key is reliance on situations that recur, but with slightly different features. Cases are instances of specific knowledge in particular situations, but that allow operationalization of tasks, methods, and applications. They permit us to in a sense capture or systematize information or data that is too complex to summarize in a general model of the sort we typically envision in traditional philosophy of science, especially when we use mechanistic physical models as our paradigms. Then what are the hallmarks of the types of science or scientific investigations that might be best served by application of a case-based reasoning approach? There are at least five main characteristics of scientific practice, one or more of which might indicate case-based approaches are explicitly or implicitly at play:

(1) focus on entities that have historical evolutionary processes underlying them or on non-static, dynamic processes;

(2) focus on entities that are in some way not accessible through traditional intervention-style experimentation, due to reasons that are ethical (e.g., medicine), practical (e.g., astrophysics), and so on;

(3) focus on entities or systems that are overly complex or variable, at least at the outset (types of immature or developing science, or what have been termed 'weak-theory domains');

(4) focus on entities that are unique or nearly unique, such as science that examines a discrete phenomenon such as a particular rock formation; and/or

(5) the need for or the prevalence of applications of the science as the primary examples within a field.

This way of viewing the prospects of case-based reasoning then includes in its potential purview everything from much of medicine and biology to paleontology, geology, meteorology, materials science, and so on. But it also takes us to model organisms, in as much as they are characterized at least by the first three of these characteristics, and helps to explain much about their privileged epistemic status in recent years both in the practice of the biomedical sciences and in the philosophy and history of biology.

## 4. Model Organism as Cases

Recent life science is faced with dilemmas similar to that confronted by medicine in its practice. There is a desire to get to the fundamental biological characteristics shared by all living things, be they biochemical, genetic, developmental, or neurobiological processes. At the same time, biologists are aware that any model system selected may be problematic and atypical, particularly inasmuch as such systems are proving to be complex in ways previously unanticipated. My previous research on the nematode worm *Caenorhabditis elegans* as a model organism has shown various ways in which the organism as studied by biologists is an abstract entity, in effect the sum of a series of cases about genetics, development, and neurobiology which allows construction of a generalization that can be worked with, understood, modified, and applied to further investigations [1]. This strategy of using the organism as a case serves as a means of control of complexity, a way to create an appropriately simplistic yet descriptively rich basis for future studies and more traditional hypothesis testing, experimentation, and explanation. This is part of a larger story about how a lowly, soil-dwelling worm has become what some consider to be the most thoroughly investigated organism on earth, but more importantly how such an object has been transformed from a natural entity to take on epistemological significance. In other words, how has the worm as a model organism become a thing 'to know with'?

An overview of the stages involved in the life of a model organism is required to answer this question. I focus here on those stages that are most relevant for more general questions about modeling and case-based reasoning (and as a result, do not discuss stage 3, generation of explanations, in detail), and also necessarily simplify the relevant biological details. Note that not all of these are *necessary* conditions for something to be a model organism, nor do I wish to argue that together they form *sufficient* conditions, although they have been significant and most of them arguably have been essential to the historical development of this particular model organism. Note also that the ordering of the stages is *not* meant to indicate causal relations (with the previous stages in any sense producing subsequent ones), though there are some temporal dependencies inherent in the ordering of the process. For instance, standardization of an organism typically precedes development of descriptive models, or at least successful attempts to do so. In this particular case study, the stages as outlined here are chronological, with each corresponding roughly to a particular decade with the first stage starting at around 1964, the beginning of the worm as we know it today.

Stage 1, choice and standardization, is a familiar theme in much historical work on model organisms. For our purposes, we need only note that the story of the standardization of *C. elegans* differs in a significant way from most of these histories (see [3]; also cf. [14]). The researchers involved set out with the goal of trying to 'optimize' an organism in large part through making a careful *organismal choice* to begin with, rather than focusing on inbreeding and other typical standardization techniques. In other words, most standardization occurred due to the choice of a strain of an organism already known to be relatively invariant in a number of its biological features, a strategy clearly derived from bacteria and phage work where inbreeding is fairly trivial.

Stage 2, the development of descriptive models, is crucial to the life of a model organism (for a fuller defense of this claim, see [1] from which the following summary is drawn). Work in the philosophy of science that places emphasis on models as theories fails to provide an adequate conceptual framework for scientific research occurring in the early, protoexplanatory stages of a scientific program, where models are being developed but generalized explanations or theories are not yet being generated. This is not to say that biological work with model organisms may not sometimes involve theoretical modeling or indeed modeling of other kinds. Instead, the claim is that whether or not other sorts of modeling occur, there must be a protoexplanatory phase in which a descriptive model of the experimental organism is developed. The biological material itself is abandoned as the basis of research in favor of these descriptive models, for instance of the nervous system (in the form of a wiring diagram), the genetics of the organism (through the genomic sequence), or the developmental processes (as captured by cell lineage diagrams).

One of the predominant uses of *C. elegans* as a model organism relies on the argument that a particular part (or combination of parts that form particular structures) of the organism can be used as a model to which particular empirical instantiations of these parts can be compared. In the case of neurobiology, this claim holds that general principles upon which the structure of the nervous system are built can be determined by establishing a descriptive model, against which observations (e.g., the structure of worms that are variant or abnormal in neural patterns) can be compared in order to assess the range of the applicability of the model and of more general theories about neural structure.

Some of the details of the research methodology used in the construction of the wiring diagram [38] raise general issues relevant to what constitutes a descriptive model, which in turn serves as the starting point for the process of case-based reasoning. First, these diagrams were actually a mosaic of the nervous systems of four individual worms, but were presented as a so-called 'canonical nervous system.' Second, the experimental techniques sometimes resulted in micrographs that were difficult to interpret without some interpretive assumptions. Finally there are other technical problems associated with low-level variability and qualitative variations in neural patterns between animals even with the same genotype (for a review of some problems associated with the reconstructions see [6]). However, given that the logic behind the reconstructions was extremely simple and rather noncontroversial, the authors were able to conclude that they were "reasonably confident that the structure that we present is substantially correct and gives a *reasonable picture* of the organization of the nervous system in a *typical C. elegans* hermaphrodite" ([38], 7). In this conclusion, it can be seen that *C. elegans* is viewed at this stage as an abstract, idealized descriptive model, specifically of the nervous system of the wild type worm.

This example of modeling structures in neurobiology has several important characteristics that motivate the concept of a descriptive model. First, *C. elegans* as a model organism for structures in neurobiology is an abstract entity rather than a natural one (though of course it is based on natural, living worms). The wiring diagram is based on an abstract model of the worm in terms of the typical or usual neural connections exhibited not by any one specimen alone or by numerous individual organisms, but by a more abstract construct hybridized from a few individual specimens. In some sense, this worm represented an *ur*-worm, idealized not only in the sense of being a wild type worm but also because individual differences among wild type worms were eliminated in favor of describing the most commonly occurring structures, a canonical form or consensus state of the worm. A target object thus was constructed which it was hoped would allow comparison between it as normal and the object of interest, some individual worm (e.g., one abnormal in phenotype).

Furthermore, in an important sense, even the natural entity *C. elegans* is extremely idealized for investigations of the general principles of neural wiring because of its relative invariance in contrast to other metazoans. Thus the wiring diagram is a descriptive model not only of the nervous system of *C. elegans,* but more generally of a simple metazoan nervous system. It is taken as a prototype which represents one likely state of basic structures (neural patterns and connections) that may also be found in other organisms, thus allowing comparison of empirical evidence regarding these other organisms in order to assess isomorphism, the limits and applicability of the descriptive model.

Given various concerns about experimental approaches and tractability (involving technical details not reviewed here), the wiring diagram can be viewed as representing a descriptive model that requires alteration and tweaking as further scientific investigations reveal its limitations. At the extreme, alterations in the descriptive model may result in alterations to the idealized presuppositions behind the description (for instance in this case that the *functional* properties of a nervous system are determined directly by its *structure).* In the closing of the magnum opus describing the wiring diagram, the authors proposed that it provides a framework for posing more specific questions about how neurons and their interconnections develop and become organized, as well as how the neural network functions, using abnormal worms: "A knowledge of the detailed structure of a nematode's nervous system does not in itself provide any answers to these questions, but it does at least provide a framework within which it is possible to pose rather more specific questions" ([38], 58).

There are at least two important points implicit in this account which support the claim that descriptive models serve as the basis for case-based reasoning processes using model organisms. First, the idea that model organisms are in fact abstractions as outlined has resonance with the construction of epistemologic entities elsewhere in the sciences, for instance of the 'average man' in medical science, going back to the work of Quetelet ([29], 2:267, as quoted in [12], 12): "The consideration of the average man is so important in the medical sciences that it is almost impossible to judge the state of an individual without comparing him to a fictive being that one regards as being the normal state and who is nothing but the [average man]." Rich, descriptive idealizations thus are the starting point for case-based reasoning, as some baseline case must be provided to initiate the reasoning process. But these idealized cases are necessarily fictitious, as is the nervous system of the so-called 'canonical worm,' at the same time as they are essential tools for developing an understanding of the actual organism.

Second, note that as with medical case reports, usually there is no explicit (or implicit) testing of a hypothesis or theory, or other 'typical' scientific behaviors. Instead, the process proceeds by the proffering of observations and detailed descriptions, which may well point to testable hypotheses and explanations, particularly if they are to have an impact on the development of theory or on practice (see [34]). Thus there is a creation of epistemological space (or framework, to borrow a term from the scientists quoted above) within which to ask questions. However, as bluntly stated by a commentator on medical reasoning, "with higher organisms, and especially with patients, it becomes hopeless to attempt to create complete descriptions. . . This is a kind of epistemologic surrender and consists in simply ignoring many of the things that could be truthfully said *in order to say what must be said*" ([7], 848, my emphasis). Both in medicine and in biological reasoning from model organisms, complexity and completeness are sacrificed in favor of selective construction of manageable material with which scientists can work.

To return to the stages of model organism development, the stage 3 is the generation of explanations.[3] Note that in these sciences, and particularly model organism work, this stage is at first relatively basic, with its content relying heavily on the descriptions developed in the previous one. In fact much of the time this stage is short-shrifted at least in the initial pass through each of the stages in favor of further refinement of the descriptions through case based reasoning.

Stage 4, case-based reasoning using descriptive models, is a key step in model organism research, and depends on a double feedback loop between base cases and cases of interest. As with the medical example outlined previously, a set of base cases are developed, for instance composed of descriptions of genetic sequences in organisms. The paradigm begins with a descriptive model of the organism that is established as being 'normal' in phenotype, for which the genomic sequence is identified. This sequence can then be compared to that of organisms that are abnormal in phenotype in order to draw out the functional properties of the genomic sequence within a particular model organism. The reasoning relies on a prediction that determining the sequence in various model organisms will reveal conserved genetic regions in these organisms, which in turn will allow investigation of the same sequences in the normal human genome and prove fruitful for understanding the functional properties of these sequences. Finally, the eventual goal is to understand the higher level, phenotypic results of abnormal, human genomic sequences found to be similar to the paradigmatic 'abnormal' sequences in model organisms, based on a correlation between higher level properties such as disease conditions or abnormalities.

What is most important to notice in this analysis of the use of this form of reasoning is that answering the question of whether a model organism will in fact prove to be a useful model, for instance for human genome sequencing, requires that researchers not only work on sequencing on the model organism but that this sequencing occur in tandem with sequencing in the object of interest, the human genome, and other comparative genomic work (for the biological details of this argument with regard to developmental biology, see [10]). This conclusion points to an important, but easily overlooked, aspect of modelling (one with which we are quite familiar from [23]): in order for models to actually function well as models there must be ongoing refinement of the original descriptive model, as well as constant interplay between the original descriptive model and the subject being modeled, and continuous development of the positive analogies between them (and identification of the relevant disanalogies and their import). Much rhetoric surrounding model organism research unconstructively obscures this interplay and hence misrepresents the potential limitations of even good models. In other words, providing a model requires an interaction between the model and the object of interest being modeled, including construction of similarity relations, which are impossible to devise without a detailed description of the process to be modeled, including in this case the functional properties of the sequence.

## 5. Reflections and Conclusions

In addition to the specific conclusions about the use of case-based reasoning with model organisms summarized above, there is an important metalevel issue implicit in this examination, namely the epistemic basis for the fruitfulness of studying model organisms in history and philosophy of science (HPS). To return to the quote offered previously, since it is hopeless to offer complete descriptions, as historians and philosophers we too are faced with the necessity of 'epistemic surrender,' that is, of saying not all that could truthfully be said, but rather that which must be said. Historians and philosophers of biology who are studying organisms that serve as model organisms have focused on them because they provide a means of access to the complexities in the practices of the life sciences and of the actual biology itself (cf. [17]). Model organisms now serve as cases for the history/philosophy of biology, as can be noted in the following features: first, historical work on model organisms has emphasized that the focus of study is not static, but in fact is an organismal entity which is the result of complex evolutionary processes, and not only those processes that are natural and biological, but as the result of the actions of the researchers including organismal choice, standardization, and so on. Second, HPS research on model organisms has revealed features of scientific practice that it can be argued were not easily accessible through much traditional HPS focused primarily on theory or on individual practitioners. Model organisms have proven extremely fruitful cases for HPS because they force us not look just at theories in biology, but much more deeply at practices, conceptual assumptions, and research groups as communities focused on a single organism from which they derive much of their self-identity and direction.

But as discussed previously, cases are problematic, in that they are oftentimes used in order to focus on entities that are unique or nearly unique, and hence it is difficult to know what kind of generalizations can be drawn from the particular cases. Just as statistics caused a great deal of excitement well before its methods were powerful enough to produce reliable results [12], so too is there a danger that HPS scholars are overestimating our abilities to work through the complexities of scientific practice by using cases particularly in these messy domains. Furthermore, case studies and medical anecdotes are traditionally held in disrepute, a trend which is associated with the historical shift away from single cases to more scientific, reductionistic explanations of disease [24]. So, too, do our cases of model organisms potentially suffer from their uniqueness and individuality, from the quirkiness of any particular organism, its development as a model organism, and the scientists who work with it. However, again as in medicine and other sciences, model organisms as cases have an important pedagogical and heuristic use: they help not only as exemplars for learning by novices but also can point to new domains for investigation. Out of the details of these cases, patterns begin to emerge in the practices of modern life science that arguably have not been visible but for viewing the biological practice through a lens focused primarily on the organism itself, both as an individual and as representative of a larger class of what have become the classic model organisms.

While striving for generalized explanations, model organism work clearly begins as a science of particulars, based on descriptive models which come to serve as cases, and whether it ultimately can achieve unified, broader explanations or theories is a topic for continued biological and philosophical investigation. Model organism work exemplifies the recognition and development of means for managing complexity in a domain plagued with variability and particularity. Case-based reasoning is a process that results initially not in unified theories or mechanistic explanations, but in a form of scientific understanding (perhaps of a weaker sort than our traditional theories and explanations) which is constantly evolving, incomplete, and uncertain, but nonetheless has the status of knowledge for its practitioners.

## Notes

[1] Acknowledgements: I wish to thank the many individuals who have discussed the issues explored in this paper with me, both in formal and more informal settings, including the Workshop on Model Systems, Cases, and Exemplary Narratives in Science and History, Department of History, Princeton University, 1999 (and the Davis Center at Princeton which hosted me in 1999–2000), especially Mary Morgan, Angela Creager, and Jane Hubbard; the Center for the Philosophy of Science and Program in Science and Technology Studies, University of Minnesota, 2000; the History of Science Society Annual Meeting 2002, particularly my session co-convenors Karen Rader and Judy Johns Schoegel; the Model Systems Strategic Network working group sponsored by the Stem Cell Network, a member of the Network of Centres of Excellence, Canada; and not least of all, the International Congress for Logic, Method-

ology and Philosophy of Science in Oviedo, 2003. I also am grateful to Fiona
Mackenzie for research assistance.

² On viewing cases more generally as a style of reasoning in the human sciences,
see also Forrester [16]. A notable exception to this gap in the literature is
Bogen's work [8] on neurobiology including so-called 'n of 1' cases and poorly
replicated evidence.

³ For a useful discussion of genetic explanations of behavior resulting from
model organism work, see [33].

## REFERENCES

1. R. A. Ankeny. Fashioning descriptive models in biology: Of worms and wiring
   diagrams, *Philosophy of Science*, 67(Proc.):S260–S272, 2000.
2. R. A. Ankeny. Model organisms as cases: Understanding the 'lingua
   franca' at the heart of the Human Genome Project, *Philosophy of Science*,
   68(Proc.):S251–S261, 2001a.
3. R. A. Ankeny. The natural history of *C. elegans* Research, *Nature Reviews
   Genetics*, 2:474–478, 2001b.
4. R. A. Ankeny. Reconceptualizing Reduction: Cystic Fibrosis as a Paradigm
   Case for Molecular Medicine. In: L. S. Parker and R. A. Ankeny (eds.), *Mu-
   tating Concepts, Evolving Disciplines: Genetics, Medicine, and Society*, Dordrecht: Kluwer Academic Publishers, 127–142, 2002.
5. R. A. Ankeny. Wormy Logic: Model Organisms as Case-Based Reasoning. In:
   A. N. H. Creager, E. Lunbeck, and M. N. Wise (eds.), *Science without Laws:
   Model Systems, Cases, Exemplary Narratives*. Chapel Hill: Duke University
   Press (forthcoming).
6. C. Bargmann. Genetic and cellular analysis of behavior in *C. elegans*, *Annual
   Review of Neurosciences*, 16:47–71, 1993.
7. M. S. Blois. Medicine and the nature of vertical reasoning, *New England Journal
   of Medicine*, 318:847–851, 1988.
8. J. Bogen. 'Two as good as a hundred': Poorly replicated evidence in some
   nineteenth-century neuroscientific research, *Studies in History and Philosophy
   of Science Part C: Studies in History and Philosophy of Biological and Biomedical Sciences*, 32:491–533, 2001.
9. J. Bogen and J. Woodward. Saving the phenomena, *The Philosophical Review*,
   47:303–352, 1988.
10. J. A. Bolker. Model systems in developmental biology, *BioEssays*, 17:451–455,
    1995.
11. N. Cartwright. *How the Laws of Physics Lie*, New York: Oxford University
    Press, 1983.
12. J. Cole. The chaos of particular facts: Statistics, medicine and the social body
    in early 19th-century France, *History of the Human Sciences* 7:1–27, 1994.
13. L. Daston. *Classical Probability in the Enlightenment*. Princeton: Princeton
    University Press, 1988.

14. S. de Chadarevian. Of worms and programmes: *Caenorhabditis elegans* and the study of development, *Studies in the History and Philosophy of Biological and Biomedical Sciences*, 29:81–105, 1998.

15. A. S. Elstein, L. S. Shulman and S. A. Sprafka, *Medical Problem Solving: An Analysis of Clinical Reasoning*. Cambridge: Harvard University Press, 1978.

16. J. Forrester. If *p*, then what? Thinking in cases, *History of the Human Sciences*, 9:1–25, 1996.

17. G. L. Geison and M. D. Laubichler. The varied lives of organisms: Variation in the historiography of the biological sciences, *Studies in the History and Philosophy of the Biological and Biomedical Sciences*, 32:1–29, 2001.

18. G. Gigerenzer et al. *The Empire of Chance: How Probability Changed Science and Everyday Life*. Cambridge: Cambridge University Press, 1989.

19. S. Gorovitz and A. MacIntyre. Toward a theory of medical fallibility, *Journal of Medicine and Philosophy*, 1:51–71, 1976.

20. I. Hacking. *The Emergence of Probability*. Cambridge: Cambridge University Press, 1975.

21. I. Hacking. *The Taming of Chance*. Cambridge: Cambridge University Press, 1990.

22. J. Hamel. *Case Study Methods*, with Stéphane Dufour and Dominic Fortin, Newbury Park, CA: Sage Publications, 1993.

23. M. B. Hesse. *Models and Analogies in Science*, London: Sheed and Ward, 1963.

24. K. M. Hunter. 'There was this one guy...': The uses of anecdotes in medicine, *Perspectives in Biology and Medicine*, 29:619–630, 1986.

25. K. M. Hunter. An N of 1: Syndrome letters in The New England Journal of Medicine, *Perspectives in Biology and Medicine*, 33:237–251, 1990.

26. K. M. Hunter. 'Don't think zebras': Uncertainty, interpretation, and the place of paradox in clinical education, *Theoretical Medicine*, 17:225–241, 1996.

27. J. Kolodner. *Case-Based Reasoning*. San Mateo: Morgan Kaufmann Publishers, Inc., 1993.

28. T. Porter. *The Rise of Statistical Thinking, 1820–1900*. Princeton: Princeton University Press, 1986.

29. A. Quetelet. *Sur l'homme et le developpement de ses facultes, ou Essai de Physique Sociale*. Paris: Bachelier, 1835.

30. C. K. Riesbeck and R. C. Schank. *Inside Case-Based Reasoning*, Mahwah, NJ: Lawrence Erlbaum Associates, Inc., 1989.

31. K. F. Schaffner. Exemplar reasoning about biological models and diseases: A relation between the philosophy of medicine and the philosophy of science, *The Journal of Medicine and Philosophy*, 11:63–80, 1986.

32. K. F. Schaffner. *Discovery and Explanation in Biology and Medicine*, Chicago: University of Chicago Press, 1993.

33. K. F. Schaffner. Genetic explanation of behavior: Of worms, flies, and men. In: D. Wasserman and R. Wachbroit, *Genetics and Criminal Behavior*, Cambridge: Cambridge University Press, 79–116, 2001.

34. R. J. Simpson and T. R. Griggs. Case reports and medical progress, *Perspectives in Biology and Medicine*, 28:402–406, 1985.

35. S. Toulmin and A. R. Jonsen. *The Abuse of Casuistry: A History of Moral Reasoning*. Berkeley: University of California Press, 1988.

36. M. Wartofsky. How to begin again: Medical therapies for the philosophy of science. In: F. Suppe and P. D. Asquith (eds.), *PSA 1976*, 2:109–122, 1977.
37. C. Whitbeck. The relevance of philosophy of medicine for the philosophy of science. In: F. Suppe and P. D. Asquith (eds.), *PSA 1976*, 2:123–135, 1977.
38. J. G. White, E. Southgate, J. N. Thomson and S. Brenner. The structure of the nervous system of the nematode *Caenorhabditis elegans:* The mind of a worm, *Philosophical Transactions of the Royal Society of London: B. Biological Sciences*, 314:1–340, 1986.
39. J. Woodward. Data and phenomena, *Synthese*, 79:393–472, 1989.
40. R. K. Yin, *Applications of Case Study Research*, Newbury Park, CA: Sage Publications, 1993.

# Models, Theories and Phenomena

Daniela M. Bailer-Jones

*Department of Philosophy, University of Bonn*
*daniela@bailer-jones.de*

**Abstract.** I address the relationship between phenomena, models and theories. A phenomenon starts off as something that raises interest and therefore becomes subject to investigation. As research into a phenomenon proceeds, what we take the phenomenon to be is increasingly influenced by the way in which we model it. Models are designed to 'describe' and interpret phenomena. While models are about concrete phenomena, theories are abstract in the sense that they do not account for some of the concrete properties of a phenomenon which, in turn, makes them more general than many models. Correspondingly, theories are applied to phenomena only via models; they provide constraints for the model construction.

## 1. Introduction

Let me start with the question of what models are models of. Models can be models of things or processes, or models of data, or models of a theory in the model-theoretical sense. I am not concerned with models as the term may be used in mathematics. I am, however, interested in all those models that are models of 'things' in nature. I call these things in nature 'phenomena' and will spell out in Section 2 what I mean by a phenomenon. Very broadly construed, the subject of science is to deal with phenomena. 'Dealing with' such phenomena can, for instance, mean describing them or explaining them. If a model deals with a phenomenon, meaning that a models *describes* or *explains* a phenomenon, then a relationship is established between the model and 'the world' to which the phenomenon belongs. I shall outline what I take models of phenomena to be in Section 3. Given the history of the philosophy of science, there is no way that one can discuss models without also addressing theories. So, in Section 4 I consider what the difference between models and theories is. My concern in Section 5 is whether what we take a certain phenomenon to be changes in the course of the phenomenon being modelled. Section 6 contains my conclusions.

## 2. What is a phenomenon?

In my understanding of phenomenon, I largely follow Jim Bogen and Jim Woodward [3,4]. A phenomenon is a fact or event in nature, such as bees dancing, rain falling or stars radiating light. A phenomenon is not necessarily something as it is observed; Bogen's and Woodward's point is precisely to distinguish data about a phenomenon from the phenomenon. A phenomenon may be something that is originally picked up by observation and then raises certain questions. Observing a bees' dance – and even calling it that – may bring about the conjecture that there is something systematic about the bees' movements which warrants further investigation. To conjecture thus is not to take the movements of the bees as something happening entirely at random. It is *treating* what is observed as a phenomenon. So, in the very first instance, a phenomenon is something that is taken to be a subject to be researched. At this stage, it is not strictly known whether there really is a distinguishable fact or event to be found, even if one has an inkling that this is so. Similarly, if we observe rain falling, asking what causes the rain is the first step of turning this observation into something that constitutes a phenomenon. This seems to suggest that picking out a phenomenon has something to do with distinguishing the causal processes that make up that phenomenon.

Some may say that the phenomenon exists, even if it is not recognized. As Jim Bogen puts it (private communication), Jacksonian epileptic seizures occurred long before Jackson began to study them and were not changed by his investigation of them. Bogen and Woodward have been criticised for their 'static' understanding of 'phenomenon' [12,10]. Glymour presents Bogen's and Woodward's position quite poignantly ([10], p. 30):

> "To say that a scientist is wrong about the data she reports is necessarily to say that she did not in fact see what she claims to have seen, while to say that a scientist is wrong about the phenomena she reports need only be to say that she has drawn incorrect inferences from what she indisputably did see. The two differ ontologically in that phenomena are stable, repeatable features of the natural world, while data are not. Phenomena are ineliminable, bedrock elements of the furniture of the world."

Bogen and Woodward [3] are adamant that data must not be identified with the phenomenon itself. How they are different can be illustrated with an example. Bogen and Woodward [3] adopted this example from Ernest Nagel's ([15], p. 79) *Structure of Science*. The topic is the melting point of lead. This can be measured and found to be 327 degrees centigrade. What does it mean, however, to find the melting point to be this precise temperature? To establish the melting point of lead, data are collected: a whole series of measurements are carried out. The temperature at which lead melts is typically not measured only once, but many times, in order to take account of measuring errors that are expected. In principle, it can happen that, during a whole series of measurements, the precise value of 327 centigrade is never once read off the thermometer. Instead, the average of the measured values is taken, the measurement error calculated and the result of this data analysis declared the melting point of lead. Because measuring errors occur in exploring nature, measurements have to be repeated many times. As a consequence, large amounts of data are produced that need to be interpreted and analysed in a way that allows scientists to extract a definite empirical finding about a phenomenon, i.e. one value for the melting point of lead. Moreover, the same phenomenon, e.g. the melting point of lead, can be examined with many different experiments. The experiments can be varied, while the phenomena remain fixed. Bogen and Woodward consider them as natural kinds.

McAllister [12] and Glymour [10] have different reasons for questioning that phenomena can be quite so static and unchangeable. McAllister thinks that scientists need to add criteria, other than simply looking for patterns in the data, in order to make out phenomena. Glymour, in turn, thinks that the concept of a phenomenon is unnecessary because causal relations could be hypothesized directly from statistical correlations in a data sample suitably analysed. My own line of thought on the problem is the following: Yes, in some sense Jacksonian epileptic seizures may have existed before Jackson discovered them, but there is no way that the phenomenon existed *for us* before their discovery or recognition. In my view, before their recognition, Jacksonian epileptic seizures were not a phenomenon because they were not a subject of study that raised our curiosity. Only *after* our curiosity is raised about what is happening and how it is brought about do the seizures become noticed and eventually established *as* a phenomenon.

Interestingly, certain phenomena would not even be recognized without at least some basic research into them. An example for such a phenomenon is the order of acquisition of prepositions in children. It takes considerable observation and experimentation to establish that children acquire the meaning of IN before ON and finally UNDER. Once this is established as a phenomenon in some languages (e.g. English and German), it becomes possible to examine whether this finding is more universal and holds for other, structurally different languages (e.g. Polish) too ([16]). It will matter for the constitution of this phenomenon whether this prepositional order is only found in English and German, say, or whether research shows that this order of acquisition of prepositions holds for *all* (or most) languages (in a way appropriate to the structures of the languages). In the case of the latter, we would consider the phenomenon not as one resulting from language and/or culture, but the phenomenon would appear to be cognitively universal. Whether or not the order of acquisition of IN, ON and UNDER is found in languages other than English and German will have an impact on what this phenomenon *is*. In fact, for a phenomenon like this, without prior examination it is not even obvious that anyone could recognize it *as a phenomenon*. Sensual perception is certainly not in all instances enough to identify and establish a phenomenon. Correspondingly, James Bogen and James Woodward acknowledge ([3], p. 352):

> "It is overly optimistic, and biologically unrealistic, to think that our senses and instruments are so finely attuned to nature that they must be capable of registering in a relatively transparent and noiseless way all phenomena of scientific interest, without any further need for complex techniques of experimental design and data analysis."

Thus, that something is identified as a phenomenon is, in many instances, already the result of research, i.e. of systematic data acquisition on a subject.

Let me now illustrate the point about a phenomenon 'changing definition' in the course of being examined. To start with, the facts or events that come to be considered as a phenomenon may not be clearly defined (although sufficiently defined to be treated as a phenomenon). At this stage, a phenomenon is something about which one wants to know more. Then, in the process of learning and discovering more about a phenomenon, what the phenomenon is taken to be changes. Take gold. Gold is a material which was originally identified probably by its colour and some of its properties. In a larger theoretical context it is the element on the periodic table that has the proton number 79. Gold can be involved in various physical processes constituting phenomena, such as chemical reactions, and it is in the context of these reactions that the proton number of gold receives its significance. It turns out that this proton number is inseparably linked to how gold is involved in natural processes that constitute certain phenomena. This means, for instance, that gold behaves and reacts in a way that is comparable to other elements of the same main group of the periodic table, such as copper and silver. One phenomenon is constituted by the fact that metals of this group do not corrode as easily as other metals, such as iron. So, the study of gold which places gold in a certain theoretical context changes how one would delineate phenomena involving gold.

Of course, even if the data are to be distinguished from phenomena, as Bogen and Woodward argue, one expects that a phenomenon manifests itself empirically somehow – that phenomena can become noticed in the empirical world, even if they also get to be captured at a different level. Let me now look again at the phenomenon of lead melting at a certain temperature[1] which, as it is claimed, differs significantly from data about the melting point. This phenomenon is also about the factors that make up that melting point. One can, for instance, ask why the melting point is as high or as low as it is. At a theoretical level, this question has to do with the forces that hold together the atoms and molecules and so influence the melting point. There are different forces and correspondingly different models of chemical binding. For different chemical elements, different forces are pronounced, e.g. the London forces in crystals, hydrogen bridge binding in water or ion binding in metals. The melting point of a specific element depends on the binding forces that act in the case of that element. One has to ask which binding force is most central in determining the melting point. For this, the comparison with similar elements, those of the same main group of the periodic table, is relevant. In the case of lead, this is the fourth main group of the periodic table, carbon, silicon, germanium, tin and lead. In this group, covalent binding decreases and metallic binding increases (Figure 1). Correspondingly, lead is subject to stronger metallic binding than tin which is why it has a higher melting point, whereas both have much lower melting points than carbon, silicon and germanium because these latter elements have much stronger covalent binding. Although these theoretical considerations may not suffice to calculate the melting point of lead, they nonetheless allow us to appreciate what constitutes the phenomenon of the melting point of lead. Part of this is to be able to explain and predict a certain behaviour of lead, something which is not possible purely on the basis of measuring the melting point. Such systematic explanation of facts about phenomena is precisely what can be achieved with models. The phenomenon of the melting point of lead is thus conceptually different from data about lead which are produced when one tries to establish the melting point of lead experimentally. Different ramifications apply to the data analysis and to the theoretical description of the phenomenon in a model.

---

[1]For help with this example, I am very grateful to Rüdiger Stumpf.

| Element | | | Melting Temperature |
|---|---|---|---|
| C | | | 3550°C |
| Si | metallic | covalent | 1414°C |
| Ge | binding | binding | 938.25°C |
| Sn | increases | decreases | 231.93°C |
| Pb | | | 327.46°C |

Figure 1.

## 3. What is a scientific model?

Let me now introduce my notion of a scientific model. I consider the following as the core idea of what a scientific model is: A model is an interpretative description of a phenomenon that facilitates access to that phenomenon. This access can be perceptual as well as intellectual.[2] Interpretative descriptions may rely, for instance, on idealisations or simplifications or on analogies to interpretative descriptions of other phenomena. Facilitating access usually involves focusing on specific aspects of a phenomenon, sometimes deliberately disregarding others [2]. As a result, models tend to be partial descriptions only. Models can range from being objects, such as a toy aeroplane, to being theoretical, abstract entities, such as the Standard Model of the structure of matter and its fundamental particles. As regards the former, scale models facilitate looking at something by enlarging it (e.g. a plastic model of a snow flake) or shrinking it (e.g. a globe as a model of the earth). This can involve making explicit features which are not directly observable (e.g. the structure of DNA or chemical elements contained in a star). The majority of scientific models are, however, a far cry from consisting of anything material like the rods and balls of molecular models used for teaching; they are highly theoretical. They often rely on abstract ideas and concepts, frequently employing a mathematical formalism (as in the big bang model, for example), but always with the intention to provide access to aspects of a phenomenon that are considered to be essential. Bohr's model of the atom informs us about the configurations of the electrons and the nucleus in an atom, and the forces acting between them; or modelling the heart as a pump gives us a clue about how the heart functions. The means by which scientific models are expressed range from the concrete to the abstract: sketches, diagrams, ordinary text, graphs, mathematical equations, to name just some. All these forms of expression serve the purpose of providing intellectual access to the relevant ideas that the model describes. Some of these forms of expression are non-propositional. Providing access means giving information and interpreting it and expressing it efficiently to those who share in a specific intellectual pursuit. In this sense, scientific models are about empirical phenomena, whether these are how metals bend and break or how man has evolved.

---

[2]If access is not perceptual, it is often facilitated by visualisation, though this need not be the case.

## 4. What is a theory, in contrast to a model?

Nancy Cartwright [8] characterizes theories as abstract. She does so in the context of the analogies between models and fables ('theories are like morals of fables'). Fables have a moral which is abstract and they tell a concrete story that instantiates that moral, or 'fits out' that moral. A moral of a fable may be 'the weaker is prey to the stronger', and a way to 'fit out' (Cartwright's formulation) this abstract claim is to tell the story of concrete events of the marten eating the grouse, the fox throttling the marten, and so on. Similarly, an abstract physical law, such as Newton's force law, $F=ma$, can be fitted out by different more concrete situations: a block being pulled by a rope across a flat surface, the displacement of a spring from the equilibrium position, the gravitational attraction between two masses. Thus, Newton's law may be fitted out by 'different stories of concrete events'. Drawing from the analogy between models and fables, models are about concrete things; they are about concrete empirical phenomena. The contrast between models and theories is not that theories are abstract and models are concrete. Rather, models are about concrete phenomena, whereas theories are not about concrete phenomena. If at all, theories are about concrete phenomena only in a very derivative sense. 'Force', which is a theoretical concept and belongs to the realm of the abstract, does not manifest itself outside concrete empirical situations. Cartwright's everyday example for this relationship is 'work': The abstract concept of 'work' may be filled out by washing the dishes and writing a grant proposal, and this does not mean that a person washed the dishes and wrote a grant proposal, *and* worked – working does not constitute a separate activity – since working consists in just those activities. Force is a factor in and contributing to empirical phenomena ([8], p. 65)[3]:

> "*Force* – and various other abstract physics' terms as well – is not a concrete term in the way that a color predicate is. It is, rather, abstract, on the model of *working*, or *being weaker than*; and to say that it is abstract is to point out that it always piggy-backs on more concrete descriptions. In the case of *force*, the more concrete descriptions are ones that use the traditional mechanical concepts, such as *position*, *extension*, *motion*, and *mass*. Force then, on my account, is abstract relative to mechanics; and being abstract, it can only exist in particular mechanical models."

I will concentrate on one aspect only of theories being abstract. This aspect is that theories, being abstract, are not directly about empirical phenomena.[4] Abstractness is opposite to concreteness. The phenomena that are explored by modelling are *concrete* in the sense that they are (or have to do with) real things – things such as stars, genes, electrons, chemical substances, and so on.

---

[3]Cartwright seems to imply here that position, extension, motion and mass are concrete concepts, or at least more concrete than force. This seems like a claim hard to defend, but I will leave this issue here.

[4]Cartwright discusses idealisation and abstractness in Chapter 5 of *Nature's Capacities and their Measurement* [7]. There the notions of abstractness and idealisation are expected to do work in the context of the concept of capacities and of causality, but this is a somewhat different context from *theories* being abstract.

Wanting to say that models are about concrete phenomena, while theories are not, brings with it still another problem, however. Often, the subject of models is a class of phenomena, rather than a specific individual phenomenon. Of most phenomena we can find many specimens in the world; these phenomena belong to the same class.[5] Modelling a star, there are many different individual stars that could serve as a prototype.[6] One tries to model, however, not any odd specimen of a phenomenon, but a typical one. Often this involves imagining the object of consideration as having 'average' or 'typical' properties, and this 'prototypical' object or phenomenon may not even exist in the real world. The point is that it could typically exist in just this way and that there exist many very much like it. So, the prototype is selected or 'distilled' from a class of objects. The prototype has all the properties of the real phenomenon; it is merely that the properties are selected such that they do not deviate from a 'typical' case of the phenomenon. It is then this prototype that is addressed in the modelling effort. The assumption behind this process of prototype formation is nonetheless that the model is not only a model of the prototype, but one of the real phenomenon, including specimens that display a certain amount of deviation from the norm. Correspondingly, modelling the human brain is not about modelling the brain of a specific person, but that, roughly, of all 'typical' people. For my purposes, the prototype of a phenomenon still counts as concrete, because it has all the properties of the real phenomenon and *could* exist in just this manner. The target of the examination remains an empirical phenomenon, even if members of the class of phenomena that belong to a certain type can come in different shapes and variants. This prototype-forming procedure is often needed in order to grasp and to define a phenomenon and to highlight what it is that one wants to model. The important point here is that despite prototype formation, the phenomenon is not in any way stripped off any of its properties.

Phenomena have properties. Abstraction I take to be a process where properties are taken away from a phenomenon, and are not replaced by another property.[7] That which is abstract lacks certain properties that belong to any real, concrete phenomenon. To put it very crudely, something concrete becomes abstract when certain properties, that belong to the 'real thing' (and that make it concrete), are taken away from it.[8] Not all concepts, principles or theories that are called 'abstract' are abstract in the same way, but I think the notion of taking away some of those properties that make something concrete can still serve as a guideline. It is important to recognize that no theory is conceivable without the concrete in-

---

[5]There are exceptions to this. For some phenomena that are modelled there exists only one specimen that is taken into account, e.g. the earth.

[6]I am aware that the term 'prototype' has some connotations that are counterintuitive to my use of it, but for want of a better alternative I introduce it here as a technical term to be used in the way described in the following.

[7]Idealisation, in contrast, means that properties are *changed* rather than omitted.

[8]The *Oxford English Dictionary* gives for 'abstract': "Withdrawn or separated from matter, from material embodiment, from practice, or from particular examples. Opposed to *concrete*", besides older uses. Cartwright ([7], p. 197 and 213f.) identifies this as the Aristotelian notion of abstraction. She recounts: "For Aristotle we begin with a concrete particular complete with all its properties. We then strip away – in our imagination – all that is irrelevant to the concerns of this moment to focus on some single property or set of properties, 'as if they were separate'' (p. 197). See also [9], (p. 327f.).

stantiations from which the theory has been abstracted. We need to go through different example problems in order to understand how $F=ma$ is instantiated in different models. The theory is that which has been distilled from several more concrete instantiations. In this sense, the abstract theory is not directly about concrete phenomena in the world. The properties that are missing in such an abstract formulation as $F=ma$ are how the force makes itself noticed in different individual situations. Think again of a block being pulled across a flat surface, or the displacement of a spring from the equilibrium position, or the gravitational attraction between two masses. It depends on the situation what the force or the acceleration consists in (deceleration due to friction, the repulsion of a spring, or acceleration due to gravitation). Moreover, for each *concrete* situation one would have to establish what the body is like whose mass features in the physical system. Correspondingly, force, acceleration and mass can be associated with different properties in different physical systems. Force, abstractly speaking, can be something that applies to an object or system, but force alone, without an object or a system, is not something about which we can say anything, nor know the properties of. To establish a theory we need models that tell us how the theory is relevant with regard to the phenomenon or process modelled.[9]

Consider the example of the pendulum. To think about the force in this particular case, it is necessary to take into account the specifics of this system, first of all the geometry of the system, involving the displacement angle, the length of the string and the gravitational attraction of the earth. The treatment of the ideal pendulum then usually continues assuming that the displacement angle of the pendulum is small, because this allows us to replace the sine of that angle with that angle itself. Obviously, physicists modelling pendula are fully aware of the idealisations they have introduced into their model. In reality,

- the string is not weightless;

- the string is not inextensible;

- the mass of the pendulum bob is not located in one point.

It is for a good reason that they specifically talk about the 'ideal' or 'mathematical' pendulum when they refer to this particular model involving the specified idealisations. This is why they also consider the physical pendulum. This is supposed to be a model that is nearer to some real pendula. This kind of pendulum is taken to be a rigid body of any arbitrary shape, pivoted about a fixed horizontal axis. In this case, the centre of mass is treated as if it were the pendulum bob and the moment of inertia about the axis of rotation plays a role when calculating the restoring force. It is perfectly possible to make the model of a pendulum 'more real' and to correct for, e.g., the frictional forces of the air resistance, the buoyancy of the pendulum bob (that the apparent weight of the bob is reduced by the weight of the displaced air) and the gravitational field of the earth not being uniform, etc. (cf. [14], p. 48-51). It is clear that, in order to make these corrections to the model, theories that go beyond Newton's Law are employed and customized to the

---

[9]For a case study supporting this kind of 'division of labour' between models and theories, see Suárez [17].

problem in hand.[10] Of course, in this case they all fall under the reign of classical mechanics, but this need not be so. Morrison comments ([14], p. 51):

> "We know the ways in which the model departs from the real pendulum, hence we know the ways in which the model needs to be corrected; but the *ability* to make those corrections results from the richness of the background theoretical structure."

Let me go back to the earlier example of the melting point of lead. There can be phenomenological laws that are merely generalisations of concrete instances, e.g. "the melting point of lead is 327 degrees Centigrade" which is presumably true of all lead. This is not abstract. An abstract law would be one that told us, for instance, how to infer the melting point of quite different metals. It would cover the general differences between London forces, hydrogen bridge binding and ion binding. These are not specific to any element, but can apply in all sorts of different elements, more in some than in others, depending on the element. For a law that simply states the melting point of lead, be it right or wrong, i.e. for phenomenological laws, we do not need a model in order to apply it to the world. Such a law does not apply to a range of different instances from which it is abstracted; such a law applies only to one kind of instance (rather than, perhaps, to all metals in a certain main group of the period table). This makes the law not theoretical.[11]

Theories can become general because they are abstract; they are free of the properties that are typical of certain individual instances where the theory might apply, or the properties that are typical of different prototypes. In order to model a phenomenon, abstract theory needs to be made more concrete, taking into account the specifications of the phenomenon that is modelled and inserting the ramifications and boundary conditions of that phenomenon (or the prototype thereof). To see how the theory holds in a model, we need to fill in the concrete detail that is not part of the theory because, being abstract, the theory has been stripped precisely of those details.

## 5. What happens to a phenomenon in the course of being modelled?

The thesis I promote in this section is that what we take a phenomenon to be is shaped by how we model this phenomenon. I said earlier that, in order to be recognized as a phenomenon, something about this phenomenon needs to raise a question. Think of the bees' dance. A model is some kind of answer to the question raised about a phenomenon, and the answer to this question will influence how we think about the phenomenon.

---

[10]Interestingly, Suárez [17] employs just this example of the pendulum to argue that using idealisation, and subsequently de-idealisation, in order to apply theory to phenomena makes models appear superfluous in comparison to theories. I tried to illustrate here, in contrast, that modelling a real pendulum involves more than merely de-idealisation: fitting bits of different theories together to approach a fairly realistic account of the phenomenon.

[11]Some laws have the status of theories, but not all do. Some sciences may be hard-pressed to formulate theories or principles that are abstract enough to apply quite generally, although an effort is often made. In other words, there can be sciences which only employ models and do not have theories.

Interpreting data without having a phenomenon in mind is sometimes hardly possible. Prajit Basu [1] presents a nice case study illustrating just this point that "observations, when transformed into evidence for a hypothesis, phenomena, or a theory, are theory infected" ([1], p. 356). Basu considers a case where two researchers perfectly agree on the data, but they take it to be evidence for different phenomena. The two researchers are Antoine-Laurent Lavoisier (1743-1794) and Joseph Priestley (1733-1804). Lavoisier wanted to argue that water is a compound. One indication was that hydrogen and oxygen react together to form water. The other indication was that iron and steam (water) react to iron oxide and hydrogen (i.e. in this reaction, water is split up into oxygen and hydrogen).[12] While Priestley had no doubts concerning the observational side of this reaction, he doubted that the black powder which Lavoisier took to be iron oxide was in fact iron oxide (see also [13], p. 208). The lesson Basu draws from this example is the following ([1], p. 357):

> "A piece of evidence for (or against) a theory is a construction in the context of that theory from (raw) data. In this construction, a set of auxiliary assumptions is employed. These auxiliaries may themselves be theoretical in character. From the same (raw) data it is possible to construct different evidence for (or against) different theories since the auxiliaries employed in connection with different theories can be different. Finally, although the (raw) data are expressed in a language which is acceptable to partisans of competing theories, the evidence constructed from the same (raw) data is often expressed in the partisans' differing theoretical languages."

That the chemical reaction described above resulted into a black powder (raw data) both Priestley and Lavoisier could agree upon, but not on what that black powder would be taken evidence for. Priestley claimed that in addition to the black powder a gas was formed during the reaction. Lavoisier accepted the principle of the conservation of mass, that the weights of what went into the reaction were the same as the weights of what came out. On the basis of this (theoretical) principle Lavoisier had reason to argue that there was no gas in addition to the black powder. (This is evidence rather than data.) Basu even identifies a number of levels of evidence. For instance, even to establish that the sample of iron is pure requires a test that is based on certain theoretical assumptions (the Stahlian theses). At this lower level, Priestley agreed with Lavoisier, but obviously evidence could be questioned at any level. Agreement may be required even to establish to which level something counts as raw data. Basu then concludes ([1], p. 364):

> "To the extent that these (raw) data are transformed into evidence, and for any evidential bearing these data might have on a particular theory and hence any bearing they might have on theory resolution, the evidence is theory-laden."

---

[12]For an account of these experiments, see Carrier ([6], to appear).

So, in a way, depending on their theoretical assumptions, Priestley and Lavoisier could have taken the same data, on which they agreed (namely black powder), as evidence for different phenomena.[13] There may not be any problem with raw data, but using raw data as evidence for a phenomenon is difficult without having the particular phenomenon in mind.

It is a long way from data to phenomenon, or sometimes from suspecting that there is a phenomenon to systematically collecting data about it in order to establish facts about the phenomenon (that can then be organized in a model). A phenomenon is experimentally or observationally examined. In the first instance, a phenomenon may be an object encountered in nature or the human environment. Yet, how to capture this phenomenon also increasingly depends on its empirical examination and theoretical description to this point, i.e. on one or more existing theoretical models of the phenomenon. The theoretical model is an attempt to capture the phenomenon by providing a description of it that is as complete as possible. This includes highlighting those factors that are relevant for constituting the phenomenon and may require omitting others that are considered more accidental. Data about a phenomenon can be produced by a whole range of different experimental procedures. They therefore present individual or isolated evidence about the phenomenon, while the phenomenon itself is expected to display a certain robustness in the face of the different experimental situations. Bogen and Woodward are right about this point, but this robustness does not automatically make phenomena into natural kinds or "bedrock elements of the furniture of the world", as Glymour put it. Raw data cannot serve to confirm a theoretical model about a phenomenon, but have to undergo procedures of data analysis and be put into the form of a data model in order to be usable for an empirical test. Thus, empirical confirmation takes place between the analysed data and the theoretical model, not between data and phenomenon.

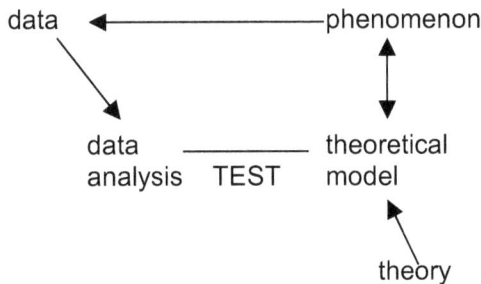

Figure 2.

Figure 2 shows that phenomenon and theoretical model remain closely connected, but the test of the model for a phenomenon takes a 'detour' via data

---

[13]The aim here is specifically not to express a preference for Lavoisier's chemical interpretation over Priestley's. At the time, such a preference would not have been empirically warranted. See, for instance, [13] who emphasizes the overlap between Lavoisier's and Priestley's methodological and ontological practices. See also Carrier [5].

generation and data analysis. When the phenomenon is examined experimentally or observationally, data about the phenomenon are produced. To compare the data with the theoretical models, data are required, but what the data are taken to be evidence for can depend on the phenomenon one has in mind and which one takes to be one's subject of investigation.

## 6. Conclusions

To summarize,

- phenomena are facts or events of nature that are subject to investigation;

- models are interpretative descriptions of phenomena that facilitate access to phenomena;

- theory in science is not that which tells us what the world is like, but that to which we (sometimes) resort when we try to describe what the world is like by developing models.

- 'Abstract', said of theories, means having been stripped of specific properties of concrete phenomena in order to apply to more and different domains.

- Models, in turn, are about concrete phenomena (or prototypes thereof) that have all the properties that real things have.

- Theories are applied to real phenomena only via models – by filling in the properties of concrete phenomena.

Being abstract and therefore not directly about empirical phenomena does not, however, render theories worthless or unimportant. Theories and models have to prove themselves at different levels, models by matching empirical phenomena and theories by being applicable in models of a whole range of different phenomena (or prototypes thereof).

A scientific model and its phenomenon are closely connected from the start in that the model is designed as a model *of* the phenomenon. Modelling involves judgements regarding which properties of a phenomenon need to be covered in a model in order to capture the central features of the phenomenon, and which properties count as accidental and can be omitted. What is taken to be the phenomenon becomes somewhat reconstructed in the course of the modelling process. The modelled phenomenon may depart somewhat from the physical reality, due to idealisations, both of the phenomenon (causal idealisation) and of the model (construct idealisation) [11]. However, despite this air of constructivism, it is the phenomenon the investigation of which results into data about the phenomenon. The link of the model to empirical evidence is required to be strong. While there is an empirical link, how we delineate and describe a phenomenon is invariable linked to the way we have learned to model it. What we take a phenomenon to be and how we model it develop together over time.

**REFERENCES**

1. P. K. Basu. Theory-ladenness of evidence: a case study from history of chemistry, *Studies in History and Philosophy of Science*, 34:351-368, 2003.
2. D. M. Bailer-Jones. Modelling Extended Extragalactic Radio Sources, *Studies in History and Philosophy of Modern Physics*, 31B:49-74, 2000.
3. J. Bogen and J. Woodward. Saving the Phenomena, *The Philosophical Review*, 97:303-352, 1988.
4. J. Bogen and J. Woodward. Observations, theories and the evolution of the human spirit, *Philosophy of Science*, 59:590-611, 1992.
5. M. Carrier. Cavendishs Version der Phlogistonchemie oder: Über den empirischen Erfolg unzutreffender theoretischer Ansätze. In: Mittelstraß, J., and Stock, G. (eds.), *Chemie und Geisteswissenschaften. Versuch einer Annäherung*, Berlin: Akademie Verlag, 35-52, 1992.
6. M. Carrier. Antoine L. Lavoisier und die Chemische Revolution, in: Leich, P. (ed.), *Leitfossilien naturwissenschaftlichen Denkens*, Würzburg: Königshausen & Neumann, 2004, to appear.
7. N. Cartwright. *Nature's Capacities and their Measurement*, Oxford: Clarendon Press, 1989.
8. N. Cartwright. Fables and models, Proceedings of the Aristotelian Society, Suppl., 65:55-68, 1991.
9. A. Chakravartty. The semantic or model-theoretic view of theories and scientific realism, *Synthese*, 127:325-345, 2001.
10. B. Glymour. Data and phenomena: A distinction reconsidered, *Erkenntnis*, 52:29-37, 2000.
11. E. McMullin. Galilean idealization, *Studies in History and Philosophy of Science*, 16:247–273, 1985.
12. J. W. McAllister. Phenomena and patterns in data sets, *Erkenntnis*, 47:217-228, 1997.
13. J. McEvoy. Continuity and discontinuity in the chemical revolution, *Osiris* ($2^{nd}$ series), 4:195-213, 1988.
14. M. C. Morrison. Models as Autonomous Agents, in: M. Morgan and M. Morrison (eds.), *Models as Mediators*. Cambridge: Cambridge University Press, 38-65, 1999.
15. E. Nagel. *The Structure of Science*, Indianapolis: Hackett Publishing Company, 1979.
16. K. J. Rohlfing. UNDERstanding. How infants acquire the meaning of UNDER and other spatial relational terms. Dissertation [on-line]. Bielefeld University, 2002. URL: http://archiv.ub.uni-bielefeld.de/disshabi/2002/0026/_index.htm
17. M. Suárez. The role of models in the application of scientific theories: epistemological implications. In: M. Morgan and M. Morrison (eds.), *Models as Mediators*. Cambridge: Cambridge University Press, 168-196, 1999.

# Truth versus Precision

Marcel Boumans

*University of Amsterdam*
*m.j.boumans@uva.nl*

**Abstract.** A typical difference between social science and natural science is the degree in which control is possible. Strategies in both sciences to obtain true facts are consequently different. Measurement errors are due to background noise. Laboratories are environments in which background conditions can be controlled. As a result, accurate observations – measurement results close to the true values of the measurands – can only be obtained in laboratories. Therefore, measuring instruments are built such that they function as mini laboratories. However, observations in social science are usually passive, in the sense that control of background conditions is impossible. Models are built to solve this problem of (lack of) control. They function as nonmaterial laboratories by aiming at precision, that is reducing the spread of the measurement errors. The application of models as measuring instruments necessitates a shift of the requirement of accuracy to the requirement of precision, which is a feature of the instrument and not of the environment.

> **Salviati** ... in our time it has pleased God to concede to human ingenuity an invention so wonderful as to have the power of increasing vision four, six, ten, twenty, thirty, and forty times, and an infinite number of objects which were invisible, either because of distance or extreme minuteness, have become visible by means of the telescope.
>
> **Simplicio** Everything that Salviati is presently setting forth is truly new to me. Frankly, I had no interest in reading those books, nor up till now have I put any faith in the newly introduced optical device. Instead, following in the footsteps of other Peripatetic philosophers of my group, I have considered as fallacies and deceptions of the lenses those things which other people have admired as stupendous achievements.
> (Galileo Galilei, *Dialogue Concerning the Two Chief World Systems – Ptolemaic & Copernican*, [8], 335-6)

## 1. Introduction

I share Dan Hausman's reason for being interested in a methodology of social science: "I would like to understand better how people manage to learn about the social world around them" ([11],4). This practice of managing to learn about the social world is dominated by model building. Therefore, to understand this practice we must try to apprehend how models function in empirical research. The kinds of models discussed in this paper are the mathematical models built and used in empirical social and economic research. These models are designed as quantitative representations of our social world. Their function is to generate numbers to inform us about social and economic aspects of the world. The central problem of this article is the assessment of the reliability of these bodies of knowledge.

To understand their specific function in empirical research, models should be distinguished from theories. They are not theories about the world but instruments through which we can see the world and so gain some understanding of it. Models are the social scientist's instruments of investigation, just as the microscope and the telescope are tools of the biologist and the astronomer. In a textbook on optical instruments, we find the following description that can easily be projected on models:

> The primary function of a lens or lens system will usually be that of making a pictorial representation or record of some object or other, and this record will usually be much more suitable for the purpose for which it is required than the original object. ([2], 15)

In the same way, models can be used to function as instruments to perform a particular kind of observation, namely, measurement. Models as measuring instruments generate numerical representations of the phenomena under investigation, which is often the kind of information needed for the purpose of policy deliberations.

Although mathematical models are not material, they function as though they are physical instruments. Therefore, standard methodology, traditionally focused on theories, is not suitable. Standard accounts define models in terms of their logical or semantic connections with theories, and methodology is traditionally seen as a way to appraise theories. Instruments (models) are not theories and therefore should be assessed differently. A separate methodology needs to be developed that is able to assess how mathematical models function as instruments. The aim of this paper is to indicate in which directions one could construct and refine such a methodology.

My starting point is Margaret Morrison and Mary Morgan's [18] account of models. According to their account, models must be considered as one of the critical instruments of modern science. Morrison and Morgan demonstrate that models function as instruments of investigation helping us to learn more about theories and the real world, because they are autonomous agents: that is to say, they are partially independent of both theories and the real world. We can learn from models because they can represent either some aspect of the world, or some aspect of a theory. The domain of inquiry here is limited to empirical models, that is, those models that inform us about the world.

To understand the function and nature of mathematical models they should not be considered as linguistic entities. Ronald Giere [9] also takes models as tools for representing the world. In his account, models represent aspects of the world similar to the way maps represent parts of the world. His example is a standard tourist map. He mentions two properties of maps that are relevant (p. 44): First, maps are not linguistic entities. They are physical objects. Therefore, it does not make sense to ask whether a map is true or false. Secondly, maps are not usually thought of as instantiations of any linguistic forms. Such an interpretation plays no role in understanding the nature or function of maps. Giere emphasizes that models represent by being similar – and not isomorphic – to the phenomena they represent: Models only represent partially, only some aspects of the phenomenon in question are represented, and which aspects are represented is context dependent.

One usually associates the word instrument with a physical device, such as a thermometer, microscope or telescope. However, the instruments of social science are not material objects, they are mathematical objects. Nevertheless, the mathematics fulfills the same role as metal, glass, cords and pulleys do in thermometers and clocks. Mathematics is the stuff non-material models are made of. Therefore, as for physical instruments like maps, thermometers and clocks, for mathematical models it also holds that instrument and representation coincide.

Modeling is a process of committing oneself to how aspects of the phenomenon should mathematically be represented and at the same time being constrained by the selected mathematical forms. Moreover, not every element in the mathematical model necessarily is connected to the phenomenon in question. To represent a phenomenon, sometimes, elements of convenience or fiction have to be introduced.[1] Because mathematics functions as material from which the instrument is built, possible flaws or blind spots in the use of mathematics could, like aberrations of the lenses in an optical system, be held responsible for possible artifacts.

Despite the fact that models function as physical instruments in empirical research, they cannot be assessed as such. The absence of materiality means that the physical methods used to test material instruments, such as control and insulation, cannot be applied to models. So, we cannot easily borrow from the philosophy of technology, which is geared to physical objects. Models, being 'quasi-material' objects belonging to a world in between the immaterial world of theoretical ideas and the material world of physical objects, require an alternative methodology.

Several case studies of model building in economics have led to the analogy of the process of model building with the process of baking a cake [1]. Models are built by fitting together ingredients from disparate sources. These ingredients are theoretical ideas, mathematical concepts and techniques, metaphors and analogies, facts about phenomena and empirical data. Integration takes place by translating the ingredients into a mathematical form and merging them into one framework. As a result, the context of discovery and the context of justification cannot be separated: justification is the successful integration of all required ingredients, including facts about the phenomenon and empirical data. Models built in this way are not assessed after they are built, by ex post empirical testing; models are assessed by whether they achieve their purpose, and, because in the model building process one works towards this goal, integration and justification are two sides of the same coin. A well-known sayings tells us that 'the proof of the pudding is in the eating', but if one prepares a pudding, tasting and smelling are essential steps to obtain a satisfactory result.

This paper will outline a few strategies of a methodology of models as instruments. The approach to understand the nature and function of models in the practice of empirical social research is similar to the way the process of model building is understood. The theoretical ingredients are bits from philosophy of science and bits from metrology. The analogy to understand the function and nature of models are physical instruments, like the thermometer, clock, microscope and filter. The

---

[1] A similar view is developed by Nancy Cartwright [3] in her simulacrum account of models.

empirical input comes from case studies of empirical research practices in social science. The justification of this approach hinges on a successful integration of all these ingredients.

## 2. Making Phenomena Visible

The models discussed here are built and used to produce numbers to inform us about social phenomena. Although phenomena are investigated by using observed data, the facts about them are not directly observable. To 'see' them we need instruments, and to obtain numerical facts about the phenomena in particular we need measuring instruments. This view is a consequence of James Woodward's [22] account of the distinction between phenomena and data. According to Woodward, phenomena are relatively stable and general features of the world and therefore suited as objects of explanation and prediction. Data, that is, the observations playing the role of evidence for claims about phenomena, on the other hand involve observational mistakes, are idiosyncratic and reflect the operation of many different causal factors and are therefore unsuited for any systematic and generalizing treatment.

Woodward characterizes the contrast between data and phenomena in three ways. In the first place, the difference between data and phenomena can be indicated in terms of the notions of error applicable to each. In the case of data the notion of error involves observational mistakes, while in the case of phenomena one worries whether one is detecting a real fact rather than an artifact produced by the peculiarities of the instrument. A second contrast between data and phenomena is that phenomena are more 'widespread' and less idiosyncratic, less closely tied to the details of a particular instrument or detection procedure. A third way of thinking about the contrast between data and phenomena is that scientific investigation is typically carried on in a noisy environment, an environment in which the observations reflect the operation of many different causal factors.

> The problem of detecting a phenomenon is the problem of detecting a signal in this sea of noise, of identifying a *relatively stable and invariant* pattern of some simplicity and generality with recurrent features – a pattern which is not just an *artifact* of the particular detection techniques we employ or the local environment in which we operate. Problems of experimental design, of *controlling* for bias or error, of selecting appropriate techniques for measurement and of data analysis are, in effect, problems of tuning, of learning how to separate signal and noise in a *reliable* way. ([22], 396-397; italics are mine)

Underlying the contrast between data and phenomena is the idea that theories do not explain data, which typically will reflect the presence of a great deal of noise. Rather, an investigator first subjects the data to analysis and processing, or alters the experimental design or detection technique, in an effort to separate out the phenomenon of interest from extraneous background factors: "It is this extracted signal rather than the data itself which is then regarded as a potential object of explanation by theory" (ibid. 397).

Theories are incomplete, in the sense that they do not provide information about facts about the phenomena. Though theories explain phenomena, they often (particularly in social science) do not have built-in application rules for mathematizing the phenomena. Moreover, theories do not have built-in rules for measuring the phenomena. For example, theories tell us that metals melt at a certain temperature, but not at which temperature (Woodward's example); or they tell us that capitalist economies give rise to business cycles, but not the duration of recovery. In practice, by mediating between theories and the data, models may overcome this incompleteness of theories. As a result, models that function as measuring instruments are located on the theory-world axis mediating between 'data' and 'facts about the phenomenon'.

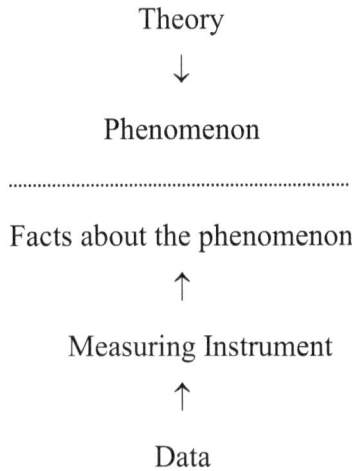

Theory

↓

Phenomenon

...............................................................

Facts about the phenomenon

↑

Measuring Instrument

↑

Data

Figure 1.

Instruments located between 'data' and 'facts about the phenomenon' on the theory-world axis are not assessed as rendered in the standard account of testing, namely by confronting the output of a model with data. The output of a model are 'facts about the phenomenon'. They are previously unobservables made visible by the detection instrument. As we will see, for the assessment of any instrument we need another instrument to compare outputs.

The italicized terms in the quotation above indicate the problems that will now be explored further:

– the problem of invariance

– the problem of artifacts

– the problem of control

– the problem of reliability

## 3. The Problem of Invariance

The dominant measurement theory of today is the representational theory of measurement (cf. [15]). The core of this theory is that measurement is a process of assigning numbers to attributes of the empirical world in such a way that the relevant qualitative empirical relations among these attributes or characteristics are reflected in the numbers themselves as well as in important properties of the numbers system. The problem, however, is that the representational theory of measurement has turned too much into a pure mathematical discipline, leaving out the question of how the mathematical structures gain their empirical significance in actual practical measurement. The representational theory lacks concrete measurement procedures and devices.

This problem of empirical significance is discussed by Michael Heidelberger [13], who argues for giving the representational theory a 'correlative interpretation', based on Gustav Fechner's principle of measurement. Fechner had argued that

> the measurement of any attribute $x$ generally presupposes a second, directly observable attribute $y$ and a measurement apparatus $A$ that can represent variable values of $y$ in correlation to values of $x$. ... Normally, we try to construct (or find) a measurement apparatus which realizes a 1:1 correlation between the values of $x$ and the values of $y$ so that we can take the values of $y$ as a direct representation of the value of $x$. [12][2]

To realize a correlation between observables and measurand we need an apparatus.

Thus, measurement is based on a correlative relation, represented by $f$, between the true values of the measurand, $x$, and the associated observable quantity $y$:

$$y_i = f(x) + \varepsilon_i \tag{1}$$

where $y_i$ is the $i$th observation ($i = 1, \ldots, N$) and $\varepsilon_i$ the corresponding observational error. To gain a better understanding of measurement we must have a closer look at the nature of correlative relations and how it is linked to instruments.

$y$ can only be an observation of $x$ if $x$ influences the performance of $y$. The observational errors, the noise, are caused by the operation of many different background conditions, also called 'other circumstances', symbolized by $OC$:

$$y = f(x, OC) \tag{2}$$

To clarify the correlation as a causal relationship, it is represented as a differential equation:

$$\Delta y = \frac{\partial f}{\partial x} \Delta x + \frac{\partial f}{\partial OC} \Delta OC \tag{3}$$

---

[2] $Q$ and $R$ in the original text are replaced by $x$ and $y$, respectively, to make the discussion in this article uniform.

A necessary requirement for correlations is that they are invariant, that is that the differential coefficient of $x$, $\frac{\partial f}{\partial x}$, is invariant for (a certain range of) variation in $x$, $\Delta x$, and (a certain range of) variations of the background conditions, $\Delta OC$ (see [23]).

## 4. The Problem of Reliability

In the correlational theory of measurement, the correlation and the measurement apparatus are connected. So to consider the nature of correlations we can use theories of instruments in which requirements of an instrument's performance are discussed. In his *Measuring Instruments: Tools of Knowledge and Control,* Sydenham emphasizes that

> all measurements are always imperfect. They are subject to a whole host of errors sources. These decide the degree of confidence that can be placed upon the measurements made. Creation and application of measuring technology is basically a case of creating a system that has optimal compromises of features so that it is just slightly better than the task requires. ([21], 46)

Two basic terms are used to describe the performance of measuring instruments. These terms are precision and accuracy, and will be explicated before we continue discussing correlations.

According to the 'International Vocabulary of Basic and General Terms in Metrology'[3], accuracy is defined as the "closeness of the agreement between the result of a measurement and a true value of the measurand" ([14], Definition 3.5, 24). A note to this definition explicitly mentions that accuracy is a qualitative concept. The distance between a measurement result, $\hat{x}$, and the true value, $x$, can be expressed by the measurement error $\hat{\varepsilon}$:

$$\hat{x} = x + \hat{\varepsilon} \tag{4}$$

The true value is unknown, so for the assessment of the instrument's accuracy we need a theory of the measurand, that is, of the phenomenon to be measured.

According to a dictionary of statistics [6], precision is defined as the likely spread of estimates. It should be noted that precision is not defined in the International Vocabulary of Metrology, only closely related concepts like 'repeatability' and 'reproducibility'. For the assessment of the instrument's precision we need a theory of error. Accuracy is a statement about closeness which implies judgments about location and spread. We might require of a measurement procedure that it have

---

[3]The 'International Vocabulary of Basic and General Terms in Metrology' is the standard dictionary of metrology that has been prepared simultaneously by a joint working group consisting of experts appointed by: BIPM (International Bureau of Weights and Measures), IEC (International Electrotechnical Commission), IFCC (International Federation of Clinical Chemistry), ISO (International Organization for Standardization), IUPAC (International Union of Pure and Applied Chemistry), IUPAP (International Union of Pure and Applied Physics), and OIML (International Organization of Legal Metrology).

its mean located near or equal to the true value – in statistics referred to as unbiasedness – and to have small spread, that is the measurement results are as precise as possible. To illustrate the difference between the concepts of precision and accuracy, Sydenham [21] compares measurement with rifle shooting. A precise group of shots lie close together. A group of shots is unbiased when the mean lies in the bull's-eye. The mean of a precise instrument can lie quite eccentric, far off from the bull's eye, due to a deficiency of the instrument, e.g. one of its pointers is bent. So a precise instrument can produce an artifact, that is, the aim for preciseness does not prevent errors due to the instrument being used.

## 5. The Problem of Control

Accuracy is, of course, the ultimate goal of any measurement. Truth is what we aim at. The prime strategy to obtain accurate measurements is to take measurements in a controlled experiment. In a laboratory, we artificially isolate a selected factor $x$ from other influences, in other words we take care that *ceteris paribus* (*CP*) conditions are imposed: $\Delta OC = 0$ or $OC = 0$, see equation (3), so that the remaining factor $x$ can be varied in a systematic way to gain knowledge about the relation between $x$ and $y$:

$$\frac{\partial f}{\partial x} = \frac{\Delta y_{CP}}{\Delta x} \tag{5}$$

If the ratio of the variation of $y_{CP}$ and the variation of $x$ appears to be stable, the correlation is an invariant relationship and can thus be used for measurement aims.

So, an observation in a controlled experiment is an accurate measurement because of the elimination of background noise (noise $\varepsilon = 0$).

$$y_{CP} = f(x) \tag{6}$$

A measuring instrument functions as an accurate instrument if it is built as a mini-laboratory.

$$\hat{x} = y_{CP} \tag{7}$$

A nice example of such a mini-laboratory is the precision balance, invented by J.J. Magellan: the actual balance is enclosed inside of a glass box with all of the necessary equipment to perform experiments.

However, in social science we can only control the environment to a certain limited extent. Moreover, even in a laboratory, there are always circumstances one cannot control. Fortunately, a measuring instrument is also accurate when it is designed, fabricated and used in such a way that the influences of all these uncontrollable circumstances are negligible. For example, a gas thermometer is more accurate than a mercury thermometer, because the expansion of glass is negligible compared with the expansion of gas (see also [5]).

To avoid the problem of the lack of control is to design and use measuring instruments in such a way that the influences of all the uncontrollable circumstances are negligible; in other words, a measuring device should be constructed and used in such a way that it fulfills the *ceteris neglectis* condition. To clarify this condition, let us suppose that we care about the correlative relation between a property $x$ to be measured and the associated quantity $y$, see equation (3). The instrument should be constructed and used so that it is sensitive to changes in $x$ and at the same time insensitive to changes in the other circumstances ($OC$):

$$\frac{\partial f}{\partial OC} \approx 0 \tag{8}$$

If we can construct the instrument fulfilling the *ceteris neglectis* condition, we do not have to worry about the extent to which the other circumstances are changing. They do not have to be controlled as is assumed by the conventional *ceteris paribus* requirements.

A measurement formula must be a representation of a lawful, that is, invariant, relationship. According to Nancy Cartwright [4], for lawful relationships we need stable environments: nomological machines. A nomological machine is

> a fixed (enough) arrangement of components, or factors, with stable (enough) capacities that in the right sort of stable (enough) environment will, with repeated operation, give rise to the kind of regular behaviour that we represent in our scientific laws. ([4], 50)

So, measuring instruments can only fulfill their measurement task when they are nomological machines. To build them we must be able to control the circumstances which is highly problematic in social science. However, invariant relationships are not always the result of *ceteris paribus* environments but could also occur because the influence of the environment is negligible, in other words invariant relationships could also be *ceteris neglectis* regularities.

Though we can be more liberal about nomological machines – we can relax the requirement of stable environments – in social science we are 'passive' observers, that is, not being able to carry out controlled experiments, and thus fully dependent on the existence of nomological machines in the real world. Unfortunately, they are rare. God did only create a few.

Even, if we observe passively ($PO$) stable relationships (and we often do of course)

$$\Delta y_{PO} = \frac{\partial f}{\partial x} \Delta x \quad \text{and} \quad \frac{\partial f}{\partial OC} \Delta OC = 0 \tag{9}$$

we are never certain whether this stability is due to the fact that the other influences are negligible or whether the background conditions were stable for the data set used to arrive at this relationship:

$$\frac{\partial f}{\partial OC} \approx 0 \quad \text{or} \quad \Delta OC = 0 \tag{10}$$

Each empirical relationship is a representation of a specific data set. So for each data set it is not clear whether potential causal factors are negligible or only dormant. In econometrics, this is the so-called problem of passive observation and is discussed in length and detail by Haalvemo [10] in his revolutionary paper 'The Probability Approach in Econometrics'.

This problem is dealt with by the strategy of 'realisticness'[4] and it works as follows (see [20]): when a relationship appears to be inaccurate, this is an indication that a potential relevant factor is omitted. As long as the resulting relationship is inaccurate, potential relevant factors should be added. The expectation is that this strategy will result in the fulfillment of two requirements: 1) the resulting model captures a complete list of factors that exert large and systematic influences; 2) all remaining influences can be treated as a small noise component. The problem of passive observation is solved by accumulation of data sets: the expectation is that we converge bit by bit to a closer approximation to the complete model, as all the most important factors reveal their influence.

This strategy however is not applicable in cases when there are influences that we cannot measure, proxy, or control for, but which exert a large and systematic influence on the outcomes. Confronted with the inability of control, social scientists deal with the problem of invariance and accuracy by using models as virtual laboratories. Mary Morgan [17] discusses the differences between 'material experiments' and 'mathematical models as experiments'. In a mathematical model, control is not materialized but assumed. As a result, accuracy has to be obtained in a different way.

To measure $x$, a model, $M$, is specified of which the $y_i$'s function as input and where the $\alpha_k$'s are the parameters of the model:

$$\hat{x} = M[y_i(i = 1, \cdots, N), \alpha_k(k = 1, \cdots, K)] \tag{11}$$

If one substitutes equation (1) into model $M$, one can derive that, assuming $M$ is a linear operator (usually the case):

$$\hat{x} = M[f(x) + \varepsilon_i, \alpha_k] = M_x[x, \alpha_k] + M_\varepsilon[\varepsilon_i, \alpha_k] \tag{12}$$

A necessary condition for $\hat{x}$ to be a measurement of $x$ is that a model $M$ must involve a representation of the phenomenon, $M_x$, and a specification of the error term, $M_\varepsilon$. As we have seen, to obtain an accurate observation in a laboratory the observational error is reduced by controlling the environment. To obtain an accurate measurement result with an immaterial mathematical model, the model parameters are adjusted.

## 6. The Problem of Artifacts

So, tuning, that is separating signal and noise, is done by adjusting the parameter values. However, a true signal can only be obtained by a perfect measurement, and

---

[4]This term is introduced by Uskali Mäki [16] to separate it from 'realism', which refers to a philosophical view of scientific theories.

so is by nature indeterminate. To deal with this problem we split the measurement error in two parts:

$$\hat{\varepsilon} = \hat{x} - x = \hat{x} - M_x + M_x - x = M_\varepsilon + (M_x - x) \tag{13}$$

$(M_x - x)$ is the part of the error term that reflects the location problem and like true value cannot be completely known. Therefore, in practice we aim at reducing the error term $M_\varepsilon$ as much as possible by reducing the spread of the error terms, in other words by aiming at precision.

The aim at precision does not prevent the problem of artifacts. Accuracy and precision determine the performance of an instrument and thus decide the degree of confidence that can be placed upon the measurements made. Often the improvement of one feature implies a reduction of the other, and then we have to balance between accuracy and precision. An important strategy to deal with the problem of artifacts is calibration.

In metrology, calibration is defined in terms of a comparison of the measuring instrument and a standard ([14], Definition 6.11, 48). A standard is an instrument chosen as reference ([14], Definition 6.1, 45). We label this kind of calibration 'external assessment' because it is the comparison of (the output of) an instrument with (the output of) another instrument, chosen to be a standard. Though both lie probably close together, the standard value and true value do not necessarily coincide.

If there is no standard available because there are no other models available that produce the same facts about the phenomenon in question, in other words if the instrument is unique – which is often the case in social science – then the assessment is carried out by investigating the inner workings of the instrument. Allan Franklin [7] discusses nine epistemological strategies to distinguish between a valid observation and an artifact. One of these strategies is calibration, which he defines as "the use of a surrogate signal to standardize an instrument":

> If an apparatus reproduces known phenomena, then we legitimately strengthen our belief that the apparatus is working properly and that the experimental results produced with that apparatus are reliable. ([7], 31)

We need known facts about phenomena to calibrate an instrument, but how are these facts about phenomena known: by another instrument or by theory? Moreover, these facts have to be stable across measurement environments, but which facts satisfy this stability requirement?

Assessment by calibration runs the danger of circularity. Even in the case of calibration as an internal assessment, we need knowledge generated by other instruments to avoid too much convention. For example, in economics, models are calibrated using so-called 'stylized facts', or they are called 'natural rates' but these are disputed as empirical facts.

For calibration we need stable and general features of the world. In natural science, these are the so-called physical constants, but what are they in social science? A usual response of economists to the accusation that economic models do not meet standard (read natural) scientific requirements is that the economic world is not as stable as the physical world. Margaret Schabas [19] undermines this defense of economists by showing that "economists may study a world of much greater constancy than they are traditionally willing to concede" (195). It might be true that when we look at the world of the natural scientist, things appear to be fairly stable, but what matters is the world that we know and perceive. The phenomena being studied in natural science have drastically altered within one century, think of (Schabas' examples) radioactive decay, black holes, isotopes, or chromosomes. In contrast to this, for the economist, once capitalism had taken hold several centuries ago, all of the key elements have remained more or less the same: factor markets, central banking, joint stock companies, insurance brokers, government bonds. The economic phenomena are all products of these human agencies and insofar as such agencies remain fairly constant, so do the phenomena that are generated. Schabas mentions striking examples of stable facts about these phenomena, which support her account: the exchange ratio of gold to silver hovering in the vicinity of 1:16, the periods of business cycles of roughly ten years, and the real interest rate for peacetime developed economies in the vicinity of 3 percent. When these known facts are available why not to use them to calibrate economic models?

## 7. Conclusions

A separate methodology of models that does justice to the idea that models function as autonomous instruments of investigation has to reconsider and consequently redefine central methodological concepts of instruments: To see how models acquire reliability, testing of models has to be reevaluated in terms of calibration. Models as measuring instruments induce faith if they are precise. However the aim for precision does not prevent the instrument from generating artifacts. Whether outcomes of a model are facts or artifacts is assessed by calibration: the evaluation whether the model is able to reproduce known stable (enough) facts.

## REFERENCES

1. M. Boumans. Built-in Justification. In: *Models as Mediators*, eds. Mary S. Morgan and Margaret Morrison, Cambridge: Cambridge University Press, 66-96, 1999.
2. R. J. Bracey. *The Technique of Optical Instrument Design*. London: The English University Press, 1960.
3. N. Cartwright. *How the Laws of Physics Lie*. Oxford: Clarendon Press, 1983.
4. N. Cartwright. *The Dappled World. A Study of the Boundaries of Science*. Cambridge: Cambridge University Press, 1999.
5. H. Chang. Spirit, Air, and Quicksilver: The Search for the "Real" Scale of Temperature. *Historical Studies in the Physical and Biological Sciences*, 31.2: 249-284, 2001.
6. B. S. Everitt. *The Cambridge Dictionary of Statistics*. Cambridge: Cambridge University Press, 1998.
7. A. Franklin. Calibration. *Perspectives on Science*, 5: 31-80, 1997.

8.  G. Galilei. *Dialogue Concerning the Two Chief World Systems – Ptolemaic & Copernican*, $2^{nd}$ revised edition, translated by Stillman Drake. Berkeley, Los Angeles, London: University of California Press, 1967.

9.  R. N. Giere. Using Models to Represent Reality. In *Model-Based Reasoning in Scientific Discovery*, eds. Lorenzo Magnani, Nancy J. Nersessian and Paul Thagard. New York, e.a.: Kluwer Academic/Plenum Publishers, 41-57, 1999.

10. T. Haavelmo. The Probability Approach in Econometrics. Supplement to *Econometrica*, 12, 1944.

11. D. M. Hausman. *The Inexact and Separate Science of Economics*. Cambridge: Cambridge University Press, 1992.

12. M. Heidelberger. Fechner's Impact for Measurement Theory. *Behavioral and Brain Sciences*, 16.1:146-148, 1993.

13. M. Heidelberger. Alternative Interpretationen der Repräsentationstheorie der Messung. In: *Proceedings of the 1st Conference "Perspectives in Analytical Philosophy"*, eds. G. Meggle and U. Wessels. Berlin and New York: Walter de Gruyter, 1994.

14. *International Vocabulary of Basic and General Terms in Metrology*. International Organization for Standardization, 1993.

15. D. H. Krantz, R. D. Luce, P. Suppes and A. Tversky. *Foundations of Measurement*. 3 Vols. New York: Academic Press, 1971, 1989, 1990.

16. U. Mäki. On the Problem of Realism in Economics. *Fundamenta Scientiae*, 9:353-373, 1988.

17. M. S. Morgan. Experiments without Material Intervention: Model Experiments, Virtual Experiments, and Virtually Experiments. In: *The Philosophy of Scientific Experimentation*, ed. Hans Radder. Pittsburgh: University of Pittsburgh Press, 216-235, 2003.

18. M. Morrison and M. S. Morgan. Models as Mediating Instruments. In *Models as Mediators*, eds. Mary S. Morgan and Margaret Morrison. Cambridge: Cambridge University Press, 10-37, 1999.

19. M. Schabas. Parmenides and the Cliometricians. In *On the Reliability of Economic Models*, ed. Daniel Little. Boston: Kluwer Academic Press, 183-202, 1995.

20. J. Sutton. *Marshall's Tendencies: What Can Economists Know?* Leuven and Cambridge, MA: Leuven University Press and The MIT Press, 2000.

21. P. H. Sydenham. *Measuring Instruments: Tools of Knowledge and Control*. London: Peter Peregrinus, 1979.

22. J. Woodward. Data and Phenomena. *Synthese*, 79:393-472, 1989.

23. J. Woodward. Explanation and Invariance in the Special Sciences. *The British Journal for the Philosophy of Science*, 51:197-254, 2000.

# Humean Effective Strategies

Carl Hoefer

*ICREA (Institució Catalana de Recerca i Estudis Avançats, Passeig Lluís Companys 23, Barcelona, 08010 (Spain)*
*carl.hoefer@uab.es*

**Abstract.** In a now-classic paper, Nancy Cartwright argued that the Humean conception of causation as mere regular co-occurrence is too weak to make sense of our everyday and scientific practices. Specifically she claimed that in order to understand our reasoning about, and uses of, effective strategies, we need a metaphysically stronger notion of causation and causal laws than Humeanism allows.

Cartwright's arguments were formulated in the framework of probabilistic causation, and it is precisely in the domain of (objective) probabilities that I am interested in defending a form of Humeanism. In this paper I will unpack some examples of effective strategies and discuss how well they fit the framework of causal laws and criteria such as *CC* from Cartwright's and others' works on probabilistic causality. As part of this discussion, I will also consider the concept or concepts of objective probability presupposed in these works. I will argue that Cartwright's notion of a *nomological machine*, or a *mechanism* as defined by Stuart Glennan, is better suited for making sense of effective strategies, and therefore that a metaphysically primitive notion of causal law (or singular causation, or capacity, as Cartwright argues in (1989)) is not – here, at least – needed. These conclusions, as well as the concept of objective probabilities I defend, are largely in harmony with claims Cartwright defends in *The Dappled World*. My discussion aims, thus, to bring out into the open how far Cartwright's current views are from a radically anti-Humean, causal-fundamentalist picture.

## Introduction.

Throughout her career, Nancy Cartwright has consistently argued against the Humean prejudices of her logical empiricist predecessors, at least in the areas of causality and the epistemology of science. The first assault in her campaign was the classic paper "Causal Laws and Effective Strategies" (1979, 1983 ch. 1). This paper argues for two main theses. First, that there is no way to reduce facts about causation to facts about probabilistic relations; and second, that in order to understand the effective strategies we use to achieve desired results, we need to invoke a strong notion of *causal laws*. When we know that it is a causal law that C brings about E (or raises the level of E, or makes E more probable, ...), then we have an effective strategy for E. It is only the second thesis that I will be attempting to undermine in this paper, by showing that the talk of causal laws, and the implicit picture of Cartwright's (1979) paper, have some serious faults.

What I will do is focus on aspects of the problematic that Cartwright glosses over relatively briefly, and try to show how a slightly different way of thinking about things can work equally well – perhaps better – at uncovering and describing our effective strategies. The point will be to show that this different perspective is wholly compatible with *Humeanism about (real, or objective) probabilities*, and with *agnosticism about causation* as a primitive relation (i.e., causal agnosticism as opposed to causal fundamentalism). The goal is to show how we can account for our effective strategies, without buying wholesale into an ontology of causal laws, singular causation, and capacities. In this sense, I will defend the spirit of Humeanism about causation, at least in a small way. I will not, however, try to argue for a view effective strategies that is purged of any taint of causal talk.

In elaborating a different view of effective strategies, I will borrow heavily from some leading ideas of Cartwright's latest book, *The Dappled World*.[1] This paper is therefore an attempt both to defend much of the perspective offered in *The Dappled World*, and to show that it contrasts strongly with the causal-fundamentalist picture to be found in some of Cartwright's earlier works.

## 1. Cartwright against Humean probabilistic causation.

In this section I will describe the main points of Cartwright's (1979) paper, including the famous criterion *CC* and three key examples of effective strategies (Malaria, TIAA-CREF, heart disease). Cartwright uses her examples to argue very convincingly that a Humean account of causation that seeks to reduce causal facts to facts about probability relations is doomed to fail. What she offers in its stead is a species of what might be called "causal fundamentalism",[2] namely a view that takes "causal laws" to be fundamental facts of our universe.

> "If indeed, it *isn't true that* buying a TIAA policy is an effective way to lengthen one's life, but stopping smoking is, the difference between the two depends on the causal laws of our universe, and on nothing weaker." (1983), p. 22.

I have never felt I understood what a "causal law" is, and in (1983) Cartwright does not give us an explicit definition. However, we do get an implicit definition: At least, the true statements "C ↪ E" that pass the test of principle *CC* should be counted as causal laws. Later we will come back to the issue of what constitutes a causal law.

The basic idea of a Humean reductive theory of (probabilistic) causation is that C causes E if $P(E|C) > P(E|\text{-}C)$ and some other conditions (all of which should be non-causal, i.e. compatible with whatever Humeanism is in play) are satisfied as well. C and E should be event types, not particulars, at least in the class of theories of interest to us here. Since there are many well-known examples of factors that increase the probability of an effect E *via* spurious correlation rather than causation, the real content of the theory will naturally be in the extra conditions. For example, since the probability of lightning is greatly increased by the presence, less than ten seconds later, of thunder, we need a condition that helps our theory rule out thunder being a cause of lightning. Suppes' (1970) theory of probabilistic causation, one of the first, took a sensible approach to cases like this: insist that the cause C must occur before the effect E. Unfortunately, this by no means finishes the task of eliminating spurious correlations. There are ubiquitous cases of effects of a common cause (D $\hookrightarrow$ C and D $\hookrightarrow$ E), where C regularly happens before E and is strongly positively correlated with E, but is not a cause of E. Ruling out cases of this form, and a variety of more complicated forms, is a job that has never proved possible, at least within the strictures of Humeanism.

Cartwright offers cases with the probability structure known as Simpson's paradox to illustrate her general argument against the Humean approach to probabilistic causation. Let's take the smoking/heart disease thought example (the probabilities we will posit are by no means true of any actual populations). Suppose that it were found that, in the statistics for the whole adult population, P(HD|SM) < P(HD|-SM). This could happen even if smoking is in fact a cause of heart disease and not a preventer of it. How? Well, suppose that regular exercising is a strong preventer of heart disease, and that as it happens, the frequency of regular exercising is much higher in the smoking population than in the non-smoking population. Then the probability relation mentioned above could hold, yet when we partition the population into exercisers and non-exercisers (and "hold fixed" this factor, conditionalize on it), the probabilistic significance of smoking reverses: P(HD|SM & EX) > P(HD|-SM & EX) and P(HD|SM & -EX) > P(HD|-SM & -EX). And these probabilities, we are to take it, reflect the true causal facts, that smoking does cause heart disease.

The point Cartwright makes with these examples is simple but devastating: if there are other causal factors relevant to an effect E (positively or negatively) that may induce misleading probabilities, we have to hold them fixed in order for the probability of E given C to genuinely reflect the fact that C [causes/prevents] E. So if there are five other genuine causes/preventers of a given E, $C_i$ for i = 1 to 5, then in order to judge whether C causes E what we need to look at is the probabilistic relevance of C for E, in each of the subpopulations where each of these five factors $C_i$ is held fixed (positively or negatively), e.g., P(E|C & $C_1$ & -$C_2$ & $C_3$ & $C_4$ & -$C_5$). Formalizing this notion Cartwright gets *CC*:

> "*CC*: C $\hookrightarrow$ E iff Prob(E|C & $K_j$) > Prob(E|$K_j$) for all state descriptions $K_j$ over the set $\{C_i\}$, where $\{C_i\}$satisfies
>
> (i) $C_i \in \{C_i\} \Rightarrow C_i \hookrightarrow$ +/- E
>
> (ii) C $\notin \{C_i\}$
>
> (iii) $\forall$D (D $\hookrightarrow$ +/- E $\Rightarrow$ D = C or D $\in \{C_i\}$)
>
> (iv) $C_i \in \{C_i\} \Rightarrow \neg(C \hookrightarrow C_i)$." [3]

This is not, of course, an analysis or definition of the causal relation $\hookrightarrow$ in terms of probability, because the relation occurs on both sides of the *iff*. It is, rather, as Cartwright puts it, "... the strongest connection that can be drawn between causal laws and laws of association."[4]

It is also a disaster for the basic Humean programme of reducing causation to probabilistic relations. Viewing it for the moment as an epistemic recipe, what *CC* says is that in order to infer that C is a cause of E from probabilities, one has to first know *all the other causes* of E, and examine the effect of C on E in *each* of the subpopulations holding fixed a combination +/- of these other causal factors. *CC*'s truth (if it is true) does not logically preclude a successful reduction of causation to probabilistic facts. But it does make it look rather unlikely, and makes it more natural to see the logical relationship going in the other direction: causal facts are (logically, or ontologically) prior, and give rise to the probabilistic facts. And that is part of Cartwright's causal fundamentalist view: causal relations *give rise to* probabilistic relations by their operation, and the latter are at best a dubious tool to be used in trying to infer the existence of the former.

So much for causal laws, as implicitly defined by *CC*. The application to effective strategies is straightforward: If C $\hookrightarrow$ E, then introducing (or augmenting, increasing, ...) C is an effective strategy for bringing about E, in all circumstances. Joining TIAA is not an effective strategy for extending ones' life, because there is (presumably) no causal law that joining TIAA $\hookrightarrow$ longer life. And spraying oil on swamps is an effective strategy for preventing malaria, while burning the blankets of the sick is not, because of the (presence/absence) of the corresponding causal laws linking these event types.

Before we take a critical look at the examples of causal laws and effective strategies used by Cartwright, we need to pause for a moment to think about the *objective probabilities* being used in the discussion.[5] She discusses the question in section 2.2 of CLES, and insists that they must be understood as simply sufficiently-stable [actual] frequencies. She does not want to make a stronger linkage between probability and causation possible by going metaphysical, opting for some primitive notion of propensity or a translation into counterfactuals. And with this I am in full agreement: there is no call to ruin a perfectly good notion like objective probability, just because we can't make it link up nicely with facts about causation.

> "Probabilities serve many other concerns than causal reasoning and it is best to keep the two as separate as possible. In his *Grammar of Science* Karl Pearson taught that probabilities should be theory free, and I agree." (1983), p. 39.

As we will see in section 3, however, this simple frequentist view of objective probabilities lands *CC* and the associated view of causal laws in great difficulties. But before we get to these, in the next section I want to lay out the elements of a different – but equally empiricist, equally Humean – account of objective chance.

## 2. Humean objective chance.

The need for a Humean account of objective probabilities (or chance, as I will usually say) different from simple actual frequentism is not hard to see. Though not all – perhaps not even most – of the traditional complaints against actual frequentism are sound[6], still there are some glaring problems that have made the view a nearly extinct species in recent decades. First of all, we expect the actual frequencies of things to at best come close to the real probabilities, an expectation that is weaker the smaller the number of actual cases involved. For example, the proportion of heads among well-flipped coins, in the history of the world, is no doubt near 0.5, but it is also no doubt *not exactly* 0.5. Yet it would be nice to have a way to say that 0.5 is in fact the correct value of the probability. The example can be strengthened by considering similar cases where the numbers are much smaller. Suppose that a proper roulette wheel with exactly 25 slots was only built once in history, and used just briefly in an obscure French casino. And suppose that the ball only fell into the 00 slot on that wheel in 2.333 percent of the spins. It would nevertheless be nice to be able to say that the probability was in fact 4%, without going off the metaphysical deep end in order to do so.

A second problem, perhaps worse, is that statistics and frequencies are ubiquitous, but not all of them should be thought of as probabilities. The frequency of men with only silver coins in their pockets on Tuesdays that are their birthdays, in the whole population of such men (on such days) is a statistic of no meaning and no utility. Notice that this sort of statistic does not support a temporal reading such as to potentially guide expectations. It is *not* to be identified with the probability that, on my next Tuesday birthday, I will end up having only silver coins in my pocket. It can be converted to a probability if we instead read it as the probability of *getting someone who only has silver coins in his pocket*, if one *randomly selects a man on one of his Tuesday birthdays* from the entire pool of such individuals over all history. The latter sort of gloss allows one to turn any mere statistic into a genuine objective probability, but this does not justify the former sort of reading, which is what we mostly would like to have.[7]

It is commonplace now to insist, following Ian Hacking, that objective probabilities can only be associated to proper *chance setups*. Frequentism does not build in this restriction, even if we add a requirement of stability. (The method of translating any statistic into a probability in footnote 7 is, in effect, a method of building the statistic into a proper chance setup.) But what sort of a thing is a chance setup, and how should we motivate the distinction between proper objective chances and mere statistics?

The account I will briefly sketch reflects the trajectory of my own interest in probabilities, which began with an interest in David Lewis' (1994) account and grew off in a different direction from there. Cartwright has never had much sympathy for Lewis-style Humean programmes, and her interest in objective probabilities has always been closer to the needs and practices of ordinary science. Despite this difference, I think the account I will sketch here is very close to the account developed in Cartwright (1999), chapter 7.[8]

## 2.1. What Chances are For.

If you know that something is the case, or you know that some other thing is definitely going to happen, then you are all set; knowing the probability of those things is then at most of academic interest to you. But often we have to work in circumstances of ignorance. I don't know whether it will rain tomorrow, so knowing the objective probability that it will (if such a thing exists) would be very useful to me. If it is less than 20% I will wear my new shoes and not take an umbrella, but if it is more than 80% I will dress warmly, wear old shoes, and carry my big umbrella. Objective probability is, in the now well-known phrase, a "guide to life". That is its nature or essence, if you like, and this role is neatly captured in David Lewis' "Principal Principle", which says roughly:

> *PP:* If you have no background knowledge relevant to whether or not it

is (or will be) the case that $A$, other than perhaps background knowledge concerning the objective chance of $A$, then if you come to believe that the objective chance of $A$ is in fact $x$, your subjective degree of belief in the truth of (or coming to pass that) $A$, should also be $x$.[9]

PP is meant to be a rather obvious principle of rationality: if you don't follow it, you are either being perverse in some way, or falling short of logical coherence, or you simply don't understand the concept of objective chance. *The point of saying that* the probability of 6 upon rolling a fair die is $1/6$ is precisely to indicate what a rational degree of belief (hence rational/fair betting behavior, etc.) in that outcome is. It is not just a shorthand way of saying what the actual frequency of 6's is, in the past or even in all of history, though we do expect chances and frequencies to be numerically close, in most cases. Nor is it a way of saying that there are 6 possible ways for a die to land and that we are indifferent between them. Not only are there other cases where we can "be indifferent" in two or more ways, yielding contradictory prescriptions for the probabilities, but moreover mere indifference is no grounds for saying what objective chances are. If you know *nothing* about a die that someone hands you, then you certainly don't know what its chance of landing 6 is! On the other hand if you know that it is a perfect cube (with rounded edges), has uniform density and is not magnetic, etc., then you may indeed have grounds for saying that the chance of heads is $1/6$, but these grounds are not best thought of as a matter of "indifference".

Finally, to say that the chance of heads is $1/6$ is not to attribute a mysterious causal power to the die that "necessitates" a roll of 6 – but only to the degree $1/6$. Whatever that might mean. There are too many varieties of propensity theories of chance to try to survey them here, but what I want to emphasize is that whatever objective chances are, they are certainly compatible with the reign of determinism at the level of physical law (contrary to what at least many propensity theorists claim).[10] One of the claims I will argue for below is that there may be fewer objective chances out there than some people assume. But I would argue strenuously that it can't be the case that there are *none* (or none whose value is neither zero nor one). Objective probabilities are the kinds of features of reality displayed *par excellence* in gambling devices and coin flippings, and presumably radium decays and many other phenomena.[11] A view which says that there are no objective chances if the world turns out to be at bottom deterministic, is in my view just changing the topic of conversation. Even if the world is deterministic, we (in our ignorance) still need all the guides to life we can get. There are indeed features of reality that we can see will serve to play the role of guide as specified in *PP* – such as the features of a fair die mentioned earlier, plus what we know about how people throw dice and how they bounce, etc. – so there are objective probabilities in the world.

Like David Lewis, I claim that *what chances are for* (as expressed in PP) is our best guide to what chances *are*. Objective chances must, at the least, be facts that entail the rationality or correctness, in some sense, of the Principal Principle. Now I will sketch a Humean view of objective chance that is meant to satisfy this constraint.

## 2.2. What chances are.

A proper Humean empiricist will insist that objective probabilities, whatever they may be, must at least supervene on the sum total of actual events in world history. They are not some mysterious or hidden springs lurking underneath (as some views take the laws of nature to be) and forcing the world's events to be the way they are. Instead they are patterns that can be discerned in the vast panoply of events occurring in the world. What kind of patterns? Finite frequentism answers the question in a simple way: relative frequencies. Or perhaps: relative frequencies meeting certain tests of stability and distribution. But there are too many of these relative frequencies, and that undermines the sensibility of PP. The chance of rain tomorrow in Castelldefels should be defined as the relative frequency of rain-the-next-day in a reference class of preceding-days "like today". But – like today, in which respects? If we specify too many respects, we whittle our reference class down to nothing, or nearly-nothing, in which case it would seem wrong to let the frequency guide our credence. (If the chance of rain does exist, I am certain that it is neither 0 nor 1.0!)[12] On the other hand, there may be no good reason (from the perspective of simple Humean frequentism) to choose one set of attributes that days "like today" share, over another set; and the other set will likely give different frequencies. This is why it is better to let go of frequentism,[13] and move to a more sophisticated Humean account based on the two key notions of *best systems* and *nomological machines*.

Lewis (1994) offers a package account of laws and objective chances together, one that in effect says objective chances (if they exist) are dictated by laws of nature. It is called a *best systems* account because it meets the demands of Humeanism by defining the laws of nature as a set of axioms that systematize the patterns in actual occurrent events, obtaining a "best" combination of simplicity and strength. In our world, it may be that the best system of axioms we can have does not deterministically specify what *will be* the case, always and everywhere, but rather tells us the objective probability of various occurrences. These then are the objective chances.

There is no space here to go into the details of Lewis' account and the many ways in which (I believe) it goes astray. What I do wish to keep from his account is the idea of chances supervening on actual occurrences, and the idea of *systematic patterns* to be discerned in those occurrences, patterns that may be something more than just actual frequencies, and which can sensibly play the role of chance defined in PP. Just to give the simplest example of how this may work: The overall pattern of events may exhibit the kind of behavior patterns known as Newtonian mechanics (for middle-sized objects in certain circumstances). That fact, plus the symmetry of objects like coins and dice, gets us almost all the pattern-facts we need to see that the chance of heads on flipping a coin is 1/2 (and 1/6, respectively, for the die). The further fact we need is an aspect of the overall pattern of events that is truly crucial to the existence of objective chances. We might call it the "micro-stochasticity of events". In the case of coin flips, what this refers to is the fact that there is a nice random-looking distribution in the size, angle, etc. of the initial impulses given to coins in ordinary coin flips. If coin flipping is basically a Newtonian phenomenon, then it is the random-looking distribution of these initial impulses that makes coin flips display the approximate 50/50 distribution we rely on.[14]

The stochastic-lookingness of initial conditions, boundary conditions, influences from outside, etc., is such an important aspect of the overall Humean pattern of actual events that it deserves a title, and I propose to call it the Stochasticity Postulate. I call it a "postulate" because we don't *know*, for a guaranteed fact, that we can rely on it everywhere and at all times.[15] But it is as well-confirmed as anything in our scientific world-picture, and we rely on it to make many of our machines – nomological or otherwise – function predictably and reliably. It is not restricted to microphysics; for the purposes of economics, the car-buying decisions of consumers may supply the micro-stochasticity that is needed for an efficient model of new-car-delivery to work adequately well. Exaggerating only slightly, we might put the Stochasticity Postulate like this: all over the place, at all sorts of levels, events are nicely random-looking. It is this fact, above all else, that grounds the existence of Humean objective chances.

But the stochastic-lookingness of events does not, by itself, give us stable and reliable objective chances of the kind that could (ideally) serve to guide belief as per PP. We need in addition a stable structure or set of conditions that utilize this stochasticity, in constrained ways, to generate stable probabilities. As I said earlier, we need proper chance set-ups. Generally I will follow Cartwright (1999)'s terminology and describe these setups as probability-generating *nomological machines*. A nomological machine is a stable arrangement of things, with appropriate shielding as needed, that generates a regularity. A probability-generating machine (or *stochastic nomological machine*, SNM, as I propose to call them) is a well-defined setup or arrangement of things that produces outcomes with a well-defined probability. It is thus something over and above mere superficial Humean "laws of association" (i.e., actual frequencies), and will therefore violate Pearson's admonition to avoid entanglement with theory – but only, I think, to an extent that is both harmless and unavoidable.

The best examples of SNMs are, naturally, classical gambling devices, so let us look at a few of them to illustrate the main points.

*1. The coin flipper.* Not every flip of a coin is an instantiation of the SNM we implicitly assume is responsible for the fair 50/50 odds of getting heads or tails when we flip coins for certain purposes. Young children's flips often turn the coin only one time; flips where the coin lands on a grooved floor frequently fail to yield either heads or tails; Persi Diaconis was alleged to be able to reliably achieve statistics far from 50/50 when flipping a coin in an apparently normal way (and he is, no doubt, not the first person to achieve this); and so on. Yet there is a wide range of circumstances that do instantiate the SNM of a fair coin flip, and we might characterize the machine roughly as follows:

i.   The coin is given a goodly upward impulse, so that it travels at least a foot upward and at least a foot downward before being caught or bouncing;

ii.  The coin rotates while in the air, at a decent rate and a goodly number of times;

iii. The coin is a reasonable approximation to a perfect disc, with reasonably uniform density and uniform magnetic properties (if any);

iv.  The coin is either caught by someone not trying to achieve any particular outcome, or is allowed to bounce and come to rest on a fairly flat surface without interference

v. If multiple flips are undertaken, the initial impulses should be distributed randomly over a decent range of values so that both the height achieved and the rate of spin do not cluster tightly around any particular value.

Two points about this SNM are worth mentioning right off. First, the characterization is obviously vague. This is not a defect. If you try to characterize what is an *automobile*, you will generate a description with similar vagueness at many points. This does not mean that there are no automobiles in reality. Second, the last clause refers to a "random distribution" in the initial impulses, and this might seem to be cheating, or creating some sort of vicious circularity. But in fact this is not the case. "Random" here simply means "random-enough looking" and has nothing to do with a mysterious "process-randomness" that fails to supervene on the actual happenings. For example, we might instantiate our SNM with a very tightly calibrated flipping machine that chooses (a) the size of the initial impulse, and (b) the distance and angle off-center of the impulse, by selecting the values from a pseudo-random number generating algorithm. In "the wild", of course, the reliability of nicely randomly-distributed initial conditions for coin flips is an aspect of the Stochasticity Postulate.

*2. The biased coin flipper.* Here I will describe a proper machine, and not worry whether Persi Diaconis or other practitioners of legerdemain fit the description. Suppose we take the tightly-calibrated coin flipper (and "fair" coin) mentioned above, and: make sure that the coins land on a very flat and smooth, but very mushy surface (so that they never, or almost never, bounce); try various inputs for the initial impulses until we find one that regularly has the coin landing heads when started heads-up, as long as nothing disturbs the machine; and finally, shield the machine from outside disturbances. Such a machine can no doubt be built (probably has been built, I would guess), and with enough engineering sweat can be made to yield as close to chance = 1.0 of heads as we wish.

This is just as good an SNM as the ordinary coin flipper, if perhaps harder to achieve in practice. Both yield a regularity, namely a determinate objective probability of the outcome heads. But it is interesting to note the differences in the kinds of "shielding" required in the two cases. In the first, what we need is shielding from conditions that bias the results (intentional or not). Conditions i, ii, iv and v are all, in part at least, shielding conditions. But in the biased coin flipper the shielding we need is of the more prosaic sort that many of our finely tuned and sensitive machines need: protection from bumps, wind, vibration, etc. Yet, unless we are aiming at a chance of heads of precisely 1.0, we cannot shield out these micro-stochastic influences completely! This machine makes use of the micro-stochasticity of events, but a more delicate and refined use. We can confidently predict that the machine would be harder to make and keep stable, than an ordinary 50/50 -generating machine. There would be a tendency of the frequencies to slide towards 1.0 (if the shielding works too well), or back toward 0.5 (if it lets in too much from outside).

*3. The radium atom decay.* Nothing much needs to be said here, as current scientific theory says that this is a SNM with no moving parts and no need of shielding. In this respect it is an unusual SNM, and some will wish for some explanation of the reliability of the machine. Whether we can have one or not remains to be seen.

In each of these cases we are able to describe a repeatable set of conditions that constitute the chance setup or SNM, and give at least some reasons for expecting it to yield a fairly reliable regularity. Sometimes the reasons may be expressed in causal terms; I think Cartwright expects this to be the case most, if not all the time. The reasons may also be grounded partly or wholly in what we take to be laws of nature, as is the case in the biased flipper (presumably modellable decently well with classical mechanics) and the radium atom (where the decay half-life follows from laws of quantum mechanics). This may seem to undermine the Humean credentials of objective probabilities. But there are two responses to this worry. First, there is an ineradicable link (or constraint) between the chances and the actual outcomes, at least when the numbers are high enough: had 99% of all coin flips in history landed heads despite the apparent satisfaction (in a huge variety of different ways) of conditions i - v, we would have to say that the objective chance of heads is 0.99, not 0.5.[16] The objective chances may be different from the actual frequencies, to some extent, in light of features of the chance setup (such as physical symmetry, presumed random-looking distribution of initial and boundary conditions, and so on), but not *greatly* different, at least not when the numbers are high. This constraint arises automatically from the need to satisfy PP. If chance is to be a good guide to belief, and 0.99 of all coin flips in world history land heads, then the chance had better be 0.99 too, or very close to it.

Second, while we may need to use causal and/or law-talk in describing our reasons for believing in the reliability of an SNM, we are not committed to any *particular* metaphysical account of these notions. Lewis, for example, offers accounts both of laws and of causation that satisfy his view of Humean supervenience. While I do not subscribe to those accounts, the point remains that this Humean account of objective probabilities leaves it an open question what account of causation or of laws is best (if any is needed at all, in the end). The notion of a SNM does not come loaded with any particular anti-Humean notion of probabilistic propensity. Indeed, in most of the cases I can think of, causal talk covers mainly the "deterministic" part of the workings of an SNM (e.g., for coin flips: what goes up, comes down because of gravity), while the part of the description that justifies the stochasticity, and the expectation of a stable probability, adverts mainly to the "randomness" of the inputs to the SNM from outside (force of the impulse on the coin or roulette ball, disturbing effects from random wind forces, etc.). And that is all just part of the Humean-acceptable pattern of actual events.

*4. Inflation > 6% in the UK economy.* What is the probability that inflation will exceed 6%, next year, in the UK? This example, as well as 1. and 2. above, is discussed in Cartwright (1999).[17] But this is an example of something that is *not* a proper chance set up, not a SNM. Why? It simply has none of the elements of one: no repeatable structure that it is reasonable to expect to generate a stable probability. If we were to correctly ascribe an objective probability, it would have to be based on a stable, enduring structure whose properties make it reasonable to expect it reliably to yield that probability. But over the years, both the meaning and structure of these notions (UK, inflation) changes greatly. There is no reason to think that any SNM is out there, waiting to be discerned by economists, that in fact grounds an objective $P(I>6\%|UK)$.[18] If the UK economy lasts a few hundred years more and if we *could* see the statistics for all years, my guess is that there would probably not be any stable regularity discernible in them (e.g., inflation $> 6\%$ in approximately 3 out of every 18 years). Certainly, we cannot discern any reason why there *should* be such a regularity. There is no chance-generating nomological machine here, and so there is no objective chance.

Unfortunately, the vast majority of statistics that we can gather in economics, medicine, and other sciences – even statistics that we feel are important, and that we wish to understand and control – will be like this example, and not like the first three. There are many more statistics in the world than objective chances. For statistics can be seen everywhere, but genuine nomological machines – stochastic or otherwise – are much more rare.

Now we can return to the topic of causal laws and effective strategies.

### 3. Re-thinking the examples.

How does spraying oil on swamps prevent malaria? We know the answer very well:[19] Particles of oil kill mosquitos when ingested (or when they land on larvae, perhaps); mosquitos are the carriers of the malaria virus; when the swamps are sprayed, some mosquitos should be killed (or larvae killed); so there should be fewer mosquitos around afterward; hence fewer mosquito bites; hence fewer bites by malaria-carrying mosquitos; hence fewer cases of malaria. Each of these steps makes common-sense causal sense; but each is also merely probabilistic, in some sense. The oil may kill more or fewer mosquitos, but is unlikely to kill all; fewer mosquitos should mean fewer bites, though of course that depends on how active the remaining mosquitos are; fewer bites should mean fewer bites transmitting malaria, though again it depends on precisely how active the malaria-carrying mosquitos are, and how frequently their bites do in fact transmit the virus; and so on. In fact, at any of these stages if things don't *happen* to go the way one expects (due to chance, or unusual initial conditions if one prefers to think of it deterministically), then the oil-spraying may *fail* to reduce the rate of malarial infection. The problem for Cartwright's (1979) picture is this: this mooted failure of the "right" statistical relation to obtain is not due to any "missing" causal factors for malaria that we have failed to hold fixed. "Bad luck with which mosquitos survived" and "Bad luck with which mosquitos bit more" are not causes that Cartwright can recognize, or declare to be part of "the causal laws of the universe".

This sort of bad luck might have turned out to be universal in the whole reference class of oil-sprayed-swamps. More realistically, it might just happen in a few cases, leading to (say) the probability of malaria going up in one or more of the reference classes $^\wedge K_i$ mentioned in *CC*. Let's suppose this happens for the class in which we hold fixed {Don't drink quinine, use bug repellent, European ancestry, etc.} Then contextual unanimity (Dupre's term for the demand that the probability change in the same direction in *all* reference classes $^\wedge K_i$) would fail, contrary to what Cartwright thinks is possible in (1979).[20]

But we need to dig deeper into several aspects of this case, especially the probabilities. There would be an obvious response to this example, if Cartwright were supposing the probabilities in *CC* to be the "true" probabilities, identical to the "real" propensities of systems of such-and-so type. The response would be: well, these statistics just don't count; they don't reflect the *true* probabilities. *CC* is still true, but only of the true probabilities.

But as we noted earlier, Cartwright does not hold with such things (which are in effect chances-as-metaphysical-propensities), and she is right not to.[21] Instead, as we saw, for Cartwright in 1979/1983, probabilities are just actual frequencies meeting certain tests. And that's not bad, from my perspective: it is better than invoking mythical propensities or hypothetical frequencies, and some such actual frequencies are indeed objective probabilities. But not all, by any means! And surely the frequencies of malaria infection, *in the tiny populations in which all these K-factors are held fixed*, are not – or are not all – genuine Humean OC probabilities. In fact, they are unlikely even to meet Cartwright's criteria of stability and so on. Nor, I would guess, will they in general meet the criteria for a chance-generating SNM.[22]

One might think that, once all the causal factors are held fixed and the situations of the classes $^\wedge K_i$ clearly defined, then a stable SNM must surely be the result. Unfortunately, this is just not so, in general. The patterns among events at the macro- and micro-levels, relative to mosquito bites and so on, may not display any systematic regularity that entails, e.g., that in a particular homogeneous reference class, the chance of infection should be .046 rather than .054. The chance is supposed to reflect what *would* occur if the reference classes were sufficiently large; but the actual patterns of events simply are not enough to dictate an answer to this question. This is in contrast with systems such as gambling devices, where the physical symmetries and the even distribution of initial and boundary conditions found all over the place in nature do dictate well-defined chances.

What if there really are no probabilities out there, in the reference classes holding fixed all the causally relevant factors? Then the project of deriving causal laws from probabilistic data is impossible to even begin. That is no big loss, however, since we already knew that we had to know, ahead of time, what all the other causally relevant factors are (to hold them fixed), in order to prove that a given factor is indeed a cause. Our knowledge of "the causal laws" has to be nearly complete anyway, on Cartwright's (1979) view, before we can look to probabilistic data to help complete it. As Cartwright has often stressed, the situation may not be so bad: if we feel we know enough about the causal structure(s) at issue, we may be able to use randomized controlled trials to discover whether C raises the probability of E in various subpopulations that we can't examine individually. But this only salvages the utility of the CC-based method on the assumption that the method is applicable in the first place, i.e., assuming that *all the statistics we need to look at correspond to genuine objective chances.* [23]

Unfortunately, this will not be true in general, at least if we understand objective chances in the way that I advocate here, or that Cartwright advocates in *Dappled World*.

What is it, for there to be a causal law of our universe that C causes E? *CC* cannot now be taken as part of the answer to this question. Contextual unanimity may fail (in actual statistics) for non-causal reasons, as well as the causal reasons recognized by writers on causation; and, much more importantly, the objective probabilities invoked in *CC* may simply fail to exist.[24] Instead, we must fall back on the answer from *Nature's Capacities and Their Measurement* (1989): it is a causal law that C causes E just in case some *c*'s do, on some occasions, by virtue of being *c*'s, cause *e*'s.[25] And this singular causation concept is one that notoriously resists all attempts at further definition (or analysis), though I think we can say two things about Cartwright's views on the matter, based on her later writings. First, at least in some cases the common-sense counterfactual is true: "If *c* had not occurred, *e* would not have occurred." Second, at least sometimes, C will be part of an INUS condition for E, hence a particular *c* may occur along with the rest of a set of circumstances instantiating an INUS condition, and jointly necessitate *e*. But since we know we can't really count on these explications being right all the time, basically we are down to this: that *c* singularly causes *e* is a primitive relation that we know how to recognize *sometimes*, and thank goodness, for without it we could never get science started.

While this last part is hard to deny – it is the core of Cartwright's view of how science works, and it is the most true-to-life account anyone has yet offered – still it is possible to chip at it around the edges, and I think it is important to do so. Because again we seem to be retreating to a black-box perspective on causation, and again there are at least some cases where we know a lot about the internal mechanism. We could look back at the malaria case now, since we do know a lot about the internal mechanism. But it is an awkward example to use (and, to some extent, this in fact undercuts Cartwright's story), because it is hard to take seriously the singular-causation version of the mooted causal law. "Some oil-sprayings do sometimes prevent malaria infections." No doubt they do, but it is hard to read this in a singular-causation (prevention) way: *whose* malaria infections were prevented? Or at least, how many? Nor is an INUS condition reading very easy to put on the case, given that, as we saw, the effect is in some sense *likely*, but by no means *necessitated* to happen.

So instead let's take Cartwright's favourite example: some aspirin-takings do relieve headaches. Here too, we know a lot now about the mechanisms inside the black box causal statement.[26] Aspirin molecules float around in the stomach and get absorbed into the bloodstream. There, they mix thoroughly into the blood, and so some get pumped toward the brain. Because the molecules are small enough, they pass through the blood-brain barrier. The molecules then interact with the swollen vein and artery walls in the head, causing (by a yet-more-microscopic mechanism, which we will skip over) reduction in the swelling. The reduced swelling relieves the pressure that causes the pain.

That's an awful lot of structure, hidden underneath a black-box-style singular causation statement. In fact it can be considered quite analogous to the oil-swamp-malaria case. At any of several stages of the story, the process relies on what are essentially statistical regularities (*not* brute cause-effect relationships, *not* necessitations supported by the (non-probabilistic) laws of nature): how many aspirin particles and of what size pass into the blood; how many pass into the relevant area of the brain; how many of these get into interactions that help reduce swelling; and so on. At each of these stages there is presumably a wide numerical distribution of the relevant events that may result, even when things go "normally". And as in the malaria case, only perhaps more plausibly here, sometimes not enough reduction in swelling will result to cause headache-relief. And this will happen by mere chance, we may say, or by "hap", or "just as a matter of random bad luck".

When this occurs for the reasons just posited, we may advert to a useful metaphor and say that the cause "failed to fire" as a purely chance matter of fact. This can be misleading, though, in two respects. First, the metaphor calls to mind the (apparently) irreducible failure-to-decay that may be demonstrated, in a given stretch of time, by a radioactive atom. That is not a good comparison, since here we can in principle understand the failure to fire. The aspirin does what it always does, it is just that the micro-movements of its particles after swallowing happen not to be good enough to relieve the headache. We have a lot to say about what may have occurred, and none of it is black-box or irreducible.[27] Second, there is a temptation to assimilate this failure to fire to genuine "probabilistic causation", thereby implying that there is some objective probability for this failure to occur (in a given population) at all times. But for the kinds of reasons already discussed above, there may in fact be no such objective probability.

The other way of thinking of the aspirin's failure, consonant with the typical discussions of mixed capacities and interactions, would be to suppose that some *cause* prevents the aspirin from curing the headache. This may be wrong-headed as well. The aspirin may well not, on such an occasion, have been "prevented" from relieving the headache by any well-characterized factor whose causal power goes in the opposite direction, so to speak. Maybe it sometimes is, maybe even *most times* when it fails, it is. But it need not always be viewable that way. That is what my description was meant to highlight: the aspirin doesn't necessarily fail because some more-powerful-headache-causer or aspirin-action-preventer wrestles it to the ground, but rather because at the micro-level, things just don't happen to go the way they normally do.

When we look inside the black boxes of probabilistic causation, at least sometimes – and perhaps *every* time – we find a lot of stuff going on that is best described as a sequence of "causal" steps that rely on statistical regularities. Like the coin-flipping SNM, we can think of them as based around (fairly-)reliable statistical regularities that are treated either as unexplained, or as arising from the result of initial and boundary conditions given underlying natural laws. We may like to say that taking aspirin is an effective strategy for getting rid of a headache, because of aspirin's "causal capacity" to relieve headaches. But underneath the metaphors of powers struggling and capacities firing, what's really going on is the existence of some regularities that are stable and repeatable (-enough), which we exploit cleverly for our own ends.

Now let's turn to effective strategies. In the malaria case, or the aspirin-taking case, I have been arguing that the $CC$-based causal law story breaks down upon close examination. *A fortiori*, it would seem, we can't claim these are effective strategies on the basis of the truth of some causal laws. But that does not mean that these are *not* effective strategies! They probably are, in many or most circumstances. But in explicating why they are, we should avoid both talk of causal laws, *and* talk of specific objective chances at work in the strategies (either at the gross, desired-outcome level, or at the level of the underlying steps in the mechanism).

Oil spraying and aspirin-taking are effective strategies not because there is an SNM (or NM) to be discerned in their working, but rather (merely) a *mechanism* that can be expected to work at least sometimes – perhaps often, if we are lucky.

Here, then, we see one big difference betwen NM's and mechanisms in general: a mechanism need not give rise to a stable regularity. It simply has the potential, by virtue of its structure, to give rise to a certain outcome (or output) – when things go right. How often and how reliably they do go right is a separate question. Aspirin-taking is an effective strategy for curing a headache just because there exists the mechanism described above, that *can* (and does) function sometimes.

Before we say more about this view of effective strategies, I want to finish criticizing the causal law-based picture by looking at one more of Cartwright's central cases. Why isn't joining TIAA-CREF an effective strategy for extending your life? Well, actually, it might be, as Cartwright herself notes; and only a little imagination is needed to work out reasons why it could be. But let's suppose that on the whole it is not, in fact. Cartwright's (1979) story about why it is not goes like this: There is in the overall population a correlation between belonging to TIAA and having longer-than-average life. But once we partition the population into sub-classes in which we hold fixed the true causes of longevity (exercise? wealth? happiness? good genes? good diet? . . .) the correlation disappears. And at the level of singular causation, we can note: joining TIAA just never does cause longer life, in any individual case.

It is now clear what's deeply wrong about this story. First, the list of things that might be thought to affect longevity is too big, open-ended, and ill-defined for *CC* (or its strategies-directed correlate, from section 2 of Cartwright (1983)) to be useful. And contrary to the "singular causes first" view of *Nature's Capacities and Their Measurement*, I would argue that there is no fact of the matter about whether, for example, 1 hour of hard exercise in the hot sun increases, decreases, or fails to affect my longevity. And the same could be said for a myriad of factors that may, *statistically*, be positively or negatively associated with lifespan.[28] But even at a non-singular level, the same problem arises: does exercising in the hot sun regularly or eating yoghurt daily cause greater longevity? There are reasons for answering yes, others for answering no, and still others for saying that there's no fact of the matter. (No SNM.) If we did manage to agree on a list of causes, and we partitioned the whole population up by homogeneity in these causes, our subpopulations would be too small to support genuine probabilities, on either my account or on Cartwright's early account of these. So the CC story just fails to make sense here.

Returning to TIAA at the level of singular causation: as in the malaria case, only much more so here, the mooted cause is so far removed from its effect, that (a) the notion of singular causing hardly seems decently applicable, but (b) if it is, then it is highly implausible that for *no-one* does joining TIAA actually increase their life expectancy. We can think of myriads of causal-counterfactual chains leading from joining TIAA to changes in lifestyle that *are* causes of increased life expectancy (to the extent anything is), and can imagine a person instantiating such changes. (Imagine a hard-drinking, smoking grad student who joins TIAA on getting her first academic job. She is sent a folio of information about how TIAA can help you get healthy by paying for your nicotine patches, subsidizing your health club membership, etc. . . . .). But if the singular-causing story holds even once, then contrary to our starting assumption, it *is* a "causal law" that joining TIAA increases life expectancy – given the reading of *Nature's Capacities and Their Measurement.*

Finally, we should see how the pieces may fit together to offer a different, arguably Humean, picture of effective strategies.

## 4. Humean effective strategies.

Cartwright's early way of talking about effective strategies may work well in a lot of cases, but in many others it tends to fall apart, as we have seen. The remedy, it seems to me, is to be even more stringent than *CC* in thinking about effective strategies, but stringent in somewhat different directions. Instead of looking for the complete sets of causal factors for a given effect, what we need to do is look for *mechanisms* or *nomological machines* – probabilistic or deterministic – that "produce" the effect. Where we can create, or discern in nature, a mechanism or a NM for a given effect, there we have a strategy for bringing it about. Where we can't find one, there we don't have an effective strategy, at least not one we have reason to think we can rely on.[29] We may have statistical regularities, and we may follow our temptation to base an effective strategy on the regularities. It may even work successfully in some cases. But that is just getting lucky; without a mechanism or NM, we are shooting in the dark.[30]

By contrast, if you have a mechanism or NM for producing an effect, you don't need to know *all* the causes and preventatives of the given effect. Instead, the mechanism/NM builds in "shielding" from interference, of two kinds. First, overt shielding from known disturbances, about which I have nothing in particular to say. Second, shielding by random initial and boundary conditions: the NM relies on nature's own fortuitous tendency to distribute uniformly the microscopic factors that might skew the results in undesired ways. This is of course analogous to the way in which human experimenters try to control for unknown skewing factors by randomized controlled experiments. But at a relatively microscopic level of description, Nature usually takes care of the randomizing for us, and that – part of the Humean supervenience pattern in the actual events – is a key fact around which many of our mechanisms and NMs are based.

A good example of such an NM to illustrate the role of nature's randomizing is the classical statistical-mechanical model of something like an ice cube being used to cool down a tepid drink. The model may not correspond to reality – it doesn't have to, to serve its illustrative purposes. But it may well so correspond, in its salient features.

What could be a more effective strategy for cooling down a tepid drink, than dropping a couple of ice cubes in it? Few things in this world are so reliable. But according to the classical stat-mech model, the strategy works not because of iron deterministic law, nor because of primitive causal powers of ice cubes to cool. Instead, the micro-motions of the liquid and the ice cubes are going to be *almost always* such that the future evolution of the system (ignoring outside influences) involves approach to equilibrium, with the equilibrium temperature being of course cooler than the initial temp of the liquid. This is the story, ignoring outside influences (and many other complications). But we should not ignore the environment: for this to be a good nomological machine for cooling drinks, it must be adequately shielded from outside heating. It must also be shielded from *coincidentally unfortunate boundary conditions (BCs)*, i.e., bumps from the outside that just happen, by bad luck, to be such as to keep the liquid + ice mixture moving *away* from equilibrium rather than toward it. But *we* don't provide this second kind of shielding; nature does that for us, via the reliable *typicality* and *randomness* of ICs and BCs to be found at the (relative) micro-level. Like any NM, it may on some occasion fail, but this one is a pretty good one compared to most that we devise. And notice one key point: the randomness (random-lookingness) of the micro-movements of molecules that is a key aspect of the pattern of actual events for a Humean account of objective chance is also the crucial to the functioning of this NM.

Laws and initial conditions underlie this SNM, not causes or capacities. Of course, this model of the situation relies on an ontological picture (billiard-ball style molecules interacting by action-at-a-distance forces, under Newtonian mechanical laws) that Cartwright would find incredible. And it may indeed be nothing more than a fiction. But if it is, it is a fiction that still works remarkably well at modelling one of nature's most reliable regularities. In light of it, and other examples that we could multiply indefinitely (the coin-flipping machine being another, for example), the claim that we should resign ourselves to causal fundamentalism in understanding our NMs seems premature.

Let me illustrate the NM-based view of effective strategies with a final case, the infamous heart-disease and exercising example, to point up how it is true to what we actually do when looking for real mechanisms in nature. The initially observed correlation between smoking and having less heart disease does not prompt us to immediately seize on smoking as an effective strategy for reducing heart disease (though it might be, if somehow smoking induces people to exercise who otherwise wouldn't). Rather it induces us to look to see if there might be a NM or mechanism linking smoking to reduced heart disease. There are two sides to this task. First, we may conduct further statistical studies to try to verify whether there really is such a mechanism at work, tests that give evidence *that* such a linkage exists without doing much to reveal *what* the mechanism is. For such tests the danger of misleading correlations is always there, and the implicit advice of *CC* – hold fixed known causally relevant factors as much as you can, and randomize – is of course correct as far as it goes. Second, we may directly test possible NM mechanisms *via* the hypothetico-deductive and other methods. We can no doubt immediately think of several ideas to test out: for example, nicotine might enter the bloodstream and have the effect of dissolving small clots inside the arteries, making the blood run more freely. Testing this might be more or less tricky, and depending on how it was done, might have more or less risk of deception via the exercise-heart disease link (or other correlations). But some tests of potential NMs might be fairly easy to do, and not have to rely on inferences made from mechanism-blind statistical studies.

The search for a NM or mechanism linking smoking to heart disease (or its prevention) is not a search for a causal law (as implicitly defined by *CC*), nor is it a search for singular causings. In these senses, I would say that the search is Humean-neutral: it does not imply causal fundamentalism at the level of the relationship under study, nor does it imply that a non-Humean notion of causation is *not* needed, at the lower level where we describe the workings of mechanisms or NMs.

## 5. Conclusion.

What I hope to have shown in this rambling discussion is that the framework of Cartwright's early discussions of causal laws and effective strategies is in many ways fragile, and that a different view of effective strategies is possible that makes use of her later concept of a nomological machine (and/or Glennan's concept of a mechanism). This view fits nicely with the Humean approach to objective probabilities that I advocate, which is agnostic about causation. The alternate view of effective strategies based on NMs is not meant to be Humean-sanitized, *vis a vis* causation: (i) I have not tried to revive the Humean project of defining "causal" facts purely in terms of statistical relations, a doomed project; (ii) as we dug into the various mechanisms by which causes such as aspirins effectively produce effects such as headache-relief, we had causal talk popping up frequently at the lower levels of description. But this doesn't mean that regularities (statistical, law-like or merely universal) are not enough to reconstruct what is going on, or that we need to fall back on some notion of causal capacity or causal law as a primitive, at the lower level. It means that the question is left open, and we can remain agnostic. My personal suspicion is that talk of causal capacities and causal laws can be replaced by Humean NM's all the way down to a level where all that is left are iron deterministic physical laws and fortunate accidental regularities.[31] *Lots* of fortunate regularities, which underlie at least as much of the predictability and stability of nature that we count on as the iron laws do (as we see in the ice-melting example). But this is not something I claim to have shown. Instead, what I hope my discussion has shown is that, *at the level of the original "causal laws"* that Cartwright wished us to accept, we can reject the need for any such things as primitives, and also reject their correlate singular-causings (taken, again, as primitives), and thereby make room, *at the level of these causal laws, relations and effective strategies*, for a more Humean approach to succeed.

Finally, let me stress that most of the points I have tried to make here are in harmony with Cartwright's most recent work on causation and probability, in *The Dappled World* (especially chapters 4, 5 and 7). The lessons I would wish to draw might be put this way: talk of causal laws should perhaps be avoided where possible, and the fact that causal capacities exist because of underlying mechanisms deserves more emphasis and investigation. Or more bluntly: it is better not to be too much of a causal fundamentalist.

A further result of these considerations seems to me worth mentioning (one that has been implicit in Cartwright's work from *How the Laws of Physics Lie* onward, but especially strongly in *The Dappled World*): the proposed methodology of trying to read off *useful* causal conclusions (hence effective strategies) from purely statistical data is *really hopeless*. In the first book, it proved hopeless because to decide that C was a cause of E (and hence a handle for increasing the level of

E, at least in principle), you had to know *all the other causes of E first.* The methods of Spirtes, Glymour and Scheines (SGS, 1993) are meant to help one partially circumvent that problem, and they build in all sorts of idealistic features to their causal graph-systems to try to make it work (e.g. CMC, faithfulness). In *Nature's Capacities and Their Measurement* and *The Dappled World*, Cartwright argues very effectively that these assumptions are implausible, for the real world in general. That is bad enough already. But perhaps the worst problem of all is one she doesn't sufficiently highlight: *most of the objective probabilities one needs as input simply don't exist.* There are SNMs in the world, including some we don't ourselves make. But they are hardly ubiquitous. And where they don't exist, the methods of Pearl and SGS may be literally inapplicable. Unfortunately, such methods are most likely to be needed and desired in precisely the sorts of fields (like macroeconomics) where it is extremely implausible that all the probabilities needed really exist. In those areas, what we have are at best what I would call *mere statistics*, not probabilities.

The difference is crucial. When you have a set of variables that are all connected by NM-like stable structures (or, using the SGS terminology, whose values are *generated by a causal graph*), there is at least some *prima facie* plausibility to the claim that the data will conform to the causal Markov condition and to faithfulness. But for the messy domains of mere statistics, what sort of arguments can be given for these conditions? You can only argue for their holding *after* you know that the statistics were generated by a real causal structure (i.e., a set of NMs and/or SNMs). But the SGS methods are supposed to start with mere statistical data, and *search out* a causal structure hidden underneath. Evidently, this can only be justified if one assumes that *all* sets of statistical data we may get hold of, arise out of some causal structure or other involving just those variables. This is unlikely not only because we will often latch onto irrelevant variables (and leave out relevant ones), but also for the reason stressed by Nancy Cartwright: much of what happens may occur just "by hap". What this means for our purposes here is: much of what happens is not appropriately thought of as "arising from a causal structure among event-types" or "happening because of the causal laws of the universe."

The question the causal modellers need to address is this: can any argument for the potential utility of such methods be mounted, given that we must largely work with mere statistics that we *know* are not generally objective probabilities? Instead of further proofs of how we can get true conclusions from ideally perfect probabilities assuming very strong conditions such as CMC and faithfulness, what is needed is some exploration of this difficult question.[32]

## Acknowledgments

I would like to thank Henrik Zinkernagel, Jordi Cat, Mauricio Suarez, José Díez, Paul Teller, and Stuart Glennan for helpful comments on earlier drafts of this paper. Special thanks go to Nancy Cartwright, whose extensive comments tried to set me straight about a number of issues and led to important improvements in the text. Needless to say, none of the above endorse anything written here. The research for this paper was partly funded by grant BFF2002-01552 from the Spanish MCYT.

# Notes

1   After I presented the main contents of this article in Oviedo (LMPS '03), Paul Teller pointed out to me that Stuart Glennan has written articles defending a mechanism-based view of causality that is very close to some of the ideas I advocate here. See Glennan (1996, 1997, 2002).

2   Turnabout is fair play: Cartwright's philosophical opponents who believe in fundamental laws of nature may deserve the epithet "fundamentalist", but she often seems to be no less a fundamentalist about causation. John Norton (2003) used this term first, I think, and I gladly borrow it from him.

3   Cartwright (1983), p. 26. Condition (iv) is needed to handle problems that would occur if one held fixed causes of E that sometimes are intermediate steps between C and E. CC is not itself immune to counterexamples and problems; see for example Otte (1985).

4   Cartwright (1983), p. 26.

5   Cartwright is explicit that these probabilities must be objective, not subjective, and indeed the reason is obvious: my (or anyone, including an "ideal rational agent") having certain degrees of belief cannot make it the case that C causes E, nor that C is an effective strategy for bringing about E.

6   For a compendium of these arguments, see Hajek (1997).

7   Generally when we use a statistic, we would like to give it a sort of causal (or perhaps better, expectation-guiding) reading. For example, we would like to use the statistics concerning incidence of breast cancer in a certain population as though it gave us the probability that a person who is about to enter that population group contracts breast cancer while a member of the group. But it is no such thing (at least, on the face of things). It is only a genuine objective probability if it is read as the probability of obtaining a person who has breast cancer, if one randomly samples one person from the population. And *this* objective probability is, unfortunately, rarely of use or interest to us.

8   See Hoefer (1997), (2003). My views have developed mainly out of a desire to correct and perfect the Lewisian approach to chance, but have certainly been influenced also by reading Cartwright's works and discussing many issues with her, during the years 1998 - 2002.

9   I eschew the usual mathematical formulations of *PP* here in order to make its common sense nature more clear.

10  And also contrary to David Lewis. See Hoefer (2003) for discussion of why linking objective chance to indeterminism is a mistake.

11  But not necessarily all the phenomena that we typically pretend have objective chances. For example, it is far from clear to me that there is an objective chance of rain tomorrow, in Castelldefels (Spain). (In Europe, unlike the U.S., weather forecasters rarely give numerical probabilities in their forecasts.)

[12] Here I am presupposing that the frequencies in *past cases* determine the frequentist objective probability. If instead all past and future cases were included, then narrowing in on the frequency in the reference class of days-like-today with only one member (namely, today) would yield a "frequency" of either 0 (if in fact it doesn't rain tomorrow), or 1 (if it does) that is splendid for guiding credence about rain tomorrow. But nobody wants to salvage PP by making the concept of chance degenerate into that of truth/falsity.

[13] I have not discussed so-called "hypothetical frequentism" because it seems to me that such accounts usually amount to propensity theories, once they are fully spelled out. What a Humean wants is to identify chances with some actual facts – aspects or patterns, of some sort, in the huge panoply of actual events, able to play the chance role as specified in PP. If, *after* identifying the chances as something actual, one wishes to go on and assert that, in addition, they inform us about *what limiting frequencies would result if the antecedent conditions could be repeated infinitely*, that is one's own business. I personally don't see the need for this metaphysical extravagance.

[14] If coin flipping is not best thought of as a deterministic Newtonian process (e.g., if quantum interactions between coin and air molecules play an important role), then *other* sources of micro-stochasticity may be involved. But either way, it is the random-lookingness of influences at the micro-level (relative to the coin) that account for the actual statistical behaviours of coins.

[15] See the film "Rosenkrantz and Guildenstern are Dead" for a lovely example of the breakdown of this postulate.

[16] In Hoefer (1997) I argue that any Humean approach to chance is obliged to take this stance, denying the possibility of radically improbable outcomes for large sets of chance events such that the actual frequencies diverge strongly from the alleged objective chances.

[17] Cartwright (1999) borrows the first two examples from a discussion by Mary Morgan and David Hendry (1995).

[18] This does not mean that there are no SNMs in economics generally. And with some work, we can imagine a fictional setup for the UK economy that might constitute a genuine SNM for inflation of a certain level. But the actual world is not such a setup.

[19] Actually, I am making this up, I do not *know* how this process works. But I assume some people do. More importantly, the true story will have a number of stages, like my possibly-fictional reconstruction here.

[20] In Dupré and Cartwright (1988) and Cartwright (1989) she does allow that failures of contextual unanimity may occur, and she does not endorse it except where it reflects the presence of a stable causal capacity. But the reason for its failure in these works is "mixed" causal powers on the part of some causes, or interaction, rather than statistical bad luck.

21 Some philosophers think of the "true probabilities" not as metaphysical propensities, but rather as parts of scientific models of certain situations. This is not the place to discuss the virtues of such a proposal, and how it may differ from the Humean account I favor; what matters here is that we are looking at situations for which we have *no* model, nor any reason to think that (in a non-trivial sense) we can have one.

22 In Cartwright (1989) probabilities are no longer actual frequencies, and instead are something more idealized. She does not give an overt account of what they are, but the perspective of *Nature's Capacities* may be seen as moving toward the view adopted in *Dappled World*.

23 By "correspond" here I just mean that two conditions are fulfilled: (a) the objective chances do, in fact, exist; and (b) the statistics being looked at are appropriately close to them. Typically causal searchers hoping to infer causal relationships from statistical data only consider condition (b), and deal with it by making it an unabashed, optimistic starting assumption.

24 Of course, not every theorist of probabilistic causation defends the kind of contextual unanimity found in CC, and there are alternatives to CC that weaken the requirement. They do so, however, to handle cases like the "mixed causal capacity" of birth control pills to both cause and prevent thrombosis. That is not the sort of problem we are looking at here. The problems arising from either non-existence of objective probabilities, or (if one takes the probabilities to be by definition the actual statistics) accidentally misleading statistics, affect these other versions of probabilistic causation just as much as they do the views of Cartwright (1983).

25 "A generic claim, such as 'Aspirins relieve headaches', is best seen as a modalized singular claim: 'An aspirin can relieve a headache'; and the surest sign that an aspirin can do so is that sometimes one does so." (1989), p. 95.

26 As before, I am making this up, and as before the details do not matter for the philosophical points being illustrated.

27 Until, perhaps, we get down to the micro-chemical molecular interactions, which are in some sense quantum-mechanical – what matters here is not whether irreducible causation enters the picture somewhere deep down, but rather whether it is present at the level we start with.

28 See Glennan (1997) for extended criticism of the singular-causation perspective on capacities found in *Nature's Capacities and Their Measurement*.

29 Here I am joining NM's and mechanisms, which are closely related things but not the same. The difference: a mechanism brings about a result fairly reliably, but the result need not be a *regularity*, whether statistical or not. A NM generates a regularity (fairly reliably).

30 It might be thought that this is too strong, and that surely if we have a positive statistical relationship between C and E (perhaps holding fixed some

possibly-relevant and easy-to-measure further variables), then we have *prima facie* evidence that increasing C is an effective strategy for producing more E. I would deny even this *prima facie* claim. There are myriads of positive statistical relationships out there in the raw data, even ones that obtain given the stipulated constraints. Few of these will we ever measure, but they are there, and few of *them* correspond to genuine effective strategies. If, in practical experience, the kinds of variables we *do* measure and test in these ways turn out often to reflect causal connections, that is because we had good reason to suspect, prior to doing the statistical tests, that such a relationship might obtain. And such suspicions most often come from common sense and antecedent causal/mechanical knowledge (i.e, from suspecting there is the right sort of NM or mechanism to be found), not from noticing a statistical correlation.

<sup>31</sup> Stuart Glennan seems to have a similar suspicion in his discussion of the mechanism-based view of causation. See (1996) section 4.

<sup>32</sup> And, as Cartwright has stressed, there is another topic that might be more useful to address: the nomological machines that do exist out there in the world. "[A] causal structure arises from a nomological machine and holds only conditional on the proper running of the machine; and the methods for studying nomological machines are different from those we use to study the structures they give rise to. Unfortunately these methods do not yet have the kind of careful articulation and defence that Spirties, Glymour and Scheines and the Pearl group have developed for treating causal structures." (1999), pp. 134 – 5.

**REFERENCES**

1. Cartwright, N. *How the Laws of Physics Lie* (Oxford University Press, 1983).
2. Cartwright, N. *Nature's Capacities and their Measurement* (Oxford University Press, 1989).
3. Cartwright, N. *The Dappled World: A Study of the Boundaries of Science* (Cambridge University Press, 1999).
4. Dupré, J. & Cartwright, N. "Probability and Causality: Why Hume and Indeterminism Don't Mix", *Noûs* **22**, 521 - 36.
5. Glennan, S. (1996) "Mechanisms and the Nature of Causation", *Erkenntnis* **44**, 49 - 71.
6. Glennan, S. (1997)"Capacities, Universality, and Singularity", *Philosophy of Science* **64**, 605 - 626.
7. Glennan, S. (2002) "Contextual Unanimity and the Units of Selection Problem", *Philosophy of Science* **69**, 2002.
8. Hajek, A. (1997) "'Mises Redux' – Redux: Fifteen Arguments Against Finite
9. Frequentism", *Erkenntnis*, Vol. 45, 209-227.
10. Hoefer, C. (1997) "On Lewis' Objective Chance: *Humean Supervenience Debugged*", *Mind* **106** no. 422.
11. Hoefer, C. (2003) "The Third Way on Objective Chance", unpublished manuscript.

12. Norton, J. (2003) "Causation as FolkScience",
http://www.philosophersimprint.org/003004/
13. Otte, R. (1985) "Probabilistic Causality and Simpson's Paradox", *Philosophy of Science* **52**, 110 - 125.
14. Spirtes, P., Glymour, C., and Scheines, R. *Causation, Prediction and Search* (Springer-Verlag, 1993).

# Inductive Logic, Verisimilitude, and Machine Learning

Ilkka Niiniluoto

*Department of Philosophy, P.O. Box 9, 00014 University of Helsinki, Finland*
*ilkka.niiniluoto@helsinki.fi*

**Abstract.** This paper starts by summarizing work that philosophers have done in the fields of inductive logic since 1950s and truth approximation since 1970s. It then proceeds to interpret and critically evaluate the studies on machine learning within artificial intelligence since 1980s. Parallels are drawn between identifiability results within formal learning theory and convergence results within Hintikka's inductive logic. Another comparison is made between the PAC-learning of concepts and the notion of probable approximate truth.

One of the major debates in the twentieth century philosophy of science was the Carnap – Popper controversy about the possibility of inductive logic. An outcome of this sometimes bitter but eventually fruitful debate was the development of systems of inductive probabilities or rational degrees of belief, designed for the analysis of non-demonstrative inferences (such as singular induction, inductive generalization, and analogical reasoning). As spin off products, these systems helped to create accounts of "epistemic utilities", such as semantic information and explanatory power, which turned out to cover some important aspects of Popper's falsificationism and Peirce's abduction. Another outcome was the non-probabilistic definition of the notion of truthlikeness or verisimilitude, expressing how "close to the truth" a hypothetical statement or theory is. As a kind of reconciliation of the Carnapian and Popperian approaches, scientific inference can be conceptualized as the maximization of expected verisimilitude, whereas ordinary Bayesianism involves the maximization of expected truth value. Other ways of combining the notions of probability and closeness-to-the-truth include probable approximate truth and probable verisimilitude.

After illustrating the power of these philosophical notions, this paper proceeds to compare them with some parallel work within artificial intelligence (AI), especially formal theories of machine learning. The success and convergence theorems of inductive inference are related to results expressing what kinds of hypotheses are effectively identifiable or decidable "in the limit" or "gradually". The notion of PAC (probable approximate correctness), applied in AI for concept learning, is discussed within the framework of inductive logic and verisimilitude. These results give new light to the debate whether recent studies in AI support inductivism or Popperian anti-inductivism.

## 1. CARNAP VS. POPPER

Both Rudolf Carnap and Karl Popper learnt in 1935 from Alfred Tarski that truth is a semantic conception. Carnap immediately presented a paper where he made a clear distinction between the time-dependent and evidence-relative notion of confirmation and the timeless notion of truth (see [39]). In the 1940s, Carnap started to develop a system of *inductive logic* (see [2]). It started in the Cambridge tradition of logical probability (J.M. Keynes, Harold Jeffreys), where probability statements are logically or analytically true, but developed into a branch of Bayesianism: inductive probabilities of the form $P(h/e)$ are degrees of belief in the sense that $P(h/e) = 1$ iff h is certain on e. In the 1960s, Jaakko Hintikka showed how the Carnapian approach can be modified so that genuine universal generalizations receive non-zero probabilities.

Popper had argued already in his *Logik der Forschung* in 1934 that science does not need induction. Carnap's turn to inductive probabilities was a great disappointment to Popper, who produced a series of arguments, collected as appendices to *The Logic of Scientific Discovery* (1959), against the possibility of inductive logic. Popper attacked "inductivism" as the view that there is a mechanical inductive method of discovery of scientific hypotheses. But his main point against induction as a method of justification was that probability in fact measures the logical weakness of a hypothesis, while in science one should seek bold and informative theories. If posterior probability $P(h/e)$ is understood as the degree of *confirmation* of h by e, then Popper's thesis is justified:

(1) If h is logically at least as strong as g (i.e, $h \vdash g$), then $P(h/e) \leq P(g/e)$.

However, Popper's [42] own proposals for a measure of *corroboration*, indicating how well a hypothesis has stood up in serious tests, were comparable to other definitions of degrees of confirmation like the relevance measure $P(h/e) - P(h)$ or the ratio measure $P(h/e)/P(h)$, which combine the ideas of high posterior probability $P(h/e)$ and high information content $\text{cont}(h) = 1 - P(h)$ (cf. [2]). Both of these definitions satisfy the *Positive Relevance* criterion: e confirms h relative to b if and only if $P(h/e\&b) > P(h/b)$. As Hintikka and Isaac Levi showed in the late 1960s, Bayesianism can incorporate many Popperian elements, if science is conceptualized as the maximization of expected epistemic utilities like information content and explanatory power (see [13,15,25], cf. [11]).

As a new move in the debate, Popper proposed a comparative notion of truthlikeness in 1960 (see [43]). The maximum value of *truthlikeness* is obtained when theory T is equivalent to the whole truth, while the weaker notion of *approximate truth* has its maximum value for all true theories. Against the probabilistic weakness ordering (1), true theories should be ordered with respect to their logical strength:

(2) If h is logically stronger than g, and both h and g are true, then h is more truthlike than g.

However, for false theories it should not be the case that the stronger theory is always more truthlike. Further, some false theories may be so close to the truth that they are better than trivial tautologies. Popper's notion of truthlikeness thus combines the ideas of truth and information. For example, tautologies (logical truths) are true and, hence, approximately true, but they are not highly truthlike, as they do not give much information about the true state of affairs. Even though Popper's own definition failed, these basic intuitions of Popper's can be saved within a theory of truthlikeness which combines the semantic Tarskian notion of truth with the notion of similarity (as applied to states of affairs or canonical statements expressing such states of affairs).

Even though truthlikeness is a logical notion, and thus is independent of all epistemic notions, it is possible to combine it with the Bayesian idea that science aims at *high expected verisimilitude*, where the expectation is calculated by means of inductive probabilities (see [29]). But the notion of expected verisimilitude ver(h/e) of hypothesis h on the basis of evidence e differs clearly from posterior probability $P(h/e)$. For example, ver(h/e) should not be high for a tautology h, even though $P(h/e) = 1$ when h is logically true or $e \vdash h$. Further, ver should not satisfy the principle that the probability or confirmation of h receives its minimal value 0 if e contradicts (falsifies) h. Instead, in some cases we may know that h is false but still estimate that h is close to the truth.

## 2. INDUCTIVE PROBABILITIES

In inductive logic, inductive probabilities are at least partly determined by symmetry assumptions concerning the underlying language ([2,15,40]).Carnap's one-time favorite measure $c^*$ is a generalization of Laplace's rule of succession. But in Carnap's $\lambda$-continuum the probabilities depend on a free parameter $\lambda$ which indicates the weight given to logical or language-dependent factors over and above purely empirical factors (observed frequencies), and in Hintikka's 1965 system one further parameter $\alpha$ is added to regulate the speed in which positive instances increase the probability of a generalization.

More precisely, let $Q_1, \ldots, Q_K$ be a K-fold classification system with mutually exclusive predicates, so that every individual in the universe U has to satisfy one and only one Q-predicate. A typical way of creating such a classification system is to assume that our monadic language L contains k basic predicates $M_1, \ldots, M_k$, and each *Q-predicate* is defined by a k-fold conjunction of positive or negative occurrences of the M-predicates: $(\pm)M_1 x \,\&\, \ldots \,\&\, (\pm)M_k x$. Then $K = 2^k$. Each predicate expressible in language L is definable as a finite disjunction of Q-predicates. Carnap generalized this approach to the case where the dichotomies $\{M_j, \sim M_j\}$ are replaced by families of mutually exclusive predicates $M_j = \{M_{j1}, \ldots, M_{jm}\}$, and each Q-predicate is defined by choosing one element from each family $M_j$ (see [18]). For example, one family could be defined by color predicates, another by a quantity taking discrete values (e.g., age).

A *state description* relative to individuals $a_1, \ldots, a_m$ tells for each $a_i$ which Q-predicate it satisfies in universe U. A *structure description* tells how many individuals in U satisfy each Q-predicate. Every sentence within this first-order monadic framework L can be expressed as a disjunction of state descriptions. Let e describe a sample of n individuals in terms of the Q-predicates, and let $n_i \geq 0$ be the observed number of individuals in cell $Q_i$ (so that $n_1 + \ldots + n_K = n$). Carnap's

$\lambda$-continuum takes the posterior probability $P(Q_i(a_{n+1})/e)$ that the next individual $a_{n+1}$ will be of kind $Q_i$ to be

(3) $(n_i + \lambda/K)/(n + \lambda)$.

The choice $\lambda = K$ gives Carnap's measure $c^*$, which allocates probability evenly to all structure descriptions. The choice $\lambda = 0$ gives Reichenbach's Straight Rule. The choice $\lambda = \infty$ would give the range measure proposed in Wittgenstein's *Tractatus*, which divides probability evenly to state descriptions, but it makes the inductive probability (3) equal to $1/K$ which is independent of the evidence e and, hence, does not allow for the learning from experience.

If the universe U is potentially infinite, so that n may grow without limit, all measures of Carnap's $\lambda$-continuum assign the probability zero to universal generalizations on evidence e. Hintikka's $\lambda$-$\alpha$-system solves the problem of universal generalization by dividing probability to constituents. A *constituent* $C^w$ tells which Q-predicates are non-empty and which empty in universe U. The number w of non-empty Q-predicates is called the width of $C^w$. Each generalization h in L (i.e., a quantificational sentence without individual names) can be expressed as a finite disjunction of constituents. When $\alpha$ grows without limit, Hintikka's measures approach in the limit the Carnapian values. When $\alpha$ is small, the posterior probability of universal generalizations grows rapidly. In this sense, the choice of a small $\alpha$ is an index of boldness of the investigator, or a regularity assumption about the law-likeness of the relevant universe U. If background assumptions are expressible by a theory T, possibly in a language that is richer than L, then posterior probabilities of the form $P(h/e\&T)$ can be calculated relative to observational and theoretical evidence (see [40]).

In Hintikka's system, there is one and only one constituent $C^c$ which has asymptotically the probability one when the size n of the sample e grows without limit. This is the constituent $C^c$ which states that the universe U instantiates precisely those c Q-predicates which are exemplified in the sample e:

(4) $P(C^c/e) \rightarrow 1$, if $n \rightarrow \infty$ and c is fixed.

It follows from (4) that a constituent which claims some unexemplified Q-predicates to be instantiat in U will asymptotically receive the probability zero. It is natural to assume with Hintikka that the prior probabilities of constituents $C^w$ are proportional to their width, so that the information content of $C^w$ decreases with w. (This holds for $\alpha > 0$; if $\alpha = 0$, all constituents are equally probable.) Then the result (4) means that inductive evidence e asymptotically favors the most informative generalization compatible with e.

It should be added that high posterior probability alone is not sufficient to make a generalization h acceptable, as this probability may result from the logical weakness of h (cf. (1)). But in Hintikka's system one may calculate for the size n of the sample e a threshold value $n_0$ which guarantees that the informative constituent $C^c$ has a probability exceeding a fixed value $1 - \epsilon$:

(5) Let $n_0$ be the value such that $P(C^c/e) \geq 1 - \epsilon$ if and only if $n \geq n_0$. Then, given evidence e, accept $C^c$ on e iff $n \geq n_0$.

(See [12]).

The Carnap-Kemeny axiomatization of Carnap's $\lambda$-continuum was generalized by Hintikka and Niiniluoto in 1974, who allowed that the inductive probability (3) of the next case being of type $Q_i$ depends on the observed relative frequency of kind $Q_i$ and on the number c of different kinds of individuals in the sample e. The latter factor expresses the variety of evidence e, and it also indicates how many universal generalizations e has already falsified. Carnap's systems turns out to be biased in the sense it assigns a priori the probability one to the constituent $C^K$ that claims all Q-predicates to be instantiated in universe U. In this axiomatic way, a system of inductive probability measures is obtained where Carnap's $\lambda$-continuum is the only special case with zero probabilities for universal generalizations (see [18,23]).

Further developments of inductive logic include its modification to problems concerning *analogical reasoning* where the distances between Q-predicates play a significant role in inference. In such cases, the probability (3) depends on the distance of $Q_i$ from the cells exemplified in e (see the papers of Kuipers and Niiniluoto in [10]). Another kind of extension allows uncertain evidence with observational errors (see [36]).

## 3. TRUTHLIKENESS AND ITS ESTIMATION

The similarity approach to truthlikeness has been developed, basically in a uniform manner, in two different contexts (see [32,37]). First, degrees of truthlikeness can be defined for qualitative statements or theories in first-order languages. Secondly, degrees of similarity may take advantage of the mathematical structures used in the definition of quantitative problems. In both cases, truthlikeness is defined relative to a *cognitive problem* that can be represented by a finite or infinite set of self-consistent statements B = $\{h_i| i \in I\}$, where the elements of B are mutually exclusive and jointly exhaustive. For each cognitive problem B there is an associated language L, where the statements $h_i$ are expressed. If L is an interpreted and semantically determinate language, it describes a unique fragment $w_L$ of the actual world. There is one and only one element h* of B which is true (in the sense of the Tarskian model theory) in the structure $w_L$. This unknown h* is the *target* of the problem B: which of the elements of B is true? The statements $h_i$ in B are the *complete potential answers* to this problem. The *true* complete answer is h* itself. The *partial* potential answers are non-empty disjunctions of complete answers; their set is denoted by D(B). For example, a tautology is a true partial answer, but it represents complete ignorance about the target, as it corresponds to the disjunction of all elements $h_i$ in B.

The basic step of the similarity approach is the introduction of a real-valued function $\Delta$: BxB→**R** which expresses the *distance* $\Delta(h_i,h_j) = \Delta_{ij}$ between the elements of B. Here $0 \leq \Delta_{ij} \leq 1$, and $\Delta_{ij}=0$ iff i=j. This distance function $\Delta$ has to be specified for each cognitive problem B separately, but there are canonical ways of doing this for special types of problems. First, $\Delta$ may be directly definable by using the metric in the structure of B (e.g., B may be a subclass of the K-dimensional Euclidean space $\mathbf{R}^K$). Secondly, if B is the set of state descriptions, the set of structure descriptions, or the set of constituents of a first-order language L, the distance $\Delta$ can be defined by counting the differences in the standard syntactical form of the elements of B. For example, a monadic constituent tells that certain kinds of individuals (given by Q-predicates) exist and others do not exist; the simplest distance between monadic constituents is the relative number of their diverging claims about the Q-predicates. If a monadic constituent $C_i$ is characterized by the class $CT_i$ of Q-predicates that are non-empty by $C_i$, then the *Clifford distance* between $C_i$ and $C_j$ is the size of the symmetric difference between $CT_i$ and $CT_j$:

(6) $|CT_i \Delta CT_j|/K$.

The next step is the extension of $\Delta$ to a function $BxD(B) \rightarrow \mathbf{R}$, so that $\Delta(h_i,g)$ expresses the distance of $g \in D(B)$ from $h_i \in B$. Let $\Delta_{min}(h_i,g)$ be the minimum of the distances $\Delta_{ij}$ of the disjuncts $h_j \in B$ in g. Then g is *approximately true* if $\Delta_{min}(h^*,g)$ is sufficiently small. Degrees of approximate truth can now be defined by

(7) $AT(g,h^*) = 1 - \Delta_{min}(h^*,g)$.

Truthlikeness should include a factor which tells how effectively a statement is able to exclude falsities. This can be expressed by the relativized sum-measure $\Delta_{sum}(h_i,g)$ which includes a penalty for each mistake that g allows, and weights this mistake by its distance from the target. At the same time, a truthlike statement should preserve truth as closely as possible. As a sufficient condition, we might now suggest that a partial answer $g'$ is more truthlike than another partial answer g if $g'$ is closer to the target $h^*$ than g with respect to both the minimum distance and the sum distance, but only few answers in $D(B)$ would be comparable by this criterion. Full comparability is achieved by the *min-sum* measure $\Delta_{ms}$, where the weights $\gamma$ and $\gamma'$ indicate our cognitive desire of finding truth and avoiding error, respectively:

(8) $\Delta_{ms}(h_i,g) = \gamma\Delta_{min}(h_i,g) + \gamma'\Delta_{sum}(h_i,g)$ $(\gamma > 0, \gamma' > 0)$.

Then a partial answer g is *truthlike* if its min-sum distance from the target $h^*$ is sufficiently small. One partial answer $g'$ is *more truthlike* than another partial answer g if $\Delta_{ms}(h^*,g') < \Delta_{ms}(h^*,g)$. The *degree of truthlikeness* $Tr(g,h^*)$ of $g \in D(B)$ (relative to the target $h^*$ in B) is now defined by

(9) $Tr(g,h^*) = 1 - \Delta_{ms}(h^*,g)$.

If the distance function $\Delta$ on B is trivial, i.e., $\Delta_{ij}=1$ for all $i \neq j$, then $Tr(g,h^*)$ reduces to a special case of Levi's [25] definition of epistemic utility.

If the cognitive problem B constitutes a subset of $\mathbf{R}^K$, then the partial answers may be taken include connected regions in $\mathbf{R}^K$. For example, if $B \subseteq \mathbf{R}$, then the complete answers are point estimates of some unknown parameter $\theta^*$, and the partial answers J in $D(B)$ are interval estimates. The distance between point estimates x and y can be defined by $\Delta(x,y) = |x - y|$, and the min-sum distance (8) can be reformulated as

(10) $\Delta_{ms}(x,J) = \gamma\Delta_{min}(x,J)^2 + \gamma' \int_J \Delta(x,y)dy$.

As the target h* is unknown, the value of Tr(g,h*) cannot be directly calculated by our formulas (8) and (9). However, there is a method of making rational comparative judgments about verisimilitude, if we have – instead of certain knowledge about the truth – rational degrees of belief about the location of truth. Thus, to *estimate* the degree Tr(g,h*), where h* is unknown, assume that there is an epistemic probability measure P defined on B, so that $P(h_i/e)$ is the rational degree of belief in the truth of $h_i$ given evidence e. The *expected degree of verisimilitude* of $g \in D(B)$ given evidence e is then defined by

(11) $\text{ver}(g/e) = \sum_{i \in I} P(h_i/e) \text{Tr}(g,h_i).$

(11) gives us a comparative notion of estimated verisimilitude: g′ *seems more truthlike* than g on evidence e, if and only if ver(g/e) < ver(g′/e).

If B is a continuous space, P is replaced by a probabilistic density function p on B, and the expected distance of a partial answer J from the truth, given evidence e, is

(12) $\text{dist}(J/e) = \int_{B} p(x/e) \Delta_{ms}(x,J)\, dx.$

For example, let $\theta^* \in \mathbf{R}$ be the unknown value of a real-valued parameter. Then maximizing the expected verisimilitude of a point estimate $\theta_o$ is equivalent to the "Bayes rule" of the Bayesian statisticians: minimize the posterior loss

$$\int_{R} p(\theta/e)(\theta - \theta_o)^2 d\theta.$$

This value is equal to

$$D^2[p(\theta/e)] + (E[p(\theta/e)] - \theta_o)^2$$

(see [4]). By this criterion, and choice of the quadratic loss function, the best point estimate is the mean of the posterior distribution $p(\theta/e)$. A similar theory of Bayesian interval estimation can be based upon the posterior loss function (10) (see [31,32]).

The relation between the functions Tr and ver is analogous to the relation between truth value tv (1 for true, 0 for false) and probability P, i.e.,

(13) Tr:ver = tv:P.

This can be seen from the fact that the posterior probability P(g/e) equals the *expected truth value* of g on e:

$$\sum_{i \in I} P(h_i/e) tv(g,h_i) = \sum_{i \in Ig} P(h_i/e) = P(g/e).$$

By (13), expected verisimilitude ver(g/e) is an estimate of real truthlikeness Tr(g,h*) in the same sense in which posterior epistemic probability is an estimate of truth value. The standard form of Bayesianism which evaluates hypotheses on the basis of their posterior probability can thus be understood as an attempt to maximize expected truth values, while ver replaces this goal by the maximization of expected truthlikeness.

In Hintikka's system of inductive logic, the result (4) guarantees that asymptotically it is precisely the boldest constituent compatible with the evidence that will have the largest degree of estimated verisimilitude:

(14) $\text{ver}(g/e) \to \text{Tr}(g, C^c)$, when $n \to \infty$ and c is fixed.

(15) $\text{ver}(g/e) \to 1$ iff $\vdash g \equiv C^c$, when $n \to \infty$ and c is fixed.

## 4. PROBABLE APPROXIMATE TRUTH

Expected verisimilitude ver is not the only way of combining the notions of epistemic probability and closeness to the truth. It is also possible to define the concepts of *probable verisimilitude* (i.e., the probability given e that g is truthlike at least to a given degree) and *probable approximate truth* (i.e., the probability given e that g is approximately true within a given degree). (See [32,35].)

Let g in D(B) be a partial answer, and $\epsilon > 0$ a small real number. Define

(16) $V_\epsilon(g) = \{ h_i \text{ in } B \mid \Delta_{\min}(h_i, g) \leq \epsilon \}$.

Denote by $g^\epsilon$ the "blurred" version of g which contains as disjuncts all the members of the neighborhood $V_\epsilon(g)$. Then $g \vdash g^\epsilon$, and g is approximately true (within degree $\epsilon$) if and only if $g^\epsilon$ is true. The probability that the minimum distance of g from the truth $h^*$ is not larger than $\epsilon$, given evidence e, defines at the same time the posterior probability that the degree of approximate truth $AT(g, h^*)$ of g is at least $1 - \epsilon$:

(17) $\text{PAT}_{1-\epsilon}(g/e) = P(h^* \in V_\epsilon(g)/e) = \sum_{h_i \in V_\epsilon(g)} P(h_i/e)$.

PAT defined by (17) is thus a measure of *probable approximate truth*. Clearly we have always $P(g/e) \leq \text{PAT}_{1-\epsilon}(g/e)$. When $\epsilon$ decreases toward zero, in the limit we have $\text{PAT}_1(g/e) = P(g/e)$. Further, $\text{PAT}_{1-\epsilon}(g/e) > 0$ if and only if $P(g^\epsilon) > 0$. Unlike ver, PAT shares with P the property (1) that logically weaker answers will have higher PAT-values than stronger ones.

In comparison with the result (13), we may note that the probable approximate truth of g equals the expected approximate truth value $\text{atv}_\epsilon(g)$, where $\text{atv}_\epsilon(g)$ is one if g is approximately true (within degree $\epsilon$) and zero otherwise.

An important feature of probable approximate truth is that its value can be non-zero even for hypotheses with a zero probability on evidence: it is possible that $\text{PAT}_{1-\epsilon}(g/e) > 0$ even though $P(g/e) = 0$. This suggests that PAT-measure gives an alternative to Abner Shimony's [44] "tempered personalism" where all seriously entertained hypotheses (even points on a real line) are assigned non-zero probabilities.

Let us still illustrate these notions in the context of statistical point estimation of real-valued parameters (cf. [32]). We shall see that they are in harmony with the statement of Bayesian statisticians in a well-known textbook: "The only realistic expectation from a statistical analysis is that the conclusions will provide a good enough *approximation* to the truth" ([1]). Here we apply the formula (17) to define PAT, and a point estimate y of the unknown parameter $\theta* \in \mathbf{R}$ is treated as a degenerate interval [y,y]. Choosing $\gamma = 1$ in (10), we obtain $\Delta_{ms}(x,[y,y]) = \Delta_{min}(x,[y,y])^2 = (x - y)^2$ .

Assume that we are making repeated independent trials with an unknown constant probability r of success. Then the number of successes in n trials is a random variable $\underline{x}$ which is distributed binomially Bin(r,n), i.e.,

$$P(\underline{x} = k \ /r) = \binom{n}{k} r^k (1 - r)^{n-k} \text{ for all } k = 0, \ldots, n.$$

The expected mean of $\underline{x}$ is $E\underline{x} = nr$, and its variance is $D^2\underline{x} = nr(1 - r)$. If $\underline{z} = \underline{x}/n$ is the relative frequency of successes in n trials, then $E\underline{z} = r$ and $D^2\underline{z} = r(1 - r)/n$. Assume that the prior distribution g(r) of the parameter $r \in [0,1]$ is a Beta distribution B(a,b) with parameters $a > -1$ and $b > -1$:

$$B(a,b) = [(a + b + 1)!/a!b!] \ r^a (1 - r)^b.$$

The mean of this distribution is $(a + 1)/(a + b + 2)$, and the variance is $(a + 1)(b +1)/(a + b + 2)^2(a + b +3)$. Then the posterior distribution g(r/k) of r given the observed number k of successes is Beta(a + k, b + n - k) (see [26]). The choice of the parameters $a = b = 0$ gives a uniform prior distribution B(0,0), i.e., g(r) = 1 for all r (see [4], p. 60; cf. [3]). The choice $a = b = -1/2$ is recommended by Box and Tiao [1], p. 35, as expressing a "noninformative prior" for r. The choice $a = b = -1$ gives an "improper" prior for r [26], but its posterior distribution is Beta(k - 1, n - k - 1) with the mean k/n and the variance $k(n - k)/n^2(n + 1)$. As long as a and b are small in relation to both k and n, the general behavior of g(r/k) can be illustrated by the uniform prior B(0,0). Its mean is (k + 1)/(n + 2), and its variance is

$$(k + 1)(n - k + 1)/(n + 2)^2(n + 3).$$

We see (allowing that k/n approximates some constant for large values of n) that the variance of the posterior distribution of r decreases to zero when n grows without limit. If we now choose the observed relative frequency z = k/n as our point estimate of the unknown parameter r, then the approximate truth of the hypothesis z = r can be calculated from the posterior distribution g(r/z):

$$PAT_{1-\epsilon}([z,z]/z) = \int_{z-\epsilon}^{z+\epsilon} g(x/z)dx = (n+1) \int_{z-\epsilon}^{z+\epsilon} \binom{n}{k} x^k(1 - x)^{n-k} \ dx.$$

This value approaches one, when n grows without limit. For the expected verisimilitude we have a similar result:

$$dist([z,z]/z) = \int_0^1 g(x/z)(x - z)^2 \ dx = D^2[g(r/z)] \to 0, \text{ when } n \to \infty.$$

## 5. FORMAL LEARNING THEORY

Parallel to philosophical and statistical work on probabilistic inference, there has been a growing interest in similar topics in computer science. Following in the footsteps of Alan Turing, since the 1960s experts of "machine intelligence" have studied processes of perception, pattern recognition, learning, and reasoning. A special feature of such work in AI has been the goal of developing algorithms that can be effectively implemented in computers - and thereby investigating the limits that computability may put on methods of inference. For example, studies in inductive logic programming have treated the problem of inductive generalization by devising systems that derive from data general conditions expressible in the Algol programming language. In cognitive science such formal systems are combined with psychological knowledge about human thinking, in order to find models of the actual processes of reasoning and problem-solving (see [16]).

Sophisticated work has been done within AI on topics like machine learning, scientific discovery, abduction, analogical inference, non-monotonic reasoning, reasoning from uncertain data, data mining, neural computing, and Bayesian networks (see, e.g., [24,10,19,6]). The journal *Machine Learning* started to appear in 1986. The so-called formal learning theory was inspired by Hilary Putnam's critical assessment of Carnap's inductive logic in 1963. But, in many cases, the studies in AI have been original and independent contributions without links to the logical systems that philosophers have developed, and equally often philosophers have been ignorant of these parallel developments. It certainly would be desirable to increase co-operation and mutual understanding of philosophers and computer scientists in these areas. (For a good starter, see [8].)

In the remaining two main sections of this paper, I shall restrict my comments to two topics. First, the success theorems of formal learning theory can be compared to the convergence results about induction and truth-approximation. Secondly, the AI study on concept learning, especially the model of PAC-learning, has interesting relations to our treatment of verisimilitude and related notions like PAT.

Formal learning theory is a study of the possibility of finding reliable solutions to learning problems (see [20,22]). The goal of learning may concern concepts, statements, or whole languages, but we shall first consider the case of learning new information about the environment or world w. An important class of such learning problems includes inputs e and an algorithmic learning rule F that attempts to decide on the basis of e whether sentences in some first-order language L are true or false in w. The inputs e are infinite *data streams* which consists of singular data true in w. A finite initial segment of e of length n is denoted by $e_n$. The success of the F depends on its performance with respect to all data streams generated by w, and the reliability of F depends on its success in various environments w. The simplest statements in L are such quantifier-free *empirical propositions* that are decided to be true or false in w by every input data e after a finite number of observations. A statement h is *verifiable with certainty* if for every input data e the learning rule halts with 'true' if h is true, but otherwise may say 'false' without halting. A statement h is *refutable with certainty* if for every input data e the learning rule halts with 'false' if h is false, but otherwise may say 'true' without halting. If statements in L are classified by their quantificational structure (in the prenex normal form), then existential statements (E-statements) are verifiable with certainty and universal statements (A-statements) are refutable with certainty.

Formal learning theory also appeals to the notion of identification *in the limit*: the rule F decides (verifies, refutes) a statement h in the limit if for each infinite input data e there is a finite n such that after $e_n$ rule F stabilizes to the truth value (true, false) of h when h is true or false in w. Here stabilization is a special case of "convergence" in the sense that F produces the same truth value for all $m \geq n$. Then EA-sentences are verifiable in the limit, and AE-sentences refutable in the limit. If truth-functional combinations of E- and A-statements are called *verifutable*, then in formal learning theory it is possible to show that a reliable learning rule in the limit (over all models of language L) exists for each verifutable sentence h in L ([3], p. 214). But this means that most hypotheses in the Borel hierarchy (with more and more complex quantifier structures) are not effectively decidable in the limit.

A broader notion is *gradual* verification and refutation, which employs the ordinary mathematical notion of convergence: for any given small number $\epsilon > 0$, there is a value n such that after $e_n$ rule F produces outcomes within the distance $\epsilon$ from the goal. Here the goal may be the truth value (1 for true, 0 for false) of a hypothesis. Similarly, Carnap's predictive probability (3) satisfies "Reichenbach's axiom" in the sense that (3) converges towards the limit of the relative frequency $n_i/n$ when n increases.

Kelly suggests that Peirce's fallibilism introduced the goal of solving all scientific problem in the limit: "for each clear proposition, there is a time at which its belief status is permanently settled by an ongoing inquiry" ([21], p. 186). But Peirce's own statements appeal to the fact that induction "pursues a method which, if duly persisted in, must, in the very nature of things, lead to a result indefinitely approximating to the truth in the long run" (*CP* 2.781). This certainly sounds more like gradual identification in Kelly's sense. What is more, Peirce seemed at least occasionally in his later work to realize that such convergence toward the truth will take place at best *with probability one* (*CP* 4.547n; see [30], p. 82).

Kelly proves the strong result that, given a countably additive probability measure P, each Borel hypothesis is decidable in the limit with probability one ([20], p. 317). But he emphasizes that such a probabilistic approach with "almost sure" decidability has to pay the prize that failures are allowed in a non-empty class of measure zero, while the logical approach of formal learning theory requires success with respect to all data streams (*ibid.*, p. 320). So the question remains whether this is an adequate answer of a scientific realist to an inductive sceptic.

Popper's falsificationism (at least in its naive form) is based on the observation that universal statements (like typical laws of nature formulated in scientific theories) are refutable with certainty, but not verifiable with certainty by any finite data $e_n$. The traditional problem of inductive generalization (in the simplest case) concerns the possibility of learning A-statements by observational data. In what sense is the generalization 'All ravens are black' decidable in the limit? The results of formal learning theory, treated as a logic of discovery, give one answer: after a finite sample of black ravens, we may guess that all ravens are black, and we will not give up this hypothetical belief until (if ever) a non-black raven is found. In the same way, in Hintikka's system the rule of proposing constituent $C^c$ after a sample e with n individuals of c kinds is an effective hypothesis generator: if $C^c$ is false, some falsifying instances will eventually be found, but if $C^c$ is true, it will not be falsified by future data.

This consideration does not yet solve the problem of justification or confirmation: can we approach certainty about the color of ravens when our sample becomes larger and larger? But it serves to reveal an important presupposition in Kevin Kelly's [20] suggestive illustration. In the simplest case "worlds can be identified with the unique data streams they produce" (*ibid.*, p. 38): the world w is conceived as an infinite tape or array of individuals, and each infinite data stream e goes through this tape step by step in some order. The same holds when worlds are structures for a first-order language and truth is defined by Tarski's model theory: "the evidence true of a world may arrive in order whatsoever", and each event or object in the world is "eventually described in the evidence" (*ibid.*, p. 270). Thus, if there is non-black raven, it will be found in a finite number of steps. But if all ravens are black, we could verify this claim only by the complete infinite sequence e. On this account, Hume's problem of induction arises from the limitation to "a finitely bounded perspective", since "the empirical scientist can see only a finite initial segment of an infinite data stream at any given time" ([22], p. 160; [20], p. 138). Science has to face the problem of "local underdetermination": no finite time is sufficient to decide the truth value of general hypotheses, even though "a demigod who can see the entire future at once" (i.e, an infinite evidence in the limit) could decide the truth value of 'All ravens are black'.

This account about "convergence to the truth" can be compared to our best results about induction in the earlier sections. Recall that in Hintikka's system the posterior probability $P(C^c/e)$ approaches one when c is fixed and n grows without limit. But this result (4) states only that our degrees of belief about $C^c$ *converge to certainty* on the basis of inductive evidence. It does not yet guarantee that $C^c$ is identical with the *true* constituent $C^*$. Similarly, by (15) we know that the expected verisimilitude $\mathrm{ver}(C^c/e)$ converges to one when c is fixed and n grows without limit. Again this does not guarantee that the "real" truthlikeness $\mathrm{Tr}(C^c, h^*)$ of $C^c$ is maximal, i.e., that $C^c$ is true. For these stronger results an additional *evidential success condition* is needed:

(SC) Evidence e is true and fully informative about the variety of the world w.

SC means that e is exhaustive in the sense that it exhibits (relative to the expressive power of the given language L) all the kinds of individuals that exist in the world (see [32], p. 276). With SC we can reformulate our results so that they concern *convergence to the truth*:

(4′) When SC holds, $P(C^*/e) \to 1$, when c is fixed and $n \to \infty$.

(14′) When SC holds, $\mathrm{ver}(g/e) \to \mathrm{Tr}(g, C^*)$, when $n \to \infty$ and c is fixed.

(15′) When SC holds, $\mathrm{ver}(g/e) \to 1$ iff $\vdash g \equiv C^*$, when $n \to \infty$ and c is fixed.

Similar modifications can be made in the convergence results about probable approximate truth.

Now we can see that the convergence results of formal learning theory are not stronger than those of probabilistic approaches, even though they demand success with respect to all data streams, since they *presuppose* something like the success condition SC: decidability in the limit assumes that the data streams are "compelete in that they exhaust the relevant evidence" ([3], p. 210) or "perfect" in that "all true data are presented and no false datum is presented" and all objects are eventually described ([20], p. 270). Without SC even "global underdetermination" cannot be avoided (*ibid.*, p. 17), since we cannot be certain that even an infinite sample of swans refutes the false generalization 'All swans are white': it logically possible that an infinite stream of white swans is picked out from a world containing white and black swans.

A more formal way of expressing these conclusions is to note that the learner of Hintikka's system uses *epistemic* probabilities in her prior distribution $P(C^w)$ and likelihoods $P(e/C^w)$. The result (4) as such needs no assumption that behind these likelihoods there are some objective conditions concerning the sampling method. The same observation can be made about the famous results of Bruno de Finetti and L. J. Savage about the convergence of opinions in the long run, when the learning agents start from different non-dogmatic priors ([17,3]; cf [30], p. 102). But it is possible combine a system of inductive logic with the assumption that the evidence arises from a *fair* sampling procedure which gives each kind of individual an objective non-zero chance of appearing the evidence e [23], where such chance is defined by a physical probability or *propensity*. As such propensities do not satisfy the notorious Principle of Plenitude, claiming that all possibilities will sometimes be realized, they do not exclude infinite sequences which violate SC (see [34]). But such sequences are extremely improbable by the convergence theorems of probability calculus (cf. [4], p. 76).

Suppose that we draw with replacement a fair sample of objects from an urn w. Let r be the proportion of objects of kind A in w. Then the objective probability of picking out an A is also r. The Strong Law of Large Numbers states now that the observed relative frequency k/n converges *with probability one* to the unknown value of r. Such "almost sure" convergence is weaker than convergence in the ordinary sense. The reason for using this notion of convergence is that there are no logical reasons for excluding such non-typical sequences of observations that violate SC – even though their measure among all possible sequences is zero.

Formal learning theory and probabilistic theories of induction, as plausible attempts to describe scientific inquiry, are in the same boat with respect to the crucial success conditions: SC is precisely the reason why inductive inference is always non-demonstrative or fallible even in the ideal limit, since there are no logical reasons for excluding the possibility that SC might be incorrect. The conclusion to be drawn from these considerations can be stated as follows: the best results for a fallibilist "convergent realist" do not claim decidability in the limit or even gradual decidability, but rather convergence to the truth with probability one.

## 6. CONCEPTS AND PAC-LEARNING

An important branch of machine learning is the study of *concept learning*. Tom Mitchell [28] characterizes this as "acquiring the definition of a general category given a sample of positive and negative training examples of the category". Each element in the set of *instances* X is defined conjunctively by the values of a finite set of finitely valued attributes (e.g., color, shape, size) (see also [9]). Thus, an attribute-based domain is the same as Carnap's family of predicates, and each instance corresponds to a Q-predicate of a classification system (see Section 3). A concept G is a subset of the instance space X (see also [7]), so that in the Carnapian terminology it is a disjunction of Q-predicates. The learner has some positive and negative *training examples* E of the unknown target concept G, where the source of information is an "oracle" (e.g., nature or teacher). The task is to find a hypothesis H $\subseteq$ X which fits the training data E and is identical to G. Usually only hypotheses consistent with the training data E are considered, but in choosing the hypothesis space **H** additional restrictions on the logical form of H may be added. Such restrictions are known as the "inductive bias" of the learning algorithm ([28], p. 43). Also "noisy data" E may be allowed. The "inductive learning hypothesis" then states that a hypothesis H which approximates the target concept well over a sufficiently large set E of training examples will also approximate the target concept G well over other unobserved examples (*ibid.*, p. 23).

For philosophers, this account of "concept learning" may seem a little perplexing. What they might expect from such a study is an analysis of two important methods of concept formation. First, a child learns many basic descriptive terms (like 'red') by *ostension*: some paradigm cases of red objects are presented, and the term 'red' is then applied to all objects that are sufficiently similar to these paradigms. Secondly, concepts are introduced by *definitions* (e.g., 'bachelor' means the same as 'unmarried man') or by *theories* (e.g., 'mass' is what satisfies the postulates of Newton's mechanics). Giving sharper definitions for the vague terms of the ordinary language is known as the method of *explication*.

The AI approach to concept learning has some similarity to ostension. But the AI framework does not always make clear distinctions – cherished by philosophers – between propositions, concepts, and individuals.

For example, Valiant's [46] seminal paper defines instances X as "vectors" which are essentially constituents of a propositional logic, and "concepts" G are sets of such constituents. A typical result tells that the class of expressions in the conjunctive normal is effectively learnable by the introduction and identification of such vectors. If each atomic proposition $p_i$ is replaced by a formula $M_i x$, where $M_i$ is a basic predicate of a monadic first-order language L, then Valiant's total vectors would correspond to Q-predicates, and a set of such "vectors" would correspond to a predicate in L, definable as a disjunction of Q-predicates in L. In this interpretation, concept learning methods would involve logical reasoning which proceeds from a disjunction of Q-predicates to its shorter expression in L. For example, if the predicate family *Age* has three values {young, middle-aged, old}, then from the Q-predicates 'male&young', 'male&middle-aged', and 'male&old' we may derive their disjunction which is equivalent simply to 'male'. While such logical tasks are no doubt interesting and useful for some purposes, and may be effectively taught to computers, they do not involve anything like induction. What is more, they can hardly be described as examples where a new concept has been learned: the outcome predicate 'male' was already assumed to be available and known in the descriptions of the instances.

One way of formulating the concept learning model in terms familiar from inductive logic would be to take the instances in X to be the individuals $a_1, \ldots, a_K$ of a monadic language L, where the target concept G is the only predicate in L. Hence, the Q-predicates of L are simply G and $\sim$G. Then the evidence E contains n singular statements of the form $(\pm)G(a_i)$, where n < K, and each hypothesis H states for all individuals $a_1, \ldots, a_K$ whether they belong to G or not. In other words, each H is a Carnapian state description in L. Carnap's and Hintikka's systems are now available for calculating the probabilities $P(H/E)$. If $\lambda < \infty$, all hypotheses are not equally probable, and learning from evidence E is possible. The measures of truthlikeness Tr and expected verisimilitude ver can also be applied to this situation (see [32]). However, while this account shows what is inductive in concept learning, it does not relate the target concept G to other concepts.

Still another formulation accepts a literal interpretation of the statement that the positive and negative examples in E themselves are instances in space X (e.g., [9], p. 645), but treats instances in X as individual objects rather than Q-predicates: a Q-predicate is only a description of an object in L, and there may be many different objects that satisfy the same Q-predicate. This is the typical situation in Carnapian inductive logic. It is reinforced by Mitchell's examples where a concept or category is taken to describe "some subset of objects or events defined over a larger set (e.g., the subset of animals that constitute birds)" ([28], p. 20), or X is chosen as the set of all people, each described by some attributes (*ibid.*, p. 203). In this case the repetition of instances of the same kind would be relevant, as they correspond to the observed relative frequencies of the Q-cells in a sample, but the practice of Mitchell is to list the instances as if they were the Q-predicates (see e.g. Table 2.6), and even to calculate relative probabilities on the basis of such tables (see Section 6.9.1). Mitchell mentions the "m-estimate of probability" (formula 6.22), which is in fact the same as Carnap's $\lambda$-continuum, but his application of Bayesian methods to concept learning (Section 6.3) makes the assumptions that all hypotheses H are a priori equally probable and the probability of training data E given H is 1 if D is consistent with H and 0 otherwise. This leads to the disappointing conclusion that "the brute-force Bayesian algorithm" recommends all consistent hypotheses as a posteriori equally probable.

Consider typical examples: classify days according to whether some will play tennis by using attributes weather outlook, temperature, humidity, and wind (Table 3.2); characterize people who are skiers in terms of their sex, age, and wealth, and generate instances by observing persons who walk out of the largest sports store in Switzerland (*ibid.*, p. 203). The term "concept learning" appears to be a misnomer for such tasks, since the goal is neither to learn some conceptual or analytic truths (such as definitions of concepts) nor to learn something about the actual use of words in the colloquial language. Rather the goal is to find *factual truths* about questions like 'On what kinds of days do people practice tennis?' and 'What kinds of people do practice downhill skiing?'. The task of characterizing swans in terms of their size, neck, aquatic, and color might produce something that could be called a "definition" of swans, but yet it is better to describe the goal as "a correct classification law for swans" ([8], p. 32).

It would be misleading to formulate such a classification law as a generalization of the form 'For all people x, x is a skier if and only if x is an old rich man or x is a young rich man or ...)', since some rich old men do not practice skiing. A better formulation is to tell which classes defined by 'skier, male, old, rich', 'skier, male, young, rich', 'non-skier, female, young, rich' etc. are occupied and which are not. This means that the relevant goal is a *constituent* $C^w$ in the appropriate language L, i.e., a statement of the form $(\pm)(\exists x)Q_1 x \& \ldots \& (\pm)(\exists x)Q_K x$. The Q-predicates, which are non-empty by $C^w$ and contain a positive occurrence of the term 'skier', constitute then the target concept G. The rival hypotheses H are also constituents in L. It follows that Hintikka's inductive logic is directly applicable to "concept learning" tasks. The training data E, containing positive and negative instances of G, include now information about the number of observed individuals which exemplify different Q-predicates of L, and Hintikka's system gives immediately posterior probabilities of hypotheses H on the basis of evidence E. In harmony with Mitchell's inductive learning hypothesis, for a sufficiently large evidence E, the posterior probability P(H/E) will be close to one when H is the constituent $C^c$ which claims that precisely the Q-predicates so far exemplified in E are instantiated in the world (see (4)). As a modification of this approach, predictive inductive probabilities about the next individuals could be allowed to include an analogy factor, so that the posterior probability $P(Q_i(a_{n+1})/E)$ of finding instances in a so far unexemplified cell $Q_i$ depends on the similarity of $Q_i$ with cells already exemplified in E (see [34]).

The AI literature has a special interest in the number of training examples needed to learn a concept. A learning method is clearly unrealistic if it requires that all possible instances in X are known in E. An interesting new approach in AI was proposed by Valiant [46]: allow that the learner does not output a zero-error hypothesis H, but require that the error of H is bounded by some given small constant $\epsilon > 0$, and require further that the probability of failure in producing such a hypothesis is bounded by some small constant $\delta > 0$. In other words, the learner is expected to probably learn a hypothesis that is approximately correct, so that this model is called *probably approximate correct* learning or *PAC learning* for short. A class $\Gamma$ of concepts is PAC($\epsilon, \delta$)-learnable if for each concept G in $\Gamma$ the learner is able to produce with probability at least $1 - \delta$ a hypothesis H such that the error of H is not more than $\epsilon$, in time that is polynomial in $1/\epsilon$, $1/\delta$, the size |X| of the instance space X, and the size |G| of G (see [28], p. 206).

In the PAC-learning model, it is assumed that instances are drawn with replacement from the instance space X by a probability distribution D, and the main results should hold for any D. (In the so called PAC-Bayesian models, there is in addition to D also information provided by a prior probability measure on the concept space. See [27].) The *error* of hypothesis H is the D-probability that a randomly drawn instance will fall in the region where H differs from the target concept G. If all instances in X are equally probable, then the error of H is simply "the fraction of all instances on which the hypothesis and target concept disagree" ([9], p. 649; [28], p. 205). This is familiar to us as the Clifford distance (6) between the constituents corresponding to H and G. Then a typical result tells that, for a finite hypothesis space $\mathbf{H}$, the value $[1/\epsilon]/[\ln(|\mathbf{H}|) + \ln(1/\delta)]$ gives a lower bound to the number n of training examples to assure that any consistent hypothesis H in $\mathbf{H}$ will be with probability $1 - \delta$ approximately correct within error $\epsilon$ ([28], p. 209).

In Hintikka's system, we may calculate the minimum number $n_0$ of observations needed to make $C^c$ highly probable (see (5)). Note that if $\alpha = n$, then all non-falsified constituents will have the same posterior probability $1/2^{K-c}$, so that $n_0 > \alpha$. Similarly, given a small number $\epsilon > 0$, we may calculate the degree of probable approximate truth $PAT_{1-\epsilon}(H/E)$ for a hypothesis H given evidence E. By (17), this value is equal to the posterior probability of the blurred version $H^\epsilon$ of H. If H is the constituent $C^c$, then its $\epsilon$-blurred version is the disjunction of all constituents which are at most at the distance $\epsilon$ from $C^c$. We may then calculate the minimum number $n_1$ (smaller than $n_0$ above) such that $C^c$ is highly probably approximately true within error $\epsilon$. In this way, the main ideas of PAC-learning can be translated to the framework of inductive logic with respect to monadic languages. The same is true of our earlier results about PAT in connection with statistical estimation problems. By the same token, the notions of expected verisimilitude and probable verisimilitude can be applied to problems of machine learning. For example, in analogy with PAC-learnability, the notions of PRt-learnability and ver-learnability could be introduced.

## 7. CONCLUSION

Donald Gillies [8] has considered the question whether recent work on AI supports inductivism or Popperian anti-inductivism. He concludes that at least some of Popper's more extreme claims against induction cannot be maintained in the light of results on machine learning. Guglielmo Tamburrini [45] in turn argues that "AI investigations on learning systems do not compel one to relinquish radical scepticism toward induction": machine learning can be accounted by Popperian trial and error correction processes and thus need not appeal to principles of induction. It seems that this discussion involves several different ideas about induction.

Gillies chooses Francis Bacon to represent inductivism: Bacon believed that induction can be a mechanical method of discovery. Popper denied this, but Carnap too rejected the possibility of "inductive machines" [2]. Formal learning theory and AI studies in machine discovery certainly give interesting new material and insights to this debate. So does Peirce's account of abduction (not mentioned by Gillies), even though abduction and its AI versions do not provide a completely mechanical heuristics.

Bacon's method was based upon systematic tables of positive and negative instances which serve - he hoped - eventually to refute all but one lawful statement. In this sense, Bacon's method has been called *eliminative induction*, since it allows the learning from mistakes. It is no wonder that Gillies finds many points of contact between Baconian induction and the Popperian method of conjectures and refutations.

But there is another controversy about "inductivism" which I have studied in this paper: the Carnap - Popper debate about the viability of *enumerative induction* as a fallible method of justification. With respect to this issue, I have argued already in Niiniluoto and Tuomela [40] that a Hintikka-type inductive logic can be "non-inductivist": its probabilities and rules of acceptance may be relative to background theories (as emphasized also by Gillies [8]), its principles of confirmation may favor bold and informative hypotheses, and the validity or reliability of its methods may be relative to contextual presuppositions. The logical problem of inductive generalization has not disappeared from cognitive science, where induction is seen to be relative to the choice of a language with projectible predicates (see [7], Ch. 6) or to tacit assumptions about population distributions (see [16], Ch. 8). Hintikka's system, as a theory within "logical pragmatics", does not deny the relevance of such contextual factors, but expresses them in a condensed form in the parameter $\alpha$.

Probabilistic inductive inference in this broad Bayesian sense is always hypothetical and tentative, and its conclusions may be at any time revised in the light of new evidence or conceptual and theoretical innovations. Still, its systematic use will lead, under appropriate success and convergence conditions, to probably true, probable approximately true, or probably truthlike results. We have thus argued that fallibilism rather than scepticism is the proper attitude towards non-demonstrative inferences like induction and abduction. Further, a non-inductivist inductive logic, combined with suitable notions of approximate truth and verisimilitude, has been seen to provide a flexible framework for interpreting and reconstructing the AI approach to concept learning.

## REFERENCES

1. G. Box and G. Tiao. *Bayesian Inference in Statistical Analysis*. Reading, MA: Addison-Wesley, 1973.
2. R. Carnap. *The Logical Foundations of Probability*, 2nd ed. Chicago: The University of Chicago Press, 1962.
3. J. Earman. *Bayes or Bust? A Critical Examination of Bayesian Confirmation Theory*. Cambridge, MA: The MIT Press, 1992.
4. R. Festa. *Optimum Inductive Methods*. Dordrecht: Kluwer, 1993.
5. R. Festa. Bayesian Confirmation. In: Galavotti, M. and Pagnini, A. (eds.), *Experience, Reality, and Scientific Explanation*. Dordrecht, Kluwer, 55-87, 1999.
6. P. Flach and A. Kakas (eds.). *Abduction and Induction: Essays on their Relation and Integration*. Dordrech: Kluwer, 2000.
7. P. Gärdenfors. *Conceptual Spaces: The Geometry of Thought*. Cambridge, MA: The MIT Press, 2000.
8. D. Gillies. *Artificial Intelligence and Scientific Method*. Oxford: Oxford University Press, 1996.
9. D. Haussler. Applying Valiant's Learning Framework to Concept-Learning Problems. In: Kodratoff, Y. and Michalski, R.S. (eds.), *Machine Learning: An Artificial Intelligence Approach III*. San Mateo, CA: Morgan Kaufmann, 641-669, 1990.
10. D. Helman (ed.). *Analogical Reasoning*. Dordrecht: D. Reidel, 1988.
11. C. G. Hempel. *Aspects of Scientific Explanation*. New York: The Free Press, 1965.

12. R. Hilpinen. *Rules of Acceptance and Inductive Logic.* Amsterdam: North-Holland, 1968.
13. J. Hintikka. The Varieties of Information and Scientific Explanation, in: van Rootselaar, B. and Staal, J.F. (eds.), *Logic, Methodology, and Philosophy of Science III.* Amsterdam: North-Holland, 151-171, 1968.
14. J. Hintikka and P. Suppes (eds.). *Aspects of Inductive Logic.* Amsterdam: North-Holland, 1966.
15. J. Hintikka and P. Suppes (eds.). *Information and Inference.* Dordrecht: D. Reidel, 1970.
16. J. J. Holland, K. J. Holyoak, R. E. Nisbett and P. R. Thagard. *Induction: Processes of Inference, Learning, and Discovery.* Cambridge, MA: The MIT Press, 1986.
17. C. Howson and P. Urbach. *Scientific Reasoning: The Bayesian Approach.* La Salle: Open Court, 1989.
18. R. Jeffrey (ed.). *Studies in Inductive Logic and Probability*, vol. 2. Berkeley: University of California Press, 1980.
19. J. R. Josephson and S. G. Josephson (eds.). *Abductive Inference.* Cambridge: Cambridge University Press, 1994.
20. K. Kelly. *The Logic of Reliable Inquiry.* New York: Oxford University Press, 1996.
21. K. Kelly and C. Glymour. Convergence to the Truth and Nothing but the Truth. In: *Philosophy of Science*, 56:185-220, 1989.
22. K. Kelly and O. Schulte. Church's Thesis and Hume's Problem. In: Dalla Chiara, M. L. et al. (eds.), *Logic and Scientific Methods.* Dordrecht: Kluwer, 159-177, 1997.
23. T. Kuipers. *Studies in Inductive Logic and Rational Expectation.* Dordrecht: D. Reidel, 1977.
24. P. Langley, H. A. Simon, G. L. Bradshaw and J. M. Zytkow. *Scientific Discovery: Computational Explorations of the Creative Processes.* Cambridge, MA: The MIT Press, 1987.
25. I. Levi. *Gambling with Truth.* New York: Alfred A. Knopf, 1967.
26. D. Lindley. *Introduction to Probability and Statistics from a Bayesian Viewpoint, Part 2: Inference.* Cambridge: Cambridge University Press, 1965.
27. D. McAllester. Some PAC-Bayesian Theorems. In: *Machine Learning*, 37:355-263, 1999.
28. T. M. Mitchell. *Machine Learning.* New York: McGraw-Hill, 1997.
29. I. Niiniluoto. On the Truthlikeness of Generalizations. In: Butts, R. E. and Hintikka, J. (eds.), *Basic Problems in Methodology and Linguistics.* Dordrecht: D. Reidel, 121-147, 1977.
30. I. Niiniluoto. *Is Science Progressive?.* Dordrecht: D. Reidel, 1984.
31. I. Niiniluoto. Truthlikeness and Bayesian Estimation. In: *Synthese*, 67:321-346, 1986.
32. I. Niiniluoto. *Truthlikeness.* Dordrecht: D. Reidel, 1987.
33. I. Niiniluoto. Analogy and Similarity in Inductive Logic. In: Helman, 271-298, 1988.
34. I. Niiniluoto. Probability, Possibility, and Plenitude. In: Fetzer, J. (ed.), *Probability and Causality.* Dordrecht: D. Reidel, 91-108, 1988.

35. I. Niiniluoto. Corroboration, Verisimilitude, and the Success of Science, in: Gavroglu, K., Goudaroulis, Y. and Nicolacopoulos, P. (eds.), *Imre Lakatos and Theories of Scientific Change*. Dordrecht: Kluwer, 229-243, 1989.

36. I. Niiniluoto. Inductive Logic, Atomism, and Observational Error. In: Sintonen, M. (ed.), *Knowledge and Inquiry*. Amsterdam, Rodopi, 117-131, 1997.

37. I. Niiniluoto. Verisimilitude: The Third Period. In: *The British Journal for the Philosophy of Science*, 49:1-29, 1998.

38. I. Niiniluoto. *Critical Scientific Realism*. Oxford: Oxford University Press, 1999.

39. I. Niiniluoto. Theories of Truth: Vienna, Berlin, and Warsaw. In: Woleski, J. and Köhler, E. (eds.), *Alfred Tarski and the Vienna Circle*. Dordrecht: Kluwer, 17-26, 1999b.

40. I. Niiniluoto and R. Tuomela. *Theoretical Concepts and Hypothetico-Inductive Inference*. Dordrecht: D. Reidel, 1973.

41. C. S. Peirce. *Collected Papers* 1-6, ed. by C. Hartshorne and P. Weiss, 7-8, ed. by A. Burks. Cambridge, MA: Harvard University Press, 1931-35, 1958.

42. K. R. Popper. *The Logic of Scientific Discovery*. London: Hutchinson, 1959.

43. K. R. Popper. *Conjectures and Refutations*. London: Hutchinson, 1963.

44. A. Shimony. Scientific Inference. In: Colodny, R.C. (ed.), *The Nature and Function of Scientific Theories*. Pittsburgh: The University of Pittsburgh Press, 79-172, 1970.

45. G. Tamburrini. Artificial Intelligence and Popper's Solution to the Problem of Induction, a paper presented in the Popper Centennial Conference, Vienna 2002.

46. L. G. Valiant. A Theory of the Learnable. In: *Communications of the ACM*, 27:1134-1142, 1984.

# Conditional Probability, Conditional Events, and Single-Case Propensities

Peter Milne

*School of Philosophy, Psychology and Language Sciences, University of Edinburgh,*
*David Hume Tower, George Square, Edinburgh EH8 9JX, United Kingdom*
*Peter.Milne@ed.ac.uk*

**Abstract.** What some probability textbooks say about conditional probabilities doesn't make much sense. What are conditional probabilities? Attempts to wriggle out of Lewis's triviality results notwithstanding, they're not probabilities of ordinary events or propositions. One way—and it is only one among others—to address the question is to start with another: What sort of things could be "the bearers of conditional probabilities"? That is, what sort of more-or-less event-like things or proposition-like things could conditional probabilities be probabilities of? We can work this out in the abstract, and obtain identity conditions and a natural logico-algebraic structure, from fairly obvious constraints drawn in analogy with absolute probabilities but we still have some work to do to make any real sense of the answer. A formal connection with rough sets suggests one way to do this.

Is there a point to this exercise? I maintain that, contrary to some recent claims in the literature defending it from previous attack by Paul Humphreys and myself, the single-case propensity interpretation still faces a difficulty in accommodating conditional probabilities. Given that difficulty and granted the formal possibility previously elaborated, one way to go might be to think of these conditional probabilities as propensities relating to conditional events. Working through this suggestion well illustrates the difficulties involved in making sense of talk of conditional events and conditional assertions.

What I want to do here is to draw together material from various things I have been working on and working out over the past few years and indicate how this might be brought to bear on a problem that first bothered me some time ago. The topics are those of my title and I'll introduce them in the order they are given there. Sometimes I think the different strands in this work all converge nicely. At other times, it seems to me that they do not fit neatly together. At present I am more inclined to the latter view, but it's interesting to see where the cracks show up (if, in fact, they do).

## 1. What some textbooks say

We begin with what some textbooks have to say about conditional probability. There's nothing special about my choice of texts or editions; the examples are drawn from what was available on a quick perusal of the shelves of the University of Edinburgh's Main Library.

In Emanuel Parzen's [29] (p. 60, my emphasis) we find:

> Given two events, $A$ and $B$, by the conditional probability of the event
> $B$, given the event $A$, $P[B|A]$, we mean intuitively the probability that
> $B$ will occur, under the assumption that $A$ has occurred. In other
> words, $P[B|A]$ represents our re-evaluation of the probability of $B$ *in
> the light of the information that* $A$ *has occurred.*

William Feller is perhaps more circumspect. He gives the standard formal account
then goes on to say [4] (p. 115):

> The quantity so defined will be called the conditional probability of $A$
> on the hypothesis that $H$ (or for given $H$). [...] Though the symbol
> $\mathbf{P}\{A|H\}$ itself is practical, its phrasing in words is so unwieldy that in
> practice less formal descriptions are used. [...] Often the phrase "on
> the hypothesis $H$" is replaced by "if it is known that $H$ occurred".

Confirming that last assertion, here's a potted extract from Boris Gnedenko's [6]
(pp. 51-2):

> [I]n a number of cases it is necessary to find the probability of events,
> given the supplementary condition that a certain event $B$ has occurred.
> We shall call such probabilities *conditional* and denote them by the
> symbol $\mathbf{P}(A|B)$; this signifies the probability of event $A$ on condition
> that event $B$ has occurred. [...]
>
> **Example 1.** Two dice are thrown. What is the probability that
> the sum 8 comes up (event $A$) if it is known that this sum is an even
> number (event $B$)?

That's a Soviet probabilist writing here, one whose fellow-countryman and fellow-
probabilist Alexander Yakovlevich Khinchin engaged in the struggle against ideal-
ism in the calculus of probability [12]. Given that none of these authors is advo-
cating a degree of belief interpretation of probability, we might well ask, What has
knowledge (or information) got to do with it? To which the short answer should
be, Nothing at all.

More importantly for what is to follow, the other thing to notice about these,
and other accounts—*e.g.* Moode & Graybill's [26] (p. 35, my emphasis):

> Let $A$ and $B$ be two events in a sample space $S$ such that $P(B) > 0$.
> The conditional probability of the event $A$, given that $B$ *has happened,*
> which is written $P(A|B)$, is

$$P(A|B) = \frac{P(A, B)}{P(B)}$$

—is the use of the past tense expressions 'has occurred', 'has happened', 'occurred'.
They pretty much invite one to read $P(b|a) = r$ as:

> if $a$ has occurred then the probability of $b$ is $r$.

Read literally, this leads straight to contradiction. If $a$ and $c$ have both occurred then $a$ has occurred, but $P(b|a)$ may differ from $P(b|a\&c)$.

That's my quick survey. While what these texts say does not bear close scrutiny, they do—and this should be emphasised—they do have the merit of seeing that the reader should be told something about how to think about conditional probabilities, unlike some other texts which just introduce conditional probabilities formally, as induced probabilities over a reduced space of events, without giving the least hint as to why this might be a useful or interesting thing to do.

## 2. Triviality

Conditional probabilities are not probabilities of any ordinary events or propositions. That's what David Lewis showed in 1972 [15]. Here's a variation on Lewis's argument (taken from [23]).

Karl Popper, it seems (see [2]), first noticed that the probability of the material conditional, $a \supset c$, is never less than the conditional probability $P(c|a)$ and but for exceptional cases exceeds it:

When $P(a) > 0$,

$$\begin{aligned} P(a \supset c) &= P(\sim a \lor (a\&c)) \\ &= P(\sim a) + P(a\&c) = P(\sim a) + P(c|a)P(a) \\ &\geq P(c|a)P(\sim a) + P(c|a)P(a) = P(c|a), \end{aligned}$$

with equality if, and only if, $P(a) = 1$ or $P(c|a) = 1$.

Now, suppose that $P(c|a)$ is the probability of some event or proposition $a \Rightarrow c$ and see what happens:

When $P(a\&c) > 0$,

$$\begin{aligned} P(a \Rightarrow c) &\geq P((a \Rightarrow c)\&(a \supset c)) = P(a \Rightarrow c|a \supset c)P(a \supset c) \\ &= P(c|a\&(a \supset c))P(a \supset c) = P(c|a\&c)P(a \supset c) \\ &= 1.P(a \supset c) = P(a \supset c). \end{aligned}$$

Putting the pieces together,

$P(.|a)$ is two-valued when $0 < P(a) < 1$.

This is equivalent to Lewis's own triviality results.

## 3. Modus tollens

Lewis's argument assumes that the *conditional construal of conditional probability*, *CCCP* as Hájek and Hall call it, is preserved under conditionalization, and that one can form logical combinations of these conditionals with other elements. While Lewis's original result has been considerably strengthened—see Hájek and Hall [9], in efforts to circumvent Lewis's results various authors set about side-stepping these assumptions. Ernest Adams denied that we know what we are about in forming logical combinations involving indicative conditionals [1]; Bas van Fraassen [33] suggested that what a conditional utterance means depends on the degrees of belief of the utterer, which allows one to maintain, formally at least, the preservation of $CCCP$ under conditionalization, but disallows appeal to the identity

$$P((a \Rightarrow c)\&(a \supset c)) = P(a \Rightarrow c|a \supset c)P(a \supset c)$$

because we may not identify the $a \Rightarrow c$ in $P((a \Rightarrow c)\&(a \supset c))$ with the $a \Rightarrow c$ in $P(a \Rightarrow c|a \supset c)$, $P$ and $P(.|a \supset c)$ being different distributions of degrees of belief.

What I present immediately below is an "all purpose" argument that avoids those assumptions that some would say take the edge off Lewis's proof. (Most of what follows in this section is taken from [25].)

Let's think first about *modus ponens* inferences:

If $P(a) = 1$ and $P(a \Rightarrow c) = P(c|a)$ then $P(a \Rightarrow c) = P(c)$.

The conclusion of the *modus ponens* inference

$a \Rightarrow c$, $a$ therefore $c$

is just as probable as its major premise given certainty about the minor premise. On reflection, that ought, I think, to strike you as just fine. (By the way, it even holds good of the material conditional.)

Now for *modus tollens* inferences:

If $P(\sim c) = 1$ and $P(a \Rightarrow c) = P(c|a)$ then either $P(a \Rightarrow c) = P(c| \sim c\&a) = 0$ or $P(a) = P(a\&\sim c) = 0$.

When the minor premise of the *modus tollens* inference

$a \Rightarrow c$, $\sim c$ therefore $\sim a$

is certain, either the major premise is certainly false[1] or the conclusion is certainly true. On even minimal reflection that ought not to strike you as anything close to fine. It implies that $a \Rightarrow c$ doesn't behave anything like a natural language indicative "if ... then ..." statement. To see why not, suppose, for example, that you think it likely but not certain that if Jim arrived yesterday, he arrived on the three o'clock bus, and you're not certain that he didn't arrive yesterday by some other means. Now you learn, and learn nothing more than, that Jim didn't arrive on the three o'clock bus. So now your degrees of belief are represented by a probability distribution in which $P(\sim c) = 1$. All of a sudden, you're either sure that Jim didn't arrive yesterday, or sure that the conditional 'if Jim arrived yesterday, he arrived on the three o'clock bus' is to be rejected out of hand. If you're not yet convinced that he didn't arrive by some other means—and you haven't learned anything that obviously makes that thought untenable—then, no matter how confident, short of certainty, you previously were of the conditional, on learning that Jim didn't arrive on the three o'clock bus you become absolutely convinced that the conditional is to be rejected.

I contend that we just don't use conditionals that way. Being sure that $c$ is false isn't always a way of being sure that $a$ is false or sure that 'if $a$ then $c$' is to be rejected. One way to look at this goes like this: when you're sure that $c$ is false, $\sim a$ and 'if $a$ then $c$' are ways of saying pretty much the same thing, so you should be, more or less, *as* sure about one as about the other, but this doesn't mean that you have to be sure, full stop.[2]

## 4. The bearers of conditional probabilities

Having cleared indicative conditionals out of the way, a new vista opens up. Let us suppose—we don't *have* to do this, nothing *forces* us to do this, but we can, so—let us suppose conditional probabilities *are* probabilities, sort of, anyway, of "things", that $P(b|a)$ is the probability of some entity we shall denote by '$b{:}a$', read '$b$ given $a$'. We know that $b{:}a$ isn't very much like an ordinary language indicative conditional. We know, following Lewis, that it isn't an ordinary event or proposition. The question is: what is it like?

No entity without identity, says Quine. We must discriminate these things, whatever they are, at least as finely as probability theory allows, but need discriminate no more finely. Taking a domain, a boolean algebra, **B**, of ordinary events or propositions as given, and taking the standard Kolmogorov axiomatization as given, we may stipulate that $b{:}a = d{:}c$ when (and only when)

(i) for all probability distributions $P$ defined on **B**, $P(a) > 0$ if, and only if, $P(c) > 0$,

and

(ii) for all probability distributions $P$ defined on **B**, if $P(a) > 0$ and $P(c) > 0$ then $P(b|a) = P(d|c)$.

If we are to identify $b{:}a$ with $d{:}c$ then either the probabilities assigned both are fixed by the probabilities of boolean elements or neither is, which is what (i) guarantees, and if they are so fixed then they must be the same, which is what (ii) says.

Granted the Boolean Prime Ideal Theorem, there's a non-probabilistic way to reformulate this criterion:

$b{:}a = d{:}c$ if, and only if, $a = c$ and $a \wedge b = c \wedge d$.

This is our criterion of identity for the entities ("things") we take to be "the bearers of conditional probabilities".

## 5. Logico-algebraic structure

There are a number of ways to impose logico-algebraic structure on this range of entities. I'll give two ways of getting at one very well-behaved structure. (For more details see [24].)

### (i) The Mazurkiewicz (–Koopman) approach

Given any partially ordered set $\mathbf{T} = \,<T,\leq_T>$ we can form the set $\mathbf{I}(T)$ of intervals of $\mathbf{T}$, *i.e.*, of all sets of the form $\{c \in \mathbf{T}: a \leq_T c \leq_T b\}$ where $a \leq_T b$. There are two natural orderings on intervals. One is the subset (or inclusion) ordering, under which intervals form an upper semi-lattice. The other is more interesting for present purposes:

The natural induced partial ordering on $\mathbf{I}(T)$ is this:

$[a,b] \leq [c,d]$ iff $a \leq_T c$ and $b \leq_T d$.

We let $\mathbf{I}(\mathbf{T})$ denote the partially ordered set $<\mathbf{I}(T), \leq>$. The structure of $\mathbf{I}(\mathbf{T})$ is fully determined by the structure of $\mathbf{T}$ and *vice versa*. (In fact, $\mathbf{T}$ is copied into $\mathbf{I}(\mathbf{T})$ by the degenerate intervals of the form $[a,a]$.) In the case of a boolean algebra we find that the algebra of intervals over a boolean algebra is what is variously known as: a centred Łukasiewicz algebra of order three; a regular double Stone algebra; a semi-simple Nelson algebra.

For fixed $a$, consider the equivalence relation

$b \approx_a c$ iff $b : a = c : a$,
*i.e.*, $b \approx_a c$ iff $a \wedge b = a \wedge c$.

The resulting equivalence classes are *intervals* over the underlying boolean algebra. That is,

$[b]_a^{\approx} = \{c \in B : a \wedge c = a \wedge b\} = \{c \in B : a \wedge b \leq_B c \leq_B \sim a \vee b\}$.

Conversely, every interval is an equivalence class:

where $a \leq_B b$, $[a,b] = [a]_{a \vee \sim b}^{\approx}$.

Following a well worn mathematical path, we may as well take the object $b{:}a$ to *be* the equivalence class $[b]_a^{\approx}$, as Stanislaw Mazurkiewicz was, it seems, the first to suggest [16].[3] Even without that move, we can impose the natural ordering of intervals on these conditional entities. We have

$b{:}a \leq d{:}c$ iff $a \wedge b \leq_B c \wedge d$ and $\sim a \vee b \leq_B \sim c \vee d$.

### (ii) The betting order

Suppose that you are offered the choice between betting to win $X$ if $a$ occurs and to lose $Y$ if it doesn't, and betting, for the same gains and losses, on $b$, $X$ and $Y$ both positive. An easy expected utility calculation tells us that, irrespective of the values of $X$ and $Y$, you do not prefer the bet on $a$ to the bet on $b$ if $P(b) \geq P(a)$, and so you *CANNOT* prefer the bet on $a$ to the bet on $b$ if, and only if, the probability of $a$ *cannot* exceed the probability of $b$, which is the case just in case $a \leq_B b$.

Now consider conditional bets: You are offered the choice between betting to win $X$ if $a$ and $b$ both occur and to lose $Y$ if $a$ occurs but $b$ doesn't, the bet being called off if $a$ doesn't occur, and betting to win $X$ if $c$ and $d$ both occur and to lose $Y$ if $c$ occurs but $d$ doesn't, the bet being called off if $c$ doesn't occur, $X$ and $Y$ both positive. A slightly messier expected utility calculation tells us that you *CANNOT* prefer the bet on $b$ conditional on $a$ to the bet on $d$ conditional on $c$, irrespective of the values of $X$ and $Y$, if, and only if, the probability of $a \wedge b$ *cannot* exceed the probability of $c \wedge d$ and the probability of $\sim a \vee b$ *cannot* exceed the probability of $\sim c \vee d$, which is the case just in case $a \wedge b \leq_B c \wedge d$ and $\sim a \vee b \leq_B \sim c \vee d$.

The betting order *coincides with* the interval ordering; and this ordering fits exactly with our identity criterion:

$$b : a \leq d : c \text{ and } d : c \leq b : a \text{ iff } b : a = d : c.$$

## 6. A step towards interpretation

That has all been very formal. We're still not much closer to really knowing what these conditional entities are. Surprisingly, perhaps, a couple of formal results will help us.

The Boolean Prime Ideal Theorem (= Ultrafilter Theorem) tells us that for any boolean algebra $\mathbf{B} = \; < B, \wedge, \vee, ^c, \mathbf{0_B}, \mathbf{1_B} >$, if $c \nleq_B d$ then there is a function $v : B \to \{0, 1\}$ with these properties:

(i) $v(a \wedge b) = min\{v(a), v(b)\}$;

(ii) $v(a \vee b) = max\{v(a), v(b)\}$;

(iii) $v(a^c) = 1 - v(a)$;

(iv) $v(\mathbf{1_B}) = 1$;

and such that

$$v(c) > v(d).$$

There's always a two-valued evaluation. That's what we expect: events occur or they don't; propositions are true or they're false; properties or outcome-types are instantiated or not.

The Boolean Prime Ideal Theorem also tells us that for any centred Łukasiewicz algebra of order three $\mathbf{L} = \; < L, \wedge, \vee, ^+, \to, \mathbf{u}, \mathbf{0_L}, \mathbf{1_L} >$, if $C \nleq_L D$ then there is a function $v : L \to \{0, \frac{1}{2}, 1\}$ with these properties:

(i°) $v(A \wedge B) = min\{v(A), v(B)\}$;

(ii°) $v(A \vee B) = max\{v(A), v(B)\}$;

(iii°) $v(A^+) = 1 - v(A)$;

(iv°) $v(A \to B) = 1$, if $v(A) \leq v(B)$, and $v(A \to B) = v(B)$, if $v(A) > v(B)$;

($v^\circ$) $v(\mathbf{1}_L) = 1$;

($vi^\circ$) $v(\mathbf{u}) = \frac{1}{2}$;

and such that

$$v(C) > v(D).$$

There's always a three-valued valuation.

Objects of the form $b{:}a \vee \sim a$ behave "classically" and take only the values 0 and 1. We can identify the conditional object $b{:}a \vee \sim a$ with the original boolean element $b$. We find:

For any $a$ and $b$ in $B$, the domain of the underlying boolean algebra,

$v(b{:}a) = 1$, when $v(a) = v(b) = 1$;
$v(b{:}a) = 0$, when $v(a) = 1$ and $v(b) = 0$;
$v(b{:}a) = \frac{1}{2}$, when $v(a) = 0$.

So, going at this bluntly, we arrive at this list of conditional entities:

| | |
|---|---|
| *Conditional propositions*: | $b{:}a$ is *true* when $a$ and $b$ are both *true*; $b{:}a$ is *false* when $a$ is *true* and $b$ is *false*; $b{:}a$ takes the third value and so is neither *true* nor *false*, when $a$ is *false*. |
| *Conditional events*: | $b{:}a$ *occurs* when $a$ and $b$ both *occur*; $b{:}a$ *fails to occur* when $a$ *occurs* and $b$ *fails to occur*; $b{:}a$ neither *occurs* nor *fails to occur*, when $a$ *fails to occur*. |
| *Conditional properties and outcomes*: | $b{:}a$ is *instantiated* when $a$ and $b$ are both *instantiated*; $b{:}a$ *fails to be instantiated* when $a$ is *instantiated* and $b$ *fails to be so*; $b{:}a$ is neither *instantiated* nor *fails to be instantiated*, when $a$ *fails to be instantiated*. |

The account here of conditional propositions was, to the best of my knowledge, first proposed, as an analysis of the indicative conditional of natural languages, by Joseph Schächter ([31], p. 144). Arne Naess reports that Schächter, a 'prominent young member' of the Vienna Circle, 'had difficulties with "proper" appreciation of Russell and his "material implication"' [27].[4] The analysis has been revived from time to time, perhaps most intriguingly by Peter Wason in response to what we now call the Wason selection task ([34], p. 146). The point of our §3 is that it does not offer a viable account of the indicative conditional of natural language: indicative conditionals are not our just introduced "conditional propositions".

What we have shown so far is this: that given a probability distribution over the elements of a boolean algebra, we can always construe the conditional probabilities determined by this distribution as probabilities *in some extended sense* of elements of a larger domain, the domain of conditional events determined by the underlying boolean algebra. In this respect we can say that, at a formal level, conditional probabilities do have an interpretation as probabilities, or, perhaps better, probability-like quantities, assigned to conditional entities. However, this does not give us much of a grip on what these conditional entities are.

The language we use in speaking of probabilities of propositions, of events, of outcomes is hopelessly wedded to binary classifications: ordinary probabilities are probabilities of truth of propositions, of occurrence of events, of instantiation of outcomes and propositions are true or false, events occur or fail to, outcomes are instantiated or not. The domain of conditional propositions/events/outcomes is wider than that of ordinary propositions/events/outcomes; ordinary propositions/events/outcomes are those conditional entities conditioned by a condition that cannot but be satisfied. When we turn to genuinely conditional entities there is a third possibility. So unaccustomed are we to thinking in these terms that there is no easy way to speak of this.

## 7. A better interpretation?

We shall briefly explore another way to read the formal results. It is prompted by a technical result [24]:

Every atomic, centred, Łukasiewicz algebra of order three is isomorphic to the algebra of rough sets over an approximation space.

We can think of rough sets as a crude way of dealing with vagueness. Some things are definitely red; some things definitely are not; where fuzzy set theory says the rest are on a sliding scale from 1 to 0, rough set theory puts the rest in a single box: the not-definitely-red-and-not-definitely-not-red box. Of course, there are other problems with vagueness, notably the matter of tolerance: something that looks enough like something that's red is red. Rough sets, because they, in their original formulation by Zdzisław Pawlak, ignore the issue of tolerance, perhaps do not provide a good analysis of vagueness, a fact that does not matter for present purposes.[5] A better application is to location on a large-scale map. Some squares in the grid-reference system will be wholly contained in, let's say, a lake; some will be wholly separate from it; and, unless the lake is of a very unlikely and regular shape, some squares contain both parts of the lake and of its margins. Asked to locate the lake using only the squares on the grid, the best one can do, the most informative one can be, is to give a list of the squares wholly contained in the lake, thereby providing a "lower bound", *and* give a list of all the squares into which the lake intrudes but which are not exhausted by parts of the lake, the "boundary set", which, when added to the lower bound yields an "upper bound". (Another example: saying where the coast-line is on a map of sufficient scale to distinguish high- and low-tide margins; in places renowned for the mud-flats the difference can be significant.)

Putting vagueness to one side, as they do, and giving no thought to the sort of location example we have just mentioned, philosophers of language have been seduced by the appeal of necessary and sufficient conditions. Clear-cut examples give us sufficient conditions; clear-cut counter-instances give us necessary conditions. Mostly, outside of mathematics, we don't have much reason to suppose that these are exhaustive, that we have necessary and sufficient conditions. (If you don't believe that, try giving necessary and sufficient conditions for the application of that universal term of approbation in contemporary British English, *cool*.)

By analogy, taking the formal result seriously, we arrive at a possibly more palatable account of the various sorts of conditional entities a probabilist might sanction:

| | |
|---|---|
| *Conditional propositions*: | *b*:*a* is *definitely true* when *a* and *b* are both *true*; *b*:*a* is *definitely false* when *a* is *true* and *b* is *false*; *b*:*a* is neither *definitely true* nor *definitely false* when *a* is *false*. |
| *Conditional events*: | *b*:*a* *definitely occurs* when *a* and *b* both occur; *b*:*a* *definitely fails to occur* when *a* occurs and *b* *fails to occur*; *b*:*a* neither *definitely occurs* nor *definitely fails to occur*, when *a* *fails to occur*. |
| *Conditional properties and outcomes*: | *b*:*a* is *definitely instantiated* when *a* and *b* are *both instantiated*; *b*:*a* *definitely fails to be instantiated* when *a* is *instantiated* and *b* *fails to be so*; *b*:*a* is neither *definitely instantiated* nor *definitely fails to be instantiated*, when *a* *fails to be instantiated*. |

'Determinately' might do as well as 'definitely' here.

## 8. Conditional propensities: the problem

In 1985 Paul Humphreys published an article posing problems for the single-case propensity interpretation of probability, his 'Why Propensities Cannot be Probabilities'. The following year, independently, I published an article with the less categorical title, 'Can There Be a Realist Single-Case Interpretation of Probability?'. Our arguments were different but the theme was the same: the failure, as we both saw it, of the accounts of probability in question to give an adequate interpretation of conditional probabilities in conformity with orthodox mathematical probability theory.

The important issue is this: how are conditional probabilities to be understood in application to the single case? Single-case probabilities are what one has in mind when one talks of "the probability of this coin here to land heads on this throw" or "the chance of this photon passing through this half-silvered mirror". Realistically construed, *i.e.*, not taken epistemically, single-case probabilities are often called propensities or objective chances; they are held to be objective features of physical reality—Popper compares them to Newtonian forces. Single-case propensities are widely held only to take values other than zero and one if the universe is indeterministic. The problem that Humphreys and I focused on is how to construe conditional probabilities in this setting.[6]

I used a couple of examples to indicate why there is, or at least seemed to me to be, a problem.[7] The examples are familiar but not usually thought of as genuinely indeterministic. Just pretend that they are indeterministic and that the various events have the probabilities that you would find ascribed to them in any textbook on probability. Consider, first, tossing two (fair) coins consecutively. There doesn't seem to be much of a problem concerning the conditional probability of heads with the second given tails with the first but now think about the conditional probability of tails with the first given heads with the second: by the time you get heads with the second, the first coin is tossed and the outcome of tossing it is no longer a matter of chance. Similarly, if you toss a single die once, the conditional probability of a six given that the die has produced an even number face up is not $\frac{1}{3}$ but either 0 or 1, because in order to obtain an even number on the uppermost face either a two or a four or a six appears and it is no longer any matter of chance which.

Those are the examples that led me to the thought that there cannot be a realist single-case interpretation of the *standard* probability calculus *including conditional probabilities*. Subsequently I have been taken to task by two commentators: firstly by David Miller in his book *Critical Rationalism: A Restatement and Defence*, next by Christopher McCurdy in an article that appeared in *Synthese* in 1996, then by Miller again in his contribution to the British Academy *Bayes's Theorem* volume [20]. Both complain that I overlooked the temporal aspect of propensities, that propensities are possessed by chance set-ups at times and that if one carefully attends to the times relative to which one is ascribing the propensities then my argument is hopelessly confused and worthless. (Neither puts the matter that strongly but that's the gist of it.)

Let's take the coin-tossing example and go through it more carefully: at $t_1$ there's a chance set-up comprising a mechanism and two coins about to be tossed, one after the other; at $t_2$ the first coin has been tossed and has landed, the second one has yet to be tossed; at $t_3$ the experiment is over, the second coin has been tossed and come down to earth. At $t_1$ there are four possible future outcomes and the set-up has, we suppose, a propensity of strength $\frac{1}{4}$ to produce each one: $Pt_1(Ht_2Ht_3) = Pt_1(Ht_2Tt_3) = Pt_1(Tt_2Ht_3) = Pt_1(Tt_2Tt_3) = \frac{1}{4}$. At $t_2$ we have a quite different set-up; the world has changed, certain futures that were possible at $t_1$ are no longer so. We have one of two set-ups, we are in one of two states, depending on whether the first coin landed heads or tails: at $t_2$ we have that $Pt_2(Ht_3) = Pt_2(Tt_3) = \frac{1}{2}$. And at $t_3$ everything is over and done with and we have no chance set-up at all.

McCurdy commends the following "updating rule" which, in the case of this example, may well strike the reader as very natural [17], (p. 112):

If the first coin falls heads up then $Pt_2(Ht_3) = Pt_1(Ht_3|Ht_2)$; if the first coin falls tails up then $Pt_2(Ht_3) = Pt_1(Ht_3|Tt_2)$.

As it happens, the two conditional probabilities both take the value $\frac{1}{2}$ in our example—this is, of course, merely an accident of the example. A word of warning— the updating rule is to be read as follows:

the *magnitude* of the propensity of the system we have at time $t_2$ if the first coin falls heads up to produce the outcome heads at $t_3$ is the same as the *magnitude* of the (conditional) propensity of the original set-up at time $t_1$ to produce heads at time $t_3$ given that it produces heads at $t_2$.

The updating rule says that the two propensities are equal in magnitude; it does *not* say that the propensities themselves are identical, and *a fortiori* it does not say that the conditional propensity at $t_1$ is to be interpreted as being the unconditional propensity at $t_2$. That clarified, two problems face us:

(i) What *is* the (conditional) propensity of the original set-up at time $t_1$ to produce heads at time $t_3$ given that it produces heads at $t_2$? Not, What is its magnitude (value)? but, quite literally, What *is* a propensity to produce outcome $a$ conditional upon producing outcome $b$?

(ii) Why *is* the updating rule correct? *I.e.*, why *should* the magnitude of the propensity of the system we have at time $t_2$ to produce the outcome $Ht_3$ when the first coin falls heads up equal the magnitude of the (conditional) propensity of the original set-up at time $t_1$ to produce heads at time $t_3$ given that it produces heads at $t_2$?

In my 1986 paper I solved these two problems in one blow by doing what Miller and McCurdy say I must not: in effect I took the updating rule as providing a single-case propensity interpretation for conditional probabilities and simply identified the (conditional) propensity of the original set-up to produce heads at time $t_3$ given that it produces heads at $t_2$ with the propensity of the system that results *when* heads is the outcome at $t_2$ to produce heads with the second coin at time $t_3$. Having done that it's no surprise that I thought inverse conditional probabilities—probabilities such as the conditional probability of heads at $t_2$ *given* heads at $t_3$—do not receive a satisfactory propensity interpretation.

You will have noticed, I trust, that I interpreted conditional propensities just as we saw some textbooks introducing conditional probabilities: the past-tense reading. Not surprisingly, that leads to trouble when the conditioning event occurs after the conditioned event. Why should I have done that? I did it because I thought I had a fair idea of what the propensity of the system that results *when* heads is the outcome at $t_2$ to produce heads with the second coin at time $t_3$ is. I hadn't a clue what else the propensity at $t_1$ to produce the outcome heads at $t_3$ given heads at $t_2$ might be. I took conditional propensities given, say, that heads *is* produced at $t_2$ to be propensities when heads *has been* produced at $t_2$.

Interpreting single case unconditional probabilities as propensities of a chance set-up to produce outcomes is fine as far as it goes but it tells us nothing about how conditional probabilities are to be understood. With that lacuna in mind, McCurdy's updating rule may well seem to confer an interpretation *on some but not all* conditional probabilities. Moreover, they are interpreted in the same way as ordinary probabilities—as propensities of a system to produce outcomes. But this is exactly what one should not do if one is to understand conditional probabilities as propensities of the system in play at the earlier time $t_1$. McCurdy and Miller set themselves the task of spelling out the details of such an understanding. As it happens, I don't think either succeeds, but that is not important here. What I want to do here is to make use of what we have been over above, exploring whether we can make sense of what should be a quite distinct possibility, namely, understanding conditional propensities as propensities (more or less) to produce conditional events.

## 9. A solution?

What are single-case conditional propensities? — They are propensities, relative, as all propensities are, to chance set-ups or the state of the world at a time, of (single-case) conditional events.

We know that this general form of response to the interpretive question is open to us (provided that the propensity interpretation of absolute probabilities makes sense). To see whether the response makes sense in the present case, we must consider more closely what the propensities of possible outcomes are, and how, if at all, that notion may be extended to include conditional outcomes.

Here is my crude conception of propensities à la Popper: a propensity at $t_1$ to produce an outcome at a later time $t_2$ is, speaking loosely, the oomph in the system/state of the world pushing towards (or in favour of) those possible futures in the garden of forking paths in which that outcome is realized. Now, the conditional propensity at $t_1$ for $a$ at $t_2$ given $b$ at $t_3$ is *not* the oomph pushing towards futures realizing the conditional event $a$ at $t_2$ given $b$ at $t_3$, for this is (determinately) realized only by the joint occurrence of $a$ at $t_2$ and $b$ at $t_3$. Nor is it the oomph pushing towards futures in which the conditional event does not determinately fail to occur, for these are futures in which $a$ occurs at $t_2$ or $b$ fails to occur at $t_3$. Conditional events (as conditional assertions, properties, and outcome-types) take us beyond binary classifications; degenerate cases aside, we are dealing with the essentially three-valued. The conditional propensity is not a force or oomph pushing towards futures realizing *any* particular event: the conditional event $at_2:bt_3$ does not pick out a class of futures, for that would take us back to a binary classification. There is no more to say than that the conditional propensity at $t_1$ for $a$ at $t_2$ given $b$ at $t_3$ is the oomph in the system/state of the world at $t_1$, pushing towards $a$ (at $t_2$) given $b$ (at $t_3$), equivalently, the oomph in the system/state of the world at $t_1$, pushing towards the conditional event $at_2:bt_3$'s determinately occurring given that it either determinately occurs or determinately fails to occur, but to say that is to say too little. I, for one, am none the wiser as to what that might be. Propensities are propensities to produce events/outcomes and in the case of conditional events we just do not seem to have the events to produce.

Is there a way out of this snorl? Perhaps. Let's go back to rough sets and the example of the lake.

The lake is included in an artillery firing range. You have a lot of information giving the locations at which artillery shells land, a lot of information presented in terms of the map's grid reference system: so many shells in this square, so many in that. Told *only* the lower and upper bounds for the lake's location on the grid system, how do you use this information to estimate the probability of a shell's landing in the lake? The probability of a shell's landing within the lower bound— *prob*(definitely in the lake) —is too low; the probability of a shell's landing within the upper bound— 1 - *prob*(definitely not in the lake) —is too high. The ratio

$$\frac{prob(\text{definitely in the lake})}{prob(\text{definitely in the lake}) + prob(\text{definitely not in the lake})}$$

which is the conditional probability

*prob*(definitely in the lake|definitely in the lake or definitely notin the lake),

may be about right, and at least is appropriately sensitive to the probability of a shell landing in the boundary set of squares, which probability is given by

*prob*(not definitely not in the lake) $-$ *prob*(definitely in thelake).[8]

Returning to propensities, what, it seems, we have to say, is this: $Pt_1(at_2|bt_3)$, the conditional propensity at $t_1$ for $a$ at $t_2$ given $b$ at $t_3$, is the oomph in the system at $t_1$ towards $a$-at-$t_2$-given-$b$-at-$t_3$ futures, a possible future being determinately (or definitely) an $a$-at-$t_2$-given-$b$-at-$t_3$ future if, and only if, $a$ occurs at $t_2$ and $b$ occurs at $t_3$ in it, and being determinately (or definitely) *not* an $a$-at-$t_2$-given-$b$-at-$t_3$ future if, and only if, $b$ occurs at $t_3$ and $a$ does not occur in it at $t_2$. *And that is all there is to say.*

Perhaps this is just what we have to say once we take seriously the ternary nature of classification induced by conditional events. And perhaps it is just its unfamiliarity that makes this feel an uncomfortable position to be in. So perhaps we have arrived at an account of what conditional propensities are. Perhaps.

## Notes

[1] Given that some—see in this regard especially Edgington 1989—have denied that the indicative conditionals of natural language have truth-conditions, strictly I should say 'certainly to be rejected' or 'most strongly disbelieved' or some such instead of 'certainly false'.

[2] We should note two facts. One is that *modus tollens* is valid according to Ernest Adams' probabilistic criterion: the sum of the uncertainties of the premises in a *modus tollens* inference is always at least as great as the uncertainty of the conclusion, the uncertainty of $a$ being given by $P(\neg a)$ for ordinary propositions and by $P(\neg b|a)$ for the indicative conditional 'if $a$ then $b$'. That is one fact, the other is that Adams has noticed this oddity concerning *modus tollens*. He takes the line that when one discovers its consequent to be false one will stop asserting an indicative conditional and affirm, in its place, the corresponding counterfactual. 'It must be stressed', he says, 'that this "finding the consequent to be false" type of situation is not one in which the indicative conditional is found to be *false* while the counterfactual is *true*, but rather one in which the *probability* of the indicative conditional becomes low as a result of learning new evidence (that its consequent is false), while presumably the probability of the counterfactual is high or becomes high' ([1], p. 105). This is all rather odd. What evidence favours the counterfactual? — Bar falsity of its consequent, the same evidence as previously supported the indicative conditional, and when asserting a conditional one usually allows for the possibility that its consequent and antecedent may be false. So why

should your degree of belief in the conditional drop so dramatically when you discover something that all along you recognized as very much a live possibility?

If Adams is right that as a matter of linguistic practice we do assert the counterfactual rather than the indicative conditional when we know the consequent to be false, this is, I think, best explained by Gricean considerations. Rules of conversation mean that use of the indicative indicates ignorance as to the truth or falsity of antecedent and consequent. The counterfactual conversationally implicates falsity of both. (Falsity of the antecedent is conversationally implicated by the form 'If *a* were the case, *b* would still be the case', but in this case truth of the consequent is entailed, not just implicated.) None of this shows that the speaker has given up commitment to truth of the indicative conditional.

³ B. O. Koopman came close ([13] and [14]). (Knowing no Polish, I take it on authority that Mazurkiewicz did make the stated suggestion.) For elements of the history of conditional event algebra, including its anticipation in the work of Boole, see [5], pp. 33-35, [28], [7], [8], and [22].

⁴ For more biographical information on Schächter see [32].

⁵ For a brief introduction to rough sets and to an adaptation of them to the analysis of vagueness, along with suggestions for further reading, see [30], pp. 194-202.

⁶ Humphreys returns to these issues, giving a very fastidious reckoning of how various propensity theories fare in the face of the various problem cases, in Humphreys [11].

⁷ No problem of this kind arises in a propensity theory like Hugh Mellor's; there the propensities ascribed to chance set-ups are propensities to produce a probability distribution over the range of possible outcomes (including thereamong conditional probabilities, presumably), not propensities to produce particular outcomes. See [18], Ch. 4.

⁸ If the lake only just makes it into most of the boundary squares, *prob*(definitely in the lake) is a better estimate; if the lake almost fills up most of the boundary squares, *prob*(not definitely not in the lake) is a better estimate; but, of course, information of this sort is exactly what you do not have when you are given only the lower (definitely in) and upper (not definitely not in) bounds with no indication of how fully the lake fills the boundary squares.

## REFERENCES

1. E. Adams. *The Logic of Conditionals*, Dordrecht: Reidel, 1975.
2. G. Dorn. Popper's Laws of the Excess of the Probability of the Conditional over the Conditional Probability, *Conceptus*, 26:3–61, 1992/93.
3. D. Edgington. Do Conditionals Have Truth Conditions?, *Crítica: Revista Hispanoamericana de Filosofía*, XVIII, 52:3-30, 1986; reprinted in *Conditionals*, ed. F. Jackson, Oxford: Oxford University Press, 176-201, 1991.

4. W. Feller. *An Introduction to Probability Theory and Its Applications*, second edition, Volume I, New York and London: John Wiley & Sons, 1957.

5. I. R. Goodman, H. T. Nguyen and E. Walker. *Conditional Inference and Logic for Intelligent Systems: A Theory of Measure-Free Conditioning*, Amsterdam: North-Holland, 1991.

6. B. Gnedenko. *The Theory of Probability*, trans. G. Yankovsky, Moscow: Mir Publishers, 1969.

7. T. Halperin. *Boole's Logic and Probability*, Amsterdam: North-Holland, 1st edition, 1976; 2nd edition, 1986.

8. T. Hailperin. *Sentential Probability Logic: Origins, Development, Current Status, and Technical Applications*, Bethlehem PA: Lehigh University Press, 1996.

9. A. Hájek and N. Hall. The Hypothesis of the Conditional Construal of Conditional Probability. In: Ellery Eells and Brian Skyrms (eds.), *Probability and Conditionals: Belief Revision and Rational Decision*, Cambridge: Cambridge University Press, 75-111, 1994.

10. P. Humphreys. Why Propensities Cannot be Probabilities, *The Philosophical Review*, XCIV, 557-70, 1985.

11. P. Humphreys. Some Considerations on Conditional Chance, *British Journal for the Philosophy of Science*, 2004, to appear.

12. A. Y. Khinchin. Die Methode der willkürlichen Funktionen und der Kampf gegen den Idealismus in die Warscheinlichkeitsrechnung, *Sowjetwissenschaft Naturwissenschafftliche Abteilung* (German translation of Russian original), 7:261-273, 1954.

13. B. O. Koopman. The Axioms and Algebra of Intuitive Probability, *Annals of Mathematics*, 41:269-92, 1940.

14. B. O. Koopman. The Bases of Probability, *Bulletin of the American Mathematical Society*, 46:763-74, 1940; reprinted in H. Kyburg and H. Smokler (eds.), *Studies in Subjective Probability* (second edition), Krieger: Huntington NY, 117-31, 1980.

15. D. K. Lewis. Probabilities of Conditionals and Conditional Probabilities, *Philosophical Review* 85:297–315, 1976. Reprinted with 'Postscript' in his *Philosophical Papers*, Volume II, New York: Oxford University Press, 133–156, 1986 and in *Conditionals*, ed. F. Jackson, Oxford: Oxford University Press, 76–101, 1991.

16. S. Mazurkiewicz. *Podstawy Rachunka Prawdopodobienstwa*, Warsaw: Państowe Wydawnictwo Naukawe, 1956.

17. C. S. I. McCurdy. Humphrey's Paradox and the Interpretation of Inverse Conditional Probabilities, *Synthese*, 108:105-25, 1996.

18. D. H. Mellor. *The Mater of Chance*, Cambridge: Cambridge University Press, 1971.

19. D. Miller. *Critical Rationalism: A Restatement and Defence*, La Salle IL: Open Court, 1994.

20. D. Miller. Propensies May Satisfy Bayes's Theorem. In: R. Swinburne (ed.), *Bayes's Theorem* (*Proceedings of the British Academy, 113*), Oxford: Oxford University Press, 111-116, 2002.

21. P. Milne. Can There Be A Realist Single-Case Interpretation of Probability?, *Erkenntnis*, 25:129-32, 1986.

22. P. Milne. Bruno de Finetti and the Logic of Conditional Events, *British Journal for the Philosophy of Science*, 48:195-232, 1997.
23. P. Milne. The simplest Lewis-style triviality proof yet?, *Analysis*, 63:300-303, 2003.
24. P. Milne. Algebras of Intervals and a Logic of Conditional Assertions, *Journal of Philosophical Logic*, to appear.
25. P. Milne. The Ramsey Test, conditional probabilities, and *modus tollens*, submitted for publication.
26. A. McF. Mood and F. A. Graybill. *Introduction to the Theory of Statistics*, second edition, New York: McGraw-Hill, 1963.
27. A. Naess. Logical Empiricism and the Uniqueness of the Schlick Seminar: A Personal Experience with Consequences. In: F. Stadler (ed.), *Scientific Philosophy: Origins and Developments* (*Vienna Circle Institute Yearbook, 1*), Dordrecht, Kluwer, 11-25, 1993.
28. H. T. Nguyen and E. Walker. A History and Introduction to the Algebra of Conditional Events and Probability Logic, *IEEE Transactions on Systems, Man, and Cybernetics* (Special Issue *Conditional Event Algebra*, Dubois, Goodman and Calabrese (guest eds.)), 24(12):1671-5, 1994.
29. E. Parzen. *Modern Probability Theory and Its Applications*, New York and London: John Wiley & Sons, 1960.
30. S. Read. *Thinking about Logic: An Introduction to the Philosophy of Logic*, Oxford and New York: Oxford University Press, 1994.
31. J. Schächter. *Prolegomena zu einer kritischen Grammatik*, Springer, Vienna, 1935; trans. *Prolegomena to a Critical Grammar*, Reidel, Dordrecht, 1973. Page reference to the English translation.
32. F. Stadler. *Studien zum Wiener Kreis: Ursprung, Entwicklung und Wirkung des Logischen Empirismus im Kontext*, Vienna: Springer, 1997; revised edition trans. by C. Nielsen as *The Vienna Circle: Studies in the Origins, Development, and Influence of Logical Empiricism*, Vienna: Springer, 2001.
33. B. C. van Fraassen. Probabilities of Conditionals. In: W. Harper and C. A. Hooker (ed.), *Foundations of Probability Theory, Statistical Inference, and Statistical Theories of Science*, Vol. I, *Foundations and Philosophy of Epistemic Applications of Probability Theory*, Dordrecht: Reidel, 261-308, 1980.
34. P. C. Wason. Reasoning. In: B.M. Foss (ed.), *New Horizons in Psychology*, Harmondsworth: Penguin, vol. I, 135-151, 1966.

.

# Prospects for a Manipulability Account of Causation

Jim Woodward

*California Institute of Technology Department of Humanities and Social Sciences 12000 East California Boulevard, Pasadena, California, 91125 USA, jfw@hss.caltech.edu*

**Abstract.** This paper defends a manipulability account of causation. Causal claims are counterfactual claims about what would happen to effects under interventions on their causes. An intervention is a special sort of causal process that is a generalized and abstract version of an ideal experimental manipulation, purged of its anthropomorphic elements. A number of different causal concepts – total cause, direct cause, and contributing cause – are distinguished and each is shown to correspond to a different interventionist counterfactual. The resulting theory is then compared with David Lewis' well-known counterfactual theory of causation.

My aim in this paper is to sketch a version of a manipulability account of causation and to highlight some of its attractions. Section 1 provides some background and preliminary motivation. Section 2 distinguishes among several different causal notions and attempts to characterize each by appealing to the notion of an ideal manipulation or intervention. Section 3 characterizes the notion of an intervention more precisely and compares my characterization with some alternatives. Section 4 briefly compares the manipulability account with David Lewis' counterfactual theory.

## 1.

A very natural idea about causation is that causes are means or handles for manipulating their effects: if $Y$ would change under a manipulation of $X$, then $X$ causes $Y$ and if, under some manipulation of $X$, $Y$ would change, then $X$ causes $Y$. This idea has a number of attractive features. First, it provides a natural account of the difference between causal and merely correlational claims. The claim that $X$ is correlated with $Y$ does not imply that manipulating $X$ is a way of changing $Y$, while the claim that $X$ causes $Y$ does have this implication. And given the connection between causation and manipulation, there is no mystery about why we should care about knowledge of causal as opposed to merely correlational relationships. Second, a manipulationist account of causation fits very naturally with the way such claims are understood and tested in many of the so-called special sciences – particularly biology and the social and behavioral sciences – and with a substantial methodological tradition in statistics, econometrics and experimental design[1].

---

[1] See, for example, [5,4].

Despite this, recent philosophical discussion has been quite unfavorable to manipulationist accounts. First, such accounts have been rejected as "circular", on the grounds that manipulation is itself a causal concept and hence cannot be used to elucidate what it is for a relationship to be causal. Second, manipulation is seen as an anthropomorphic notion and hence unsuitable for the elucidation of causal relationships in contexts in which human action is not involved.

The second objection may be addressed, at least in part, by the strategy of (i) formulating an account of what an "ideal manipulation" (or, as I will call it, an *intervention*) involves that is causal in character but makes no reference to human action (cf. section 3) and (ii) formulating the manipulability theory as a counterfactual theory – that is, in terms of claims about what *would happen* to the effect if an intervention on the cause *were* to be performed. When formulated in this way, the manipulability account does not require that when $X$ causes $Y$, an intervention on $X$ must actually occur, but only the truth of appropriate counterfactual claims about what would happen were such intervention to occur.

With respect to the first objection about circularity, I think that it must be conceded that any notion of intervention suitable for formulating a manipulability account of causation must be causal in character. Indeed, as we shall see, this is so not just in the obvious sense that to speak of an intervention on $X$ implies that $X$ has been caused (by the intervention process or event) to have some value but also in a number of more subtle ways as well. In particular, to characterize the notion of an intervention $I$ on $X$ for the purpose of assessing whether $X$ causes $Y$ (or, as I will say below, with respect to $Y$) one must build into $I$ constraints having to do with the relationship between $I$ and other causes (besides $X$) of $Y$. This has the immediate consequence that a manipulability account of causation will not be a reductionist account in the sense that it explains causal notions just in terms of other notions (e.g., correlation, spatio-temporal contiguity) that are non-causal in character. On the other hand, it does not follow that a manipulability theory must be viciously circular in the sense that it is completely trivial or unilluminating or that it presupposes exactly what it is that we are trying to explain.

One possibility that I will explore below is this: although the characterization of what it is for $I$ to be an intervention on $X$ with respect to $Y$ must be causal in character, it need not involve reference to the existence or non-existence of a causal relationship *between X and Y*. Instead the characterization will only involve reference to *other* causal and correlational information – for example, information about the correlation, if any, between $X$ and $Y$ under interventions on $X$, information about the correlation between $X$ and certain other possible causes of $Y$ and so on. While not reductionist, such a characterization would not presuppose a prior grasp of the very thing we are trying to understand – what it is for there to be a causal relationship between $X$ and $Y$.

This possibility fits with a sort of Neurath's raft picture of causal inference and causal understanding. According to this picture, one begins in causal inference with some causal truths and then uses these, typically in conjunction with other sorts of information to infer to other causal truths. Similarly, one can understand the content of any particular causal claim by connecting it, in the way described by the manipulability account, to the content of other causal claims and to non-causal information, even though one cannot translate the original causal claim into a non-causal claim without remainder. Non-reductionist patterns of connection of this general sort are proposed by a number of other theories of causation[2]. They are also familiar from other subject areas, where they are generally regarded as unproblematic. For example, the claim that a coin has probability 0.5 of coming up heads when tossed is not translatable into facts about relative frequencies, but when this probability claim is combined with other probabilistic assumptions (e. g., that successive tosses of the coin are independent and identically distributed), it follows that certain outcomes, expressible as facts about frequencies, are highly probable. We can use such connections between probabilities and frequencies to get some purchase on what probabilistic claims mean, although the connections are not reductive.

Even if it turns out that we cannot characterize what it is for there to be an intervention on $X$ with respect to $Y$ in a way that does not build in assumptions about the existence or non-existence of a causal relationship between $X$ and $Y$, a manipulability account of causation can still be illuminating. For one thing, there are, as we shall see, a variety of different causal concepts. Each of these is characterizable in interventionist terms and, given such characterizations, we can investigate the interrelationships between these concepts. In addition, once we have a manipulability account of causation, we can ask about the relationship between the conditions imposed by that account and whether various other conditions imposed on causation in the philosophical literature – for example, conditions having to do with spatio-temporal continuity, unanimous probability increase across all background contexts, and so on. We can ask whether these additional conditions follow from the manipulability account or have a natural motivation in terms of it.

## 2.

I turn now to the task of providing a somewhat more detailed statement of a manipulability account. First some preliminaries. Although it is possible to provide a treatment of token causation with a manipulability framework[3], my focus in this paper will be on type causation and on capturing a broad notion of causal relevance that corresponds to the idea of one factor being positively, negatively, or of mixed casual significance for another. Within a manipulability framework, it is most natural to think of causation as a relationship between variables, where the mark of a variable is that it is capable of taking more than one value. The standard assumption within the philosophical literature that causation is a relationship between events or event types can be readily captured within this framework in terms of indicator or two valued variables corresponding to the occurrence or non-occurrence of the events of interest. Thus we may express the causal claim that short circuits cause fires in terms of a relationship between two variables $S$ and $F$, with $S$ taking two possible values corresponding to the occurrence or non-occurrence of a short circuit, and $F$ taking possible values corresponding to the

---

[2]For example, a number of versions of probabilistic theories of causation such as [2] claim that $C$ causes $E$ if and only if $C$ raises the probability of $E$ across all background contexts, where these contexts are characterized by reference to other causes of $E$ besides $C$. Such accounts are non-reductive but are thought to provide some purchase on what it is for $C$ to cause $E$.

[3]See, for example, [8] and [20].

occurrence or non-occurrence of a fire.

The causal notion I will aim at characterizing is relatively weak and uninforma-
tive. It corresponds to the following question: Is $X$ causally relevant to $Y$ at all –
that is, is there some change in the value of $X$ which will change the value of $Y$ or
the probability distribution of $Y$? We are of course also interested in more precise
causal claims having to do with the exact way in which $X$ is causally relevant to $Y$
– that is, which changes in $X$ will be associated with which changes in $Y$ and under
what conditions? These also may be captured within a manipulability framework
by extending the characterizations below in obvious ways.

As a point of departure, consider the following naïve proposal giving necessary
and sufficient conditions for "$X$ causes $Y$":

> **Sufficient Condition (SC):** If (i) there are possible interven-
> tions (ideal manipulations) that change the value of $X$ such that (ii)
> under such interventions (and no others) $X$ and $Y$ are correlated, then
> $X$ causes $Y$.

> **Necessary Condition (NC):** If $X$ causes $Y$ then (i) there are
> possible interventions that change the value of $X$ such that (ii) under
> such interventions (and no other interventions) $X$ and $Y$ are correlated.

The question of what "possible" in clause (i) means is a complex one which I
have explored elsewhere [20][4]. Here I will just note that it cannot mean anything
like "within the present technological powers of human beings", given that there are
true causal claims about the past, about large scale cosmological events such as the
expansion of the universe, and so on. On the other hand, interventions on $X$ must
at least be conceptually well-defined in the sense that $X$ must be capable of taking
more than one value or of undergoing a change[5]. The motivation for the restriction
in clause (ii) of **SC** (and **NC**) to a single intervention on $X$ is that without this
restriction, correlated interventions on $X$ and $Y$ will produce correlated changes in
these variables even if they are causally unrelated[6]

Finally, we should note that if **SC** is to be even prima–facie plausible, we need to
impose restrictions on the sorts of changes in $X$ that count as interventions. Consider a system in which $A$ = atmospheric pressure is a common cause of the reading

---

[4]It may be tempting to suppose that the difficulty of explaining what "possible" in (i) means
can be avoided by simply dropping (i) and formulating a manipulability theory just in terms of
(ii)–i. e., by opting for a conditional rather than a conjunctive formulation of the theory. (This
option is sympathetically explored, but not fully endorsed, by Ernest Sosa and Michael Tooley
in the introduction to their [18]) This strategy strikes me as unhelpful. To avoid trivialization,
the conditional in (ii) cannot be understood as a material or strict conditional but must instead
be understood as a counterfactual of some kind. An illuminating version of the manipulability
theory thus needs to make it clear how such counterfactuals are to be understood and what their
truth conditions are. In particular, we need to know just what sort of possibility we should be
envisioning when we envision the antecedent of a conditional along the lines of (ii). This in turn
means that we cannot duck the question of what (i) means.

[5]As an illustration, suppose that there is no well-defined notion of changing a raven into a non-
raven or vice versa. Then interventions on the property or magnitude of being a raven will not
be "possible" in the relevant sense and being a raven will not be an acceptable candidate for a
cause of anything. This restriction of bona-fide causes to what can be manipulated "in principle"
if not in practice is emphasized by statisticians like [15] and [9].

[6][3] uses examples of this sort to object to a proposal about the connection and intervention in
[7].

$B$ of a barometer and a variable $S$ corresponding to the occurrence/non-occurrence of a storm, but in which $B$ does not cause $S$ or vice-versa. If we manipulate the value of $B$ by manipulating the value of $A$, then the value of $S$ will change even though, in contravention of **SC**, $B$ does not cause $S$. Intuitively, an experiment in which $B$ is manipulated in this way is a badly designed experiment for the purposes of determining whether $B$ causes $S$. We need to formulate conditions that restrict the allowable ways of changing $B$ so as to rule out possibilities of this sort. Informally, the constraint we want is that the change in $B$ should be of such a character that any change in $S$ (if it occurs at all) can only come about through the change in $B$. Operationally, this might be accomplished by, for example, employing a randomizing device which is causally independent of $A$ and $B$ and then, depending on the output of this device, experimentally imposing (or "setting") $B$ to some particular value. Under such interventions, the value of $S$ will no longer be correlated with the value of $B$ and **SC** will not judge that $B$ causes $S$.

Provided that the notion of an intervention is understood in the appropriate way, I believe that **SC** is extremely plausible. It says, in effect, that if it is possible to manipulate $Y$ by intervening on $X$, then we may conclude that $X$ causes $Y$, regardless of whether the relationship between $X$ and $Y$ lacks various other features standardly regarded as necessary for causation. Thus cases of "double prevention" (Hall, 2000) or "causation by disconnection" [17] are (I believe correctly) counted as causal by **SC** even though the cause is not connected to its effect via a spatio-temporally continuous process and even though there is no transfer of energy.

What about **NC**? Consider the causal structure represented by means of the equations

(2.1) $Y = aX + cZ$
(2.2) $Z = bX$

and by the associated directed graph

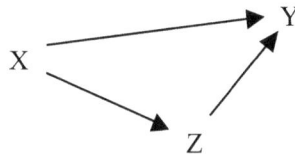

Figure 1.

Here the convention is that variables that appear on the right side of an equation are "direct causes" (a notion that will be elucidated below) of those that are on the left. Similarly, an arrow drawn from $X$ to $Y$ means that $X$ is a direct cause of $Y$.

If $a = -bc$, the direct causal influence of $X$ on $Y$ will be exactly canceled out by the indirect influence of $X$ on $Y$ that is mediated through $Z$. If it is correct to think that $X$ (in some relevant sense) causes $Y$, then **NC** will be false, since there are no interventions on $X$ alone that will change $Y$[7].

I take this example to show that we need to distinguish between two notions of "cause"[8]. Let us say that $X$ is a [insert/indexes and tables/entry/item] *total cause* of $Y$ if and only if it has a non-null total effect on $Y$ – that is, if and only if there is some intervention on $X$ alone (and no other variables) such that for some value of other variables besides $X$, this intervention on $X$ will change the value of $Y$. The notion of a total cause contrasts with the notion of a [insert/indexes and tables/entry/item] *contributing cause* which is intended to capture the intuitive idea of $X$ influencing $Y$ along some route or directed path even if, because of cancellation, $X$ has no total effect on $Y$. While both **SC** and **NC** are plausible if "cause" is interpreted as "total cause", **NC** is not correct if "cause" is interpreted as "contributing cause", although **SC** remains plausible under this interpretation.

Can we capture the notion of a contributing cause within a manipulability framework? The strategy I will follow is to first formulate a necessary and sufficient condition for $X$ to be a [insert/indexes and tables/entry/item] *direct cause* of $Y$ and then to use this formulation to arrive at a necessary and sufficient condition for $X$ to be a contributing cause of $Y$. I propose the following characterization of direct causation:

> **Direct Cause (DC):** A necessary and sufficient condition for $X$ to be a direct cause of $Y$ with respect to some variable set **V** is that there is a possible intervention on $X$ that will change $Y$ (or the probability distribution of $Y$) when all other variables in **V** besides $X$ and $Y$ are held fixed at some value by interventions.

Using **DC** we may formulate a necessary condition (**NC\***), expressed in terms of claims about the outcomes of hypothetical interventions, for $X$ to be a contributing, (type-level) cause of $Y$ as follows:

> (**NC\***) If $X$ is a [insert/indexes and tables/entry/item] contributing cause of $Y$ with respect to the variable set V then there is a directed path from $X$ to $Y$ such that each link in this path is a direct causal relationship–i.e., if $Z_1 \ldots Z_n$ are intermediate variables along this path, $X$ is a direct cause of $Z_1$ which is a direct cause of $Z_2$ which is a direct cause of. . . . $Z_n$ which is a direct cause of $Y$. Put differently, if $X$ causes $Y$ then $X$ must either be a direct cause of $Y$ or there must be a causal chain, each link of which involves a relationship of direct causation, extending from $X$ to $Y$.

Note that **NC\*** does *not* require the assumption that contributing causation is transitive. Assumptions about transitivity involve sufficient conditions for causation and **NC\*** purports to provide only a necessary condition. The proper role for transitivity and related requirements, if they are assumed at all, is in the statement of a sufficient condition for contributing causation.

---

[7] [19] call this a failure of "faithfulness".
[8] For a similar view, see [14] and [8].

If it were justifiable to assume [insert/indexes and tables/entry/item] transitivity, we could simply replace the "if"s in (**NC\***) with "if and only if"s and we would have a sufficient as well as a necessary condition for contributing causation. But, as a number of examples show, such assumptions are dubious. Consider the following case, which is due to [13]. A dog bites off my right forefinger. The next day I detonate a bomb by using my left forefinger. If I had not lost my right finger, I would have used it instead to detonate the bomb. The bite causes me to use my left finger which causes the bomb to explode but (it seems) the bite does not cause the bomb to explode.

This example has a natural treatment within a manipulability framework[9]. Let $B$ specify whether a bite occurs, $L$ take one of *three* values, according as to whether the left hand is used, the right hand is used, or the button is not pushed at all, and $E$ specify whether the bomb explodes. An intervention that changes whether or not a bite occurs changes whether I use my left or right hand to detonate the bomb. An intervention that changes the situation from one in which I use my left finger to detonate the bomb to a situation in which I do not detonate the bomb at all (with either hand) changes whether the bomb explodes. However, changing whether I use my left or right hand to detonate the bomb does not change whether the bomb explodes. Although there are changes in $L$ (whether I use my right or left hand) which are sensitive to changes in $B$, and *other* changes in $L$ (whether I use my left hand rather than not pressing the button at all) to which the value of $E$ is sensitive, there is no set of changes in the value of $L$ which fulfill both these roles. In other words, the function $F$ linking $B$ to $L$ and the function $G$ linking $L$ to $E$ compose in such a way that the composite function $E = G(F(B))$ assigns the same value of $E$ to both values of $B$, so that there is no intervention on $B$ which changes $E$.

If this analysis is correct, then one natural way to deal with failures of transitivity of the sort under discussion is to require that for $X$ to be a contributing cause of $Y$ not only must there be at least one chain of direct causal relationships (a directed path or route) from $X$ to $Y$ but it must also be the case that the value of $Y$ is sensitive along that path to some interventions that change the value of $X$ – that is, it must be the case that there is a directed path from $X$ to $Y$ such that an intervention on $X$ will change $Y$ when all variables that are not on this path, including intermediate variables on *other* paths between $X$ and $Y$ and variables along any other paths leading into $Y$ are fixed at some value. This leads to the following proposal for an interventionist account of contributing causation:

> (**M**) A necessary and sufficient condition for $X$ to be a (type-level)
> *contributing cause* of $Y$ with respect to variable set V is that (i) there
> is a directed path from $X$ to $Y$ such that each link in this path is a
> direct casual relationship – i.e., a set of variables $Z_1 \ldots Z_n$ such that $X$
> is a direct cause of $Z_1$ which is in turn a direct cause of $Z_2$ which is a
> direct cause of.... $Z_n$ which is a direct cause of $Y$ and that (ii) there
> is some intervention on $X$ that will change $Y$ when all other variables
> in V that are not on this path are fixed at some value. If there is only
> one path $P$ from $X$ to $Y$ or if the only alternative path from $X$ to $Y$

[9]See Hitchcock, 2001 and Woodward, 2003.

besides $P$ contains no intermediate variables (i.e., is direct) then $X$ is a contributing cause of $Y$ along $P$ as long as there is some intervention on $X$ that will change the value of $Y$, for some values of the other variables in V.

## 3.

I turn now to the task of formulating a notion of intervention that fits with the project of providing truth conditions for claims about total, direct, contributing causes (as characterized above) by appealing to facts about what would happen under interventions. I emphasize that this is a semantic or interpretive project that is very different from the project of specifying the full range of conditions under which causal claims can be inferred from statistical information Note, in particular, that it is obviously false that only when an experimental manipulation meets the conditions that follow that one can reach reliable causal conclusions.

The basic strategy I will follow is to assume that if $X$ and $Y$ are correlated under manipulations of $X$, this correlation must have some causal explanation. I will then attempt to characterize an intervention on $X$ with respect to Y so that all *other* ways (in addition to $X$s causing $Y$) that changes in $X$ might be associated with changes in $Y$ are ruled out. In effect, I will begin by assuming that a version of the common cause principle:

If $X$ and $Y$ are correlated then $X$ causes $Y$ or $Y$ causes $X$ or $X$ and $Y$ have a common cause

holds in the special context in which there are interventions on $X$, but not necessarily in other contexts in which no interventions are involved. One advantage of this strategy, to which I will return below, is that we don't need to assume, in the characterization of an intervention, the full common cause principle or such stronger principles as the Causal Markov Condition[10]. The characterization of an [insert/indexes and tables/entry/item] intervention I propose is the following:

**IN**: Let $X$ and $Y$ be variables, with the different values of $X$ and $Y$ representing different and incompatible properties possessed by the unit $u$,the intent being to determine whether some intervention on $X$ produces changes in $Y$. Then$I$is an intervention on $X$ (an intervention variable for$X$) with respect to $Y$ if and only if

I1. $I$ causes $X$.

I2. $I$ acts as a switch for all the other variables that cause $X$. That is, certain values of $I$ are such that when $I$ attains those values, $X$ ceases to depend upon the values of other variables that cause $X$ and instead only depends on the value taken by $I$.

I3. Any directed path from $I$ to $Y$ goes through $X$. That is $I$ does not directly cause $Y$ and is not a cause of any causes of $Y$ that are distinct from $X$ except,

---

[10][insert/indexes and tables/entry/item] The Causal Markov Condition says that conditional on its direct causes, every variable is independent of every other, except possibly for its effects. It is a generalization of the screening off conditions often assumed by philosophers. See Spirtes, Glymour, and Scheines 1993.

of course, for those causes of $Y$, if any, that are built into the $I - X - Y$ connection itself; that is, except for (a) any causes of $Y$ that are effects of $X$ (i.e., variables that are causally between $X$ and $Y$) and (b) any causes of $Y$ that are between $I$ and $X$ and have no effect on $Y$ independently of $X$.

I4. $I$ is independent of any variable $Z$ that is causally relevant to $Y$ and that is on a directed path from $I$ to $Y$ that does not go through $X$.

All of these conditions may be thought of as conditions on ideal experimental manipulations of $X$ in a context in which the object is to determine whether $X$ causes $Y$. For illustration and motivation, consider an experiment designed to determine whether treatment with a certain drug causes recovery from a disease. Suppose that each subject $u_i$ in a group, all of whom have the disease, is assigned a treatment which may be represented by a binary variable $T$, the values of which depend on whether $u_i$ takes or does not take the drug. The characteristics of the process that determines the value of $T$ for each $u_i$ will determine whether or not this process qualifies as an intervention.

Suppose that prior to the setting up of the experiment, whether or not subjects take the drug depends on the value of some "endogenous" variable $Z$ ($Z$ might represent access to medical care or some personality variable like willingness to seek medical help). Then any effects that $Z$ may have on recovery which are independent of the action of the drug will be confounded with the effect of the drug on recovery. IN1 and IN2 are designed to rule out this sort of possibility. They require that to qualify as an intervention the treatment assignment must "break" or "turn off" any such pre-existing endogenous causal connection between $Z$ and recovery, so whether a subject receives the drug is now set entirely "exogenously" by the experimental design and is no longer influenced by $Z$– subjects are no longer allowed to determine on their own whether they will take the drug etc. Operationally, this might be accomplished by allowing some randomizing process to determine who does and who does not receive the drug. Graphically, IN2 corresponds to the idea that the intervention $I$ should "break" any pre-existing arrows directed into the variable $X$ intervened on, so that the value of $X$ is determined entirely by $I$, an idea that is developed in [19] and [14].

Conditions I3 and I4 are designed to rule out possibilities like the following: If the experimenter's manipulations $I$ are correlated with certain other causes of recovery ($R$) besides $T$ (whether because $I$ is caused by these other causes of recovery or for any other reason), this will undermine the reliability of the experiment. This would happen, for example, if those patients who receive the drug have, on average, stronger immune systems than those who do not receive it. However, it would be too strong to require that $I$ (or $T$) be uncorrelated with *all* other causes of $R$. As long as $T$ is efficacious, $I$ and $T$ will be correlated with any other causes of $R$ that are themselves caused by $I$ or by $T$. For example, if treatment by the drug does cause recovery and does so by killing ($K$) a certain sort of bacterium, then it will be no threat to the validity of the experiment if the experimenters' interventions $I$ are correlated with $K$, even though $K$ causally affects $R$. What we need to rule out is the possibility that there are causes of $R$ that are correlated with $I$, and that affect $R$ independently of the $I \rightarrow T \rightarrow K \rightarrow R$ causal chain. Relatedly, we need to rule out the possibility that $I$ affects $R$ directly via a route that does not go through

$T$. This would happen if, for example, administration of the drug was by shot gun blast (assuming that this would directly adversely affect recovery). Less fancifully, it would be violated if the subjects learn whether they have been assigned to the treatment group or the control group, and this makes those in the treatment group more hopeful and those in the control group more discouraged and this in turn has an effect on whether they recover which is independent of any effects of the drug per se. I3 and I4 are intended to rule out such possibilities.

Note that as claimed above, **IN** is not anthropomorphic – the characterization is entirely in terms of causal and correlational notions and makes no reference to human beings or their activities. Note also that **IN** is not circular in the sense of building into the characterization of an intervention on $X$ with respect to $Y$, information about the existence or non-existence of a causal relationship between $X$ and $Y$.

How does **IN** compare with other notions of [insert/indexes and tables/entry/ /item] intervention in the literature? **IN** is similar in a number of respects to the characterization that [19], give of a "manipulation", but their characterization assumes that the directed graph representing the manipulation variable(s) and the system in which the manipulation occurs come with an associated probability distribution that satisfies the Causal Markov Condition (**CM**). By contrast, **IN** does not assume that the system intervened on satisfies **CM**. As far as the project of providing an interventionist interpretation of causal claims goes, this seems to me to be an advantage for several reasons. First, we want the notion of an intervention and causal notions generally to be applicable to systems that do not satisfy **CM**[11]. Second, if possible, one would like to have some insight into when and why **CM** holds and for this purpose characterizations of causation and intervention that do not directly build in the truth of CM are obviously preferable[12].

In his [14] Judea Pearl characterizes one notion of intervention as follows:

> (**PI**) The simplest type of external intervention is one in which a single variable, say $X_i$, is forced to take on some fixed value $x_i$. Such an intervention, which we call "atomic" amounts to lifting $X_i$ from the influence of the old functional mechanism $x_i = f_i(pa_i, u_i)$ and placing it under the influence of a new mechanism that sets the value $x_i$ while leaving all other mechanisms unperturbed. Formally, this atomic intervention, which we denote by do $(X_i = x_i)$ or do $(x_i)$ for short, amounts to removing the equation $x_i = f_i(pa_i, u_i)$ from the model and substituting

---

[11] Consider a system like that described in [16] in which the collision of a cue ball with the eight ball raises the probability that the eight ball will go into the pocket. Let us suppose that at the relatively coarse-grained level of description associated with variables like {collision, no collision}, {eight ball falls into the pocket, does not fall into the pocket} the behavior of the eight ball is not determined just by whether the collision occurs but that, because the system is governed by conservation laws, the probability that the eight ball falls given that there is a collision *and* the cue ball goes into a second pocket is one. Then with respect to these variables **CM** fails. Nonetheless, **IN** gives us a well-fined notion of an intervention for such a system and since an intervention on the position of the cue ball after the collision will not alter the probability that the eight ball goes into the pocket, the combination of **NC** and **NC\*** correctly yield the judgment that the sinking of the cue ball does not cause the sinking of the eight ball.

[12] Hausman and Woodward [7] and forthcoming, attempt to show how one can derive **CM** from a manipulability conception of causation and certain other assumptions.

$X_i = x_i$ in the remaining equations. [14]

Pearl takes the notion of a causal mechanism, as expressed by a functional equation relating an effect variable to its parents (direct causes) as primitive and then characterizes the notion of an intervention in terms of this primitive. The requirement that an intervention on $X$ must leave *all* other mechanisms (besides the mechanism linking $X$ to its parents) undisturbed has the consequence that any mechanism linking $X$ to its putative effect $Y$ must be left undisturbed. In contrast to **IN**, Pearl thus builds reference to the causal relationship (if any) between $X$ and $Y$ into the characterization of an intervention on $X$. In proceeding in this way, it looks as though he loses any possibility of using the notion of an intervention to explain the notion of causal mechanism or direct causal relationship. This is *not* a criticism of Pearl who, as I read him, is not (at least primarily) interested in providing an interventionist interpretation of causation, but is rather interested in formulating a notion that fits well with various calculational purposes. However, if one's purpose is to provide such an interventionist interpretation, a characterization along the lines of **IN** seems preferable.

**4.**

The manipulability account sketched above is a counterfactual theory of causation – it connects causal claims to claims about what would happen if an intervention were to occur. What is the relationship between this account and the well-known version of a [insert/indexes and tables/entry/item] counterfactual theory developed by David Lewis? [10–12]. Lewis defines causation as the ancestral of counterfactual dependence: $c$ causes $e$ if and only if $c$ and $e$ occur and there is a chain of counterfactual dependence between $c$ and $e$–that is, a sequence of events $c_1, c_2, \ldots, c_n$ such that $e$ is counterfactually dependent on $c_n$, $c_n$ is counterfactually dependent on $c_{n-1}, \ldots$, and $c_1$ is counterfactually dependent on $c$. One obvious difference is that Lewis' theory is an account of causation between particular events while the account sketched above is intended to capture a type causal notion. I will abstract away from this difference in what follows and attempt to compare a natural extension of Lewis' theory to type causal claims with the intervention based approach.

As is well-known, Lewis understands the truth conditions for counterfactuals in terms of similarity relationships among possible worlds. The criteria for evaluating "similarity" are as follows: ([12], p. 47)

(S1) It is of the first importance to avoid big, widespread, diverse violations of law.

(S2) It is of the second importance to maximize the spatio-temporal region throughout which perfect match of particular fact prevails.

(S3) It is of the third importance to avoid even small, localized simple violations of law.

(S4) It is of little or no importance to secure approximate similarity of particular fact, even in matters that concern us greatly.

As Lewis explains ([12], Appendix B, p. 56), "big, widespread, diverse violations of law" are events that

> consist of many little miracles together, preferably not all alike. What makes a big miracle more of a miracle is not that it breaks more laws; but rather that it is divisible into many and varied parts, any one of which is on a par with a little miracle.

Let me begin with an example designed to illustrate the broad similarities between Lewis' account and the manipulability theory. Consider the counterfactual

(4.1) If $e_1$ had not occurred, then $e_2$ would not have occurred

evaluated with reference to a deterministic causal structure in which $c$ is the common cause of two joint effects, $e_1$ and $e_2$, neither of which causes the other. Since counterfactual dependence is sufficient for causation on Lewis' theory, this counterfactual should come out false. Lewis' criteria achieve this result since the most similar world to the actual world is a world (world 1) which matches the actual world exactly up to just before $e_1$ occurs, at which point a small localized miracle occurs which results in the non-occurrence of $e_1$. In this world both $c$ and $e_2$ will still occur and hence the counterfactual (4.1) is false. By contrast, in a world (world 2) in which the non-occurrence of $e_1$ is achieved through the earlier non-occurrence of $c$, as the result of a small miracle just before the time $c$ would have occurred, we still require a miracle and there is a less extensive region of perfect match of particular fact with the actual world than is the case with world 1, since divergence from the actual world begins earlier, with the non-occurrence of $c$. Accounting for the non-occurrence of $e_1$ by introducing a miracle at some still earlier time would produce an even less extensive region of perfect match. In this way, Lewis' theory arrives at the result that so-called backtracking counterfactuals such as

(4.2) If $e_1$ had not occurred, then $c$ would not have occurred are false and, moreover, that counterfactuals like

(4.3) If $e_1$ had not occurred, then $c$ still would have occurred are, as one intuitively expects, true.

The description of this example should make it clear that Lewis' similarity criteria, and in particular, the "small localized miracles" which they require often function in broadly the same way as the notion of an intervention. Again abstracting from the type/token difference, on both approaches, when we evaluate a counterfactual of form " if $C$ had not occurred, then $E$ would not have occurred " with respect to a world in which $C$ does occur, we think of the antecedent of this counterfactual as made true by some exogenous source of change – an intervention or a localized miracle – which breaks whatever endogenous casual relationships are at work in the actual world in producing $C$ (Recall condition IN2 in the characterization **IN**). This gives the non-occurrence of $C$ an independent causal history, and if the miracle has the right character (in particular if it has no other effects on $E$ except those that occur through the non-occurrence of $C$), and if it is inserted in the right place (see below for the significance of both of these qualifications), it follows that any change in $E$ will be an effect of just the change from a situation in which $C$ occurs to one in which it does not, and not the result of a change in some other factor.

In other words, the invocation of "miracles" in Lewis' framework works, to the extent that it does, because, like the notion of an intervention, it requires us to consider a counterfactual the antecedent of which differs from the actual world only in the non-occurrence of $C$. That is, nothing else changes that might have an effect on $E$ independently of $C$–thus insuring that if $E$ does change this can only be because it is an effect of $C$. In opposition to those (e. g., [1]) who claim that a theory of counterfactuals should countenance only a single interpretation of counterfactuals that permits backtracking and is appropriate for both causal and non-causal contexts, the manipulability account agrees with Lewis in holding that there is a fundamental distinction between those similarity criteria that are appropriate for the counterfactuals that may be used to analyze causal claims and which should not permit backtracking, and those criteria that are appropriate for backtracking counterfactuals.

Despite these similarities, there are deep differences between Lewis' theory and the version of the manipulability theory that I have been defending. One of these concerns the question of reduction. Lewis' theory is avowedly reductionist in aspiration: the idea is to define the notion of causation in terms of a more general notion of counterfactual dependence that does not itself presuppose causal notions. By contrast, as explained above, the manipulability account does not purport to provide such a reduction. According to the manipulability account, given that $C$ causes $E$, which counterfactual claims involving $C$ and $E$ are true will always depend on which other *causal* claims involving other variables besides $C$ and $E$ are true in the situation under discussion. For example, it will depend on whether other causes of $E$ besides $C$ are present. Thus in Figure 1 (Section 2) with $a = -bc$, the causal relationship between $X$ and $Y$ fails to reveal itself in a straightforward Lewisian pattern of counterfactual dependence of $Y$ on $X$ because of the presence of another cause $Z$ of $Y$. To reveal the direct causal dependence of $Y$ on $X$ we must invoke a more complex counterfactual than any that figures in Lewis' account: a counterfactual about how the value of $Y$ would change if an intervention were to hold $Z$ fixed and if at the same time another intervention were to change the value of $X$. This sort of counterfactual, with its reference to *two* interventions, one of which is on the putative cause $X$ but the other of which is on the variable $Z$ which is off the direct route from $X$ to $Y$, has no direct analog in Lewis' system. Thus which counterfactuals are appropriate for capturing the counterfactual dependence of $Y$ on $X$ when $X$ is a direct cause of $Y$ will depend on the causal features, including the causal route structure, of the larger system in which $X$ and $Y$ are embedded: the specification of which additional variables should be held fixed in the antecedent of the counterfactual relating $X$ to $Y$, requires reference to causal facts about the directed path structure of the larger system in which $X$ and $Y$ figure. The manipulability account does not assume that (or try to show how) reference to such causal facts can be eliminated in favor of purely non-causal counterfactual claims.

The reductive character of Lewis' theory might be regarded as an advantage if the application of the similarity criteria (S1-S4) always correctly captured the connection between causal and counterfactual claims but in fact they do not, as the following example shows:

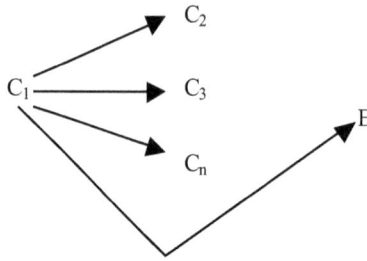

Figure 2.

Here $C_1$ is a direct cause of each of $C_2 \ldots C_n$ . In addition there is a direct causal link from $C_1$ to $E$. There are no other casual links. Assume that the occurrence of each cause is sufficient for its effect to occur.

Now consider the counterfactual.

> (4.4) If $C_2$ and $C_3$ and $\ldots C_n$ had not occurred, then $E$ would not have occurred.

On the usual understanding of the connection between causation and counterfactuals, (4.4) is false since $C_1$ is a deterministic cause of $E$ and even if $C_2 \ldots C_n$ had not occurred, $C_1$ would have occurred and would have caused $E$.

On the Lewisian non-backtracking interpretation of (4.4), we consider a possible world which diverges from the actual world in that $C_2 \ldots C_n$ do not occur. This requires the insertion of a miracle someplace. There seem to be two possibilities:

> (World 3): $C_1$ occurs but then $n - 1$ miracles occur, one for each of $C_2$ through $C_n$ in such a way that each of $C_2 \ldots C_n$ does not occur. In other words, each of the links between $C_1$ and each of $C_2 \ldots C_n$ is broken.

However, this requires many distinct miracles (a "big" miracle) and S1 tells us to give the greatest weight to avoiding this. (We may suppose that $n$ is large and that the links from $C_1$ to each of $C_2, \ldots C_n$ are different from each other)
Another possibility is:

> (World 4): A single miracle occurs just before the occurrence of $C_1$, so that $C_1$ does not occur, and in this way the non-occurrence of $C_2 \ldots C_n$ is ensured, as the antecedent of (4) requires.

This involves a somewhat less extensive region of match with the actual world than under world 3, since $C_1$ occurs in world 3 and not in world 4, but we only have to introduce one miracle as opposed to the many required in world 3.

Since Lewis gives a higher priority to avoiding postulating a number of different miracles than to maximizing match, he views world 4 as the more similar world to the actual world. However, if world 4 is closest to the actual world, (4.4) is true. Moreover, the following counterfactual is also true:

(4.5) If $C_2$... $C_n$ had not occurred, $C_1$ would not have occurred.

Hence, since on Lewis' theory, counterfactual dependence is sufficient for causation, it follows that $C_2 \ldots C_n$ cause $C_1$. But, intuitively, the correct world to look at in evaluating (4.4) is not world 4 but rather world 3 in which (4.4) comes out false, despite the fact that world 3 requires many more miracles than world 4.

In contrast to Lewis' theory, the manipulability account correctly tells us that in evaluating the counterfactual (4.4) we should consider world 3 rather than world 4. As explained above, **IN** commits us to an "arrow-breaking" conception of interventions, according to which, in considering a counterfactual like (4.4), we should imagine that all arrows directed into $C_2 \ldots C_n$ are broken. All other arrows, including the arrow from $C_1$ to $E$ and any arrows directed into $C_1$ are preserved. A process that made the antecedent of (4.4) true by removing $C_1$ would not be an intervention on $C_2 \ldots C_n$ because it would violate the requirement that an intervention must not change other causes of the putative effect variable $E$ except those causes that are causally between $C_2 \ldots C_n$ and $E$. ($C_1$ is a cause of $E$ that is not causally between $C_2, \ldots, C_n$ and $E$.) On the manipulationist account, both the counterfactuals (4.4) and (4.5) are false, as they should be.

The reason why the manipulability account inserts the needed miracle or intervention in the right place is that the characterization of interventions is framed in causal language. This allows us to take account of the relationship between the intervention and the details of the causal structure of the system in which the intervention occurs and to insure that the intervention changes only the putative cause variables and those variables, if any, that are caused by these and that only causal links directed into the variable intervened on are broken. The price of doing this is that we lose the possibility of a reduction, but as the example illustrates, the attempt to get by with a non-causal theory leads to the insertion of miracles at the wrong points.

## REFERENCES

1. J. Bennett. Counterfactuals and Temporal Direction, *The Philosophical Review*, 93:57-91, 1984.
2. N. Cartwright. *How the Laws of Physics Lie*, Oxford: Clarendon Press, 1983.
3. N. Cartwright. Against Modularity, the Causal Markov Condition and Any Link Between the Two: Comments on Hausman and Woodward, British Journal for the Philosophy of 53:411-453, 2002.
4. T. Cook and D. Campbell. *Quasi-Experimentation: Design and Analysis Issues for Field Settings*, Boston: Houghton Miflin Company, 1979.

5. T. Haavelmo. The Probability Approach in Econometrics. *Econometrica*, 12 (Supplement):1-118, 1944.

6. N. Hall. Causation and the Price of Transitivity, *Journal of Philosophy*, 97:198-222, 2000.

7. D. Hausman and J. Woodward. Manipulation and the Causal Markov Condition, *PSA 2002*, vol. 2, forthcoming.

8. C. Hitchcock. The Intransitivity of Causation Revealed in Equations and Graphs. *Journal of Philosophy*, 98:273-99, 2001.

9. P. Holland. Statistics and Causal Inference. *Journal of the American Statistical Association*, 81:945-960, 1986.

10. D. Lewis. Causation. *Journal of Philosophy*, 70:556-67, 1973. Reprinted with Postscripts in [12], 159-213.

11. D. Lewis. Counterfactual Dependence and Time's Arrow, *Noûs*, 13:455-476, 1979. Reprinted with Postscripts in [12], 32-66.

12. D. Lewis. *Philosophical Papers, Volume II,* Oxford: Oxford University Press, 1986.

13. M. McDermott. Redundant Causation, *The British Journal for the Philosophy of Science*, 46:523-544, 1995.

14. J. Pearl. *Causality: Models, Reasoning and Inference.* Cambridge: Cambridge University, 2000.

15. D. Rubin. Comment: Which Ifs Have Causal Answers?, *Journal of the American Statistical Association*, 81:961-962, 1986.

16. W. Salmon. *Scientific Explanation and the Causal Structure of the World.* Princeton: Princeton University Press, 1984.

17. J. Schaffer. Causation by Disconnection. *Philosophy of Science*, 67:285-300, 2000.

18. E. Sosa and M. Tooley (eds.). *Causation.* Oxford: Oxford University Press, 1993.

19. P. Spirtes, C. Glymour and R. Scheines. *Causation, Prediction and Search.* New York: Springer-Verlag. Second Edition, 2000. Cambridge: MIT Press, 1993.

20. J. Woodward. *Making Things Happen: A Theory of Causal Explanation.* New York: Oxford University Press, 2003.

# Section C:

# Philosophical Issues
# of Particular Sciences

# Some Thoughts and a Proposal in the Philosophy of Mathematics

Haim Gaifman

*Philosophy Department, Columbia University, NY 10027, USA*
*hg17@columbia.edu*

**Abstract.** The paper outlines a project in the philosophy of mathematics based on a proposed view of the nature of mathematical reasoning. It also contains a brief evaluative overview of the discipline and some historical observations; here it points out and illustrates the division between the philosophical dimension, where questions of realism and the status of mathematics are treated, and the more descriptive and looser dimension of epistemic efficiency, which has to do with ways of organizing the mathematical material. The paper's concern is with the first. The grand tradition in the philosophy of mathematics goes back to the foundational debates at the end of the $19^{th}$ and the first decades of the $20^{th}$ century. Logicism went together with a realistic view of actual infinities; rejection of, or skepticism about actual infinities derived from conceptions that were Kantian in spirit. Yet questions about the nature of mathematical reasoning should be distinguished from questions about realism (the extent of objective knowledge–independent mathematical truth). Logicism is now dead. Recent attempts to revive it are based on a redefinition of "logic", which exploits the flexibility of the concept; they yield no interesting insight into the nature of mathematics. A conception of mathematical reasoning, broadly speaking along Kantian lines, need not imply anti–realism and can be pursued and investigated, leaving questions of realism open. Using some concrete examples of non–formal mathematical proofs, the paper proposes that mathematics is the study of forms of organization—-a concept that should be taken as primitive, rather than interpreted in terms of set–theoretic structures. For set theory itself is a study of a particular form of organization, albeit one that provides a modeling for the other known mathematical systems. In a nutshell: *"We come to know mathematical truths through becoming aware of the properties of some of the organizational forms that underlie our world. This is possible, due to a capacity we have: to reflect on some of our own practices and the ways of organizing our world, and to realize what they imply. In this respect all mathematical knowledge is meta-knowledge; mathematics is a meta-activity par excellence."* This of course requires analysis and development, hence the project. The paper also discusses briefly the axiomatic method and formalized proofs in light of the proposed view.

## 1. Different Dimensions in the Philosophy of Mathematics

There is a grand tradition in the philosophy of mathematics, stemming from the first three or four decades of the last century. At that time key figures, such as Frege, Russell, Whitehead, Hilbert, Poincaré, Weyl, Brouwer, to mention some, worked out and debated major foundational projects in mathematics. The competing positions are often characterized by the terms 'logicism', 'intuitionism' (or 'constructivism') and 'formalism', a somewhat misleading division, as I shall later explain. The classification has become a cliché, and as the debates have lost much of their original vitality, resistance has arisen to conceiving the discipline along these lines. There have been calls for more "down to earth" approaches, which focus on the way mathematics is actually practiced, and on historical and sociological aspects. Lakatos – who treats mathematics within the general framework of scientific research-programs – has been a forerunner in this "non-metaphysical" trend, which has gathered some momentum in the last ten years. To the extent that this tendency has broadened the perspective in the philosophy of mathematics – serving as a corrective to the neglect of epistemic, historical and sociological aspects – it should be welcome. But it cannot be taken seriously as a philosophical position, when it marches under the "anti-foundational" banner, that is, as a rejection of direct philosophical questions about the nature mathematics and mathematical truth. By 'foundational', I do not mean an approach whose goal is to provide mathematics with "safe foundations", but an inquiry into the nature of mathematical reasoning, its validity and the kind of truth it yields. The positions mentioned above are foundational in this respect; but so is Mill who construed mathematical truth as empirical (in the same sense that physics is empirical), and so is Wittgenstein, who denied the factual status of mathematical statements. Paying attention to epistemic, historical and sociological aspects in the practice of mathematics is complementary to, not a rival of foundational approaches. When it aspires to become a rival it results in bad philosophy, as well as a wrong phenomenal picture of the very practices it aims to describe. It leads to superficial positions that dismiss basic questions about the nature of mathematics, in favor of certain external descriptions of the activity.

Informative and interesting accounts can be provided by intelligent external observers; an atheist may comment knowledgeably on disputes concerning the divinity of Christ. But the analogy, for philosophy of mathematics, is misleading. What does "atheism" here mean? Anti-foundationalism is not a position that denies existence of mathematical facts; or, to put it in a different way, denies that mathematical statements are, independently of our knowledge, true or false. For such a denial is by itself a substantial foundational position, and its adherents are drawn into foundational debates when they are called upon to provide satisfactory accounts of their own. Only Wittgenstein, as far as I can tell, accepted the full implications of such a denial, across the board; he ended with an interesting, provocative, though untenable picture of mathematics. But Wittgenstein is a singularity in our story, and is not my present concern.

"Anti-foundational" positions merely reflect a certain tiredness of the old debates, a reaction to the overemphasis on formal logic, and an awareness (in itself true and valuable) of the gap between the formalized systems and the actual practice. It is therefore useful to remind ourselves how direct and natural basic questions are in the philosophy of mathematics. Quite aside from choosing a program or following a trend, these questions face us *qua* philosophers. Elementary questions about realism are a good example:

7 is a prime number, 6 is not; these are such trivialities that one is tempted to regard them as tautological consequences of a convention, rather than "facts".

That 113 is a prime and that 111 is not is less obvious, but still trivial. Mersenne believed, in 1644, that $2^{67} - 1$ is prime, but he was proven wrong by Cole in 1903. It has been now verified that $2^{24,036,583}-1$ (a number with about 724,000 digits) is prime. What kind of truths are these? It is remarkable that a claim such as the last can be easily understood by anyone with the most elementary arithmetical knowledge (a seventh-grader say), who has no idea how it was proved.[1] The gap between grasping the claim and understanding anything about its proof is even more striking in the case of Fermat's last theorem. "Everyone" can understand what the theorem says, but only an expert in this particular area can access the proof. "Everyone" moreover will initially find no wiggle place: either there are four non-zero numbers, $x$, $y$, $z$, $n$ such that $n > 2$ and $x^n + y^n = z^n$, or there are none. A host of extremely simple questions, accessible to anyone with elementary knowledge, constitute open problems. It is not known, as I write this, whether every even number greater than 2 is a sum of two primes (Goldbach's conjecture); or whether there are odd perfect numbers (a perfect number is a natural number that is equal to the sum of its proper divisors, e.g., 6=1+2+3, or 28 = 1+ 2 + 4 + 7 + 14); or whether there is an infinite number of pairs of twin primes (a pair of twin primes is a pair of primes that differ by 2, such as 11, 13, or 17, 19; the conjecture that there is an infinite number of such pairs is the twin-prime conjecture). Do these mathematical questions have true answers, independently of our state of knowledge and methods of proof?

Consider, for comparison, a toss of a fair coin, whose outcome is unobserved and will never be known, because it is followed immediately by a second toss, or because the coin is destroyed. Did the coin land heads or tails? The answer is a hard fact beyond our knowledge. One might try to mitigate the gap between the fact and its knowledge by invoking a counterfactual: we *could have* observed the outcome and we *would have* then known the answer. But it is not clear that our belief in the objective evidence-independent outcome of the toss derives from our belief in the counterfactual, or vice versa: we accept the counterfactual because we consider the outcome an objective fact. In any case, the counterfactual and the fact-of-the-matter view reinforce each other. (I remark in passing that this connection has been utilized in modal interpretations of the theory of natural numbers.) As we move away from everyday scenarios, the appeal to counterfactuals becomes more problematic. Physical theory itself may imply the existence of certain events, which, in principle, cannot be known by us. Without going into this any further I can only say that our realistic conception of various parts of our world – from everyday events to events in the centers of stars – rests on a far-reaching web of theories, practices and beliefs. And at least in certain sectors of our framework the realistic conception is constitutive of the meaning of the concepts in question.[2]

In the case of mathematics – as exemplified by arithmetic – the situation differs in a crucial respect. The framework is, so to speak, transparent, and our thorough

---

[1] In principle, we can check every smaller number greater than 1, whether it is a divisor of this number. This, and similar brute-force methods (the sieve of Eratosthenes) would have taken more time than the life span of our sun.

[2] It is constitutive of the meaning of 'tree' that the proverbial tree that falls in the forest, falls independently of being observed. This by itself is not an answer to a radical skeptic who might argue that, in principle, we might be misguided in applying 'tree' in its ordinary meaning.

understanding of it does not seem to leave place for contingency: we may not know how things are, but whatever they are, they cannot be otherwise. The empirical factor, the opacity that endows everyday truths and the truths of science with the nature of hard facts, is lacking. The mathematical setup seems to be a creation of the human mind, or a game whose possibilities have been determined by certain rules. This picture might encourage an attempt to reduce questions of truth to questions of provability, or to deny that 'truth' applies here at all. Yet the statement that the twin-prime conjecture is true and the statement that it is provable have altogether different meanings. The first is a clear sharp mathematical statement. The second depends on how we read 'provable'; if it means having a proof like those accepted in current mathematical literature, then the second statement is not a mathematical one. It is a clear mathematical statement only if 'provable' means provable in a given formal deductive system. In any case, our seven grader, who understands perfectly what the twin-prime conjecture is, has hardly a clear idea what it means for it to be provable (surely, one can understand the twin-prime conjecture without knowing anything about formal deductive systems). This still leaves open the possibility that, the difference of meaning notwithstanding, there is some deductive system, such that simple arithmetical statements of the kinds exemplified above are true just when they are provable. But this possibility is ruled out by the incompleteness and undecidability results. These results have revealed an unbridgeable gap between provability and truth. They make it extremely plausible that any system of mathematical reasoning accessible to humans, will fail to decide certain simple elementary questions. Further results (lower bounds in complexity theory) show that something like this holds even for simple statements about particular large numbers – statements which can, in principle, be decided, but whose minimal verification will necessitate a number of steps that puts it beyond human reach (say, more than the estimated life, in nanoseconds, of the sun, or of the galaxy, or of the universe). We cannot, of course, say of a particular problem that humans will not be able to solve it, because we do not have foreknowledge of what future deductive systems – in particular, what kind of axioms – will be used. But any formal system whose proofs can be effectively recognized will fail to decide some elementary statements, like the existence of a solution to a given system of diophantine equations.

Moreover, if a formalist appeals to a well-defined notion of 'proof', then the question, is there a proof for such and such a sentence? is itself a mathematical question, which, in principle, is not different from the question about the existence of a number with certain properties. For proofs can be encoded into natural numbers.

These observations are no more than opening moves in the lengthy analysis of realism in mathematics, moves that can be followed and responded to according to different strategies. In general, anti-realism with regard to actual infinities is expressed by rejecting the appeal to "either A or not-A", when this involves quantification over an infinite domain (e.g., either there is an odd perfect number, or there is no odd perfect number). An intuitionist will base the rejection of the excluded middle on a non-standard interpretation of the logical particles ('not', 'or', 'if...then...'), which leads to a different logic. The resulting theory will be weaker than the classical one, in the case of first-order arithmetic, but incomparable with the classical system in the case of real-number theory. A host of theories, motivated by different conceptions of real numbers and other higher-order entities, have been looked into.

In general, the extent of realism that one subscribes to is indicated by those statements one considers objectively true or false, independently of our state of knowledge. Anti-realism is therefore indicated by excluding certain statements from this class. It may or may not go with a different logic. For example, one can be a realist with regard to natural numbers, but reject the realistic picture for second-order logic on grounds of predicativity, as Feferman does. This position is expressed by replacing set theory with the weaker predicative version; it does not involve a change in classical logic.

A realistic conception rests, of course, on the *intended interpretation* (or intended model) of the language in question. It signifies an attitude that treats that language as sufficiently clear and univocal, so that questions of truth and falsity are completely settled. Thus, our understanding of the natural numbers seems quite clear, sufficient for convincing the great majority of mathematicians (and seven-graders) that the truth or falsity of the twin-prime conjecture is objectively determined by the interpretation of the numeric terms – the so-called standard model. Much more should and can be said, concerning the grounds for this conviction and concerning the fact that the situation is quite different with regard to set theory, where there is a broad spectrum of positions: from skepticism about the set of all reals, to full scale Platonism with respect to Cantor's universe. But this is not my subject here.

I conclude these brief observations about realism by pointing out that even constructivists must adopt a certain realistic attitude when it comes to quantification over finite but very large domains. Statements about finite domains can be decided, in principle, through finite checking; on the usual constructivist views, the excluded middle applies in these cases. But the number of steps can be so large as to render the checking practically unfeasible. Gödel's construction of a sentence that "says of itself" that it is not provable can be modified, so that it yields a sentence that "says of itself" that it is not provable in less than $k$ steps, where $k$ is a very large number. Using this technique we can get relatively short sentences whose minimal proofs are extremely long. Underlying this phenomenon is the fact that short names can denote very large numbers. Already in decimal notation numbers can be exponentially related to the length of their names. Using an exponentiation symbol, we can, furthermore, have numerical terms such as '$10^{100,000}$' or '$10^{10^{100,000}}$'. The exponentiation symbol can be eliminated by using the inductive definition of exponentiation in terms of addition and multiplication; the elimination of each occurrence will increase the length of the name by some constant but not more. Let $t_n$ be a term of this kind, denoting the number $n$. Given a deductive system, e.g., ZFC (Zermelo-Fraenkel set theory with the axiom of choice), there is an arithmetical wff, $\varphi(x)$, involving only bounded quantifiers, such that: for all $n$, $\forall x < 2^{t_n} \varphi(x)$, is provable in the system, but every proof contains no less than $2^n/c \cdot |t_n|$, steps, where $c$ is some constant and $|t_n|$ is the length of $t_n$. Since all quantifiers in the sentence are bounded, the excluded middle applies to it in intuitionistic arithmetic. There is no practical way to prove the sentence in the given system (unless the system is inconsistent).

Realism is not the subject of this paper; I focused on it, since it provides an example of an elementary question that calls for philosophical analysis. It can also provide an illuminating contrast to another dimension in the philosophy of mathematics, the dimension of efficiency and epistemic fruitfulness. Consider: (i) The extension of the system of positive numbers, 1,2,..., to the system 0,1,2, ..., of natural numbers obtained by adding 0, (ii) The extension of the system of natural numbers to the system of all integers, obtained by adding negative numbers, (iii) The extension of the system of integers to that of rational numbers, (iv) The extension of the system of real numbers to the system of complex numbers.[3] All of these might appear as "ontological enrichments", since they consist in "adding objects". But this conception is wrong. As far mathematical reality is concerned (the statements that have objective truth-values) there is no more to the natural numbers with 0 than there is to the non-zero ones, no more to the integers than to the natural numbers, and so on. Whatever is described in the system of natural numbers can be also described in terms of the non-zero ones. The addition of 0 is a formal move that rounds up the system and makes for a very efficient notation, but the enlarged system can be reinterpreted in terms of the original system: Let 1 play the role of 0, 2 – the role of 1, and so on; define addition, multiplication and any other function accordingly (i.e., for non-zero $m$, $n$, put $m +' n = ((m - 1) + (n - 1)) + 1$, $m \cdot' n = (m - 1) \cdot (n - 1) + 1$). Statements about the larger system can be then rephrased as statements about the smaller one. The same goes for the addition of negatives, which can be accomplished by considering all pairs $(i, n)$, where $i$ is either 0 or 1, and $n$ is a natural number that is strictly positive if $i > 0$; our previous natural numbers are now identified with the pairs $(0,n)$ and the negative numbers with the pairs $(1, n)$. Rational numbers can be defined in the well-known way as pairs of integers (under the congruence that identifies fractions of equal value), and complex numbers as pairs of reals. These easy reductions lead to obvious translations of statements from one system into the other.[4] (The move from the rationals to the reals is, of course, quite another matter; there is all the difference in the world between the two.)

Yet, these extensions constitute momentous developments in the history of mathematics. It is only with the hindsight of the 19th and the 20th centuries that we can describe them as we just did. They constitute advances in several respects: notational, algorithmic, and epistemic. The so-called "discovery of 0" and the importance of '0' for arithmetical notation is a well-researched story that needs no further comment. But the inclusion of 0 as a natural number was far from obvious

---

[3]I use 'system' in a broad sense: an underlying class of objects with various mathematical concepts that are associated with them. In the case of numbers, these will include various functions and relations. I assume that the ordering is included so that we can recover the non-negative integers from the integers, via '$x \geq 0$', without appeal to non-trivial theorems. Similarly, I assume that we have in the system of rational numbers a predicate '$x$ is an integer' and that in the system of complex numbers we have '$x$ is real'.

[4]This type of semantically-based translation leaves the logic intact. It should be distinguished from the
    syntactic type that preserve provability relations, exemplified by the translations between classical and intuitionistic first-order arithmetic. The second type does not amount to an "ontological reduction". If we want to get an "ontological reduction" using a syntactic approach we should also require that the translation should commute with the logical connectives. This blocks the translation from classical to intuitionistic systems.

even at the end of the $19^{th}$ century. In Dedekind's system (from 1888), the natural numbers start with 1, as they do in Peano's earlier work (from 1890); Cantor begins his ordinals with 1, though he sometimes recognizes the technical advantage of adding an additional zero-element. Negative and imaginary numbers have made their way into mathematical practice gradually during hundreds of years. First they were used as computational props – auxiliary symbols, which do not stand for numbers but are treated *as if* they did. They were therefore regarded with suspicion and were granted first-class citizenship only in the $19^{th}$ century, when it was realized that they can be modeled as pairs of *bona fide* numbers, as indicated above. Their incorporation had far reaching consequences, far beyond the computational aspect. It amounted to a restructuring of the mathematical space: the way a mathematician organizes his or her material. Although there is no more to the integers than there is to the natural numbers, viewing the natural numbers as part of the integers is altogether different from viewing them by themselves. The difference in epistemic organization is even more striking in the case of complex numbers. Although complex numbers are no more than pairs of reals, complex analysis – with its underlying two-dimensional geometric interpretation – is a different subject than real analysis, with different heuristics, different natural questions, and different techniques.

Philosophy of mathematics should address both of these dimensions: the ontological – as it is revealed by questions about realism, the reductions of systems, and the relative strength of various theories – and the epistemic, which has to do with ways of organizing the mathematical space. The first can be handled with the help of precise technical tools, and here mathematical logic is indispensable. This is a kind of philosophy that merges with its subject matter, since metamathematics itself is being treated mathematically. A realistic view can sometimes be stated with precision that is impossible in any other philosophical investigation, e.g., "I think that all first-order statements about natural numbers have objective truth-values, but not all higher-order statements do." The second dimension is far less technical and far less precise. Naturally, it helps itself more to phenomenal descriptions. The great difficulty here is in finding the right questions and the right parameters that will make possible a systematic approach.

The philosophical question about the nature of mathematics goes back at least to Kant, who posed it explicitly and gave an elaborate account. The question of realism, on the other hand, is relatively a new comer, an outcome of the foundational debates at the turn of the last century.[5] The two questions are, of course, related, but not to the extent that an answer to one must always determine an answer to the other. For example, Frege was a realist with regard to arithmetic as well as

---

[5]Questions about actual, versus potential infinity and skepticism with regard to the former are at least as old as Galileo. But only at the turn of the last century did the debates take place, which later crystallize into well-defined positions in terms of realism. The present day use of 'Platonism' in the philosophy of mathematics stems from these debates, whatever its affinity is with Plato's original philosophy. Finally, mathematical realism has to be distinguished from the subject of the old Newton-Leibniz debate regarding the reality of space. The question whether statements in pure geometry have objectively determined truth-values did not arise. I think both Newton and Leibniz would have answered it positively. The debate was rather about the physical interpretation of geometry. With hindsight, it can be rephrased as being about the factual status of certain statements; but these statements involve reference to physical bodies.

geometry; but he thought that arithmetical truth is logical and geometrical truth is not.[6] In the other direction, one's view regarding the nature of mathematics may leave undetermined which parts of set theory one conceives realistically. On the whole, the question about the nature of mathematical truth and the source of its validity is looser and less direct than the question of realism, since it depends more on other presupposed debatable categories. Thus, Kant used the double distinction of *a priori* versus *a posteriori* and analytic versus synthetic, as well as other conceptual tools in his philosophical arsenal.

## 2. The Grand Tradition

In Kant's account mathematics derives from a sort of intuition, a form of grasping that cannot be reduced to logic. At the turn of the last century the main opposite view was logicism, which claimed that mathematics (or, at least, arithmetic) *is* reducible to logic. This was one of the main points of contention, and was viewed as such by the people involved.[7] The other, more obvious issue of contention was infinity. Those who, broadly speaking, toed the Kantian line accepted potential infinity as legitimate in mathematics, but rejected, or were suspicious of actual infinities; to this group belonged, among others, Hilbert, Brouwer and Poincaré. Logicists, on the other hand, accepted the infinities provided by Cantorian set theory (in this or that version of it) as fully meaningful and legitimate. They thought, wrongly as it turned out, that set theory can be reduced to pure logic; their acceptance of actual infinity fitted nicely within the logicist view.

Subsequently, the issue was given a precise form: Can we use quantification over the natural numbers in forming statements that are objectively true or false? The first point of contention was thus about the nature of mathematics, the second was about realism.

On both issues Hilbert should be grouped with Poincaré and Brouwer. But unlike the intuitionists he did not propose a revision of logic but a different kind of restriction: finitism. Hilbert did not present his conception formally, but it is not difficult to see what he was aiming at. PRA (Primitive Recursive Arithmetic), or something like it, is a plausible candidate; this is a very simple system, more restrictive than intuitionistic arithmetic, which uses classical logic but disallows the usual quantification over the natural numbers. Hilbert is often associated with "formalism". But the upshot of his position was that his proposed formalization was a tool in the service of finitism. His idea was ingenious in its simplicity: Given a mathematical system that uses actual infinities, formalize it. If the resulting formal system is consistent, then general finitistic equalities that are provable in it, $f(x, y, \dots) = g(x, y, \dots)$ – where $f$ and $g$ are say, primitive recursive and '$x$', '$y$'... range over the natural numbers – must be valid; it cannot have a counterexample, $f(m, n, \dots) \neq g(m, n, \dots)$, because any particular inequality can be verified and this would lead to contradiction. Hence the consistency of the formal system means that we can use it safely to derive valid finitistic results. If, moreover,

---

[6]He declared himself a Kantian with respect to geometry, though it is not clear that this was an accurate description of his view of geometry, as revealed in his debates with Hilbert.

[7]In particular, Poincaré, in his *Science et méthode* represents the dispute as a momentous struggle between Kantians (in a broad sense of the term) and logicists.

we have a consistency proof that satisfies finitistic standards, then we have also an effective way of transforming non-finitistic proofs of general finitistic equalities into finitistic ones. This will show that, in principle, actual infinities can be eliminated; their value is instrumental.

Roughly speaking, the constructivist views (among which I include Hilbert's position) viewed infinity as reflecting an unbounded process of iterative constructions, grounded in an intuitive grasp à la Kant. The principle of predicativity belongs here as well. Predicative set theory uses classical two-valued logic and is realistic about the natural numbers: it accepts the standard model. But it conceives the subsets of the natural numbers not as pre-existing things, but as entities that are constructed in an ongoing process of definition. The constructive process proceeds bottom-up, in a well-ordered sequence, in which earlier sets can be used in constructing later ones. Russell subscribed to predicativity and incorporated it into the basic setup of the *Principia*; but this undermined logicism, since the system is then insufficient for the purpose of reconstructing arithmetic. Therefore an additional axiomatic scheme, the so called reducibility axiom, was included to undo the limitations imposed by predicativity. Russell frankly admitted that the axiom was not logical and hoped to eliminate it.

The views characterized by the three standard terms mentioned at the beginning of the paper had different fates. Russell's attempt to eliminate his reducibility axiom failed; the axiom cannot be dropped (this was first proved by Myhill). More generally, one can also see that Gödel's results undermine logicism, but I shall not go into this here. In recent years there have been attempts to resuscitate defunct logicism through artificial respiration – a broadening of what was originally conceived as logic. An assortment of arguments – some bad, others with some appeal perhaps – has been invoked in order to classify under "logic" substantial portions of set-theory. This line provides no insight about mathematics, but is rather a thesis about logic, which arguably redefines "logic" in a rather uninteresting way. If anything, the attempt only shows how pliant a concept can become under the expert massaging of some skilful philosophers.

While Hilbert's project has been proven unfeasible by Gödel's results, finitism remains a philosophical option. This is the view that actual infinities are no more than useful fiction for deriving valid finitistic theorems. The thesis that the use of actual infinities leads to valid finitistic theorems is equivalent to the claim that the use of actual infinities produces no contradiction, and this, by Gödel's results, must remain an unprovable belief. Finitists are called upon to give some account for the prevalence of this belief – the belief, which presumably they share, that systems used in current mathematics, such as Peano arithmetic, or analysis, or set theory are consistent. The obvious gesture at "inductive confirmation" (no contradiction found so far) is not convincing. Past experience is, to be sure, crucially relevant, but is not sufficient by itself. We are not concerned here with a black box whose output, so far, has been consistent. We know how this "black box" functions. Whatever philosophical view we espouse, we understand quite well the reasoning used in the standard systems and our convictions about consistency stem from this understanding.

The various positions that come under "constructivism" remain, on the whole, viable positions. I do not intend to discuss them at any detail. The problem of accounting for the general belief in the consistency of classical systems arises here, as it arises for finitism, but is less of a challenge – since these systems are richer. In certain cases the belief can be fully accounted for by relative consistency proofs; e.g., any proof of a contradiction in classical first-order arithmetic can be effectively converted to a proof of a contradiction in the corresponding intuitionistic system. An intuitionist can therefore justify, from his own point of view, the belief that the classical system is consistent.

In the first two decades of the last century, constructivists had extremely high expectations of transforming mathematical practice. These hopes petered out. Some constructivist programs continue to produce active research; also intuitionistic mathematics has found new uses in the context of theoretical computer science. But the impact of constructivism on mathematical practice as a whole is hardly noticeable. The reason for this is not hard to discover: the use of actual infinities, within the framework of classical logic, makes for a simpler more convenient organization of the mathematical space, and is therefore more effective. When classical results can be reproduced within constructivist systems the proofs are usually less transparent, and when the systems diverge, as in the case of intuitionistic analysis, the intuitionistic one is far more complicated. Abraham Robinson, who, as a philosopher, rejected actual infinities altogether, had a considerable corpus of mathematical works in diverse fields, all within the classical systems. He regarded actual infinities as useful fiction, but was not willing to give the fiction up. Other, less philosophically inclined mathematicians do not bother with the problem.

In its grand tradition period, philosophy of mathematics was produced by researchers with comprehensive foundational projects, who saw themselves responding to the discipline's needs, continuing the overhaul that took place in the $19^{th}$ century. Most were active mathematicians whose philosophical insights derived from an intimate acquaintance with their subject. The combination of philosophy and mathematical technique is also represented by the more philosophical figures of Frege and Russell, who produced ground breaking technical innovations. In the second half of the last century this is no longer the case. There are mathematical logicians, who produce technical work – either motivated by philosophical questions, or with clear philosophical implications – and there is a community of less-technical, or non-technical philosophers who address their subject from some general philosophical perspective. Works that impinge on the question of realism – and here I include the various constructivists positions – belong mostly to the first group. To this group belong also works that measure mathematical theorems by the logical theories within which they can be derived, this is the Friedman-Simpson project of reverse mathematics. In the second group we find works that explain and clarify historical positions, as well as works that try to tell some sort of philosophical story about current mathematics. Many of the latter adopt a deliberate non-critical position, intending to give a philosophically palatable description of the reigning practice. We get valuable insights, as in some versions of structuralism, or useful proposals, such as the modal approach (which can also involve technical work). But sometimes one cannot avoid the impression of philosophical decorative annotations, or, in other cases, of ideas moving in a closed circle of a philosophical game that adds little to our understanding. This danger is hardly avoidable in philosophy, where opinions might vary almost as they vary in art criticism.

## 3. The Proposal

I propose that, using concrete simple examples, we take a hard look at the nature of mathematical reasoning and the source of its validity. At least provisionally, the subject can be treated separately from questions concerning realism. At the turn of the last century, those who conceived mathematics on, broadly speaking, Kantian lines were also those who adopted an anti-realist position with regard to actual infinities. But this is not at all necessary. Suppose that our reasoning in arithmetic derives from an intuitive grasp of finite strings of abstract strokes, arranged in an unending sequence (Hilbert's position). Does this imply that statements in first-order classical logic, with quantifiers ranging over the natural numbers, do not have objective truth-values? It does not. It also does not imply that they do. Either decision is an additional step. Gödel explained our mathematical knowledge by appealing to some sort of intuition, or perception. This did not prevent him from being an extreme realist with regard to the set-theoretical universe. One's position regarding realism will, as a rule, be sensitive to developments in mathematical logic, and, possibly, even to physical theory. There are two kinds of possible investigations. One can have a definite conviction about what should and should not be conceived realistically, that is, about the sentences that have objective truth values and those that do not; one can develop a line of argument for this view, which conceivably can also involve some technical work. Or one may suspend judgment and contend oneself with a systematic analysis of possible positions. I myself find questions of realism extremely hard and I opt for the second approach. Realism however is not the subject of the paper.

I shall proceed with some simple examples of mathematical reasoning, which come mostly in the form of mathematical puzzles. They are easily understood by "everyone" (our proverbial seventh-grader); their solutions are difficult, but, once given, are obvious and utterly convincing ("how neat, why didn't I think of it before?"). They can teach us something significant about mathematical reasoning. There are scores of such problems. Here are three.

### 1. A Tiling Problem

A domino-tiling of an area, made up of non-overlapping squares of equal size, is a covering of the area by rectangular domino pieces, in which every domino covers two adjacent squares; there are no overlapping dominos, the whole area is covered and nothing else. From now on 'tiling' refers to domino-tilings. Consider a division of a square by a grid, into 8×8 squares: 8 rows and 8 columns; call this a "standard board" ( fig. 1). Trivially, it can be tiled by using 4 dominos to tile each row. An example of a more complicated tiling is given in fig. 2.

Consider now boards obtained by removing squares from the standard board. Since each domino covers two squares, a board that can be tiled must have an even number of squares. The removal of a single square yields a board with 63 squares, which cannot be tiled. Suppose we remove two: the bottom-left corner and the top-right corner (fig. 3) can the remaining board, which consists of 62 squares, be tiled? This is the puzzle.

The solution uses an ingenious idea. Color the squares of the standard board black and white, alternating the colors as in a board of chess. Our removed squares

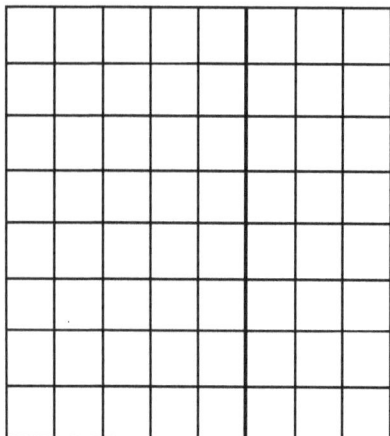

Figure 1

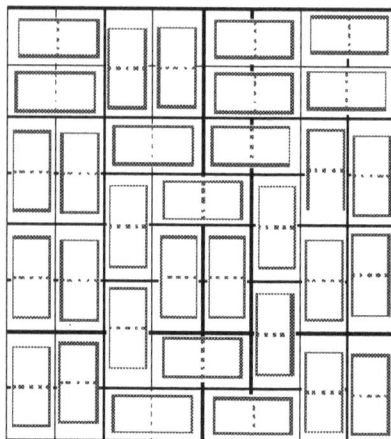

Figure 2

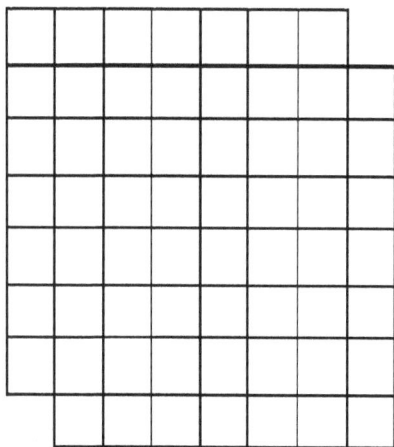

Figure 3

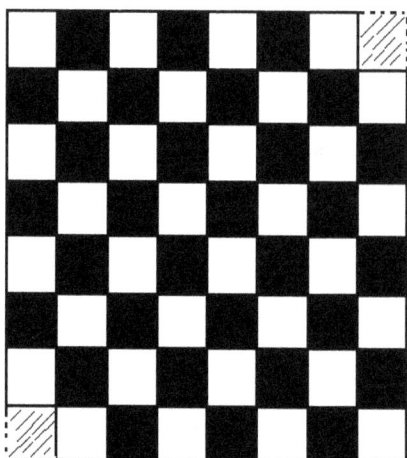

Figure 4

are the endpoints of a main diagonal; they have the same color, say they are black (fig. 4). Then the mutilated board consists of more white squares than black. Since each domino covers adjacent squares, and adjacent squares have different colors, the covered area must have equal numbers of white and black squares. Hence the mutilated board cannot be tiled.

As fig. 5 shows, there are boards obtained by removal of two squares, quite apart from each other, which can be tiled. The removed squares must of course have different colors. Is this condition sufficient for the existence of a tiling? It can be shown that the answer is yes (it holds moreover for any standard board with an even number of rows). Our ingenious trick is therefore more than a trick;

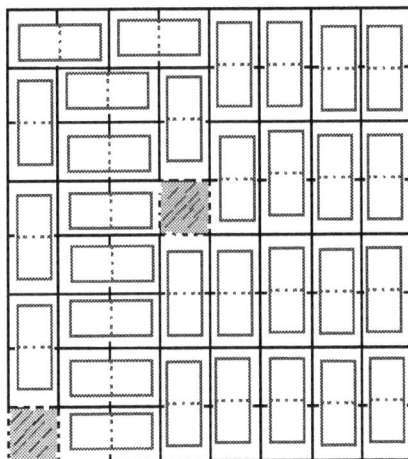

Figure 5

it gives us a necessary and sufficient condition for the existence of a tiling after a removal of two squares. I note this in order to put our trick in wider perspective; it is not part of the example and is not needed for the philosophical points I am going to make, but it shows something about the way mathematics poses questions and progresses.

I do not know the history of this puzzle; some books present it as a question about chess boards, providing thus the major step in the solution and making it easy and not very interesting (an 8×8 board may give a hint already, a 10×10 board is preferable). For the purpose of the discussion let us ignore the history and treat it as a mathematical question. The following are noteworthy features:

1. It is an elementary problem whose understanding hardly requires mathematical training. Tiling games are easily taught to first-graders (and below) and it does not take much to understand the goal of tiling a given board. Understanding the impossibility proof requires more, but should pose no problem to anyone who has minimal grasp of chess-coloring and of concepts such as *equal numbers of black and white squares*.

2. The proof exemplifies the methodology of solving a problem by adding structural elements not found in the given description. In this it resembles some famous impossibility proofs, e.g., the impossibility of a ruler-and-compass trisection of an angle. There, the added structure consists in automorphism groups, here it consists in the coloring.

3. The trick works like magic. The problem now seems easy, but make no mistake , this is a hard problem! It takes considerable ingenuity to invent such a trick (unless one has seen already similar devices).

I think that the problem is as clear, and the proof as certain as we can ever get in mathematics. As far as clarity and validity are concerned we have hit rock bottom.

A more formal proof, which requires a more formal rephrasing of the question, will add nothing in this respect. Although "everything" can, in principle, be recast in set theoretic terms, the route from the question, as stated above, to some formal rephrasing is far from clear. Let us try. The first suggestion is to represent boards as relational structures consisting of (i) a finite set (whose members are the squares), (ii) the adjacency relation, consisting of all pairs of adjacent squares. A tiling is then a partition of this set into a family of disjoint sets, each of which consists of two adjacent members. When the standard board is thus viewed, we loose any insight into the situation; even the existence of trivial tilings becomes a fact in need of careful checking. The impossibility proof will then involve the following steps: We divide the given set into two disjoint subsets (the "white squares" and the "black squares"), for which we show that adjacent elements cannot belong to the same division member. We also show that the two deleted squares belong to the same division member. From this we can derive in set theory the impossibility claim. But the definition of the two-fold division and the alleged truth of the claims are complex affairs into which we have little insight. Note that our non-formal proof generalizes trivially to any even-sized board and yields a general statement for all boards of $2n \times 2n$. In the suggested formal version this is altogether obscure.

We can do better by representing our standard boards not as arbitrary sets but as Cartesian products, $A \times A$, where the set $A$ is also provided with a "neighboring relation". Adjacent members of $A \times A$ are pairs $(a, b)$, $(a', b')$, such that either $a = a'$ and $b$ and $b'$ are neighbors, or vice versa. We can improve further by letting the set $A$ be ordered; yet the coloring remains a complicated affair. The way that leads to a comprehensible set-theoretic proof (where "set-theory" includes also arithmetic) is the following. The board is $A \times A$, where $A = \{1,\ldots,8\}$, or in general $\{1,\ldots,2n\}$. Two elements, $(i, j)$, $(i', j')$, are adjacent if $i = i'$ and $|j - j'| = 1$, or vice versa. We partition the squares, $(i, j)$, into two subsets, the odd ones and the even ones, according as $i + j$ is odd or even. Modeled thus, the original proof can be recast as a formal one, which is comprehensible.

Even the comprehensible formal version may give pause to non-mathematicians, who have no problem with our original solution. It also takes some mathematical training to realize that the formal version is a faithful translation of the original into a different setting. Yet, our original argument needs no translation. It is as valid a piece of mathematical reasoning as any.

What is then the value of a translation into set theory, or into some other accepted mathematical framework? Before we consider this, let us consider a more fundamental question: From where does our original grasp of the mathematical problem derive? The required understanding of concepts such as *area made of non-overlapping squares,* and *tiling* can be acquired through simple games, or – in the case of grown ups – through explanations backed by a couple of examples. The number of games and examples is finite, yet we find no difficulty in generalizing definite features, from very few cases to a potentially infinite collection; this makes the reasoning that proves theorems possible. Wittgenstein characterizes the phenomenon in terms of rule following: somehow we acquire the ability to follow the "right" rule. On his view it would be wrong to say that we recognize certain patterns, since rule-following generates the very patterns we come to recognize; rule following is an irreducible primitive that determines arbitrarily what we consider as "being of the same kind". I have no objection to speaking in terms of rule follow-ing, as long as this leaves place for genuine discoveries, the discovery of non-trivial facts that are implied by the rules we follow. The way we organize our conceptual space – having a world that is informed, among other things, by areas, squares, and tilings – may be arbitrary (let us grant this); yet, *in a world thus organized* the mutilated board has no tiling, and *that* is not arbitrary, but a substantial truth. Wittgenstein denies this possibility because he denies that we can discover truths by reflecting on some of our rules; this kind of meta-level perspective has no place in his account. As he sees it, the impossibility of a ruler-and-compass trisection is not a discovery of a truth, which is implied by our geometric conception, but a kind of choice that we made after seeing the impossibility proof.[8] The choice can be "good" or "natural", but it is still a choice. The non-existence of a tiling is a miniature example of the same type, which shares the important feature (ii); I do not see how one can accept it as a "good choice" rather than a necessary truth. Consider, for illustration the game of chess. The rules that define it are perhaps a "good choice", since they yield an enjoyable intriguing game. They are nonetheless arbitrary. But, *given that these are the rules*, some far from obvious statements follow, which are not arbitrary at all; e.g., generally, it is impossible to mate with a king and two knights against a king. Some claims of this type are known to be true only through the use of computers. Sometimes we know that one of several alternatives holds, without knowing which: either white has a winning strategy, or black has one, or both have strategies that guarantee at least a draw. It would be a major discovery to find which of these is the case. In this respect mathe-matics is like chess. Wittgenstein was right in seeing the rules of mathematics as rules that determine meaning. He was wrong in denying the factual character of mathematical truths.

I am going to suggest that mathematics is the study of forms of organization (or patterns, or structures, or kinds of configurations), and what they imply. Here 'implication' marks a semantic notion that goes with a notion of truth, which cannot be reduced to provability in some deductive system. Let me consider first two other examples.

## 2. Euler Graphs

This is a famous historical case, known as the problem of the seven bridges of Königsberg. It concerns seven bridges connecting four land areas that are separated by a river (Pregel). A known puzzle of the time asked whether it is possible to cross in a single walk all the bridges, so that every bridge is crossed no more than once. Figure 6 represents the configuration of the areas and bridges as a graph, in which the areas appear as vertices, marked with capital letters, and the bridges as edges (connecting lines). In 1735 Euler gave a lecture to the Russian Academy in St. Petersburg in which he proved that this is impossible. His proof established a necessary condition for the existence of such a walk in similar configurations, a condition that can be easily checked and which obviously fails in the case of the

---

[8] *Lectures on the Foundations of Mathematics, Cambridge 1939*, edited by C. Diamond, University of Chicago Press, p. 56 (paperback edition). Cf. also the discussion in lectures V and VI. Wittgenstein's paradigmatic example is the impossibility of a ruler-and-compass construction of a heptagon.

Königsberg bridges. In a paper published a year later, he noted that some people thought that the walk was impossible, others were doubtful, and none claimed that it can be done. His proof thus confirmed a general suspicion. He also notes that, in principle, the answer can be found by checking all possible combinations, but that this is an involved task and when there are more bridges it is unfeasible.

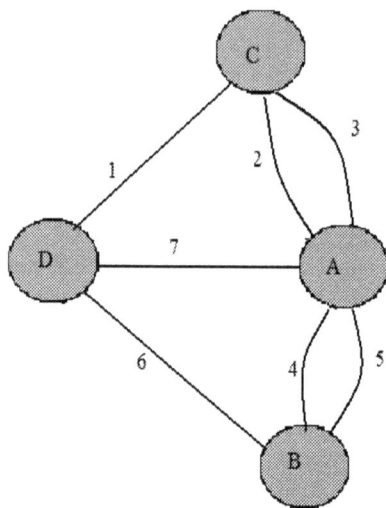

Figure 6

The idea of a graph – a system of vertices and edges, where each edge has two vertices, pictured as its "end points" – is easily explained with the help of a few examples (such as fig. 6). An edge is said to be *incident* on each of its vertices, and each of its vertices is *incident* on it. The *degree* of a given vertex is defined as the number of edges incident on it. In fig. 6, the degree of A is 5, and of each of C, D, B has degree 3.

A general "walk" on a graph can be given as a sequence of alternating vertices and edges that starts and ends with vertices (the beginning and the end of the walk), $v_1$ , $e_1$ , $v_2$, ..., $e_{k-1}$, $v_k$, such that for any edge of the sequence, $e_j$ , its two vertices are its two sequence-neighbors, $v_j$, $v_{j+1}$. Such a sequence is also referred to as a *path*. The condition that every bridge be crossed exactly once means that every edge of the graph occurs in the sequence one time exactly; such a path is called an Euler path, and a graph that has it is called an Euler graph. The argument in Euler's paper involves unnecessary details. The following is a proof that "everyone" can easily grasp.

Suppose we start the walk in vertex $v'$ and end it in vertex $v''$, where the two may or may not be the same vertex. Consider any vertex, v, different from the end points. Each time we pass through v, there is an edge through which we arrive and an edge through which we depart. Since every edge incident on v is traversed exactly once, the number of edges incident on v is twice the number of times that we pass through it (which is the number of times v occurs in the sequence). This implies that except for the end points, $v'$ , $v''$, every vertex has an even degree.

The same counting applies to the end points, except that, for v′, the first edge of our walk is unpaired, and, for v″, the last edge is unpaired. If v′ ≠ v″, this implies that the degrees of the end points are odd. But if v′ = v″ then the first outgoing edge is paired with the last ingoing edge, hence all the degrees are even. Therefore a necessary condition for the existence of an Euler path is: either (i) all vertices have even degrees, or (ii) all vertices except two have even degrees. In fig. 6 all vertices have odd degrees. There is no Euler path. If we delete edge 7 between A and D, then A has degree 4, B and C have degree 3 and D has degree 2. In this case there is an Euler path:

B 4 A 5 B 6 D 1 C 2 A 3 C .

We can also see that (i) is necessary for the existence of an Euler *cycle* ( an Euler path that starts and ends in the same vertex), while (ii) is necessary for the existence of an Euler path that is not a cycle.

Euler claimed that his condition is also sufficient; this claim is true, provided that the graph is connected. His argument for the sufficiency is incomplete; the full proof of sufficiency is more involved than the argument for necessity and I omit it. Figure 7 is based on an illustration given by Euler in the same paper, with 16 bridges (including the dotted bridge between C and D) and 6 land areas marked by capital letters. The degrees of all vertices except D and E are even; indeed, there is an Euler path. If we omit the dotted edge, then also C and F have odd degree, and there is no Euler path.

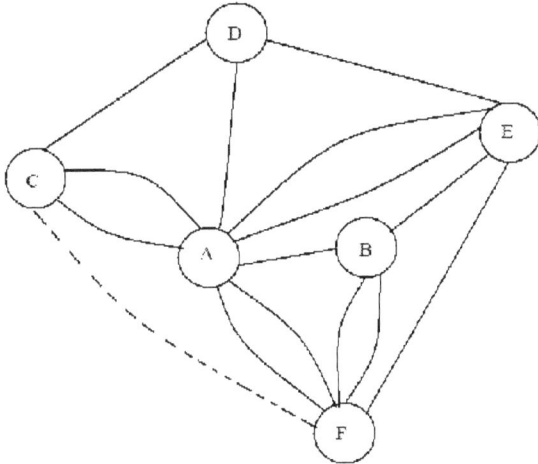

Figure 7

Unlike the tiling problem, the present example is based on a concept (that of a graph) which is not fully specified in the statement of the problem. The drawings used to explain the concept are bound to be *planar*: graphs that can be embedded in the plane, with vertices appearing as points and edges as continuous arcs that do not intersect except at their end points. But evidently the argument does not depend on this restriction. Although our explanations leave the concept somewhat open, I claim that the proof is as valid as it can be. It works in the more restrictive interpretation of 'graph'; and it works, exactly in the same way, under other less restrictive interpretations. Actually the proof depends only on an abstract conception, under which a graph consists of two finite disjoint sets – a set of vertices and a set of edges – and a correlation that associates with every edge two vertices.

The third example goes further in this direction: a valid argument that proves a claim that involves an open concept.

## 3. The Coin Placing Game

Two players take turns in placing coins of equal size on a perfectly round table. The coins can touch, but should not overlap and should not extend beyond the table's perimeter. The player who cannot place a coin loses. (The table is of course much larger than a coin, say its diameter is at least six times that of a coin.) Does player I (who goes first) have a winning strategy, or does player II have one? Since the game must terminate after finite number of moves (a number that is smaller than the ratio of areas of the table and the coin) and one of the players must lose, a well-known theorem says that one of the players has a winning strategy. 'Winning strategy' is a precise, well-defined term of game theory. Our riddle however is addressed to the mathematically non-sophisticated. 'Strategy' therefore means no more than a prescription (in some intuitive loosely understood sense) for playing the game, and a winning strategy is a prescription that leads to a win. No appeal is made to game theory. In any case, the theory does not help us towards a solution. Again, the solution is not easy, if you have not seen similar devices before; but, once given, is completely obvious.

Player I starts by placing his coin at the center of the table. Then, whatever player II does, player I responds symmetrically in the diametrical opposite place: if player II places her coin at point $x$, player I places his at $x'$, which is on the diameter through $x$ on the opposite side of, and at the same distance from the center; cf. figure 8.

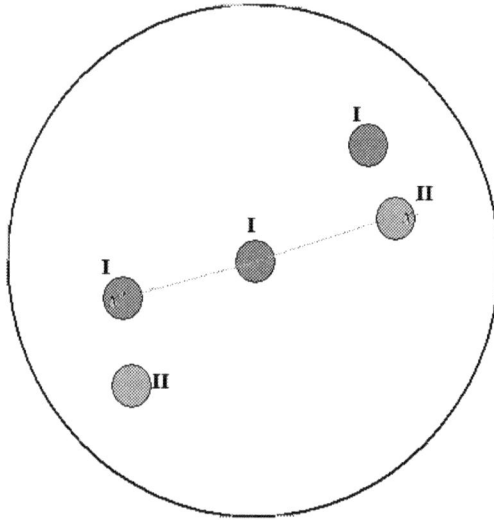

Figure 8

After each move of player I the coin configuration has radial symmetry. If player II has a place for her coin, player I has the diametrically opposite place for his. Hence, the first to run out of places must be player II.

Playing according to this prescription, player I ensures a win. Here again we have hit rock-bottom as far as validity is concerned. The solution is a well-defined particular prescription; it does not matter that a general definition of 'winning strategy' is still pending.

Each of our three examples involves geometry, but geometry plays in them different roles. The tiling problem is purely combinatorial, but it makes little sense without the geometric organization that helps to define it. The graph theoretic problem is introduced in a geometric setting, but the geometry can be easily dispensed with. The last example differs in that geometry enters essentially, since the metric is crucial. Hence the last example may give rise to the standard old observations about the imprecision of the real world: a table is never a perfect circle, a player can measure distances with limited accuracy only, etc. Evidently, the game is supposed to be *mathematical*: the table and coins are ideally circular and places are determined with perfect precision. It does not take mathematical sophistication to appreciate this aspect of the problem.

The examples illustrate and support my main point to which I now return: Mathematics studies patterns, structures, kinds of configurations, what I shall refer to as *forms of organization*. I take this notion as primitive. When a formal language is used in characterizing a system, the underlying form of organization is indicated by the structures that are considered as possible interpretations of the formal language.[9] But in this context 'structure' is a technical, or semi technical term of set theory – a theory that presupposes already a kind of "structure" of a more basic, non-technical kind. This generic concept is marked by 'form of organization'.

The view advocated here has an obvious affinity with Kant, who, speaking broadly, saw mathematics as the discipline that studies time and space, where these are forms of perception – the way we organize sense data. But my proposal avoids the Kantian metaphysics of raw sense data that are organized into experience. It is concerned with general forms of organization that underlie our practices and views of the world. This is as broad a category as what comes under Wittgenstein's rule following. We come to know mathematical truths through becoming aware of the properties of some of the organizational forms that underlie our world. This is possible, due to a capacity we have: to reflect on some of our own practices and the ways of organizing our world, and to realize what they imply. In this respect all mathematical knowledge is meta-knowledge; mathematics is a meta-activity *par excellence*. The meta-reflection need not be explicit, or deliberate. The ancient Egyptians and Babylonians used particular examples, involving the adding, subtracting and dividing of quantities of merchandize, not for their

---

[9]As a rule, mathematical theories are considered in contexts that include at least arithmetic (if not analysis and more). Hence, as a rule, the structures that are considered as a possible interpretations should include the standard model of natural numbers as a component. For example, group theory can be characterized by first-order axioms; but when we speak of finite groups, or groups generated by a finite number of generators, etc., the structure involves, besides the group itself, the natural numbers.

own sake, but as *generic*: to illustrate how similar problems are to be solved; this marks already the move to the meta-level.

In mathematics the study of organizational forms yields clear–cut, necessary, and far from trivial truths. This is a special combination of features; when one of them is missing we do not have mathematics. The late Jerry Katz claimed that, underlying the semantics of natural languages, there is a theory of *sense*, which is a sort of mathematics. I do not know why there should not be such a system, but the brute fact is that (so far) there is not. The semantics of natural languages is an empirical discipline, which did not breed any new mathematical system.

Mathematics is continuously expanding by deriving new mathematical systems from areas of human cognition and practice. A domain may be thus "mathema-tized". The ongoing list is numerous. Graph theory, initiated by Euler's work (our example 2) is a nice illustration; game theory, which grew in the context of economics, is another. Many new systems originate within mathematical practice itself, or mathematical practice supplemented by concrete examples (e.g., the the-ory of knots). Probability theory and mathematical logic are two grand examples whose significance can be hardly overestimated. In mathematical logic, which in-cludes the study of formal languages, aspects of mathematical activities become themselves the subject of mathematical investigations.

Each of the points made above needs elaboration and further philosophical work. In particular, further explanations are due concerning the key notion of "form of organization". I am arguing here for a line of investigation, rather than for a worked out account. Let me only note that a form of organization is related to what is known as a structure, but the link is not rigid. A form of organization need not correspond to a single standard interpretation of a given mathematical language, it can correspond to a family of interpretations, which share some basic features. Note also that this is not a mathematical concept. Philosophy must have recourse to a looser conceptual apparatus than the mathematics it studies.

The story I told so far is quite partial. The interconnections of mathematical disciplines, the reductions, the translations, the unifying grand systems, the ax-iomatic method, all these are missing. I focused on certain examples in order to bring to the fore essential features of mathematical reasoning. They should not mislead us into picturing mathematics as a dispersed collection of puzzle-solving techniques. How does the analysis offered so far fit into the bigger picture?

First, note that proofs of the kind exemplified above have limited use only. As a rule, difficult theorems are established through derivations from a relatively small number of axioms, because they cannot be proved by ingenious immediate devices. Even when a theorem seems evident (as is the case with quite a few geometrical claims), its derivation from a small number of axioms is useful in the interest of a streamlined system, one that gives us a good overview of the area. It is highly desirable to keep one's working desk uncluttered. The system serves also as a common framework for a wide community of researchers. It facilitates communication by establishing a common terminology, sets up accepted standards, and reduces the possibilities of misunderstanding.

There is also, as in any cognitive enterprise, the danger of error. In reasoning about our forms of organization, we can make mathematical errors. There is no lack of mistakes in the history of mathematics. The axiomatic method does not guarantee an error-free activity but it provides us with an error-correcting method-ology. We can go again, and again, over the steps in a proof, and we can make it more formal and detailed as we need; faced with contradicting claims of different researchers, we can sort out the conflict and deliver a verdict. The uniqueness of mathematics does not consist in its being error-free – which it is not – but in its error -correcting mechanisms.

All this is compatible with my previous claim that the solutions of our three examples are as valid as they can be. Nothing will be added by recasting them in some acceptable system and deriving them formally from axioms. If I am asked, how I can be sure that I have not overlooked something in the tiling proof, I respond: how can we be sure that 59 is a prime number? we have checked and rechecked, but perhaps we have overlooked something? At a certain stage doubt must come to an end. In each of the above examples that stage is reached with the solution – the one given above, not some formalized version. Writing the tiling-proof formally and having a computer verify all the steps (every modus ponens) is of no help. First there is a possibility that we have not formalized the problem correctly, second, there might have been an error in the proof-checking program, third, the system's software, even its hardware, could have been faulty.

Finally, the axiomatic method is crucial in uncovering hidden assumptions. This is a process of making explicit various features of an organizational form, which, unaware, we took for granted. In geometry the process spanned long stretches of history, culminating in the full axiomatization, at end of the $19^{th}$ century, where everything was made explicit; it also inspired the development of pure formalisms. The uncovering of a hidden assumption opens the possibility of modifying the form of organization, either by omitting the assumption or by modifying it. This is not a mere syntactic replacement of an uninterpreted axiom by another, but a change of meaning, where different structures are considered as possible interpretations of the language.[10] Thus, the non-derivability of the parallel postulate from the other geometric axioms was not sufficient for the emergence of non-Euclidean geometry. Quite a few mathematicians were convinced that the postulate was not derivable, but Non-Euclidean geometry emerged when Gauss, Bolyai, and Lobachevsky real-ized the possibility of a different geometric structure; and the theory was developed before any consistency proof was considered.

What then of set-theory? Like other mathematical systems set theory studies a certain form of organization. It is a form that arose within mathematics itself, when Cantor saw how basic patterns of reasoning about collections of real numbers can be generalized, extended and made into a self-standing discipline. That form of organization was developed by focusing on some basic concepts and by estab-lishing some non-trivial properties (e.g., the Cantor-Bernstein theorem); later it was recast axiomatically and, still later, was crystallized into what is now known as the iterative concept of sets. It is an elegant system, which can be grasped after

---

[10] In situations where the completeness theorem applies, the change of possible interpretations can be fully characterized syntactically. The passage from commutative to non-commutative fields is captured in a purely formal way, as long as we are interested in the first-order theory of fields. But such theories are, as a rule, considered in wider contexts that include arithmetic (perhaps also analysis and more), cf. footnote 9. Once we throw in the standard model of natural numbers, we must appeal to a semantic, not a syntactic, consequence relation.

some elementary training. Its unique position is due to the fact that other current mathematical systems can be modeled in it, and then the theorems of current mathematics can be derived from its axioms. The last feature underlies the significance of some set-theoretic independence results. What this tells us about the philosophical status of set theory is a difficult question and a matter of debate. It should be noted that, in principle, other systems can have this universal character. Indeed, category theory is a contender – though category theory, I am told, must help itself to a modicum of set theory, since it has to rely on the distinction between sets and proper classes. The view proposed in this paper and the line suggested by it, can be pursued independently of one's conclusion regarding set theory.

# The Iterative Conception of Sets from a Cantorian Perspective

Ignacio Jané

*Departament de Lògica, Universitat de Barcelona,*
*Baldiri Reixac s/n, E-08028 Barcelona, Spain.*
*jane@ub.edu*

**Abstract.** According to the iterative conception, a set is, in Gödel's words, something obtainable from some well-defined objects by iterated application of the operation "set of". Such a conception, then, depends on the prior concepts of power set (Gödel's *set of*) and iteration, and we develop it by first dealing with them. After introducing the opposition between the relative and the general view of sets, we discuss the idea of the totality of all (relative) sets of a given domain. Then we turn to the iteration, which should rest on some version of the ordinal sequence that can be accounted for before developing set theory proper. This we find in the notion of transfinite number as it occurs in Cantor's *Grundlagen*, but which we free from any existence assumptions. Principles about the existence of ordinals are then explicitly introduced in order to guarantee that the axioms of Zermelo-Fraenkel set theory hold of the iterative sets. All these principles are found in Cantor's *Grundlagen*. After a brief remark on the status of the axiom of choice, we end with some philosophical reflections on the nature of set theory prompted by our construal of the iterative conception.

There is a theorem of ZF which is often suggestively read as asserting that the set-theoretic universe can be divided into a well-ordered chain of layers in such a way that each layer contains exactly those sets all of whose members belong to some previous layer. Since, as follows from this description, every layer extends all the preceding ones, this well-ordered chain is called the *cumulative hierarchy*. In the presence of the other axioms of ZF, this theorem is equivalent to the axiom of foundation, which asserts that the membership relation is well-founded. This axiom, and thus this theorem, is usually disregarded in the uses of set theory as a tool for the rest of mathematics, but it plays a fundamental role in the development of set theory as an autonomous mathematical discipline.

This theorem has also been the subject of considerable attention from a philosophical perspective, in so far as it both suggests and embodies the so-called *iterative conception of sets*. According to this conception, a set is, in Gödel's words, "something obtainable from... some... well-defined objects by iterated application of the operation 'set of'."(Gödel [17], 259). The iterative conception—which was put forward only when a substantial amount of set theory was available—is meant to provide an account of the universe of sets from which the usual set-theoretical axioms can be justified.

On the face of it, the very statement of the conception depends on the concepts of *power set operation* (Gödel's *set of*) and *iteration*. Since the iteration is to be carried out beyond the finite, this in turn depends on some extension of the natural number sequence into the transfinite along which to iterate. Such an extension is provided by the ordinal numbers, which are usually described as the order-types of well-ordered sets or are defined as sets of a special kind—so that the notion of an ordinal is made to rest on those of set and well-order. Because of this dependence, the iterative conception appears to be too set-theoretically laden to be a suitable basis for set theory.

Accordingly, the iterative conception of sets is usually presented without explicit recourse to ordinals. Moreover, as George Boolos, the author of an influential presentation of this conception (Boolos [1]), has stressed, one should admit that there actually is no iteration, no generation of sets—this being only a way of speaking, a description of how things really are with a narrative flavor (Boolos [2], 90-1). The difficulty with such a course, however, is that any conviction that the iterative conception may carry is made to depend on metaphorical details that are dismissed as inessential to it. If there is no generation, why must sets be hierarchically distributed?, why must the membership relation be well-founded?, why cannot the totality of all sets be a set?

The naturalness of the iterative conception is apparent only if we conceive of sets as built by iteration of the power set operation from a domain of individuals (the empty domain if we restrict, as we shall do, to hereditary sets). Now, for this direct approach to the hierarchy to work at all, some notion of ordinal is needed which does not rest on set-theoretical ground. This, we submit, is the original notion of transfinite number introduced by Cantor in his 1883 *Foundations of a General Theory of Manifolds* (Cantor [7]), *Grundlagen* for short, which is neither extracted from well-orders nor rests on any idea of well-ordering any set.

We must also have recourse to some form of the power set operation before setting up the iterative conception. This is an important point that is often obscured and whose neglect might lure us into believing that the power set axiom of ZF simply follows from the idea of iteration. The reason given for the validity of this axiom is that if a set lies on a layer, so do all its subsets, and therefore the set of all of them lies on the next layer. One question about this way of presenting the matter is what is meant by "all subsets" of a set $a$. Perhaps from the standpoint that the iterative conception only describes how the world of sets is actually structured there is really no question to be asked (for if we can resort to the universe of sets, there is no difficulty in saying what are all subsets of $a$; they are just those sets all of whose members are members of $a$). But if we want to account for the set-theoretic universe as built by iterated application of the power set operation, such an explanation is of no use whatever. Since we cannot turn to the result of the iteration to tell what to do at each step, the notion of *all subsets* of a given set cannot be taken for granted, but must be clarified at the outset.

This is a description of the contents of the paper. In section 1 we bring in the general concept of a definite totality, or a *domain*. One basic feature of a domain is that any plurality of objects in it gives rise to a set. The concept of a domain is central to our account of the iterative conception, but, although many domains can be exhibited (as the empty domain, and the finite domains), none will be assumed to exist unless asserted in a principle. The first principle of existence of domains is a conditional one: to every domain a power domain corresponds, containing the sets of objects of the domain. As we shall argue in section 2, the notion of power domain involves a great deal of idealization and goes beyond our evidence regarding sets. As a result, the power domain operation becomes partly opaque to our understanding, and its opacity extends to the infinite levels of the cumulative hierarchy. Nevertheless, the description of the iterative conception is unaffected by the details of the power domain operation, which is one reason why we may feel that it describes a unique structure. The concept of ordinal needed for the iteration of the power domain operation will be introduced in section 3 following Cantor's steps in *Grundlagen*, and the iteration itself will be defined in section 4. Our treatment of the generation of ordinals and of the iteration along them is purely formal, in so far as they yield the general scheme of the ordinal sequence and of the iteration stages, but no ordinal and no iterative set can be inferred to exist from them (iterative sets being the objects produced in the iteration). This shows that, by itself, the iterative conception is unable to account for the existence of even the empty set. This, we think, is a virtue of our presentation, which makes the strength of the conception dependent on explicit principles of ordinal existence. The principles we need to ensure that the iterative sets satisfy the axioms of ZF, all found in Cantor's *Grundlagen*, are introduced, motivated and used in sections 5-8. After a brief note on the axiom of choice in section 9, we close the paper in section 10 with some philosophical considerations on set theory.

One last word about the reference to Cantor in the title of the paper. There is no evidence that the iterative conception can be traced back to Cantor. Even if we accept that he took the members of a set to be prior to the set itself, there is no hint that he conceived of all sets as fitting a pattern remotely similar to the cumulative hierarchy. Accordingly, we do not claim to be developing a view implicit in Cantor. But we do show how the iterative conception can be developed with the tools that Cantor had when he wrote *Grundlagen*. Although the particular operation to be iterated, namely power set, was not discussed by Cantor at the time, the notion of transfinite iteration and the means to carry it out are unquestionably due to him, as are the principles needed to ensure that the result is strong enough to deliver all the axioms of ZF.

## 1. Domains

The notion of set involved in the iterative conception—Gödel's *set of*—is a *restricted* or *relative* one, in the sense that a set is always a set of objects of some fixed (perhaps implicitly) domain—Gödel's *well-defined objects*. This relative notion stands opposed to the *general* notion of set, according to which sets are conceived as collections of any objects whatsoever. Both notions occur in Cantor's work. Before *Grundlagen*, he dealt only with restricted sets, the restriction being to what he called a "conceptual sphere"(Begriffssphäre). From *Grundlagen* onwards, he dealt with general sets (see Tait [22], 271-2).

In *Grundlagen*, Cantor gave the first explanation of the notion of a general set as "a plurality that can be thought as a one",[1] thus implying that not every

---

[1]He added: "i.e., every totality of definite elements which can be united to a whole by means of a law." [Unter einer 'Mannigfaltigkeit' oder 'Menge' verstehe ich nämlich allgemein jedes Viele, welches sich als Eines denken läßt, d.h. jeden Inbegriff bestimmter Elemente, welcher durch ein

plurality of objects is the collection of members of a set. He also gave the first example of such a plurality: what he called the absolutely infinite sequence of the transfinite numbers. But not long before *Grundlagen*, Cantor had all but explicitly stated that every definite condition on the objects of a conceptual sphere singles out a set. He said that a set is *well-defined* (by which we propose to understand that the defining condition does indeed define a set) "if on the grounds of its definition and as a consequence of the principle of the excluded third, it is internally determined whether any object of the same conceptual sphere belongs to the set as an element."[2] He didn't require that the plurality of objects should be seen as a unity. That, whatever it exactly meant, was an idle requirement, because their belonging to a domain warranted it.

Every definite condition on the objects of a domain gives rise to a set, and the characteristic of a domain that ensures that this is so is that of being a definite totality, as opposed to an open-ended plurality. Thus we understand by a domain a plurality of definite extent, meaning that it is fully determinate of what objects a domain consists, i.e., which objects are all the objects of the domain.

This notion of domain is very close to Cantor's idea of a general set—as opposed to what he called inconsistent, incomplete or absolutely infinite multiplicities.[3] Unlike domains and general sets, incomplete multiplicities are *quantitatively indeterminate*—to borrow a phrase from Cantor's review of Frege's *Grundlagen der Arithmetik*.[4]

How can a multiplicity fail to be quantitatively determinate? Not by being too big to have a number, but rather by not being definite enough to sustain one. As to what this lack of definiteness consists in, Cantor suggested that a multiplicity is not definite enough to be a set just in case it has a merely potential existence. A multiplicity, Cantor told Hilbert in 1897, is a completed one (is a set)

> if it is possible without contradiction to think of *all its elements as*

---

Gesetz zu einem ganzen verbunden werden kann.] (Cantor [7], 204, note 1; Ewald [15], 916.)

[2] "wenn auf Grund ihrer Definition und infolge des logischen Prinzips vom ausgeschlossenen Dritten es als *inner bestimmt* angesehen werden muß, ... ob irgendein derselbe Begriffssphäre angehöriges Objeckt zu der gedachten Mannigfaltigkeit als Element gehört oder nicht" (Cantor [6], 150). The inner determination, as he went on to explain, stands opposed to the external, or factual determination, which depends on the improvement of the resources at hand.

[3] We don't make any distinction between multiplicities and pluralities. We normally use "plurality", but prefer "multiplicity" as a translation of Cantor's "Vielheit".

[4] Cantor says that Frege "completely overlooks that ... only in certain cases the 'extension of a concept' is quantitatively determinate" [er übersieht ganz, daß der "Umfang eines Begriffs" quantitativ im allgemeinen etwas völlig Unbestimmtes ist; nur in gewissen Fällen ist der "Umfang eines Begriffs" quantitativ bestimmt] (Cantor [9], 440). Taken literally, Cantor's complaint against Frege's recourse to extensions of concepts is misplaced. Frege used them only to define the *number corresponding to a concept F* as the extension of the concept *concept equinumerous with F*. As Zermelo notes in his comment to Cantor's review, whether this extension is quantitatively determinate or not is irrelevant, because it is not to this extension, but to the concept *F* itself, that the number is assigned (Cantor [9], 441-2). Nevertheless, Cantor's words that only in certain cases the extension of a concept is quantitatively determinate make it very plausible that his objection was really aimed at Frege's assumption that to each concept a number corresponds. In any event, Zermelo's assertion that "in fact, Frege understands by 'number' exactly the same as Cantor understands by 'cardinal number', namely the invariant, that which all mutually equivalent (Frege says 'equinumerous') sets (Frege says 'concepts') have in common" ([9], 441) is especially unfortunate in this context.

*existing together,* ...; or (in other words) if it is *possible* to conceive the set as *actually existing* with the totality of its elements. So, the "transfinite" coincides with what has since antiquity been called "the actual infinite" (Purkert-Ilgauds [21], 226-7; translated in Ewald [15], 927-8).

By implication, then, an incomplete multiplicity, being infinite but not transfinite, is something potential (but it should not be described as being "potentially infinite", since it includes actually infinite totalities as parts). A merely potential multiplicity lacks the closed character of a collection. That is why, Cantor tells Hilbert, he defined 'set' in *Beiträge* as a collection [Zusammenfassung], and he adds that "collecting together is only possible when *'existing together'* is *possible*".[5] More explicitly still, in a letter to Jourdain of 9 July 1904, in which he explained why his conception is immune to Russell's Paradox, Cantor wrote:

> As elements of a multiplicity only complete things can be taken, only sets, but not inconsistent multiplicities, as it is essential to these that they can never be thought of as completed and actually existing.[6]

Let's come back to the iterative conception. According to it, sets are all relative to domains. We view the iterative conception as accounting for an ever increasing sequence of domains so that every set (every *iterative set*) is a set relative to one of them. Cantor's distinction between sets and incomplete multiplicities is easily explained in the iterative setting, since a multiplicity is incomplete just in case no domain of this sequence encompasses all its objects. This explanation becomes both mathematically efficient and philosophically satisfying once the sequence of domains is properly introduced and the set-theoretical axioms are shown to hold of the iterative sets.

Domains are basic to our account of the iterative conception. Thus, our likening of domains to Cantor's general sets might led one to suspect that the very articulation of the iterative conception depends on having a clear account of Cantor's distinction between sets and inconsistent multiplicities. If this were so, our project would be doomed, since the attempt to clarify Cantor's distinction from scratch, without the help of a previously developed set theory, seems hopeless. Fortunately, this is not our situation, for in order to articulate the iterative conception we don't

---

[5]In Purkert-Ilgauds [21], 226-7; translated in Ewald [15], 927-8. Cantor is dealing here with the multiplicity of the alephs. He agrees with Hilbert that it is determined of each given thing whether it is an aleph or not, but this, he claims, is not enough to guarantee that the alephs form a completed set. In fact, they don't, because they cannot all coexist. The reason of this impossibility, Cantor adds, is the absolute boundlessness [die absolute Grenzenlosigkeit] of the multiplicity of all the alephs (Cantor to Hilbert, 6.10.1898. In Cantor [10], 393-5.) In this and in the previously mentioned letter, Cantor speaks of the "set of all alephs", and of "inconsistent sets". However, later he used "set" as we have used it here. As he wrote to Hilbert in 9.5.1899: "I am now used to call 'consistent' what before I called 'complete', but I don't know whether this terminology deserves to be maintained. 'Sets' will now be 'consistent multiplicities'."(Cantor [10], 399).

[6]"Zu *Elementen* einer Vielheit, können nur *fertige Dinge* genommen werden, nur *Mengen,* nicht aber *inconsistente Vielheiten,* in deren Wesen es liegt, daß sie nie als *fertig* und *actuell existierend* gedacht werden kann" (Grattan-Guinness [19], 119). Cantor's late view of incomplete multiplicities as merely potential is discussed in Jané [20], where it is opposed to his early conception of absolute infinity as a quantitative maximum.

need any general theory of domains. In order to carry it out, we only need to be able to recognize some particular domains as such and accept that some specific operations on some specific domains yield new domains. For this, our inchoate idea of a domain as a definite totality will hopefully suffice.

One enlightening feature of the iterative conception as we develop it is that it sheds light on the nature of all incomplete multiplicities by referring them to just one whose incomplete nature is reasonably clear, namely that of Cantor's ordinals as introduced in *Grundlagen*. Moreover, the obscure component in Cantor's explanation of inconsistent multiplicities (lack of coexistence of their objects, mere potential existence, absolute boundlessness) can be tamed somewhat in the case of the transfinite ordinals. Their generation endows them with a natural order, of which it can be proven that it is a well-order and that it is absolutely unbounded in a rather clear sense. In fact, the absolute unboundedness of the ordinal sequence amounts to the circumstance that the generating rules can have no closure. Thus, the incomplete nature of the ordinal sequence is just the statement that no domain (independently of what exactly counts as a domain) can encompass all ordinals. And the satisfying thing is that this follows immediately from the very definition of the ordinals, as Cantor saw and emphasized. It doesn't strike one as surprising or in any way paradoxical, as will be apparent when Cantor's ordinal generation is discussed.

## 2. Power domains

Now we deal with the component *set of* in Gödel's mention of the iterative conception. This is the operation that assigns to any given domain $D$ a new domain, $D^*$, the *power domain* of $D$, which consists of the sets of objects in $D$. We refer to the objects in $D$ as the "$D$-*objects*", and to sets of $D$-objects as "$D$-*sets*". Of course, $D$-sets are sets in the relative or restricted sense.

Let's assume that we are given a domain $D$. Our aim is to secure the domain of $D$-sets. As a first attempt, we view $D$-sets as mere pluralities of $D$-objects, i.e., we take a $D$-set to be the plurality of its elements. This provisional decision is innocuous since, sets being extensional, each set is, as a set, uniquely determined by its elements. For us, such a decision means simply that we will only care about what $D$-objects a $D$-set contains—and not about what kind of objects $D$-sets are or which particular object an individual $D$-set is.

Even without being clear on what pluralities are, we see that they suffice to account for $D$-sets, when $D$ is a finite domain, for in this case we may think of a plurality as a selection, i.e., as the end result of a selecting process. If we are given a finite domain $D$ of $n$ elements, not only can we tell that there are exactly $2^n$ distinct selections of $D$-objects, but we also know how to describe them explicitly (in terms of any given enumeration of $D$) in an orderly manner. Moreover, we can decide from their descriptions which $D$-objects belong to each selection, and thus how each selection can be identified. This being so, we may treat the selections as objects of some sort and see them as forming a domain. This means that, in the finite case, $D$-sets can be taken to be these reified pluralities, and $D^*$ can be described in full as their totality.

No such procedure works for infinite domains. But there we may resort to a reasonably clear general notion of a set of $D$-objects, namely that of a plurality of $D$-objects which is specifiable with certain means. This general notion encompasses a variety of particular ones, each obtained by fixing the means allowed for specification—for example the notion of a set of natural numbers definable in first-order arithmetic (assuming, of course, that the natural numbers form a domain). Each particular choice of means yields a definite notion of $D$-set, so that the $D$-sets specifiable with these particular means form a domain. But, if $D$ is infinite, no specific choice of means yields a power domain rich enough to meet the demands of set theory. Thus, given any list of means for specifying sets of natural numbers, Cantor's diagonal method will allow us to specify, in terms of the list, a set which is not specifiable with those means.

On the other hand, placing no restrictions on the means allowed for specification, i.e., taking a $D$-set to be (or to correspond to) a plurality of $D$-objects specifiable with any means whatever, will not do, because there will be no guarantee that the plurality of sets so described forms a domain. If it is not determinate—which it isn't—what counts as a possible means of specification, how would we argue that it is determinate what specifiable pluralities there are?

The customary proposal to overcome all limitations is to get rid of specifications altogether and to introduce the idea of a *combinatorial $D$-set*, that is, of a plurality selected by arbitrarily and independently deciding for every object in the domain whether to select it or not.

While this proposal can be taken at face value for finite domains, in the infinite case it is at most a metaphor. Besides, for infinite domains, the combinatorial approach is arguably less effective than the previous one. For even if the pluralities of $D$-objects specifiable with arbitrary means do not form a domain, we certainly know what we mean by some such plurality and, in favorable cases (as when dealing with the domain of the natural numbers), we can produce many examples of them. However, as to what a combinatorial set is we have nothing but a hint—and no instructions about how to follow it.

Moreover, the combinatorial approach to $D$-sets is meaningful and will single out a domain only under the assumption that such a domain exists (which is what we are supposed to argue for). Under this assumption, there is no need to be precise about what a combinatorial $D$-set is in order to succeed in referring to them. For, if the combinatorial $D$-sets, whatever they be, all exist, a mere clue to what they are may suffice to identify them by marking them off from other entities—even if we are unable to characterize them with any accuracy. But as an explanation of "$D$-set", the combinatorial proposal is helpless.

In view of the impossibility to give an explanation of what a $D$-set is which is rich enough for the needs of set theory and from which it can be justified that the $D$-sets form a domain, we stop looking for it and propose to invert the priority relation between $D$-sets and the domain $D^*$. That is, we don't describe $D^*$ as the totality of all $D$-sets. Instead, *we postulate the existence of a domain called $D^*$*—together with a binary membership relation $\in_D$ between $D$-objects and $D^*$-objects—and *we define a $D$-set to be an object in $D^*$*. All we require of $D^*$ is that it be *maximally extensional* over $D$. That $D^*$ is extensional means that for any two distinct $D$-sets $a$ and $b$ there is some $D$-object $x$ such that $x \in_D a$ iff $x \notin_D b$. That $D^*$ is *maximally* extensional means that $D^*$ (and $\in_D$) cannot be extended without loss of extensionality.

We want to emphasize that maximality cannot be fully rendered as a mathematical condition. To make clear where the hindrance lies, it may prove useful to compare the intended maximality of $D^*$ with that of a filter on a boolean algebra. A filter $F$ in a boolean algebra $B$ is maximal if it is proper but no longer remains proper by the adjunction of any new element of $B$. Similarly, $D^*$ is maximally extensional if it is extensional over $D$ but no longer remains extensional by the addition of any new object (with an account of what $D$-objects relate to it by a suitable extension of the $\in_D$-relation). The essential difference between the two situations is that in the case of a filter on a boolean algebra we are given from the outset all the objects to be considered for adjunction, namely, the elements of the algebra, while no such supply of objects is available in the case of the power domain.

Strictly speaking, then, we don't know what all $D$-sets are and we don't know what $D^*$ is. That is why we have to posit it. Nevertheless, we can reason about $D^*$, we can define some $D$-sets, and we can argue for the existence of $D$-sets with certain properties (this means: *we can argue from our understanding of the maximality condition that such $D$-sets exist*). Thus, no matter what plurality of $D$-objects we would ever acknowledge, there should be a $D$-set corresponding to it.[7] In particular, in $D^*$ there is a set corresponding to each plurality of $D$-objects which we know how to specify in some given context, as there are $D$-sets corresponding to those pluralities specifiable in terms of other members of $D^*$. $D^*$ is thus conceived as being closed under various operations, some of them inspired by the suggestion of combinatorial sets. In a sense, we can think of $D^*$ as the ideal completion of the open-ended range of specifiable pluralities of $D$-objects.

We lay down two principles on $D$-sets, or, equivalently, on $D^*$: the **principle of extensionality**, according to which *$D$-sets with exactly the same members are identical*, and the **specification principle**, which says that *whenever $C$ is a definite condition on $D$-objects, there is a $D$-set whose members are exactly the $D$-objects which satisfy $C$*.[8] These two principles, which spring from the maximal extensionality of $D^*$, are all we use regarding $D$-sets. In particular, nowhere in the development of the iterative conception shall we appeal to the meaning of the maximality condition in order to justify some particular step. We will resort to it only in section 9, where we introduce a selection principle related to the axiom of choice.

We end our discussion of power domains by making two simplifying assumptions. We said nothing about the nature of $D$-sets, and, in fact, we want to say as little as possible, since what matters about $D$-sets is not what particular objects they are, but what members they have. In order to keep with this maxim, when dealing with two domains $D_1$ and $D_2$ in a common setting we shall assume that $D_1$-sets

---

[7] A $D$-set corresponds to a plurality of $D$-objects if the members of the set are just the objects in the plurality.

[8] A condition on $D$-objects is definite if, for every $D$-object $a$, the question whether $a$ satisfies it has a definite *yes* or *no* answer. The specification principle is a loose one in that it leaves open how conditions are to be articulated. We don't want to confine ourselves to conditions formulated in any particular language. The intended notion of condition is open-ended. As such, this principle is unsuitable as a mathematical axiom, and we don't intend it to be one.

and $D_2$-sets with the same members coincide. In other words, we shall assume that if $a_1$ is a $D_1$-set and $a_2$ is a $D_2$-set and the $\in_{D_1}$-members of $a_1$ are exactly the $\in_{D_2}$-members of $a_2$, then $a_1 = a_2$. In particular, if $D_1$ is a subdomain of $D_2$, then $D_1^*$ is a subdomain of $D_2^*$ and $\in_{D_1}$ is a subrelation of $\in_{D_2}$. This assumption allows us to simplify statements of membership about $D$-objects and $D$-sets for diverse $D$ by dropping the reference to $D$ in the membership relation.

Here is the second simplifying assumption. If $D$ is a domain and $X$ is a $D$-set, then the members of $X$ form a domain, say $D_X$. Since all that matters about $X$ is what members it has, and all that matters about $D_X$ is what objects it consists of, no ill effects will ensue from failing to distinguish $X$ and $D_X$. So, we shall treat, where convenient, a $D$-set as a domain, and, for any $D$-object $a$, we shall express with "$a \in X$" both that $a$ is a member of the $D$-set $X$ and that $a$ is a $D_X$-object. Similarly, since to $D$ itself a $D$-set corresponds, we may view $D$ as a $D$-set and use "$a \in D$" to express that $a$ is a $D$-object.

## 3. The generation of ordinals

For the tools needed to iterate the power domain operation we turn to Cantor, who first brought in the transfinite numbers in [5]—before he recognized them as numbers—as a means to carry out long iterations. As is usual nowadays, we shall refer to Cantor's numbers as "ordinals", but we must keep in mind that in *Grundlagen*, where Cantor introduced them, he didn't define them as order-types of well-ordered sets (see Tait [22], 273-4).

Cantor conceived the ordinals (the *real whole numbers* [die realen ganzen Zahlen], as he called them) as being obtained from a first ordinal (which we take to be 0, although he began with 1) with the help of two generating principles: the first one giving, for any generated ordinal $\alpha$, its immediate successor $\alpha + 1$; the second one applying to any available definite sequence of generated ordinals to yield its limit, i.e., the least ordinal largest than all ordinals in the sequence.

We fuse Cantor's two principles into one single rule and deprive them of existential import. Our only rule $\Gamma$ will yield the form of the generation, so to speak, and the requirements for the existence of ordinals will be introduced independently as explicit existence principles.

Ordinals, to which we refer by Greek lower-case letters, are meant to be generated by rule $\Gamma$. The generation of ordinals induces the *generating order* $<$ among them. Each ordinal is generated from a *segment*, i.e., from a domain $X$ such that whenever $\alpha \in X$ and $\beta < \alpha$, also $\beta \in X$. Applied to a segment $X$, $\Gamma$ yields $\Gamma(X)$, the least ordinal larger than all the ordinals in $X$.

We summarize the import of rule $\Gamma$, plus the assumption that each ordinal is generated by it, in the following four **principles of ordinal generation**:

($\Gamma$1)  *If $X$ is a segment, $\Gamma(X)$ is an ordinal not in $X$.*

($\Gamma$2)  *If $\alpha$ is an ordinal, there is some segment $X$ such that $\Gamma(X) = \alpha$.*

($\Gamma$3)  *If $X$ is a segment and $\alpha$ is an ordinal, $\alpha < \Gamma(X)$ iff $\alpha \in X$.*

($\Gamma$4)  *If $X$ is a segment, $\alpha$ is an ordinal, and every ordinal in $X$ is $< \alpha$, then $\Gamma(X) \leq \alpha$.*

What ordinals (or equivalently, what segments) exist will be discussed when principles of ordinal existence are introduced. For the time being no segment is assumed to exist. Since, by definition, each segment is a domain, we can apply the specification principle to obtain sets of ordinals in it.

From these four principles we can show that $<$ well-orders the generated ordinals. More precisely, we can derive (a) – (d):

(a) $<$ *is irreflexive.*

Let $\alpha$ be any ordinal. By $(\Gamma 2)$, let $X$ be a segment such that $\Gamma(X) = \alpha$. By $(\Gamma 1)$, $\alpha \notin X$, so that, by $(\Gamma 3)$, $\alpha \not< \alpha$.

(b) $<$ *is transitive.*

Let $\alpha < \beta < \gamma$ be any ordinals. By $(\Gamma 2)$, let $X$ be a segment such that $\gamma = \Gamma(X)$. By $(\Gamma 3)$, $\beta \in X$. Since $X$ is a segment and $\alpha < \beta$, $\alpha \in X$. By $(\Gamma 3)$ again, $\alpha < \gamma$.

(c) $<$ *is connected.*

Let $\alpha$ and $\beta$ be any ordinals. We must show that $\alpha \leq \beta$ or $\beta \leq \alpha$. By $(\Gamma 2)$, let $X$ and $Y$ be segments such that $\alpha = \Gamma(X)$ and $\beta = \Gamma(Y)$. Let $Z = X \cap Y$ ($Z$ is an $X$-set, which exists by *specification*). $Z$ is a segment. Let, by $(\Gamma 1)$, $\gamma = \Gamma(Z)$. By $(\Gamma 3)$, every ordinal in $Z$ is less both than $\alpha$ and $\beta$, so that, by $(\Gamma 4)$, $\gamma \leq \alpha$ and $\gamma \leq \beta$. By $(\Gamma 1)$, $\gamma \notin Z$, and thus $\gamma \notin X$ or $\gamma \notin Y$, that is, by $(\Gamma 3)$, $\gamma \not< \alpha$ or $\gamma \not< \beta$. If the former, $\gamma = \alpha$, and hence $\alpha \leq \beta$; if the latter, $\gamma = \beta$ and $\beta \leq \alpha$.

(d) *Every non-empty set of ordinals in a segment has a least member.*

Let $X$ be a segment and let $Y$ be a non-empty $X$-set. By *specification*, there exists the $X$-set $Z$ of all ordinals in $X$ which precede ($<$) all ordinals in $Y$. Since $Z$ is a segment, by $(\Gamma 1)$ there is an ordinal $\gamma$ such that $\gamma = \Gamma(Z)$. We claim that $\gamma$ is the minimum ordinal in $Y$. First, we see that if $\eta$ is an ordinal in $Y$, then $\gamma \leq \eta$. This is so by $(\Gamma 4)$, since every ordinal in $Z$ is $< \eta$. Thus, since $X$ is a segment and $Y$ is a non-empty $X$-set, $\gamma \in X$. It remains to show that $\gamma \in Y$. But if not, $\gamma$ would be an ordinal in $X$ less than all ordinals in $Y$, i.e., $\gamma \in Z$, in contradiction to $(\Gamma 1)$.

It also follows that

(e) *No domain of ordinals contains all ordinals.*

A domain containing all ordinals would be a segment, but, by $(\Gamma 1)$, no segment contains all ordinals.[9]

No matter how unclear the idea of ordinal generation may be, we see that it implies that, whatever ordinals be and whatever domains of ordinals there be, there is no domain to which all ordinals belong. This was a basic tenet of Cantor's in *Grundlagen*, who thus knew (and said it in clear enough terms, before any

---

[9] Besides the four principles of ordinal generation, only specification has been used in the proof of (a)-(e): to form the $X$-set $X \cap Y$ in the proof of (c), and to define $Z$ in the proof of (d). In the definition of $Z$ as an $X$-set, all quantification is over ordinals in $Y$, thus in $X$.

paradox arose) that not every condition determines a set. That the ordinals do not form a domain is a purely formal result which holds regardless of whether there is anything answering to the words "ordinal" or "domain". It is the expression that the rule $\Gamma$ can have no closure.

Our four assumptions on $\Gamma$ do not guarantee the existence of a single ordinal. They hold if there are neither ordinals nor segments. They hold also if the only segment that exists is the empty domain and the only ordinal is zero. For our basic principles to yield any ordinals we need to allow for the existence either of ordinals or of segments thereof—which by principles ($\Gamma 1$) and ($\Gamma 2$) is essentially the same.

## 4. Iteration

Now that we have the required ingredients, we can formally describe the iteration of the power domain operation along the ordinals starting from the empty domain. There is no doubt that this is a domain, since it is fully determined what its members are, namely none. Our description, however, will not depend on the assumption that there is a domain, not even an empty one. That any exists should be derived from principles of existence of ordinals—and we have none yet. Our present description of the iteration will be purely formal.

To every ordinal $\alpha$ the iteration assigns a domain $R_\alpha$. $R_\alpha$ is to be the $\alpha$-th iterate, with $R_0$ the empty domain (if zero exists). Cantor never considered this particular iteration, but we can imitate his description of the iteration of the derivative operation on pointsets;[10] indeed he introduced the ordinals (first as mere indices or *symbols of infinity* in Cantor [5]) to iterate this operation beyond the finite. Iterating along the ordinals is certainly a Cantorian notion. The iteration is regulated by the **iteration principle**, according to which, *for each ordinal $\alpha$, the $R_\alpha$-objects are just all $R_\beta$-sets, for the ordinals $\beta < \alpha$*:

$$x \in R_\alpha \leftrightarrow (\exists \beta < \alpha)\,(x \in R_\beta^*).$$

Regardless of any assumption about the existence of ordinals we can prove that for any ordinal $\alpha$: (1) if $\beta < \alpha$, all $R_\beta$-objects are also $R_\alpha$-objects, (2) the members of $R_\alpha$-objects are also $R_\alpha$-objects ($R_\alpha$ is transitive),[11] and (3) if $X$ is a non-empty $R_\alpha$-set, there is a least ordinal $\beta$ such that $X$ has some member in $R_\beta$. These three consequences of our assumptions are of help in the proof that the set-theoretical axioms hold of the iterative sets, which we now define.

## 5. Z for iterative sets.

We define an **iterative set** to be an object in some iterate $R_\alpha$. Since every iterate is a domain and, by the iteration principle, each object in $R_\alpha$ is a member of some $R_\beta^*$, all iterative sets are indeed sets in the relative sense. Moreover $R_\alpha$ itself is an iterative set if there is an ordinal larger than $\alpha$. Finally, since each

---

[10]This is the operation that assigns to every set of points $P$ in a topological space the set $P'$ of all its accumulation points, i.e., of all those points every neighborhood of which has infinite intersection with $P$.

[11]It makes sense to speak of members of $R_\alpha$-objects, since by the iteration principle an $R_\alpha$-object is an $R_\beta$-set, for some $\beta < \alpha$.

iterate is transitive, the members of iterative sets are also iterative sets. Ordinals are not assumed to be sets, iterative or otherwise. We examine what we need to assume regarding the existence of ordinals in order to secure all the axioms of ZF for the iterative sets.

Without assuming the existence of any ordinal, we can prove that the axioms of *extensionality, foundation, separation* and *union* hold of the iterative sets.

For the other axioms of ZF some assumptions on ordinal existence are needed. We only consider those that can be found in Cantor's *Grundlagen*. The ones we bring in first, namely (1) *there exists a limit ordinal* (the **principle of infinity**), and (2) *there is no largest ordinal* are certainly there.

With the help of these two principles we can prove that the axioms of *empty set, pair, power set* and *infinity* hold of the iterative sets. In other words, granted (1) and (2), all the axioms of the so-called Zermelo set theory Z (the axioms of ZF save for replacement) are satisfied by the iterative sets.

## 6. Two Cantorian principles of ordinal existence

We extract two principles of ordinal existence from Cantor's *Grundlagen* from which we derive the axiom of replacement for iterative sets. The first is just a version of Cantor's second generating principle, which is to be applied when "any definite sequence of generated whole numbers is available, none of which is the largest," and yields a number larger than all of them.[12]

The question is, when is a definite sequence of (generated) ordinals available so that the second principle can be applied to it? From Cantor's explicit examples of the generation of countable ordinals, we infer that such a sequence of ordinals is certainly available when we have a definite ordinal $\alpha$ and an assignment to each ordinal $\xi$ less than $\alpha$ of some definite ordinal $\alpha_\xi$.[13] This is how Cantor uses the principle to show the existence of larger and larger countable limit ordinals (as limits of $\omega$-sequences).

The principle we draw from Cantor's work we call the **boundedness principle**: *for any ordinal $\alpha$, every assignment $\xi \mapsto \alpha_\xi$ of a definite ordinal $\alpha_\xi$ to each ordinal $\xi < \alpha$ is bounded, i.e., there is an ordinal $\beta$ such that $\alpha_\xi < \beta$ for all $\xi < \alpha$.*

The boundedness principle is not enough to yield replacement for iterative sets, as we can see by noticing that this principle, but not replacement, holds in $R_{\omega_1}$. Moreover, it seems clear that whatever additional principle we use to get replacement for iterative sets, it must relate sets to ordinals. One such principle was stated and used by Cantor in *Grundlagen*—first to define the arithmetical operations on the ordinals. We call it the **isomorphism principle**: *every well-ordered set is isomorphic to the segment of predecessors of some ordinal.*

Cantor stated the isomorphism principle in *Grundlagen*, immediately after the introduction of the ordinals by means of the generating rules and the subsequent definition of the concept of well-order. From Cantor's perspective at the time it

---

[12]This is the complete statement of the second principle: "wenn irgendeine bestimmte Sukzession definierter ganzer Zahlen vorliegt, von denen keine größte existiert, auf grund dieses zweiten Erzeugungsprinzips eine neue Zahl geschaffen wird, welche als *Grenze* jener Zahlen gedacht, d.h. als die ihnen allen nächst größere Zahl definiert wird." (Cantor [7], 196; Ewald [15], 907-8.)

[13]Of course, every segment is also an available sequence of generated ordinals. But the import of the second principle for segments is already embodied in rule $\Gamma$.

was a rather obvious remark, since any given well-order can be taken as directing the generation of a segment of ordinals.

The definition of well-order that Cantor gave in the second section of *Grundlagen* is a replica of his principles of generation of ordinals. Since we have condensed his two principles into one single rule Γ, we can suitably, but faithfully, render Cantor's definition thus: *A linear order < on a set A is a well-order if and only if every proper initial segment of A with respect to < has an immediate successor.*[14] From this definition it is apparent that the generating order on the set of predecessors of an ordinal is a well-order.

Just after the definition, Cantor introduced the concept of *Anzahl* (later, order-type) of a well-order by declaring that two well-orders have the same *Anzahl* if and only if they are isomorphic. He then asserted that the isomorphism, if it exists, is unique, "and since in the extended number sequence (*Zahlen*reihe) there is always one and only one number α such that the numbers preceding it" have the same *Anzahl*, then "one must set" the *Anzahl* of both well-orders equal to α (Cantor [7], 168; Ewald [15], 885).[15]

If we think in Cantor's terms, we also find that the assertion that any well-order can be measured by an ordinal is immediately clear, because in order to show that a certain ordinal exists it is enough to describe the steps of its generating process, and any well-order can be used to describe the process of generation of some ordinal. In other words, the *Anzahl* of the well-order—its form, so to speak—exhibits the pattern to be followed in order to generate the ordinal that measures it.

Cantor didn't view the isomorphism principle as a principle of ordinal existence at all. No well-order extraneous to the ordinals is needed to generate an ordinal. The existence of a well-order is good evidence that the corresponding ordinal is generable, thus exists —but the existence of the ordinal doesn't depend on that of the well-order. In principle, it could happen that no well-order existed whose length were that of some large generated ordinal—it could, that is, if we don't take into consideration well-orders made out of ordinals.

## 7. The axiom of replacement

With the principles at hand we are able to show that the axiom of replacement holds of the iterative sets. As we did with all our principles, we state the axiom of replacement informally as:

> *Whenever a is a set and $x \mapsto b_x$ is an assignment of a definite set $b_x$ to each $x \in a$, there is a set c such that $b_x \in c$ for all $x \in a$.*

We begin by remarking that the axiom of replacement for iterative sets can be easily proven from the assumption that every assignment of a definite ordinal to each element of an iterative set is bounded, i.e., from:

---

[14] An initial segment of $A$ is a subset $B$ of $A$ such that if $x \in B$ and $y < x$, then $y \in B$. It is proper if it is different from $A$. Thus, if $A$ is non-empty, the empty set is a proper initial segment of $A$.

[15] Since Cantor's first number was 1 (not 0), this assignment of a number to a well-order has to be corrected if the set is finite. If a finite well-ordered set is isomorphic to the set of predecessors of $\alpha$, Cantor assigned the ordinal $\alpha-1$ to it.

($\star$) *If $a$ is an iterative set and $x \mapsto \alpha_x$ is an assignment of a definite ordinal $\alpha_x$
to each $x \in a$, there is an ordinal $\beta$ such that $\alpha_x < \beta$ for all $x \in a$.*

For suppose that ($\star$) holds. Let $a$ be an iterative set and let $b_x$, for each $x \in a$, be
an iterative set. We must conclude that there is an iterative set $c$ such that $b_x \in c$
for all $x \in a$. To this end we assign to each $x \in a$ the least ordinal $\alpha$ such that
$b_x \in R_\alpha$. Call this ordinal $\alpha_x$. By ($\star$), there is $\beta$ such that $\alpha_x < \beta$ for all $x \in a$. It
follows that each $b_x$ is a member of $R_\beta$, which is thus the set $c$ we were looking for.

In *Grundlagen*, Cantor introduced the *well-ordering principle*, according to which
every set can be well-ordered, and took it to be a fundamental law of general validity (Cantor [7], 169; Ewald [15], 886). If we are willing to use the well-ordering
principle as well, we can get ($\star$), and thus replacement, quite easily. For let $a$ be
an iterative set an let $\alpha_x$ be a definite ordinal for each $x$ in $a$. By the well-ordering
principle, $a$ admits a well-order, which, by the isomorphism principle, is isomorphic to the set of predecessors of some ordinal $\gamma$. The inverse of this isomorphism
is a one-to-one function $f$ on the set of predecessors of $\gamma$ onto $a$. Hence, we can
define the assignment $\xi \mapsto \alpha_{f(\xi)}$ (for $\xi < \gamma$) which, by the boundedness principle
is bounded below some ordinal $\beta$. But then $\alpha_x < \beta$, for all $x \in a$.

Although present in *Grundlagen*, we don't want to admit the well-ordering principle among our principles. Accordingly, we now prove ($\star$) without recourse to it.
Assume we are given an iterative set $a$ and an assignment $x \mapsto \alpha_x$ of ordinals to
the members of $a$. We must conclude that there is an ordinal larger than all the
$\alpha_x$, for $x \in a$. We define the equivalence relation $\equiv$ on $a$ so that $x \equiv y$ iff $\alpha_x = \alpha_y$.
Since Z holds of iterative sets, the quotient $b$ of $a$ by $\equiv$ is an iterative set which is
well-ordered by the relation $\prec$ defined by:

$$u \prec v \leftrightarrow (\forall x \in u)\,(\forall y \in v)\,(\alpha_x < \alpha_y),$$

so that, for $x$, $y$ in $a$,

$$[x] \prec [y] \leftrightarrow \alpha_x < \alpha_y,$$

where $[x]$ and $[y]$ are the equivalence classes of $x$ and $y$, respectively. As above,
by the isomorphism principle, there is an ordinal $\gamma$ and a function $f$ on the set of
predecessors of $\gamma$ onto $b$. Now we can assign an ordinal $\alpha_\xi$ to each ordinal $\xi < \gamma$
in such a way that if $f(\xi) = [x]$, then $\alpha_\xi = \alpha_x$. By the boundedness principle,
the ordinals assigned are bounded below some ordinal $\beta$. But then $\alpha_x < \beta$, for all
$x \in a$.

## 8. The list of principles

These are the principles we have used to define the iterative sets and to show
that the axioms of ZF hold of them:

(1) **Principles on power domains.**    For any domain $D$ there is a domain
$D^*$ and a relation $\in_D$ between $D$-objects and $D^*$-objects (i.e., $D$-sets) which
satisfy:

(1a) The principle of extensionality.

(1b) The specification principle.

**(2) Principles of ordinal generation.** Ordinals are generated from segments thereof according to rule $\Gamma$ in such a way that:

(2a) If $X$ is a segment, $\Gamma(X)$ is an ordinal not in $X$.

(2b) If $\alpha$ is an ordinal, there is some segment $X$ such that $\Gamma(X) = \alpha$.

(2c) If $X$ is a segment and $\alpha$ is an ordinal, $\alpha < \Gamma(X)$ iff $\alpha \in X$.

(2d) If $X$ is a segment, $\alpha$ is an ordinal and every ordinal in $X$ is $< \alpha$, then $\Gamma(X) \leq \alpha$.

**(3) Iteration principle.** To each ordinal $\alpha$ a domain $R_\alpha$ is assigned such that:

(3a) $\forall x \, \forall \alpha \, [x \in R_\alpha \leftrightarrow (\exists \beta < \alpha) \, (x \in R_\beta^*)]$.

**(4) Principles of ordinal existence.**

(4a) The principle of infinity.

(4b) The boundedness principle.

(4c) The isomorphism principle.

We have also assumed that there is no largest ordinal. This, however, need not be explicitly postulated, since it follows from the other principles. For let $\alpha$ be any ordinal. If $\alpha = 0$, then any limit ordinal is larger that $\alpha$, whereas if $\alpha \neq 0$, by the boundedness principle the constant map $\xi \mapsto \alpha$, for $\xi < \alpha$, is bounded below some ordinal $\beta$, which is thus larger than $\alpha$.

## 9. On the axiom of choice

Whether the axiom of choice holds of the iterative sets does not depend on the iterative aspect of the conception, but on the intended content of the power domains, thus on how the idea of maximal extensionality is unfolded. Consequently, some principle related to the axiom of choice must be assumed for power domains in order to ensure the axiom of choice for iterative sets.

The axiom of choice—say in Zermelo's version that to every family of pairwise disjoint non-empty sets there is a *selection set*, i.e., a set having exactly one member in common with every set of the family—is usually justified or motivated in terms of the combinatorial view of sets. As a matter of fact, the existence of selection sets is taken to be self-evident if our view of sets is the combinatorial one.

Although we propose to understand the power domain of a domain $D$ in terms of maximal extensionality (and not as the alleged totality of all combinatorial $D$-sets), the combinatorial idea can be used to argue for the existence of a selection set for any disjoint family. For the mere possibility of a selecting set for a particular family of $D$-sets counts against the maximality of a power domain containing none. Nevertheless, the stronger reason for the acceptance of a principle of selection for $D$-sets comes from the mathematical development of set theory. By reflecting on the use of the axiom of choice we can endow the sketchy notion of maximality with some more specific content. This, by the way, is also true of the combinatorial notion of set. It is not a notion that preceded the mathematical development of a theory of sets and that led to the acceptance of the axiom of choice; on the contrary, the metaphoric view of combinatorial sets has been proposed as a conceptual, or philosophical, articulation of the way sets are dealt with in mathematics—and the use of the axiom of choice has played an important part in it. We don't mean to suggest that there is something wrong with the attempt to intuitively justify an axiom from a notion of set that has been built in part from the use of that very axiom. In this and in other cases, our view of what sets are (or of what sets are to be) is shaped by the actual development of set theory and does not precede it. A case in point is the iterative conception of sets.

In order to derive the axiom of choice for iterative sets, we may assume the following **selection principle** for $D$-sets: *If $\mathcal{F}$ is a function on $D$ to $D$, there is a $D$-set $a$ such that* (1) *$\mathcal{F}$ is one-to-one on $a$, and* (2) *for any $D$-object $x$ there is $y \in_D a$ such that $\mathcal{F}(x) = \mathcal{F}(y)$.*

The reason for choosing this particular form is that it is a principle about $D^*$, involving only $D$-objects and $D$-sets (the function $\mathcal{F}$ is to be understood as a specification of an assignment, not as a set of pairs. It has the same status as the condition $\mathcal{C}$ in the specification principle).

## 10. Two aspects of set theory

The iterative conception of sets has a schematic character, in that the cumulative hierarchy it purports to describe is not a particular structure, but rather a schema with two under-determined parameters: the power domain operation and the ordinal sequence.

The under-determination of the power operation as applied to an infinite domain $D$ has to do with the hazy gap between the open-ended range of specifiable pluralities of $D$-objects and the posited closed domain $D^*$ of all $D$-sets—a gap of whose contents we have no full description, since we don't know how to fill it. The positing of power domains is not the result of any probing into the concept of relative set, but rests on external requirements. As a matter of historical fact, power domains were first brought in (they were implicitly assumed) in order to secure a notion of real number robust enough for the requirements of analysis—and, as we know, the definiteness of the concept of real number is equivalent to that of an arbitrary set of natural numbers.[16]

The under-determination of the ordinal sequence concerns the import of absolute infinity. The ordinal sequence is to be absolutely infinite, but how to spell this out? Failing to be a domain is clearly not enough, for, as we saw, we can prove from the very definition of the ordinals that they do not form a domain, no matter how few of them there are. We unfold Cantor's idea of absolute infinity by means of principles of ordinal existence—but we don't want to claim at all that the ones we introduced exhaust its content. However, even the principles that we stated and

---

[16]The construction of the real numbers in Cantor [4] assumes the totality (the domain) of all infinite sequences of rational numbers as given, while that of Dedekind in [11] presupposes the power set of the rational numbers. Both Cantor's and Dedekind's essays were published in 1872, at the dawn of set theory. It wouldn't be inappropriate to say that the positing of power domains is (both historically and conceptually) the first set-theoretical act.

used are less specific than it might seem, in particular the isomorphism principle: since it is formulated in terms of sets, its strength depends on what sets there are, hence on the contents of power domains. Besides, because of the isomorphism principle, the generation of the ordinals, although defined prior to the iteration, is affected by the very iteration that the ordinals are intended to direct.[17]

The Cantorian ordinal sequence cannot be taken to be a domain, and neither can the set-theoretical universe, which consists of the iterative sets. Both the concept of ordinal and that of iterative set are paradigmatic examples of what Michael Dummett calls "indefinitely extensible concepts" (Dummett [13], 316 and [14], 22). Adapting Dummett's definition to our setting,[18] a concept is *indefinitely extensible* if from any domain of objects falling under the concept a larger domain can be defined all of whose objects fall under the concept as well. According to Dummett, the right logic for the statements involving quantification over the objects falling under such a concept is not classical, but intuitionistic—from which he concludes that the acceptance that the concept of set is indefinitely extensible "entails a revision of mathematical practice in accordance with constructivist principles" (Dummett [13], 319).

For us, a philosophical approach to set theory which is not faithful to mathematical practice is unsatisfactory, since our aim is to understand set theory, not to change it. But we claim that to acknowledge that the set-theoretic universe is open-ended, as entailed by the iterative conception, does not force us to take a revisionist standpoint. For we have to distinguish two aspects in set theory, which for want of better names we call the *conceptual* aspect and the strictly *mathematical* aspect. The iterative description of the set-theoretical universe belongs in the conceptual aspect. When involved in it, we start from some more or less precise ideas about what sets are (or rather, about what we want sets to be). These ideas are rooted in our activities of counting, collecting, selecting, iterating, etc., both in everyday situations and, above all, in mathematics. We elaborate these more or less inchoate ideas by positing the existence of objects (ordinals, sets) in order to develop them more efficiently. In this way we outline a structure that will play the role of the universe about which the mathematical theory of sets is meant to be about. The notions of ordinal generation, of a domain, of absolute infinity, which are so important from this conceptual standpoint, do not have a place in the mathematical theory, but in them we may find the motivation behind some particular set-theoretical assumptions.

Even though the generation of ordinals (and the subsequent iteration of the power domain operation) cannot be taken at face value, it has a sort of formal meaning which is rich enough to suggest a list of principles from which to obtain ZF. Moreover, the generating aspect of the ordinal sequence has a compelling character that Cantor himself acknowledged. Thus, when (about three years before he set up the generating principles) Cantor introduced the symbols of infinity as indices of the derivative iterates beyond the finite, he insisted on how inevitable

---

[17]This paragraph was prompted by a comment of Haim Gaifman.

[18] "An indefinitely extensible concept is one such that, if we can form a definite conception of a totality all of whose members fall under that concept, we can, by reference to that totality, characterize a larger totality all of whose members fall under it." (Dummett [14], 22).

and "free from any arbitariness" the procedure was.[19] And in the first section of *Grundlagen*, he expressed his firm conviction that the extension of the sequence of the positive integers into the infinite, effected by the generating rules, would be regarded with time as "thoroughly simple, proper, and natural" (Cantor [7], 165; Ewald [15], 882). Moreover, this ideal generation had clear mathematical effects, to begin with, the Cantor-Bendixson theorem, a weak version of which was arguably the inducement to step from the symbols of infinity to the fully-fledged ordinals (see Ferreirós [16]).

If we look at the mathematical theory of sets without taking into account this conceptual aspect (or rather, without seeing it as a separate feature), we may be tempted to say, in view of the use of classical logic and the recourse to non-constructive methods of proof, that set theorists treat the set-theoretical universe as a definite, complete totality, thus as a domain. The temptation becomes particularly pressing if, instead of ZF, we consider the theory NBG (von Neumann-Bernays-Gödel), which allows us to talk directly about $V$, the class of all sets, and about OR, the class of all ordinals. In NBG, $V$ and OR are taken to be objects as definite as any individual set or ordinal (both are values of the variables and are denoted by singular terms). This temptation, however, has to be resisted, because its attractiveness rests on the confusion of the two aspects of set theory under consideration—has to be resisted, that is, if we take the words "definite totality", "complete totality", "domain", as having the same meaning in this context as they had when we dealt with the conceptual aspect of set theory. These words, we should emphasize, do not belong to the vocabulary of mathematical set theory (besides, they are too imprecise to be of mathematical use). If, from the way set theory is practiced, we concluded that $V$ is a domain, what would preclude it to have its power domain? Why, if OR is a complete totality, can it not be an argument of the generating rule $\Gamma$? Questions like these, which lie behind the so-called paradoxes of set theory, become suspect as soon as the two aspects of set theory are distinguished. When we go from the conceptual to the mathematical aspect, we change our perspective and our basic notions. In mathematics proper there is no room for the ideal closure of an open plurality of relative sets, as there isn't either for the generating rules or for the notion of an indefinitely extensible concept.

The gist of the relation between the two aspects of set theory is not hard to convey. As we saw, our description of the iterative hierarchy in terms of domains and generated ordinals is sketchy and under-determined, it is more a draft or a suggestion of a structure than a full account of one. Now, in order to deal with this schematic product with mathematical means we *do as if* the two basic parameters (the extent of power-domains and the length of the ordinal generation) were fixed. We know that they aren't, but our taking them to be fixed will have no ill effects, because, on the one hand, we never say (how could we?) what value the parameters take, and, on the other hand, as soon as we turn mathematical we leave behind the generating outlook. Nevertheless, the features that the ordinal sequence or the set-theoretical universe possess in virtue of their being open-ended or indefinitely extensible are not forgotten—they translate into the mathematical aspect of set

---

[19] "wir sehen hier eine dialektische Begriffserzeugung, welche immer weiter führt und dabei frei von jeglicher Willkür in sich notwendig und konsequent bleibt" (Cantor [5], 148).

theory in terms of restrictions about proper classes. If we don't set apart the two aspects of set theory, we feel that the distinction between sets and proper classes lacks intuitive support—we don't even see how there could be proper classes at all. Only if we pay attention to the distinction between the two aspects of set theory can we see that talk of proper classes (in NBG) is innocuous, and that their mathematical treatment is dictated by conceptual considerations.[20]

In Cantor's work, the conceptual and the mathematical aspects often occur intertwined, but as the theory is being consolidated, the conceptual component is left behind. This regress can be seen in the published writings of Cantor himself. Just compare his *Grundlagen*, which bears the subtitle of *A mathematico-philosophical investigation into the theory of the infinite*, with the systematic *Beiträge* (*Contributions to the founding of transfinite set theory*). In the former, the ordinals are introduced by means of the generating principles; in the latter as order-types of well-orders.

Cantor's own description of why, only a few months after *Grundlagen*, he changed his definition of the ordinals, is rather telling. In September 1883 he attended a scientific meeting in Freiburg, where, although he had no intention to talk about his work, he felt compelled to do so in order to dispel the skepticism of some mathematicians about his *conceptual constructions* [Begriffssbildungen]. So he spoke about the infinite numbers. Writes Cantor:

> I limited myself to treat the matter in a purely mathematical way. I started from my concept of well-ordered set and showed how the task of establishing the various *types* of well-ordered sets leads with necessity to the whole numbers, both the finite and the infinite. The ... whole numbers are nothing more than "signs" for the diverse "types" of well-ordered sets.
>
> The operations with these numbers arise from this foundation with no further ado in the simplest and clearest way, and this path is *purely mathematical*[21] (Letter to Mittag-Leffler of 23 September 1883; in Cantor [10], 130).

---

[20]The theory MK of Morse-Kelley, with the axiom of non-predicative comprehension for classes, cannot be so understood. The reason why it cannot is related to the impossibility of accounting from below for the power domain of an infinite domain. From the conceptual standpoint, a proper class corresponds to a plurality of iterative sets of unbounded rank, but, for any such merely potential plurality, to exist is to be specifiable. At the mathematical level, this prompts the acceptation of the axiom of predicative comprehension, which asserts the existence of classes which are definable in the language of the theory with quantification restricted to sets. In order to motivate non-predicative comprehension from the iterative conception, we should be able to account at the conceptual level for a fixed inclusive stock of something like *arbitrary pluralities*. This should be at least as hard to obtain as the power domain of an infinite domain, which, as we argued, we can only get by postulation. But such a move would be preposterous here, for how could we consistently entertain a full domain of pluralities of sets if there is no domain comprising all sets? This is not to deny that MK can be given a justification. It can as the theory of a sufficiently closed iterate $R_\alpha$, the proper classes being understood as the sets in $R_{\alpha+1}$. This requires stronger principles of ordinal existence which go beyond Cantor's, but which can be justified by resorting to Cantor's idea of absolute infinity.

[21] "Ich beschränkte mich darauf, die Sache rein mathematisch zu behandeln; ging von meinem Begriff der wohlgeordneten Menge aus und zeigte wie die Aufgabe: "die verschiedenen *Typen* wohlgeordneten Mengen aufzustellen" ebensowohl auf die endlichen wie auf die unendlichen ganzen Zahlen mit Notwendigkeit führt. Die ... ganzen Zahlen sind nichts anderes als "Zeichen" für

Perhaps this new path was purely mathematical, but it couldn't lead to the insight on absolute infinity that the generating rules aimed to convey. With no strong principles of set existence at his disposal,[22] the order-type approach to ordinals was utterly unable to provide Cantor even with a clue to the existence of infinitely many cardinals—let alone with the vision that they form an absolute infinite sequence, as he confidently asserted in *Grundlagen* (Cantor [7], 205 note 2; Ewald [15], 916-7).

**Acknowledgements.** I am indebted to David Asperó, Joan Bagaria, Frederik Muller, and Gabriel Uzquiano for helpful comments and suggestions. This paper is based on the lecture given at the 12th. International Congress of LMPS. Earlier versions were presented at the Logic Seminar of the University of Barcelona, April 2003, and at the Abstraction Weekend at Arché, University of St. Andrews, May 2003. The research leading to it has been partially supported by the Spanish MCYT under grant BFM2002-03236.

## REFERENCES

1.  Boolos, G. (1971). The Iterative Conception of Sets. In Boolos [3], 13-29.
2.  Boolos, G. (1989). Iteration Again. In [3], 88-104.
3.  Boolos, G. (1991). *Logic, Logic, and Logic*. Harvard University Press, Cambridge, Mass.
4.  Cantor, G. (1872). Über die Ausdehnung eines Satzes aus der Theorie der trigonometrischen Reihen, in Cantor [9], 92-102.
5.  Cantor, G. (1880). Über unendliche lineare Punktmannigfaltigkeiten. Part II, in Cantor [9], 145-8.
6.  Cantor, G. (1882). Über unendliche lineare Punktmannigfaltigkeiten. Part III, in Cantor [9], 149-57.
7.  Cantor, G. (1883). *Grundlagen einer allgmeinen Mannigfaltigkeitslehre. Ein mathematisch-philosophischer Versuch in der Lehre des Unendlichen*, in Cantor [9], 165-208. English translation in [15], 878-920.
8.  Cantor, G. (1895, 1897). Beiträge zur Begründung der transfiniten Mengenlehre, parts I and II. In [9], 282-356.
9.  Cantor, G. (1932). *Gesammelte Abhandlungen mathematischen und philosophischen Inhalts*, edited by E. Zermelo. Springer-Verlag, Berlin.
10. Cantor, G. (1991). *Briefe*, edited by H. Meschkowski and W. Nilson. Springer-Verlag, Berlin.
11. Dedekind, R. (1872). Stetigkeit und irrationale Zahlen, in Dedekind [12], 315-334. English translation in [15], 765-779.
12. Dedekind, R. (1932). *Gesammelte mathematische Werke*, vol. 3, edited by R. Fricke, E. Noether, and Ö. Ore. Vieweg, Braunschweig.
13. Dummett, M. (1992). *Frege. Philosophy of Mathematics*. Duckworth, London.
14. Dummett, M. (1994). What is Mathematics About? In George, A. (ed.) *Mathematics and Mind*. Oxford University Press, New York, pp. 11-26.

---

verschiedene "Typen" von . . . wohlgeordneten Mengen. Die Operationen mit den Zahlen ergeben sich auf dieser Grundlage ohne weiteres in der einfachsten und übersichtlichsten Weise und dieser Gang is *rein mathematisch*."

[22]The diagonal argument was more than six years into the future.

15. Ewald, W. (1996). *From Kant to Hilbert. A Source Book in the Foundations of mathematics*, vol. II. Oxford University Press, Oxford, ney York.

16. Ferreirós, J. (1995). "What Fermented in Me for Years": Cantor's discovery of Transfinite Numbers. *Historia Mathematica*, 22, 33-42.

17. Gödel, K. (1964). What is Cantor's continuum problem? In Gödel [18], 254-270.

18. Gödel, K. (1990). *Collected Works*, volume II. Oxford University Press, New York.

19. Grattan-Guinness, I. (1971). The correspondence between Georg Cantor and Philip Jourdain, *Jahresbericht der Deutschen Mathematiker-Vereinigung*, 73, 111-30.

20. Jané, I. (1995). The role of the absolute infinite in Cantor's conception of set. *Erkenntnis*, 42, 375-402.

21. Purkert, W. and Ilgauds, J. (1987). *Georg Cantor*, Birkhäuser, Basel.

22. Tait, W. (2000). Cantor's *Grundlagen* and the Paradoxes of Set Theory. In Sher, G. and Tieszen, R. (eds.) *Between Logic and Intuition*. Cambridge University Press, Cambridge, UK, pp. 269-90.

# Linguistic Invariants and Language Variation

Edward L. Keenan, Edward P. Stabler

*University of California, Los Angeles, Department of Linguistics*
*keenan@humnet.ucla.edu, stabler@ucla.edu*

**Abstract.** We illustrate a novel conception of linguistic invariant which applies to grammars of different natural languages (English, Korean,...) even though they may use different categories and have different rules. We illustrate formally how semantically defined notions, such as "is an anaphor" may be invariant in all linguistically motivated grammars (the issue is an empirical one), and we show that individual morphemes, such as case markers, may be invariant in grammars that have them in exactly the same sense in which properties, such as "is a Verb Phrase" or relations such as "is a constituent of" are invariant. Finally we distinguish "stable" invariants from "logical" ones, arguing that they reflect empirically based linguistic symmetries.

Since the publication of Noam Chomsky's field founding *Syntactic Structures* in 1957, generative grammarians have been formulating and studying the grammars of particular languages to extract from them what is general across languages. The idea is that properties which all languages have will give us some insight into the nature of mind. A widely acknowledged problem to which this work has led is how to reconcile the goal of generalization with language specific phenomena and the cross language variation they induce. Good science requires that cross linguistically valid generalizations be based on accurate, precise and thorough descriptions of particular languages. But such work on any given language increasingly leads us to describe language specific phenomena: irregular verbs, exceptions to paradigms, lexically conditioned rules, etc. So this work and cross language generalization seem to pull in opposite directions.

Here we propose an approach in which these two forces are reconciled. Our solution, presented in greater depth in *Bare Grammar* [5], is built on the notion of linguistic invariant. On our approach different languages do have non-trivially different grammars: their grammatical categories are defined internal to the language and may fail to be comparable to ones used for other languages. Their rules, ways of building complex expressions from simpler ones, may also fail to be isomorphic across languages. So languages differ. Nonetheless certain properties and relations may be invariant in all natural language grammars, as we will see below. And it is to these linguistic invariants that we should look for properties of mind.

Our approach contrasts with that of the most widely adopted linguistic theories, where the dominant idea is that *there is only one grammar*, the grammars of particular languages being, somehow, special cases. This has led to a mode of description in which grammars of particular languages are given in a notationally uniform way: the grammatical categories of all languages are drawn from a fixed universal set,[1] as are the rules characterizing complex expressions in terms of their components. It has also led to the postulation of a level of unobservable structure ("LF", suggesting "Logical Form"), where structural properties of observable expressions may be changed in important ways. So this allows that structural generalizations which appear to be false on the basis of observable expressions may be true at LF where structural properties have been modified. We shall be concerned with one such case in this paper.

## 1. Linguistic Invariants

Consider the minimally complex expressions in (1):

(1)    a.  Casper coughed

       b.  Carson sneezed

Different linguistic theories - GB/Minimalism [4], HPSG [6], LFG [2], Relational Grammar and Arc-Pair Grammar [1] - differ with regard to the structure they attribute to (1a), and of course the notation they use to express that structure. But each of these theories would assign the same structure to (1a) and (1b). And it is this latter type of judgment - Under what conditions do X and Y have the same structure? - that forms the basis of the Bare Grammar (BG) approach.

Consider how we might argue pretheoretically that (1a,b) have the same structure. We agree that replacing 'Caspar' by 'Carson' in (1a) yielding *Carson coughed* does not change structure. And then replacing 'coughed' by 'sneezed' deriving thus (1b) does not change structure. So the intuition is that expressions X and Y have the same structure if each can be derived from the other by a succession of structure preserving transformations.

Here is a more explicit statement, leading up to our definition of *invariant*. We think of a grammar as a way of defining (and semantically interpreting) a class of expressions. Specifically the syntax of a grammar G is primarily a pair (Lex$_G$, Rule$_G$), where, omitting subscripts, Lex is a (normally) finite set of expressions, called *lexical items*, and Rule is a set of functions, called *generating* or *structure building functions*. L$_G$, the *language generated by* G, is the set of all expressions you can build starting with those in Lex and applying the structure building functions finitely many times.

Lexical items on our view do present some internal structure. Like the expressions in L$_G$ in general, they are partitioned into classes by grammatical categories. So we represent an expression, and in particular a lexical item, as an ordered pair (s, C) where s is a string over the vocabulary V$_G$ of G and C is an element of the set Cat$_G$ of category symbols of G. For any expression e = (s, C), Cat(e) =$_{df}$ C, its second coordinate. Slightly more formally:

**Definition 1.** *A* bare grammar *G is a four-tuple,* $\langle V_G, Cat_G, \text{Lex}_G, \text{Rule}_G \rangle$, *where* Lex $\subseteq$ V $\times$ Cat, *and* Rule *is a set of partial functions from* (V$^*$ $\times$ Cat)$^+$ *into* V$^*$ $\times$ Cat. V$^*$ $\times$ Cat *is the set of* possible expressions *over* G, *and the* language generated by G, L$_G$, *is the closure of* Lex *under* Rule.

For any set K we can find a grammar G as above such that K is the set of strings of expressions in L$_G$. So any universal properties of natural language will have to be given explicitly as axioms (or consequences of other axioms), they do not follow from the mere formalism we use to express the grammar.

**Definition 2.** *An* automorphism *of a grammar* G *is a bijection* h : $L_G \to L_G$ *which fixes each* F *in* Rule, *that is,* h(F) = F. *This just means that* F *maps a tuple* $\langle s_1, \ldots, s_n \rangle$ *to* $s_{n+1}$ *iff* F *maps* $\langle h(s_1), \ldots, h(s_n) \rangle$ *to* $h(s_{n+1})$.

**Fact 1.** $id_{L_G}$, the identity map on $L_G$, is in $Aut_G$, the set of automorphisms of G; so is $h^{-1}$ whenever h is, and so is $g \circ h$ whenever g and h are. So $Aut_G$ is a group, as expected.

**Definition 3.** *For all* s, t $\in L_G$, s *is isomorphic to* t, *noted* s $\simeq$ t, *iff* h(s) = t *for some* h $\in Aut_G$. *We write* [s] *for* {t $\in L_G|$ s $\simeq$ t}.

We may, when useful, treat [s] as the "structure" of s. In practice we have not found this very useful; $\simeq$, however, is a very useful relation.

**Fact 2.** For each G, $\simeq$ is an equivalence relation partitioning $L_G$ into blocks {[s]| s $\in L_G$}.

Now, leading up to our definition of invariant, observe that whenever g is a function from a set A to a set B we can canonically lift g to a map $P_g$ from $\wp(A)$, the power set of A, into $\wp(B)$ by setting $P_g(K) = \{g(x)|$ x $\in K\}$. We usually just write $g(K)$ instead of $P_g(K)$. Similarly we can extend g to a map $g^*$ from $A^*$, the set of finite sequences of elements of A, into $B^*$ by setting $g^*(a_1, \ldots, a_n) = (g(a_1), \ldots, g(a_n))$. Again we usually write g for $g^*$ here.

**Definition 4.** *The* invariants *of a grammar G are the expressions, properties (sets) of expressions, relations between expressions,. . . that are fixed, mapped to themselves, by all the automorphisms of G.*

So given a grammar, its (logical) invariants are those linguistic objects (expressions, properties of expressions, relations between expressions, functions from expressions to expressions,. . . ) which cannot be changed without changing structure.

Later we introduce the notion of a *stable automorphism* and define the linguistic invariants of a grammar G to be those linguistic objects fixed by all stable automorphisms. But first let us learn to use the more general notion (and in any event in our initial examples of grammars the automorphisms and the stable automorphisms coincide).

## 2. Eng, an illustrative grammar for a fragment of English

We present a very simple grammar Eng in order to illustrate in a concrete way the notions of grammar and invariant defined above. It has some proper nouns, like *John* and *Bill*, some one place predicate symbols (P1s), like *laughed* and *cried*, some two place predicate symbols (P2s), like *praised* and *criticized*. We also have some conjunctions, *and* and *or* which form boolean compounds of expressions in a fairly obvious way. Finally, Eng has a reflexive pronoun *himself* that combines with P2s to form P1s, but does not combine with P1s to form anything. Eng has just two rules: Merge, which combines nominal elements with Pn+1s to form Pn's (we use P0 where many use 'S' for 'sentence'), and Coord which forms boolean compounds with *and* and *or*. Formally, Eng=$\langle$V, Cat, Lex, Rule$\rangle$, where these are given as follows:

| | |
|---|---|
| **V:** | laugh, cry, sneeze, praise, criticize, see, |
| | John, Bill, Sam, himself, and, or, both, either |
| **Cat:** | P0, P1, P2, P01/P12, P1/P2, CONJ |
| **Lex:** | P1      laughed, cried, sneezed |
| | P2      praised, criticized, interviewed |
| | P01/P12    John, Bill, Sam |
| | P1/P2     himself |
| | CONJ     and, or |
| **Rule:** | Merge and Coord, defined below. |

| Domain | Merge | Value | Conditions |
|---|---|---|---|
| s   t<br>A   B | $\longmapsto$ | s⌢t<br>P0 | $A = P01/P12, B = P1$ |
| s   t<br>A   B | $\longmapsto$ | t⌢s<br>P1 | $A \in \{P1/P2, P01/P12\}, B = P2$ |

So the domain of Merge is the set of pairs $\langle(s, A), (t, B)\rangle$, for any s, t in $V^*$ and any A,B in Cat meeting the specified conditions. We summarize the argument that (John laughed, P0) is in $L_{Eng}$ using a Function-Argument (FA) tree in which mother nodes are labeled with the values of generating functions applied to the labels on the daughter nodes:

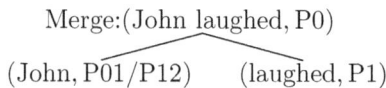

$$\text{Merge:(John laughed, P0)}$$
$$\text{(John, P01/P12)} \quad \text{(laughed, P1)}$$

Linguists more often represent this derivation with slightly less explicit "standard" trees like the following:

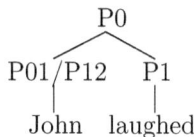

$$P0$$
$$P01/P12 \quad P1$$
$$\text{John} \quad \text{laughed}$$

Letting the set of coordinable categories $cC_{Eng} = Cat - \{CONJ\}$ and the class of nominal categories $nC_{Eng} = \{P1/P2, P01/P12\}$, we define the other generating function Coord as follows:

| Domain | | | Coord | Value | Conditions |
|---|---|---|---|---|---|
| and | s | t | $\longmapsto$ | both⌢s ⌢and⌢t | $C \in cC_{Eng}$ |
| CONJ | C | C | | C | |
| or | s | t | $\longmapsto$ | either⌢s ⌢or⌢t | $C \in cC_{Eng}$ |
| CONJ | C | C | | C | |
| and | s | t | $\longmapsto$ | both⌢s ⌢and⌢t | $C \neq C' \in nC_{Eng}$ |
| CONJ | C | C' | | P1/P2 | |
| or | s | t | $\longmapsto$ | either⌢s ⌢or⌢t | $C \neq C' \in nC_{Eng}$ |
| CONJ | C | C' | | P1/P2 | |

This rule is used in the derivation of (John criticized both himself and Bill, P0), as we see in the following FA derivation tree:

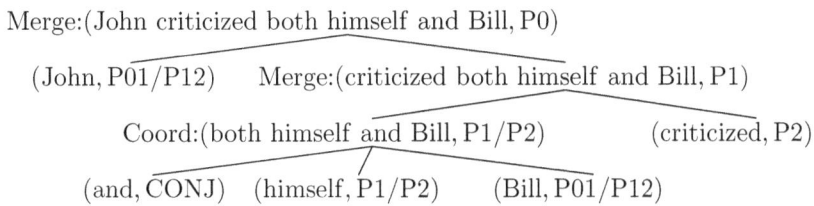

Merge:(John criticized both himself and Bill, P0)

    (John, P01/P12)    Merge:(criticized both himself and Bill, P1)

        Coord:(both himself and Bill, P1/P2)        (criticized, P2)

      (and, CONJ)  (himself, P1/P2)    (Bill, P01/P12)

## 3. Some invariants of Eng

E1. At the lowest level, the only expression that is invariant is (himself, P1/P2). The reason is that it has a unique distribution. It is the only lexical item that combines with P2s to form P1s but does not combine with P1s to form P0s.

E2. At the level of properties, we find several interesting invariants. First, the property of being a lexical item is invariant. That is, for all automorphisms h of Eng, $h(Lex_{Eng}) = Lex_{Eng}$. Indeed one might think that the property of being a lexical item was invariant in all G, but this is not the case.

E3. For each category C of Eng, the property of being an expression of category C is invariant. That is, for all $h \in Aut_{Eng}$, $h(PH(C)) = PH(C)$, where $PH(C) =_{df} \{s \in L_G | s = (t, C)$ for some string $t\}$. This also is not a universal invariant, as we see explicitly later.

E4. A more interesting invariant property in $L_{Eng}$ is: the property of being an anaphor. Informally anaphors are expressions like *himself, both himself and Bill*, etc. which are obligatorily interpreted as referentially dependent in a certain way. (Below we provide a properly semantic, language independent, definition of 'anaphor'.) We can show that the (infinite) set of expressions in $L_{Eng}$ which have this property is fixed by all the automorphisms of Eng.

E5. At the level of relations and functions, the binary relation *is a constituent of* ($CON_{Eng}$) is invariant, but this is universally invariant in the sense that for all G, $CON_G$ is invariant (as explained in the next section). Also invariant, but not universally so, is the three place relation *s is a possible antecedent of an anaphor t in u*. To illustrate the intuition behind this relation consider that in the expressions below *himself* may be understood as referentially dependent as the underlined nominals in the expression, and if there is none it is ungrammatical (indicated by the asterisk):

(2)  a. John thought that <u>the duke</u> defended himself well
     b. *John thought that Mary defended himself well
     c. <u>John</u> protected <u>Bill</u> from himself

E6. And lastly, as an example of an invariant (partial) function on $L_{Eng}$ consider $SUBJ_{Eng}$, which maps a P0 to its subject it if has one: for any $s \in L_{Eng}$,

$$Domain(SUBJ_{Eng}) = Range(Merge) \cap PH(P0)$$
$$SUBJ_{Eng}(s) = t \text{ iff for some u of category P1}, s = Merge(t, u).$$

So $SUBJ_{Eng}$(both John and Bill praised Sam, P0) = (both John and Bill, P01/P12). But (Either John laughed or Bill cried, P0) is not mapped to anything by this function, since it is not in the range of Merge.

## 4. Universal invariants

We referred above to invariants as universal if they are invariant in all G, no matter how implausible G might be considered as a grammar for a natural language. So these are invariants that follow from our definition of a grammar plus that of invariant. But linguistically our interest lies primarily in properties, relations, etc. which are empirically invariant – they hold for all motivated grammars of natural language but admit of formal counterexamples. We shall argue that *is an anaphor* and *is a possible antecedent of* are two such cases. But first, let us list some universal invariants, since they place boundary conditions on empirical invariants and they are very useful in showing that one or another property of a particular grammar G is invariant. In our statements we use 'structural' and 'structurally definable' as synonyms of 'invariant'. We have the following, for all grammars G:

U1. $L_G$ is invariant. That is, the property of being grammatical in G is structural.

U2. For any $F \in Rule_G$, F is invariant (trivially), as is its domain and range. So the property of being derived by any given $F \in Rule_G$ is structural.

U3. If $Lex_G$ is invariant then for all n, $Lex_n$ is invariant, where we define the complexity hierarchy $Lex_n$ by: $Lex_0 = Lex_G$ and for all n, $Lex_{n+1} = Lex_n \cup \{F(t)| F \in Rule_G,\ t \in Lex_n^* \cap Domain(F)\}$.
    Note that $L_G = \bigcup_n Lex_n$ and if for all $F \in Rule_G, Range(F) \cap Lex_G = \emptyset$ then $Lex_G$ is invariant.

U4. If G is category functional and each Lex(C) is invariant then each PH(C) is invariant, where $Lex(C) =_{df} PH(C) \cap Lex$ and G is *category functional* iff for all $F \in Rule$ and all n-tuples $u, v \in Domain(F)$, if $Cat(u_i) = Cat(v_i)$ all $1 \le i \le n$ then $Cat(F(u)) = Cat(F(v))$.

U5. The set of invariant subsets of $L_G$ is closed under relative complement and arbitrary intersections and unions, and thus forms a complete atomic boolean algebra (with atoms [s]). So conjunctions, disjunctions, and negations of invariant properties are themselves invariant properties. Comparable claims hold for $R \subseteq (L_G)^n$, for all n. Equally, cross products of invariant sets are invariant.

So if the property of being a feminine noun is invariant, and the property of being a plural noun is invariant then the property of being a feminine plural noun is invariant, as is that of being a feminine non-plural noun, etc.

U6. The is a constituent of relation, CON, is invariant, as are PCON (is a proper constituent of) and ICON (is an immediate constituent of), where for all $s, t \in L_G$, we define:

  a. $sICONt$ iff for some $u_1, \ldots, u_n \in L_G$ and some $F \in Rule_G$, $t = F(u_1, \ldots, u_n)$ and $s = u_i$, some $1 \leq i \leq n$.

  b. $sPCONt$ iff for some $n \geq 2$ there is a sequence $v = \langle v_1, \ldots, v_n \rangle$ of elements of $L_G$ with $v_1 = s$, $v_n = t$ and for each $1 \leq i < n$, $v_i ICON v_{i+1}$.

  c. $sCONt$ iff $s = t$ or $sPCONt$

U7. The *sister of* relation is invariant, where, s sister of t in u iff some $F(v_1, \ldots, v_n)$ is a constituent of u and for some $i \neq j$, $s = v_i$ and $t = v_j$.

U8. CC, *c-commands*, is invariant, where, sCCt in u iff for some constituent v of u, s is a sister of v in u and t is a constituent of v.

U6-U8 define linguistic notions on expressions, not, as is more usual, on derivations or tree-like structures representing derivations. We give the definitions more generally than usual because there are a variety of linguistic phenomena that are not naturally representable with standard trees and in which constituency is not recoverable by merely segmenting the derived string. Examples are reduplication, second position placement of Latin -*que* 'and', and the Dutch crossing verb dependencies (see Keenan and Stabler 2003, Chapter 3).

## 5. Empirical invariants: Anaphor-Antecedent relations

For illustrative purposes we limit ourselves to the simplest environment in which non-trivial anaphora obtains: that between the two arguments of a binary relation denoting expression (e.g. a transitive verb). Consider the data pattern in English below, where the intended antecedent of the anaphor *himself* is underlined, and constituency is indicated by brackets for later reference:

(3)  a. [Every student [criticized himself]]

  b. *[Himself [criticized every student]]

A first attempt to describe these data might use left-right order: "X is a possible antecedent of an anaphor Y iff X and Y are co-arguments and X precedes Y". This claim works surprisingly well for quite a range of fairly simple sentences in English. But it is cross linguistically not valid. Languages such as Malagasy (Austronesian; Madagascar) and Tzotzil (Mayan; Mexico) which use Verb+Patient+Agent as a pragmatically neutral order in simple sentences, (4a,b), naturally present anaphors before their antecedents (5a,b)

(4)  a. Namono ny akoho  Rabe              Malagasy
     Killed  the chicken Rabe

  'Rabe killed the chicken'

    b. **ʔ**i-s-poxta Xun li   j**ʔ** ilol-e        Tzotzil (Aissen 1987:90)
       Asp-3-care Xun the shaman-clitic

      'The shaman treated Xun'

(5)  a.  Namono tena Rabe                 Malagasy
       Killed    self  Rabe

       Rabe killed himself


    b. **ʔ**i-s-poxta s-ba  li   Xun-e         Tzotzil
       Asp-3-care 3-self art Xun-clitic

      'Xun treated himself'

A more comprehensive proposal, accepted by many linguists as valid for natural languages in general, would replace "X precedes Y" with "X c-commands Y". This characterization of the AA (Anaphor-Antecedent) relation is consistent with the Tzotzil and Malagasy data above. But again it seems insufficiently general to account for a quite widespread language type: the verb is peripheral (usually final) and the arguments of the verb carry morphological markings, *case markers*, which identify the arguments. In the verb final case, illustrated below by Korean, the relative order of arguments is often rather free. We give the examples directly with the anaphors, but non-anaphoric nominals may replace them without change.

(6)  [Caki-casin-ul [motun haksayng+tul-i piphanhayssta]]   Korean
    Self-emph-acc all      student+pl-nom criticized

    'All the students criticized themselves'

(7)  [[Sinampal ng babae] ang sarili niya]                 Tagalog
    slap+GF   gen woman top self   3poss

    'The woman slapped herself'

There is reasonable evidence in these cases that the antecedent of the anaphor does not c-command it; indeed the anaphor seems to asymmetrically c-command its antecedent. But the important structural regularity here concerns the case markers. They cannot be interchanged preserving grammaticality:

(8)  *[Caki-casin-i [motun haksayng+tul-ul piphanhayssta]]   Korean
    Self-emph-nom all      student+pl-acc   criticized

    'All the students criticized themselves'

(9)  *[[Sinampal ang babae] ng   sarili niya]                Tagalog
    slap+GF    top woman gen self   3poss

    'The woman slapped herself'

The c-command relations have not changed, but the case marking has, resulting in ungrammaticality. So case marking plays a structurally important role in these languages, and in our models is provably invariant.

    The appropriate generalization for Korean then is: in simple sentences, X is a possible antecedent for an anaphor Y iff X and Y are co-arguments and X is *-i* marked and Y is *-ul* marked.[2] In Tagalog X is *ng* marked and Y is *ang* marked.

Based on the Korean data we exhibit a mini-grammar for a verb final case marking language in which case relations determine the distribution of anaphors. We
provide a compositional semantic interpretation, including a semantic, language
independent, definition of anaphor, thereby establishing that the expressions we
call anaphors are indeed interpreted as anaphors. But first let us give the language
independent definition of anaphor (for the restricted class of contexts considered).

## 6. A semantic definition of 'anaphor'

For each domain E we interpret P2s as binary relations over E, represented as
functions from E into $[E \rightarrow \{0,1\}]$. Anaphors and ordinary NPs, such *John, most
of John's friends*, etc. map P2 denotations into $[E \rightarrow \{0,1\}]$. The difference in the
two cases concerns what the values of the functions depend on. Compare:

(10)   a.  Sam criticized most of John's students

      b.  Sam criticized himself

In (10a) whether the denotation of *criticized most of John's students* holds of Sam is
decided just by checking the set of objects that Sam criticized. If that set includes
a majority of John's students the whole S is true. We don't need to know who Sam
is. If Bill praised exactly the people that Sam criticized then (10a) and *Bill praised
most of John's students* must have the same truth value. In contrast it might be
that the individuals Sam criticized are just those that Bill praised but (10b) and
*Bill praised himself* have different truth values. Formally,

**Definition 5.** *Given a domain* E, *a binary relation* R *over* E, *and* $x \in E$,

$$xR =_{df} \{y \in E|\ (R(y))(x) = 1\}.$$

*So in set notation,* $xR = \{y \in E|\ (x,y) \in R\}$.
    *Let* F *map binary relations to properties. Then* F *satisfies the Extensions Con-
dition* (EC) *iff for all* $a, b \in E$, *all binary relations* R, S *over* E,

if $aR = bS$ then $F(R)(a) = F(S)(b)$.

    *And* F *satisfies the Anaphor Condition* (AC) *iff for all* $a \in E$, *all binary relations*
R, S *over* E,

if $aR = aS$ then $F(R)(a) = F(S)(a)$.

    *Let* D *combine with P2s to form P1s. Then* D *is an* anaphor *iff all non-trivial* [3]
*interpretations of* D *satisfy the AC but fail the EC.* [4]

So for example, for E with at least two members, the function SELF from binary
relations to sets given by: $SELF(R)(x) = R(x)(x)$ is easily seen to fail the EC but
satisfy the AC.

## 7. Kor, a verb final case marking language

Consider the following language Kor, inspired by Korean:

| | |
|---|---|
| **V:** | laughed, cried, sneezed, praised, criticized, saw, -nom, -acc, John, Bill, Sam, himself, and, or, nor, both, either, neither |
| **Cat:** | NP, $NP_{refl}$, Ka, Kn, KPa, KPn, P0, P1a, P1n, P2, CONJ |
| **Lex:** | Kn     -nom |
| | Ka     -acc |
| | P1n    laughed, cried, sneezed |
| | P2     praised, criticized, interviewed |
| | NP     John, Bill, Sam |
| | $NP_{refl}$   himself |
| | CONJ   and, or, nor |
| **Rule:** | CM (case mark), PA (predicate-argument) and Coord, as follows. |

| Domain | | CM | Value | Conditions |
|---|---|---|---|---|
| -nom | t | $\longmapsto$ | t⌢-nom | $t \neq$ himself |
| Kn | NP | | KPn | |
| -acc | t | $\longmapsto$ | t⌢-acc | none |
| Ka | NP | | KPa | |

| Domain | | PA | Value |
|---|---|---|---|
| s | t | $\longmapsto$ | s⌢t |
| KPn | P1n | | S |
| s | t | $\longmapsto$ | s⌢t |
| KPa | P1a | | S |
| s | t | $\longmapsto$ | s⌢t |
| KPn | P2 | | P1a |
| s | t | $\longmapsto$ | s⌢t |
| KPa | P2 | | P1n |

Letting the coordinable, "boolean" categories be

$$cC_{Kor} =_{df} Cat - \{CONJ, Ka, Kn, KPa, KPn\}$$

and the nominal categories be

$$nC_{Kor} =_{df} \{NP, NP_{refl}\},$$

we define a coordination rule as follows:[5]

| Domain | | | Coord | Value | Conditions |
|---|---|---|---|---|---|
| and | s | t | $\longmapsto$ | both$^\frown$s $^\frown$and$^\frown$t | $C \in cC_{Kor}$ |
| CONJ | C | C | | C | |
| or | s | t | $\longmapsto$ | either$^\frown$s $^\frown$or$^\frown$t | $C \in cC_{Kor}$ |
| CONJ | C | C | | C | |
| nor | s | t | $\longmapsto$ | neither$^\frown$s $^\frown$nor$^\frown$t | $C \in cC_{Kor}$ |
| CONJ | C | C | | C | |
| and | s | t | $\longmapsto$ | both$^\frown$s $^\frown$and$^\frown$t | $C \neq C' \in nC_{Kor}$ |
| CONJ | C | C' | | $NP_{refl}$ | |
| or | s | t | $\longmapsto$ | either$^\frown$s $^\frown$or$^\frown$t | $C \neq C' \in nC_{Kor}$ |
| CONJ | C | C' | | $NP_{refl}$ | |
| nor | s | t | $\longmapsto$ | neither$^\frown$s $^\frown$or$^\frown$t | $C \neq C' \in nC_{Kor}$ |
| CONJ | C | C' | | $NP_{refl}$ | |

The following tree represents the argument that (himself-acc John-nom praised, P0)∈L(Kor).

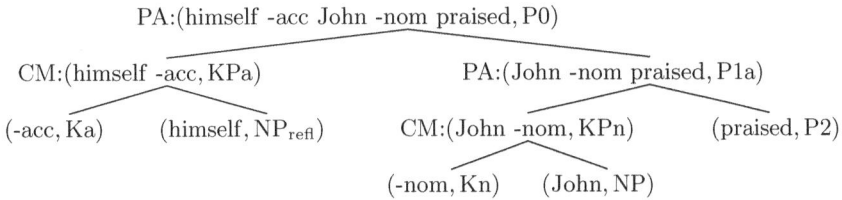

```
                PA:(himself -acc John -nom praised, P0)
         ┌────────────────────┴──────────────────────┐
CM:(himself -acc, KPa)                    PA:(John -nom praised, P1a)
    ┌────────┴────────┐                   ┌──────────┴──────────┐
(-acc, Ka)   (himself, NP_refl)   CM:(John -nom, KPn)      (praised, P2)
                                   ┌──────┴──────┐
                            (-nom, Kn)      (John, NP)
```

This is the only derivation of this expression, and so, in this expression, (himself,NP$_{refl}$) c-commands and is not c-commanded by (John-nom,KPn).

## 8. Some invariants of Kor

**K1.** The set Lex is invariant. So by U3, Lex$_n$ is invariant for each n.

**K2.** The expressions (-nom,Kn) and (-acc,Ka) are both invariants.

Pretheoretically case markers are grammatical formatives, so the fact that they are provably invariants in Kor supports that our formal notion of invariant identifies expressions independently judged to be grammatical in nature. So no automorphism can interchange (-nom,Kn) and (-acc,Ka).

**K3.** The expression (himself,NP$_{refl}$) is invariant, but (Bill,NP) is not.

**K4.** For all C ∈ Cat, the set PH(C) of expressions of that category is invariant.

**K5.** The *co-argument* relation is invariant, defined by: s co-argument t in u iff for some v of category P2, either PA(s,PA (t,v)) or PA(t,PA(s,v)) is a constituent of u.

## 9. Semantic interpretation for Kor

This section provides L(Kor) with a compositional semantics which shows that sentences with reflexives are interpreted correctly in all cases. Those willing to take our word for this can move directly to the next section. We assume a modest familiarity with a model theoretic semantics and boolean lattices.

**Definition 6.** *Given a non-empty universe* E, *we let* $R_0 =_{df} \{0, 1\}$, *regarded as the boolean lattice 2 where the* $\leq$ *relation coincides with the numerical one. In general* $R_{n+1}$ *is* $[E \to R_n]$, *regarded as a boolean lattice with* $\leq$ *understood pointwise:* $f \leq g$ *iff for* $x \in E$, $f(x) \leq g(x)$.

*Type 1 is the set of functions from n+1-ary relations to n-ary ones, for all n:*

$$\{f \in [\bigcup R_{n+1} \to \bigcup R_n] | \text{ for all n, all } r \in R_{n+1}, \ f(r) \in R_n\}.$$

**Definition 7.** *A model for* L(Kor) *is a pair* $M = \langle E, m \rangle$, E *a non-empty domain and* m *a function mapping elements* $\langle v, C \rangle$ *of* Lex *into* $Den_E(C)$, *the set of possible denotations of expressions of category* C *in* M, *defined as follows. Note in particular the definition of NOM(f); its value at properties determines its value at relations.*

$Den_E(NP_{refl})$ = $\{f \in$ Type 1$|$ if nontrivial, f satisfies AC *and fails EC*$\}$

$Den_E(P0)$ = $R_0$

$Den_E(P1n)$ = $R_1$

$Den_E(P2)$ = $R_2$

$Den_E(NP)$ = Type1

$Den_E(KPa)$ = Type1

$Den_E(P1a)$ = $[Type1 \to R_1]$

$Den_E(CONJ)$ = $\{\wedge_C, \vee_C\}$, *where* $\wedge_C$ *is the greatest lower bound operator in* $Den_E(C)$ *and* $\vee_C$ *is the least upper bound operator*

$Den_E(KPn)$ = $\{NOM(f)| f \in$ Type1$\}$, *where for any* $f \in$ Type1, *NOM is the function with domain* $R_1 \cup R_2$ *such that for* $P \in R_1$, $NOM(f)(P) = f(P)$ *and for* $R \in R_2, h \in$ Type1, $NOM(f)(R)(h) = f(h(R))$

1. m *at elements of* Lex *satisfies the following conditions:*

   a. *for all* $s \in$ Lex(NP), $m(s) \in \{I_b| \ b \in E\}$, *where for all* $R \in R_{n+1}$, $I_b(R) = R(b)$

   b. $m(-acc, Ka)$, *noted* ACC, *is the identity map on Type 1.*

   c. $m(-nom, Kn) = NOM$, *defined above*

   d. $m(himself, NP_{refl}) = SELF$, *that map from R2 to R1 defined earlier*

   e. *for all* $x, y \in Den_E(C)$, C *boolean,*

   $$m(and, CONJ) = \wedge_C \qquad and \qquad m(or, CONJ) = \vee_C$$

2. m *extends to a function* $m^*$ *on* L(Kor), *called an interpretation of* L(Kor) *relative to* M, *by:*

   a. $m^*(CM(s, t)) = m(s)(m^*(t))$

   b. $m^*(PA(s, t)) = \begin{cases} m^*(s)(m^*(t)) & \text{unless} \\ m^*(t)(m^*(s)) & Cat(t) = P1a \text{ and} \\ & Cat(s) = KPa \end{cases}$

c.  $m^*(\mathrm{Coord}(s,t,u)) = m(s)(m^*(t), m^*(u))$

Using these definitions one computes that (11a,b) are logically equivalent (always interpreted the same): for all models $\mathcal{M} = (E, m)$, $m^*(11a) = m^*(11b)$.

(11)  a.  (John-nom Bill-acc praised, P0)

   b.  (Bill-acc John-nom praised, P0)

PA:(John -nom Bill -acc praised, P0)

CM:(John -nom, KPn)          PA:(Bill -acc praised, P1n)

(-nom, Kn)   (John, NP)   CM:(Bill -acc, KPa)      (praised, P2)

(-acc, Ka)      (Bill, NP)

PA:(Bill -acc John -nom praised, P0)

CM:(Bill -acc, KPa)          PA:(John -nom praised, P1a)

(-acc, Ka)      (Bill, NP)   CM:(John -nom, KPn)      (praised, P2)

(-nom, Kn)      (John, NP)

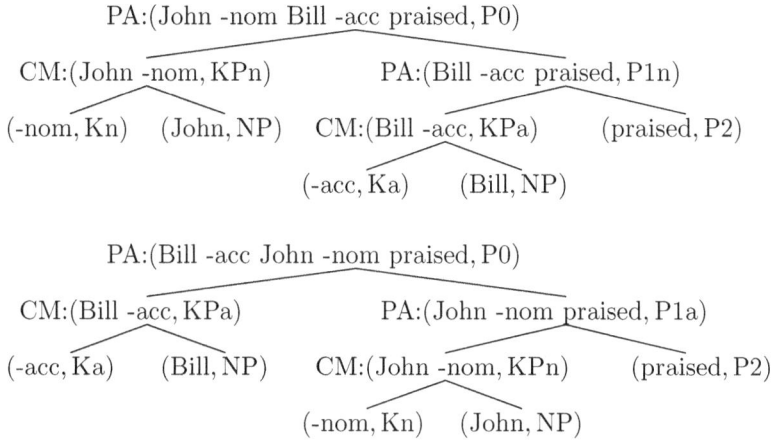

The logical equivalence of these sentences relies on the interpretation of (-nom,Kn). When the nominative KP looks at a P2, in effect, it knows to wait until the next KP denotation comes along. So the interpretation of bound morphology here is critical. Moreover the same reasoning shows that the result of replacing (Bill,NP) by (himself,NP$_{\mathrm{refl}}$) in (11a,b) are also logically equivalent:

$m^*$(John-nom himself-acc criticized, P0)
$= m^*$(himself-acc John-nom criticized, P0)

Thus the interpretation of *himself* as an anaphor does not depend on it being c-commanded by its antecedent. We note that these sentences, like (11a,b), have isomorphic derivation trees (standard or FA). But the expressions are not isomorphic in L(Kor) since automorphisms can't map KPn's to KPa's, P1n's to P1a's, etc.

## 10. Two further invariants of Kor

Now we are in a position to state invariants that involve semantic notions.

K6.  The property of being an anaphor is invariant, where the expressions interpreted as anaphors following Definition 5 are precisely those in PH(P1/P2).

K7.  The Anaphor-Antecedent relation is invariant in Kor, where we define:

s AA t in u iff t is an anaphor and s co-argument t in u

(AA is invariant because it is defined as a boolean compound of invariants).

## 11. Concluding remarks on Kor

It is unproblematic that anaphors asymmetrically c-command their antecedents. The interpretation of case markers guarantees the right semantic interpretation (sentence internally) independent of c-command. We also note that a compositional interpretation of L(Eng) is even easier than of L(Kor), and that *himself* in Eng denotes SELF, just as *himself* in Kor does. So our claims about anaphors are claims about expressions with the same denotation.

*Morphology is structural*, independent of c-command relations within the clause. The case markers, (-nom,Kn) and (-acc,Ka), are invariant even though the KPs they build do not have fixed structural positions. Specifically a KPa does not always combine with a P2 to form a P1; it also combines with P1s to form P0s.

Our formulation of Kor abstracts away from the conditioned variants of the case markers: *-i/-ka* for -nom and *-ul/-lul* for -acc. This seems reasonable when our concern is syntax and semantics, as these differences in form are phonologically conditioned.

Still, an interesting option arises when we do distinguish two categories of NP in Lex, say NPc and NPv (according as the string coordinate ends in a consonant or a vowel). So Lex would contain (John,NPc) and (Joe,NPv) of different categories, but ones that had the same distribution except for the choice of case marker: *-i*, *-ul* in the first case, *-ka, -lul* in the second. And we would then find that if the cardinalities of the lexical NPv's and NPc's were the same (permitting a bijection between them) we could design an automorphism that would map all NPv's to NPc's and conversely. It would also interchange (-i,Kn) with (-ka,Kn) and (-ul,Ka) and (-lul,Ka). The resulting grammar would be one in which not all PH(C) were invariant.

## 12. Categorial symmetry and stable automorphisms

The case of conditioned variants noted above for Korean has much more extensive and systematic manifestations in other grammatical subsystems. In BG for example we present a grammar, Span (Spanish), illustrating basic adjective and determiner agreement with masculine (m) and feminine (f) nouns. The Lexicon arbitrarily distinguishes Nm's and Nf's, and when adjectives and determiners combine with them they get marked with an *-o* or an *-a*, of category Agr(m) and Agr(f) respectively. The m/f distinction is inherited by NPs built from the Nm's and Nf's, and then the P1s show predicate agreement with them.

And analogous to the Korean case, if we design the grammar so that the number of lexical Nm's and Nf's is the same then we can find an automorphism of Span which interchanges PH(Nm) and PH(Nf), as well as the derived masculine and feminine adjectives, NPs and P1s. So again not all PH(C) are invariant in Span. However the automorphisms that can effect this category swapping are unstable in that slight additions to the Lexicon rule out their existence. Thus if we add just one new feminine noun, say (poet,Nf) making no other changes then no automorphism changes category and all PH(C) are invariant since then the lexical Nm's and the lexical Nf's would have different cardinalities, so there could be no bijection between them.

The possibility of category changing automorphisms above reveals a categorial symmetry present, in principle, in natural language. Noun classes partition a subset of the expressions in such a way that the blocks of the partition can be structurally interchanged. This possibility is "unstable" in the sense that many "minor" changes in the language, ones we agree are insignificant, such as adding new lexical items, result in languages in which these blocks cannot be interchanged.

Ignoring this accidental possibility would be, we feel, a mistake. A grammar with unequal numbers of lexical Nm's and Nf's could always be extended by adding new lexical items to one in which the numbers evened out again, permitting category changing automorphisms. And the ability to add new content words freely is a basic property of a NL. More generally various types of allomorphy present a similar phenomenon. In English we might distinguish classes of Nouns according to how their plural is formed: with /z/ as in *dog/dogs*, with /s/ as in *cat/cats*, with /əz/ in *judge/judges*, /f/→/vz/ as in *leaf/leaves*, -*on*→-*a*, as in *phenomenon/phenomena*, no change as in *sheep→sheep*, etc.

We will treat agreement and allomorphy by distinguishing among automorphisms according as they remain stable under such changes. Informally, an automorphism is stable if it remains an automorphism after the addition of new expressions isomorphic to old ones. "New" means not inducing new derivations of expressions in the original language (thanks to Greg Kobele for this formulation, and thanks to Philippe Schlenker for forcing us to treat allomorphy):

**Definition 8.** *For* $G = \langle V, \mathrm{Cat}, \mathrm{Lex}, \mathrm{Rule}\rangle$ *and* $S \subseteq_{\mathrm{finite}} V \times \mathrm{Cat}$,

a. $G[S] =_{\mathrm{df}} \langle V, \mathrm{Cat}, \mathrm{Lex} \cup S, \mathrm{Rule}\rangle$. *Write* $G[s]$ *or* $G_s$ *for* $G[\{s\}]$, $s \in V \times \mathrm{Cat}$. *So* $G_s$ *results from adding* $s$ *to* $\mathrm{Lex}_G$ *with no changes in* $\mathrm{Cat}$ *or* $\mathrm{Rule}$.

b. $G$ *is free for* $s$ *in* $V \times \mathrm{Cat}$ *iff*

   i. *for all* $t \in L(G_s)$, *if* $t \in L_G$ *then* $\neg(s\mathrm{CONt})$, *and*

   ii. *For some* $h \in \mathrm{Aut}_{G_s}$ *and some* $t \in \mathrm{Lex}_G$, $h$ *interchanges* $s$ *and* $t$ *and fixes all other elements of* $\mathrm{Lex}_{G_s}$.

   iii. $G$ *is free for* $S$ *iff for all* $s \in S$, $G$ *is free for* $s$ *and* $G_s$ *is free for* $S - \{s\}$. *(Note that all* $G$ *are free for* $\emptyset$.)

So (b.i) blocks adding as new lexical items expressions that are already in $L_G$.

**Definition 9.** $h \in \mathrm{Aut}_G$ *is stable iff* $h$ *extends to an* $h' \in \mathrm{Aut}_{G[S]}$, *all finite* $S$ *for which* $G$ *is free*.

*An expression, a property of expressions,... over* $G$ *is a linguistic invariant iff it is fixed by all stable automorphisms.*

Of course all logical invariants of a grammar are linguistic invariants since an object fixed by all automorphisms is a fortiori fixed by all stable automorphisms. But the converse may fail. In Kor enriched with the phonologically conditioned case markers PH(NPv) is a linguistic invariant but not a logical one. Equally each case marker (-i, Kn), (-lul,Ka), etc. is a linguistic invariant (but not a logical one). And in Span PH(Nm) is a linguistic invariant but not an logical invariant, as is each agreement marker (-o,Agr(m)), (-a,Agr(f)).

## 13. Conclusion

We have provided a way of establishing invariants of natural languages while countenancing that different languages may have quite different grammars. Our specific claims, that *is an anaphor* or *is a possible antecedent of* are invariant in all natural languages, are empirical, not mathematical, and further empirical research could show them false.

In addition our approach has led us to formulate several conceptually new generalizations about natural language. Here are two, of somewhat different sorts:

**Stable Categories**  In adequate natural language grammars G, each PH(C) is a linguistic invariant.

**Thesis**  Grammatical Formatives are linguistically invariant lexical items.

The Thesis above offers a characterization of those expressions linguists variously call "function words" or "grammatical formatives". To our knowledge this is the first non-stipulative characterization of these objects. In contrast, Stable Categories is offered as an axiom of a theory of language structure. It provides a principled account of how the expressions of a language may be partitioned into grammatical categories. They are sets of expressions fixed by all stable automorphisms.

## Notes

[1]  Advocates of this approach intend more than the claim that we use the same notation for grammatical categories in different languages but it is quite unclear what this "more" is.

[2]  In more detail, an expression is -nom marked iff it is suffixed with *-i* if it is consonant final and with *-ka* if it is vowel final. It is -acc marked iff it is suffixed with *-ul* if consonant final and *-lul* if vowel final. In addition either argument (but not both) can have their -nom/-acc suffixes replaced with a topic marker *-un/-nun* preserving the pattern of antecedence. Then a more accurate statement of the AA relation would be: "...X is -nom marked and Y is -acc marked or topic marked, or X is -nom marked or topic marked and Y is -acc marked". The important point remains: the relevant factor governing the distribution of anaphor and antecedent in simple sentences concerns their morphological marking, not their left-right order or c-command relations.

[3]  It is assumed here that the universe E of interpretation always has at least two elements. The non-triviality condition is intended for cases like at least two of the ten students besides himself, which requires for non-triviality that the E contain exactly ten students.

[4]  The definition of EC and AC and hence of anaphor generalizes directly to maps from n+1-ary relations to n-ary ones just by interpreting a and b as n-tuples rather than "1-tuples".

[5]  In head initial languages (Verb initial, or SVO as in English) framing coordinations follow the English pattern *(both X and Y, either X or Y, neither X nor Y)*, though the more typical case is where the conjunctive morphemes

are the same, as in French: *et Jean et Marie, ou Jean ou Marie, ni Jean ni Marie.* A case can be made that in verb final languages the order is *X and Y and, X or Y or,* etc. though in our examples from Korean we did not find such framing expressions, only infix coordinators. We include the framing construction to avoid semantic ambiguities with iterated coordinations. We are not really studying either coordination or ambiguity here, but we include coordination so that many categories of expression will have infinitely many members, forcing us to avoid non-general definitions by listing cases.

## REFERENCES

1. J. Aissen. *Tzotzil Clause Structure.* Reidel, Dordrecth, 1987.
2. J. Bresnan. *Lexical-Functional Syntax.* Blackwell, Oxford, 2001.
3. N. Chomsky. *Syntactic Structures.* Mouton, The Hague, 1957.
4. N. Hornstein. *Logical Form: From GB to Minimalism.* Basil Blackwell, Oxford, 1995.
5. E. L. Keenan and E. P. Stabler. *Bare Grammar.* CSLI Publications, Stanford, California, 2003.
6. C. Pollard and I. Sag. *Head-driven Phrase Structure Grammar.* The University of Chicago Press, Chicago, 1994.

# The Measurement of Quality of Life in Medicine

Alain Leplège

*Department of Philosophy and Social Sciences, University of Picardie, Campus, Chemin du Thil, 80025 Amiens CEDEX, France and Institute of History and Philosophy of Sciences and Techniques (UMR 8590 CNRS/U.Paris1), 13 rue du Four, 75006 Paris, France*

*alain.leplege@wanadoo.fr*

**Abstract.** For several years, medicine has experienced a quasi exponential surge of quality of life measurements. This field offers us an example of a technical innovation that contributes to the production of new scientific knowledge whose progress can be observed almost on a day to day basis and can be participated in situ. The development and implementation of standardized questionnaires that allow theses measurements and the integration of their results in decisions affecting patients raise numerous theoretical, methodological and practical problems, which makes them a fruitful object for epistemological studies. This paper reviews some of them : First of all, conceptual questions: What is the object being measured? How to conceive it? How to define it? What are its distinctive characteristics? Then, questions related to the epistemology of measurement. For example, the study of some probabilistic measurement models of recent use in the development of standardized questionnaires (i.e. the family of models identified by Georg Rasch) lead to questions regarding : the way the requirement of invariance is operationalized the relationship with the classical theory of tests and representational theory, the methodological consequences (relationship model- empirical data) and the impact of these models on the demarcation problem in the Social Sciences. Finally, good practices of experimental science in the health field necessarily entail an ethical aspect whose analysis cannot be dissociated from the conceptual and method- ological interrogations. Quality of life and well-being measurements are developed explicitly in a decisional prospect. They thus entail a normative dimension. Classic problems of individual and social ethic are then raised, which primarily fall within an ethic of Good. These questionings meet contemporary reflections on rationality, as well as applied ethic, as for instance norms in scientific research or the analysis of conflicts of interest likely to influence these measurements.

## 1. Introduction

For several years, medicine has experienced a quasi exponential surge of quality of life measurements. It was originated in the medical field, in 1970 with a paper authored by Fanshell and Bush (1970) published in a journal called Operation Research. Since then, the number of studies and articles dedicated to this topic increases regularly (Number of new reference per year in Medline with 'quality of life' as key word: 1970: 4 references; 1980: 284 references; 1990: 1399 references; 2000: 4495 references) and several scientific journals are entirely devoted to this topic, for example, the International and Interdisciplinary Journal for Quality of Life Measurement, and in the medical area, Quality of Life Research.

The general objective is pragmatic. It is to clarify the terms of a rational debate, based on empirical observations and measurements. In this regard, the development of quality of life measurements takes part in a wide ranging research program called 'evidence based medicine' whose objective is to increase the proportion of medical decisions based on verified or scientific information [34]. Ultimately, the purpose of these measurements is to improve patients' quality of life and to contribute to improve general population's well-being and satisfaction regarding health care by increasing relevance in medical interventions. In order to meet these objectives, ever since the seventies, psychologists, economists and researchers in the public health field have developed instruments and methods that allow to measure health interventions effects in terms of health status, quality of life or health preferences [36].

Quality of life measurements are based on the evaluation by the subjects themselves of the consequences of their health condition through standardized questionnaires, developed according to psychometric methods. These subjects are asked whether they are satisfied with their health condition or with their lives and to what extend their life has been modified by their illness or by the medical interventions they have undergone.

Indicators of "quality of life" have ranged from the purely physiological and the ability to return to work to complex series of questionnaires on social activities and psychological problems. The first measurements used in the health field were devised to quantify the health status of subjects ; accordingly, a number of valid and reliable tools developed in both USA and Britain were designed to measure, for example, perceived distress, the impact of illness [5], physical functional capacity, and life satisfaction [17].

There is today a large number of instruments of varying types for measuring health status or health-related quality-of-life [25,8]. What these instruments have in common is a certain way of viewing results of medical acts, which brings out at least some of the aspects of what is generally understood by "health" or "absence of illness" or quality of life. For example the World Health Organisation Quality of Life group stated that: *Quality of life is defined as individual's perceptions of their position in life in the context of the culture and value systems in which they live and in relation to their goals, expectations, standards, and concerns* [35]. In most cases, what is involved is the elaboration of very precise questionnaires based on statements relating to health conditions [31]. Most often these statements are grouped in sets of discrete dimensions (e.g. see, table 1). The different statements (and/or dimensions) are often explicitly weighted on the basis of classification or subjective preferences collected from a certain number of 'judges'. The metric properties of these instruments are carefully studied [6]. These instruments differ from one another in the way they are constructed, in the nature of the questions asked to the subject, and in their overall purpose. There are many uses for quality of life measurements in medicine and public health [27]. They can be classified in two main categories:

a) A descriptive use with a primarily cognitive prospect: questionnaires can be incorporated into epidemiological studies to complement the knowledge of diseases' history with the observation of symptoms' subjective consequences. Measurements can also participate to the observation of individuals' or groups' health condition.

b) A normative use: evaluation of new therapies or technologies, clinical research, economical evaluation within cost-efficiency studies. Quality of life instruments can contribute to numerous medical decisions in order to increase their relevance. By objectifying a difference of level in Quality of Life condition, these measurements also allow to legitimate new requests for health care or assistance and to identify patients' needs in order to improve their quality of life. These instruments participate also to more complex processes of negotiation and information to multiple participants, whether individual or collective, public or private. In that way, they contribute to some kind of rationalization of the debate in the area of medical decision making.

Table 1
WHOQOL-BREF domains

| Domain | Facets incorporated within domains (1 item per facet) |
|---|---|
| 0. General quality of life (2 items) | |
| 1. Physical health (7 items) | Activities of daily living<br>Dependence on medicinal substances and medical aids<br>Energy and fatigue<br>Mobility<br>Pain and discomfort<br>Sleep and rest |
| 2. Psychological (6 items) | Bodily image and appearance<br>Negative feelings<br>Positive feelings<br>Self-esteem Spirituality / Religion / Personal beliefs<br>Thinking, learning, memory and concentration |
| 3. Social relationships (3 items) | Personal relationships<br>Social support<br>Sexual activity |
| 4. Environment (8 items) | Financial resources<br>Freedom, physical safety and security<br>Health and social care: accessibility and quality<br>Home environment<br>Opportunities for acquiring new information and skills<br>Participation in and opportunities for recreation / leisure activities |

From the point of view of public decision-making regarding the allocation of an overall health budget, one of the main questions that needs to be asked is whether these instruments enable comparisons from one medical care procedure to another and from one patient population to another. Certain instruments measuring health status regard specific diseases or conditions, and thus cannot be used to compare programs aimed at different pathologies. Other "generic" instruments are intended for comparisons of this nature. They can, for example, assess the relative importance to be given to a renal dialysis program and a hip replacement program.

For a philosophical perspective, the quality of life measurement field offers an example of a technical innovation that contributes to the production of new scientific knowledge whose progress can be observed almost on a day to day basis and can be participated to and studied in situ. However, despite quality of life questionnaires can be conceived as technical objects, I will not address, in this general presentation, specific Philosophy of Technique issues as meant for example by Simondon [32]. I will merely review some of the current conceptual, methodological and ethical problem that are associated with the development of quality of life measurement and their actual and future use in medical and public health decision making.

## 2. CURRENT EMPIRICAL AND EPISTEMOLOGICAL RESEARCH

As an object of research, quality of life measurements are as relevant to economists as they are to other investigators in the social sciences or to physicians.

Amongst other empirical research issues raised by quality of life measurement development, let's mention, on one side, difficulties in taking into account individual variability and time perspective in evaluations as well as issues related to establishing intercultural equivalence and translating instruments. Current empirical research primarily tends to develop specific instruments for particular health issues, to internationalize evaluations (how to test the invariance of the instruments being used) and to model individual behaviors. Finally, the objective of numerous research teams is to contribute to the production and publication of reference literature that allows to understand and interpret the signification of the measurements obtained. However, methodological and technical research has been of greater importance than theoretical and conceptual research in the studies that have been carried out so far.

The development and implementation of standardized questionnaires that allow theses measurements and the integration of their results in decisions affecting patients raise numerous theoretical, methodological and practical problems, which makes them a fruitful object for epistemological studies.

First of all, conceptual questions: What is the object being measured? How to conceive it? How to define it? What are its distinctive characteristics?

Then, questions related to the epistemology of measurement. For example, at the Institute of History and Philosophy of Sciences and Techniques in Paris, we currently study the measurement conceptions underlying certain probabilistic measurement models of recent use in the development of standardized questionnaires (i.e. the family of models identified by Georg Rasch. See [30]), the way the requirement of invariance is operationalized the relationship with the classical theory of tests and representational theory [20], the methodological consequences (relationship model- empirical data) and the impact of these models on the demarcation problem in the social sciences.

Finally, good practices of experimental science in the health field necessarily entail an ethical aspect whose analysis cannot be dissociated from the conceptual and methodological interrogations that are central to our reflection. Quality of life and well-being measurements are developed explicitly in a decisional prospect. They thus entail a normative dimension. Classic problems of individual and social ethic are then raised, which primarily fall within an ethic of Good. These questionings meet contemporary reflections on rationality, as well as applied ethic, as for instance norms in scientific research or the analysis of conflicts of interest likely to influence these measurements.

Philosophy was initially introduced into this reflection, through the analysis of ethical issues raised by the prospect of using quality of life measurement in medical and public health decisions [12,11]). Fewer works addressed the standard problems in normative ethic raised by interpersonal comparisons and the aggregation of individual preferences.

The epistemological reflection is relatively less advanced so far. Answers that are given (or not given) to each one of those points may have repercussions on the methods used to shape and validate measurement instruments and eventually to justify the uses that are (or will be) made of them.

## 3. THE NEED FOR CONCEPTUAL AND THEORETICAL CLARIFICATION

As any area of technical innovation, this new field of knowledge is highly influenced by the inter-disciplinarity of collaborations. These collaborations involve scientists from various fields including: sociology, psychology, economics, public health or health service research. The measurement operation that consists in assigning a number to a concept or to a quality of life area implies a theoretical stand point, and even if users and analysts do not explicitly refer to a specific theory (or a group of theories) when they select concepts and quality of life dimensions to be quantified, their choices are in line with the culture in which they live and were educated.

As a result, inter-disciplinarity goes along with some theoretical confusion: references are made to methodological individualism, to psychological functionalism, to the theory of needs or to the utility theory. Yet, the part theories play in the conception and the development of quality of life measurement instruments is often underestimated and certainly understudied. Unfortunately, it doesn't seem that conceptual clarification has been a priority for researchers who contribute to the emergence of this new field [14,22]. One noticeable exception is the book edited by Lennard Nordenfeld which addresses explicitely conceptual and theoretical issues [26].

For example, when the demand for 'quality of life' measures arose, there were, in fact, no such measures available since this had not been the focus of intellectual or research effort in the health field. The focus has been on health status measurement [6]. The term 'health-related quality of life' was coined apparently as a way of justifying the use of such measures as were available under a new banner. The rationale was that such questionnaires, since they focused upon those aspects of existence which were affected by ill health, must also give some indication of the impact of illness on quality of life. This view does not acknowledge the inter-connectedness of health status with other aspects of existence such as changes in income, work status, personal relationships, coping strategies, responsibilities, self-image and customary modes of being. In other words, the concept of 'health related quality of life' can be accused of being an empirical nonsense since it implies that people can divide quality of life into its health and non health related component. It is puzzling that this has rarely been discussed by those who work in the field of health measurement.

The consequence of this confusion is that it is often difficult to know exactly what is being measured and the rationale for the inclusion of particular measures is often unclear. This non satisfying situation is detrimental to the field of quality of life measurement. It is difficult to make any scientific advances in any field and to assess its progress if there is no shared definition of the concept or phenomenon under study.

In addition, because of theoretical and conceptual differences, showing the diversity of investigators' scientific cultures, different measurement strategies can be adopted, potentially leading to different results and therefore to different conclusions. In a medical context, these differences can incur the risk of influencing the decisions based on these measurements and in some cases of altering patients' health, which would undoubtedly raise ethical issues.

## 4. DIFFICULTIES IN TAKING INTO ACCOUNT THE PATIENTS'POINT OF VIEW

It can be argued that with quality of life measurement instruments, medicine has undertaken a change of perspective which conduced to substitute its traditional object (health) for patients' quality of life. This evolution was, no doubt, anticipated by the definition of health the W.H.O. adopted at the time of its constitution as "a state of complete physical, psychological and social well-being and not only as the absence of disease" [35] and which has changed somewhat the objective of medical action.

Since then, researchers have tried to figure out how to develop quality of life indicators that actually reflect patients' point of view.

What really seems new in this approach is not as much the interest health professionals take in their patients' quality of life which is, after all, a traditional task of the physicians as their wish, first to use questionnaires in order to quantify this quality of life and second to base these quantifications on patients' answers in order to take into account their point of view.

With quality of life measurements, we always find ourselves in a situation with a double point of view: from the person who observes (the physician for instance) and from the person who is being observed (the patient). The physician observes the patient, the patient is seen through the physician's point of view but this object of the physician's observation is also, from his side, a subject with his own point of view.

The interest that clinicians and health service researchers take in quality of life measurements can be explained in part by the recent awareness in the medical community that patients' preferences and perceptions have to be taken directly into account when it comes to health decisions. Even if physicians are the one who prescribe health treatments, the patients are those who decide whether to consult a physician, to follow their prescriptions and recommendations, or to seek other ways to fulfil their own needs. In other words, the patients' views on their own health and quality of life have a direct influence on the demand for care and on their adherence or compliance to medical treatments. These two elements are obviously of professional interest to the physicians. In addition to the recognition of the importance of the patients own views, numerous studies have shown that physicians' opinions on the outcome of care, for example on treatments' efficiency, differ notably from patients' opinions or from their next of kin's. Indeed, each of those three groups adopt different evaluation criteria: Physicians focus primarily on clinical signs and symptoms, patients on what they feel and on their capacity to fulfil their needs when their next of kin tend to value behaviors and attitudes. These studies entail that in order to access the patients perceptions one has to question directly the patients [33].

In other terms, the idea is that in order to give better health care to their patients, physicians (medicine) who, until then only considered them from their medical stand point, should also try and consider them through their own point of view on themselves. In this respect, the purpose of quality of life measurements instruments is undoubtedly to help physicians take patients' point of view into account. In the prospect of a normative use of quality of life measurement instruments, other problems arise from the complexity, inherent to the project, to take patients' point of view into account. Naturally, the fact that patients answer questions formulated by experts does not ensure that their answers will indeed reflect their own point of view (the questions' content, their hierarchy of health conditions, may or may not reflect patients' preoccupations or own hierarchy of values).

As a matter of fact, there has been some confusion between questionnaires which are completed by patients and those which reflect the concerns of patients. For example, current questionnaires designed to measure quality of life in epilepsy tend to have a content which focuses on frequency and severity of seizures, physical functioning and paid employment. However, qualitative studies of people with epilepsy have noted that their principle concern is with being labeled "epileptic" and experiencing 'felt stigma' leading to a need to conceal their condition and avoid situations which might reveal it. Severity and frequency of seizures are of much less concern [18]. In other words, although lip service has been paid to the primacy of the patient's viewpoint in matters which might have a bearing on "quality of life", the resistance to relinquish a medical model remains.

Yet, quality of life measurements that aim to reflect patients' point of view can only be developed and implemented by experts, who also have their own point of view. Numerous decisions are taken based on the judgments of experts involved in the development and the implementation of the instruments. When it comes to quality of life measurements we are, therefore, in a paradoxical situation. The paradox resides in a fundamental methodological difficulty: quality of life measurements whose objective is to reflect patients' point of view can only be developed and implemented by experts (physicians, methodologists etc.) who have their own point of view which is well known to be significantly different from patients'. Because the trust being put in the impartiality and the legitimacy of quality of life measurements depends almost exclusively on the trust being put in the judgment and integrity of the experts who developed and implement them, this question goes beyond a mere technical point. Given the institutional and financial context of these instruments' development, this question also regards investigators' deontology and justifies an analysis in terms of conflicts of interest [7,29]. What is, indeed, threatened by a conflict of interest is the trust put in an expert's judgment which in this case has some consequences in terms of the impartiality of the measures, in their normative use.

In the case of clinical medicine, the general framework of medical deontology, designed by a series of regulations, was intended to limit the negative consequences conflicts of interest may have on professional practices in order to protect patients. The same can be observed, more recently, in the case of biomedical research. The idea supported here is that a reflection on conflicts of interest should be initiated, in the clinical research field in general, and more particularly in the field of quality of life evaluations. I suggest that some institutional safeguards, similar to those implemented in order to minimize unethical consequences in clinical trials, should be considered in order to enable the development of trustworthy quality of life measurement.

## 5. MEASUREMENT INVARIANCE IN CROSS-CULTURAL EVALUATIONS

These issues are compounded by studies which combine data from more than one country. In relation to cross cultural work, a definition of quality of life is inevitably bound up with the issues of translation, meaning and conceptual equivalence in different cultures.

It is fair to assume that the processes of disease are likely to be culture free, with clinical measurements having an almost universal acceptance. However, a notion such as quality of life is not at all independent of cultural norms or indigenous patterns of behavior and expectations. Cultural forces can be expected to influence such variables as the type of activities engaged in by individuals, including preferred ways of spending time; the relative values placed upon these activities and on relationships, material possessions, physical strength, health and independence, expectations of what it means to feel good or to be healthy or to be ill and conventions about seeking health care; the form of language which is used to refer to personal experiences and the expressions with which people describe their feelings together with conventions governing the communication of these feelings to other people.

Although cross-cultural and anthropological issues are extremely significant, they are rarely considered in this type of research (Guyatt, 1993). However, all projects of developing invariant measurement instruments are not necessarily pointless since some similarity exist between cultures, regarding certain values related to health and the medical field. From a psychometric point of view, the question in the case of developing instruments in a trans-cultural context is the equivalence of measurement collected in different linguistic contexts [9]. The issue is then to identify the best way to empirically test the invariance between several linguistic version of a same measurement instrument and to formulate recommendations.

While methods for developing and validating instruments within a given culture are numerous, there are only a few articles and debates regarding the actual concepts, the criteria and the methods for developing instruments that work invariantly across cultures [1]. This literature suggests methodological principles and steps that have been adopted by leading research projects [9]. These principles can be summarized as follow:

1/ Formulation of a standardized methodology of development and of psychometric analysis of each version of the instrument;

2/ Publication of each national development;

3/ Comparison of the contents of the translations;

4/ Comparison of the values attributed to the modalities of answers;

5/ Comparison between the quality of the data and the hypotheses related to the establishment of the scores;

6/ Comparison of the given structures gathered/collected by the intervention of different versions, according to several methods (factorial analysis, structured equations models);

7/ Comparison between the average scores of the general population data collected according to the similar methods in each participating centre.

This approach, although quite systematic, poses several problems: On one hand, it does not make it possible to clearly test the hypothesis of invariance. On the other hand, it does not directly answer the question: are the different linguistic versions of the same instrument invariant when compared with each other? That is to say, do they yield measurements which are reasonably comparable and that can be aggregated at an international level? Moreover, it is difficult to identify and to specify the cases where the invariance of the instruments of measurement has not been obtained -for such cases must exist- [28]. Finally, our difficulties are explained by the fact that the concept of invariance used so far is itself quite inaccurate. For example, most authors use equivalence and invariance interchangeably confusing, i.e., the equivalence of the measured concept and the instrument of measurement that is expected to be invariant whatever the measured object is.

In an empirical project, aiming at comparing the scalar value of the item of five linguistic versions of the WHOQOL-Bref, we were specifically concerned with one central aspect of invariance, namely the invariance of the relative intensity of the items of each of the WHOQOL-Bref subscales across groups ([35]; see tables 2, 3, 4). This example illustrates how epistemological reflections may interact with the development of research methodologies.

This kind of invariance is a requirement for quantitative comparisons and for the aggregation of the data: the observed differences should not result because the

Table 2: Sex characteristic of the sample

| | Argentina | | France | | GB | | Hong-Kong | | Spain | | USA | | Total | |
|---|---|---|---|---|---|---|---|---|---|---|---|---|---|---|
| | N | % | N | % | N | % | N | % | N | % | N | % | N | % |
| Male | 170 | 40.4 | 143 | 48.3 | 138 | 39.1 | 425 | 50.2 | 162 | 53.3 | 203 | 45.8 | 1241 | 46.6 |
| Female | 251 | 59.6 | 151 | 51 | 197 | 55.8 | 415 | 49 | 136 | 44.7 | 240 | 54.2 | 1390 | 52.2 |
| MD | 0 | 0 | 2 | 0.7 | 18 | 5.1 | 7 | 0.8 | 6 | 2.0% | 0 | 0.0% | 33 | 1.2% |
| Total | 421 | 100 | 296 | 100 | 353 | 100 | 847 | 100 | 304 | 100% | 443 | 100% | 2664 | 100% |

Table 3: Age distribution

| Country | Argentina | France | GB | Hong-Kong | Spain | USA | Total |
|---|---|---|---|---|---|---|---|
| N | 421 | 298 | 330 | 828 | 298 | 441 | 2601 |
| Mean age | 47.1 | 44 | 48 | 45.2 | 42.6 | 44.2 | 45.3 |
| Std Dev | 14.6 | 15.4 | 15.7 | 16.1 | 14 | 15.7 | 15.5 |
| Minimum | 20 | 17 | 9 | 12 | 19 | 20 | 9 |
| Maximum | 80 | 81 | 85 | 92 | 80 | 90 | 92 |

Table 4: broadly defined health status

| | Argentina | | France | | GB | | Hong-Kong | | Spain | | USA | Total | |
|---|---|---|---|---|---|---|---|---|---|---|---|---|---|
| | N | % | N | % | N | % | N | % | N | % | Unknown | N | % |
| Healthy | 240 | 57 | 47 | 15.9 | 114 | 32.3 | 155 | 18.3 | 194 | 63.8 | | 750 | 33.8 |
| Ill | 181 | 43 | 249 | 84.1 | 239 | 67.1 | 690 | 81.5 | 110 | 36.2 | | 1469 | 66.1 |
| MD | 0 | 0 | 0 | 0 | 0 | 0 | 2 | 0.2 | 0 | 0.0 | | 2 | 0.1 |
| Total | 421 | 100 | 296 | 100 | 353 | 100 | 847 | 100 | 304 | 100.0 | | 2221 | 100 |

relative intensities of the items are different in the different groups. By analogy, to compare the heights of men and women, an essential requirement is that the locations of the marks on the ruler (scale) remain invariant when applied to men and to women and that they are not affected by whether or not the males or females are taller on the average in the groups. This requirement leads to modern test theory where the study of items that are not invariant is referred to as the study of differential item functioning -DIF- [23] although such studies were originally referred to as item bias (Ironson, 1976). It has been shown formally [2,3] that the requirement of invariance of the locations of the items on a scale can be tested directly from the responses of the persons whose attitudes are to be measured using a measurement model called the Rasch model [30]. A powerful feature of this approach to measurement is that, if the data fit the model, the relative scale value of the items of a scale are estimated independently of the location value of the persons who answer the items.

For example, two items of the social dimension of WHOQOL-BREF are *"How satisfied are you with your personal relationships?"* (item # F13.3) and *"How satisfied are you with the support you get from your friends?"* (item # F14.4). It is required that the relative intensities of these items on the scale, that is their locations on the scale, should be similar, irrespective of which cultural group responds to them and irrespective of whether one cultural group tends to agree strongly with these items and another tends to disagree with them. To stress, this invariance of the locations of the items on the scale irrespective of which groups respond, holds only if the data conform to the model. If they do hold, then quantitative comparisons can be made across different cultural groups. If the data do not conform to the model, then there is an opportunity to understand qualitatively the source of the lack of invariance. Therefore, central to any analysis is the test of fit between the data and the model regarding the criterion of invariance across relevant groups.

In this analysis we assumed that the minimum level of invariance is that an item is invariant with respect to just one pair of countries. In addition to pair-wise invariance, the retained items should be transitively invariant across at least three different cultural (centre/language) versions. By transitivity, we mean that if a given items in language A is equivalent to the same in language B and this item in language B is equivalent to this item in language C then it should follow that the language A version should be equivalent to the language C version of this item.

This provides an extra check over an above the usual methods for checking the stability of a questionnaire across translations. Typically, in large projects that involve many translations of one questionnaire, the equivalence of any translated questionnaire is checked -against the source questionnaire and never against each another. Empirical transitivity of invariance is an intuitive, albeit rigorous, requirement for any measuring instrument that is to be used in cross cultural studies.

The results of this analysis are fairly intuitive: invariance is obtained in a majority of cases, due to the state of the art methodology which led to the development of the various versions of the WHOQOL-BREF, but not in all (table 5).

Often, two or more versions of a given item have the same value along the measurement continuum, but, in quite a number of cases, the item location is different in each version. This result is in principle compatible with the work of medical anthropologists. Of course, such knowledge is critical when it comes to aggregating datasets and comparing the results of studies undertaken in different centers/cultural setting.

## 6. QUALITY OF LIFE MEASUREMENT AND MEDICAL ETHICS

The concept of quality of life is close to happiness as understood by numerous moralists and philosophers since Aristotle [26]. The main ethical point in favor of quality of life measurements rests on the principle that their uses are likely to contribute to increase subjects' quality of life [24]. Quality of life measurement, at least in certain cases, rests on the same postulate as the hedonist calculus. In that way, these studies fit within a broad utilitarian framework [15].

Moreover, as we have seen, the purpose of quality of life measurement instruments is undoubtedly to help physicians to take patients' point of view into account. Overall these measurements are compatible with the double ethical imperative of contemporary medicine: to do good and to respect patients' autonomy [4]. However, it is uneasy to empirically discriminate the best policies and the measurement of health policies consequences holds many ambiguities. Let's mention standard problems in normative ethic raised by interpersonal comparisons and the aggregation of individual preferences. For this reason, we must make sure that the uses of these measurements produce effects with a reasonable chance to be qualified as "good". This question regards each area of application of quality of life measurements: clinical research, clinical medicine and public health.

The different measurement instruments we have mentioned so far apply to health status or quality-of-life at a given moment. If they are combined with survival statistics, they can then measure "Quality Adjusted Life Years" (QALYs). Here each year of life considered is given a coefficient between 0 and 1. A certain number of QALYs is associated with each type of intervention. The beneficial medical act can then be defined as that which produces a positive number of QALYs [10].

The underlying idea is this: if the choice is offered, a rational person may perfectly well prefer a life that is shorter but coupled with a satisfactory state of health, to a longer life with a considerable handicap or serious discomfort. Preferences of this nature are however totally subjective, and the aggregation of results from questionnaires for the purpose of public policy-making would be open to very serious objections if it resulted in penalizing certain individuals whose psychological profile is not catered for. Individuals psychological characteristics are contingent and all should be considered equally legitimate. One could then go on to define an efficient intervention as that for which the cost per QALY is low. At this point, it is important to note that this approach is neither "technical" nor completely "economic". It is rather the result of a certain partisan approach, aiming to reduce public expenditure or budget deficit [21].

This problem of aggregation, combined with the complexity of the issues of "efficiency", make discussion of the relationship between equity and efficiency in the context of health policies a rather delicate matter. To approach the issue, we can distinguish two problems, which will be designated respectively the "first application" and the "second application".

The first application concerns allocation decisions on procedures or treatment applying to patients in the same situation from a medical point of view. For a given

Table 5: item/country invariance and location of the items

| Social | Loc. | Arg. | Spain | Fr. | GB | Hong--Kong | USA |
|---|---|---|---|---|---|---|---|
| F13.3 How satisfied are you with your personal relationships? | -0.343 | X | X | X | X | | |
| F14.4 How satisfied are you with the support you get from your friends? | -0.191 | X | X | X | | X | |
| F13.3 How satisfied are you with your personal relationships? | -0.117 | | | | | X | X |
| F15.3 How satisfied are you with your sex life? | 0.651 | X | X | | X | | X |
| **Psychological** | | Arg. | Spain | Fr. | GB | H-K | USA |
| F7.1 Are you able to accept your bodily appearance? | -1.3 | X | | | X | | X |
| F7.1 Are you able to accept your bodily appearance? | -0.585 | | X | X | | X | |
| F4.1 How much do you enjoy life? | -0.35 | | | X | X | | X |
| F6.3 How satisfied are you with yourself? | 0.097 | | X | | | X | |
| F24.2 To what extent do you feel your life to be meaningful? | 0.107 | | | X | X | X | X |
| F8.1 How often do you have negative feelings, such as blue mood, despair, anxiety, and depression? | 0.167 | | X | | | X | X |
| F5.3 How well are you able to concentrate? | 0.214 | | | X | X | X | |
| F4.1 How much do you enjoy life? | 0.481 | X | X | | | X | |
| **Physical** | | Arg. | Spain | Fr. | GB | H-K | USA |
| F9.1 How well are you able to get around? | -0.369 | | X | X | | X | X |
| F10.3 How satisfied are you with your ability to perform your daily living activities? | -0.192 | | | X | | | X |

| | | Arg. | Spain | Fr. | GB | H-K | USA |
|---|---|---|---|---|---|---|---|
| F2.1 Do you have enough energy for everyday life? | -0.088 | | X | X | | | X |
| F11.3 How much do you need any medical treatment to function in your daily life? | -0.058 | | X | X | X | X | X |
| F2.1 Do you have enough energy for everyday life? | -0.022 | X | | | X | X | |
| F1.4 To what extent do you feel that (physical) pain prevents you from doing what you need to do? | 0.194 | | X | X | X | | |
| F12.4 How satisfied are you with your capacity for work? | 0.238 | | X | X | X | X | |
| F3.3 How satisfied are you with your sleep? | 0.297 | X | X | X | X | X | |
| **Environment** | | Arg. | Spain | Fr. | GB | H-K | USA |
| F19.3 How satisfied are you with your access to health services? | -0.236 | X | X | | X | | X |
| F23.3 How satisfied are you with your transport? | -0.209 | | X | X | X | | X |
| F17.3 How satisfied are you with the conditions of your living place? | -0.152 | X | X | X | | X | X |
| F20.1 How available to you is the information that you need in your day-to-day life? | -0.095 | | X | X | X | X | |
| F16.1 How safe do you feel in your daily life? | -0.091 | | X | X | X | X | X |
| F22.1 How healthy is your physical environment? | 0.124 | X | X | X | X | | X |
| F21.1 To what extent do you have the opportunity for leisure activities? | 0.309 | X | X | X | | X | |
| F18.1 Have you enough money to meet your needs? | 0.349 | X | X | X | X | X | |

disease, several modes of treatment may be available: which should be preferred? Generally speaking, most people agree that it is reasonable to seek, among available treatments with identical clinical benefit for a given patient, the treatment which is associated with the better quality of life. However, the systematic use of quality of life measurement as choice criteria may deprive the practitioner and the patient of their respective rights of advice and choice, which may conflict with the traditional medical obligation to do whatever is judge best for the patient's good. The patient in turn may be deprived of this expert role in what concerns his or her own quality of life and hierarchy of values. In particular, if someone is an outlier, i.e. his/her values are not fairly different from the mean values and his responses to the questionnaires accordingly different from the mean, and if the responses are used normatively to decide which treatment to prescribe, this patient's values may not been respected.

The second application concerns allocation choices among procedures applying to patients who are not in the same medical situation. Inevitably, an "application" of this nature implies a choice favoring certain groups of individuals, and not others. For example, declaring on the basis of a QALY calculus that total hip replacement should have a more favorable financial allocation than renal dialysis, is not as much as making a pronounced judgment on the value of the two medical procedures, as to favor one group of patients at the expense of the other. It is always regrettable, and it can also be tragic, to have to make this sort of choice.

Other moral problems are originated in the propension of quality of life measures to put on the same metric, life saving interventions and quality of life improvement. One consequence may be that no attention is paid to the urgency of some perceived needs. When some cares seem more efficient than others in terms of quality of life gains, even if the later are increasing survival, should we prefer (i.e. fund in priority) without further discussion the former? Here again, the mechanical application of a criteria which would lead to choose the treatment that maximize quality of life score does not necessarily fit with our common moral intuitions.

## 7. CONCLUSION

Quality of life cannot be identified to happiness but, like well-being, it falls within its necessary conditions. It can be defined in relation to the level of satisfaction of a series of needs and desires, determined from a standpoint which, idealistically, should be the proper subjects' point of view.

Subjects' quality of life must be evaluated through methods capable of reflecting the preoccupation of individuals suffering diseases.

Since it is not possible to envision all aspects of existence nor to design a specific questionnaire per subject, quality of life, in order to be measured, must be reduced to some of its dimensions. This reduction is legitimate considering that the loss of reality implied is compensated by the objectivation of reality and by the prospects of improving the patients well-being.

The debate over the conceptual and methodological standpoints underlying these measurements should be wished for, given the human importance of the decisions, whether individual or collective, that can be made based on quality of life measurements.

These debates should aim at clarifying: 1/ the concept to be measured, 2/ the underlying theories, 3/ the methodology of taking into account the patient's view point, 4/ the identification of which view point is taken into account by currently used instruments and finally 5/ which measurement models are being used.

In conclusion, the quality of life measurement project in the health field, as ambitious as it may seem, is no less plausible and respectable, both from a scientific point of view and for the improvement of the health system. Particularly, the idea according to which patients and clinicians' perspectives need to be equally sought, when it comes to evaluate diseases' consequences, must not be discarded. The increasing sophistication of research allows to consider designing instruments able to gradually measure this concept and to make it reasonable to think that a new field of scientific enquiries has emerged.

Epistemology and Philosophy of technique can help to understand how these measurements interact with our culture in addition to contribute to the advancement of knowledge provided by these instruments, to the definition of rules for their good use and to the development of research methodologies.

## REFERENCES

1. R. Anderson, N. K. Aaronson, A. Leplège and D. Wilkin. Review of the International Assessment of Health Related Quality of Life. In: Spilker B. ed. *Quality of Life, and Pharmacoeconomics in Clinical Trials*, Raven Press, 613-632, 1996.
2. D. Andrich. A rating formulation for ordered response categories, *Psychometrika*, 43:357-374, 1978.
3. D. Andrich. *Rasch models for measurement*, Sage University Paper Series on Quantitative Applications in the Social Sciences vol serie no 07-068, Beverly Hills, CA, Sage Publications, 1988.
4. T. Beauchamp and J. F. Childress. *Principles of Biomedical Ethics*, 2nd ed 1983, Oxford University Press, 1979.
5. M. B. R. Bergner, S. Kressel, M. E. Pollard. The Sickness Impact Profile: Conceptual formulation and methodology for the development of a health status measure, *International Journal of Health Services*, 6:393-415, 1976.
6. M. Bergner. Quality of life, health status and clinical research, *Medical Care*, 27(3 suppl):S148-S156, 1989.
7. D. Blumenthal. Ethics issues in academic - industry relationships in the life sciences: the continuing debate, *Academic Medecine*, 71(12):1291-1296, 1996.
8. A. Bowling. *Measuring Health: a review of quality of life measurement scales*, Open University Press, 1991.
9. M. Bullinger. Ensuring International Equivalence of Quality of Life Measures: Problems and Approaches to Solutions, in J Orley and W Kuyken (eds) *Quality of life Assessments: International Perspectives*, Springer Verlag Berlin, 33-40, 1994.
10. R. A. Carr-Hill. Background material for the workshop on QALYs: Assumptions of the QALY Procedure, *Social Science and Medicine*, 29:469-77, 1989.
11. R. Faden, A. Leplège. Assessing quality of life. Moral implications for clinical practice. *Medical Care*. 30(Sup 1992):MS166-175.
12. A. Fagot-Largeault. *Réflexions sur la notion de qualité de la vie Archives de philosophie du droit*, 36:138-139, 1991.

13. S. Fanshel, J. W. A. Bush. Health-Status Index and its Application to Health-Services, *Outcomes Operation Research* 18:1021-1066, 1970.
14. T. M. Gill, A. R. Feinstein. A critical appraisal of the quality of quality-of-life measurements. *JAMA*, 272:619-626, 1994.
15. J. Griffin. Well-*being, its Meaning, Measurement and Moral Importance*, Oxford University Press, 1988.
16. G. H. Guyatt. The philosophy of health-related quality of life translation, *Qual Life Res*; 2(6):461-5, 1993.
17. S. M. Hunt, J. McEwen & S. P. McKenna. *Measuring Health Status* Croom Helm. London, 1986.
18. S. M. Hunt, S. P. McKenna. The measurement of quality of life of people with epilepsy. In: Epilepsy 2nd edition, A Hopkins et al (eds) Chapman & Hall, London, 1995.
19. G. H. Ironson and M. J. Subkoviak. A comparison of several methods of assessing item bias, Journal of Educational Measurement, 16:209-225, 1979.
20. D. H. Krantz, R. D. Luce, P. Suppes, A. Tversky. *Foundations of Measurement*, Vol 1, New-York: Academic Press, 1971.
21. A. Leplège and E. Picavet. Optimalité sociale, efficacité économique et bien-être mesurable dans l'évaluation des politiques de santé, *Information sur les sciences sociales*, 36(1):159-197, 1997.
22. A. Leplège and S. Hunt. The problem of quality of life in medicine, *Journal of the American Medical Association*, 1(278):47-50, July 2, 1997.
23. A. Leplège. Emmanuel Ecosse, and the WHOQOL Rasch project scientific committee, the analysis of four WHOQOL -100 data sets (Argentina, France, Hong-Kong, UK), *Journal of Applied Measurement*, 1(4):389-418, 2000.
24. R. J. Levine. Quality of Life Assessment in Clinical Trials: an Ethical Perspective. In: Spilker B. (editor), *Quality of Life and Pharmacoeconomics in Clinical Trials*, 2nd ed Lippincott-Raven, 489-95, 1996.
25. McDowell and C. Newell. *Measuring health: a guide to rating scales and questionnaire*, New York, Oxford University Press, 1987.
26. Nordenfelt et *Concept and Measurement of Quality of Life in Health Care*, Kluwer Academic Publishers, 1994.
27. D. L. Patrick and P. Erickson. *Health Status and Health Policy, Quality of Life in Health Care Evaluation and Resource Allocation*, Oxford University Press, New-York, 1993.
28. T. Perneger, A. Leplège and J. F. Etter. Cross cultural adaptation of a psychometric instrument: two methods compared, *J Clin Epidemiol*, 52(11):1037-046, 1999.
29. M. Pritchard. Conflicts of interest: conceptual and normative issues, *Academic Medecine*, 71(12), December 1996.
30. G. Rasch. *Probabilistic models for some intelligence and attainment tests* Danish Institute of Educational Research, 1960; University of Chicago Press, 1980; MESA Press, 1993.
31. N. Sartorius. WHO Method for Assessment of Health-related Quality of Life, In: SR Walker and RM Rosser (Eds), *Quality of Life Assessment: Key Issues in the 1990s* Dordrecht, Netherlands: Kluwer Academic Publishers, 1993.

32. G. Simondon. *Du mode d'existence des objets techniques*, Aubier, 1958.
33. M. L. Slevin, H. Plant, D. Lynch et al. Who should measure quality of life, the doctor or the patient? *British Journal of Cancer*, 57:109-12, 1988.
34. The Evidence-Based Medicine Working Group. Evidence-based medicine. A new approach to teaching the practice of medicine, *JAMA*, 268(17):2420-2425, Nov 4, 1992.
35. The WHOQOL Group. The Development of the World Health Organization Quality of Life assessment instrument (the WHOQOL). In: Orley, J. and Kuyken, W. (Eds, 1994) *Quality of Life Assessment in Health Care Settings* Heidelberg: Springer-Verlag, 1994 WHO Chron 1947.
36. D. Wilkin, L. Hallam and M. A. Doggett. *Measures of need and outcome for primary health care*, Oxford Medical Publications, 1992.

# What is Life? A New Look at an Old Question

Michel Morange

*Centre Cavaillès d'Histoire et de Philosophie des sciences Ens, 45 rue d'Ulm, 75230 Paris Cedex 05, France*
*morange@wotan.ens.fr*

**Abstract.** For molecular biologists, the question "What is Life?" disappeared in the 1960s to reemerge recently. The reasons for this reemergence will be analysed: they tell us much about the recent transformations of biology, and its present state. This question can be considered as a thermometer, which measures the balance between reductionist vs. holist explanations in biology: when the question disappears, reductionist approaches are dominant; when the question reappears, the reductionist vision is challenged.

In this contribution, I do not intend to provide a personal answer to this question or to relate all the answers put forward since the time of Aristotle. Such a commentary on the rich philosophical tradition associated with this question would require much more space than I have at my disposal.

My aims are more modest. I will limit myself to the answers to this question proposed by scientists, biologists for the most part, since the 1940's. This period has seen considerable changes both in the nature of the answers given and in the status of the question itself, which ceased to be posed at one point and has only recently re-emerged.

Changes in the answer to this question have been observed before. In *Les mots et les choses*, Michel Foucault stated that the notion of "Life" was invented during the 18th Century ([8], p. 139; trans. [9], p. 127-128). This point of view has been heavily criticised by philosophers of science. Personally, I prefer the vision of Georges Canguilhem. In his article "Life" published in *Encyclopaedia Universalis*, Georges Canguilhem pointed out that although the question "What is Life?" was not new, the nature of the answers and the urgency of the response required were highly variable [1]. This question became increasingly important at the end of the 18th Century, in parallel with the development of vitalism, and remained central to biology throughout the 19th Century. It continued to be posed in the 1940's by the founders of molecular biology and those influential in its development. It also served as the title of a famous book written by Schrödinger in 1944 [26].

At the beginning of the 1970's, Georges Canguilhem pointed out in his article for *Encyclopaedia Universalis* that this question was scarcely dealt with in the writings of contemporary biologists. He quoted the famous aphorism of François Jacob: "Life is no longer questioned in laboratories" ([10], p. 320). The translation of Jacob's sentence is difficult, because its meaning is ambiguous in French. It may mean that the existence of life, as a phenomenon distinct from physico-chemistry, is no longer considered seriously or that the question as to the nature of life is no longer posed. Several years previously, in 1962, the biologist Ernest Kahane published a book with the provocative title *La vie n'existe pas* (Life Does Not Exist) [13].

My aim here is to follow in the footsteps of Georges Canguilhem. I will show that this question has increased in importance again in the last few years and will try to identify the reasons for the decline and recrudescence of interest in this question. Finally, I will briefly show that different answers, corresponding to different scientific traditions, are currently in vogue and will outline two major points of divergence between biologists.

## 1. The disappearance of the question "What is Life?" in the 1960's

How is it possible to account for the disappearance of the question "What is Life?" from prominence between 1960 and the end of the 1980's? The simplest answer is that the question was no longer considered important because an answer had been found. This year (2003) marks the 50th anniversary of the discovery of the double-helix structure of DNA. When Jim Watson showed Francis Crick the structure he had deduced for DNA, Crick quickly realised that it was entirely consistent with what was known about the DNA molecule and that it provided a very simple model for the replication of the genetic material. In a state of some excitement, Crick dashed into his local pub, the Eagle, exclaiming that they had discovered "the secret of Life". Seventeen years later, Monod came to much the same conclusion in *Chance and Necessity*: the riddle of Life had been solved ([20], p. 12 of the paperback edition).

The question "What is Life?" was therefore no longer raised because a satisfactory answer had been found. The secret of Life lays in the existence of a genetic code, genetic information, and a genetic programme. Life was considered to be the possession of genetic information and of a genetic code allowing this information to be translated. The appearance of Life on the primitive Earth was thought to coincide with the appearance of this genetic code. This answer fits into the tradition founded by the geneticist Hermann Muller, who, in the 1920's, linked the appearance of genes with the appearance of the first organisms [23]. The probability of this event was considered very low: the formation of the first living cell not only required the formation of two different macromolecules, DNA and protein, but also of a precise relationship – the genetic code – linking them. Monod felt that this event was so unlikely that the emergence of organisms was probably a unique event in the history of the Universe, limited to our planet [20].

However, there was probably another, more subtle reason for which the question "What is Life?" ceased to be posed. The question itself became taboo. Merely posing this question indicated a lack of acceptance of the answer provided by the most active branch of biology, molecular biology. It suggested that the person asking the question considered Life to be something different, not the simple result of physics and chemistry, that could not be reduced to the properties of macromolecules. This question thus became a sign of heresy.

The rise of molecular biology therefore had two different consequences. The first was to provide an answer to the question "What is Life?". The second was to provoke what the biologist Stanley Shostak called *"The Death of Life"* [27]. For the most reductionists in the molecular biologist camp, Life was no more than the sum of the properties of the macromolecules present in organisms.

## 2. The reemergence of "Life"

Nonetheless, "What is Life?" has re-emerged as a valid question and we should now consider the reasons for which the "Death of Life" proved to be only temporary.

The question "What is Life?" has clearly become fashionable again in the United States, although this effect is less clear in Europe. Proof that this question has again increased in importance can be found in its use in book titles ([18]), and in the titles of chapters within books ([4], trans. [5], chapter 1). This question is frequently raised at the beginning of articles considering the origin of Life on Earth or the possible existence of extra-terrestrial forms of Life[1]. However, although this question is frequently posed, answers are not always given. In most cases, the authors remark that the question is difficult to answer and that a number of different answers have been provided. Following these introductory remarks, they abandon such vague, philosophical questions in favour of real, hard science. It is not an easy task to find an answer![2] Nevertheless, this interest, albeit limited, in the question "What is Life?" is really something new. The question is no longer taboo, even for the molecular biologists and molecular geneticists who considered it meaningless a few decades ago. The new element is not that the question is being asked — it continued to be asked by heterodox scientists who considered the answers of molecular biology to be unsatisfactory — but that this question is now being posed by scientists from mainstream biological research. An interesting example is that of Craig Venter: his project to create new, artificial forms of Life is for him a way to answer this question.

How can we account for this dramatic turnaround in the last few decades? The answer to this question probably lies both in changes in biology itself and in external influences. The simplest way to interpret the re-emergence of the question "What is Life?" is that it corresponds to the abandonment of the previous answers: the existence of genetic information and of a genetic code.

No observer of biology can fail to note that the informational vision of Life is no longer as dominant as it was in the 1960's. For the sake of simplicity, I will consider only three series of works and experiments that undermined, for totally different reasons, this genetic, informational vision.

The notion of a "genetic programme" was proposed independently by Ernst Mayr [19], and by François Jacob and Jacques Monod in 1961 [11]. It was very fashionable in the 1970's, and taken very seriously by biologists like Sydney Brenner, who spent more than a year learning computer programming [6]. It was also heavily criticised by biologists and philosophers of science. This notion is still used today, but in a metaphoric sense: no modern biologist would accept the idea that the genome has a programme resembling that of a computer, which was extremely widespread among molecular biologists in the 1960's and 1970's.

---

[1] "What is Life? Natural science has never found a satisfactory definition", [24]; " No broadly accepted definition of life exists ", [2].

[2] see for instance the story recalled by Daniel Koshland, [16].

Similar changes are apparent if we consider, more generally, the notions of "gene action" and "gene function". What did molecular biologists think about gene action in the 1960's, and what do they think now, at the beginning of the 21st Century? Most of the first molecular biologists thought that the complex structures and functions of organisms could be accounted for directly by the action of a limited number of genes. These genes were the *bearers* of the information required for these complex structures and functions. A precise knowledge of the sequences of these genes was expected to provide immediate clues as to their role in these complex functions. This vision could be called preformationist, as these structures and functions were seen as being somehow contained in these genes.

Nowadays, every complex structure and function of the organism is thought to result from the action of hundreds or thousands of molecular components, each encoded by a different gene, organised in pathways and networks. No one molecular component is considered more responsible than any other for these structures and functions. Each molecular component is involved in various pathways and networks, in the accomplishment of various tasks, and the construction of various structures in the organism. Characterising these components provides no immediate clue as to the general functions they fulfil [15,21].

This change in the concept of "gene action" is the result of many experiments. We have space here to deal with only a few of these studies. For example, the search for genes involved in animal behaviour, in *Drosophila* in particular, demonstrated that a large number of genes are involved but that no one gene is specific for any particular type of behaviour. Characterisation of the intercellular and intracellular signalling pathways demonstrated that there are extremely large numbers of molecular components involved in these pathways and networks. Gene inactivation or "knockout" experiments have turned up many surprises in terms of the pleiotropic action of genes, and the redundancy of molecular components.

All these experiments followed on from the development of genetic engineering technology in the 1970's. All have delivered, one way or another, the same message: no single gene is solely responsible for complex tasks and structures. The notion of a "gene for" a particular process or structure is therefore absurd. Each gene encodes a single molecular component that participates, together with many other components, in the achievement of particular functions and the formation of certain structures.

Genetic information can provide clues as to the function of individual molecules, but not to the ways in which complex structures and functions are generated by combinations of these molecular components.

The initiation of the Human Genome Sequencing Programme at the end of the 1980's was emblematic of the previous informational vision of molecular biology: it was thought that deciphering the genome would immediately reveal the characteristics of human beings, and the origin of our specificity. In contrast, the results officially announced at the beginning of 2001 were emblematic of the new vision: the human genome contains only about 30,000 genes, a number not very different to that in *Drosophila* or in nematodes. Furthermore, the nature of the genes present in these species seems to be similar. Simply reading the genetic information tells us nothing about the origin of human complexity. This complexity seems instead to originate in the way in which these molecular components co-operate to generate structures and functions.

These recent changes in our view of gene action are particularly important for two reasons. Firstly, they came about as a result of numerous different experiments rather than a single study, making them much harder to detect. Secondly, they have helped to lead to the gradual abandonment of the idea that the secret of Life lies in the existence of genetic information. Genetic information is essential for production of the building blocks of organisms, but it cannot tell us in itself anything about the way in which these building blocks are assembled. The secret of Life lies to a much greater extent in this process of assembly than in the simple direct properties of the molecular components, although the assembly process in itself depends on the nature of the building blocks.

Results obtained in the 1980's have also demonstrated the limitations of the earlier identification of genetic information, genetic code and Life. Since the 1960's, molecular biologists have often pointed out that it was very difficult to explain the origin of Life in the informational vision. According to this vision, the appearance of Life depended on the highly unlikely simultaneous emergence of two macromolecules — DNA and protein — and of a precise relationship between them (the genetic code). As we have seen previously, some molecular biologists, including Jacques Monod, have concluded from this that the appearance of life is likely to have been a unique event and that there is little hope of finding any extra-terrestrial forms of Life.

At the beginning of the 1980's, it was discovered that RNA molecules could have catalytic activity, which was previously considered to be an exclusive property of proteins. Results obtained since this initial discovery have confirmed that RNA molecules exist with most, if not all of the known catalytic capabilities of proteins.

Based on the problems associated with the simultaneous appearance of DNA and proteins and the discovery that RNA, which is structurally related to DNA, can fulfil catalytic functions, Walter Gilbert suggested, in 1986, that an RNA-based form of life probably preceded the DNA- and protein-based form of life we know today. This suggestion has been followed by many others, proposing primitive forms of life based on different kinds of macromolecular structures.

The most important take-home message from these new models on the origin of Life is that the appearance of Life is no longer associated with the appearance of a genetic code, or even the formation of genetic information. This second part of the sentence is particularly important. It has often been suggested that RNA-based life is not very different from DNA- and protein-based life, being the only difference the macromolecule bearing the genetic information: DNA today, RNA in the past. In fact, the difference is much more important. In the absence of any relationship between two kinds of macromolecule, the meaning of the expression "genetic information" would be very different in an RNA-based world than in our present DNA and protein world. Therefore, the existence of genetic information – with the meaning we give to this term – and of a genetic code are not necessarily linked with the origin of Life. Life may have existed before the genetic code and genetic information came into existence, with these elements being late consequences of the appearance of life rather than its origins. If the secret of life and the answer to the question "What is Life?" do not lie in the existence of a genetic code and genetic information, they must be sought elsewhere.

The question "What is Life?" has therefore become important once more because previous answers are no longer considered satisfactory. Two other events are also responsible for this return of a forgotten, taboo question.

The first is the transformation of biological research linked to the development of genome sequencing programmes and, more recently, of post-genomic technologies. The accumulation of data and their treatment have required the collaboration of specialists in computer science. More recently, physicists and mathematicians have also become involved in efforts to treat these data, and to develop models from this knowledge. For bioinformatics specialists, physicists and mathematicians, organisms are complex physical systems. These researchers do not have the same attachment to the informational concept of Life as molecular biologists, whose views stem largely from the informational context of the 1950's and 1960's. For physicists, organisms are simply "systems of organized complexity", to quote Warren Weaver in an article published in *American Scientist* in 1948 [30]. Most physicists today would probably still consider this to be the best description of organisms.

The second is the development in the United States, and then in Europe, of astrobiology programmes, formerly known as exobiology programmes. These programmes search for current or past signs of Life on other planets, within the Solar System or around other stars.

Most of the projects currently underway in the domain of astrobiology focus on finding life-forms similar or identical to those on Earth, on other planets. The reasons for this limited ambition on the part of astrobiologists are complex and diverse. Firstly, there are chemical constraints limiting the form that Life could adopt: organisms must theoretically contain water, carbon — the only abundant chemical element able to generate such a diversity of bonds — amino acids, which spontaneously form in very different environmental conditions, and so on. Some astrobiologists even think that all organisms, regardless of their site of origin in the Universe, are likely to have a genetic code similar to that of terrestrial organisms [25].

Most do not share the same restrictive vision of what Life is and might be. Nevertheless, they support current projects for the simple pragmatic reason that it is simpler to look for something we already know, than for something that we cannot even begin to imagine.

One major incentive to the development of these astrobiology programmes was the discovery, in 1995, of giant planets outside the Solar System. Although these planets were not found to be suitable for the development of Life, their discovery has increased hopes of finding planets similar to Earth. However, the motivation behind these programmes was probably more political than scientific: making the search for extra-terrestrial Life one of the major objectives of NASA was probably the most efficient way to attract interest and funds from the members of the American Congress, or simply to reduce planned decreases in these funds. Regardless of the true motives behind the development of the astrobiology programme, the search for extra-terrestrial Life clearly provides a fresh impetus for considering the question "What is Life?"

## 3. The present answers

Here are three examples of answers to the question "What is Life?": 1. "Life is a potentially self-perpetuating open system of linked organic reactions, catalysed stepwise and almost isothermally by complex and specific organic catalysts, which are themselves produced by the system", (M. Perrett, quoted by [22]); 2. "Life is a self-sustained chemical system capable of undergoing Darwinian evolution" [12]; 3. "Life is a name we give to certain emergent processes of complex systems" [3]. This is only a limited sample of the possible answers, but the diversity of responses is nonetheless clear. To these explicit answers, we should also add the implicit or "subliminal" answers. Scientists investigating the beginnings of Life on Earth, or designing experiments to be run on spacecrafts with the aim of searching for traces of past or present Life on Mars or on the satellites of Jupiter, have an implicit concept of "Life" and of its hallmark characteristics. When studying early Life on Earth, or looking for signs of Life on other planets, they look at the presence or development of these characteristics, which are considered to be necessarily linked with Life.

Three characteristics are currently thought to be indicative of Life and the presence of organisms: 1. Living beings rapidly exchange matter and energy with their environment: they are very active chemical reactors, and have what biologists call a metabolism; 2. They possess highly complex molecular structures; 3. They are able to reproduce with imperfections, opening up the possibility of evolution by natural selection.

Surprisingly, most biologists, chemists and astrobiologists seem to agree that these three characteristics are necessary for Life — not only as we know it on Earth, but also in all its possible forms anywhere in the Universe.

However, this consensus masks strong divergences and differences in opinion. The first problem concerns the ranking of these three characteristics of Life and even the issue of whether such ranking is necessary. Are the presence of an active metabolism, of highly complex molecules and the capacity of reproduction equally important in the definition of an organism, or is one of these characteristics more important than the others, determining the other characteristics in a specific environment?

The scientists who believe that there is only one essential characteristic of Life fall into two camps. One of these two groups sees reproduction as the essential property of Life, and the possession of an active metabolism and highly structured molecules only as the conditions required for reproduction to occur. Those holding this view often consider other life-forms, based on different material supports, as possible; for example, they would see artificial forms of Life based on computers as new forms of Life. This view is supported by those working with C. Langton in Santa Fe, who carried out research on artificial forms of Life [17]. The other group considers an active metabolism and the possession of highly complex molecules to be at the core of Life, with reproduction being no more than a consequence of this chemical activity. This view is promoted by supporters of the autopoietic models of Humberto Maturana and Francisco Varela [29]. Such diametrically opposite viewpoints are not new among specialists investigating the origin of Life: Harmke

Kamminga, a historian, described such divisions years ago [14]. Those who consider that the acquisition of reproductive power is essential for Life clearly adhere to the tradition of genetics, molecular genetics and molecular biology, whereas those considering that the formation of active chemical systems and complex macromolecules was the essential step in the formation of Life belong to a biochemical tradition.

The second point on which specialists are divided concerns the need to appeal to new laws, new principles, to explain the specific characteristics of Life. This division has already existed in the past. In the 1940's, one of the founders of molecular biology, Max Delbrück, hoped to discover new physical principles at the origin of the capacity of living beings – and their genes – to reproduce ([7]). He was disappointed when the self-replicative power of genes was explained by the double-helix structure of DNA. Many other attempts have been made to explain the characteristics of living beings in terms of specific physical laws. This approach today takes a specific form: some biologists and chemists consider organisms to be complex systems, the specific characteristics of which will be explained in the near future by the development of theories of complexity ([28]).

Most biologists would agree that organisms are complex systems: they have thousands of molecular components, each of which has a complex structure, interacting one with each other in a dynamic way. Most biologists and chemists would also agree that increasing our knowledge of the behaviour of complex systems should enable us to better understand the structures and functions of organisms. However, there is a divergence between those who consider that the new laws of complexity that will emerge from studies currently underway will tell us what Life is and those who think that they will only help us to understand some of the characteristics of organisms, without providing fundamental answers to the question "What is Life?". For the latters, the best answer lies in the co-existence of the three characteristics described above.

I have described recent developments concerning the question "What is Life?" in the last few decades and the reasons for these developments. This question, which was taboo some decades ago, has become fashionable again, principally due to the failing of the informational vision of molecular biology. Diverse, and in some cases opposing answers to this question are put forward today. I have tried to outline some of the points of conflict, and the scientific traditions underlying the opposing views. It is obvious that both current answers and the points of divergence are embedded in the very rich philosophical tradition of answers to the question "What is Life?"

The re-emergence of the question "What is Life?" is a sign that biological research has reached a crossroads and is presently in a metastable state. We can also imagine the impact of the discovery of other forms of Life, or simply of remnants of them, on Mars or other planets. Such discoveries would probably have an even greater impact than the discovery of America had, some centuries ago, on Western civilisation.

## REFERENCES

1. Canguilhem, G. 1989. article «vie», *Encyclopaedia Universalis*, **23**:546–553.
2. Chyba, C. and C. Phillips 2001. Possible ecosystems and the search for life on Europa. *Proc. Natl. Acad. Sci. USA* **98**:801–804.
3. Cohen, J. and I. Stewart 2001. Where are the dolphins? *Nature* **409**:1119–1122.
4. Danchin, A. 1998. *La barque de Delphes.* Paris: Odile Jacob.
5. Danchin, A. 2002. *The Delphic Boat.* Cambridge Mass.: Harvard University Press.
6. de Chadarevian, S. 1998. Of worms and programmes: *Caenorhabditis elegans* and the study of development. *Studies in History and Philosophy of Biological and Biomedical Sciences* **29**:81–105.
7. Fischer, E. P. and C. Lipson 1988. *Thinking about science: Max Delbrück and the origins of molecular biology.* New York: W.W. Norton and Co.
8. Foucault, M. 1966. *Les mots et les choses.* Paris: Gallimard.
9. Foucault, M. 1970. *The Order of Things: An Archaeology of the Human Sciences.* London: Tavistock.
10. Jacob, F. 1970. *La logique du vivant.* Paris: Gallimard.
11. Jacob, F. and J. Monod 1961. Genetic regulatory mechanisms in the synthesis of proteins. *J. Mol. Biol.* **3**:318–356.
12. Joyce, G. 1992. Exobiology: discipline science plan. internal NASA document.
13. Kahane, E. 1962. *La vie n'existe pas.* Paris: Editions rationalistes.
14. Kamminga, H. 1988. Historical perspective: the problem of the origin of life in the context of developments in biology. *Origins of Life and Evolution of the Biosphere* **18**:1–11.
15. Keller, E. F. 2000. *The century of the gene.* Cambridge Mass.: Harvard University Press.
16. Koshland Jr., D. E. 2002. The seven pillars of life. *Science* **295**:2215–2216.
17. Langton, C. G. (ed.) 1989. *Santa Fe studies in sciences of compexity;* vol. 6: *Artificial Life* Reading Mass.: Addison Wesley.
18. Margulis, L. and D. Sagan 1995. *What is Life?* University of California Press.
19. Mayr, E. 1961. Cause and effect in biology. *Science* **134**:1501–1506.
20. Monod, J. 1970. *Le hasard et la nécessité.* Paris: Le Seuil.
21. Morange, M. 2001. *The misunderstood gene.* Cambridge Mass.: Harvard University Press.
22. Morowitz, H. J. 1992. *Beginnings of Cellular Life.* New Haven: Yale University Press.
23. Muller, H. J. 1929. The gene as the basis of life. In *Proceedings of the 4th International Congress of plant science* (Ithaca) 1, 879–921.
24. Nisbet, E. and N. Sleep 2001. The habitat and nature of early life. *Nature* **409**:1083–1109.
25. Pace, N. R. 2001. The universal nature of biochemistry. *Proc. Natl. Acad. Sci. USA* **98**:805–808.
26. Schrödinger, E. 1944. *What is Life?* Cambridge: Cambridge University Press.
27. Shostak, S. 1998. *Death of Life: the legacy of molecular biology.* London: MacMillan Press.

28. Solé, R. and B. Goodwin 2000. *Signs of Life: How complexity pervades biology.* New York: Basic Books (Perseus Books).

29. Varela, F., H. Maturana, and R. Uribe 1974. Autopoiesis: The organization of living systems, its characterization and a model. *Biosystems* **5**:187–196.

30. Weaver, W. 1948. Science and complexity. *American Scientist* **36**:536–544.

# Anthropic Explanations in Cosmology

Jesús Mosterín*

*Instituto de Filosofía, CSIC (Madrid)*

**Abstract.** The claims of some authors to have introduced a new type of explanation in cosmology, based on the anthropic principle, are examined and found wanting. The weak anthropic principle is neither anthropic nor a principle. Either in its direct or in its Bayesian form, it is a mere tautology lacking explanatory force and unable to yield any prediction of previously unknown results. It is a pattern of inference, not of explanation. The strong anthropic principle is a gratuitous speculation with no other support than previous religious commitment or the assumption of an actual infinity of universes, for which there is no the slightest empirical hint. But even assuming so much, it does not work. In particular, the assumption of an infinity of different universes is no guarantee of finding among them one like this one. The loose anthropic way of reasoning does not stand up to the usual methodological standards of empirical science. And it does not signal any anthropocentric turn in contemporary science.

In the last century cosmology has ceased to be a dormant branch of speculative philosophy and has become a vibrant part of physics, constantly invigorated by new empirical inputs from a legion of new terrestrial and outer space detectors. Nevertheless cosmology continues to be relevant to our philosophical world view, and some conceptual and methodological issues arising in cosmology are in need of epistemological analysis. In particular, in the last decades extraordinary claims have been repeatedly voiced for an alleged new type of scientific reasoning and explanation, based on a so-called "anthropic principle". "The anthropic principle is a remarkable device. It eschews the normal methods of science as they have been practiced for centuries, and instead elevates humanity's existence to the status of a principle of understanding" ([34], p. 47). Steven Weinberg has taken it seriously at some stage, while many physicists and philosophers of science dismiss it out of hand. The whole issue deserves a detailed critical analysis. Let us begin with a historical survey.

## 1. History of the anthropic principle

After Herman Weyl's remark of 1919 on the dimensionless numbers in physics, several eminent British physicists engaged in numerological or aprioristic speculations in the 1920's and 1930's (see [3]). Arthur Eddington [26] calculated the number of protons and electrons in the universe (Eddington's number, N) and found it to be around $10^{79}$. He noticed the coincidence between $N^{1/2}$ and the ratio of the electromagnetic to gravitational forces between a proton and an electron: $e^2/Gm_e m_p \approx N^{1/2} \approx 10^{39}$. He also tried to explain the value of the fine structure

---
*I thank Jeremy Butterfield, Carl Hoefer and Terry Jones for useful comments.

constant $\alpha = e^2/\hbar c$ through numerological reasonings which were obscure and unconvincing to other physicists. In his relentless but sloppy search for ratios and numerical coincidences, Klee [44] sees Eddington "attempting to extract numerological revenge on behalf of Pythagoras." In the 1930s Edward Milne developed a "kinematic theory of relativity", based on philosophical ideas, such as the cosmological principle. He advocated the idea that the "constants" of physics, like the gravitational constant $G$, changed over the life of the universe. These predictions proved unfounded, as did the kinematic theory of relativity.

After a most distinguished career in quantum mechanics, Paul Dirac came under the spell of Eddington and Milne and in 1937 became also involved in numerology. As already mentioned, it was well known that the ratio of the electrostatic attraction between the proton and the electron in the hydrogen atom to the gravitational force between the same two particles is about $10^{39}$. Dirac found other combinations of fundamental constants with a somehow similar value. If we take as unit of time the time it takes light to travel a distance equal to the classical electron diameter, then the current age of the universe (estimated at that time to be just about two billion years) is about $6 \times 10^{39}$ of those units. So, again the order of magnitude $10^{39}$! Dirac suggested that this coincidence should be explained by looking for some link between the fundamental constants and the age of the universe. Since the age of the universe increases with time, the fundamental constants of physics also have to change in time, in order to keep that relation. Specifically, the value of the gravitational constant $G$ would decrease with time. Later data from our solar system, from space probes and from the binary pulsar discovered by Hulse and Taylor in 1974, allow us to exclude that $G$ is weakening at even a hundredth of the rate assumed by Dirac.

The scientific community soon became sick of these speculations. Already in 1931 Beck, Hans Bethe and Riezler [6] spoofed Eddigton's numerology in a parody they managed to get published in *Naturwissenschaften*. It was a curious precursor of Alan Sokal's 1996 'hoax' paper . In 1937 Herbert Dingle denounced in *Nature* the whole speculative approach: "This combination of paralysis of the reason with intoxication of the fancy is shown, if possible, even more strongly in Prof. Dirac's letter in *Nature* ... in which he, too, appears victim of the great 'Universe'-mania ... Milne and Dirac ... plunge headlong into an ocean of principles' of their own making ... The criterion for distinguishing sense from nonsense has to a large extent been lost ... "

In 1961 Robert Dicke published in *Nature* a short paper entitled "Dirac's cosmology and Mach's principle" [18]. Dicke rejected Dirac's speculation about the change of $G$ in time and found a simpler explanation in the selection effect (on possible values of the constants) of the fact that we, humans, are here. So the Hubble time $T$ elapsed since the Big Bang (the age of the universe) "is not a 'random choice' from a wide range of possible choices, but is limited by the criteria for the existence of physicists." So the values of $T$ are constrained by the requirement "that the universe, hence galaxy, shall have aged sufficiently for there to exist elements other than hydrogen. It is well known that carbon is required to make physicists." Dirac published a short reply to Dicke, saying that Dicke's analysis was sound, but that he (Dirac) preferred his own argument because it allowed for the possibility that planets "could exist indefinitely in the future and life need never end." Dicke was a practical man, more interested in observation than in speculation. After his 1961 paper, he did not dwell on that piece of anthropic reasoning nor did he show any further interest in the matter.

The carbon of which Dicke spoke is produced by nuclear fusion of helium inside red giant stars. This process takes several billion years (in small or medium-size stars) or a few million years (in large stars), after which period the star can explode as a supernova, scattering the newly formed elements throughout space, where they can eventually become part of a planet, on which life could evolve. So, in order to be able to produce carbon-based life, the universe must be at least several million years old. On the other hand, it can not be too old (older, let us say, than $10^{12}$ years), because if it was, all the stellar processes would have already concluded and there would be no life-sustaining radiation energy around. This is the reason for the coincidence remarked by Dirac, and there is no need to go to the length of postulating a variable gravitational constant. In any case, the "prediction" is extremely vague, and the range of time that allows carbon atoms or planets to exist extremely broad: from a few million to a trillion years.

In 1973 C. B. Collins and Steven Hawking noticed that only a narrow range of initial conditions (out of all the possible values of the physical constants) could give rise to the observed isotropy of the actual universe. They found this result unsatisfactory, because current theory did not offer any explanation for the fact that the universe turned out this way rather than another. Collins and Hawking reasoned along anthropic lines to discuss the flatness problem. Starting with the assumption that galaxies and stars are necessary for life, they argued that a universe beginning with too much gravitational energy would recollapse before it could form stars, and a universe with too little of it would never allow gravitational condensation of galaxies and stars. (Notice that they were talking of galaxies, and the assumption that galaxies are indicators of life does no real work in the argument). Thus, out of many different possible initial values of $\Omega$ (the ratio of the actual average density of the universe to the critical density), only in a universe where the initial value of $\Omega$ was almost precisely 1 could we have existed. This would explain why $\Omega$ is so near to 1.

In 1974 Brandon Carter published "Large Number Coincidences and the Anthropic Principle in Cosmology" [14], in which he presented his ideas, previously exposed in oral form. In this article Carter baptized the type of reasoning already present in Dicke's paper as the *anthropic principle*. He distinguished two versions of it, the weak and the strong. The weak anthropic principle says that "what we can expect to observe must be restricted by the conditions necessary for our presence as observers". This true but trivial version is very different from the strong anthropic principle, which says that "the universe (and hence the fundamental parameters on which it depends) must be such as to admit the creation of observers within it at some stage". Others have formulated the strong anthropic principle as saying that it is a law of nature that life or intelligent life has to evolve.

In 1979 Bernard Carr and Martin Rees pointed to many alleged "cosmic coincidences" [13], numerical relations among physical magnitudes that, if allowed to change (keeping everything else in the theoretical structure constant), would make carbon-based life impossible. Carr [12] and others began to speak of a fine tuning of the physical constants to make life possible.

All these speculative developments culminated in 1986 in the book of John Barrow and Frank Tipler, *The Anthropic Cosmological Principle* [4]. This 700 page book exhaustively traced the history of teleological ideas and cataloged the alleged applications of the anthropic principle to lots of coincidences and contingencies in the initial conditions of the universe and in the fundamental constants of physics. For example, the strength of the fundamental forces of nature (gravitation, electromagnetism, weak and strong interaction) as given by their corresponding fine structure constants (dimensionless numbers which are ratios of fundamental constants, like $c$, $h$, $G$, $e$, $m_p$, $m_e$) is found to be so well proportioned and fine tuned, that any tinkering with their values or ratios would make life impossible. Other speculations, for example on (and against) extraterrestrial intelligence, were also extensively dealt with. The scientific reception of the book was rather negative. In his review in *Nature* [66], astrophysicist William Press even wrote that "there is some fundamental intellectual dishonesty here, some snake oil to be peddled". Nevertheless the book popularized the "anthropic" talk. The anthropic principle made its way into the popular science literature and even popped out (in a loose and redundant way) in some serious technical papers.

Some outstanding physicists like John A. Wheeler, Hawking and Weinberg have at some stage appealed to the anthropic principle as a desperate way out of their difficulties. Rees has been promoting the anthropic principle in a continuous stream of popular science books. In 1990 Shaposhnikov and Tkachev tried to estimate the mass of the Higgs boson by anthropic considerations. From 1987 on, Weinberg has tried to find an anthropic bound to the cosmological constant. More recently, he has become more skeptical: "This sort of reasoning is called anthropic, and it has a bad name among physicists. Although I have used such arguments myself in some of my own work on the problem of the vacuum energy, I am not that fond of anthropic reasoning." ([85], p. 173). In 1998 Hawking and Neil Turok used the Hawking-Hartle wave function for the universe, coupled with the anthropic principle, as a way of achieving an open universe in a broadly inflationary scenario (without false vacuum). Shortly thereafter, the new distance measurements of type Ia supernovae seemed to favor a flat universe again, and at least Turok does not wish to appeal to the anthropic principle any longer. Still in 2003 Leonard Susskind invoked the anthropic principle as a desperate way out of the huge multiplicity of solutions plaguing string theory. Most physicists are appalled at the introduction of these loose ways of reasoning in science. As commented by Peter Mittelstaedt in 2000, the anthropic principle is not a problem in philosophy of science, but in psychology of science: how could competent physicists take such a thing seriously?

## 2. Cosmic coincidences and fine tuning

As we saw, the numerological speculations of Eddington, Milne and Dirac were at the origin of the anthropic thinking. Numerology is the resort to obscure and far-fetched explanations for numerical coincidences. As a matter of fact, and as documented by Klee [44], the authors in the anthropic tradition have had a rather sloppy and cavalier way of seeing astonishing coincidences in numbers different by several (even by six) orders of magnitude. If we look for broad numerical coincidences, we will find them everywhere. The number of neurons in our brain seems to be of the same order of magnitude as the number of stars in our galaxy, about $10^{11}$. And so what? The numerologist would be tempted to look for hidden designs behind this harmless coincidence.

Let us consider the following six fundamental physical constants: the gravitational constant, the speed of light, Planck's constant, the electric charge of the electron and the proton, the rest mass of the proton, and the rest mass of the electron ($G$, $c$, $h$, $e$, $m_p$, $m_e$). Let us consider a 6-dimensional space, each of whose 6 coordinates coincides with the set of all possible values of one of those 6 fundamental physical constants. Each vector or point of this space represents a possible combination of values for the six physical constants considered, or, if you prefer, each point represents a (logically) possible universe. In most of these possible universes there would have been no galaxies, no long lasting main sequence stars, no life, no intelligence, no scientists. Only in a small subset of the set of all possible universes can all these things exist. So, if we already know that there are scientists, or humans, or rabbits, or stones, we can infer from this item of information that the actual universe is a point of the restricted subset which allows for the existence of such things. This inference rule has been called the (weak) anthropic principle.

The anthropic speculations often focus on the fact that most points in the possibility space would represent universes unfit for life, on the many coincidences which are necessary for life to arise and on the alleged evidence of fine tuning provided by these coincidences. For example, if the charge of the proton had been (in absolute value) different from the charge of the electron, no stable objects could have formed. Every two atoms would repel each other, every star, planet or organism would explode. "If we modify the value of one of the fundamental constants, something invariably goes wrong, leading to a universe that is inhospitable to life as we know it" [35].

Carr and Rees [13] reviewed the many "anthropic coincidences", the many cases where the values of constants are in the narrow ranges compatible with life. They concluded that "nature does exhibit remarkable coincidences and these do warrant some explanation. ...The anthropic explanation is the only candidate and the discovery of every extra anthropic coincidence increases the *post hoc* evidence for it."

Carr and others continued to elaborate the idea and they began to speak of fine tuning. For example, it is well known that the density of the universe is very close to the critical density, which would make the space flat, marking the frontier between an open and a closed spacetime. Now the actual density deviates from the critical density by at most an order of magnitude (a factor of ten). In the past the deviation was much smaller: it was only one part in $10^{16}$ one second after the Big Bang, and still smaller before. These are the type of densities which allow the universe to expand at the adequate rate for the formation of chemical elements like carbon and the evolution of life.

There is no known physical reason why the initial expansion rate should have been what it was, so one is led to speculate why this should be. One suggestion is that we could not be here if things were otherwise. On the one hand, if the expansion rate were slightly too low, the universe would recollapse before life had time to arise; on the other hand, if the expansion rate were slightly too high, life could not arise either because galaxies could not have formed amid the general expansion [12].

All this talk of coincidences and fine tuning is rather muddled and careless. As pointed out by Ernan McMullin [59], the large-number coincidences have nothing to do with the "fine tuning" of the constants, and this has nothing to do with the laws. The same applies to the alleged improbability of the actual world. No one wants to buy a lottery ticket with the number 5555555. It seems very improbable that such a number will win. As a matter of fact, that number is no less probable than any other, let us say 3405175, which looks less peculiar. A repetition of draws would erase any surprising coincidences in the long run, but in a single draw any result (however full of coincidences) is as likely as any other. Already Rémi Hakim [37] rejected as lacking any foundation the sense of "probability" used in talk of multiple universes and arguments of fine tuning: "The notion of probability issues from (and implies) the real possibility of repeating the same random experiment a large number of times in an independent way. However the choice of universe is, for us, not only impossible to repeat, but even to realize at a single time".

The universe (as far as we can know it) is something unique ('uni-verse' and 'uni-que' come from the same root 'uni', one). We can learn a posteriori how the universe is, but it makes no sense to speculate on how it should be on the basis of a priori statistical considerations. This is the reason why John Leslie's [47] firing squad argument is flawed. He compared our existence to the survival of a sentenced man, because each of the guns in his execution squad misfires. Has someone (God?) tinkered with the guns beforehand? Of course, there have been lots of firing squads and seldom have all the guns misfired. There is a grim statistics of firing squads. But the universe is a unique historical fact. There are no statistics of universes. Besides, the components of the firing squad are people with the intention of shooting, but there are no intentions in the fabric of the universe. At least in usual language, fine tuning implies intentionality and multiplicity of cases. The question of fine tuning does not arise in unintentional one-element sets.

## 3. The weak anthropic principle

Brandon Carter [14] introduced the weak anthropic principle with the words: "what we can expect to observe must be restricted to the conditions necessary for our presence as observers." Barrow and Tipler [4] formulate the same principle in this way: "The observed values of all physical and cosmological quantities are not equally probable, but they take on values restricted by the requirement that there exist sites where carbon-based life can evolve and by the requirement that the Universe be old enough for it to have already done so." Roberto Torretti [81] expressed perplexity at the usage of the probability notions by Barrow and Tipler: "What does it mean to say that 'the observed values of all the physical ... quantities are not equally probable'? ... In fact, any correctly observed value, just because of having been observed, has probability 1."

In a nutshell: The fact that we exist implies that the universe satisfies all the necessary conditions for our existence. Or, in probabilistic garments: The conditional probability of the real universe being in the restricted region of the possible-universes space where life is possible, given the fact that we exist, is different (and much higher) than the absolute probability would be in an a priori probability distribution which did not take into account the fact that such things as people and rabbits actually exist.

Gale [31] stressed that the weak anthropic principle, "even if acceptable, ... appears *so* weak as to be meaningless. At first glance, it looks either trivial, or tautological or transcendental, or all three at once." Nevertheless, he thought it could function as a heuristic device. Earman [24], after careful examination, concluded that in the weak anthropic principle "it is hard to find anything stronger than a tautology".

The (weak) anthropic principle is a valid rule of inference, but it is not a physical (or non-physical) explanation of anything. If the physical constants being what they are is a necessary condition for the existence of humans (or cockroaches), and there are humans (and cockroaches), then, we can conclude, the physical constants are what they are, ie they are in the limited region of the possibility space which make humans and cockroaches possible. This is not a principle of physics, but an application of a trivial theorem of logic: whenever $A$ is a necessary condition for $B$ and $B$ obtains, $A$ must also obtain. This is just an equivalent reformulation of the good old inference rule of *modus ponens*: from (if $A$ then $B$) and $A$ you can infer $B$. Remember that 'if A then B' is equivalent to '$B$ is a necessary condition for $A$'.

So-called anthropic reasonings are very often just indirect reasonings (by *reductio ad absurdum*), in which, knowing already that $A$, we prove that $B$ by showing that if not $B$ then not $A$. So, if we already know (as we do) that there are people, or stones, and that if protons and electrons had different electrical charges, there would be neither people nor stones, we can conclude that protons and electrons have to have the same electric charge (in absolute value). This type of reasoning is again an application of another old inference rule, *modus tollens*.

Far from representing any breakthrough in scientific reasoning, the (weak) anthropic principle is just the restatement of an elementary rule of logic, already known in the Middle Ages and even by the ancient Stoics. It is valid with the barren and trivial validity of tautologies. It only allows us to infer what we already knew (that the constants have the values we know they have), but it does not allow us to explain anything. Neither does it lead to any new prediction.

## 4. No predictions from the anthropic principle

It is usually agreed that the anthropic principle has never led to any genuine scientific prediction (i.e. to any prediction of something previously unknown). So, for example, Carr and Rees at the end of their 1979 sympathetic review, aknowledge that the anthropic principle "is entirely *post hoc*: it has not yet been used to predict any feature of the Universe." Nevertheless Barrow and Tipler ([4], p. 252-253) pretended that there was one case of anthropic prediction: Fred Hoyle's prediction in 1953 of an excited state of the carbon isotope $^{12}C$ at 7.6 MeV above the ground state. This contention was rejected in several reviews of the book. So, Helge Kragh [46]:

> Barrow and Tipler claim that Fred Hoyle's remarkable 1953 prediction of the resonance energy level of $^{12}C$ was based on the anthropic principle. What Hoyle showed was that only if there exists a certain carbon

resonance can astrophysical theory be consistent with the present existence of carbon ... But Hoyle's prediction is not anthropic since it does
not refer to the existence of human beings but only to carbon atoms
...

Barrow and Tipler's pretension was acritically accepted by some later writers,
like Yuri Balashov [2]: 'Many authors argue that the anthropic principle... is absolutely incapable of predicting anything new. One must remember, however, that
in 1953 Hoyle *predicted*, based on what we now call anthropic arguments, the *unknown* excited resonance level in $^{12}$C." Some popular science books have made
the view notorious, even if the comments often went in opposite directions. So,
in connection with the resonances between helium, beryllium and carbon, Gribbin
and Rees [35] write: "This combination of coincidences, just right for resonance
in carbon-12, just wrong in oxygen-16, is indeed remarkable. There is no better
evidence to support the argument that the universe has been designed for our benefit – tailor-made for man". On the contrary, Greenstein [34] comments: "Those
resonances really are coincidences. They are genuinely remarkable strokes of luck.
The anthropic principle provides no explanation for anything, and no amount of
anthropic reasoning can explain these coincidences."

According to our actual understanding of nucleosynthesis, most of the atomic
nuclei of hydrogen (protons, H), deuterium ($^{2}$H), helium ($^{3}$He and $^{4}$He) and the
isotope lithium-7 ($^{7}$Li) were formed shortly after the Big Bang. The rest of the
atomic nuclei were cooked later in the core of red giant stars. Hydrogen and helium
continue to be by far the most abundant nuclei. After hydrogen and helium, carbon
and oxygen (in the form of the isotopes $^{12}$C and $^{16}$O) are the two most abundant
nuclei in the visible universe.

In his classic 1954 paper "On nuclear reactions occurring in very hot stars:
The synthesis of elements from carbon to nickel", Hoyle gave the first satisfactory
account of the production of carbon, oxygen and neon in the interior of red giant
stars. It was already well understood that main sequence stars burn hydrogen into
helium. Once the hydrogen supply is exhausted, they leave the main sequence,
expand dramatically, become red giant stars and begin to burn helium. The burning
or fusing of helium to produce first carbon and then oxygen and neon, was still
not understood. Once carbon was available, oxygen could be formed by fusing
carbon with helium and neon could then be formed by fusing oxygen with helium.
The main difficulty laid in the production of carbon from helium in the first place.
Edwin Salpeter suggested originally that the fusion of three "alpha particles" ($^{4}$He
nuclei) to $^{12}$C would occur by a simultaneous collision.

$$3\ ^{4}\text{He} \rightarrow\ ^{12}\text{C} + \gamma$$

Calculations showed that the three-collisions were too rare and the burning rate
of helium would be too slow. Taking into account that two-collisions were much
more frequent than three-collisions, Salpeter then made the further suggestion that
helium burning occurred in a two-step process: first two helium nuclei collided to
form the isotope beryllium-8 and then this beryllium nucleus collided with an
helium nucleus to form carbon.

$2 \ ^4\text{He} \rightarrow ^8\text{Be}$
$^4\text{He} + \ ^8\text{Be} \rightarrow ^{12}\text{C} + \gamma$

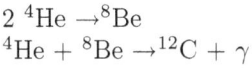

The problem is that $^8$Be is highly unstable, it bursts apart in $10^{-15}$ s, making the encounter with an helium nucleus too improbable. If it nevertheless happened, it would have to be due to the existence of a certain resonance in $^{12}$C. Hoyle accurately calculated the resonance, an excited state of the $^{12}$C nucleus at about 7.6 MeV above the ground state. Subsequently this resonance was experimentally detected in the exact predicted energy range, to the astonishment of the scientists (Cook, Fowler and others) who performed the experiment. Hoyle's calculation is considered a tour de force of modern astrophysics. Hoyle was able to derive the properties of the excited energy levels of $^{12}$C and $^{16}$O from the astronomical facts then known, like the cosmic abundances of carbon and oxygen. In the whole 1954 paper by Hoyle there is of course no reference to the still unborn anthropic principle, but neither is there any reference to humans or life or observers. Taking the known facts of astronomy and chemistry into account to formulate a bright hypothesis that solves a previous puzzle and gets confirmed by experiment is an example of the standard scientific method at its best. It has nothing to do with humans or with any specifically anthropic way of reasoning.

The anthropic authors have often underlined the very precise fine tuning needed for the famous resonant excited energy level of $^{12}$C at 7.644 MeV, which is "just" (in fact, 277.3 keV) above the sum of the combined energies of a $^8$Be and a $^4$He nucleus. Mario Livio and his fellow astrophysicists Hollowell, Weiss and Truran have run systematic computer simulations of the consequences for carbon production in stars of increasing or decreasing that value of 7.644 MeV. In 1989 they published their results in *Nature*. By lowering the level of 7.644 MeV by 60 keV, the production of carbon was four times higher than normal. "It appears that carbon production is not strongly favored by nature, because a small reduction in the energy difference would lead to a relatively much greater increase in carbon abundance ..." An increase of 60 keV would not significantly alter the level of carbon production. So carbon abundances like the observed or still higher are compatible with a window (of the resonant excited energy level of $^{12}$C) of 120 keV, equivalent to a temperature window of 1.39 billion K. As Klee [44] asks: "How can a temperature window *that* wide within which the resonant energies can fall count as a case of 'fine tuning' that results in energy levels that are 'just barely' resonant?"

## 5. Misnomer

"Anthropic principle" is a complete misnomer. First of all, and as already remarked by McMullin [59] and acknowledged even by Rees [70], it is not a principle at all. More importantly, it does not deserve the adjective "anthropic" (relative to humans), as there is nothing specifically human or about humans in the type of reasoning it refers to. It could also have been called the rabbit principle or the cockroach principle. There cannot be rabbits or cockroaches without heavy chemical elements having been formed in the interior of massive stars and scattered around in supernova explosions. But there are rabbits and cockroaches. So (we can conclude by the rabbit principle or the cockroach principle) the fundamental physical constants must be in the narrow margin that allows for heavy chemical elements to be formed in stars and scattered in supernova explosions. Perhaps we should rather talk about the beetle principle, for, as observed (tongue in cheek) by Haldane, God loves beetles above anything else, as shown by the many species of beetles (more than 300,000) He has created. Neither is there anything specific about life or living organisms in the principle. It could also be called the washing machine principle, or the limestone principle. Of course, there could not be any washing machines or limestones without heavy chemical elements having been produced in the interior of massive stars and scattered around in supernova explosions.

That the weak anthropic principle has no more to do with humans than with beetles or uranium atoms was soon recognized. According to Earman [24], the motivation force of the weak anthropic principle "does not derive from any consideration about Man, Consciousness, or Observership. The weak anthropic principle, as used by Dicke and Carter, is in fact nothing but a corollary of a truism of confirmation theory. Nor does the application of the corollary have to rely on life or minds, for the selection function is served just as well by the existence of stars and planetary systems supporting a carbon-based chemistry but not life forms". Helge Kragh [46] remarks that "in virtually all the examples of the (weak) Anthropic principle mentioned by Barrow and Tipler the existence of human beings, or even life, is in fact irrelevant. Most of the so-called anthropic arguments can be reduced to standard scientific arguments of the retroductive form in which it is asked which constraints have to be put on nature in order to make it consistent with current theory and observation". The same point was raised by Wilson [91] and McMullin [60].

The anthropic authors frequently switch between carbon atoms and human consciousness (or observership), as if they were somehow equivalent, or at least as if the first was a precondition of the second. So Alexander Vilenkin talks about consciousness, but only to say that galaxies are good tracers of life and consciousness and to confine his arguments to galaxies. In general, either heavy chemical elements or galaxies are all that is needed for "anthropic" arguments. Only in the very particular version of Wheeler's participatory anthropic principle does consciousness or observership as such play any role. For the rest, it is just carbon atoms what we are talking about. Besides, in traditional religious thought, observers, minds or intelligences need not be made out of carbon or other heavy chemical elements. God(s), angels and other alleged spiritual beings were supposed to be minds and to be observing and even watching us all the time.

## 6. Privileged position

Carter [14] wrote: "We must be prepared to take account of the fact that our location in the universe is *necessarily* privileged to the extent of being compatible with our existence as observers". But *any* position is privileged to the extent of being compatible with the existence of whatever exists at that position. Such an alleged privilege is so universal as to constitute no privilege at all.

The special conditions of temperature that prevailed at the very early time when the universe was a quark soup made for a privileged position for the quark soup. Indeed as soon as the universe expanded and cooled, the conditions were not right any more, and the quark soup lost its privileged position and disappeared, transformed into something else. Black holes are also privileged in the sense of having very special conditions in their corner of the world. Anything is privileged when it is in a cosmic position which satisfies the conditions for its existence.

Aerobic creatures (like us) are privileged because they are in an oxygen-filled medium like the earth's atmosphere, while anaerobes like the bacteria in the guts of mammals are privileged due to the lack of oxygen in their environment. The sulfate-reducing bacteria *Thiopneutes* require sulfur for their respiration and are quickly poisoned by exposure to oxygen. So there is little wonder that they enjoy the privilege of living in media rich in sulfates and lacking free oxygen, like certain muds or soils of geothermal regions. Even neutrons bound in atomic nuclei and neutron stars are in a privileged situation, which allows them to exist. Free neutrons decay into protons and electrons in about ten minutes.

Carter [15] pretended to react "against exaggerated subservience to the 'Copernican principle'". He called Copernican principle "the assumption that our own situation in the Universe is not in any way privileged, but is typically representative in a Universe that is entirely homogeneous apart from minor local fluctuations". He identified the modern version of the Copernican principle with the "perfect cosmological principle" of Hermann Bondi and Thomas Gold.

The "cosmological principle" is the name some people give to the assumption that the universe is spatially homogeneous and isotropic, i.e., that the 3-dimensional slices (hypersurfaces) of constant time of the 4-dimensional spacetime are symmetric. This assumption is necessary for the application of the FRW metric, which is the component of the standard Big Bang model that allows us to solve Einstein's field equations. (As is well known, the non linear equations of general relativity cannot be solved in general, but only in some extremely simple cases, like the Schwarzschild or the FRW metrics.) The so-called perfect cosmological principle also postulates the temporal symmetry and homogeneity and is thus incompatible with the Big Bang model, which represents a dynamically evolving universe. It is the basis of the steady state model.

In the day-dreaming world of anthropic speculation, nothing is what it appears. The anthropic principle is not anthropic and the Copernican principle is not Copernican. Actually Copernicus never defended anything like the perfect cosmological principle, not even the cosmological principle. He did not think that the positions of the Sun and the Earth were homogenous. He just thought that the "honorific" central position was reserved for the Sun, not the Earth. His point was that the Earth was not at the center, but was a planet circling the central Sun. McMullin [59] comments that Carter

> begins from what he calls the "Copernican principle" to the effect that the cosmic abode of man is in *no* way privileged. (Copernicus would, I suspect, be astonished to have this taken to be an inference from his theory!) By "privileged", he does not mean to have an honorific status or advantage attached, rather that the human abode has *no* special features associated with it that would mark it off from any other part of the cosmos. He calls it a "dogma" of earlier cosmologists. In the face of it, it is *obviously* false, since the Earth is a planet, has an atmosphere,

and has many other features that *do* mark it off from empty space, for example. What he seems to have in mind is the much more limited claim that the earth, the human abode, is not "privileged" ... in its overall spatial or temporal *location*. ... The reference to humans here (rather than, say, to beetles) comes only from the expectations engendered by religious or philosophical traditions that led people to expect that the human abode *would* be privileged in its cosmic location, privileged in the proper sense of being at the center and not just of being different.

## 7. Bayesian argument

Carter [15], Garret & Cole (1992) and others have pointed out that the weak anthropic principle can be viewed as an application of Bayes' theorem.

The Bayesian approach to inductive inference is based on the assignment of prior probabilities to competing hypotheses (using the principle of maximum entropy or otherwise), and the reassignment of posterior probabilities, once new data are available, as a function of the likelihood of those data, assuming the truth of the different hypotheses. The hypothesis which gives the maximum probability to the data is favored.

The posterior probability of the hypothesis $H$ in view of the new data $D$ and the background knowledge $K$, $P(H|D \wedge K)$, is given by Bayes' theorem in terms of the prior probability $P(H|K)$ (assigned using the maximum entropy principle) and the likelihoods $P(D|H \wedge K)$ and $P(D|K)$:

$$P(H|D \wedge K) = \frac{P(H|K) \cdot P(D|H \wedge K)}{P(D|K)}$$

From here follows the inference principle:

$$P(D|H \wedge K) > P(D|K) \Rightarrow P(H|D \wedge K) > P(H|K)$$

which says that the likelihood of the data, given the hypothesis, gives support to the hypothesis.

Now, if we take the hypothesis to be $R$ (that the values of the constants and parameters of the standard cosmological model are in the restricted range that allows – via the formation of stars and planets and heavy chemical elements – for the evolution of carbon-based life), the datum to be $L$ (that there is carbon-based life on Earth) and the background knowledge to be $M$ (the standard cosmological model), then we can reformulate the last formula as:

$$P(L|R \wedge M) > P(L|M) \Rightarrow P(R|L \wedge M) > P(R|M)$$

The likelihood of there being carbon-based life, assuming the standard cosmological model with its constants and parameters in the narrowly restricted range which allows for the evolution of life, is much greater than the likelihood of finding carbon-based life, assuming only the standard model without any specific values for the parameters (because most combinations of the values of its parameters preclude the evolution of carbon-based life). So we can conclude that the posterior probability of the parameter values being in the life-producing zone, given the datum that there is life, is much greater than its prior probability. And this can be considered to be another version of the weak anthropic principle.

The argument is formally correct. What is a little odd is to consider the standard cosmological model as background knowledge and the fact that there is life as a *new* datum. The Bayesian rule is a rule of inference, not a principle of explanation. But even as an inference principle couched in Bayesian terms, the argument only leads to already known results, and remains conspicuously sterile.

## 8. Difference between inference and explanation

We don't need to assume that all syntactically correct why questions make sense. How many hairs do I have on my head now? Why that many? Why is today Tuesday? Why is water $H_2O$? Why does the Earth have only one natural satellite? Why does the Sun have nine planets? Why is there anything? As remarked by Sylvain Bromberger [11], "Why questions, unlike other wh-questions [what, how, when, where, ...], can be obscure ...It is the obscurity of not knowing what, if anything, controls whether it has an answer at all. ...I don't know why there are nine large planets. I don't know whether there is an answer to that why-question. ...I don't know whether there being only nine planets isn't simply a brute fact".

Perhaps the values of the fundamental constants of physics are brute facts. Perhaps there is nothing to explain about them. Or perhaps there is an explanation, but we do not know whether we will ever find it. That explanation could only be provided by some sort of overarching theory, which would allow us to deduce such values from more general principles, instead of measuring them empirically. The existence of such a theory is just a hope. Anyway, the fact that we exist does not explain why the constants of physics and the parameters of the standard cosmological model have the values they have. Our life and existence do not explain the initial conditions of the universe or the value of the fine structure constants. The alleged anthropic explanation follows a pattern to which the following "explanations" also belong:

Why is it raining? Because I have opened my umbrella. Why did I get the bacterial infection? Because I am taking antibiotics. Why did he commit a crime? Because he is now in prison. Why did they get married in the first place? Because they later got divorced. Why did he smoke? Because he later gave up smoking. Why did he die yesterday? Because he was alive one week ago. Humankind developed the linguistic ability, because I am writing today. Writing was invented, because I am writing today. There is still hydrogen left in the sun's core because I am writing today. There is something rather than nothing, because I am writing here today. My grandmother did not die a virgin, because I am writing today.

Why is there oxygen in the Earth's atmosphere? Because we humans breathe oxygen. Of course this "anthropic explanation" is no explanation at all. The oxygen in the atmosphere antedates our arrival on Earth by two billion years. On the contrary, that the atmosphere contains oxygen is a precondition of the existence of aerobic creatures like us. Philip Gasper (1991, in a context independent of any concern for the anthropic principle) gave this as a text-book example of derivation without explanation: "For instance, from the laws of biology together with the fact that there are mammals on the earth, we can deduce that there is oxygen in the atmosphere. We clearly have not explained the presence of oxygen in this way ... "

In whatever account of explanation, the *explanans* contains initial conditions which are temporally prior (or at most simultaneous) to the facts to be explained (*explanandum*). The only point that all people who talk about causes and explanations agree on is that you cannot explain a cause by its effects. You cannot explain the initial conditions of the universe fifteen billion years ago by our existence now. Carr and Rees [13] remarked: "From a physical point of view, the anthropic "explanation" of the various coincidences in nature is unsatisfactory, in three respects. First, it is entirely *post hoc*: it has not yet been used to predict any feature of the Universe ... Second, the concept is based on what may be an unduly anthropocentric concept of the observer. ... Third, the anthropic principle does not explain the exact values of the various coupling constants and mass-ratios, only their order of magnitude." Gale [30] gave this analysis of Dicke's initial reasoning: "In general arbitrariness has been eliminated by showing that a phenomenon can be predicted or that a theory can be deduced from some more fundamental premise. Dicke's technique is quite different. Deductive or predictive logic proceeds from a fundamental assumption to a derived result: the future is deduced from the past. The temporal flow of Dicke's argument is in the opposite direction. He cites a present condition (man's existence) as the explanation of a phenomenon grounded in the past (the age of the universe)."

The blunder of anthropic explanations is so obvious that it does not even escape sympathetic theologians. Commenting Collins and Hawking's [16] explanation, "...the isotropy of the universe is a consequence of our existence", William Craig [17] wrote "literally taken, such an answer would require some form of backward causation whereby the conditions of the early universe were brought about by us acting as efficient causes merely by our observing the heavens". And Richard Swinburne [76] added: "The suggestion might seem to be that our existence is in some way the *cause* of the laws of nature and boundary conditions being the way they are (because if they were not that way we wouldn't be able to observe them). That suggestion is nonsense. The laws of nature and boundary conditions cause our existence; we do not cause theirs."

## 9. Anthropocentrism and design

According to the Jewish, Christian and Islamic religious tradition the world was made by God for the sake of man, and man was made for singing the praises of God. Everything, from the nightly skies to the minute bugs, bears witness to God's precise craftsmanship and intelligent design. God designed an anthropocentric universe, in which the Earth, humans' abode, occupied center stage, surrounded by the atmosphere, the seven skies (of the sun, the moon and the five known planets) and the sphere of the fixed stars, above which God and the rest of the heavenly Court dwelt. God and the angels were always looking down at us, humans, watching us, caring about us. The world was like a theater and we were the protagonists of the cosmic drama.

All these tenets are deeply alien to modern science. The whole history of the Scientific Revolution has been the story of the continuous and increasing abdication by man of any anthropocentric pretensions to occupy a privileged cosmic position in a universe designed for him. This epic story is well known. Copernicus demoted the Earth from the center of the cosmos to a mere planet circling the sun. Bruno demoted the sun to just one more among myriads of similar stars. Still in 1920 most astronomers doubted the existence of any galaxies outside our own Milky Way, as shown in the public confrontation between Shapley and Curtis at the Washington meeting of the National Academy of Science that year. Since then, we have become aware that not only our sun is just one unimpressive star among many billions of other stars of our galaxy, but that also our own galaxy is just one among many billions of other galaxies. The isotropy inferred from the cosmic background radiation is the most radical denial of any local anthropocentrism. And, as emphasized by cosmologist Joel Primack [67], the fact that most of the matter in the universe seems to be dark matter, matter of a different kind of the usual one we are acquainted with, "is the ultimate Copernican revolution. ... Not only will the Earth no longer be the center of the universe, it won't even be made of the same sort of stuff". Even Weinberg ([85], p. 46) rejects emphatically any form of anthropocentrism: "Nothing that scientists have discovered suggests to me that human beings have any special place in the laws of physics or in the initial conditions of the universe." And teleology has disappeared from modern biology.

Against this background it is surprising that some authors have tried to reopen the debate on cosmic design and anthropocentrism under the banner of the (strong) anthropic principle. Their popular writings have often led to rather muddled misrepresentations of modern science results.

## 10. The strong anthropic principle

The weak anthropic principle is a tautological inference rule. The strong anthropic principle, on the contrary, is a substantial metaphysical assertion, committing its upholders to a thoroughly anthropocentric view of the universe. The whole cosmic evolution is seen as a gigantic plot to produce people (or life). Some hint of it was already in Dyson (1971): "As we ... identify the many accidents of physics and astronomy that have worked together to our benefit, it almost seems as if the Universe must in some sense have known that we are coming". Its name and its first formulation are due to Carter [14]: "the Universe (and hence the fundamental parameters on which it depends) must be such as to admit the creation of observers within it at some stage". In Barrow and Tipler's [4] words: "The Universe must have those properties which allow life to develop within it at some stage in its history". According to Greenstein [34]: "The weak principle states that *humanity can exist only in a habitable environment* [i.e., humanity can exist only where it can exist]. But the strong Anthropic principle goes further. It states that *a habitable environment must exist*. It states that ... a planet must be found wrapped in gasses of precisely the composition required by humans". Remark the unconditional "must" in all three formulations. The universe *must* be fit for human life. Human life is not a welcome chance result of cosmic evolution, but instead its aim and ultimate reason. In the last author's words: "Life obeys the laws of physics –

this much is a truism. What is new ... is that ... the reverse seems also to be true – that the laws of physics conform themselves to life. ... How did it come to pass that against all odds the cosmos succeeded in bringing forth life? It had to. It had to in order to exist".

The strong anthropic principle is an acknowledgedly metaphysical speculation, with no base whatsoever on logic or physics. No deduction from it can be considered a scientific explanation of anything.

Even Carter [14] conceded: "It is of course always philosophically possible – as a last resort, when no stronger physical argument is available" to take resource to the anthropic principle. "I would personally be happier with explanations of the values of the fundamental coupling constants etc. based on a deeper mathematical structure (in which they would no longer be fundamental but would be derived)". Carr and Rees ([13], at the end of their sympathetic review) declared: "The anthropic principle ... may never aspire to being much more than a philosophical curiosity. One day, we may have a more physical explanation for some of the relationships discussed here that now seem genuine coincidences. For example, the coincidence $\alpha_G = (m_e/m_w)$, which is essential for nucleogenesis, may eventually be subsumed as a consequence of some presently unformulated unified physical theory". Alan Guth (1990) commented: "...the anthropic principle kind of rubs me the wrong way. ...Obviously, there are some anthropic statements you can make that are true. If we weren't here, then we wouldn't be here. ...I find it hard to believe that anybody would ever use the anthropic principle if he had a better explanation for something. ...the physical constants are determined by physical laws that we can't understand now. ...I don't think the laws were contrived in order to allow life to exist". Kane, Perry and Zytkow [41], in a paper significantly entitled "The beginning of the end of the anthropic principle", argue that the success of the string theory program would preclude any anthropic considerations. Murray Gell-Mann [32] has written that the strong Anthropic principle "would supposedly apply to the dynamics of the elementary particles and the initial condition of the universe, somehow shaping those fundamental laws so as to produce human beings. That idea seems to me so ridiculous as to merit no further discussion." Weinberg ([85], p. 238) has asserted that this type of anthropic reasoning "just amounts to an assertion that the laws of nature are what they are so that we can exist, without further explanation. This seems to me to be little more than mystical mumbo jumbo."

The most desirable epistemic situation would be that the values of the fundamental constants of physics that we now put by hand (through experimental measurement) could one day be derived from some fundamental physical theory. In the mean time (and this mean time can last forever) they have to be accepted as brute facts.

Biologists have not been less severe on the strong anthropic principle than physicists. John Maynard Smith (1996) commented: "How can this curious claim be understood? The simplest interpretation is that the Universe was designed by a creator who intended that intelligent life should evolve. This interpretation lies outside science. ... as biologists, we are unhappy with the anthropic principle because, faced with a need for historical explanation, it seems to be a cop-out". In biology we explain, for example, the historical origin of the eukaryotes by the symbiotic theory, that is supported by empirical evidence, like the presence of specific DNA and a bacteria-like translating machinery in the mitochondria of the eukaryotic cells. "It would be unsatisfactory to argue that, because eukaryotes are in fact here, the many accidents, however unlikely, needed to give rise to them must have happened."

## 11. Religion in disguise

As the strong anthropic principle cannot be justified in terms of physics, some authors have given it a theological interpretation: the Universe must produce life, because God created it with that intention. The fact of our existence would allegedly explain why the physical constants have the (life-compatible) values they have only if complemented by the assumption that either there is a personal omnipotent God intent on producing humans or all the possible worlds do in fact exist. "The conditions in our universe really do seem to be uniquely suitable for life forms like ourselves ... But the question remains – *is* the universe tailor-made for man? Or is it ... more a case that there is a whole variety of universes to "choose" from, and that by our existence we have selected, off the peg as it were, the one that happens to fit?" [35].

The metaphysical speculations on the strong anthropic principle have trickled down through popular science books into the hands of theologians and moralists. So the moral philosopher Derek Parfit began his 1993 Harvard course on ethics and metaphysics with the words: "Modern physics tells us that either God exists (and wants to produce humans) or all possible worlds actually exist (and so this one, fit for us)." In published words of Parfit [65]: "Of these different global possibilities, one must obtain, and only one can obtain. ... That is why, rather than believing that the Big Bang merely happened to be right for life, we should believe either in God or many worlds."

Reviewing Barrow and Tipler's book, Helge Kragh [46] commented: "Under the cover of the authority of science and hundreds of references Barrow and Tipler ... contribute to a questionable ... mystification of the social and spiritual consequences of modern science. This kind of escapistic physics, also cultivated by authors like Wheeler, ... and Dyson, appeals to the religious instinct of man in a scientific age." He also underlines "the speculative, quasi-religious nature of this kind of science writing". William Press [66] in his review of the same book in *Nature* made unusually harsh remarks: "The authors badly want to be the founding doctrinal theorists of a 'new' resurgence of teleological belief in science." Their "end is nothing less than the fusion of matters of science with matters of individual faith and belief... It has taken us a long way to separate these matters... We should not lightly allow them to become once again jumbled, least of all by a book ... whose extra-scientific agenda most of us will, ultimately, wish to reject" ... "There is some fundamental intellectual dishonesty here, some snake oil to be peddled. The authors ... do not always play fair with their readers. ... There is too a distressing amount of what seems to be mathematical flim-flam, that is, quotations of precise results in a manner designed to mislead less-mathematical readers and cause them to jump to the author's desired (usually non-mathematical) conclusion."

Several theologians and moralists, like Swinburne, Craig or Parfit, have heard the talk about fine tuning of initial conditions, and infer that we are confronted with a choice between only two alternatives: either God exists and is keen on producing humans, or many (all possible) worlds exist, and our existence has the selection effect of picking up one of the few compatible with life. Some theologians use the obvious difficulties of the many worlds hypothesis as grist for their own theistic mill. "We appear then to be confronted with two alternatives: posit either a cosmic Designer or an exhaustively random, infinite number of other worlds. Faced with these options, is not theism just as rational a choice as multiple worlds?" [17].

Some cosmologists of strong religious convictions are ready to go to still greater lengths than the theologians in defense of their faith. Frank Tipler (a Tulane University physicist and co-author with Barrow of the classical book on the anthropic principle) later went on to make propaganda for the Christian doctrine of resurrection in the TV program *Soul* (1992), in which he predicted that life will take over the whole physical universe in the future, and resurrection of all the dead people in mind and body will take place, via a gigantic processing of information under the pushing will of God. "What I mean by resurrection is exactly the same thing as is taught in our churches" – said Tipler in the program. Life is information processing and the future supercomputer will run the complete brain-program of everyone who ever lived. A background voice summarizes the message: "Theology has become a branch of physics." In 1994 Tipler published a large book, *The Physics of Immortality*, in which he pretended to "prove" that all creatures who ever lived will be resurrected at the end of time. Its reception in the scientific press was unanimously devastating. All reviewers coincided that the book is a hoax.

Well, theology has not yet quite become a branch of physics. Neither has the anthropic principle. It has not even become a part of theology. The traditional theological idea pictures God as an absolute and omnipotent monarch, who can directly satisfy any of His desires. The "anthropic" view would demote God to the ceremonial role of a mere constitutional monarch, who has to work within the limits of a constitution (the laws of physics and the standard model of cosmology) that He is unable to change, His power reduced to the fixing of some specific details. It is as if He was unable to create people straight away, but had to go the long and roundabout way of tinkering with the values of the fundamental constants and fine tuning them, so that after many billions of years in some insignificant speck of the universe His wish could finally be fulfilled. Such a God would not resemble the God of the Bible. Little wonder that some theologians [17] are less than enthusiastic about the anthropic principle.

## 12. Participatory anthropic principle

Wheeler [87] took a hint from a version of the Copenhagen interpretation of quantum mechanics implying that the quantum characteristics of the observed system are created by the act of observation. He asked: "Is the architecture of existence such that only through 'observership' does the universe have a way to come into being?" He did not advocate this line of thought. "It is too frail a reed to stand either advocacy or criticism." But, as remarked by Earman [24], "excessive caution is not one of the faults of anthropic theorists". Barrow (1982) and Barrow and Tipler [4] elevated the speculation on the alleged observer-dependency of the universe to the status of a principle, the participatory anthropic principle: "Observers are necessary to bring the universe into being".

Other authors were more critical. So Gale [30]: "With his hypothesis Wheeler has carried the anthropic principle far beyond the domain of the logic of explanation; he has crossed the threshold of metaphysics: Few scientists or philosophers of science would be comfortable with his vision". And in 1986 he added:

> Wheeler ... has incorporated the strongest possible teleology and anthropocentrism into his cosmos. Wheeler takes the ensemble view, and adds a further bit of spice from quantum mechanics. ... Observed events become actual only in the very event of being observed. Although this proposition ... has roots going back to Bishop Berkeley, ... Wheeler, however, goes even further than these ontologies by coupling together the three notions of quantum reality, ensemble universe, and anthropic principle. According to his view, that universe comes into existence which, through the participation of intelligent observers, can come into existence via the act of observation itself. On this view, the observer and the observed are linked together in a self-excited loop of self-causation. ... Here a creation drama rather than a salvation drama would be the focus of history and evolution, but still of totally central focus would be the human (or other) intelligence which would serve as the reason for it all.

Earman [24] uncovered a crucial misunderstanding in Wheeler's participatory anthropic speculation: "even if one opts for a dualistic Process 1 – Process 2 [Copenhagen-style] interpretation of QM, with conscious observers playing a central role in the former, it does not follow that without conscious observers the world would not have being, existence, reality, or actuality, but only that certain kind of changes would not take place in it. After a Process 1 [reduction or collapse by observation] change the world is no more real or actual than before; and the QM state after measurement contains just as many (though different) unactualized possibilities as before." It is difficult not to subscribe to Earman's conclusion: "Failing to find any firm ground in physics for PAP, ... my concern is with attempts to wrap PAP in the cloth of scientific respectability. These attempts amount to no more than hand waving. As a scientific principle, the participatory anthropic principle has a peculiarly apt acronym."

## 13. Many worlds

Some authors adhere to the strong anthropic principle, but reject religious explanations based on the will of God. They try to square the circle by appealing to a plurality of universes. It is doubtful that the trick works.

At the end of his classic 1974 paper, Carter wrote:

> It is of course always philosophically possible – as a last resort, when no stronger physical argument is available – to promote a *prediction* based on the strong anthropic principle to the status of an *explanation*

by thinking in terms of a "world ensemble". By this I mean an ensemble of universes characterized by all conceivable combinations of initial conditions and fundamental constants.

He offered a "world ensemble explanation of the weakness of the gravitational constant". (A stronger $G$ would be incompatible with planets and so with observers). Then he tried to relate his idea of many universes to Everett's interpretation of quantum mechanics:

> Although the idea that there may exist many universes, of which only one can be known to us, may at first sight seem philosophically undesirable, it does not really go very much further than the Everett doctrine (...) to which one is virtually forced by the internal logic of quantum theory. According to the Everett doctrine the Universe, or more precisely the state vector of the Universe, has many branches of which only one can be known to any well-defined observer (although all are equally "real"). This doctrine would fit very naturally with the world ensemble philosophy that I have tried to describe.

There are several confusions here. As pointed out by McMullin [59], not only is there no warrant whatsoever for the many-universe concept, but

> Carter cites the Everett "branching worlds" model in quantum theory "to which one is virtually forced by the internal logic of quantum theory". But one is *not* virtually forced to it; indeed, it has found little support among quantum theorists. But more important, Everett's branching worlds do not provide the range of alternative initial conditions or alternative physical laws that this version of an anthropic explanation of the initial parameter constraint would require.

The assumption of infinitely many universes and the application of the anthropic principle to "select" one suitable for life is supposed to explain the "cosmic coincidences" that allow carbon atoms to be available for life and ultimately for humans to exist. One philosopher convinced by this line of argument is the neo-Platonist Leslie, who has devoted an entire book [47] and several articles to the defense of this position. His most recent stand [49] tends to build on an assumed metaphysical principle (another one!) of "ethical requiredness" that is supposed to be creatively effective. The other convinced philosopher is Nick Bostrom [9], whose anthropic reasoning emphasizes the observation selection effects. Of course statisticians have to be careful in the selection of their samples and mindful of distortions introduced in their data by unintended selection effects of their procedures. If you catch fish in a pond with a net that only retains fish larger than 15 cm, you cannot conclude that all or most of the fish in the pond are larger than 15 cm. Your result is merely an artifact of your sampling method and it doesn't need to reflect the actual situation in the pond. But it is difficult to follow Bostrom and Leslie when they try to apply statistical considerations to the one and only universe available to empirical science.

Roger White [89] argues that even if the hypothesis of there being many universes increases the probability that some universe will be life-permitting, it does not increase the probability that *our* universe is life-permitting. The hypothesis is that the initial conditions and constants of each universe are chosen randomly and independently of the other universes. The choices are like independent rolls of a die. So, the appeal to many worlds does nothing to explain why this one allows for the existence of life.

## 14. Multiverses galore

Authors fond of many universes talk about them in a variety of incompatible ways. The totality of the many universes accepted by an author forms the multiverse for that author. There are at least as many multiverses as authors talking about them; in fact, there are more, as some authors have several multiverses to offer. Let us recall the main proposals for multiple universes. The first was due to a philosopher; the rest to physicists.

(1) David Lewis [50] put forward an impressive and tightly argued theory of the plurality of worlds, called modal realism. Possible worlds and possible individuals occupy the whole logical space. "There are so many other worlds, in fact, that absolutely every way that a world could possibly be is a way that some world is." (p. 2). All possible worlds and possible individuals exist in a full and concrete sense, in the same sense in which we and our own "worldmates" exist. Our actual world is just one world among uncountably many, the world where we happen to be, but it is no more real or concrete than other worlds, which are possible for us but actual for their inhabitants (if any). Nevertheless there are no spatiotemporal or causal relations of any kind between worlds. Each world is isolated. Only of (logically) impossible worlds and impossible individuals can we say that they do not exist.

Lewis thought he could define the modal notions of necessity, possibility, predetermination, supervenience, counterfactuals and natural properties by quantification over possible worlds and possible individuals. For example, necessity is truth at all possible worlds; possibility amounts to existential quantification over the worlds. He called his theory of modal realism "a philosopher's paradise" and was ready to pay the price for it: an overblown ontology. But one philosopher's paradise is another philosopher's nightmare. He himself complained of "the incredulous stare" often meeting the exposition of his ideas. But even if, for the sake of argument, we accepted such other worlds, totally isolated from ours, how could we say anything about them? Lewis insisted that we can gain information about other worlds by relying on "our abundant modal knowledge". He thought obvious that there are uncountable infinities of donkeys. We have "necessary knowledge that there are donkeys at some worlds – even talking donkeys, donkeys with dragons as worldmates ..." (p. 112).

If the condition for the existence of a possible world is just the logical consistency of its description, why don't we let consistency do the work instead, as in mathematics? Lewis compared his acceptance of many worlds with the acceptance in mathematics of many sets. Both are useful and allow for a simple theory. But, as Cantor remarked, mathematics is the realm of freedom. Physics and metaphysics are not. Lewis denied that possible worlds are just mathematical structures. He wanted them to be concrete and physical, like ours.

His ontological flamboyance notwithstanding, Lewis gave a sober assessment of the pretences at explanation of other multiversalists. "There may indeed be a sense in which modal realism makes us more comfortable with the arbitrary, brute facts of the world ... but I insist that they remain arbitrary and they remain unexplained." (p. 128). "Of course, any inhabitant of a world will find that his world is a habitable one. That is only to be expected. It does not cry out for further explanation." And he did not see any merit in so-called anthropic explanations:

> It's all very well to invoke the anthropic principle when the remarkable habitability of our world seems to cry out for explanation. But I do not think that this invoking of the anthropic principle is *itself* an explanation. ... It is not an explanation because it gives no information about the causal or nomological ways of our world. It tells us nothing about how any event was caused; it does nothing to subsume laws under still more unified and general laws. (p. 132-133)

(2) George Ellis and G. Brundrit [28] suggested the existence of many unconnected domains beyond each other's horizon and inside an open infinite FRW universe. This is the most modest of the proposals and it might be plausible, even if the conclusions drawn by the authors rely on a misleading argument about the infinite (more on this later).

(3) Wheeler [87] suggested an oscillatory or cyclic universe. (It has nothing to do with Wheeler's participatory anthropic principle). An infinite sequence of alternatively crunching and expanding universes goes on for ever. Each big crunch "rebounds" in a new Big Bang. The successive universes are reborn with entirely new initial conditions. So all (?) Big Bang models are realized and (no wonder) a few of them are hospitable to life. There are obvious difficulties with this proposal. If the expansion rate of a cycle is sufficiently great (if the universe of that cycle is open or flat), the recollapse and crunch will not take place. The universe will continue to expand forever and the cyclic scenario will destroy itself. Besides and more fundamentally, as pointed by Earman [24], a causal curve approaching a Big Bang singularity cannot be continuously extended through the singularity. Ian Hacking [36] contended that the alleged cosmic fine tuning does not allow inference to a cyclical ensemble of universes that come into being one after the other with no memory of previous universes being carried into subsequent ones. In this last case, the argument would be as fallacious as assuming that a good poker hand is evidence that a long series of poor hands has been previously dealt. Hacking calls it "the inverse gambler's fallacy."

(4) Everett's [29] many-worlds interpretation of quantum mechanics. Actually, this proposal has nothing to do with the alleged cosmological multiverse needed in connection with the anthropic principle. Everett's many worlds of QM correspond to the various states (vectors) in a superposition of states. They are introduced in order to avoid the reduction or collapse of the wave function at the act of measurement. The many worlds of inflation and other cosmological contexts have nothing to do with them. It is an obvious mistake to confuse one with the other. This confusion (already present in Carter) has been denounced by McMullin, Peter Mittelstaedt and others. Earman [24] uncovered a crucial flaw:

Anthropic theorists are not above some double dealing. ...they appeal to the Everett many-world interpretation of QM to generate an actual ensemble of worlds; but recall that the main motivation for this interpretation was to avoid process 1 [the reduction of a superposition to an eigenstate of the observable being measured] changes altogether, whether such changes are induced by conscious observers or otherwise. This fact is conveniently ignored when it does not suit.

The confusion of Everett's many worlds with the alleged cosmological multiverse is too much even for a staunch advocate of many universes like Rees. Indeed there are different scenarios for the many universes. "However one of them, at most, can be correct. Quite possibly none is: there are alternative theories that would lead to just one universe" ([70], p. 171).

(5) Vilenkin [82] and Andrei Linde [53] proposed a different kind of multiverse, based on quantum cosmology (a speculative offshoot of quantum mechanics) and the inflationary universe scenario. According to them, quantum fluctuations in nothing (where "nothing" means not only the absence of matter, but also of space and time) produce a multiplicity (infinity) of nucleations, each of which leads to a different eternally inflating superuniverse. Each of these superuniverses is totally (spatially, causally, informationally) disconnected from the others. Each superuniverse of eternal inflation produces a never-ending series of different bubble-universes, but all of them are parts of the same spacetime (even if cosmic time is only well defined inside a single bubble), and you can define hypersurfaces cutting across universes. Our observable universe is just a region of our bubble universe, a particular region flattened by inflation (always according to this eternal inflation scenario). Vilenkin applies the anthropic principle very crucially in his quantum cosmology to constrain the immense set of initial conditions. The probabilities are weighted by the probability of producing observers or civilizations. But he considers galaxies as good tracers of observers and civilizations, and so he restricts his attention just to the conditions for galaxy formation. So the anthropic principle is here a galactic principle.

Eternal inflation seems to be almost unavoidable in the inflationary scenario. The false vacuum energy does not get exhausted with the inflationary bubble, because the expansion of the bubble proceeds more rapidly than the decay of the false vacuum. If there is inflation (a big if), then it is plausible that eternal inflation obtains. The inflationary superuniverse goes on inflating forever (outside our bubble). Notice that if you accept both quantum cosmology and eternal inflation, you get an infinity of infinities of bubble universes.

(6) Lee Smolin [74] proposed a highly speculative evolutionary cosmological model: whenever a black hole is formed, processes deep inside it might trigger the creation of another universe into a space disjoint from our own. The so created "baby-universe" inherits some of the properties of its parent-universe, including the capacity to produce new black holes and so to spawn new generations of universes without end.

(7) There are still other conjectures that suggest a multiplicity of universes. For instance, Lisa Randall and Raman Sundrum [68,69] think of multiple universes as branes, four-dimensional subspaces of an assumed five-dimensional spacetime. The

branes are separated in extra dimensional space and there is no contact between them.

(8) Max Tegmark [78] is the ultimate multiversalist, a true believer in all kinds of proposed multiverses and oblivious of the warnings of their mutual incompatibility: "The key question is not whether the multiverse exists but rather how many levels it has." To all the previous proposals, he adds a new "Platonic" level of multiple universes, so that the whole physical (?) multiverse gets final symmetry and closure. This level IV encompasses just any thinkable structure, because anything thinkable is physically realized:

> Why was only one of the many mathematical structures singled out to describe the universe? A fundamental asymmetry appears to be built into the very heart of reality. As a way out of this conundrum, I have suggested that complete mathematical symmetry holds; that all mathematical structures exist physically as well. Every mathematical structure corresponds to a parallel universe. The elements of this multiverse do not reside in the same space but exist outside of space and time. Most of them are probably devoid of observers. This hypothesis can be viewed as a form of radical Platonism, asserting that the mathematical structures in Plato's realm of ideas ... exist in a physical sense. Level IV brings closure to the hierarchy of multiverses, because any self-consistent fundamental physical theory can be phrased as some kind of mathematical structure. [p. 40].

There is nothing wrong with grand speculations with no connection to empirical reality, but they should not be confused with empirical science. Empirical science is rooted in mistrust of mere reason and insistence on empirical checks. Some scientists forget it at their own risk. Rees [70] writes: "The multiverse concept is already part of empirical science: we may already have intimations of other universes ..." The cover of the May 2003 issue of *Scientific American* asserts: "Infinite Earths in Parallel Universes Really Exist"; in the interior, the title and banner of Max Tegmark's paper reads: "Parallel Universes: Not just a staple of science fiction, other universes are a direct implication of cosmological observations."

The hypothesis that all possible worlds exist is as difficult to understand as it is to accept. The set of all possible worlds is not at all defined with independence from our conceptual schemes and models. If we keep a certain model (with its underlying theories and mathematics) fixed, the set of the combinations of admissible values for its free parameters gives us the set of all possible worlds (relative to that model). It changes every time we introduce a new cosmological model (and we are introducing them all the time). Of course, one could propose to consider the set of all possible worlds relative to all possible models formulated in all possible languages on the basis of all possible mathematics and all possible underlying theories, but such consideration would produce more dizziness than enlightment.

In any case, there seems not to be the slightest reason for accepting an infinite (or finite, for that matter) plurality of universes different from and unconnected with the one we inhabit. Of course, they are not impossible. But neither are impossible the bizarre mythological worlds. Even supporters of the anthropic principle,

like Carr [12], are skeptical here: "Both the 'many worlds' and 'many cycles' explanations for the anthropic principle are rather bizarre and I would not recommend
that either be taken too seriously." Lightman [52] speaks about being "uncomfortable with postulating different universes. We inhabit just our one universe, and
arguments that must go outside of that universe in order to explain it may have
also gone outside science."

## 15. Infinity does not imply realization of all possibilities

A whole family of anthropic explanations proceeds from the assumption that
all physically possible universes are somehow realized in an ensemble of actually
existing universes. This ensemble would allegedly guarantee the existence of at
least a universe hospitable to life, like ours. "In an infinite ensemble, the existence
of some universes that are seemingly fine-tuned to harbor life would occasion no
surprise." ([70], p.xvii). Of course, in that case, and as stressed by Wilson (1993),
it would be the ensemble (rather than the anthropic principle) that would carry
the weight of the explanation. Unfortunately the effort is wasted, as the premise
remains arbitrary and the logical reasoning is flawed. Let us elaborate this last
point.

A frequent confusion in the anthropic literature is the notion that an infinity
of objects characterized by certain numbers or properties implies the existence
among them of objects with any combination of those numbers or characteristics.
Ellis and Grundrit [28] asserted that, in an infinite collection of universes, every
possible universe has to be realized and even that it has to be repeated an infinity
of times. The multiverse would "contain infinitely many planets with histories
almost exactly like Earth's, with infinitely many beings named G.W. Leibniz, for
instance." The same contention was repeated by Leslie [47]. Despite its falsity,
this idea keeps recurring in the anthropic and multiversalist literature to this day,
as the two following quotes show:

> If the universe were literally infinite, then anything, however improba-
> ble, could happen. Indeed, it could happen infinitely often, leading to
> replicas of our Earth, even infinitely many of them. But these clones
> would be located far beyond our own galaxy ... ([70], p. 29)

> Is there a copy of you reading this article? A person who is not you
> but who lives on a planet called Earth, ...? The life of this person has
> been identical to yours in every respect ... The idea of such an alter
> ego ... merely [assumes] that space is infinite (or at least sufficiently
> large) in size and almost uniformly filled with matter, as observation
> indicate. In infinite space, even the most unlikely events must take
> place somewhere. There are infinitely many other inhabited planets,
> including not just one but infinitely many that have people with the
> same appearance, name and memories as you ... [The] universes of
> your other selves ... are the most straightforward example of parallel
> universes. Each universe is merely a small part of a larger 'multiverse'.
> ([78], pp. 31-32).

This suggestion is mistaken. An infinity does not imply at all that *any* arrangement is present or repeated. For example, think of the trivial case of an infinite set of binary sequences $s_n$ with the $i$-th member $x_i = 1$ if $i \neq n$, and $x_i = 0$ if $i = n$:

$s_1 = 0111111111111111111\ldots$
$s_2 = 1011111111111111111\ldots$
$s_3 = 1101111111111111111\ldots$
$s_4 = 1110111111111111111\ldots$
$s_5 = 1111011111111111111\ldots$

And so on. As $n$ ranges over all natural numbers, we get an infinity of different binary sequences that are almost everywhere $= 1$, but differ in the place where they are $= 0$. This set of binary sequences is infinite, but most binary sequences you can think of (for example, any containing two or more 0's, such as $1010101010\ldots$) are not in it. And no sequence is repeated.

Of course there are much simpler counterexamples. The infinite set of the even numbers does not contain any of the odd numbers. The infinite set of the numbers greater than a trillion does not contain any of the numbers up to one trillion. In general, all infinite sets contain proper infinite subsets. This property was famously used by Dedekind to define infinity. The same happens with uncountable domains, like the n-dimensional Euclidean spaces. Any interval of the real line is an infinite set of real numbers, but does not contain all real numbers; most of them remain outside. Any straight line in the plane is an infinite set of points of the plane, but does not include all points of the plane.

If every possible world is characterized by a finite sequence of numbers (including quantum numbers, as in [78]), then every possible world can be coded or represented by a different natural number (via a Gödel numbering of all sequences). So, the assumption that all possible worlds are realized in an infinite universe is equivalent to the assertion that any infinite set of numbers contains all numbers (or at least all Gödel numbers of the sequences), which is obviously false.

Some anthropic authors are aware of these difficulties. Barrow and Tipler ([4], p. 24), discussing the alleged infinite repetition of everything in an infinite universe, remarked that "the infinity alone is not a sufficient condition for this to occur; it must be an exhaustively random infinity in order to include all possibilities". Following this observation, some authors speak of there being an "exhaustive infinity" of universes. Of course, from the assumed existence of an exhaustive infinity of universes you can tautologically infer that every possibility is realized, by definition (if it was not, the infinity would not be exhaustive). Other authors consider the universe a statistical thermodynamic system in equilibrium satisfying the ergodicity hypothesis or take at face value some far-fetched scenarios of inflationary cosmology. But in these cases infinity alone no longer carries the weight of the argument, it is just another among many unchecked assumptions being piled up in this line of reasoning.

## 16. Conclusions

In its weak version, the anthropic principle is a mere tautology, which does not allow us to explain anything or to predict anything that we did not already know. In its strong version, it is a gratuitous speculation, only sustained by previous religious faith. The attempt to secularize it, by an appeal to an infinity of universes unconnected by principle with our own, ends in failure. In McMullin's [60] words, "the weak Anthropic principle is trivial ...and the strong Anthropic principle is indefensible." Alleged anthropic explanations do not explain anything and are not needed in cosmology. And if someone still intends to revive the corpse of anthropocentrism, he will need stronger medicine than just the anthropic principle itself.

## REFERENCES

1. E. Agazzi and A. Cordero (ed.). *Philosophy and the Origin and Evolution of the Universe.* Dordrecht-Boston-London: Kluwer Academic Publishers, 1991.
2. Y. Balashov. Resource Letter AP-1: The Anthropic Principle. *American Journal of Physics,* 59:1069-76, 1991.
3. J. Barrow. The mysterious lore of large numbers. In: Bertotti, Balbinot, Bergia & Messina, 67-93, 1990.
4. J. Barrow and F. Tipler. *The Anthropic Cosmological Principle.* Clarendon Press. Oxford, 1986.
5. J. Barrow, H. Sandvik and J. Magueijo. Anthropic Reasons for Non-zero Flatness and Lambda. *Astro-ph*/0110497 (22 Oct 2001), submitted to *Physical Review D.*
6. G. Beck, H. Bethe and W. Riezler. Bemerkung zur Quantentheorie der Nullpunktstemperatur. *Naturwissenschaften,* 19:39.
7. F. Bertola and U. Curi (eds.). *The Anthropic Principle.* Cambridge University Press, 1993.
8. B. Bertotti. *Modern Cosmology in Retrospect.* Cambridge University Press, Bergia & Messina (ed.), 1990.
9. N. Bostrom. *Anthropic Bias: Observation Selection Effects in Science and Philosophy.* New York: Routledge, 2002.
10. R. Boyd. *The Philosophy of Science.* Cambridge (Mass.): MIT Press, Gasper & Trout (ed.), 1991.
11. S. Bromberger. *On What We Know We Don't Know.* The University of Chicago Press, 1992.
12. B. Carr. On the origin, evolution and purpose of the physical universe. *The Irish Astronomical* Journal, 15:237-253, 1982.
13. B. Carr and M. Rees. The anthropic principle and the structure of the physical world. *Nature,* 278:605-612, 1979.
14. B. Carter. Large number coincidences and the anthropic principle in cosmology. In: Longair (1974), 291-298, 1974.
15. B. Carter. The anthropic principle and its implications for biological evolution. *Philosophical Transactions of the Royal Society of London,* A 130:347-363, 1983.

16. C. Collins and S. Hawking. Why is the universe isotropic? *Astrophysical Journal*, 180:317-34, 1973.

17. W. Craig. Barrow and Tipler on the anthropic principle vs. divine design. *British Journal for the Philosophy of Science*, 38:389-395, 1988.

18. R. Dicke. Dirac's cosmology and Mach's principle. *Nature*, 192:440-441, 1961.

19. H. Dingle. Modern Aristotelianism. *Nature*, 139:784-786, 1937.

20. H. Dingle. Deductive and inductive methods in science: A reply. *Nature*, 139:1011-1012, 1937a.

21. P. Dirac. The cosmological constants (Letter to the Editor). *Nature*, 139:323, 1937.

22. P. Dirac. New basis for cosmology. *Proceedings of the Royal Astronomical Society of London*, A 165:199, 1938.

23. N. Dowrick and N. McDougall. Axions and the anthropic principle. *Physical Review*, D 38:3619-3624, 1988.

24. J. Earman. The SAP also rises: A critical examination of the anthropic principle. *American Philosophical Quarterly*, 24:307-317, 1987.

25. J. Earman and J. Mosterín. A Critical look at inflationary cosmology. *Philosophy of Science*, 66:1-50, 1999.

26. A. Eddington. *The Mathematical Theory of Relativity*. Cambridge University Press, 1923.

27. G. Ellis. Major themes in the relation between philosophy and cosmology. *Mem. Soc. Astronomica Italiana*, 62:553-605, 1991.

28. G. Ellis and G. Brundrit. Life in the infinite universe. *Quarterly Journal of the Royal Astronomical Society*, 20:37-41, 1979.

29. H. Everett. Relative state formulation of Quantum Mechanics,. *Reviews of Modern Physics*, 29:454-462, 1957.

30. G. Gale. The anthopic principle. *Scientific American*, 245:154-171, June 1981.

31. G. Gale. Whither cosmology: anthropic, anthopocentric, teleological? En N. Rescher (ed.): *Current Issues in Teleology*, Lanham, Md.: University Press of America, 102-110, 1986.

32. M. Gell-Mann. *The Quark and the Jaguar: Adventures in the simple and the Complex*. Little, Brown and Co., 1994.

33. J. R. Gott III. Implications of the Copernican priciple for our future prospects. *Nature*, 363:315-319, 1993.

34. G. Greenstein. *The Symbiotic Universe*. New York: William Morrow, 1988.

35. J. Gribbin and M. Rees. *Cosmic Coincidences: Dark Matter, Mankind, and Anthropic Cosmology*. New York: Bantam Books, 1989.

36. I. Hacking. The inverse gambler's fallacy: The argument from design: The anthropic principle applied to Wheeler universes. *Mind*, 96:331-340, 1987.

37. R. Hakim. The special status of cosmology in science. In: F. W. Meyerstein (ed.), *Foundations of Big Bang Cosmology*, Singapore-London: World Scientific, 85-139, 1989.

38. P. J. Hall. Anthropic explanations in cosmology. *Quaterly Journal of the Royal Astronomical Society*, 24:443-447, 1983.

39. S. Hawking and N. Turok. Open inflation without false vacua. *Phys. Lett.*, B 451:25, 1998.

40. F. Hoyle. On nuclear reactions occurring in very hot stars. *Astrophysical Journal*, Suppl, 1:121-146, 1954.
41. G. Kane, M. Perry and A. Zytkow. The beginning of the end of the anthropic principle. *Astro-ph*/0001197, 2000.
42. B. Kanitscheider. Anthropic arguments – Are they really explanations? In Bertola, F. & U. Curi (1993), 171-181, 1993.
43. P. Kirschenmann. Does the anthropic principle live up to scientific standards? *The Annals of the Japan Association for Philosophy of Science*, 8:69-96, 1992.
44. R. Klee. The revenge of Pythagoras: How a mathematical sharp practice undermines the contemporary design argument in astrophysical cosmology. *British Journal for Philosophy of Science*, 53:331-354, 2002.
45. J. Knobe, K. Olum and A. Vilenkin. Philosophical implications of inflationary cosmology. *Arxiv.org/abs/physics*/0302071, 2003.
46. H. Kragh. Review of Barrow and Tipler's *The Anthropic Cosmological Principle*. *Centaurus*, 30:191-194, 1987.
47. J. Leslie. *Universes*. London-New York: Routledge, 1989.
48. J. Leslie (ed.). *Physical Cosmology and Philosophy*. New York: Macmillan, 1990.
49. J. Leslie. Our Place in the Cosmos. *Philosophy*, 75:5-24, 2000.
50. D. Lewis. *On the Plurality of Worlds*. Oxford: Basil Blackwell, 1986.
51. A. Lightman and R. Brawer. *Origins. The Lives and Worlds of Modern Cosmologists*. Cambridge (Mass.): Harvard University Press, 1990.
52. A. Lightman. *Ancient Light. Our changing view of the universe*. Cambridge (Mass.): Harvard University Press, 1991.
53. A. Linde. Eternal chaotic inflation. *Modern Physics Letters*, A 1:81, 1986.
54. A. Linde. Inflation, quantum cosmology and the anthropic principle. In Barrow, Davies & Harper (ed.), *Science and Ultimate Reality: From Quantum to Cosmos*, Cambridge University Press, 2003.
55. M. Livio, D. Hollowell, A. Weiss and J. W. Truran. The anthropic significance of the existence of an excited state of $^{12}C$. *Nature*, 340:281-284, 1989.
56. M. Longair (ed.). *Confrontation of Cosmological Theories with Observational Data*. International Astronomical Union, 1974.
57. H. Martel, P. R. Shapiro and S. Weinberg. Likely values of the cosmological constant. *Astrophysical. Journal*, 492:29, 1998.
58. J. Maynard Smith and E. Szathmáry. On the likelihood of habitable worlds. *Nature*, 384:107, 1996.
59. E. McMullin. Indifference principle and anthropic principle in cosmology. *Studies in the History and Philosophy of Science*, 24:359-389, 1993.
60. E. McMullin. Fine-tuning the universe? In: M. Shale & G. Shields (ed.), *Science, Technology, and Religious Ideas*, Lanham: University Press of America, 1994.
61. J. Mosterín. Philosophy and cosmology. In: *Spanish Studies in the Philosophy of Science* (ed. by G. Munévar), Kluwer Academic Pubishers, Dordrecht-Boston-London, 57-89, 1996.
62. J. Mosterín. The anthropic principle in cosmology: A critical review. *Acta Institutionis Philosophiae et Aestheticae* (Tokyo, Japan), 18:111-139, 2000.

63. J. Mosterín. Examen del principio antrópico en cosmología. *Diálogos*, 79:203-236, 2002.

64. J. Mosterín and R. Torretti. *Diccionario de Lógica y Filosofía de la Ciencia.* Madrid: Alianza Editorial, 2002.

65. D. Parfit. Why does the universe exist?. *Times Literary Supplement*, 3 July, 1992.

66. W. Press. A place for teleology? *Nature*, 320:315-316, 1986.

67. J. Primack. Quote in *Time*, Jan 18, 1993.

68. L. Randall and R. Sundrum. Large mass from a small extra dimension, *Physical Review Letters*, 83:3370-3373, 1999.

69. L. Randall and R. Sundrum. An alternative to compactification, *Physical Review Letters*, 83:4690-4693, 1999.

70. M. Rees. *Our Cosmic Habitat.* Princeton Universsity Press, 2001.

71. M. Shaposhnikov and I. Tkachev. Higgs boson mass and the anthropic principle. *Modern Physics Letters*, A 5:1659-1661, 1990.

72. Q. Smith. The anthropic principle and many-worlds cosmologies. *Australasian Journal of Philosophy*, 72:371-382, 1985.

73. Q. Smith. Anthropic explanations in cosmology. *Australasian Journal of Philosophy*, 72:371-382, 1994.

74. L. Smolin. *The Life of the Cosmos.* Oxford University Press, 1997.

75. L. Susskind. The anthropic landscape in string theory. *arXiv:hep-th*/0302219, 2003.

76. R. Swinburne. Argument from the fine tuning of the universe. In: Leslie 1990, 154-173, 1990.

77. M. Tegmark. Parallel universes. In Barrow, Davies & Harper (eds.), *Science and Ultimate Reality: From Quantum to Cosmos*, Cambridge University Press, 2003.

78. M. Tegmark. Parallel universes. *Scientific American* (May 2003), 30-41, 2003.

79. F. Tipler. The anthropic principle: A primer for philosophers. *PSA 1988*, The Philosophy of Science Association, 2:27-48, 1989.

80. F. Tipler. *The Physics of Immortality.* New York: Doubleday, 1994.

81. R. Torretti. El 'observador' en la física del siglo XX. In: F. J. Ramos (ed.), *Hacer: Pensar.* Editorial de la Universidad de Puerto Rico.

82. A. Vilenkinr. Birth of inflationary universes. *Physical Review*, D 27:2848-1855, 1983.

83. S. Weinberg. Anthropic bounds on the cosmological constant. *Physical Review Letters*, 59:2607-2610, 1987.

84. S. Weinberg. A priori probability distribtion of the cosmological constant. *ArXiv:Astro-ph*/0002387, 2000.

85. S. Weinberg. *Facing Up.* Harvard University Press, 2001.

86. J. Wheeler. Beyond the end of time. In: Rees, Ruffini & Wheeler (ed.), *Black Holes, Gravitational Waves and Cosmology*, Gordon and Breach, 1974.

87. J. Wheeler. Genesis and observership. In: J. Butts & J. Hintikka (ed.), *Foundational Problems in the Special Sciences*, Dordrecht: Reidel, 3-33, 1977.

88. J. Wheeler. Beyond the black hole. In H. Woolf (ed.), *Some Strangeness in the Proportion*, Reading (Mass.): Addison-Wesley, 341-375, 1980.

89. R. White. Fine-tuning and multiple universes. *Noûs*, 34(2):260-276, 2000.

90. G. Williams. *Natural Selection: Domains, Levels, and Challenges.* New York: Oxford University Press, 1992.

91. P. Wilson. What is the explanandum of the anthropic principle? *American Philosophical Quaterly*, 28:167-173, 1991.

92. P. Wilson. Carter on anthropic principle predictions. *The British Journal for Philosophy of Science*, 45:241-253, 1994.

# Group Agency

Philip Pettit

*Princeton University, Corwin Hall, Princeton, NJ 08540, USA*
*ppettit@princeton.edu*

**Abstract.** This is a short account of some crucial conditions that individuals must satisfy if they are to be able to work together to constitute a group agent: a centre of belief, desire and decision that needs to be recognised in its own right. Those individuals will need to take active steps to identify what it is they are to be held to believe and desire, and what it is they are to be taken to intend and do. And they will have to monitor their procedures of identification to ensure that the attitudes espoused cohere with one another in the manner required for agency; no mechanical voting procedure will guard satisfactorily against the danger of inconsistency and the like.

This paper is an attempt to sketch the main elements in a story of group agency that I have elaborated more fully, and in different ways, elsewhere [11,12]. The paper is in four sections. In the first I identify two conditions that groups must arguably meet if they are to be capable of agency. In the second section I try to show how groups can meet the first of these; in the third how they can meet the second. And then in the fourth section, I outline the prospect for group-formation to which these reflections direct us.

## 1. Two conditions on group agency

Let us assume, in line with a now well-established habit of thinking, that an agent or subject — an intentional subject. [2] — is a creature that pursues a certain network of goals in a manner that is appropriate according to a system of representation that is more or less sensitive to evidence. The goals are the attractors that elicit action, the representations provide the guidance that goal-pursuit requires. The goal-seeking states constitute the agent's desires or intentions, the representations its beliefs [10].

Can a collection of human beings constitute an intentional subject that is distinct from the individual subjects who make it up? Not, for sure, if the collection is relatively unorganized. Consider the sort of collection that is made up of relatively like-minded people, with most of the members believing and desiring the same things. This is the best candidate for an unorganized collection that might be regarded as an intentional subject in its own right. But it demonstrably fails.

If a collection of people had a large number of desires and beliefs in common then it we could indeed ascribe certain attitudes to it, and the members could be expected to behave in aggregate — say, to vote — in a manner that those attitudes explained. And this means that in a deflationary sense, as Anthony Quinton [13] argues, we might regard it as a group agent.

> Groups are said to have beliefs, emotions, and attitudes and to take
> decisions and make promises. . . To ascribe mental predicates to a group
> is always an indirect way of ascribing such predicates to its members.
> . . . To say that the industrial working class is determined to resist anti-
> trade union laws is to say that all or most industrial workers are so
> minded.

But, even by Quinton's lights, the sense in which a collection of like-minded
people constitutes an intentional subject is of only marginal interest: it is 'plainly
metaphorical', as he says.

There are two problems with the claim that a collection of like-minded people
might constitute a group agent in its own right. And these point us to the two
conditions that I think groups must meet if they are to constitute agents.

The first problem is that the group's having certain attitudes ought to be capable
of explaining things that cannot be explained just by the fact that the members, or
most of the members, have those attitudes in common. Why would the attitudes
of the collection, as distinct from the attitudes of the members, be worth marking
otherwise? Under the story sketched, the attitudes ascribed to groups would not
play any explanatory role that is not already adequately discharged by the attitudes
held by members of those groups; they would be an idle wheel.

The second problem with the claim that a collection of like-minded individuals
might count as a group agent is that any such collection would be liable to invite the
attribution of an entirely inconsistent set of attitudes. The fact that a majority of
group members hold each of ten or twenty beliefs does not ensure that those beliefs
will constitute a consistent set: and this, indeed, even if each of the members of
the group is individually consistent in the things they hold. If different majorities
hold by each of the ten or twenty beliefs, then the beliefs as a whole may fail to
constitute a consistent set. The belief-set may not amount to a representation of
how the world could be, there being no inconsistent worlds.

The possibility is easily illustrated [11]. Suppose we take a group of three people,
A, B, and C, and consider three logically connected issues on which they hold
beliefs: say, to take the simplest possible case, the issues of whether it is the case
that p, whether it is the case that q, and whether it is the case that p-and-q. The
following matrix explains how each of the members of the collection may have a
consistent set of beliefs on these matters but how the group will have to be ascribed
an inconsistent set, if group beliefs are computed on a simple majoritarian basis.

|   | P? | Q? | P-and-Q? |
|---|-----|-----|----------|
| A | Yes | No  | No       |
| B | No  | Yes | No       |
| C | Yes | Yes | Yes      |

The group will have to be held to believe that p, since A and C support this;
that q, since B and C hold by that belief; and, under ordinary assumptions, that
not p-and-q, since both A and B reject p-and-q. It will have to be credited with
an inconsistent representation of things.

The point of a system of representation, as we assumed, is to enable an agent to pursue its goals in a satisfactory way. It will pursue its goals after a fashion that the representations holds out as appropriate; and it will pursue its goals effectively only so far as the representations are accurate. This means that an inconsistent system of representation cannot serve satisfactorily in a guiding role, since there is no way the world could be that answers to it; at a certain margin, indeed, the inconsistent system of representation will freeze the agent's initiatives, supporting incompatible modes of action at one and the same time. This explains why, despite the fact that we often find ourselves to be inconsistent, we always think there is reason to try to resolve any inconsistency. But the collection of people that has inconsistent beliefs on any set of issues will not necessarily be in a position even to mark the inconsistencies in those attitudes, let alone to try and resolve them. It seems quite outlandish to depict such a collection as an intentional subject.

Let us agree, then, that the mere fact that a number of people are like-minded in some respect, or have any other property in common, is not sufficient to ensure that the collection involved counts as a collective agent: an intentional subject in its own right. Quinton himself thinks that there are no serious collective subjects, believing that the collection of like-minded agents is as close as we can ever get to a group agent. But, like many others, I think he is demonstrably mistaken. There are ways in which collections of people may be organized that enable them to escape the two problems identified and to make a claim to be intentional subjects in their own right.

## 2. Satisfying the first condition on group agency

The first of the problems identified will be overcome if we require that for a collection of people to display a group-attitude two things have to be in place. First, members have to relate to one another in a way that amounts to constructing that attitude, as we might put it. And second, the existence of the attitude constructed has to explain matters that are otherwise unexplained. No group attitude, without the connivance of members in the formation of that attitude. And no group attitude without there being something that this connivance explains.

As it happens, there is a lively literature on how these conditions can be fulfilled. This body of research is due mainly to the work of Raimo Tuomela [15], Margaret Gilbert [4], and Michael Bratman [1], and continues to flourish [8]; for a useful perspective see Velleman [16,17]. The core idea shared among these writers — and of course there are many differences that divide them — is that for a collection of people to have a group-level attitude, or to perform a group-level action, they must be involved in a structure of mutual awareness and conditionalized cooperation.

Typical of the general approach is the following sort of claim about the conditions under which a number of people have a certain intention in common. First, they each intend that they collectively do something at a certain time or in a certain situation, each assuming that others are similarly disposed. Second, this is a matter of mutual awareness: each is (in a position to be) aware of the common intention and assumption, each is (in a position to be) aware that each is in (a position to be) aware of that common intention and assumption, and so on. And third, as a consequence, they are disposed jointly to perform the action intended at the appropriate time or in the appropriate situation. The jointly performed action will involve a cooperative scheme that is salient to all. It may consist in one person singing the soprano line, a second the tenor line, and the third the bass line. Or it may consist in each taking whatever part falls to them in some allocative lottery. Or it may consist in letting a designated representative act in their collective name or on their joint behalf.

Without going into detail, it should be clear that this sort of approach can in principle be extended to develop a story about how a group attitude of any kind might form. Take belief or judgment. A group of people may be said to assent to something in common, making it a matter of group belief, just so far as the following is true. They each intend that they collectively assent to it, assuming that others are similarly disposed; they are mutually aware of this pattern of intention and assumption; and this explains why as a matter of fact they do come to assent to the proposition, acting as a group in a manner that makes sense if the proposition is true or having their representatives act in that manner. They may intend to assent collectively to a particular proposition, after this pattern. Or, perhaps more plausibly, they may intend to assent to that proposition, among a given set of alternatives, that comes to be favoured by a certain authority or under a certain procedure: say, favoured in majoritarian voting.

There are many collections of people, small and large, that go about construct-ing collective attitudes of roughly the kind envisaged, allowing those attitudes to explain consequent responses. Suppose I belong to a voluntary association that is in the habit of voting on the purposes it will promote over this or that period, on the priorities that should prevail among those purposes, on the opportunities in this or that context for advancing them, and on what represents the best means of doing so in any given choice. When we members vote on any such matter, do we each intend that we jointly adopt this or that goal or judgment or plan, meeting conditions of the kind suggested in the mutual-awareness story? I believe that we do. We each acquiesce in an arrangement involving all of us that will serve to determine a goal by which we will orientate, a judgment by which we will be guided, or a plan that we will enact, whether in acting together or in acting via representatives. Such acquiescence amounts to nothing more or less than our each intending that we jointly adopt the goal or judgment or plan in question, assuming that others intend this too. And since acquiescence in the arrangement is going to be a matter of mutual awareness, and is going to affect how we in the group go on to behave, the other conditions are bound to be satisfied too. The little story sketched is clearly going to apply.

## 3. Satisfying the second condition on group agency

We said that there were two problems with the idea that a collection of like-minded people, just in virtue of being like-minded, could constitute a group subject. The first is that the attitudes ascribed to that collection — the attitudes held by most members — will not explain anything that could not be explained by reference to the members' individual attitudes. And the second is that the attitudes ascribed to the collection may constitute an irrational set. We have seen how the first problem can be overcome under the stipulation that if a collection of people is to constitute a group agent then members should actively connive in the formation of group-level attitudes. But does this get us over the second problem too? And if not, how are we to get over it?

The mutual-awareness stipulation, as we might call it, does not get us over the second problem. While that stipulation identifies constraints that must be satisfied for the formation of any particular attitude, be it an attitude of desire or goal-seeking, an attitude of belief, or an intention to act in a certain way, it does not impose constraints that are going to enforce consistency among the attitudes formed on different matters. It leaves open the possibility that the collection of people envisaged will fail to generate the consistent, unified vision that we expect any serious agent to instantiate [14].

The point can be illustrated by reference to the very example we used in raising the problem in the first place. Consider a group like the voluntary association just mentioned that goes to majority vote in determining its goals and judgments and intentions. Such a group, as we saw, will satisfy perfectly the sorts of constraints imposed in the mutual-awareness story. But consistently with meeting those constraints such a group can still find itself collectively assenting to inconsistent propositions. The members might vote in the pattern of A, B and C in our original matrix, leaving the group in the unhappy position of holding that p, that q and that not p-and-q.

|       | P?  | Q?  | P-and-Q? |
|-------|-----|-----|----------|
| A     | Yes | No  | No       |
| B     | No  | Yes | No       |
| C     | Yes | Yes | Yes      |
| A-B-C | Yes | Yes | No       |

One of the earliest thinkers to endorse something approximating the mutual awareness story of group-formation is Thomas Hobbes. He envisages the multitude of people in a society becoming a group proper so far as they each form the intention, conditionally on others doing so too, that they shall all be represented by a single person or assembly of persons. 'I authorize and give up my right of governing myself to this man, or to this assembly of men, on this condition, that thou give up they right to him, and authorize all his actions in like manner'... ([5], Ch. 17, s 13). Strikingly, Hobbes thinks that if the representative is an assembly — in particular, an odd-numbered assembly — then majority voting will work fine for the formation of attitudes; 'if the representative consist of many men, the voice of the greater number must be considered as the voice of them all' ([5], Ch. 16, s 15). But this, as we now know, is a mistake. Something over and beyond the mutual-awareness move is required if a collection of people is to constitute a group-agent proper.

What is necessary, of course, is that they adopt a mode of attitude-formation that protects them against the appearance of problems like the inconsistency identified in our matrix. And here there is an interesting social-choice result available, to the effect that impartial, mechanical methods, as we might call them, will not do the trick. Let a collection of people each have a complete and consistent set of judgments over a logically connected set of issues such as 'p', 'q' and 'p-and-q'. Is there a procedure of voting that will work for any inputs of individual sets of judgments, and that will guarantee the production of a group-level set of judgments with the same properties of completeness and consistency? The answer, roughly stated, is: not if the procedure is to treat every individual and every issue as equal; not if it is to give every individual a vote and put every issue up for resolution on the basis of voting ..()..([6,3,9,7]). If a collection of people is to constitute itself as a group agent, it must take active steps to ensure that the attitudes it adopts will form a coherent vision or representation.

What can the collection do, assuming that members want to generate an open range of attitudes: they are not content, for example, to have little or nothing to say on many questions raised? There are any number of possibilities. One would be to designate a particular individual as the dictator for the group, or at least the dictator in any case where voting leads to an irrational result. Another would be to designate certain issues as prior to others and to let the resolution of those issues always dictate the resolution of other questions — say, to let the resolution of 'p?' and 'q?' in our example dictate the resolution of 'p-and-q?'. But the first approach may not be appealing for democratic reasons and the second is not going to be easily implemented, since there may be no principled way of identifying the prior issues.

If a collection is going to ensure itself against holding by inconsistent judgments and the like, then in all likelihood members will have to play an active role in adjusting to one another. They may take a straw vote on every issue, for example, and then go to a consideration about who is willing to change his or her mind in the event of finding that the resolution of that issue is inconsistent with the line taken on other questions. The resolution they thereby identify may lead to a change of view on the issue under discussion from that which the straw vote supported. Or it may involve them changing their view on one or another issue that was previously resolved.

## 4. Group agency exemplified

We have identified conditions under which, with suitable organization, a collection of individuals might well claim to be able to constitute a group-agent. The discussion shows us how the collection might be able to get over the two problems raised earlier. And it points us at the same time towards a positive prospect for group-formation.

Suppose that there are a number of people who in the sense explained in the mutual-awareness type of story adopt certain collective purposes. Suppose that they adopt it as a policy that they and their representatives should only act for the promotion of collective purposes after a manner that is supported by the collective judgments that they adopt on relevant questions of priority, opportunity and means. Suppose that they manage to find a way of forming collective judgments that covers the issues required and that avoids problems of inconsistency. And suppose, finally, that they and their representatives are more or less faithful in pursuing collective purposes according to collective judgments. If all those conditions are fulfilled, then there is going to be every reason to recognize the group as a subject in its own right, distinct from the individuals subjects who combine to set it up.

The distinctness of the group subject constituted from the individuals who do the constituting will come out in the fact that there may not be a good match between the judgments and other attitudes that the group adopts and the judgments and attitudes that members hold individually. Thus when people act on behalf of the group, they may routinely find that they have to act on judgments that they do not individually accept; they may have to give expression, as it were, to a mind or persona — that of the group — that is distinct from their individual mind or persona.

This discontinuity may assume quite a dramatic form. On some issues, for example, the group may be led to endorse a judgment that no one member individually accepts. Suppose that members voted as follows on each of four issues.

|   | P? | Q? | R? | P-and-Q-and R? |
|---|-----|-----|-----|-----|
| A | Yes | No  | Yes | No  |
| B | No  | Yes | Yes | No  |
| C | Yes | Yes | No  | No  |
| A-B-C | Yes | Yes | Yes | No  |

In order to avoid committing the group to the inconsistent set of judgments in the bottom line of the matrix, the individuals in this group might well decide that the group as a whole has to amend the last majority vote and endorse p-and-q-and-r, despite the fact that they individually reject it. When they act on behalf of the group, then, they will have to act as if it were the case that p-and-q-and-r, though they may each individually believe that this is not the case.

Our world is populated with group agents of the type envisaged. The voluntary association in which everyone votes on everything represents one simple variety of group agent: a simple molecule, as it were, that is composed directly out of individual atoms. And that simple sort of structure is exemplified on a wide front: in the research collaboration , for example, in the small business partnership, and in the local party organization.

Most of the group agents in our world, of course, are much more complex than these simple groups. But like more complex molecules, these entities will generally be composed out of simple sub-groups, as in the corporation or church or university that is built up out coordinated chains and hierarchies of simpler groupings. Some of the sub-groups in such an organization will have the task of forming views on issues in this or that restricted domain — perhaps financial, perhaps technical, perhaps strategic; other sub-groups will play the role of mediating between such groups, providing feedback amongst them, and packaging the overall product; and under at least some patterns of organization, one sub-group, such as the board or council, will have the task of building that product into a coherent set of policies and action-plans. Our model will apply to complex organizations so far as it makes sense, one by one, of sub-groups of this kind. The molecules may be more complex but the chemistry will remain essentially the same.

This brief exposition of two elements in the theory of group-formation should give a flavour, I hope, of the work that remains to be done in this area. The literature on how mutual awareness can enable people to form joint intentions and the like has been of enormous importance and illumination. It indicates how the first condition on group agency can be met. But it needs to be supplemented by further research on how collections of people can also satisfy the second condition, sustaining collective entities that are answerable in their own right to demands such as that of consistency. As this research progresses, we should be able to reclaim the tradition of thinking under which a group can amount to more than the sum of its parts. We should be able to do this, in particular, without falling into the romantic and idealist idiom that gave that tradition such a bad name in the last century.

## REFERENCES

1. M. Bratman. *Faces of Intention: Selected Essays on Intention and Agency.* Cambridge, Cambridge University Press, 1999.
2. D. Dennett. *Brainstorms.* Brighton, Harvester Press, 1979.
3. F. Dietrich. Judgment Aggregation: (im)possibility theorems. *Group on Philosophy, Probability and Modeling.* University of Konstanz, 2003.
4. M. Gilbert. Collective Preferences, Obligations, and Rational Choice. *Economics and Philosophy,* 17:109-120, 2001.
5. T. Hobbes. *Leviathan.* Indianapolis, Hackett, 1994.
6. C. List and P. Pettit. The Aggregation of Sets of Judgments: An Impossibility Result. *Economics and Philosophy,* 18:89-110, 2002.
7. C. List and P. Pettit. Aggregating Sets of Judgments: Two Impossibility Results Compared. *Synthese,* forthcoming.
8. S. Miller. *Social Action: A Teleological Account.* Cambridge, Cambridge University Press, 2001.
9. M. Pauly and M. Van Hees. Some General Results on the Aggregation of Individual Judgments. *Dept of Computer Science.* University of Liverpool, 2003.
10. P. Pettit. *The Common Mind: An Essay on Psychology, Society and Politics, paperback edition 1996,.* New York, Oxford University Press, 1993.
11. P. Pettit. *A Theory of Freedom: From the Psychology to the Politics of Agency.* Cambridge and New York, Polity and Oxford University Press, 2001.
12. P. Pettit. Groups with Minds of their Own. *Socializing Metaphysics.* F. Schmitt. New York, Rowan and Littlefield, 2003.
13. A. Quinton. Social Objects. *Proceedings of the Aristotelian Society,* 75, 1975.
14. C. Rovane. *The Bounds of Agency: An Essay in Revisionary Metaphysics.* Princeton, NJ, Princeton University Press, 1997.
15. R. Tuomela. *The Importance of Us.* Stanford, CA, Stanford University Press, 1995.
16. D. Velleman. *The Possibility of Practical Reason.* Oxford, Oxford University Press, 2000.
17. D. Velleman. Review of Michael Bratman 'Faces of Intention'. *Philosophical Quarterly,* 51:119-21, 2001.

# Contrasting Applications of Logic in Natural Language Syntactic Description*

Geoffrey K. Pullum
*University of California, Santa Cruz*
and
Barbara C. Scholz
*San José State University*

**Abstract.** Formal syntax has hitherto worked mostly with theoretical frameworks that take grammars to be generative, in Emil Post's sense: they provide recursive enumerations of sets. This work has its origins in Post's formalization of proof theory. There is an alternative, with roots in the semantic side of logic: model-theoretic syntax (MTS). MTS takes grammars to be sets of statements of which (algebraically idealized) well-formed expressions are models. We clarify the difference between the two kinds of framework and review their separate histories, and then argue that the generative perspective has misled linguists concerning the properties of natural languages. We select two elementary facts about natural language phenomena for discussion: the gradient character of the property of being ungrammatical and the open nature of natural language lexicons. We claim that the MTS perspective on syntactic structure does much better on representing the facts in these two domains. We also examine the arguments linguists give for the infinitude of the class of all expressions in a natural language. These arguments turn out on examination to be either unsound or lacking in empirical content. We claim that infinitude is an unsupportable claim that is also unimportant. What is actually needed is a way of representing the structure of expressions in a natural language without assigning any importance to the notion of a unique set with definite cardinality that contains all and only the expressions in the language. MTS provides that.

## Introduction

For the last half century a large community of linguists has been devoted to the goal of stating grammars for natural languages, and general linguistic theories, in the form of fully explicit theories. Attainment of this goal means making fully explicit all of the consequences of grammatical statements and theoretical principles. We argue here, specifically with respect to syntax, that linguists have been led astray as a result of taking too narrow a view of what it means to construct an explicit grammar or theory. Linguists have largely restricted themselves to a class

---

*This paper presents the main content of the invited lecture given by the first author at the 12th International Congress of Logic, Methodology and Philosophy of Science (Oviedo, Spain, 2003). It was based on joint research and is here written up jointly. We are grateful to many people for stimulating questions and comments on these ideas, not only at the International Congress but also at Stanford University; the University of California, Santa Cruz; Simon Fraser University; the Technical University of Vienna; the University of Pennsylvania; and Harvard University. Thanks to Peter Alrenga, John Colby, Robert Conde, Ascander Dost, Paul Postal, Chris Potts, James Rogers, and Dag Westerstahl for helpful comments; they enabled us to avoid some errors, and bear no blame for the ones that remain.

of theoretical frameworks for syntax, what we call **generative** frameworks, which are in fact ill suited to stating theories and grammars for natural languages (though ironically, they are ideal for stating explicit definitions of the denumerable infinite sets of formulae referenced in fields like logic). In consequence, highly distinctive features of natural languages, features crucially differentiating them from invented formal languages, have been overlooked.

There is an alternative type of framework for stating explicit grammars of natural languages. We use the term **model-theoretic syntax** to refer generically to frameworks of this type. Model-theoretic syntax has great heuristic value for linguistics, in the sense of providing appropriate guidance of theoreticians' thinking with respect to just those aspects of natural languages that generative frameworks have ignored. Among these are two that we discuss here: the gradience of ungrammaticality and the lexical openness of natural languages.

We also argue that the machinery and structure of generative frameworks have misled linguists, philosophers, and psycholinguists alike into thinking that the cardinality of the set of expressions generated by a generative grammar is important theoretically, in a way that is connected with settling the question of whether natural languages are infinite. We argue that this is a particularly clear case of scientists having confused their subject matter with their theoretical toolkit, and suggest that the model-theoretic perspective permits an escape from that confusion.

## Origins of generative frameworks

In his invited address at the first International Congress of Logic, Methodology and Philosophy of Science in 1960, Noam Chomsky laid out a convincing case for the goal of stating explicit grammars. He then turned to the development of a framework for stating them, remarking:

> Clearly, a grammar must contain two basic elements: a 'syntactic component' that generates an infinite number of strings representing grammatical sentences and a 'morphophonemic component' that specifies the physical shape of each of these sentences. This is the classical model for grammar. (Chomsky 1962:539)

But such a model is not "classical" in any sense. The idea that grammars of natural languages generate infinite sets of strings was only a few years old in 1960, and the mathematics supporting it only a few decades older.

The most important source for the approach Chomsky advocated was the work of Emil Post in the period 1920–1950. Post's goal was a mathematicization of the notion of a proof in logic. In pursuit of this he developed a schema defining the general form of inference rules in terms of what he calls **productions** (see Post 1943:197). A production has a finite number of 'premises' and produces one 'conclusion'. Each of these is a finite string of symbols. A production thus maps a set of strings to a string, has the form seen in (1a), where '$\Rightarrow$' represents the relation of producing or licensing, and each $\sigma_i$ is a string of the form in (1b).[2]

(1)  a.  $\{\sigma_1, \cdots, \sigma_{n-1}\} \quad \Rightarrow \quad \sigma_n$

  b.  $x_0\ X_1\ x_1\ X_2\ x_2\ \cdots\ X_m\ x_m$

---

[2] Kozen (1997:256-7) provides a brief but useful elementary overview of Post production systems, which we have drawn on here.

Each $x_i$ in (1b) is a specified symbol string (possibly null), and each $X_i$ is a free variable over (possibly null) symbol strings. There are two associated stipulations. First, every $X_i$ in the conclusion $\sigma_n$ must appear in at least one of the premises $\{\sigma_1, \cdots, \sigma_{n-1}\}$ (intuitively because inference rules never allow arbitrary extra material to be introduced into a conclusion). Second, no assignment of strings to variables is permitted if it would lead to $\sigma_n$ being the empty string $e$ (because conclusions have to say something).

Proofs are formally represented as **derivations**. A derivation is a sequence of symbol strings in which each string in the sequence either belongs to a fixed set of strings given at the start (i.e., it is an axiom) or is derivable by means of a production from some set of the strings that precede it in the derivation (i.e., it is licensed by a rule of inference). A **production system** is a set of axioms together with a set of productions, and it **generates** the set of all and only those strings that appear as the last line of some derivation.

Post showed that production systems could generate any recursively enumerable (r. e.) set of strings, thus connecting provability to the characterization of sets of strings by Turing machines. Chomsky (1962) refers to this work just half a dozen lines below the quotation given earlier, and comments:

> A rewriting rule is a special case of a production in the sense of Post; a rule of the form ZXW → ZYW, where Z or W (or both) may be null. (Chomsky 1962:539)

The idea of defining restricted special cases of production systems without losing expressive power was already present in Post's work. Post (1943) had defined a special case of production systems by imposing the following limitations:

- $n$ in (1a) is limited to 2 (i.e., only one premise is allowed), so productions have the form $\sigma_1 \Rightarrow \sigma_2$;
- $m$ in (1b) is limited to 1 (i.e., only one variable is allowed), so $\sigma_1$ and $\sigma_2$ have the form $x_0 X_1 x_1$;
- $x_1$ in $\sigma_1$ and $x_0$ in $\sigma_2$ are both null.

This means that each production has the form '$xX \Rightarrow Xy$' for some $x$ and $y$. Post shows that this makes no difference to expressive power, provided only that the vocabulary can contain certain symbols that function in productions but do not appear in the strings derived; today these are called **nonterminals**, and the symbols appearing in generated strings are called **terminals**. The definition of 'generates' is altered slightly when nonterminals are introduced: the generated set is taken to be the set of all and only those *strings of terminals* that appear as the last line of some derivation. Post proves that any r. e. stringset over the terminal vocabulary can be generated by a production system in the restricted form. And he shows that this remains true if there is just a single axiom consisting of one symbol.

The Type 0 rules of Chomsky (1959)) are defined by a restriction very similar in spirit. Type 0 grammars are essentially Post's semi-Thue systems, deriving from earlier work by Axel Thue (see Post 1947). Again, $n$ in (1a) is limited to 2 so that productions have the form $\sigma_1 \Rightarrow \sigma_2$, but:

– $m$ in (1b) is limited to 2 (i.e., there are just two variables in each production), so $\sigma_1$ and $\sigma_2$ both have the form $x_0 X_1 x_1 X_2 x_2$; and

– in both the premise and the conclusion, $x_0$ and $x_2$ are both null.

In other words, each production has the form '$X_1 x X_2 \Rightarrow X_1 y X_2$' for some $x$ and $y$: it rewrites a string $x$ as a different string $y$ in contexts where $X_1$ precedes and $X_2$ follows. Again there is no loss of expressive power from Post's unrestricted production systems over vocabularies of terminal and nonterminal symbols (this is proved in Chomsky 1959): all (and only) the r. e. sets over the terminal vocabulary can be generated.

The idea of cutting productions completely adrift from the formalization of inference calculi, and using them instead to enumerate strings of words corresponding to natural language sentences, did not come from any classical or traditional source; it was original with Chomsky, who had read Post and acknowledges the intellectual debt (Post 1944 was Chomsky's cited source for the term 'generate'; see Chomsky 1959:137n).

Today, virtually all theoretical linguistics that aims at explicitness is based on Chomsky's development of Post. Chomsky's early transformational grammars represented an elaboration rather than a further restriction of Post systems (though expressive power is not increased). The context-sensitive, context-free, and regular grammars defined in Chomsky (1959) are special cases that have reduced expressive power so that not all r. e. sets are generable.

Categorial grammars (the origins of which antedate not only Chomsky's work but also Post's) can be seen as a kind of bottom-up special case of Post systems where instead of starting with the single-symbol axiom $S$ and iteratively rewriting it until a string over the terminal vocabulary $\Sigma$ is obtained, we start with a multiset of pairs $\langle \sigma_i, \alpha \rangle$, where $\sigma$ is a one-symbol string over $\Sigma$ and $\alpha$ is a nonterminal, and we form larger strings by iterative combination under general principles (e.g., given $\langle \sigma_1, A/B \rangle$ and $\langle \sigma_2, B \rangle$ we are allowed to form $\langle \sigma_1 \sigma_2, A \rangle$); the combination process proceeds until we obtain a pair $\langle \sigma_1 \ldots \sigma_k, S \rangle$ ($S$ being the designated category of complete or saturated expressions), which corresponds in a standard Post system to a proof that the string $\sigma_1 \ldots \sigma_k$ can be derived from $S$.

Under some formalizations which allow function composition as well as function application, categorial grammars are equivalent to tree adjoining grammars (Weir and Vijayshanker 1994); some simpler forms are equivalent to context-free grammars. The so-called 'minimalist' grammars of Chomsky's recent transformationalist work appear to be very similar to categorial systems in this regard, and if Edward Stabler's formalization of them is accepted (Stabler 1997) they are equivalent to multi-component tree-adjoining grammars (Harkema 2001; Michaelis 2001).

## Mathematical sources of model-theoretic syntax

There is an entirely distinct alternative way to state explicit grammars. With hindsight we can glimpse its mathematical beginnings in the work of Büchi (1960).[3] Büchi's results were motivated by questions of arithmetic. It took more than thirty years for their relevance to linguistics to be appreciated (e.g., by Rogers 1994,

---

[3]The recent literature contains some more elegant and accessible proofs of Büchi's theorem: see especially Thomas (1990), Engelfriet (1993), and Straubing (1994).

Rogers 1998, and Kracht 2001). The basic question concerned the expressive power of logic of a particular kind for talking about finite sequences. The logic in question was ordinary predicate logic augmented in such a way that in addition to variables ranging over individuals it has variables ranging over finite sets of individuals. It is known as **weak monadic second-order logic** (henceforth WMSOL). One very desirable property of WMSOL is that its satisfiability problem is decidable — there is an algorithm for determining whether the set of structures satisfying a WMSOL formula is empty.

When WMSOL is interpreted on finite linearly ordered structures, which we will call **string structures**, Büchi showed that the following holds:

(2)   *Büchi's theorem*
      Given any existential WMSOL formula, the set of all its string-structure models is a regular stringset, and for any regular stringset there is a WMSOL formula having that stringset as the set of all its string-structure models.

This gives us a purely model-theoretic perspective on the regular stringsets (generated by Chomsky's 'Type 3' grammars): a regular stringset is simply a set containing all and only those finite string structures that are models of a certain existential formula of WMSOL.

It took a number of years for Büchi's work (and the similar contemporaneous work of Calvin Elgot (1961)) to be appreciated, but in due course the result was generalized to structures having the form of labeled trees. Doner (1970) proved an analog of Büchi's theorem for finite, ordered, directed, acyclic, singly-rooted, non-tangled graphs with labeled nodes as used by linguists for representing syntactic structure that was of special interest in automata theory. In Rogers (1998) these results are finally applied to linguistics rather than computation. Rogers defines a WMSOL description language in which in addition to the binary relation symbol '$\prec$', interpreted by the 'left of' relation found in string structures, there is also a second binary relation symbol '$\lhd^*$', interpreted by the dominance relation. The usual exclusivity and exhaustiveness condition for dominance and precedence can be stated thus:

(3)   $\forall x \forall y[(x \lhd^* y \vee y \lhd^* x) \leftrightarrow \neg(x \prec y \vee y \prec x)]$
      A pair of nodes stands in the dominance relation or its inverse if and only if it does not stand in the precedence relation or its inverse.

The condition that says every tree has a node (the root) that is the minimum point in the weak partial dominance ordering can be stated thus:

(4)   $\exists x \forall y[x \lhd^* y]$
      There is a node that dominates every node.

The other basic axioms defining trees can be given in a similar way (for details see Rogers 1998:15-16).

Structural generalizations about trees with particular properties can also be defined (and most of the time we can do it with just the first-order fragment of our WMSOL language). For example, we can say of a tree that it has only binary branching. Let '◁' be a symbol interpreted by the immediate dominance or 'parent of' relation.[4] Then a tree is binary-branching iff it satisfies (5).

(5)      $\forall x [\exists y \exists z [x \triangleleft y \land x \triangleleft z \land y \neq z] \lor \neg \exists y [x \triangleleft y]]$
         Every node is the parent either of two distinct nodes or of none.

But WMSOL also allows quantification over finite sets of nodes, which enables further properties and relations on trees to be defined. Quantification over sets of nodes is the key to the full power of WMSOL on tree models that Doner (1970) exploited. It permits the expression of certain projections between label sets that enable the logic to define the property of being recognizable by a finite-state tree automaton, distinguishing it from the (more restrictive) property of being generable by a context-free grammar. The difference is, in effect, that a recognizable treeset may have certain dependencies between nodes in the trees that are not registered in the node labeling.[5] What Doner's theorem says is this:

(6)      *Doner's theorem*
         Given any existential WMSOL formula, the set of all its finite tree models is recognizable by a finite-state tree automaton, and thus the yield of that set of trees is a context-free stringset; and for any context-free stringset there is a WMSOL formula having that stringset as the yield of the (recognizable) set containing all and only its finite tree models.

The decidability result holds as before: WMSOL on binary trees corresponds to the WMSOL theory of arithmetic with two successor functions, which was shown by Rabin (1969) to be decidable.

We now have a characterization of both the context-free stringsets and the recognizable treesets in terms of a logic with a decidable satisfiability problem. What is important about this is that, as shown by the work of Gazdar et al. (1985), henceforth *GKPS*, a very large part of the central facts of syntax for English can be stated in terms of recognizable sets of trees. And Rogers (1997) shows that the kinds of theoretical statements made in GKPS can be restated much more simply using WMSOL to impose conditions on syntactic structure directly. For example, Rogers shows how to express the theory of feature specification defaults in a way that is vastly simpler than the cumbersome development of GKPS. A predicate $P'_f$ of sets of nodes is defined for a feature specification $f$ to characterize the property of (i) including all nodes that are free to take $f$ without violating other statements of the theory, and (ii) being closed under propagation of $f$ (Rogers 1997:739). Then the notion of being privileged with respect to feature $f$ can be defined by

$$\text{Privileged}_f(x) \equiv \forall X [P'_f(X) \rightarrow X(x)]$$

---

[4]That is, let $x \triangleleft y$ mean by definition $x \triangleleft^* y \land x \neq y \land \neg \exists z [x \triangleleft^* z \land z \triangleleft^* y \land x \neq z \land z \neq y]$.
[5]Doner gives a useful example to distinguish the two: the set of all binary-branching trees in which every node is labeled $A$ except for a unique node labeled $B$. Generating this set with a context-free grammar is impossible, yet it is easily recognizable by a finite-state tree automaton that keeps track in its state space of not just how many daughters the current node has and what the daughter labels are but also whether the unique $B$ is contained within the subtree dominated by the current node.

(Rogers 1997:740). Feature specification defaults become easily statable as simple material conditionals about trees without any special non-classical default semantics; for example,

$$\forall x [\neg \text{Privileged}_{[-\text{INV}]} \rightarrow [-\text{INV}](x)]$$

says that if a node is not privileged with respect to [−INV], then it is [−INV].

Work on other natural languages shows that they too are mostly describable in terms of recognizable sets of trees, hence by WMSOL on trees. Where natural languages do have constructions that go beyond the sets of trees (henceforth, treesets) that are recognizable in the standard sense, the expression types involved are often grammatically rather marginal. For example, English has an idiomatic adjunct type illustrated by the bracketed part of *The US will go ahead,* [*UN support or no UN support*]; the adjunct must be of the form '...*W or no W*'. But the construction is not that common, the values of *W* are typically just individual nouns or noun-noun compounds, and the syntax of the construction does not interact with central properties of clause structure at all.

There are a few central clausal constructions in other languages that go beyond the recognition power of standard tree automata, but they have so far been found only in Germanic (Shieber 1985; Miller 1991), and they seem quite rare. Work in computational linguistics has shown that if we have the power to describe the recognizable treesets, we have enough descriptive power for doing most of the work necessary in natural language processing.

Insofar as describing non-context-free constructions model-theoretically is required, Rogers (2003) shows how to do it for a very interesting class of cases. He shows that the tree adjoining treesets (and derivatively, the tree-adjoining stringsets that are their yields) can be characterized in terms of WMSOL on three-dimensional tree manifolds — roughly, trees that have trees as their nodes; and Langholm (2001) has shown that the much larger class of indexed stringsets (Aho 1968) can also be characterized with a particular kind of bounded existential quantification in WMSOL on trees with added links between the nodes.

All the work just reviewed uses model theory to characterize sets of structures. The possibility this opens for linguistics is that if we idealize expressions of natural languages as structures such as trees or similar graphs, we can formulate grammars for natural languages as sets of interpreted statements in a logic, the models being expression structures. These grammars will be fully explicit, though not generative.

### Linguistic foundations of model-theoretic frameworks

There is a sense in which McCawley (1968) might be said to have introduced the model-theoretic view to the linguistics literature, though in various ways his approach does not exactly coincide with the one we will adhere to below. McCawley pointed out that phrase structure rules can be interpreted as statements that are satisfied (or violated) by trees.[6] He noted that for context-sensitive rules there is an expressive power difference between what we could now call their generative and model-theoretic interpretations.

---

[6]This insight of McCawley's does not appear to have been influenced by the work of Büchi, and predates that of Doner. Note that McCawley (1968) does not explicitly make the link to using logic as a descriptive formalism for trees.

The generative interpretation of a rule of the form '$A_0 \rightarrow A_1 \ldots A_n/X\_Y$' is that it means "in a string containing the substring $X A_0 Y$, the $A_0$ may be replaced by $A_1 \ldots A_n$." ($A_0$ is a nonterminal, $A_1, \ldots, A_n$ are terminals or nonterminals, and $X, Y$ are strings of terminals or nonterminals.) Under this interpretation, grammars with context-sensitive rules generate all and only the context-sensitive stringsets over the terminal vocabulary (i.e., all and only the stringsets that are Turing-recognizable in linear space — a very large proper superset of the context-free stringsets).

McCawley's alternate interpretation for phrase structure rules, stated in the same formalism, was that they should be interpreted as stating sufficient conditions for well-formedness of local subtrees. The rule '$A \rightarrow A_1 \ldots A_n/W\_Y$' says that a subtree $T$ with root $A_0$ and daughters $A_1 \ldots A_n$ is legitimate provided that the sequence $W$ can be found immediately left-adjacent to $T$ in the rest of the tree, and the sequence $Y$ can be found immediately right-adjacent to $T$. A string is in the stringset defined by a grammar under this interpretation iff it is the yield of a tree in which every subtree is legitimate according to the rule set. This interpretation, surprisingly, is much more restrictive. It was shown by Peters and Ritchie (1969, 1973) that the set characterized by context-sensitive rules under the tree-admitting interpretation is the set of context-free stringsets.

A wider application of the same kind of thinking about how to state grammars can be seen in (Lakoff 1971). Lakoff proposes that the structure of a natural language expression should be idealized as a finite sequence of finite trees $\langle \Delta_0, \ldots, \Delta_k \rangle$ in which $\Delta_0$ is by definition the deep structure of the expression (under Lakoff's generative semantics view, the part that determines the semantic interpretation), and has a form determined by conditions on those structures, and $\Delta_k$ is the surface structure, determining the phonological interpretation. For each $i \geq 1$, $\Delta_i$ results from the application of a transformation to $\Delta_{i-1}$; the value of $k$ will vary, since some expressions have more complex transformational 'derivations' than others.

Crucially, Lakoff takes grammars to be sets of assertions about the structural properties of tree sequences. The analog of a transformation in Lakoff's scheme is a condition specifying the permitted differences between two adjacent trees $\Delta_{i-1}$ and $\Delta_i$ ($0 < i < k$) in a sequence. Rule-ordering conditions and global derivational constraints are claimed to be formalizable as higher-level conditions on the form of tree sequences.[7]

Lakoff did not in fact present a well-defined theory of grammar. He left many crucial matters undefined. The exact character of the models was not clarified, and no specification of the logical description language or its interpretation was given (for example, the domain of quantification was never quite clear: sometimes he seemed to be quantifying over nodes and sometimes trees). The detailed critique offered by Soames (1974) established that Lakoff's proposals did not work as stated: Soames shows convincingly that Lakoff underestimated the difficulty of making his

---

[7]Postal (1972) regards it as a fundamental clarification that under this view there is no important distinction between transformations, which filter out illicit $\langle \Delta_{i-1}, \Delta_i \rangle$ pairs from sequences, and 'global' constraints that filter out illicit tree-pairs $\langle \Delta_i, \Delta_j \rangle$ for arbitrary $i$ and $j$, since $i$ and $j$ will simply reference positions between 0 and $k$ in a finite sequence, and whether $j - i = 1$ can hardly matter very much.

proposal explicit.[8] Nonetheless, the leading idea that syntactic descriptions can be stated as *sets of well-formedness conditions on syntactic structures*, rather than procedures for generating sets, is clearly present in Lakoff's paper.

Other antecedents of model-theoretic syntax were present in the work of a group of Russian mathematicians in Moscow at around the same time. Borščev and Xomjakov (1973) take 'languages' to be collections of what they call 'texts', and idealize texts as finite models in a suitable signature, formulating grammars as sets of axioms, a grammar being interpreted as a description of those texts that satisfy it. Borščev (personal communication, May 2002) informs us that the group did intend to apply this approach to natural language description, but the applied work that got done mostly related it to the description of chemical structures.

The model-theoretic approach foreshadowed in these early works was much more fully developed by Johnson and Postal (1980), who overtly adopted the idea of treating expression structures as models of statements in a logical description language. Their structures are complex graph-like objects (actually, graphs with additional relations holding between edges), some aspects of which e.g., the linear ordering of terminals that defines the yield) are not clearly worked out, and their metalanguage is not fully defined (it may be first-order, or it may be augmented with the power to define ancestrals of relations, one cannot tell from the exposition); but without question, the leading idea of model-theoretic syntax is there, along with a clear perception of some of its metatheoretical advantages.

Virtually no linguists followed Johnson and Postal's lead in the decade that followed. Gerald Gazdar proposed a model-theoretic reconstruction of the generalized phrase structure grammar of Gazdar et al. (1985) in various lectures in 1987, but otherwise the approach was forgotten until the 1990s, when papers independently developing the idea in two directions began to appear. First, Patrick Blackburn and his colleagues, and independently Marcus Kracht, began to develop an idea from Gazdar et al. (1988): using modal logic as a description language for syntax (Blackburn et al. 1993; Kracht 1993; Blackburn and Gardent 1995; Blackburn and Meyer-Viol 1997). And second, James Rogers began his work on applying WMSOL to linguistics (Rogers 1994, 1996, 1997, 1998, 1999).

These papers from the 1990s, however, were largely preoccupied with restating particular varieties of generative grammars in model-theoretic terms, the point being to attain new insights into the character of the syntactic facts or to compare the expressive power of different classes of grammars. If model-theoretic syntax were merely a matter of stating generative grammars in a different way, it would be of only minor importance to linguistics. Our thesis is that model-theoretic syntax offers the study of natural languages not just a restatement of generative grammar but a shift of framework type with profound heuristic consequences.

---

[8]The technical difficulties could probably be overcome. It would help to link the trees in the sequence via a correspondence relation holding between the nodes of one tree in the sequence and the nodes of the next. This would permit movement or erasure of a particular node to be reconstructed in terms of where or whether a node in one tree corresponded to a node in the next. See Potts and Pullum (2002) for an application of this idea (on pairs rather than arbitrary $n$-tuples of tree-like structures) in phonological theory.

## Two key properties of natural languages

The invented languages of logic, mathematics, and computer science — henceforth, **formal languages** — are stipulated sets of strings (or other structures) defined over a finite vocabulary of symbols. Post production systems and the generative grammars that are based on them are ideally suited to stating the explicit grammars of formal languages, and were invented for exactly that purpose.

But natural languages have a number of properties that clearly differentiate them from formal languages. In this section we review two illustrative phenomena: first, the fact that being grammatically ill-formed is a matter of degree, and second, the fact that there is no fixed lexicon for a natural language. We point out that model-theoretic frameworks immediately suggest appropriate ways to describe the relevant phenomena. Generative frameworks do not.

We are not saying that it would be impossible to use a generative grammar in giving an explicit account of these phenomena. Augmentation with additional theoretical machinery is always possible. But it does appear that any such theoretical augmentations will be entirely ad hoc. The basic structure of generative grammars does not suggest them. Certain distinctive phenomena of natural language appear to have been ignored within generative grammar precisely because generative frameworks are ill suited to their description.

## Ungrammaticality is gradient

Some utterances that do not correspond to fully well-formed expressions are vastly less deviant than others. And this feature of utterances is also a feature of expressions — or rather (since it may be better to limit the term 'expression' to what is fully grammatical), those objects that are like expressions except that they are only partially well-formed. Let us call these latter QUASI-EXPRESSIONS. For example, (7a) is a quasi-expression that is clearly ungrammatical; and (7b) is clearly more ungrammatical; but neither is ungrammatical to the same degree as the utterly incomprehensible (7c).

(7)    a.  * *The growth of of spam threatens to make email useless.*

        b.  * *The growth of of the spam threatens make email useless.*

        c.  * *The of email growth make threatens spam to useless of.*

The point is that *some quasi-expressions are closer to being grammatical than others.*

No unaugmented generative grammar describes degrees of ungrammaticality, nor does it suggest a way to do so. A generative grammar generates a single set $L$ of expressions over a vocabulary $V$. $L$ is a subset of $V^*$ (the set of all strings over $V$). Given $w \in L - V^*$ and a generative grammar $G$ such that $L(G) = L$, $G$ will say absolutely nothing about $w$, since no derivation permitted by $G$ will lead to $w$.

Chomsky (1955), Chomsky (1961), and Chomsky and Miller (1963) give several basically unsuccessful attempts to augment a generative grammar so that degrees of ungrammaticality can be described. All these proposals define degrees of ungrammaticality by matching ungrammatical word sequences with lexical category strings associated with grammatical word sequences. The idea is to define a series of lexical category inventories of graded coarseness, the finest being the set of lexical categories assigned to words in a full and accurate grammar for the language, and the coarsest consisting of just the single category 'Word'. A string $w$ that is not generated by the grammar is assigned a degree of ungrammaticality according to which degree of lexical category coarseness must be used to get a match between $w$ and some string that is generated. Consider these examples:

(8)    a.    *John plays golf.* ($N_{anim}$ $V_t$[+__anim. subj.] $N_{inan}$)

       b.    *Golf plays John.* ($N_{inan}$ $V_t$[+__anim. subj.] $N_{anim}$)

       c.    * *Golf fainted John.* ($N_{inan}$ $V_i$ $N_{anim}$)

       d.    * *The of punctilious.* (D P Adj)

The example in (8a) is entirely normal; (8b) is not, but if we just ignore the requirement that *play* should have an animate subject we can say that it matches (8a) in its sequence of lexical categories;[9] (8c) is worse, because to find a match for it we have to descend to a greater level of coarseness of categorization where we ignore not only the selectional restriction that *faint* should have an animate subject but also that *faint* is syntactically required not to have a direct object (a strict subcategorization restriction); and finally (8d) is yet worse, because its sequence of lexical categories matches nothing in English unless we categorize its items at a level of coarseness where we treat it as simply 'Word Word Word'.

Proposals of this type are inadequate for describing the phenomena of ungrammaticality in natural languages. They provide too few degrees of ungrammaticality. The examples provided in the cited references yield only three levels, and it is not clear how to go beyond this. A serious problem is that the accounting system for grammatical errors is not cumulative; for example, all of the strings in (7) will be assigned exactly the same degree of ungrammaticality, clearly the wrong result.

But perhaps the most important inadequacies stem from deep theoretical failings in this kind of analysis. First, the proposal relies on an entirely nonconstructive definition of a transderivational relation between strings in various infinite sets. To determine whether the relevant relation holds between a a specific ungrammatical $w$ and a certain string of lexical categories, we have to solve this problem:

(9)         Input: an ungrammatical string $w_1 \ldots w_n$ of words categorized at coarseness level $i$ yielding a category string $K_1 \ldots K_n$.

            Output: a decision on whether there is a grammatical sentence that also has lexical category sequence $K_1 \ldots K_n$ at coarseness level $i$.

But this is undecidable for Post production systems in general, and for the transformational grammars that Chomsky developed from them. All r. e. sets have transformational grammars; but r. e. sets do not necessarily have r. e. complements. This means that although there will always be some answer to a question of the form in (9), there can be no general effective procedure for finding it.

Even more importantly, the generative grammar that derives the well-formed strings is independent of the assignment of degrees of ungrammaticality. Degree of ungrammaticality depends entirely on lexical category sequence similarities, according to this proposal; it does not depend on any aspect of syntactic structure.

---

[9] Incidentally, we would follow McCawley 1968 in taking the deviance of selectional restriction violations like *Golf plays John* as semantic rather than syntactic.

Indeed, the independence of the grammar from the assignment of degrees of un-grammaticality is so extreme that it would be the same given *any* observationally adequate grammar with the same lexicon.

By contrast, with model-theoretic grammars the same resources employed to describe the fully grammatical expressions also yield a description of the quasi-expressions. Let $\Omega$ be a domain of relevant structures of some kind and $\Gamma$ a set of constraints interpreted on structures in $\Omega$. Let $\Omega_1$ be the subset of $\Omega$ containing the structures that satisfy $\Gamma$. We note that $\Gamma$ also structures the rest of the domain, assigning a status to each member of $\Omega - \Omega_1$, depending on which of the constraints in $\Gamma$ that member fails to satisfy. And notice, the status assigned to a structure depends entirely on what structural properties it has and what the grammatical constraints in $\Gamma$ say.

That is, model-theoretic grammars do not just partition a set of structures into two subsets (the fully well-formed set and the complement of that set). Rather, even under a crude view on which structures are evaluated as wholes,[10] a set of $n$ constraints determines $2^n$ subsets, just as any set of $n$ binary attributes determines a set of $2^n$ distinct attribute-value matrices.

Which quasi-expressions are assigned which degrees of ungrammaticality will not be invariant under reaxiomatization. What the right constraints are for the gram-mar of some language has to be decided on the basis of a wide range of empirical evidence and theoretical considerations — simply selecting a framework does not settle everything. For example, if we replace a set of two or more constraints by a single constraint which is their conjunction, we get only one degree of ungrammati-cality, namely 100% ungrammaticality. So the model-theoretic account can always mimic the undesirable properties of the generative account.[11]

Under the model-theoretic view we are suggesting for consideration, therefore, a whole new range of empirical data, the class of facts about one quasi-expression being more ungrammatical than another, becomes relevant to decisions about what the constraints are, and with exactly what degree of delicacy they should be for-mulated, and in what description language.

### The lexicon is open

Take the structure for some grammatical expression in a natural language and replace one of its lexical items by a piece of phonological material unassociated with any grammatical or semantic properties (henceforth we call such a nonsense form a *pseudo-word*). For example, take a well-formed structure for (10a) and replace *fox* by the random pseudo-word *meglip*, changing nothing else in the structure, yielding a similar structure for (10b).

(10)  a.  *The quick brown fox jumps over the lazy dog*

b.  *The quick brown meglip jumps over the lazy dog*

---

[10] For a more refined view of ungrammaticality we could ask for each *node* in the structure which of the constraints is satisfied at that node; obviously, the normal interpretation strategy for modal logic would be ideal for this.

[11] Johnson and Postal (1980) actually do adopt the view that a grammar is to be stated as a single formula, the conjunction of the large set of material conditionals that state their constraints on syntactic structure, though without noticing the undesirable consequence we point out here.

What is the grammatical status of the new structure, the one with terminal string (10b)? Does it have grammatical structure?

An unaugmented generative grammar (either top-down or bottom-up), by definition, fixes a finite inventory of admissible terminal symbols. The lexicon for the grammar contains these symbols paired with certain grammatical and semantic features. In top-down generative grammars only a terminal symbol (lexical item) that appears on the right hand side of a rule can occur in a derived string, and all nonterminals must ultimately be eliminated through the application of rules if the derivation is to complete. Since non-terminating sequences are not derivations at all (by definition), no top-down generative grammar, in and of itself, allows any structure containing a pseudo-word to be in the set of grammatical sentences generated.

The failure of bottom-up generative grammars to derive the structure of expressions with pseudo-word terminals is particularly clear. Derivations start with a multiset of items (a 'numeration' in minimalist parlance) selected from some fixed, finite lexical stock, and operations of combination are performed on them. The generated language is the set of all structures that can be built using some multiset. Keenan and Stabler (1996), for example, define a bottom-up generative framework in which a grammar is a finite set Lex of lexical items together with a finite set $\mathcal{F}$ of combination operations defined on Lex. The language generated is the set of all expressions that can be derived by selecting some multiset $M$ of items from Lex and iteratively applying operations in $\mathcal{F}$ until $M$ is exhausted and a saturated expression (e.g., a sentence) has been derived. A structure containing an item not present in Lex cannot be derived.

Thus (10b) has no derivation in such grammars, and is not even classified by such grammars as a candidate for grammaticality in English, because it does not belong to the universe of strings over the appropriate set of symbols.

A distinctive and fundamental feature of natural languages is being missed by generative frameworks: the lexicons of natural languages are *not* fixed sets. Natural languages are strikingly different from formal languages in this respect. In a natural language the lexicon continuously changes, not just on an intergenerational time scale (in the sense that children do not learn exactly the same words as those their parents learned), but week by week and day by day. New brand and model names are introduced; novel personal names are given; technical terms are devised; new artifacts are dubbed; onomatopoeic terms are made up on the fly; words are borrowed from other languages; noises are imitated and used as words; and in dozens of other ways the word stock of a natural language is constantly under modification. The lexicon of a natural language is open and indefinitely extensible. In consequence, explicit grammars for natural languages, if they are to describe the phenomena, must not entail that the lexicon is some fixed, finite set.

We are not concerned here with whatever diachronic processes add new lexical items to languages during their history, nor with the psycholinguistic processes involved in the recognition of expressions containing pseudo-words or the coining or production of nonce words. Rather, we are concerned with a fundamental feature of natural languages and with how explicit grammars can be formulated in a way that is compatible with it.

A model-theoretic syntactic framework makes available a fully explicit description of lexical openness.[12] In model-theoretic terms, to say that a language has a lexical item with a certain phonological form and certain grammatical and semantic properties, is to say simply that there are constraints on that phonological form stating limits on the grammatical and semantic properties that are associated with it. That is, what it means for there to be a noun *fox* in English is that a condition in the grammar of English links the phonological representation /faks/ to the lexical category N and the property of having regular inflection and the meaning "member of the species *Vulpes vulpes crucigera*."

A pseudo-word, by definition, is phonologically well-formed but has no grammatical or semantic properties. A correct grammar will thus state no conditions on grammatical or semantic aspects of the structures in which it may occur. Thus there are no grammatical or semantic constraints in any model-theoretic grammar for a pseudo-word to violate. No expression structure will be ungrammatical solely in virtue of containing a pseudo-word at some terminal node. If the structure of (10b) is the same as that of (10a), then (10b) does not violate any constraint at all.

Introductory works on language and linguistics commonly observe that expressions containing pseudo-words are well formed. Many (see Pinker 1994:89 for a typical example) reprint the first stanza of Lewis Carroll's *Jabberwocky* ("'Twas brillig, and the slithy toves / Did gyre and gimble in the wabe; / All mimsy were the borogoves...") to make exactly this point — the point that syntactic structure is not entirely dependent on lexical inventory. The authors correctly regard the clause *all mimsy were the borogoves* as grammatical English. What they appear to miss is that this insight is implicitly at odds with the machinery of generative frameworks, which define grammars in a way that precludes the explicit description of lexical openness.

We are not claiming that no generative framework could be modified to incorporate an open lexicon. There would doubtless be some way to modify generative grammar to get the effect of a lexicon containing all possible well-formed phonological representations, the default being that phonological representations are associated with the disjunction of all sets of grammatical and semantic properties. And we are not claiming that model-theoretic frameworks entail that the lexicon is open: it is perfectly possible to close the lexicon of a model-theoretic grammar (in fact Rogers (1998:119) does this, for good reason, since he is interested in demonstrating full equivalence to a context-free generative grammar). What must be done to close the lexicon is to give an exhaustive disjunction of all the phonological realizations and all the grammatical features they can be paired with, thus disallowing for each lexical category all other realizations than the ones appearing in a finite list.

The two framework types therefore can, to some extent, simulate each other. But the contrast between them is nonetheless stark. They point in opposite directions. In a model-theoretic framework, additional content (the stipulated impermissibility of other shapes for each lexical category) has to be built into a grammar to get the

---

[12]Here, as in Pullum and Scholz (2001), we are essentially just reiterating and elaborating an important point made by Johnson and Postal 1980:675–7. It has since been discussed more fully by Postal (2004).

effect of a closed lexicon. A generative framework as standardly defined, on the other hand, would have to be modified (in a way that has so far never been worked out) in order to get the (desirable) effect of lexical openness.

## Languages, expressions, and infinity

The dominance of the generative conception of how language is to be described has led to two items of dogma: that the set of all expressions belonging to a natural language is an infinite set, and that each of those infinitely many expressions is a finite object. We now proceed to argue that both of these claims are artifactual.

## The myth that natural languages are demonstrably infinite

Contrary to popular belief, it has never been shown that natural languages have infinitely many expressions. To say this is to reject familiar arguments that are frequently repeated in introductory linguistics texts, encyclopedia articles, and other presentations. Stabler (1999:321) expresses the standard wisdom tersely by saying: "there seems to be no longest sentence, and consequently no maximally complex linguistic structure, and we can conclude that human languages are infinite." Such arguments have been much repeated over the past thirty or forty years; in (11) we offer samples, one statement from each of the last four decades.

(11)   a.   If we admit that, given any English sentence, we can concoct some way to add at least one word to the sentence and come up with a longer English sentence, then we are driven to the conclusion that the set of English sentences is (countably) infinite. (Bach 1974:24)

       b.   Is there anyone who would seriously suggest that there is a number, n, such that *n is a number* is a sentence of English and *n+1 is a number* is not a sentence of English ... ? On this basis we take it as conclusively demonstrated that the number of English sentences is infinite and, therefore, that English cannot be equated with any corpus no matter how large. (Atkinson et al. 1982:35-36)

       c.   By the same logic that shows that there are an infinite number of integers—if you ever think you have the largest integer, just add 1 to it and you will have another—there must be an infinite number of sentences. (Pinker 1994:86)

       d.   It is always possible to embed a sentence inside of a larger one. This means that Language is an infinite system. Carnie (2002:13–14)

Some of these are more carefully worded than others, but clearly the same basic argument is repeated over and over again. It assumes that natural languages have the property of **productivity** — that is, they have expression-lengthening operations that preserve well-formedness. We do not question this. A sentence as simple as (12a) illustrates that English has expressions with tautocategorial embedding — i.e., a constituent of type $\alpha$ having a proper subconstituent of type $\alpha$, as shown in (12b).

(12)   a.   *See Spot run away.*

       b.   [$_{VP}$ *see Spot* [$_{VP}$ *run away* ]]

This suggests a lengthening operation that will also permit longer expressions such as *Let Jane see Spot run away*, *Watch Dick let Jane see Spot run away*, etc., and indeed these are grammatical.

Our use of the term 'operation' here is an informal reflection of the algebraic use of the term, not the algorithmic one. In algebra, a unary operation on a set is just a function $f : A \mapsto A$, and $A$ is said to be **closed under** an operation iff $a \in A$ implies $f(a) \in A$. The authors quoted above apparently think that given any productive expression-lengthening operation it follows immediately that the set of well-formed sentences is countably infinite. It does indeed follow that the set formed by *closing* a set of expressions under a lengthening operation will be infinite. But the argument is supposed to be about natural languages such as English. What needs to be supported is the claim that (for example) English actually contains all the members of the closure of some set of English expressions under certain lengthening operations.

The illustrative quotations in (11) are attempts at providing that support. We will refer to the underlying form they all share as the Master Argument for language infinity. Let us try to restate it in more precise form. Let $E(x)$ mean '$x$ is a well-formed English expression' and let $\mu$ be a measure of expression size in integer terms — for simplicity we can let $\mu$ measure the length of the yield in words, so that $\mu(w_1 \ldots w_k) = k$. Then the argument runs like this:

(13)    The Master Argument for language infinity

      a.    There is at least one well-formed English expression that has size greater than zero:

$$\exists x[E(x) \wedge \mu(x) > 0]$$

      b.    For all $n$, if some well-formed English expression has size $n$, then some well-formed English expression has size greater than $n$:

$$\forall n[\exists x[E(x) \wedge \mu(x) = n] \rightarrow \exists x[E(x) \wedge \mu(x) > n]]$$

      c.    Therefore, for every positive integer $n$ there are well-formed English expressions with size greater than $n$ (i.e., the set of well-formed English expressions is at least countably infinite):

$$\therefore \forall n \exists x[E(x) \wedge \mu(x) > n]$$

The Master Argument fails, in one of two different ways, depending on what sort of universe we apply it to — more specifically, what set we assign as the interpretation for the predicate $E$. There are two cases.

(i) If we choose a finite set as the extension for $E$, then clearly (13b) is false, and the argument is unsound.

(ii) If we choose an infinite set as the extension for $E$, then (13b) is true, but we have assumed what we set out to show — we have begged the question.

That is, the argument fails because if English were finite one of the premises would be false, while if English were infinite the argument would be circular, and we are given no way to tell which is the case. (This point is not new; we are basically just paraphrasing Langendoen and Postal 1984:30-35.)

If some argument did show that English or some other natural language had infinitely many expressions, the fit with generative grammars would be a good one in this respect, for given nontrivial and unbounded productivity, Post production systems and all kinds of generative grammars generate countably infinite sets of expressions. But in the absence of such an argument, the cardinality of the set generated by a generative grammar is irrelevant to the choice of a framework for linguistic theory.

Under a model-theoretic framework, no claim is made about how many natural language expressions there are. Although each constraint in a model-theoretic grammar will entail claims about the structure of individual expressions, nothing in any collection of constraints need entail any claim about the size of the set of all expressions.[13] A set of constraints that accurately characterizes the syntactic properties that well-formed English expressions have in common will be satisfied by each well-formed expression no matter how many or how few there are, assuming only that no restriction stated in the grammar places a ceiling on productivity — a reasonable assumption.

The model-theoretic view enables us to distinguish two issues: (i) the existence of productive lengthening operations in natural languages, and (ii) the cardinality of the set of all expressions in a language. Our thesis is that the first of these is relevant and important to the formulation of grammars for natural languages, but the second is not.

### Natural languages and expressions of infinite length

Schiffer (1972) proposes a definition of what he calls 'mutual belief' under which what it would mean for Jones and Smith to have mutual belief of the proposition that iron rusts might be expressed in an infinite string of which the following is a representative initial subpart:

(14)   *Jones believes that iron rusts, and Smith believes that iron rusts, and Jones believes that Smith believes that iron rusts, and Smith believes that Jones believes that iron rusts, and Jones believes that Smith believes that Jones believes that iron rusts, and...*

Joshi (1982:182) actually states the mutual belief schema in terms of an infinitely long conjunctive formula of this sort. The truth conditions of the infinite conjunction are clear enough under the ordinary interpretive principles for English: each of Jones and Smith believes that iron rusts, and each has a set of beliefs about the other's beliefs that is closed under reapplying *Jones believes that* to sentences about Smith's beliefs and vice versa.

But is the infinite string of which a small finite initial subpart is seen in (14) a grammatical English sentence? If not, it is certainly not clear why, since the string

---

[13]Of course, if one added to the grammar a statement such as 'there are not more than 374 nodes', structures would be limited to a certain finite maximum size, and thus (assuming also a finite bound on the set of node labels and relations) would be finite in number. But the point is that one does not have to if one has no warrant for doing so.

is entirely Englishlike in terms of its grammatical properties. But if it is, then some English expressions have infinite length.

No generative grammar can derive the fully expanded infinite version of 14, because derivations, by definition, complete in a finite number of steps. A derivation-like sequence that goes on adding terminals to the string but in a way that always adds new nonterminals as well never generates a terminal string under the standard definitions.

Linguists seem to have been led astray by this property of Post production systems and their progeny, taking the finiteness of expressions to be a truth about natural languages rather than a stipulated fact about a class of formal systems.[14] We see no fact about natural languages here.

We therefore take it as a point in favor of model-theoretic syntax that it permits description of the structure of expressions without entailing claims any about their size. The infinite sequence suggested by (14) would appear to satisfy all the constraints for English: subjects precede predicates, clausal complements follow their licensing verbs, present-tense verbs agree with subjects, and so on. Those properties are the ones that a syntax for English should describe. The description need not say anything about whether infinite length is possible for English expressions.[15]

What seems to be wanted here is the right to remain silent concerning expression size. A model-theoretic syntax for English can provide that, defining the infinite version of (14) as not violating any grammatical constraints but not insisting that it is ipso facto a sentence.[16] It would also be possible to stipulate that the infinite version of (14) is ungrammatical by using WMSOL: assuming tree structures as above, we would state something like (15), where $X$ is a variable over sets of nodes:

(15)    a.    $\forall X [\exists x [X(x)] \rightarrow \exists x [X(x) \wedge \forall y [(y \triangleleft^* x \wedge y \neq x) \rightarrow \neg X(y)]]]$

        b.    $\forall X [\exists x [X(x)] \rightarrow \exists x [X(x) \wedge \forall y [(y \prec x) \rightarrow \neg X(y)]]]$

These statements say that every sequence of nodes ordered by domination or precedence comes to an end (Rogers 1998:22), which entails that structures are finite in size (though with no specific finite upper bound). Our point, however, is that

---

[14]The relevant definitions can easily be modified, and have been for theoretical reasons within logic and computer science (see Thomas 1990); but the standard definitions absolutely exclude infinite expressions.

[15]The observation that model-theoretic ("nonconstructive") grammars are compatible with infinite-size grammatical expressions is stressed in Langendoen and Postal (1984). What distinguishes our position from theirs is that they hold that natural languages are proper classes, closed under an operation of infinite coordinate compounding that renders them too large for the laws of set theory to apply. We have no space to discuss this position here, but we note one point. Langendoen and Postal claim that for every set $X$ of sentences in a natural language $L$ there is a coordinate sentence of $L$ having all the members of $X$ as its coordinates. This claim is not statable as an MTS constraint, because it is not interpretable on individual expressions. So under a strict construal of our position, Langendoen and Postal's closure generalization is not just unmotivated but actually unstatable.

[16]Note, though, that if a first-order logic is used as the description language, the infinite version of (14) cannot be blocked. It is a simple corollary of the compactness theorem for first-order logic that if a theory places no upper bound on the size of its finite models, it must have an infinite model. Thus any model-theoretic grammar that permits finite structures of arbitrary size (that is, any grammar that is at all plausible) *must* admit infinite structures if it is stated in a first-order language.

model-theoretic frameworks do not require this. We can state a grammar without any such stipulation, and restrict attention to an appropriate class of models as necessary for a given task. To prove Doner's theorem, we need to restrict attention to finite models; for other theoretical purposes, allowing infinite models might be appropriate. Model-theoretic syntax does not require that a decision be made once and for all on this, because model-theoretic grammars can make claims about the structural regularities expressions have without saying anything about how big expressions can be.

**Conclusion**

For nearly fifty years explicit grammars for natural languages have been formulated within generative frameworks that heuristically suggest that

- quasi-expressions have no syntactic properties at all;
- all quasi-expressions are ungrammatical to the same degree;
- the lexicons of natural languages are fixed;
- natural language expressions with pseudo-words are ungrammatical; and
- a natural language contains a countable infinity of expressions.

Our thesis is that all of these are artifacts of a view that is imposed on linguistics by generative frameworks, and leading to neglect of the actual properties of natural language.

Model-theoretic frameworks guide explicit grammar development in a notably different direction, correctly suggesting that

- quasi-expressions have a full array of syntactic properties describable by means of the same grammar that defines the grammatical expressions;
- quasi-expressions are ungrammatical in a multidimensional variety of ways and to an indefinitely large number of different degrees;
- the lexicon of a natural language is open and continually changing, often without the changes having any syntactic implications;
- natural language expressions with pseudo-words are fully grammatical as well as meaningful; and
- there is no theoretically important notion of the cardinality of the set of expressions in a natural language.

A rethinking of the mathematical and logical foundations of 20th-century theoretical syntax is in order. In 1960, when the first International Congress of Logic, Methodology and Philosophy of Science took place, linguists were rightly taken with the results of Chomsky (1959) on formal language theory, and few if any knew of the almost contemporaneous work of Büchi.[17] But generative frameworks best fit the task for which Post initially developed them: describing the syntax of artificial languages in mathematical logic. We argue for a different theoretical basis for syntax, based not on Post's formalization of proof theory but on the more recent work that has begun to forge a link between the syntax of natural languages and the semantic side of logic.

---

[17]Though Büchi was in fact present as the first Congress in 1960, and a paper of his, Büchi (1962), opens the proceedings volume.

# References

Aho, A. V.: 1968, Indexed grammars — an extension of context-free grammars, *Journal of the Association for Computing Machinery* **15**, 647–671.

Atkinson, M., Kilby, D. and Roca, I.: 1982, *Foundations of general linguistics*, George Allen and Unwin, London.

Bach, E.: 1974, *Syntactic theory*, Holt Rinehart and Winston, New York.

Blackburn, P. and Gardent, C.: 1995, A specification language for lexical functional grammars, *Proceedings of the 7th EACL*, European Association for Computational Linguistics, pp. 39–44.

Blackburn, P. and Meyer-Viol, W.: 1997, Modal logic and model-theoretic syntax, *in* M. de Rijke (ed.), *Advances in Intensional Logic*, Kluwer Academic, Dordrecht, pp. 29–60.

Blackburn, P., Gardent, C. and Meyer-Viol, W.: 1993, Talking about trees, *Proceedings of the 1993 Meeting of the European Chapter of the Association for Computational Linguistics*, pp. 21–29.

Borščev, V. B. and Xomjakov, M. V.: 1973, Axiomatic approach to a description of formalized languages and translation neighbourhood languages, *in* F. Kiefer (ed.), *Mathematical Models of Language*, Soviet Papers in Formal Linguistics, Skriptor, Stockholm, pp. 37–114.

Büchi, J. R.: 1960, Weak second-order arithmetic and finite automata, *Zeitschrift für Mathematische Logik und Grundlagen der Mathematik* **6**, 66–92.

Büchi, J. R.: 1962, On a decision method in restricted second-order arithmetic, *Proceedings of the International Congress on Logic, Methodology, and Philosophy of Science*, Stanford University Press, Stanford, CA, pp. 1–11.

Carnie, A.: 2002, *Syntax: A General Introduction*, Blackwell, Oxford.

Chomsky, N.: 1955, The Logical Structure of Linguistic Theory, Unpublished dittograph, microfilmed.

Chomsky, N.: 1959, On certain formal properties of grammars, *Information and Control* **2**, 137–167. Reprinted in *Readings in Mathematical Psychology*, Volume II, ed. by R. Duncan Luce, Robert R. Bush, and Eugene Galanter, 125–155, New York: John Wiley & Sons, 1965. [Note: The citation to the original on p. 125 of this reprinting is incorrect.].

Chomsky, N.: 1961, Some methodological remarks on generative grammar, *Word* **17**, 219–239. Section 5 republished as 'Degrees of grammaticalness' in Jerry A. Fodor and Jerrold J. Katz (eds.), *The Structure of Language: Readings in the Philosophy of Language*, 384-389 (Englewood Cliffs, NJ: Prentice-Hall).

Chomsky, N.: 1962, Explanatory models in linguistics, *in* E. Nagel, P. Suppes and A. Tarski (eds), *Logic, Methodology and Philosophy of Science: Proceedings of the 1960 International Congress*, Stanford University Press, Stanford, CA, pp. 528–550.

Chomsky, N. and Miller, G. A.: 1963, Finitary models of language users, *in* R. D. Luce, R. R. Bush and E. Galanter (eds), *Handbook of Mathematical Psychology*, Vol. II, John Wiley and Sons, New York, pp. 419–491.

Doner, J.: 1970, Tree acceptors and some of their applications, *Journal of Computer and System Sciences* **4**, 406–451.

Elgot, C. C.: 1961, Decision problems of finite automata and related arithmetics,

*Transactions of the American Mathematical Society* **98**, 21–51.

Engelfriet, J.: 1993, A regular characterization of graph languages definable in monadic second-order logic, *Theoretical Computer Science* **88**, 139–150.

Gazdar, G., Klein, E., Pullum, G. K. and Sag, I. A.: 1985, *Generalized Phrase Structure Grammar*, Basil Blackwell, Oxford.

Gazdar, G., Pullum, G. K., Carpenter, B., Klein, E., Hukari, T. E. and Levine, R. D.: 1988, Category structures, *Computational Linguistics* **14**, 1–19.

Harkema, H.: 2001, A characterization of minimalist languages, *in* P. de Groote, G. Morrill and C. Retoré (eds), *Logical Aspects of Computational Linguistics: 4th International Conference*, number 2099 in *Lecture Notes in Artificial Intelligence*, Springer Verlag, Berlin, pp. 193–211.

Johnson, D. E. and Postal, P. M.: 1980, *Arc Pair Grammar*, Princeton University Press, Princeton, NJ.

Joshi, A.: 1982, Mutual beliefs in question-answer systems, *in* N. Smith (ed.), *Mutual Knowledge*, Academic Press, London, pp. 181–197.

Keenan, E. L. and Stabler, E.: 1996, Abstract syntax, *in* A.-M. D. Sciullo (ed.), *Configurations: Essays on Structure and Interpretation*, Cascadilla Press, Somerville, MA, pp. 329–344.

Kozen, D. C.: 1997, *Automata and Computability*, Springer, Berlin.

Kracht, M.: 1993, Syntactic codes and grammar refinement, *Journal of Logic, Language and Information* **4**, 41–60.

Kracht, M.: 2001, Logic and syntax: a personal view, *in* M. Zhakharyaschev, K. Segerberg, M. de Rijke and H. Wansing (eds), *Advances in Modal Logic 2*, CSLI Publications, Stanford, pp. 355–384.

Lakoff, G.: 1971, On generative semantics, *in* D. D. Steinberg and L. A. Jakobovitz (eds), *Semantics: An Interdisciplinary Reader in Philosophy, Linguistics and Psychology*, Cambridge University Press, Cambridge, pp. 232–296.

Langendoen, T. and Postal, P. M.: 1984, *The Vastness of Natural Languages*, Basil Blackwell, Oxford.

Langholm, T.: 2001, A descriptive characterisation of indexed grammars, *Grammars* **4**, 205–262.

McCawley, J. D.: 1968, Concerning the base component of a transformational grammar, *Foundations of Language* **4**, 243–269. Reprinted in James D. McCawley, *Grammar and Meaning*, 35–58 (New York: Academic Press; Tokyo: Taishukan, 1973).

Michaelis, J.: 2001, Transforming linear context-free rewriting systems into minimalist grammars, *in* P. de Groote, G. Morrill and C. Retoré (eds), *Logical Aspects of Computational Linguistics: 4th International Conference*, number 2099 in *Lecture Notes in Artificial Intelligence*, Springer Verlag, Berlin, pp. 228–244.

Miller, P.: 1991, Scandinavian extraction phenomena revisited: Weak and strong generative capacity, *Linguistics and Philosophy* **14**, 101–113.

Peters, P. S. and Ritchie, R. W.: 1969, Context-sensitive immediate constituent analysis — context-free languages revisited, *Proceedings of the ACM Conference on the Theory of Computing*, pp. 1–8. Republished in *Mathematical Systems Theory* 6 (1973), 324-333.

Peters, P. S. and Ritchie, R. W.: 1973, Context-sensitive immediate constituent analysis — context-free languages revisited, *Mathematical Systems Theory*

**6**, 324–333.

Pinker, S.: 1994, *The Language Instinct*, William Morrow, New York.

Post, E.: 1943, Formal reductions of the general combinatory decision problem, *American Journal of Mathematics* **65**, 197–215.

Post, E.: 1944, Recursively enumerable sets of positive integers and their decision problems, *Bulletin of the American Mathematical Society* **50**, 284–316.

Post, E.: 1947, Recursive unsolvability of a problem of thue, *Journal of Symbolic Logic* **12**, 1–11.

Postal, P. M.: 1972, The best theory, *in* P. S. Peters (ed.), *Goals of Linguistic Theory*, Prentice-Hall, pp. 131–170.

Postal, P. M.: 2004, The openness of natural languages, *Skeptical Linguistic Essays*, Oxford University Press, pp. 173–201.

Potts, C. and Pullum, G. K.: 2002, Model theory and the content of OT constraints, *Phonology* **19**, 361–393.

Pullum, G. K. and Scholz, B. C.: 2001, On the distinction between model-theoretic and generative-enumerative syntactic frameworks, *in* P. de Groote, G. Morrill and C. Retoré (eds), *Logical Aspects of Computational Linguistics: 4th International Conference*, number 2099 in *Lecture Notes in Artificial Intelligence*, Springer Verlag, Berlin, pp. 17–43.

Rabin, M. O.: 1969, Decidability of second-order theories and automata on infinite trees, *Transactions of the American Mathematical Society* **141**, 1–35.

Rogers, J.: 1994, *Studies in the Logic of Trees with Applications to Grammar Formalisms*, PhD thesis, University of Delaware, Newark, DE.

Rogers, J.: 1996, A model-theoretic framework for theories of syntax, *34th Annual Meeting of the Assocation for Computational Linguistics: Proceedings of the Conference*, Morgan Kaufmann, San Francisco, CA, pp. 10–16.

Rogers, J.: 1997, "Grammarless" phrase structure grammar, *Linguistics and Philosophy* **20**, 721–746.

Rogers, J.: 1998, *A Descriptive Approach to Language-Theoretic Complexity*, CSLI Publications, Stanford, CA.

Rogers, J.: 1999, The descriptive complexity of generalized local sets, *in* H.-P. Kolb and U. Mönnich (eds), *The Mathematics of Syntactic Structure: Trees and their Logics*, number 44 in *Studies in Generative Grammar*, Mouton de Gruyter, Berlin, pp. 21–40.

Rogers, J.: 2003, wMSO theories as grammar formalisms, *Theoretical Computer Science* **293**, 291–320.

Schiffer, S. R.: 1972, *Meaning*, Clarendon Press, Oxford.

Shieber, S.: 1985, Evidence against the context-freeness of human language, *Linguistics and Philosophy* **8**, 333–343.

Soames, S.: 1974, Rule orderings, obligatory transformations, and derivational constraints, *Theoretical Linguistics* **1**, 116–138.

Stabler, E.: 1997, Derivational minimalism, *in* C. Retoré (ed.), *Logical Aspects of Computational Linguistics, LACL '96*, number 1328 in *Lecture Notes in Artificial Intelligence*, Springer Verlag, Berlin, pp. 68–95.

Stabler, E.: 1999, Formal grammars, *in* R. A. Wilson and F. C. Keil (eds), *The MIT Encyclopedia of the Cognitive Sciences*, MIT Press, Cambridge, MA, pp. 320–322.

Straubing, H.: 1994, *Finite Automata, Formal Logic, and Circuit Complexity*, Birkhäuser, Boston, MA.

Thomas, W.: 1990, Automata on infinite objects, *in* J. van Leeuwen (ed.) (ed.), *Handbook of Theoretical Computer Science*, Elsevier Science, New York, pp. 135–191.

Weir, D. J. and Vijayshanker, K.: 1994, The equivalence of four extensions of context-free grammars, *Mathematical Systems Theory* **27**, 511–545.

# The Logic of Biological Classification and the Foundations of Biomedical Ontology

Barry Smith

*Department of Philosophy, University at Buffalo and Institute for Formal Ontology and Medical Information Science, University at Leipzig*

Biomedical research is increasingly a matter of the navigation through large computerized information resources deriving from functional genomics or from the biochemistry of disease pathways. To make such navigation possible, controlled vocabularies are needed in terms of which data from different sources can be unified. One of the most influential developments in this regard is the so-called Gene Ontology, which consists of controlled vocabularies of terms used by biologists to describe cellular constituents, biological processes and molecular functions, organized into hierarchies via the relation of class subsumption. Here we seek to provide a rigorous account of the logic of classification that underlies GO and similar biomedical ontologies. Drawing on Aristotle, we develop a system of axioms and definitions for the treatment of biological classes and instances.

## 1. Introduction

In reflection of the huge amounts of data accumulating in areas such as genomics and proteomics, biology and biomedicine have come to rely increasingly on the use of computational methods in their research. One of the most impressive and influential developments in this regard is the so-called Gene Ontology (GO) [1], which is being developed as part of the effort to produce controlled vocabularies for shared use across different biological domains within the framework of the Open Biological Ontologies project.[1] We take GO as our test case in what follows, not only because it has proved so successful in serving as a common reference system for a variety of groups working at the forefront of biomedical research, but also because, as we shall see, it suffers from a series of problems which are characteristic of almost all current ontologies used in bioinformatics.

GO provides some 20,000 terms for describing gene product attributes. It is divided into three hierarchically structured networks, whose topmost nodes are, respectively: *cellular component, molecular function* and *biological process*.[2] While GO is not strictly speaking an ontology in the sense in which this term is understood by philosophers, it does go some way in this direction, in that its three constituent vocabularies are organized as hierarchies via the ontological relations of subsumption (*human being* is subsumed by *mammal*) and partonomic inclusion

---

[1] http://obo.sourceforge.net.
[2] http://www.geneontology.org/doc/GO.doc.html. We refer in what follows to the version of October 2003.

(*human heart* is included as part of *human being*). Following standard usage in GO and other similar endeavors, these relations are called '*is a*' and '*part of*' in what follows.

Here we are concerned with GO as a *classification* of biological phenomena. The classes which stand in its *is a* and *part of* relations have some obvious relation to the species and genera of more traditional biological classifications, but there are also important differences. Thus not only are classes of *objects* recognized by GO, but so too are classes of *processes* and *functions*.[3] Crucially, GO defines its three structured networks as separate ontologies, which means that no ontological relations are defined between them. In other respects, too, the GO literature provides few clues as to how the ontological correlates of its separate constituent terms are to be conceived. Thus in particular, it tells us little about how we are to understand the two central terms *biological process* and *molecular function [3]*.

As a step towards filling this gap, and in reflection of the fact that GO, like many other ontologies currently being developed for purposes of biomedical research, shuns logico-philosophical rigor, we provide here a formal account of biological and biomedical classification which is designed as a first step towards the rigorous treatment of the questions concerning classes and class-hierarchies which arise at the interface between biology and medicine on the one hand and current bioinformatics research on the other.

## 2. The Gene Ontology

For purposes of preliminary orientation, consider the two GO terms:

GO:0003673:**cell fate commitment**

GO:0045168:**cell-cell signaling involved in cell fate commitment**

The hierarchical relations between these two entries within GO's biological process ontology are shown in Figure 1[4] below.

'Is a', as it is employed in this diagram, means roughly what we would expect it to mean when interpreted as a relation of subsumption between classes (natural kinds, species, genera) in biology. Note, though, that (unlike Aristotle, and unlike Linnaeus) GO allows multiple inheritance; that is to say, it allows one and the same biological class to have two or more parent-classes (as, in the figure, *cell differentiation* has the two parents *development* and *cellular process*). In addition GO does not strive to ensure that the terms in its three hierarchies are divided into predetermined levels (analogous to the levels of kingdom, phylum, class, order, etc., in traditional biology); indeed the acceptance of multiple inheritance means that such levels cannot in any case be defined, since the notion of 'sibling' becomes indeterminate.

Multiple inheritance allows us to deal with different aspects and contexts of classification within a single network. It is thus a useful device for producing

---

[3]On the different logical frameworks needed for the treatment of objects, functions and processes see [2]

[4]The diagram is taken from the QuickGO browser: http://www.ebi.ac.uk/ego. Solid links indicate *is a* relations; broken links indicate *part of* relations.

compact networks which can facilitate computationally efficient navigation through large edifices of information.

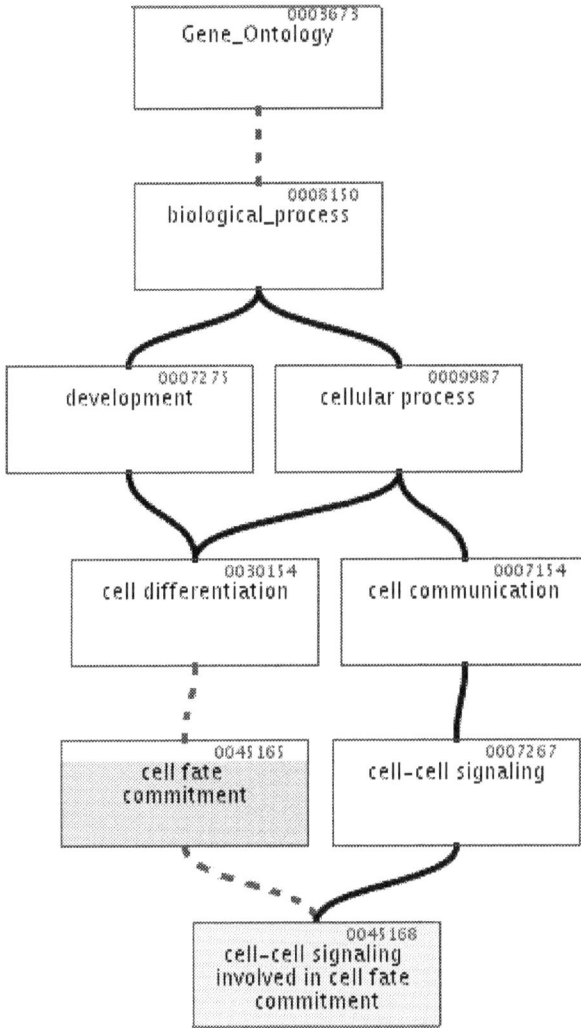

Figure 1. Example of GO Relations.

At the same time, however, multiple inheritance causes problems. These turn *inter alia* on the fact that the alignment of distinct ontologies rests crucially on the assumption that the basic ontological relations – above all relations such as *is a* and *part of*, which provide the glue which holds ontologies together – must have the same meanings in the different ontologies to be aligned. As inspection reveals, however, multiple inheritance goes hand in hand, at least in many cases, with the assignment to the *is a* relation of a variety of meanings within a single ontology. The resultant mélange makes coherent integration across ontologies achievable (at best) only under the guidance of human beings with the sorts of biological knowledge which can override the mismatches which otherwise threaten to arise. This, however, is to defeat the very purpose of constructing bioinformatics ontologies like GO as the basis for a new kind of biological and biomedical research designed to exploit the power of computers [4].

Thus for example when GO postulates

> cell differentiation *is a* cellular process

> cell differentiation *is a* development

then it means two different things by 'is a'. Only in the former case do we have to deal with a true subsumption relation between biological classes. In the latter case, rather, as is seen from the definition:

GO:0007275 **Development**

> **Definition:** Biological processes specifically aimed at the progression of an organism over time from an initial condition (e.g. a zygote, or a young adult) to a later condition (e.g. a multicellular animal or an aged adult)

the relation involved would more properly be expressed as: *contributes to the achievement of a certain end.*

When GO postulates:

> hexose biosynthesis *is a* monosaccharide biosynthesis

> hexose biosynthesis *is a* hexose metabolism,

on the other hand, then the second *is a* seems more properly to amount to a *part of* relation, since *hexose biosynthesis* is just that part of hexose metabolism in which hexose is synthesized.

And when GO postulates:

> vacuole (sensu Fungi) *is a* storage vacuole

> vacuole (sensu Fungi) *is a* lytic vacuole,

where the 'sensu' operator is introduced by GO to cope with those cases where a word or phrase has a specific meaning when applied to specific classes of organisms,[5]

---

[5]http://www.geneontology.org/doc/GO.usage.html#sensu.

then it seems that *is a* stands in neither case for a genuine subsumption relation between biological classes; rather, it signifies on the one hand the assignment of a *function* and on the other hand the assignment of special features to the entities in question.[6] The case is thus analogous to:

tank (sensu Oil Industry) *is a* storage tank

tank (sensu Oil Industry) *is a* tank with an enamel coating to prevent rust.

The term 'tank' as used in the oil industry designates in every case a tank used for storage, and all such tanks have an enamel coating to prevent rust. But in neither case do we have what should properly be represented as an *is a* relation in a well-designed ontology.

Theorists of classification have long recognized that the division into levels and the possession by every level within a classificatory hierarchy of the so-called JEPD property (for: jointly exhaustive and pairwise disjoint) represent ideals to which classifications should aspire. The feature of exhaustivity may be difficult to achieve in the realm of biological phenomena. But shortfalls from disjointness are easy to detect. The acceptance of multiple inheritance is just the rejection of the criterion of disjointness and thus also of the JEPD ideal.

We here leave open the question whether division into levels and single inheritance involving genuine *is a* relations can be achieved throughout the realm of classifications treated of by GO and similar ontologies. However, we note that, as Guarino and Welty have shown see e.g. their paper [5], methods exist which have demonstrated considerable success in removing cases of multiple inheritance from class hierarchies by distinguishing *is a* relations from ontological relations of other sorts. Using their methods, well-structured classifications can be achieved by recognizing additional relation-types (for example: *has role*, *is dependent on*, *causes*, *is involved in*, *is realized in*) and by allowing within a single ontology categories of entities of different sorts (for instance *roles, functions, qualities, processes*). GO, however, has neither of these alternatives at its disposal because of its insistence that its three constituent vocabularies represent *separate ontologies* with no relations defined between them.

## 3. Core Axioms for a Theory of Biological Classification

We shall focus, in what follows, on the logical treatment of the notion of class as a step towards building a framework within which issues of biological classification can be more rigorously addressed. One might at first suppose that the logic of classes is a matter properly to be treated on a more general level – for example as part of set theory in the mathematical sense. If, however, a class is the ontological correlate of a node in a (biological) classification, if, in other words, a class is a (biological) *natural kind*, then this means that classes must stand to their instances in

---

[6]A lytic vacuole is defined by GO as: a vacuole that is maintained at an acidic pH and which contains degradative enzymes, including a wide variety of acid hydrolases.

a relation which is quite different from the relation between a set and its members. This is because classes, but not sets, can remain identical even while undergoing a certain turnover in their instances (see [6])

Our formal theory is motivated by the theory of classes that we find in Aristotle's writings. We turn to Aristotle not only because many of his ideas still have an astonishing pertinence when it comes to laying down standards of logical rigor in the construction of classifications and in the formulation of definitions, but also because, while many Aristotelian ideas were cast aside in the wake of the Darwinian revolution in biology, his ideas on classes and classification have in recent times come to enjoy a new relevance as a result of the role of classificatory ontologies in contemporary bioinformatics.

The theory here set forth is designed as a central module to be extended and modified to deal with specific issues relating to biological classification or with specific kinds of biological classes. As we should expect, given the Aristotelian roots of the axioms presented, the theory works well when applied to the classification of organisms and of spatially extended objects (endurants, continuants, things, substances) in general. Amended versions will be needed where we are dealing with the classification of entities, such as functions and processes, in other categories.

We begin by drawing a distinction, within the realm of entities in general, between universals and particulars. We take the opposition between universals and particulars as a primitive of our theory, and introduce variables $e, f, g, \ldots$ to range over entities in general. We then adopt the axiom:

A1. $\neg\exists e(\boldsymbol{u}(e) \land \boldsymbol{p}(e))$

where $\boldsymbol{u}$ and $\boldsymbol{p}$ are primitive predicates holding of *universals* and *particulars*, respecttively. Thus A1 asserts that there is nothing that is both a universal and a particular.

Examples of particulars are: you and me, the Planet Earth, this piece of cheese. Examples of universals are: human being, enzyme, aspirin. Particulars (individuals, tokens) are simply located entities, bound to a specific (normally topologically connected) location in space and time. Universals are multiply located entities; they exist in the corresponding particulars.[7]

We introduce a primitive relational predicate *inst* to stand for the relation between an instance and a class. We then define a class as anything (any universal) that is instantiated, and an instance as anything (any particular) that instantiates some class:

D1. class(e) $=_{def} \exists f\ \boldsymbol{inst}(f, e)$

D2. instance(e) $=_{def} \exists f\ \boldsymbol{inst}(e, f)$

By admitting the predicate *inst* and treating terms for classes as logically on a par with terms for instances in this way, we can develop our theory exclusively within the framework of first order logic. We might call it: first order logic with universal terms and certain designated (relational) predicates – above all identity

---

[7]For a formal treatment of these notions see [7],

and instantiation – which have a fixed semantic evaluation in every model.[8]

Most importantly for our purposes, the realm of universals comprehends (biological) classes, i.e. what in other contexts would be called natural kinds, species, genera, and the like. We can now postulate further:

A2. $\exists e(\boldsymbol{u}(e) \wedge \neg class(e))$

There exists at least one universal which is not a class.

Examples of universals which are not classes are: *pet, adult, rational being, parent, catalyst, movement, process of development, storage vacuole*.[9] Classes are, as it were, elite entities within the realm of universals.[10] Which classes (and thus which instances) exist in a given domain is a matter for empirical research. In the macroscopic biological realm, at least, we can assume that the question as to which classes of entities exist has to do with the question as to which entities result from the coordinated expression of genes of specific sorts.

Instances, similarly, are elite entities within the realm of particulars; they are the *natural* (or standard or prototypical or canonical) exemplars of biological classes. The problems raised by non-standard instances must be dealt with in the extended version of the core module here presented, as also must the problems raised by non-standard classes, by classes in non-standard situations (for example organism species on the verge of extinction) and by the ways in which biological classes can change (evolve) over time.[11]

We need an axiom to the effect that:

A3. $\forall e \forall e\prime(\boldsymbol{inst}(e, e\prime) \rightarrow \boldsymbol{p}(e) \wedge \boldsymbol{u}(e\prime))$

We can then prove the theorems:

T1. $\forall e(class(e) \rightarrow \boldsymbol{u}(e))$

There are no classes which are not universals.

T2. $\forall e(instance(e) \rightarrow \boldsymbol{p}(e))$

There are no instances which are not particulars.

A4. $\exists e(\boldsymbol{p}(e) \wedge \neg instance(e))$

There are particulars which are not instances.

A5. $\exists e \, \boldsymbol{p}(e) \rightarrow \exists e\prime \, instance(e\prime)$

If there is a particular, then there is an instance.

As an example of a particular which is not an instance consider the mereological

---

[8] An alternative approach, which embraces a second-order logical framework, is explored in [8].

[9] In an alternative formulation of these ideas we might distinguish different *contexts of classification*. It might then be that in certain special contexts of inquiry some of these terms can indeed be held to designate classes satisfying axioms very much like the ones presented here.

[10] Our theory of classes is thus an analogue of the 'sparse theory of universals' propounded by David Lewis in [9].

[11] Some indications are provided in [10].

sum of a molecular at the end of your nose and your brother's lizard. Intuitively, every particular is such as to overlap mereologically with some instance.

We can then prove:

T3. $\neg\exists e(\text{class}(e) \land \text{instance}(e))$
Nothing can be both an instance and a class.

T4. $\forall e \, (\text{class}(e) \rightarrow \exists e\prime(\boldsymbol{inst}(e\prime, e)))$
Every class has at least one instance. (This follows trivially from D1.) This is the basic principle of Aristotelian realism as far as classes are concerned. We here leave open whether an analogous axiom holds for universals in general.

T5. $\exists e \, \boldsymbol{p}(e)$

T6. $\exists e \, \text{instance}(e)$

T7. $\exists e \, \text{class}(e)$

T8. $\exists e \, \boldsymbol{u}(e)$

There exists at least one particular; there exists at least one instance; there exists at least one class; and there exists at least one universal.

We can now introduce typed variables, $A, B, C \ldots$ to range over classes and $x, y, z, \ldots$ to range over instances, and we can postulate an axiom to the effect that at least two classes exist:

A6. $\exists A \exists B(A \neq B)$,
together with an axiom of extensionality:

A7. $\forall A \forall B \forall x((\boldsymbol{inst}(x, A) \leftrightarrow \boldsymbol{inst}(x, B)) \rightarrow A = B)$.
(We note that the relation to time must be taken into account in the extended version of the core module here presented. We should then, for example, be able to formulate principles to the effect that classes are identical if and only if they share the same instances *at the same times*.)

We can now define the *is a* relation between classes in terms of $\boldsymbol{inst}$:

D3. $A$ *is a* $B =_{def} \forall x \, (\boldsymbol{inst}(x, A) \rightarrow \boldsymbol{inst}(x, B))$.

*Is a* is thus superficially analogous to the usual set-theoretic subset relation ($\subseteq$). More perspicuously:

D3* $e$ *is a* $f =_{def} \text{class}(e) \land \text{class}(f) \land \forall x \, (\boldsymbol{inst(x, e)} \rightarrow \boldsymbol{inst}(x, f))$.

We can also define various predicates picking out special sorts of classes, as follows:

D4. $\text{genus}(A) =_{def} \text{class}(A) \land \exists B(B \, is \, a \, A \land B \neq A)$

D5. $\text{species}(A) =_{def} \text{class}(A) \land \exists B(A \, is \, a \, B \land B \neq A)$

D6. $\text{lowestspecies}(A) =_{def} \text{species}(A) \land \neg\text{genus}(A)$

D7. highestgenus$(A)$ $=_{def}$ genus$(A) \wedge \neg$species$(A)$

Aristotle uses the term 'category' as a synonym of highest genus, and we can guarantee axiomatically that at least one such highest genus exists:

A8. $\exists A$ highestgenus$(A)$

Adding:

A9. class$(A) \rightarrow$ genus$(A) \vee$ species$(A)$

we can then prove:

T9. class$(A) \rightarrow$ (genus$(A) \vee$ lowestspecies$(A)$)

T10. class$(A) \rightarrow$ (species $(A) \vee$ highestgenus$(A)$)

and also:

T11. *A is a A*                          (*is a* is reflexive)

T12. *(A is a B $\wedge$ B is a C) $\rightarrow$ A is a C*          (*is a* is transitive)

T13. *(A is a B $\wedge$ B is a A) $\rightarrow$ A = B*          (*is a* is antisymmetric)

## 4. Axioms for *Nearest Species*

When one class is immediately subsumed by another (i.e. where one is child to the other as parent in a species-genus tree) then we say that they stand in the relation of nearest species, which is defined as follows:

D8. nearestspecies*(A, B)* $=_{def}$ *A is a B $\wedge$ A $\neq$ B $\wedge \forall C$ ((A is a C $\wedge$ C is a B)* $\rightarrow$ *(C = A $\vee$ C = B))*

We can now formulate a series of axioms for biological classes which seem to come close to capturing what we mean when we say that classes are *natural* kinds. Here (following Aristotle[12] ) we focus on axioms for classes of *objects* (cells, molecules, organisms, limbs, organs, and the like), noting again that the framework will in due course need to be expanded to cope with the class-instance relations governing entities in other categories:

A10. (nearestspecies$(A, B) \wedge$ nearestspecies$(A, C)) \rightarrow B = C$

A species never has two is a parents. (This rules out cases of multiple inheritance.)

A11. lowestspecies$(A) \wedge$ lowestspecies$(B) \wedge A \neq B$
$\rightarrow \neg \exists x (\textbf{inst}(x, A) \wedge \textbf{inst}(x, B))$

Distinct lowest species never share instances.

A12. genus$(A) \wedge$ inst$(x, A) \rightarrow \exists B$ nearestspecies$(B, A) \wedge$ inst$(x, B)$

---

[12]More precisely: following Jan Berg's excellent treatment of these matters in [11].)

Every instance of a genus instantiates also some nearest species of this genus.

A13. nearestspecies(A, B) → ∃x(inst(x, B) ∧¬inst(x, A))

Each genus includes more instances than any of its nearest species.

A14. nearestspecies(B, A) → ∃C (nearestspecies(C, A) ∧ B ≠ C))

Every genus has at least two children.

A15. (nearestspecies(B, A) ∧ nearestspecies(C, A) ∧∃x(inst(x, B) ∧ inst(x, C)))
→ B = C

Species of a common genus never share instances.

A16. (genus(A) ∧ inst(x, A)) → ∃B (lowestspecies(B) ∧ B is a A ∧ inst(x, B))

Every instance also instantiates some lowest species.
The above are non-trivial. They have the following theorems as consequences:

T14. genus($A$) → ∃$B$∃$C$(nearestspecies($B$, $A$)∧ nearestspecies ($C, A$) ∧ $B ≠ C$))

Every genus has at least two nearest species.

T15. (genus($A$)∧ lowestspecies($B$) ∧ ∃$x$(***inst***($x, A$)∧ ***inst***($x, B$))) → $B$ *is a* $A$

If an instance of a lowest species instantiates some genus, then the lowest species
is subsumed by the genus.

T16. nearestspecies($A, B$)
→ ¬∃$C$(nearestspecies($A, C$)∧ nearestspecies($C, B$))

If $A$is a nearest species to $B$, then there is no path through the hierarchy from $B$to
$A$via some third class $C$.

T17. class($A$)∧ class($B$)
→(A = B ∨ A is a B ∨ B is a A ∨¬∃x(***inst***(x, A) ∧ ***inst***($x, B$)))

Distinct classes are either such that one subsumes the other or they have no instances in common.
To prove further desirable theorems we would need to add an **additional axiom**
to the effect that the universe is finite (in other words that there are only finitely
many biological classes, and only finitely many instances of such classes), a thesis
which seems intuitively plausible in the domain of biology. We could then infer:

T18. (genus($A$)∧ genus($B$) ∧ ∃$x$(***inst***($x, A$)∧ ***inst***($x, B$)))
→ ∃$C$($C$ *is a* $A$ ∧ $C$ *is a* $B$)

If two genera have a common instance then they have a common subclass.

T19. $A$ *is a* $B$ ∧ $A$ *is a* $C$ → ($B = C$∨ $B$ *is a* $C$ ∨ $C$ *is a* $B$)

Classes which share a subclass in common are either identical or one is subordinated to the other.

The system so defined implies that each class hierarchy constitutes a supremum semilattice, or in other words that every collection of classes has a least upper bound with respect to *is a*. To generate a simple model let P be any finite set. P can be, for example, a finite subset of the natural numbers. Let I be any non-empty proper subset of P and let C be any non-empty subset of $\wp(I)$ (so that C is a collection of subsets of I), with the following properties:

i) $\emptyset \notin C$

ii) $\cup C = I$

ii) if X, Y $\in$ C and X $\cap$ Y $\neq \emptyset$, then X $\subseteq$ Y or Y $\subseteq$ X

iii) if X $\in$ C, then there is some Y $\in$ C such that either X $\subset$ Y or Y $\subset$ X

iv) if X, Y $\in$ C and X $\subset$ Y,

then there are $Z_1$, ..., $Z_n \in$ C such that $X \cup Z_1 \cup ... \cup Z_n = Y$.

The particulars in these models are the members of P, universals are the members of C, and instantiation is interpreted as the set-membership relation. Instances, then, are the members of I and all members of C are not merely universals but also classes. A highest genus is a member of C that is not a proper subset of any member of C and a lowest species is a member of C that is not a proper superset of any member of C. Notice that because I is finite, there must be at least one highest genus and more than one (but only finitely many) lowest species.

## 4.1. Aristotelian Definitions

We can now, again following Berg (*op. cit.*) give an account of the Aristotelian theory of definitions. To give a definition, for Aristotle, is to say of something what it is. More precisely, a definition tells us what makes an entity of a given sort an entity of that sort. In a different terminology, an Aristotelian definition is an account of the essence or nature of something. Definitions, for Aristotle, are *real* rather than merely *nominal* definitions: thus they are not the specifications of the meanings of words.

It follows from the above that only what has an essence can be defined, and it is precisely classes, in the terms we have been using above, which satisfy this condition. More precisely, it is *species* which can be defined, via the specification in each case of the relevant nearest genus and differentia. The latter tells us what marks out instances of the species within that genus. Thus *human* is defined as *rational animal*, where *animal* is the genus and *rational* is the differentia. The differentia is also referred to in Aristotelian terms as the 'specific difference' or 'difference that makes a species'.

To specify a class is to provide an answer to a "What is it?" question. When faced with a new kind of biological phenomenon the task of the biologist is to provide the tightest possible answer to the "What is it?" question, which means: to provide the species for the phenomenon, which means also specifying the relevant nearest genus and the relevant specific difference. An Aristotelian definition must then satisfy the condition that an entity satisfies it *if and only if* it instantiates the corresponding species. Specifying a genus alone would be to provide an answer to the "What is it?" question that is not sufficiently tight, since the genus encompasses also other phenomena. Note that to specify the qualities, functions or roles of entities or to say what processes entities engage in is not to provide an answer to the "What is it?" question.

Differentia, too, are universals in the sense of this term presupposed in the above. Differentia are not *instantiated*, but rather *exemplified*, a new primitive notion which we symbolize by means of **exemp.** An Aristotelian definition then has the form:

$$\text{An A is a B which exemplifies S}$$

where the variables $S, T, \ldots$ range over differentia. We then have:

$$inst(x, A) \leftrightarrow inst(x, B) \wedge exemp(x, S)$$

and we can define what it is to be a differentia as follows:

D9. differentia(S) $=_{def} \exists$B $\exists$C (nearestspecies(B, C) $\wedge$ $\forall$x(**inst**(x, B) $\leftrightarrow$ (**inst**(x, C) $\wedge$ **exemp**(x, S))))

The genus together with the differentia of a species constitutes the essence of the corresponding species.[13]

We can then postulate axioms for differentia such as:

A17. differentia(S) $\rightarrow$ ¬class(S)

and prove theorems for example to the effect that:

T20. differentia($S$) $\rightarrow$ $\exists x$ **exemp**($x, S$)

The axioms presented above are motivated by the sorts of classifications we find in the life sciences. However, in extensions of the theory we may consider which amended versions of these axioms would be required to cope with the classification of natural kinds in non-organic domains such as chemistry, meteorology, or geomorphology [12], and also which axioms, or systems of axioms, would be needed to cope with the classification of *artefacts* of different sorts, including both physical artefacts such as drugs or drug-delivery devices, and non-physical artefacts such as medical procedures or diagnoses. We may extend the framework still further by considering what, if any, would be the analogues of the axiom systems here considered in realms such as temperature, which are marked by continuous variation, or in realms such as types of soil or types of water impurity, which are marked by combinations of factors which vary independently. We may consider what the analogues of these axioms would be for the different sorts of folk classifications carried out by human beings in different cultures and using different natural languages

---

[13]Not everything which satisfies 4.1 is a differentia, for Aristotle, who distinguished also what he called 'propria', which are properties peculiar to all the members of a given species which yet do not belong to the essence of the species – for example the property: *capable of laughing* as possessed by humans. We ignore this issue here for the sake of simplicity.

[13]. And finally, and most importantly for the realm of biomedical informatics, we may consider how to manipulate simultaneously a multiplicity of different classifications, prepared in different disciplinary contexts or for different purposes, of the same domain of phenomena in reality[14].

## 5. The Foundational Model of Anatomy

Among all existing biomedical ontologies it is the Foundational Model of Anatomy, developed at the University of Washington, Seattle as part of the Digital Anatomist Project, which comes closest to meeting the standards of formal rigor taken for granted among philosophical ontologists. The Foundational Model of Anatomy (hereafter: FMA) is a symbolic representation of the structural organization of the human body from the macromolecular to the macroscopic levels[15]. It has the goal of providing a robust and consistent scheme for classifying anatomical entities on the basis of explicit definitions of a sort which can serve as a reference ontology in biomedical informatics.

Most significant, from our present point of view, is the fact that the FMA has adopted an Aristotelian regime of definitions.

Thus definitions in FMA look like this:

**Cell** *is a* **anatomical structure** that *consists of* **cytoplasm** *surrounded by* a **plasma membrane** with or without a **cell nucleus**

**Plasma membrane** *is a* **cell part** that *surrounds* the **cytoplasm** where terms picked out in bold are nodes within the FMA classification and italicized terms signify the formal relations – including *is a* – which are defined between these nodes.

As the FMA points out, ontologies 'differ from dictionaries in both their nature and purpose [16]'. Dictionaries are prepared for human beings; their merely nominal definitions can employ the unregimented resources of natural language, can tolerate circularities and all manner of idiosyncrasy. In ontologies, however, definitions must be regimented in such a way that each reflects the position in the hierarchy to which the definiendum belongs:

> The role of definitions in an ontology is ...to specify such defining attributes in a consistent manner, thus assuring their transitive inheritance through a type hierarchy. Consistency in definitions and, therefore, in the classification, requires that a unifying viewpoint (i.e., context) be also specified for concept representation. This context should hold true for the entire ontology.[14] Provided such requirements are satisfied, the position of a concept will enrich its own definition by the definition of all of its parents within the hierarchy. Thus, unlike in a dictionary, a definition of a concept within an ontology is incomplete without that of all of its parents.[15]

This means additionally that in order to ensure transitive inheritance of essential

---

[14]The context for the FMA is: anatomical structure; this means that all the definitions in the FMA hierarchy are formulated exclusively in structural terms, which means: without appeal to normal and abnormal functions performed by the anatomical entities distinguished.

[15] *Op. cit.*

characteristics, all intermediate classes should be defined even if they have not have been explicitly identified in the scientific literature.[16] It means also that, already on the basis of its rules for the formulation of definitions, the FMA rules out multiple inheritance. And it means, finally, that the FMA, with its system of definitions, can exploit all the benefits – in terms of reliable curation, efficient error checking and information retrieval, and ease of alignment with neighboring ontologies – of *logical compositionality.*

## 6. GO Again

GO, too, like other, similar biomedical ontologies, provides not only controlled vocabularies with hierarchical structures but also definitions of its terms. Indeed part of the goal of GO, and of similar projects, is to provide a source of 'strict definitions' that can be communicated across people and applications. When we examine GO's actual practice, however, we find that its definitions are affected by a number of characteristic problems which, while perhaps not affecting their usability by human biologists, will raise severe obstacles at the point where the sort of formal rigor needed by computer applications (or by a formally rigorous biology of the future) is an issue. Consider again our two initial examples:

GO:0003673: **cell fate commitment**

> **Definition:** The commitment of cells to specific cell fates and their capacity to differentiate into particular kinds of cells.

GO:0045168: **cell-cell signaling involved in cell fate commitment**

> **Definition:** Signaling between cells that results in the commitment of a cell to a certain fate. This is often done by secretion of proteins by one cell which affects the neighboring cells and causes them to adopt a certain fate.

In both of these definitions we recognize the characteristic problem of circularity. The coarse logic of the definition of cell fate commitment is as follows:

$x$ is a cell fate commitment $=_{def} x$ is a cell fate commitment and $p$,

where $p$ is, logically speaking, a second, extraneous condition. Further problems arise in virtue of the fact that, as a result of its use of unregimented natural language and of its lack of concern for issues of logical compositionality, substitution of GO definiens for the GO terms appearing within other GO terms and definitions can be achieved, at best, only with human intervention. Thus consider:

GO:0030154: **Cell differentiation**

> **Definition:** The process whereby relatively unspecialized cells, e. g. embryonic or regenerative cells, acquire specialized structural and/or functional features that characterize the cells, tissues, or organs of the mature organism or some other relatively stable phase of the organism's life history.

GO:0007514: **Garland cell differentiation**

> **Definition:** Development of garland cells, a small group of nephrocytes which take up waste materials from the hemolymph by endocytosis.

---

[16] *Op. cit.*

In this way a number of valuable methods of inference, extrapolation of new terms, and error-checking are foreclosed.

## 7. Conclusion

The treatment of is a relations in biomedical ontologies has been thus far highly problematic. In some cases, indeed, the two relations are not clearly distinguished at all, leading to what Guarino calls 'is a overloading' [17]. The FMA defines an ontology as a 'true inheritance hierarchy [18]'. thereby drawing attention to the fact that one central reason for adopting the method of ontologies in supporting reasoning across large bodies of data is precisely the fact that this method allows the exploitation of the inheritance of properties along paths of is a relations.

When challenged with such problems, the members of the GO and associated communities standardly insist that their concerns are those of practicing biologists, and that they are thus not concerned with the sorts of scrupulousness that are important in logic. To repeat, however, if GO's adherents propose that GO should serve as a reference-platform for computer-assisted navigation between biomedical databases, then the failure to achieve consistency with standard logical principles will place considerable obstacles in the way of its efforts to achieve this end.[17]

## REFERENCES

1. The Gene Ontology Consortium, "Gene Ontology: Tool for the Unification of Biology. *Nature Genetics*, 25 (2000), 25-29. See also: http://www.geneontology.org.
2. Pierre Grenon and Barry Smith, "SNAP and SPAN: Towards Dynamic Spatial Ontology", forthcoming in *Spatial Cognition and Computation.*
3. Barry Smith, Jennifer Williams and Steffen Schulze-Kremer, 2003, "The Ontology of the Gene Ontology", in *Biomedical and Health Informatics: From Foundations to Applications*, Proceedings of the Annual Symposium of the American Medical Informatics Association, Washington DC, November 2003, 609–613.
4. Barry Smith, Jakob Köhler and Anand Kumar, "On the Application of Formal Principles to Life Science Data: A Case Study in the Gene Ontology", in *Proceedings of* DILS 2004 *(Data Integration in the Life Sciences)*, (Lecture Notes in Bioinformatics 2994), Berlin: Springer, 2004.
5. Nicola Guarino and Chris Welty, "Identity and subsumption", in R. Green, C. A. Bean, and S. Hyon Myaeng (eds.), *The Semantics of Relationships: An Interdisciplinary Perspective*, Dordrecht: Kluwer (2002), 111–126.
6. Barry Smith and Cornelius Rosse, "The Role of Foundational Relations in Biomedical Ontologies", *Proceedings of Medinfo*, 7-11 September 2004, in press.
7. Smith, "On Substances, Accidents and Universals: In Defence of a Constituent Ontology", *Philosophical Papers*, 26 (1997), 105–127.

[17]Acknowledgements: This paper was written under the auspices of the Wolfgang Paul Program of the Alexander von Humboldt Foundation. Thanks are due also to Maureen Donnelly, Kai Hauser, Ingvar Johansson, Jacob Köhler, David Mark, Carsten Pontow, Cornelius Rosse, Steffen Schulze-Kremer and Jonathan Simon.

8. Nino Cocchiarella, "On the Logic of Natural Kinds", *Philosophy of Science*, 1976; 43: 202–222.

9. David Armstrong "New Work for a Theory of Universals", *Australasian Journal of Philosophy*, 61 (1983): 343–377.

10. Thomas Bittner and Barry Smith, "A Theory of Granular Partitions", *Foundations of Geographic Information Science*, Matthew Duckham, Michael F. Goodchild and Michael F. Worboys, eds., London: Taylor & Francis, 2003, 117–151.

11. "Aristotle's Theory of Definition", *ATTI del Convegno Internazionale di Storia della Logica*, San Gimignano, 4–8 December 1982, Bologna: CLUEB, 1983, 19–30. See http://ontology.buffalo.edu/bio/berg.pdf.

12. Barry Smith and David M. Mark, "Do Mountains Exist? Towards an Ontology of Landforms", *Environment and Planning B (Planning and Design)*, 30(3), 2003, 411–427.

13. Douglas L. Medin and Scott Atran (eds.), *Folkbiology*, Cambridge, MA: MIT Press, 1999.

14. Thomas Bittner and Barry Smith, "A Theory of Granular Partitions", *Foundations of Geographic Information Science*, Matthew Duckham, Michael F. Goodchild and Michael F. Worboys, eds., London: Taylor & Francis, 2003, 117–151.

15. Cornelius Rosse and José L. V. Mejino Jr., "A Reference Ontology for Bioinformatics: The Foundational Model of Anatomy", *Journal of Biomedical Informatics*, forthcoming.

16. J. Michael, José L. V. Mejino Jr., and Cornelius Rosse "The Role of Definitions in Biomedical Concept Representation", *Proceedings of the American Medical Informatics Association Symposium*, 2001, 463–467.

17. Guarino, N. "Some Ontological Principles for Designing Upper Level Lexical Resources", in *Proceedings of the First International Conference on Language Resources and Evaluation*, Granada, 1998, 527–534.

18. Michael, Mejino, and Rosse, *op. cit.*

# Experiment and Concept Formation

Friedrich Steinle*

*Max Planck Institute for the History of Science, Berlin*
*steinle@mpiwg-berlin.mpg.de*

**Abstract.** While the recent wave on studies of experiment has shown the 'standard view' to be insufficient, it has not spelled out so far a richer perspective on the epistemic roles of experiment. In my paper, and based on historical studies, I present a specific type of experimentation, called "exploratory", that is widespread in experimental research and plays a decisive role. It often leads to forming new basic concepts in specific research fields, and it is here that its central epistemic importance is located. The delicate problem of how new concepts can be generated within an activity that itself is based on the use of concepts is in the core of my study. I also discuss the questions of how exploratory experimentation relates to concept formation in general and why it has escaped attention for so long.

Experiment has interested those who reflected on science for a long time. Francis Bacon, often cited as the champion of modern experimentation, pointed out a variety of epistemic functions of experiment, including producing new phenomena, classifying them, and deciding between competing theories and hypotheses by way of "crucial" experiments. In the $19^{th}$ century, ideas about the function of experiment developed further, with Mill's four "experimental methods" as a prominent case [31, book III, chs. 7 & 8]. In the $20^{th}$ century, however, the perspective narrowed down. In response to naïve inductivistic views, Duhem rejected any role of experiment in *generating* theories, leaving over only the *testing* of theories [10, ch. 10]. Such a view, much corroborated by Reichenbach's distinction between the contexts of discovery and justification (though in a popular, misguided interpretation), eventually became the philosophical "standard view" of experiment. Whether called the "handmaiden of theory" (Popper) or "theorizing with different means" (van Fraassen), it was believed that experiment dealt *only* with well-defined questions posed by theoreticians [50, p. 673],[37, § 30].

Only in the 1980s, the "New Experimentalism" emphasized the insufficiency of the older accounts, stimulated by Hacking's emphasis on a "Baconian variety" of experiment.[18, ch. 9] Historians of science began to focus on the local, cultural, material, rhetorical, and social aspects of experiment. While many new perspectives on experiments have thus been unearthed, one central area has remained peculiarly unanalyzed, that is, how *knowledge* is drawn from experiments, and what type of knowledge. To be sure, there have been significant attempts to bring knowledge

---

*Thanks to Giora Hon for critical comments and helpful suggestions on an earlier version, to Lilia Gurova for pointing me to an ongoing discussion, and to the Thyssen Foundation, Cologne, for supporting the research of which this paper is an outcome. Parts of this paper have been discussed in [45] and [46].

back into the picture – for example by Hacking, Burian, Gooding, Graßhoff, and Rheinberger. ([15,3,21,16], [19] emphasized that need explicitly.) Still, however, there is a striking misbalance between insights into the historical and cultural dynamics of experiments on the one hand, and their epistemic variety on the other. This is where my talk focuses.

I shall follow two guidelines here. First, concerning our analytical toolbox, it is too unspecific to simply contrast theory and experiment. Many agree now on the need to differentiate further here. I shall tackle this by focusing on the types of *questions* and of *epistemic goals* that are pursued in experimental work and, in particular, on the role of concepts and concept formation. My second guideline has a methodological character. In order to widen our philosophical view, we have to look at real science. We have to study and take seriously historical cases, in particular those that don't fit into the classical scheme and thus require us to re-think our analytical categories. Only seriously developed historical material can provide an appropriate background to grasp the epistemic complexity of experimental research. Following these guidelines, I shall first sketch out three episodes from the history of electricity. A particular approach to experiment emerges here, one that is distinctly different from the standard account. I call it "exploratory experimentation" and I shall discuss its specific details in my second section. In the third section, finally, I shall delineate some perspectives on the more general problem of concept formation.

THREE HISTORICAL CASES

*Charles Dufay, structuring electricity*

   My first case deals with Charles Dufay, an early $18^{th}$ century, brilliant academician and director of the Paris botanic garden. When he started his research in electricity in the 1730s, the field was in an unstable and incoherent state.[2] More than a hundred years of research throughout Europe had produced a multitude of different and puzzling phenomena, such as:

• Some materials could be electrified by rubbing, others sometimes, others not at all.
• Sometimes electricity acted as attraction, sometimes as repulsion.
• Sometimes sudden changes regarding attractive and repulsive effects occurred.

Dealing with those questions proved difficult, even more so since the experiments were delicate, the effects tiny, and reproducibility difficult. Dufay conducted extensive experiments, varying the procedure in many ways: he used a vast number of different materials, in individual or combined arrangements, and varied their shape, temperature, color, moisture, air pressure, and the experimental setting: two bodies in touch, in close neighborhood, in large distance, being connected by a third and so on. His work led to remarkable results and bold claims such as that *all* materials except metals could be electrified by rubbing, and *all* bodies except a flame could receive electricity by communication. But still he was left with serious questions as to when attraction and repulsion occurred, and when the one sometimes suddenly switched into the other.

---

[2][22, ch. 9] gives a brief account of Dufay's research. Dufay's eight Mémoires at the Paris Académie between 1733 and 1737 provide the main source; cf. also his own English summary [9]. Electricity here always means, of course, what we nowadays call static electricity.

Dufay continued to conduct many experiments, with electrified and/ or unelectrified bodies attracting or repelling each other, before and after touching each other. From those experiments, he extracted a regularity, which held for every two bodies: When an unelectrified body was attracted by one that was electrified, and touched it, it would suddenly repel after the contact. The sequence of attraction – contact – repulsion formed a regularity that comprised a lot of previously puzzling phenomena. But it was restricted to the interaction of two bodies of which one was electrified by the other. In other constellations it failed. Dufay was not satisfied. After additional experiments, he finally made what he called a 'bold hypothesis:' If one did not speak of electricity in general, but of two electricities, then his experimental results would suddenly made sense. The law was that similarly electrified bodies repelled each other, while dissimilarly electrified ones attracted each other. The two electricities corresponded to two classes of materials of which they were produced by rubbing, and Dufay labeled them "vitreous" and "resinous" according to the most prominent representatives of these classes. With such a concept, he could subsume hundreds of experiments under general regularities! This convinced him quickly of his proposal, and this is how the two electricities entered our scientific thinking. Within a decade, they became part of the conceptual framework, even the definition of electricity, and have been common knowledge ever since.

### Ampère on electromagnetism, looking for "general facts"

My second case is nearly a century later. In July 1820, the Danish researcher Hans Christian Oersted announced his finding of the action of a galvanic current on a magnetic needle.[3] The experimental arrangement consisted of a galvanic battery with its "closing wire" and a magnetic needle suspended as compass needle. When the wire was brought near to the needle and connected to the battery, the needle immediately deviated from its normal north-south position and returned to it as soon as the wire was disconnected. Oersted's discovery opened a new field of research: electromagnetism, and caused, moreover, fundamental puzzlement: The needle was not attracted to or repelled from the wire, but set itself somewhat across. Even more mysterious, the deflection of the needle changed when it was placed over the wire rather than below it. Such a behavior was incompatible with the notion of attractive and repulsive forces, a notion on which all reasoning on physical processes was based. It was even difficult just to formulate the experimental outcomes. The needle set itself not only across the wire, but also across established thinking.

I shall sketch an episode of the work of André-Marie Ampère, professor of mathematics at the École Polytechnique.[4] Far from being drawn towards a certain theory, as the received view suggests, his early research followed tortuous pathways, pursued various goals in parallel and had an open-ended character. First, he figured out an instrument in which the effect of terrestrial magnetism was drastically reduced: his "astatic needle." The axis of the needle was put right in the direction of

---

[3]Oersted spoke of an "electric conflict" [34], but it was still debated whether the effects of the battery could properly be called "electric" – many preferred to speak of "galvanic battery", and to avoid the talk of "current" altogether.

[4]My account on Ampère in this and the next sections is based on my more extended work [48, chs. 2-4]. For a general overview of Ampère's research see [2] and [23]. Ampère's early research has, due to a particularly bad state of the sources, not been well studied so far. For my new approach, see [47].

the magnetic dip, so the needle could not react to terrestrial effects. Ampère now varied many experimental conditions: the strength and polarity of the battery, the length and material of the needle and, most extensively, the position of the needle relative to the wire: above, below, right, left, parallel, perpendicular, etc. His aim was to find out which factors contributed to the deflection of the needle and to formulate regularities. He realized that the needle always tended to take a right angle towards the wire. But into which of the two possible positions did the North Pole move? The lack of concepts to express spatial constellations became particularly pressing here, and Ampère decided to introduce new ones. First, in order to facilitate reference to the polarity of the battery, he spoke of the "so-called galvanic current," explicitly emphasizing a merely instrumentalistic use of that notion. Furthermore, he introduced the notions of "left" and "right hand side" of the current and explained them by imagining a man with a current running through him from toe to head. If that man turned his face towards the magnetic needle, his right hand indicated the "right hand side" of the current, and the left hand accordingly. With these concepts, he was able to formulate one coherent regularity that he called "directive action" and that in older physics textbooks is still called Ampère's "swimmer-rule."[1, p. 197]

However, this was not yet sufficient. Ampère realized by chance that the battery itself exerted an action onto the magnetic needle, much like the action of the wire. In order to subsume the two cases under one regularity, he had to assign the galvanic current within the battery the opposite direction: not from the copper- to the zinc-pole as in the wire, but in reverse direction. A few days later, however, he saw an easier way. If the direction of the current was no longer referred to the poles of the battery, but rather taken as a sense of rotation, the regularity could be given a coherent and more general form. The concept of current as mere direction was much sharpened here and, this is the essential point, the battery was "conceived as forming *one single circuit* with the conducting wire" [1, p. 198]. The concept of a current circuit, comprising the battery and its connecting wire likewise, was introduced here for the first time.[5] It enabled the formulation of general laws of the electromagnetic effects, of "general facts," as Ampère called them. The concept of a current circuit quickly became part of the basic framework of electricity and came out to be most fundamental for all further research. While it was soon given a physical interpretation, it had been formed and introduced as a means to enable the formulation of regularities in a most general form.

*Ampère on electromagnetism, proving a theory*

To offer a contrast, I sketch a somewhat later episode. Parallel to his search for regularities, Ampère pursued speculations about the 'causes' of the electromagnetic interaction. By tortuous pathways, he arrived at the hypothesis that all magnetism might be caused by circular electric currents within the magnetic bodies. That was a breathtaking perspective indeed. Not only was an exceedingly wide scope envisaged, but also the possibility appeared to treat such a theory mathematically, a point of utmost importance to Paris academicians.

In looking for empirical support, Ampère considered that if circular currents

---

[5] Again the talk of "circuit" had been much older, but Ampère gave it an essentially new meaning by treating the battery and the connecting wire in exactly the same manner.

interacted with magnets and behaved like magnets, they should also interact with each other, without any iron involved. In order to test this expectation, he designed a specific experiment. The central part of the apparatus consisted of two spirals of wire, placed face to face in two parallel vertical planes. One of them was mounted on a fixed stand, while the other one was suspended like a pendulum and moved without difficulty towards the first spiral. Ampère expected that the spirals, when connected to the battery, should either attract or repel each other. But he could not obtain that effect. He suspected the failure was caused by too much friction within the apparatus due to inappropriate suspension techniques (a particularly delicate point) or by insufficient battery power. His attempts to optimize these components went so far that he spent a half of a month's salary on the strongest battery available in Paris. With that apparatus, he succeeded in obtaining the expected effect; and only a few hours later, he proudly announced the new effect in a lecture at the Paris Academy, presenting it as a "definite proof" of his hypothesis of circular currents as the cause of magnetism.

It should be noted already here that these experiments differed significantly in character from the first two sets described above. Throughout this series, the central elements of the experiment remained unchanged. What Ampère conducted was well-directed optimizing, not broad exploration. From the first idea to the final evaluation, the experiment was defined by the expectation of a hypothesis. And the result was not a broad "if – then" regularity, but was considered to be an experimental "proof" of the theory.

EXPLORATORY EXPERIMENTATION

The above cases show various uses of experiment. Obviously, the type of experimental activity in my second Ampère case closely reflects the standard view. There was a theory, which led to expecting a certain effect; the expectation led to designing and conducting an experiment; and the success of the experiment counted as support for the theory. Different philosophers such as Duhem, Popper, and van Fraassen, would agree here: yes, this is how experiment works. The other two cases, however, open a richer perspective. Indeed, they pose a challenge to the standard view. Just to mention two obvious points: No theories or hypotheses have been tested here. By contrast, laws have been established and basic concepts been shifted. These are processes the standard view has no means to account for. In order to have a shorthand label, I have labeled that type of research "exploratory experimentation" [43,44, among others]. I shall go through its several features.

*Epistemic goals and questions*

First of all, what is the epistemic goal pursued in such experiments? Exploratory experiments aim at identifying conditions, i.e. factors that have to be present for the occurrence of the phenomena in question, and at establishing regularities and laws.[6] Typically the results have the form of "if – then" propositions, where both the if- and the then-clauses refer to the empirical level. Much of this reminds us of

---

[6]Some researchers were quite explicit about this specific goal as contrasted to possible other ones. Dufay, for example, deliberately chose to focus on regularities on the level of phenomena, and to put the search for the 'hidden nature' of electricity behind, though he was well aware of the long history of speculations on that question: [8, pp. 476-7].

the four famous experimental methods described by Mill, who cast them in terms of searching for "causal relations."[7] I shall come back in a moment to where Mill's account does not suffice.

### Procedures

Second, there is the question of *procedures*: How is this epistemic goal typically pursued in laboratory work? The core issue here is to systematically vary many experimental parameters, one by one while the others are kept constant (i.e. in a *ceteris paribus* mode), and noticing the way in which the outcome changes. Of course, there arise immediately problems here. One of them is well known: it may happen that while varying one experimental parameter, one changes some others at the same time without noticing it – this is the case if the chosen parameter is not independent from others, but exactly this is what one doesn't know when conducting the experiments. This is a general problem of causal reasoning, and pragmatic ways to deal with it have been proposed.

More serious, however, is another problem. In principle, the number of possible parameters to be varied is unlimited. Neither a starting nor an end point seems to be discernible. In research practice, however, things are different. As to the starting point, previous experience in the field or in related ones provides some ideas about where to start, i.e. about what might be promising candidates for being relevant parameters and what not. In the case of electromagnetism, for example, nobody started with varying the color of the wires, since it was well known from $18^{th}$ century experiments that color did not affect electrical effects. Without providing a rigid framework, those aspects enable, in all their variability with individual backgrounds, a pragmatic entry point into the procedure.

Likewise, the question when to end is pragmatically treated. After all, the procedure of systematic variation has a definite goal: to formulate stable and ever more general empirical regularities. Once a tentatively formulated regularity comes out to be stable, that result is usually taken to indicate that the essential experimental conditions have been grasped, i.e. that the variation procedure has succeeded. Further variations are only needed when there is the intention to widen the scope of that regularity. And in this respect, it is mainly a matter of personal ambition, and of the resources and time available, how far the procedure will be driven. Dufay was quite ambitious here. Ampère, by contrast, well saw a pathway to extend further the generality of his laws, but nevertheless dropped the enterprise at an early point, in favor of a different one that would more likely resonate in the Paris academy. Thus, both the points of start and end are pragmatically shaped. In between, there is a wide space for individual preferences, biographical and cultural factors, and not the least for chance, to affect the choice of certain factors to be varied.

### Regularities and concept formation

Third, let me turn to empirical regularities and their epistemic preconditions. This is a core issue of exploratory experimentation. Scientists themselves have always been talking of empirical laws, and contrasting them to explanations by

---

[7]Indeed, with causality being understood in the sense of INUS-conditions, or in a refined version as "minimal theory" (Graßhoff), one may well put the goal in those terms.

hidden processes, from the early modern period up to this day. Philosophically, however, we know that such a distinction is not innocent. Formulating statements like "if the parameters A, B, C are present, then the effect D occurs" already requires categories A, B, C, D and so on, i.e. it requires a conceptual structure of the field in question. And this point is crucial: Where do these categories, this conceptual structure come from?

This is the point where Mill's account leaves us. He well realized that the categorization of the field had to be there before experimentation, but took it, in good empiristic tradition, that it could be achieved by a sort of contemplation.[8] And logical positivists were not much more attentive to the role of language. Later $20^{th}$ century philosophers opposed strongly to those naïve accounts, and with good reasons. Instead of really pursuing the question, however, they discredited the very notion of empirical regularities as incurably naïve. This was throwing out the baby with the bathwater, as it now becomes clear. Such a distinction *does* play an important role, and we need it for a better understanding of experiment – not by chance it has been taken up by some "New Experimentalists".[9] Rather than simply rejecting it, we should take effort to cast it in ways that avoid the shortcoming of naïve empiricism.

Again the above cases are telling. In none of them, the researcher started with the *intention* to revise the conceptual system, but rather with the definite idea to formulate regularities. The conceptual shift that eventually was proposed was the reaction to a specific epistemic constellation: At some point they realized that, in face of a wide range of experimental results, the formulation of stable regularities was not possible with the available concepts.[10] They had the choice either to give up the belief that such regularities exist – not really a favorable option – or to look for further parameters that could have an effect but went unrecognized so far, or even to open the conceptual system, at least parts of it, for possible revision. In my above cases, the last option was only taken when the second one did no longer appear promising: Dufay knew that the material was the relevant factor here, but did just not know how exactly to account for it. Likewise, Ampère had realized the decisive role of position, but had no way to express it. The criterion for alternative concepts being regarded more appropriate than the former ones was exactly that they allowed to do what formerly had not worked: to formulate the ever widening range of experimental results in stable and general regularities. It should be noted that in the process of revising newly forming and stabilizing categories and concepts, experimentation played a central role. Action and conceptualization stabilized or destabilized each other at every step.[11] Stability of concepts here is closely related to stability of acting: only reliable regularities allow a reliable handling of the effects in question, expectable outcomes under varying conditions.

---

[8][31, book III, chs. 7 and 8], see also his dispute with Whewell in bk. II, ch. 5. Among the British empiricist philosophers, it was only William Whewell to present a more sophisticated view, with much attention to the role of language.

[9][18, p. 159] or [5, p. 352], for example. Even critics of the New Experimentalism concede the need for differentiating the general notion of theory: [4].

[10]This is a subjective judgement, of course, and not all researchers perceived a specific situation in this way.

[11]The point of the flexibility and revisability of concepts within experimental acting has been overlooked even in the very stimulating account of experimentation by [16].

*Exploratory vs. standard-view experiments*

To highlight the specifities of exploratory experimentation, let me contrast it to the standard view type.

- There are different *questions* and *epistemic goals* – search for regularities vs. test of expectations.
- There are different *types of conclusions*: broad empirical laws vs. pivotal "proofs" or "refutations" of specific expectations or even theories.
- While exploratory work is open to the revision of basic concepts, nothing comparable is visible in the standard view.
- The *epistemic requirements* of exploratory work are rather low, whereas standard type experiments have at least a stable conceptual framework as precondition. In exploring, the categories and concepts by which experiments are described and ordered arise typically at the end of experimental series, as their very result, whereas standard view experiments have such an ordering – and much more: a well formulated hypothesis – as a precondition.
- Exploratory *procedures* are guided by rather unspecific methodological guidelines – mainly a *ceteris paribus* variation procedure. Standard view experiments, by contrast, focus on testing procedures that are in all essential details determined by the theory under scrutiny. Not a broad variety, but a single, elaborated arrangement is typically dealt with here.
- A further difference concerns the character of the *instruments and apparatus* used. In exploratory work, instruments have to allow for a great range of variations, and likewise be open to a large variety of outcomes, even unexpected ones. It is essential that the restrictions posed by the instrumental arrangement must not be too confining. In testing well-formulated expectations, by contrast, instruments are specifically designed and optimized for a single effect. The possibilities of variations are much restricted, and so is the openness to outcomes that are not in the range of expectation. The high specifity of the apparatus has considerable cost of flexibility and openness to unexpected results. Ampère's different experimental arrangements are a good illustration of this point. Whereas the instrument for the "directive action" allowed many variations of position and many different outcomes, the apparatus for the attraction of spirals was restricted to proving or disproving the attraction of spirals. The suspension of the moveable spiral, for example, excluded, by its very design for lowest friction, any sidewise or rotatory motions of the spiral.
- To sum up, one may characterize the main difference as a contrast of experimenting in regimes of low or high conceptual stability. This is not just a difference in degree, but points to a different focus of the epistemic enterprise: laws and concepts on the one hand, theories on the other.

*How common exploratory experimentation is*

So far, I have mentioned only two cases of exploratory work. But it is widespread indeed and was conducted in many historical periods, fields of research, and scientific traditions. Here is a list of select cases.

- The first clear and systematic division between electric and magnetic effects by William Gilbert in 1600 [14].
- In the 1790s, after Galvanis's spectacular announcement of the effects that quickly were named after him, most broad exploratory research started in many

parts of Europe. Within a short time, Humboldt and Ritter developed new and abstract means of representation of galvanic arrangements [49]

- Ampère was far from alone with his exploratory enterprise. Many, and most different researchers all over Europe, pursued a similar type of experimental activity in reaction to Oersted's discovery [36,35,40,7,41,6].
- Michael Faraday's work in electricity and magnetism (1820s to 1850s) is one of the most striking and monumental cases of exploratory experimentation. In most extensive experimental work, he developed the concept of "lines of force" (the main ingredients of field theory), and formed new basic concepts of magnetic behavior: dia- and paramagnetism.
- Julius Plücker developed in the 1860s basic concepts for dealing with electric discharges in rarefied gases.
- As a more recent case from physics, much of the research on high-temperature superconductivity has been of exploratory character ever since the effect was discovered in 1986.
- To take cases from other fields: The concept of chemical reaction was shaped and developed in the seventeenth century on the basis of broad experimentation, not by single prominent individual researchers, but rather within a community with a tight communication structure [28].
- In entering the "jungle" of organic chemistry in the 1830s, most intense and broad exploratory experimentation led to the development of new concepts and means of representation such as formulas [29].
- Biochemical research was revolutionized in the 1930s by Hans Krebs' introduction of the idea of circular reaction patterns. The experimental research on urea biosynthesis that led Krebs to develop this idea for the first time had long and essential exploratory phases [24,16].
- In yet another field, finally, Jean Brachet's research on protein biosynthesis in the 1930s and 1940s yielded new central concepts. Most of his experimental work was of the exploratory type. Richard Burian, in analyzing Brachet's work, introduced himself, and independently, the category of "exploratory" experimentation. His emphasis that "the style of exploratory experimentation ... *should* be of great historical and philosophical interest" underlines my claim of the epistemic importance of this activity [3, p. 27; original emphasis].

Disparate as these cases are: what they have in common, is the specific type of epistemic situation that I have described above.

*Experiment and theory*

Let me add here another comment on experiment and theory. My focus on exploratory experiments and their contrast to standard view experiments has been characterized as being the claim that there are "theory-free experiments" [39, p. 9, 161]. Such a characterization, however, is easily misleading (and, I should add, I have never said this). It is a typical case of the confusions that arise from an unspecified use of the notion of "theory". If theory means any use of concepts and language (this is the impression Popper provides) [37, appendix to § 30 (1968), sect. 3], the claim of "theory-ladenness" is true, but trivial. Nobody has ever claimed that experimenters do not think about their doing. If, by contrast, theory means overarching systems such as relativity, quantum, or evolutionary theories, or hypotheses about the hidden "causes" of phenomena, the claim that *all* experiments are theory-laden is evidently wrong. There are experiments, which are not driven by these types of theories, and this has indeed been my claim. The case shows how much we need to differentiate further the talk of "theory" and "theory-ladenness". Focusing on concepts is a promising option to reach a wider perspective here.

CONCEPT FORMATION IN SCIENCE

Thus let me, in my last section, sketch some general thoughts about concepts and concept formation. In contrast to classical accounts, by Carnap and Hempel, for example, a look to cases from real science will shift the focus towards the *processes* by which concepts are actually formed and stabilized in experimental research. There are some recent approaches, mostly shaped by cognitive science aspects [13,32, among others], but the role of experiment is still much underdeveloped.

*Different epistemic levels*

The talk of concepts has been notoriously unsharp and disputed. But despite ongoing debates about the exact notion of concepts [42,38,17], there are some more or less undisputed core aspects, i.e. a shared basic understanding that is sufficient for my purpose. Concepts are always understood as *elements* of our thoughts, rather than extended *systems* as theories are. Moreover, and this is a sort of corollary, it is crucial that concepts are usually just used and taken for granted, but not discussed and reflected upon. They are normally implicit, whereas theories are normally explicit. Of course, it is exactly in periods of concept formation that concepts become explicitly discussed. But these processes, I claim, have a different character from those of discussion of theories.

These aspects point to an epistemic level that is more fundamental than the level of theories. It is concepts and conceptual frameworks by which we structure our world in things, entities, categories, facts and so on; a structure that we necessarily presuppose and usually just take for granted both when debating about theories and, even more crucial, even when formulating individual empirical statements and experimental outcomes. It is exactly by focusing on this difference of epistemic levels that important aspects of the process of research appear in a new light and become understandable. One of them is the difference between exploratory and standard view experimentation, as I have explicated above. But there are more.

*How new concepts are received*

One of them concerns the ways in which new concepts find their way once they have been formed and proposed by an individual.[12] The case of the two electricities is typical. Dufay's proposal raised surprisingly little explicit discussion. Already some five years later, however, the concept showed up in articles and textbooks, not as a thing to be disputed, but as expressing a matter of fact, or even as part of the very definition of electricity.[13] The awareness that this concept had been created

---

[12]I do not claim that all new concepts follow the pattern from the individual to the communal – there might well be cases with different pathways.

[13]Even for Benjamin Franklin some years later, this would be an undisputed fact to start with, a fact for which he wanted to give a microscopic explanation with his one-fluid-theory. Not taking into account this difference has led to much confusion in the historiography of electricity. The two electricities provide an illustrative example of how facts are created, and how closely this is related to the formation of concepts.

by hard work quickly disappeared. Similar observations hold, *mutatis mutandis*, for my other cases. Such a way of filtering in is strikingly different from explicit theoretical disputes, and this difference has exactly to do with the epistemic level addressed. Concepts, categories, and language cannot be proved, be confirmed or disconfirmed, but they have to prove themselves. Stability of acting plays an essential role here. Once concepts have proved themselves, they tend to disappear from the realm of explicit discussion. They turn into parts of the conceptual framework of a research field, and are just used, not debated.

*Why exploratory experimentation disappears*

As a sort of corollary, we can understand now the puzzling observation that exploratory experimentation seems to have a tendency to disappear from the picture of scientific activity. It is often the case that scientists themselves, even when they present an account of their own research pathway, do give exploratory phases much less attention than they had actually devoted to it in their laboratory work. It was not the least this fact that made philosophers of science totally overlook exploratory experimentation so far. But why is it that exploring disappears so easily, even from scientists' own account? Again the main point is the epistemic level involved here. A change of concepts goes much deeper than a change of an explicit theoretical issue: we can easily take an "outside look" on theories, but we can not easily step outside the borders of our conceptual framework. After the conceptual structure of a research field has been revised and new concepts established, it is difficult to re-imagine the previous situation. It was a reflective practitioner of science who sharply highlighted this point: Ludwik Fleck, the Polish immunologist and philosopher of science, noted already in 1935 that

> If after years we were to look back upon a field we have worked in, we could no longer see or understand the difficulties present in that creative work. ... But how could it be any different? We can no longer express the previously incomplete thoughts with these now finished concepts [14].

Conceptual shifts tend, by their very epistemic level, to disappear from the historical picture, or to be reconstructed as the "discovery" (rather then the "formalism", as Fleck emphasizes) of "facts".

*History, theory, concepts*

There is another dimension. Concepts and conceptual frameworks enable us to think and act in the world. They shape all thinking and acting on a fundamental epistemic level – and they limit it: what lies outside our conceptual framework, and hence outside the borders of our language, can just not be thought. While this insight might well be known to philosophers of language, it has not found appropriate attention in philosophy of science. There is another insight, moreover, of which recent history of science has reminded us: Concepts have been created at a specific historical point, been formed, stabilized, established, eventually reshaped, revised, or even dismissed – all that usually in hard work. These processes have involved decisions, social acting, and cultural settings. And there is no reason

---

[14][11, p. 114], English translation [12, p. 86]

to believe that the resulting conceptual frameworks would bear no traces of their genesis. Concepts have a history, and they ,,have memories," as Hacking put it [20, p. 37].

This is not to say that concepts are reducible to, or explainable by, social, cultural and historical settings *alone*. But these settings do inevitably contribute to concept formation. New concepts are always taken out of a more or less wide reservoir that itself is culturally and historically shaped. But why some of them come out to be more appropriate than others, is not a matter of culture, but of experiment. Concept formation in experimental contexts is perhaps one of the most intriguing points of intertwining strict epistemic procedures with cultural settings of every research process.

Thus the dimension of the problem comes into view. There is something that shapes and delimits our scientific thinking and acting, but that we usually do not discuss, and, even more, we are usually not even aware of. All discussions about theories are necessarily based on a huge, but unrecognized system that is transported by the language we use. As a consequence, all philosophical talk about theory-dynamics, about verification or falsification, about corroboration, refutation, or justification, will necessarily and fundamentally be incomplete, if not inadequate, without involving a genuine historical dimension: the historic process in which the concepts used have been generated and by which they have been specifically shaped.

In a way, all this is not really new, but has already been observed from different angles. Let me again quote Ludwik Fleck who, from his own experience in laboratory research, emphasized that

> Once a field has been sufficiently worked over ... the experiments will become increasingly better defined. But they will no longer be independent, *because they are carried along by a system of earlier experiments and decisions*, which is generally the situation in physics and chemistry today. Such a system will then become self-evident know-how itself. We will no longer be aware of its application and effect.[15]

Also in the 1930s, but starting from quite different aspects, Edmund Husserl introduced the notion of "sedimentation" to point out that many concepts we use now, and never put in question, have been highly theoretical in former times.[16] Later on, Michel Foucault has pointed out the issue in a specific way, and Thomas Kuhn was, in his later career, increasingly concerned with questions of concepts [30], since it is this level of concepts and of language where problems like the ever-puzzling issue of incommensurability are located. More recently, the topic has been taken up by Ian Hacking under the label of "historical ontology" [20]. In focusing on concepts and concept formation, unexpected connexions between analytic and non-analytic traditions in philosophy of science become visible. Moreover, such an enterprise well opens new perspectives. For example, it might come out in the end

---

[15] [11], 114, English translation slightly altered from [12], 86, original emphasis.
[16] "Die Frage nach dem Ursprung der Geometrie," 1936, as attachment III to his "Krisis der europäischen Wissenschaften und die transzendentale Phänomenologie." [26], vol. 6, 365-386. It should be noted that already in 1840, William Whewell had pointed to such a process, though not in terms of concepts, but of facts: [51], pt. I, bk. I, ch. 2.

that the *fameuse* "theory-ladenness" of observation and experimentation, should better be reformulated as the "history-ladenness" of the concepts and the language in which observations and experiments are designed and their outcomes formulated.

While those issues have been addressed in psychology, social science, and the life sciences, philosophy of physics has not been affected too deeply so far.[17] Physics is often taken as "mature", with its basic concepts and procedures being stable once for all. But this is a misperception, I think. Physical research is not so categorically different from other fields. To get the issue of concept formation into closer view here, experiment may be a key issue.

*Epilogue: History and Philosophy of Science*

In order to meet that challenge, we have to look at scientific practice, as often opposed to the mere presentation of ready-made results and to afterwards self-fashioning by scientists of their procedures and pathways. We have to focus on science in its working, on the "night science", as the geneticist François Jacob put it, as contrasted to the picture it presents to the public as "day-science" [27, p. 126], to the workshop rather than to the shop window of science. This is not what philosophers are used to do. Getting an insight into the workings of science, we have to use new source material, different from published papers, and critical historiographical methods of reconstruction. These are tasks that have been done with great success by recent historians of science, though usually not with an eye on epistemological questions. Competencies have to be brought together here, either within individual persons or, even more promising, by new forms of collaboration. The challenge to better understand experiment and concept formation is at the same time a challenge to find fruitful ways to combine historical and philosophical analysis, i.e. to redefine HPS with capital letters [46]. A veritable task ahead, but one that promises rich outcomes.

## REFERENCES

1. Ampère, A.-M. 1820. Suite du mémoire sur l'action mutuelle entre deux courans électriques, entre un courant électrique et un aimant ou le globe terrestre, et entre deux aimants. *Annales de Chimie et de Physique*, 15(octobre):170–218.

2. Blondel, C. 1982. *A.-M. Ampère et la création de l'électrodynamique (1820-1827)*. Paris : Bibliothèque Nationale.

3. Burian, R. M. 1997. Exploratory experimentation and the role of histochemical techniques in the work of Jean Brachet, 1938-1952. *History and Philosophy of the Life Sciences*, 19:27–45.

4. Carrier, M. 1998. New experimentalism and the changing significance of experiments: On the shortcomings of an equipment-centered guide to history. In M. Heidelberger and F. Steinle, editors, *Experimental Essays - Versuche zum Experiment*. Interdisziplinäre Studien/ Interdisciplinary Studies, pages 175–191. Baden-Baden : Nomos Verlag.

5. Cartwright, N. 1983. *How the Laws of Physics Lie*. Oxford : Clarendon Press.

6. Configliachi 1821. Lettera del Prof. Configliachi a C. Ridolfi, etc. Lettre du Prof. Configliachi au Marquis Ridolfi sur les expériences Voltaïco-magnétique;

---

[17]Kuhn has given important impulses here, cf. [25, ch. 3.6] or [33].

communiqué au Prof. Pictet (Traduction). Paris, 26 Dec 1820. *Bibliothèque universelle*, 16(janvier):72–74.

7. Davy, H. 1821. On the magnetic phenomena produced by Electricity. *Philosophical Transactions*, 111:7–19.

8. Dufay, C. F. de Cisternai 1733. Quatrième mémoire sur l'électricité. Des l'attraction et répulsion des corps électriques. *Histoire de l'Académie Royale des Sciences, avec les Mémoires de Mathématique & de Physique pour la même année*, pages 457–476.

9. Dufay, C. F. de Cisternai 1734. A Letter from Mons. Du Fay, F. R. S. and of the Royal Academy of Sciences at Paris, to His Grace Charles Duke of Richmond and Lenox, concerning Electricity. Translated from the French by T. S. M D. *Philosophical Transactions*, 38(431):258–266.

10. Duhem, P. 1906. *La Théorie Physique: Son Objet, Sa Structure*. Paris.

11. Fleck, L. 1935[1980]. *Entstehung und Entwicklung einer wissenschaftlichen Tatsache. Einführung in die Lehre vom Denkstil und Denkkollektiv*. Frankfurt : Suhrkamp.

12. Fleck, L. 1979. *Genesis and Development of a Scientific Fact*. Chicago : University of Chicago Press.

13. Giere, R. N. 1988. *Explaining Science: a Cognitive Approach*. Chicago : University of Chicago Press.

14. Gilbert, W. 1600. *De Magnete magnetisque corporibus, et de magno magnete tellure, physiologia nova, plurimis & argumentis & experimentis demonstrata*. London : Petrus Short.

15. Gooding, D. C. 1990. *Experiment and the making of meaning: Human agency in scientific observation and experiment*. Dordrecht : Kluwer.

16. Graßhoff, G. et al. 2000. *Zur Theorie des Experimentes. Untersuchungen am Beispiel der Entdeckung des Harnstoffzyklus*. Bern.

17. Gurova, L. 2003. Philosophy of science meets cognitive science: The categorization debate. In D. Ginev, editor, *Bulgarian studies in the philosophy of science*, Boston studies in the philosophy of science, 236, pages 141–162. Dordrecht : Kluwer.

18. Hacking, I. 1983. *Representing and Intervening: Introductory topics in the philosophy of natural science*. Cambridge : Cambridge University Press.

19. Hacking, I. 1992. The self-vindication of the laboratory science. In A. Pickering, editor, *Science as Practice and Culture*. Chicago : University of Chicago Press, 29–64.

20. Hacking, I. 2002. *Historical ontology*. Cambridge : Harvard University Press.

21. Heidelberger, M. 1998. Die Erweiterung der Wirklichkeit im Experiment. In M. Heidelberger and F. Steinle, editors, *Experimental Essays - Versuche zum Experiment*, Interdisziplinäre Studien/ Interdisciplinary Studies, pages 71–92. Baden-Baden : Nomos Verlag.

22. Heilbron, J. L. 1979. *Electricity in the 17th and 18th centuries*. Berkeley : University of California Press.

23. Hofmann, J. R. 1995. *André-Marie Ampère: Enlightenment and Electrodynamics*. Oxford : Blackwell.

24. Holmes, F. L. 1993. *Hans Krebs: Architect of Intermediary Metabolism 1933-1937*. Oxford : Oxford University Press.

25. Hoyningen-Huene, P. 1993. *Reconstructing scientific revolutions: Thomas S. Kuhn's philosophy of science. With a foreword by Thomas S. Kuhn.* Chicago : University of Chicago Press.

26. Husserl, E. 1956-2002. *Gesammelte Werke, Hg. Samuel Ijsseling.* Dordrecht : Kluwer.

27. Jacob, F. 1998. *Of flies, mice and men.* Cambridge : Harvard University Press.

28. Klein, U. 1994. Origin of the concept of chemical compound. *Science in Context,* 7(2):163–204.

29. Klein, U. 1998. Paving a way through the jungle of organic chemistry - experimenting within changing systems of order. In M. Heidelberger and F. Steinle, editors, *Experimental Essays - Versuche zum Experiment,* Interdisziplinäre Studien/ Interdisciplinary Studies, pages 251–271. Baden-Baden : Nomos Verlag.

30. Kuhn, T. S. 2000. *The road since "structure".* Chicago : University of Chicago Press.

31. Mill, J. S. 1843. *A System of Logic Ratiocinative and Inductive. Being a Connected View of the Principles of Evidence and the Methods of Scientific Investigation.*

32. Nersessian, N. J. 2000. Nomic concepts, frames, and conceptual change. *Philosophy of Science,* 67:S224–S241.

33. Nersessian, N. J. 2002. Kuhn, conceptual change, and cognitive science. In T. Nickles, editor, *Thomas Kuhn,* Contemporary Philosophers in Focus, pages 129–166. Cambridge : Cambridge University Press.

34. Oersted, H. C. 1820. Experiments on the effect of a current of electricity on the magnetic needle. *Annals of Philosophy,* 16(october):273–276.

35. Oersted, H. C. 1820. New electromagnetic experiments. *Annals of Philosophy,* 16(november):375–377.

36. Pictet, M.-A. 1820. Addition des Rédacteurs (à: Experimenta circa effectum, etc. Experiences sur l'effet du conflict électrique sur l'aiguille aimantée, par Mr. J.Chr. Oersted, Prof. de physique dans l'université de Copenhague. (Traduction). *Bibliothèque universelle,* 14(août):281–284.

37. Popper, K. R. 1934. *Logik der Forschung.* Wien : Julius Springer.

38. Prinz, J. J. 2002. *Furnishing the Mind: Concepts and Their Perceptual Basis.* Cambridge, MA : MIT Press.

39. Radder, H., ed. 2002. *The Philosophy of Scientific Experimentation.* Pittsburgh : University of Pittsburgh Press.

40. Schweigger, J. S. C. 1821. Zusätze zu Oersteds elektromagnetischen Versuchen, vorgelesen in der naturforschenden Gesellschaft zu Halle den 16. September 1820. *Journal für Chemie und Physik (Schweigger),* 31(1. Heft(Januar)):1–17.

41. Seebeck, T. J. 1821. Ueber Elektromagnetismus: Auszug aus einer Abhandlung, welche in der Sitzung der Königlichen Akademie der Wissenschaften zu Berlin vom 14. Dezember 1820 vorgetragen wurde, *Journal für Chemie und Physik (Schweigger).* 32(Mai):27–38.

42. Smith, E. E. and D. L. Medin 1981. *Categories and concepts.* Cambridge, Mass : Harvard University Press.

43. Steinle, F. 1995. Looking for a "simple case": Faraday and electromagnetic rotation. *History of Science,* 33:179–202.

44. Steinle, F. 1997. Entering new fields: Exploratory uses of experimentation.

*Philosophy of Science*, 64 (Supplement):S65–S74.

45. Steinle, F. 2002. Challenging established concepts: Ampère and exploratory experimentation. *Theoria: revista de teoria, historia y fundamentos de la ciencia*, 17:291–316.

46. Steinle, F. 2003. Experiments in history and philosophy of science. *Perspectives on Science*, 10(4):408–432.

47. Steinle, F. 2003. The practice of studying practice: Analyzing research records of Ampère and Faraday. In F. L. Holmes, J. Renn and H.-J. Rheinberger, editors, *Reworking the bench: Research notebooks in the History of Science*, Archimedes, 7, pages 93–117. Dordrecht : Kluwer.

48. Steinle, F. in press. *Explorative Experimente. Ampère, Faraday und die Ursprünge der Elektrodynamik*. Stuttgart : Franz Steiner Verlag.

49. Trumpler, M. 1997. Verification and variation: Patterns of experimentation in investigations of galvanism in Germany, 1790-1800. *Philosophy of Science*, 64 (Supplement):S75–S84.

50. van Fraassen, B. C. 1981. Theory construction and experiment: An empiricist view. In P. D. Asquith and R. N. Giere, editors, *PSA 1980: Proceedings of the Biennial Meeting of the Philosophy of Science Association*, vol.2. East Lansing: Philosophy of Science Association, 663–677.

51. Whewell, W. 1840. *The Philosophy of the Inductive Sciences, founded upon their History*. London : Parker.

# Rereading Ludwig Boltzmann

Jos Uffink

*Institute for History and Foundations of Science, Utrecht University,*
*P.O. Box, 80.000, 3508 TA Utrecht, the Netherlands*
*uffink@phys.uu.nl*

**Abstract.** Boltzmann's views on statistical physics continue to play an important role in contemporary debate on the foundations of that theory. However, Boltzmann's papers are numereous, voluminous and for the most part untranslated. Accordingly, discussions of his views often rely on secondary literature. This paper aims to go back to his own writings, focusing particularly on his conceptions of probability and the role of the ergodic hypothesis in his approach.

## 1. Introduction

Ludwig Boltzmann is one of the first physicists who brought probability and statistical considerations into physics. Particularly famous is his statistical explanation of the second law of thermodynamics. However, Boltzmann's ideas on the relationship between the thermodynamical properties of macroscopic bodies and their microscopic mechanical constitution, and the role of probability in this relationship are involved and differed quite remarkably in different periods of his life. In his lifelong struggle with this problem he made use of a varying arsenal of tools and assumptions. (To mention a few: the so-called *stoßzahlansatz* (SZA), the ergodic hypothesis, ensembles, the permutational argument, the hypothesis of molecular disorder.) But the role of these assumptions, and the results he obtained from them, also shifted in the course of time. Particularly notorious are the role of the ergodic hypothesis and the status of the so-called $H$-theorem. Moreover, he used 'probability' in four different technical meanings. It is, therefore, not easy to speak of a consistent, single "Boltzmannian approach" to statistical physics. A full survey of these various approaches would fall beyond the bounds of this paper. I shall concentrate here on his view on the ergodic hypothesis and its connection to probability.

### 1.1. The Ehrenfests on the role of the ergodic hypothesis

'Statistical physics' includes at least two roughly distinguished theories: the kinetic theory of gases and statistical mechanics proper. The first theory aims to explain the properties of gases by assuming that they consist of a very large number of molecules in rapid motion. During the 1860s probability considerations were imported into this theory. The aim then became to characterize the properties of gases, in particular in thermal equilibrium, in terms of probabilities of various molecular states. This is what the Ehrenfests, in their famous Encyclopaedia article (1912) call "the older formulation of statistico-mechanical investigations" or

"kineto-statistics of the molecule". Here, molecular states, in particular their velocities, are regarded as stochastic variables, and probabilities are attached to such
molecular states of motion. These probabilities themselves are conceived of as mechanical properties of the state of the total gas system: Either they represent the
relative *number* of molecules with a particular state, or the relative *time* during
which a molecule has that state.

In the course of time a transition was made to what the Ehrenfests called a
"modern formulation of statistico-mechanical investigations" or "kineto-statistics
of the gas model", or what is nowadays known as statistical mechanics. In this
latter approach, probabilities are not attached to the state of a molecule but of the
entire gas system. Thus, the state of the gas, instead of determining the probability
distribution, now itself becomes a stochastic variable.

A merit of this latter approach is that interactions between molecules can be
taken into account. Indeed, the approach is not necessarily restricted to gases,
but might in principle also be applied to liquids or solids. (This is why the name
'gas theory' is abandoned.) The price to be paid however, is that the probabilities
become more abstract. Since probabilities are attributed to the mechanical states
of the total system, they are no longer determined by mechanical states. Instead, in
statistical mechanics, the probabilities are determined by means of an 'ensemble',
i.e. a fictitious collection of replicas of the system in question.

It is not easy to pinpoint this transition in the course of history, except to say
that in Maxwell's work in the 1860s definitely belong to the first category, and
Gibbs' book of 1902 to the second. Boltzmann's own works fall somewhere in
the middle ground. His earlier contributions clearly belong to the kinetic theory of
gases (although his 1868 paper already applies probability to an entire gas system),
while his work after 1877 is usually seen as elements in the theory of statistical
mechanics. However, Boltzmann himself never indicated a distinction between
these two theories, and it seems arbitrary to draw a demarcation at an exact
location in his work.

The transition from kinetic gas theory to statistical mechanics poses two main
foundational questions. On what grounds do we choose a particular ensemble,
or the probability distribution characterizing the ensemble? Gibbs did not enter
into a systematic discussion of this problem, but only presented special cases of
equilibrium ensembles (i.e. canonical, micro-canonical etc.) for which some special
form of the probability distribution is stipulated. A second problem is to relate the
ensemble-based probabilities with the probabilities obtained in the earlier kinetic
approach for a single gas model.

The Ehrenfests Encyclopedia paper was the first to recognize these questions,
and to provide a partial answer: Assuming a certain hypothesis of Boltzmann's,
which they dubbed the *ergodic hypothesis*, they pointed out that for an isolated
system with a fixed total energy the micro-canonical distribution is the unique
stationary probability distribution. Hence, if one demands that an ensemble of
isolated systems in thermal equilibrium must be represented by a stationary distribution, the only choice for this purpose is the micro-canonical one. Similarly, they
pointed out that under the ergodic hypothesis infinite time averages and ensemble
averages were identical. This, then, would provide a desired link between the probabilities of the older kinetic gas theory and those of statistical mechanics, at least

in equilibrium and in the infinite time limit. Yet the Ehrenfests simultaneously expressed strong doubts about the validity of the ergodic hypothesis. These doubts were soon substantiated when in 1913 Rozenthal and Plancherel proved that the hypothesis was untenable for realistic gas models.

The Ehrenfests' attempt to construct a coherent framework out of Boltzmann's work thus gave a prominent role to the ergodic hypothesis, suggesting that it played a fundamental and lasting role in his thinking. Although this view indeed succeeds in producing a coherent view of his multifaceted work, it is certainly not historically correct. Indeed, Boltzmann himself also had grave doubts about this hypothesis, and expressly avoided it whenever he could, in particular in his two great papers of [8] and [11]. Since the Ehrenfests, many other authors have presented accounts of Boltzmann's work. Particularly important are [34] and [27]. Still, there remains much confusion about what the role of the ergodic hypothesis was in his thinking. This essay aims to provide a more accurate account on this topic.

## 1.2. A concise chronography of Boltzmann's writing

Roughly, one may divide Boltzmann's work in four periods. The period 1866-1871 is more or less his formative period. In his first paper (1866) Boltzmann set himself the problem of deriving the second law from mechanics. The notion of probability does not appear in this paper. The following papers, from 1868 and 1871, were written after Boltzmann had read Maxwell's work of 1860 and 1867. Following Maxwell's example, they deal with the characterization of equilibrium, by means of a probability distribution, and explored new cases where the gas is subject to a static external force, or consists of poly-atomic molecules. He regularly switched between three conceptions of probability: sometimes defined as a time average, a particle average, or in [6] an ensemble average.

The main result of those papers is that, by adopting the SZA, one can show that the Maxwellian distribution function is stationary, and thus an appropriate candidate for the equilibrium state. In some cases Boltzmann also argued it was the unique stationary state. However, in this period he also presented a completely different method, which did not rely on the SZA but on the ergodic hypothesis. ([3] Section 3, [9], [6]) This approach led to a form of the distribution function (conceived of as the marginal of $\rho$) that, when the number of particles tends to infinity, reduces to the Maxwellian form. Ensembles would not play a prominent role in his approach until the 1880's.

The next period is that of 1872-1878, in which he wrote his two most famous papers: the *Weitere Studien* [8] and *Über die Beziehung* [11]. 1872 paper continued the approach relying on the SZA, and obtained the Boltzmann equation and the *H*-theorem. Boltzmann now claimed that the *H*-theorem provided the desired theorem corresponding to the second law. However, this claim came under a serious objection due to Loschmidt's criticism of 1876. The objection was simply that no purely mechanical theorem could ever produce a time-asymmetrical result. Boltzmann's responded to this objection in [10] and [11].

The upshot was that Boltzmann rethought the basis of his approach and in [11] produced a conceptually very different analysis (the permutational argument) of equilibrium and evolutions towards equilibrium, and the role of probability theory. The distribution function, which formerly represented the probability distribution, was now conceived of as a stochastic variable (nowadays called a macrostate) subject to a probability distribution. That probability distribution was now determined by the size of the volume in phase space corresponding to all the microstates giving rise to the same macrostate, (essentially given by calculating all permutations of the particles in a given macrostate). Equilibrium was now conceived of as the most probable macrostate– instead of a stationary macrostate. The evolution towards equilibrium could then be expressed as an evolution from less probable to more probable states.

The third period is taken up by the papers Boltzmann wrote during the 1880's. During this period, he abandoned the permutational approach, and went back to an approach that relied on a combination of the ergodic hypothesis and the use of ensembles. For a while Boltzmann worked on an application of this approach to Helmholtz's concept of monocyclic systems. However, after finding that these systems did not always provide the desired thermodynamical analogies, he abandoned this topic again.

In the 1890s the reversibility problem resurfaced again, this time in a debate in the columns of *Nature*. This time Boltzmann chose an entirely different line of counterargument than in his debate with Loschmidt. A few years later, he entered into a debate with Zermelo on the recurrence objection. The same period also saw the publication of the two volumes of his *Lectures on Gas Theory*. In the final years of his life he gave various formulations of the cosmological speculation. His last paper, the Encyclopedia article he wrote with Nabl, is also notable.

## 2. Boltzmann 1866-1871

### 1866: the mechanical meaning of the second law

Boltzmann's first paper (1866) on statistical physics is entitled "on the mechanical meaning of the second law of heat theory". Although the results of this youthful work were rather rough, the paper is still interesting since it reveals the goals that he originally set himself and the set of ideas and methods by which he hoped to achieved them.

He considers an arbitrary thermal body consisting of $N$ particles and proposes to define the temperature $T$, as the average kinetic energy of a particle. But unlike Maxwell, whose work was as yet unknown to him, he chose to average over *time* instead of over the number of particles in the gas. Thus, temperature is defined as

$$T = \frac{1}{t_2 - t_1} \int_{t_1}^{t_2} \frac{mv^2}{2} dt =: \overline{E_{kin}}$$

(In modern notation, this means he is adopting units such that the Boltzmann constant $k = 2/3$). Here the limits $t_1$ and $t_2$ are to be taken in such a way as "to obtain independence of any contingencies in the kinetic energy" (Abh. I, p. 14), i.e. they should cover a representative portion of the particle's motion. Note that temperature is thus a mechanical property of a single particle. In principle, this could lead to a different temperature being assigned to each particle. But Boltzmann seems to assume implicitly that the time average is the same for all particles.

The aim of the paper was to argue (under certain assumptions) that $T$ thus defined provides an integrating divisor for the inexact heat differential $dQ$. It is not difficult to point out defects in the argument, but that is not our concern here. Note that Boltzmann nowhere mentions probability. True, the paper talks about averages, and to some readers this might imply an implicit invocation of probability theory [40]e.g.. But I think that this overlooks the fact that the time-average of the kinetic energy of a molecule is an ordinary mechanical quantity that does not depend on any non-mechanical concepts for its definition.

### 1868: From a system of hard discs to the ergodic hypothesis

Maxwell had derived his equilibrium probability distribution

$$f(\vec{v})d^3\vec{v} = A^3 e^{-v^2/B} v^2 dv, \tag{1}$$

for two special gas models (i.e. a hard sphere gas in 1860 and a model of point particles with a central $r^5$ repulsive force acting between them in 1867). He had noticed that the distribution, once attained, will remain stationary in time (when the gas remains isolated), and had argued (but not very convincingly) that it was the *only* such stationary distribution.

In the first section of his [3], Boltzmann aims to reproduce and improve these results for a system of an infinite number of hard discs in a plane. He adopts from Maxwell the idea to characterize thermal equilibrium by a probability distribution. He regards it as obvious that this distribution should be independent of the position of the discs, and that every direction of their velocities is equally probable. It is therefore sufficient to consider the distribution over the various values of the velocity $v = \|\vec{v}\|$.

However, Boltzmann started out with a somewhat different interpretation of probability in mind than Maxwell. He introduced the probability distribution as follows:

> Let $\phi(v)dv$ be the sum of all the instants of time during which the velocity of a disc in the course of a very long time lies between $v$ and $v + dv$, and let $N$ be the number of discs which on average are located in a unit surface area, then
>
> $$N\phi(v)dv$$
>
> is the number of discs per unit surface whose velocities lie between $v$ and $v + dv$. (Abh. I, p.50)

Thus, $\phi(v)dv$ is introduced as the relative *time* during which a (given) disc has a particular velocity. But, in the same breath, this is identified with the relative *number* of discs with this velocity.

This remarkable quote shows how he connected Maxwell's interpretation of the distribution function with that of his previous paper using averages over time: simply by identifying two different meanings for the same function. This equivocation returned in different guises again and again in Boltzmann's writing.[1] Indeed, it is, I believe, the very heart of the ergodic problem, put forward so prominently by

---

[1] This is not to say that he always conflated these two interpretations of probability. Some papers

the Ehrenfests. Either way, of course, whether we average over time or particles, probabilities are defined here in strictly mechanical terms, and therefore objective properties of the gas.

Next he goes into a detailed mechanical description of a two-disc collision process. Using a two-dimensional version of the SZA, he shows that the velocity distribution is stationary if it takes the form

$$\phi(v) = 2hve^{-hv^2},\tag{2}$$

for some constant $h$. This is the two-dimensional analogue of the Maxwell distribution (1).

In the next subsections of [3], Boltzmann repeats the derivation, each time in slightly different settings: first, he goes over to the three-dimensional version of the problem, assuming a system of hard spheres, and supposes that one special sphere is acted upon by an external potential $V(\vec{x})$. He shows that if the velocities of all other spheres are distributed according to the Maxwellian distribution (1), the probability distribution of finding the special sphere at place $\vec{x}$ and velocity $\vec{v}$ is $f(\vec{v}, \vec{x}) \propto e^{-h(\frac{1}{2}mv^2 + V(\vec{x}))}$ (Abh. I, p.63). In a subsequent subsection, he replaces the spheres by material points with a short-range interaction potential and reaches a similar result. Next, he assumes that two discs are acted upon by the same external force. It is shown that mutual collisions between these two discs do not disturb the equilibrium distribution (assuming that these collisions obey the SZA). This analysis is then again transposed to three dimensions.

At the end of Section I of his paper, Boltzmann suddenly switches course. He announces (Abh. I p. 80) that all the cases treated and yet untreated follow from a much more general theorem. This theorem is the subject of the second and third Section of the paper. I will limit the discussion to the third section and rely partly on Maxwell's (1879) exposition, which is somewhat simpler than Boltzmann's own.

### 1868 section III: the ergodic hypothesis

Consider a general mechanical system of $N$ material points, each with mass $m$, subject to an arbitrary time-independent potential. In modern notation, let $(p, q) = (p_1, \ldots, p_n; q_1, \ldots, q_n)$ denote position coordinates and canonical momenta (with $n = 3N$) of the system. Its Hamiltonian is then

$$H(p, q) = \frac{1}{2m} \sum_i p_i^2 + V(q_1, \ldots, q_n)\tag{3}$$

The state of this system may be represented as a phase point $(p, q)$ in the mechanical phase space $\Gamma$. By the Hamiltonian equations of motion, the phase point evolves in time, and thus describes a trajectory $(p_t, q_t)$. This trajectory is constrained to lie on a given energy hypersurface $H(p, q) = E$. Boltzmann asks for

---

employ a clear and consistent choice for one interpretation only. But then that choice differs between papers, or even in different sections of a single paper. In fact, in [7] he even multiplied probabilities with different interpretations into one equation to obtain a joint probability. But then in [8] he conflates them again. Even in his last paper [25] we see that Boltzmann identifies the two meanings of probability with a simple-minded argument.

the probability (i.e. the fraction of time during a very long period) that the phase point lies in a region $dp_1 \cdots dq_n$, which we may write as:

$$\rho(p,q)dp_1 \cdots dq_n = f(p,q)\delta(H(p,q) - E)dp_1 \cdots dq_n$$

Boltzmann seems to assume implicitly that this distribution is stationary. This property would of course be guaranteed if the "very long period" were understood as an infinite time limit. He argues, by Liouville's theorem, that $f$ is a constant for all points on the energy hypersurface that are "possible", i.e. that are actually reached by the trajectory. For all other points $f$ vanishes. Ignoring those latter points, the function $f$ is therefore uniform over the entire energy hypersurface, and the probability density $\rho$ takes the form of the micro-canonical distribution

$$\rho_{\mathrm{mc}}(p,q) = \frac{1}{\omega(E)}\delta(H(p,q) - E) \tag{4}$$

where $\omega(E) := \int_{H(p,q)=E} dpdq$ is nowadays known as the structure function.

In particular, one can now evaluate the marginal probability density for the positions $q_1, \ldots, q_n$:

$$\begin{aligned}
\rho_{\mathrm{mc}}(q_1, \ldots, q_n) &= \int \rho_{\mathrm{mc}}(p,q)\, dp \\
&= \frac{1}{\omega(E)} \int_{p_i^2 = 2m(E - V(q))} dp_1 \cdots dp_n \\
&= \frac{(\pi)^{n/2}}{\omega(E)\Gamma(n/2)}(2m(E - V(q)))^{(n-2)/2}.
\end{aligned} \tag{5}$$

Similarly, the marginal probability density for finding the first particle with a momentum $p_1$ as well as finding the positions of all particles at $q_1, \ldots, q_n$ is

$$\begin{aligned}
\rho_{\mathrm{mc}}(p_1, q_1 \ldots, q_n) &= \int \rho_{\mathrm{mc}}(q,p)dp_2 \cdots dp_n \\
&= \frac{2\pi^{(n-1)/2}}{\omega(E)\Gamma(\frac{n-1}{2})}(2m(E - V(q) - p_1^2)^{(n-2)/2}
\end{aligned} \tag{6}$$

These two results can be conveniently presented in the form of the conditional probability that the momentum of the first particle has a value between $p_1$ and $p_1 + dp_1$, given that the positions have the values $q_1 \ldots, q_n$:

$$\rho_{\mathrm{mc}}(p_1 \,|\, q_1, \ldots, q_n)dp_1 = \frac{\sqrt{\pi}\Gamma(\frac{n}{2})}{\sqrt{2m}\Gamma(\frac{n-1}{2})}\frac{(E - V - \frac{p_1^2}{2m})^{(n-2)/2}}{(E - V)^{(n-1)/2}}dp_1 \tag{7}$$

This, in essence, is the general theorem announced.

In the limit where $n \longrightarrow \infty$, and the kinetic energy per particle $\kappa := (E - V)/n$ constant, the expression (7) approaches

$$\frac{1}{\sqrt{2\pi m\kappa}}\exp(-\frac{p_1^2}{2m\kappa})dp_1 \tag{8}$$

which takes the same form as the Maxwell distribution (1). One ought to note however, that since $V$, and therefore $\kappa$ depends on the coordinates, the condition $\kappa = constant$ is different for different values of $(q_1, \ldots, q_n)$. Some comments on this result are in order.

1. The difference between this approach and that of the first section is striking. Instead of concentrating on a gas model in which particles are assumed to move freely except for their occasional collisions, Boltzmann here assumes a much more general model with an arbitrary interaction potential $V(q_1, \ldots q_n)$. Moreover, the probability density $\rho$ is defined over phase space, instead of the space of molecular velocities. This is the first occasion where probability considerations are applied to the state of the mechanical system as whole, instead of its individual particles. If the transition between kinetic gas theory and statistical mechanics may be identified with this caesura, (as argued by the Ehrenfests and by Klein) it would seem that the transition has already been made right here. But of course, for Boltzmann this transition did not involve a major conceptual move, thanks to his conception of probability as a relative time. Thus, the probability of a particular state of the total system is still identified with the fraction of time in which that state is occupied by the system. In other words, he did not yet need ensembles or non-mechanical probabilistic assumptions.

However, one should note that the equivocation between relative times and relative number of particles, which was relatively harmless in the first section of the 1868 paper, is now no longer possible in the interpretation of $\rho$. Consequently, the marginal probability $\rho(p_1|q_1, \ldots q_n)dp_1$ gives us the relative time that the total system is in a state for which particle 1 has a momentum between $p_1$ and $p + dp_1$, for fixed values of the positions. There is no immediate route to conclude that this has anything to do with the relative number of particles with the momentum $p_1$, unless we further assume that $V(q_1, \ldots q_n)$ is permutation invariant.

2. Most importantly, the results (7,8) open up a perspective of great generality. It proves that the probability of the molecular velocities for an isolated system in a stationary state will always assume the Maxwellian form if the number of particles tends to infinity. Notably, this proof completely dispenses with any particular assumption about collisions, like the SZA, or other details of the mechanical model involved, apart from the assumption that it is Hamiltonian.

3. The main weakness of the result is its assumption that the trajectory actually visits all points on the energy hypersurface, nowadays called the *ergodic hypothesis*. Boltzmann returned to this issue on the final page of the paper ([26] I, p. 96). He notes there that exceptions to his theorem might occur if the microscopic variables would not, in the course of time, take on all values which are consistent with the conservation of energy. For example, the trajectory might be periodic. However, Boltzmann argued, such relations would be immediately annihilated by the slightest disturbance from outside, e.g. by the interaction of a single free atom that happened to be passing by. He concluded that these exceptions would thus only provide cases of unstable equilibrium.

Still, Boltzmann must have felt unsatisfied with his own argument. According to an editorial footnote in his collected works ([26] I, p. 96), Boltzmann's personal copy of the paper contains a hand-written remark in the margin stating that the point was still dubious and that it had not been proven that, even allowing the disturbance from a single external atom, the system would traverse all possible values compatible with the energy equation.

## 2.1. Subsequent doubts on the ergodic hypothesis

In any case, the problem of the validity of the ergodic hypothesis perturbed him enough to investigate the question more closely in a special case [4]. This paper focuses, "partly for illustration, but partly also for verification of the general theorem ([26] I, p. 97)" on a simple mechanical system: a material point in a plane, attracted by a central potential of the form $a/r^2 + b/r^3$. Here we also find a nice pictorial formulation of the ergodic hypothesis: if the material point were a light source, and its motion exceedingly swift, the entire energy surface would appear as homogeneously illuminated ([26] I, p. 103). Boltzmann convinces himself in this paper that the 'general theorem' in question is true in this example.

However, his doubts were still not laid to rest. His next paper on gas theory [5] returns to the study of a detailed mechanical gas model, this time consisting of polyatomic molecules, and avoids any reliance on the ergodic hypothesis. In fact, he subsequently noted twice that it had been a point of principle for him to avoid this hypothesis ([26] I, p.287; [26] II p. 573). This clearly shows that Boltzmann did not trust his (1868) theorem to be sufficiently general.

And when he did return to the ergodic hypothesis in [6], it was with much more clarity and caution. Indeed, it is here that he actually first presents the worrying assumption as an *hypothesis*, formulated as follows:

> "The great irregularity of the thermal motion and the multitude of forces that act on a body make it probable that its atoms, due to the motion we call heat, traverse all positions and velocities which are compatible with the principle of [conservation of] energy." ([26] I, p. 284)

Note that Boltzmann formulates this hypothesis for an arbitrary body, i.e. it is not restricted to gases. He also remarks, at the end of the paper, that "the proof that this hypothesis is fulfilled for thermal bodies, or even is fullfillable, has not been provided ([26] I, p. 287)".

There is a major confusion among modern commentators about the role and status of the ergodic hypothesis in Boltzmann's thinking. Indeed, the question has often been raised how Boltzmann could ever have believed that a trajectory traverses *all* points on the energy-hypersurface, since, as the Ehrenfests conjectured in 1912, and was shown almost immediately in 1913 by Plancherel and Rozenthal, this is mathematically impossible when the energy hypersurface has a dimension larger than 1.

It is a fact that both [9] [[26] I, p. 96] and [6] [[26] I, p. 284] mention external disturbances as an ingredient in the motivation for the ergodic hypothesis. This might be taken as evidence for 'interventionalism', i.e. the viewpoint that such external influences are crucial in the explanation of thermal phenomena (cf.: Blatt, Ridderbos & Redhead). Yet even though Boltzmann expressed the thought that these disturbances might help to motivate the ergodic hypothesis, he never took the idea that they are crucial very seriously. The marginal note in the 1868 paper mentioned above indicated that, even if the system is disturbed, there is still no proof of the ergodic hypothesis, and all his further investigations concerning this hypothesis assume a system that is either completely isolated from its environment or at most acted upon by a static external force. Thus, interventionalism did not play significant role in his thinking.[2]

It has also been suggested, in view of Boltzmann's later habit of discretising continuous variables, that he somehow thought of the energy-hypersurface as a discrete manifold containing only finitely many discrete cells [30]. In this reading, obviously, the no-go theorems of Rozenthal and Plancherel no longer apply. Now it is definitely true that Boltzmann developed a preference towards discretizing continuous variables (although usually adding that this procedure was fictitious and purely for purposes of illustration and more easy understanding). However, there is no evidence in the (1868) and (1871b) papers that Boltzmann implicitly assumed a discrete structure of mechanical phase space or the energy-hypersurface.

Instead, the context of his [6] makes clear enough how he intended the hypothesis.[3] Immediately preceding the section in which the hypothesis is introduced, Boltzmann gives a clear discussion of trajectories for some simple systems: he comes back to the mass point in two dimensions with potential $V(r) = a/r + b/r^2$, that he had also studied in [4] and, even more simple, a two-dimensional harmonic oscillator with potential $V(x, y) = ax^2 + by^2$. In the latter case, the mass point moves through the surface of a rectangle. (Cf. Fig. 1. See also [28].) If $a/b$ is rational, (actually: if $\sqrt{a/b}$ is rational) this motion is periodic, and Boltzmann calls $x$ and $y$ dependent: for each value of $x$ only a finite number of values of $y$ are traversed. However, if $\sqrt{a/b}$ is irrational, the trajectory will, in the course of time, traverse "almählich die ganze Fläche ([26] I, p. 271)" of the rectangle. In this case $x$ and $y$ are called *independent*, since for each values of $x$ an infinity of values for $y$ in any interval in its range are possible. The very fact that Boltzmann considers intervals for the values of $x$ and $y$ of arbitrary small sizes, and stressed the distinction between rational and irrational values, indicates that he did *not* silently presuppose that phase space was essentially discrete, where those distinctions would make no sense.

Now clearly, in modern language, one should say in the second case that the trajectory lies *dense* in the surface, but not that it covers it, or traverses all points. Boltzmann did not possess this language. In fact, he could not have been aware that the continuum contains more than a countable infinity of points. Thus, the correct statement that, in the case that $\sqrt{a/b}$ is irrational, the trajectory will traverse, for each value of $x$, an infinity of values of $y$ within any interval however small, could easily have lead him to believe (incorrectly) that *all* values of $x$ and $y$ are traversed in the course of time.

It thus seems eminently plausible, by the fact that this discussion immediately precedes the formulation of the ergodic hypothesis, that the intended statement is really what the Ehrenfests called the *quasi-ergodic hypothesis*; i.e. the assumption that the trajectory lies dense (i.e. passes arbitrarily close to every point) on the energy hypersurface; or at least some hypothesis compatible with this.[4] The

---

[2]Indeed, in the rare occasion in which he later did mention external disturbances, it was only to say that they are "not necessary" [22]. See also [23].
[3]This has also been argued by [27].
[4]As it happens, Boltzmann's two examples are compatible with the measure-theoretical hypothesis of 'metric transitivity' too.

quasi-ergodic hypothesis is not mathematically impossible for higher-dimensional phase spaces. However, the quasi-ergodic hypothesis does not imply the desired conclusion that the only stationary probability distribution over the energy surface is micro-canonical. One might then still hope that if the system is quasi-ergodic, the only continuous stationary distribution $\rho$ is microcanonical. But even this is fails in general [39].

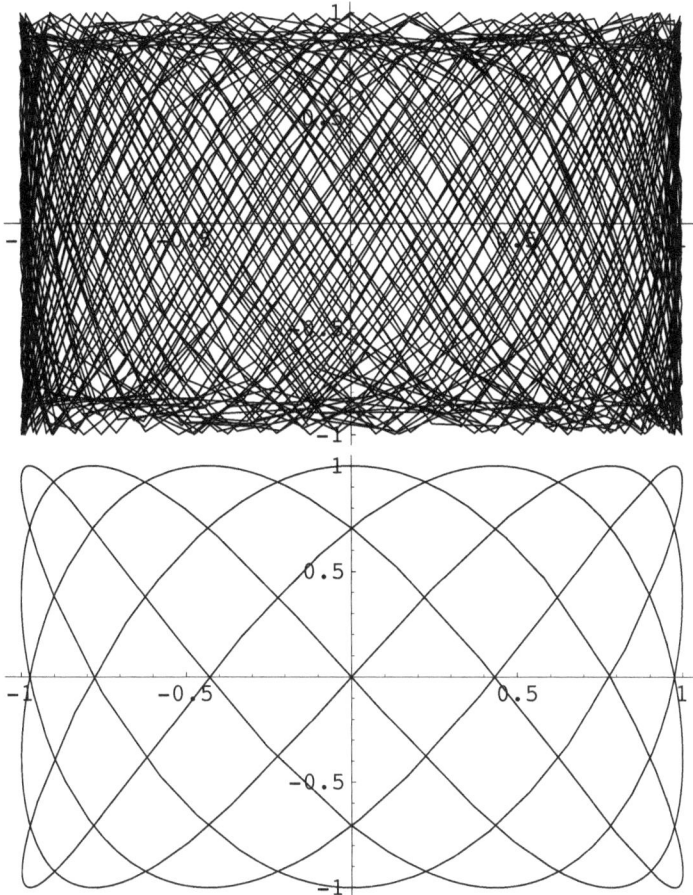

Figure 1. The trajectory in configuration space for the potential function $V(x, y) = ax^2 + by^2$, illustrating the distinction between the case where (i) $\sqrt{a/b}$ is rational (here $4/7$) and the trajectory is periodic; (ii) a fragment of the trajectory when $\sqrt{a/b}$ is irrational $(1/e)$.

Nevertheless, Boltzmann clearly remained skeptical about the validity of his hypothesis, or at least that it could be proven. For this reason, he attempted to explore different routes to his goal of characterizing thermal equilibrium in mechanics. Indeed, both the preceding [5] and his next paper [7] present alternative arguments, with the explicit recommendation that they avoid hypotheses. In fact, he did not return[5] to this hypothesis until the 1880's (stimulated by Maxwell's 1879 review of Boltzmann's 1868 paper.) At that time, perhaps feeling fortified by Maxwell's authority, he would express much more confidence in the ergodic hypothesis (see section 3).

## 3. The 1880's. Return of the ergodic hypothesis

During the 1870s, Boltzmann turned away from the ergodic hypothesis, and developed other approaches, relying on the SZA in 1872, and on the permutational argument in 1877. This latter approach introduced far-reaching conceptual shifts in the theory. Accordingly, the year 1877 is frequently seen as a watershed in Boltzmann's thinking. Concurrent with that view, one would expect his subsequent work to build on his new insights and turn away from the themes and assumptions of his earlier papers. In actual fact, however, Boltzmann's subsequent work in gas theory in the next decade and a half was predominantly concerned with technical applications of his 1872 Boltzmann equation, in particular to gas diffusion and gas friction. And when he did touch on fundamental aspects of the theory, he returned to the issues and themes raised in his 1868-1871 papers, in particular the ergodic hypothesis and the use of ensembles. Apart from two paper, i.e. [12,14], the permutational argument developed in [11] seems to leave no trace at all in his work until he was forcefully reminded of the reversibility objection in 1894. In this section, I review a selection of passages from his later work that are relevant to the the ergodic hypothesis, and the equality of different averages.

### 3.1. 1881: Boltzmann on Maxwell on Boltzmann

The next step in Boltzmann's development was again triggered by Maxwell, this time by a paper that must have pleased Boltzmann very much, since it was devoted to "Boltzmann's theorem" [38] and dealt with his work of 1868.

Maxwell, characteristically, went straight to the main point of Boltzmann's 1868 paper, i.e. the theorem discussed in its last section. He pointed out that the importance of the theorem is that it does not rely on any collision assumption. But Maxwell also made some pertinent observations along the way. He is careful and critical about stating Boltzmann's ergodic hypothesis, pointing out that "it is manifest that there are cases in which this does not take place" [38, p. 694]. Apparently, Maxwell had not noticed that Boltzmann's later papers had also expressed similar doubts. He rejected Boltzmann'a time-average view of probability and instead preferred to interpret $\rho$ as an ensemble density. Further, he remarks that any claim that the distribution function obtained was the unique stationary distribution "remained to be investigated [38, p. 722]". Boltzmann responded by

---

[5]An exceptional occasion where the hypothesis is ergodic mentioned is a passage in the Appendix of his [9], where he briefly considers some system of a single two-dimensional particle and writes: "Suppose it were possible that [the system] could traverse all states of motion that are compatible with the principle of conservation of energy, then [certain formula's from his [6]] are applicable." However, he continues: "On the other hand, if we deal with a system [...] that does not traverse all positions, velocities, and directions of velocity that are compatible with the energy principle, ...", and argues that then its characterization is more difficult. Clearly, this passage too does not indicate any confidence in the validity of the hypothesis.

writing a presentation [13] with the purpose of disseminating Maxwell's 1879 paper to the German-speaking physicists. (Curiously, he did not take the opportunity to comment upon Maxwell's critical remarks.) Maxwell's paper seems to have revived Boltzmann's interest in the ergodic hypothesis, which he had been avoiding for a decade, as well as to the ensemble view which never had received a central place in his thinking before.

## 3.2. 1884-86 Monocyclic systems

Another major influence on Boltzmann's work in the 1880s were Von Helmholtz' investigations into so-called monocyclic systems. They prompted Boltzmann to explore the same subject in three papers [15–17].

### 1884: On the properties of monocyclic systems

The opening paragraph is worth quoting:

> A most complete mechanical proof of the second law would obviously consist in showing that, for every mechanical process, relations hold which are analogous to those of the theory of heat. But, on the one hand, the second law does not seem to be valid in general and, on the other hand, the properties of the so-called atoms are unknown. Therefore the problem is rather to investigate in how far there is an analogy.

This is the first occasion (to my knowledge) where Boltzmann expressed disbelief in the validity of the Second Law. It is also noteworthy how he proposes his studies as an *analogy*; rather than an attempt to provide a mechanical theorem that could be identified with the Second Law. Here we see an influence from Helmholtz which was to last for several papers (cf. Boltzmann (1887, 1894)). The main difference between Boltzmann and Helmholtz is that Helmholtz had considered a purely mechanical problem, without any appeal to probability. These systems had to be stationary, as in a fluid moving through a ring-shaped channel, or, to quote Boltzmann's own example: Saturn's rings. Boltzmann noted that these conditions would also be fulfilled if we replace Helmholtz's mechanical system by a stationary ensemble.

A stationary ensemble is called a *monode* ([26] III, p. 129). An ensemble for which the kinetic energy is an integrating divisor is an *orthode*. Boltzmann says: "For all orthodes equations hold which are completely analogous to thermodynamics (p. 130)." Boltzmann discusses two special cases. For an ensemble of Hamiltonian systems with coordinates $q_1, \ldots, q_g$ and momenta $p_1, \ldots, p_g$ and Hamiltonian $H(p,q) = T(p) + V(q)$, let the relative number of such systems with $p, q$ within certain limits be given by:

$$\rho(p,q) \propto e^{-hH(p,p)} dp dq$$

This type of ensemble is what he calls a *holode* (i.e., a canonical ensemble). He shows (by referring to the arguments of [7]) that the ensemble average of the kinetic energy $\langle T \rangle = \frac{g}{2h}$ is an integrating divisor of the inexact heat differential $dQ$. A holode is therefore a special case of an orthode.

Another special case is an ensemble in which no integral of motion exist but the total energy. Such an ensemble is called an *ergode*. He notes that an element of this ensemble satisfies the hypothesis of traversing every state compatible with the given total energy. Indeed, being an element of an ergode and satisfaction of the hypothesis are equivalent. (This is how the hypothesis got its name.)[6] The main result of the paper is that every ergode is an orthode.

### 3.3. 1887: Mechanical analogies for the second law of thermodynamics

This paper contains the most detailed exposition of the ergodic hypothesis Boltzmann ever gave.

> "Under all purely mechanical systems, for which equations exist that are analogous to the so-called second law of the mechanical theory of heat, those which I and Maxwell have investigated ... seem to me to be by far the most important.... It is likely that thermal bodies in general are of this kind [i.e.: they obey the ergodic hypothesis]"

Again, Boltzmann discusses the trajectory of a mechanical system through state space, points out that the trajectory can be characterized by the integrals of the equation of motions, but that these integral equations might be satisfied for a finite number of combinations of the state variables or an infinity of such combinations.

To illustrate this, he returns to the two-dimensional oscillator of [6] (cf. Fig. 1). The condition $\arcsin x = A \arcsin y$ will allow for each $x$ a finite number of values of $y$, if $A$ is rational but an infinity of values if $A$ is irrational. In the latter case he calls the integral "infinitely multivalued". The integral then "looses its meaning", and "the values of the pair $x$ and $y$ that are traversed now constitute a two-dimensional manifold" ([26] III, p. 262). In general, a number of the integrals might be infinitely multi-valued, and the trajectory then becomes an even higher-dimensional manifold. In the extreme case, when all integrals except the energy become infinitely multivalued, the trajectory will thus fill the entire energy-hypersurface. Again, the only plausible reading here is not that Boltzmann is somehow assuming that the state space is discrete, but rather that he identifies a one-dimensional trajectory that lies dense in a higher-dimensional space with that space.

---

[6] The literature contains some confusion about what Boltzmann actually understood by an ergode. Brush points out correctly (in his translation of [24] and [27]) that an ergode should not be confused with an ergodic system, as used by the Ehrenfests. Indeed, an Ergode is an ensemble, not an individual system. Unfortunately, Brush construes an ergode simply as an ensemble characterized by the micro-canonical distribution (7). In his view, an ergode is just an micro-canonical ensemble. As we can see from the passage just quoted, this is still not correct. An ergode is a stationary ensemble with only a single integral of motion. As a consequence, its distribution is indeed micro-canonical, but also, every member of the ensemble is an ergodic system in the sense of the Ehrenfests (i.e., it satisfies the ergodic hypothesis).

Another dispute has emerged concerning the etymology of the term 'ergode'. The common opinion, going back at least to the Ehrenfests is that the word derived from ergos (work) and hodos (path). [30] has argued however that "undoubtedly" it derives from ergos and eidos (similar). Now one must grant Galavotti that one would expect the etymology of the suffix "ode" of ergode to be identical to that for holode, monode, orthode and planode, and that a reference to path would be somewhat unnatural in these last four cases. However, I don't believe a reference to eidos would be more natural. Moreover, it seems to me that if Boltzmann intended this etymology, he would have written "ergoide" in analogy to planetoide, ellipsoide etc. The idea that he was familiar with this usage is substantiated by him coining the term "momentoide" for momentum-like degrees of freedom (i.e. those that contribute a quadratic term to the Hamiltonian) in [19]. The argument mentioned by Cercignani (that Galavotti's father is a classicist) fails to convince me in this matter.

### 3.4. 1894: Time ensembles

Another relevant paper is Boltzmann's contributions to the 1894 BAAS meeting in Oxford, which appeared as an Appendix to a paper by Bryan. It introduced a remarkable new view on ensembles:

> To express probability by means of a number we suppose the stationary state to last for a long time, $\Theta$. Divide this time into $n$ infinitely small parts, $\theta$. We shall call the beginning of the first of these parts the time zero; the beginning of the second $t_1$, etc. After the whole time $\Theta$ has elapsed, let another series of times of length $\theta$ begin.
>
> Assume for a moment we have $n$ separate vessels, all exactly similar to the one containing the gas; that each of these $n$ vessels contains the same gas and that the motion of the gas is the same in each. The beginning, however is different. For example, let the gas in the second vessel at time zero be in the same condition in which the gas of the first vessel is at time $t_1$; in the third vessel let the gas at time zero be in exactly the same condition as it is in the first vessel at the time $t_2$, and so on.
>
> The probability may be defined in two ways. If we consider a single *vessel* containing gas, let $\tau$ be the fraction of the time during which the coordinates and momenta of a molecules lie between [certain limits], then $\tau/\Theta$ is the probability required [...]. On the other hand if we consider the above series of $n$ vessels at any single *instant* of time, we can define the probability $dw$ to be $dz/n$ where $dz$ is the number of vessels in which a molecule [has its coordinates and momenta between the given limits]. Evidently $dw$ will have different values for different values of the coordinates and momenta. [...] We may therefore put $\tau/\Theta = dz/n = dw\ldots$ [[26] III p. 520-521].

The 'first way' of defining probability mentioned here has been called a *time ensemble* by [32]. It is conceptually rather different from the usual one, presented here as the 'second way' of defining probability (cf. Fig. 3). The crucial point is that the question of the equality of ensemble averages and time averages is put into a very different perspective. In a time ensemble, *obviously* the time average and ensemble average are equal. (Disregarding niceties about discrete, continuous and infinite-limit averages (i.e. between $\frac{1}{\Theta}\sum_i A(t_i)$, $\frac{1}{\Theta}\int_0^\Theta A(t)dt$ and $\lim_{\Theta \longrightarrow \infty} \frac{1}{\Theta}\int_0^\Theta A(t)dt$). Yet, Boltzmann presents the two ensemble conceptions as two ways of defining the same thing, and simply equates them.

He does not note any explicit difference between his present conception and his earlier usages of ensembles. This might suggest that he understood ensembles also as time ensembles in these earlier cases. In any case, it reveals once again that Boltzmann did not rely on the ergodic hypothesis to obtain identity between these two kinds of averages, but rather hoped to obtain it by an appropriate choice of definitions.

### 3.5. 1898: Lectures on Gas Theory II

The second volume of the *Lectures on Gas Theory* is mostly concerned with applications and contain comparatively little material on the foundation of statistical physics. However, some sections and digressions present important passages, in particular in §30,35,41 and Chapter 7.

Chapter 3 of the book develops the principles of general mechanics needed for gas theory. This is straightforward Hamiltonian mechanics. In §26 he introduces ensembles: "Just as one can represent a curve infinitely many times, each time with a different value of a parameter, so we can represent our mechanical system infinitely often so that we obtain infinitely many mechanical systems, all of the same nature and subject to the same equations of motion but with different initial conditions."

This quote is somewhat ambiguous about what kind of ensemble is intended. However, immediately subsequent to this he considers a subset of this infinite number for which the initial conditions are specified within an infinitesimal region $dp_1 \cdots dq_n$. This suggests he has an ordinary ensemble in mind. He then proceeds to discuss to investigate the case when the ensemble is stationary. Just as in the (1871) papers, he argues that a sufficient condition for stationarity is guaranteed if the distribution function depend only on the integrals of the equation of motion (which he now calls "invariants"). But in contrast to those papers he now claims that it is also sufficient. He ends the section by saying that the most simple example of such a stationary distribution is obtained when $\rho$ is given by (7), i.e. the microcanonical distribution. He adds: I once allowed myself to call the distribution of states described by this formula [...] an ergodic one. Note that Boltzmann does not mention explicitly that his previous usage of the term 'ergodic' also assumed that there was no other integral (or invariant) than the total energy. Clearly, it is this passage that gave rise to the confusion mentioned in footnote 6.

### 3.6. 1904: Boltzmann & Nabl

The most explicit account Boltzmann ever gave about the relation between time and particles averages is to be found in his last article on statistical physics, the Encyclopedia article co-authored with Nabl.

Let $\phi(c)dc$ represent the relative number of molecules with a velocity between $c$ and $c + dc$, and let the average squared velocity be

$$\langle c^2 \rangle = \frac{1}{n} \int c^2 \phi(c)dc$$

Boltzmann an Nabl write:

> The average defined here is the so-called *statistical* average, which one obtains by forming, for every molecule, at a given time, the value $c^2$ and taking the average of all these values. On the other hand, the *historical* average is defined by considering a single molecules during a long time $t$ and forming the integral
>
> $$\frac{1}{t} \int_0^t c^2 dt.$$
>
> Yet, because on average all molecules behave equally, these two averages are the same for a stationary state [25].

This is the only occasion I know of in which Boltzmann recognizes that, in principle, these two types of averaging differ. And yet again, it is immediately claimed that they are equal for the stationary state. Not because if the ergodic hypothesis, but because "all molecules behave equally".

## 4. Conclusions

The first issue which any student of Boltzmann's work must face is that of his light-hearted identification of different meanings of probability. We have seen that this is a recurrent factor in his work; from his second paper on gas theory of 1868 to his very last in 1904. How could Boltzmann have thought that these different meanings coincided?

The Ehrenfests suggested an answer: it is because Boltzmann silently or implicitly relied on the ergodic hypothesis. Indeed they showed that if one assumes this hypothesis, then equality between different types of averages follows. Interesting as this reconstruction may be, it is clear that Boltzmann did not reason this way. He never mentioned the ergodic hypothesis as a motivation for this identification, and repeatedly expressed doubts about the validity of this hypothesis and made a point of avoiding it as much as he could. If the identification in question relied on the ergodic hypothesis, one ought assume that his distrust of the ergodic hypothesis should be accompanied by doubt about the equality of different averages. But this does not occur in Boltzmann's writing.

A (historically) more accurate explanation seems to be that Boltzmann simply believed that the interpretation of the averages or the distribution function was more or less a matter of taste which would not interfere with the mathematical results he obtained for these concepts. To be sure, this answer is not satisfactory, since this belief is wrong. But I can see no other explanation.

So what role did the ergodic hypothesis play? It seems that Boltzmann regarded the ergodic hypothesis as a special dynamical assumption that may or may not be true, depending on the nature of the system, and perhaps also on its initial state. Its role was simply to help derive a result of great generality: For any system for which the hypothesis is true, its equilibrium state is characterized by (7), which reduces in the limit $n \longrightarrow \infty$ to the Maxwell distribution, regardless of the details of the inter-particle interactions, the SZA, or indeed whether the system represented is a gas, fluid, solid or any other thermal body.

## REFERENCES

1. J. M. Blatt. *Prog. Theor. Phys.* 22, p. 745, 1959.
2. L. Boltzmann. Über die mechanische Bedeutung des zweiten Hauptsatzes der Wärmetheorie, *Wiener Berichte*, 53:195–220, 1866. In [26] Vol. I, paper 2.
3. L. Boltzmann. Studien über das Gleichgewicht der lebendigen Kraft zwischen bewegten materiellen Punkten. *Wiener Berichte*. 58:517–560, 1868. In [26] Vol. I, paper 5.
4. L. Boltzmann. Lösung eines mechanischen Problems. *Wiener Berichte*. 58:1035–1044, 1868. In [26], Vol. I, paper 6.
5. L. Boltzmann. Über das Wärmegleichgewicht zwischen mehratomigen Gas-molekülen *Wiener Berichte*. 63:397–418, 1871. In [26] Vol. I, paper 18.
6. L. Boltzmann. Einige allgemeine Sätze über Wärmegleichgewicht *Wiener Berichte*. 63:679–711, 1871. In [26] Vol. I, paper 19.
7. L. Boltzmann. Analytischer Beweis des zweiten Hauptsatzes der mechanischen Wärmetheorie aus den Sätzen über das Gleichgewicht der lebendigen Kraft. *Wiener Berichte*. 63:712–732, 1971. In [26] Vol. I, paper 20.

8. L. Boltzmann. Weitere Studien über das Wärmegleichgewicht unter Gas-molekülen *Wiener Berichte.* 66:275–370, 1872. In [26] Vol. I, paper 23.

9. L. Boltzmann. Über das Wärmegleichgewicht von Gasen, auf welche äußere Kräfte wirken. *Wiener Berichte.* 72:427-457, 1875. In [26] Vol. II, paper 32.

10. L. Boltzmann. Bermerkungen über einige Probleme der mechanische Wärmetheorie *Wiener Berichte.* 75:62-100, 1877. In [26] Vol. II, paper 39.

11. L. Bolzmann. Über die beziehung dem zweiten Haubtsatze der mechanischen wärmetheorie und der Wahrscheinlichkeitsrechnung resp. dem Sätzen über das Wärmegleichgewicht *Wiener Berichte.* 76:373–435, 1877. In [26] Vol. II, paper 42.

12. L. Boltzmann. Über einige das Wärmegleichgewicht betreffende Sätze *Wiener Berichte.* 78:7–46, 1878. In [26] Vol. II paper 44.

13. L. Boltzmann. Referat über die Abhandlung von J.C. Maxwell: "Über Boltzmann's Theorem betreffend die mittlere Verteilung der lebendige Kraft in einem System materieller Punkte" *Wied. Ann. Beiblätter.* 5:403-417, 1881. In [26] Vol. II paper 63.

14. L. Boltzmann. Über das Arbeitsquantum, welches bei chemischen Verbindungen gewonnen werden kann *Wiener Berichte.* 88:861–896, 1883. In [26] Vol. II, paper 69.

15. L. Boltzmann. Über die Eigenschaften monocyklischer und andere damit verwandter Systeme *Crelles Journal.* 98:68–94, 1885. In [26] Vol III, paper 73.

16. L. Boltzmann. Über einige Fälle, wo die lebendige Kraft nicht integrierender Nenner des Differentials der zugeführte Energie ist *Wiener Berichte.* 92:853–875, 1885. In [26] Vol III, paper 74.

17. L. Boltzmann. Neuer Beweis eines von Helmholtz aufgestellten Theorems betreffend die Eigenschaften monocyklischer Systeme *Göttinger Nachrichte.* 209–213, 1886. In [26] Vol III, paper 75.

18. L. Boltzmann. Über die mechanischen Analogien der zweiten Hauptsatzes der Thermodynamik *Crelles Journal.* 100:201–212. In [26] Vol. III, paper 82.

19. L. Boltzmann. III. Teil der Studien über Gleichgewicht der lebendigen Kraft *Münch. Ber.* 22:329–358, 1892. In [26] Vol. III, paper 97.

20. L. Boltzmann. On the Application of the Determinantal relation to the kinetic theory of gases. Appendix C to an article by G.H. Bryan on thermodynamics in *Reports of the British Association for the Advancement of Science.* 102–106, 1894. In [26] Vol. III, paper 108.

21. L. Boltzmann. On certain questions in the theory of gases, *Nature.* 51:413–415, 1895. Also in [26], Vol. III, paper 112.

22. L. Boltzmann. On the minimum theorem in the theory of gases *Nature* 52: 221, 1895. Also in [26] Vol. III, paper 114.

23. L. Boltzmann. *Vorlesungen über Gastheorie* Vol I. Leipzig, J. A. Barth, 1896. Translated, together with [24] by S. G. Brush, *Lecture on Gas Theory* Berkeley: University of California Press, 1964.

24. L. Boltzmann. *Vorlesungen über Gastheorie* Vol II. Leipzig, J. A. Barth, 1898. Translated, together with [23] by S.G. Brush, *Lecture on Gas Theory* Berkeley: University of California Press, 1964.

25. L. Boltzmann and J. Nabl. Kinetisch theorie der Materie *Encyclopädie der Mathematischen Wisenschaften,* Vol V-1, 493–557.

26. L. Boltzmann. *Wissenschaftliche Abhandlungen* Vol. I, II, and III. F. Hasenöhrl (ed.) Leipzig. Reissued New York: Chelsea, 1969.

27. S. G. Brush. *The Kind of Motion we call Heat*, Amsterdam, North Holland, 1976.

28. C. Cercignani. *Ludwig Boltzmann, the man who trusted atoms* Oxford: Oxford University Press, 1998.

29. P. Ehrenfest and T. Ehrenfest-Afanassjewa. *The conceptual Foundations of the Statistical Approach in Mechanics*, New York, Cornell University Press, 1959 (original edition 1912).

30. G. Galavotti. Ergodicity, ensembles, irreversibility in Boltzmann and beyond http: arXiv:chao-dyn/9403004. *Journal of Statistical Physics* 78:1571-1589, 1994.

31. E. Garber, S. G. Brush and C. W. F. Everitt (eds.). *Maxwell on Molecules and Gases* MIT Press, Cambridge Mass, 1986.

32. J. W. Gibbs. *Elementary Principles in Statistical Mechanics*, New York, Scribner etc., 1902.

33. H. von Helmholtz. *Wissenschaftliche Abhandlungen*. Vol. III, G. Wiedemann (ed) Leipzig, J. A. Barth, 1895.

34. M. J. Klein. The Development of Boltzmann's Statistical Ideas. In E. G. D. Cohen en W. Thirring (eds.), *The Boltzmann Equation*, Wien, Springer, 53–106, 1973.

35. J. Loschmidt. Über die Zustand des Wärmegleichgewichtes eines Systems von Körpern mit Rücksicht auf die Schwerkraft *Wiener Berichte*, (73), 128, 366, 1876, (75) 287, (76), 209, 1877.

36. J. C. Maxwell. Illustrations of the Dynamical Theory of Gases *Philosophical Magazine*. 19:19–32, 1860; 20:21–37, 1860. Also in [31].

37. J. C. Maxwell. On the dynamical theory of gases *Philosophical Transactions of the Royal Society of London*. 157:49-88, 1867. Also in [31].

38. J. C. Maxwell. On Boltzmann's theorem on the average distribution of energy in a system of material points *Trans. Cambridge Phil. Soc.* 12:547–570, 1879.

39. V. V. Nemytskii and V. V. Stepanov. *Qualitative Theory of Differential Equations* Princeton: Princeton University Press, p. 392, 1960.

40. J. von Plato. *Creating Modern Probability* Cambridge: Cambridge University Press, 1994.

41. T. M. Ridderbos and M. L. G. Redhead. *The spin-echo experiment and the second law of thermodynamics*. Found. Phys. 28: 1237-1270.

42. A. Rosenthal. *Ann. Phys.* 42:796, 1913.

43. D. Szasz. Boltzmann's Ergodic hypothesis, a conjecture for centuries?, *Stud. Sci. Math Hung.* 31:299–322, 1996.

44. E. Zermelo. Ueber einen Satz der Dynamik und die mechanische Wärmetheorie, *Annalen der Physik* 57:485–494, 1896.

# SECTION D:

# ETHICAL, SOCIAL AND HISTORICAL PERSPECTIVES ON PHILOSOPHY OF SCIENCE

# Emotion, Rationality, and Decision Making in Science

James W. McAllister

*Faculty of Philosophy, University of Leiden, The Netherlands*
*j.w.mcallister@let.leidenuniv.nl*

**Abstract.** The noncognitivist view of emotion, which has been dominant until recently, portrays emotion as inimical to cognition and rationality. This view has had a profound effect on images of science, encouraging the conclusion that scientific practice requires the suppression of emotion. Recent results in cognitive psychology have cast doubt on the noncognitivist view: emotion is now seen as a resource underpinning decision making, especially in practical contexts. The new understanding of emotion has important implications for philosophical models of science: emotion is revealed as playing an ineliminable role in scientific work. Furthermore, there are grounds for considering the appeal to emotion in science, under certain conditions, to be rationally warranted.

## 1. Introduction

What role does emotion play in science? What role should it play? The answer given until recently to both questions was "none". Science is the antithesis of emotion, it was held: scientific method is controlled and independent of standpoint, whereas emotion is unbridled and rooted in personal feelings. To practise science, one must suppress emotion.

Recent developments in the understanding of emotion oblige us to revise this view. They show that emotion plays an ineliminable role in decision making, especially in practical contexts. Furthermore, they establish that emotion is not a threat to rationality, but a faculty that assists reasoning. Since the development of science is governed by scientists' decisions, many of which are of a practical kind, and scientific practice is based on rationality, these findings have important repercussions for philosophy of science. Philosophical models of scientific practice will be enriched and made more realistic by taking account of them. The outcome will be a view of science as an activity that, as well as cognitive and rational throughout, is emotional throughout.

In this paper, I discuss the implications of the new understanding of emotion for philosophy of science. To begin, I review some basic facts about emotion. In section 3, I outline the noncognitivist view that has long dominated the study of emotion, and show how this view has influenced both popular and scholarly images of science. I turn in section 4 to some cognitivist approaches to emotion that have been developed since around 1980, and discuss the roles that, according to these approaches, emotion plays in decision making. I apply these findings to the decision making of scientists in section 5. In conclusion, I address the normative question whether and how reliance on emotion in science can be warranted.

## 2. What Emotion Is

Whereas many claims about emotion are contested, some findings are relatively uncontroversial. A brief review of these findings is useful to offer an initial approach to the phenomenon of emotion.

From a biological viewpoint, emotions constitute a class of hormonal and neural responses to external stimuli in mammals and, perhaps, some other taxa. Emotions may be triggered by sensory signals or by beliefs. For example, a threatening sound may trigger fear, and the belief that a loved one has died may trigger sadness. In virtue of the fact that they can be triggered by beliefs, emotions differ from other physiological states, such as pain, hunger, and drowsiness, which are triggered only by sensory signals ([23], pp. 263–324).

Emotions generally have two biological effects: they produce a behavioural reaction by the organism to the stimulus and a physiological change in the organism to support that reaction. Fear, for example, produces a response that generally consists of a fight reaction or a flight reaction. Simultaneously, fear produces metabolic changes in the organism to facilitate that response, such as changes in the organism's breathing rhythm, heart rate, and blood flow ([23], pp. 124–175).

From a functional viewpoint, emotions may be regarded as the outcomes of processes for detecting, evaluating, and categorizing certain events and situations encountered by the organism in its environment. In particular, emotions constitute appraisals of events and situations for the extent to which they promote or threaten the organism's wellbeing [50].

Emotions are engaged automatically and without deliberation, at least in the first instance. An organism is not necessarily conscious of its own emotions, though in many instances a human has feelings that constitute awareness of his or her emotions. To the extent to which an organism is aware of its emotions, it can acquire information from them about its internal state and surroundings ([10], pp. 127–164).

Traditional accounts identify a small number of so-called basic or primary emotion categories in humans. According to most accounts, the basic emotions are happiness, sadness, fear, anger, surprise, and disgust. Most writers believe that these basic emotions are found in all human societies. Evidence for this claim includes the observation that the facial expressions associated with these emotions are recognized panculturally [15]. Many writers hold that humans also construct conceptually more subtle and sophisticated secondary or social emotions from the vocabulary and mechanisms of the basic emotions. Most accounts identify embarrassment, jealousy, guilt, and pride as typical secondary emotions ([10], pp. 127–164; [38], pp. 112–114). Some writers identify also what they regard as specifically cognitive emotions: examples are attention, surprise, wonder, fear of the unknown, and pleasure in finding out whether something is the case ([55]; [25], pp. 100–136).

There is much discussion about the extent to which human emotions are biologically determined and the degree to which they exhibit cultural and individual differences. Most writers believe that emotions are biologically determined to a large extent, and are produced by brain mechanisms resulting from evolutionary history. Culture and learning influence the ways in which emotions are expressed, the meanings attributed to them, and the construction of higher-level emotions. The composition and dynamics of the emotional responses of each individual are also shaped by that individual's developmental and environmental history. In consequence, there is considerable individual variation in the expression of emotions [28].

## 3. Science and the Noncognitivist View of Emotion

Until recently, Western thought inclined to a noncognitivist view of emotion. On this view, emotion is opposed to cognition and conflicts with it: emotion perturbs, degrades, contaminates, disrupts, impedes, clouds, and biases cognitive processes and logical and rational thinking. This view can be retraced to Plato and Aristotle, and was given an influential modern formulation by William James. The noncognitivist view is underpinned by the empirical claim that emotion and cognition operate independently of one another and are produced by distinct neurophysiological structures: whereas subcortical structures account for affect and emotion, neocortical structures are responsible for cognitive and conceptual functions ([61], [62]).

According to the noncognitivist view, an emotionless thinker would have no difficulty in performing rational reasoning and reaching rational conclusions, but a person affected by emotion is diverted from the path of rationality. On this view, to be affected by emotion is to be in a state unfit for reasoning and decision making. Those who aspire to attain rationality must therefore restrict the influence of emotion on their reasoning as much as possible. This view was spelled out by, for example, Arthur Lefford:

> The attitude which the subject has toward the subject matter of his reasoning strongly tends to bring his reasoning into harmony with his feelings. These emotional attitudes are thus the arch-enemies of objective and clear thinking. [ ... ] To remain objective and impersonal and to follow the objective requirements of the syllogistic structure demands that the individual be able to separate the more emotional aspects of his personality from the more rational. ([39], pp. 146, 149)

The noncognitivist view of emotion, portraying rationality as an impersonal attainment that must be shielded from the body and the subject, has constituted one of the pillars of the subjectivity/objectivity polarity that dominates Western thought. Perhaps, at some point in history and under certain intellectual circumstances, the portrayal of emotion as inimical to cognition and to rationality helped to demarcate a realm of intersubjective inquiry: perhaps mistrust and rejection of emotion constituted a necessary phase in the rise of science. Today, however, the noncognitivist view of emotion no longer helps us practise and understand science. To the contrary, it hinders these activities. Its effects can be seen in images of science, both popular and scholarly.

Popular images of science—which include both the self-images of scientists and images of science produced for the general public—have traditionally drawn on two archetypes of the scientist. In one archetype, the scientist is a genius like an artist, who makes personal and irreplaceable contributions by dint of intuition, imagination, and creativity. The alternative archetype depicts the scientist as objective spokesperson of nature, unveiling facts that are the same for everyone. Whereas the former archetype leaves open the possibility that emotion plays a positive role in scientists' work, the latter suggests that subjectivity—including emotion—has no legitimate place in science. Since the Renaissance, the archetype of the scientist as intuitive genius has lost ground in popular images of science, to be replaced by that of the scientist as objective discoverer of facts ([13], [14]). This shift is consistent with, and has probably been prompted partly by, the noncognitivist view of emotion. If emotion is regarded as detrimental to cognition and rationality, anyone wishing to construct a positive image of science would attempt to play down the influence of emotion on scientists.

The noncognitivist view of emotion has similarly influenced the principal academic disciplines that take science as their object: history, social studies, and philosophy of science. Let us begin with history of science. The rise of modern science in the seventeenth century is bound up, in the minds of both early scientists and present-day historians, with the suppression of emotion. Seventeenth-century moderate skepticism encouraged the view that the passions lead us into error [32]. Partly in response to this view, the early practitioners of the new experimental science chose to propagate a rhetorical image of their procedures as impersonal and mechanical [56]. This image is somewhat at odds with many private writings of the early members of the Royal Society, which reveal a passion for inquiry that can fairly be called lusty. Nonetheless, twentieth-century accounts of the rise of modern science largely accepted the emotionless image of it proffered by its first practitioners. Robert K. Merton, whose work greatly influenced the historiography of the scientific revolution from the 1930s to the 1980s, linked the rise of modern science to the asceticism of the Puritan ethic. According to the Merton thesis, the Puritan repression of the passions fostered the pursuit of rationality and scientific inquiry ([46], [9]). This thesis is allied to Max Weber's view that ascetic Protestantism was partly responsible for the rise of capitalism [60]. On further consideration, however, these views acquire an implausible air. The functioning of science and capitalism depends on the motivations provided by the gratification of intellectual and material desires. In this light, the claim that science and capitalism arose from the suppression of emotion should be regarded with some suspicion ([20], [21], [29]).

The noncognitivist view of emotion affects the historiography of recent science too. The figure of the intellectually brilliant but emotionally stunted scientist is a staple of biographies. Helge Kragh remarks of P. A. M. Dirac: "For the most part, he concentrated his resources on theoretical physics, which acted as a substitute for human emotions and a richer social life" ([33], p. 255). Paul Hoffman describes Paul Erdös as "the man who loved only numbers", as lacking social emotions [30]. These portrayals reinforce the impression that success in science is accompanied by, and depends on, lack of emotion. There are two reasons for doubting that such portrayals are accurate and complete, however. First, social studies of science suggest that scientific achievement consists as much in making alliances as in making discoveries. Dirac actively participated in several controversies in physics, taking sides forcefully and arguing his case deftly; Erdös maintained cordial and even affectionate relations with many colleagues, with whom he coauthored hundreds of mathematical papers. Such activities require insight in social emotions. Second, even if a person lacks emotion in the social sphere, this does not necessarily entail an inability to experience emotion in scientific work. Dirac was quick to endorse and reject theories and lines of research on aesthetic grounds, even when the available empirical data were against him; Erdös had a highly intuitive and metaphorical approach in mathematics. It is reasonable to suppose that these traits are manifestations of emotional responses to hypotheses and evidence. In this light, the coupling of scientific greatness to emotional incapacity is implausible: it is more reasonable to think that even such seemingly unemotional scientists as Dirac and Erdös owe their scientific achievements partly to their ability to experience emotion.

Let us now turn to social studies of science. One might expect social scientists to be sensitive to the role of behavioural factors in science. In fact, most writers who take a sociological approach to science marginalize emotion. The marginalization began in the earliest phase of the discipline. The view of emotion that Merton puts forward in his writings in sociology of science is consistent with his historiographic thesis that science arose from the suppression of emotion. Merton acknowledges that scientists may experience emotions, such as jealousy, in the competition for priority and rewards. However, such emotions pertain to the social and institutional context of science: the cognitive content of research is free of emotion. Indeed, Merton identifies disinterestedness as one of the core norms of science, arguing that researchers ought to be and are emotionally detached from their field of study ([44], [45]). Working in the same framework, Bernard Barber describes what he regards as a fundamental value underpinning the organization of science:

> This is the value scientists set upon emotional neutrality as an instrumental condition for the achievement of rationality. Science approves of emotional neutrality not primarily for its own sake, and certainly not for all social activities, but insofar as it enlarges the scope for the exercise of rationality, and its power as well. Emotional involvement is recognized to be a good thing in science—up to a point: it is a necessary component of the moral dedication to the scientific values and methods. But in the application of those techniques of rationality, emotion is so often a subtle deceiver that strong moral disapproval is placed upon its use. ([2], p. 88)

Attention for emotion in social studies of science has also suffered from the subordination of the psychological to the social. The work of Thomas S. Kuhn provides an example. Kuhn regards both the development and the acceptance of theories as essentially social phenomena. He draws upon findings in psychology, especially pertaining to contextual influences on perception and *Gestalt* effects, in his accounts of the incommensurability of paradigms and the dynamics of paradigm switch. However, he assigns to psychology only the second-order task of accounting for individual differences in attitude within groups of scientists, whose underlying behaviour is to be explained by sociology ([34], pp. 240–241; [31], p. 59). Some more recent approaches in social studies of science, including the school known as "sociology of scientific knowledge", portray scientists as motivated by social interests, but decline to consider the possibility that these interests are grounded to some extent in emotion.

We turn lastly to philosophy of science, where the influence of the noncognitivist view of emotion has also been strong. The attitude customarily adopted by philosophers of science to emotion and other psychological factors resembles their attitude towards social factors: they regard them as inimical to rationality and essentially irrelevant to the understanding of good science ([35], pp. 196–222; [6]). Logical positivists incorporated this view in their distinction between the context of discovery and that of justification, a centrepiece of their account of science. They regarded the generation of a theory by an individual scientist as an arational act in which emotion may play a part, alongside stimuli of any other kind. The process of testing and validating theories assures the rationality and progressiveness of science, but this activity is devoid of any psychological dimension. This two-stage model, which effectively removes emotion from the picture, is similar to the approach embodied in rational choice theory, which analyzes agents' decisions in the light of their goals. A person's emotions normally play a large role in setting his or her goals, but rational choice theory pays no attention to the origin and justification of goals. The theory assumes an agent's goals as a given input, and models the agent's subsequent behaviour as a mechanical calculation of costs and benefits.

Since the decline of logical positivism, philosophers of science have reclaimed discovery as an object of analysis [49]. This development offered some prospect that philosophers would consider the emotional factors that the logical positivists had consigned to the context of discovery. Many writers, however, interpret the study of discovery as a project to automate discovery, ideally in computational terms [12]. On this view, discovery does not depend on the subjectivity of individual scientists: it could be practised just as well, or even better, by disembodied agents executing algorithms to detect patterns in data, reach inductive generalizations, and the like. Emotion is excluded from this view as securely as ever.

The dominant approaches in history, social studies, and philosophy of science, reviewed above, share one feature: they deny emotion any place in science. According to these approaches, the rise of science was fostered by the suppression of emotion; the practice of science requires emotion to be set aside; and the understanding of science should avoid reference to emotion. It is plausible to attribute this convergence of attitudes across different disciplines partly to the influence of the noncognitivist view of emotion.

A few researchers outside the disciplines mentioned above have been more willing to attribute a positive role to emotion in science. Psychological aspects of science have come under study from the late 1950s onwards, and some writers have commented on the place of emotion in scientific work ([52], [48], [41]). Unlike sociology of science, however, psychology of science has never achieved the status of academic discipline, and these efforts did not coalesce into an enduring research tradition ([24], [18]).

The influence of the noncognitivist view of emotion on academic theorizing can perhaps be dismissed as being of interest only to specialists. Of wider concern is the fact that it has affected the politics of science and the lives of scientists. In particular, it is likely that the noncognitivist view of emotion has hindered the careers of women in science.

Women have traditionally occupied a marginal place in science. Many factors, both cultural and social, have contributed to this state of affairs. One important factor is the belief that women are not suited to pursue science. This belief follows from the conjunction of three presuppositions: that women's behaviour is primarily governed by emotion, that science is rational, and that emotion is inimical to rationality.

The first of these presuppositions has pervaded Western philosophy and social thought for centuries. It is embedded in a polarity between men and women. Whereas men have the capacity systematically to pursue abstract goals and interests, women's emotions lead them to identify with and respond to their particular surroundings. A woman's sphere is thus intuitive, subjective, personal, and private, whereas a man's sphere is analytical, objective, impersonal, and public. It follows that women find it difficult to attain detachment, neutrality, and objectivity ([17]; [53], pp. 111–120; [1]; [40]). An editorial in the *New York Times* in 1921, for example, in the wake of a visit of Marie Curie to the United States, explained that there would always be more men than women in science, because more men

> have the power—a necessary qualification for any real achievement in science—of viewing facts abstractly rather than relationally, without overestimating them because they harmonize with previously accepted theories or justify established tastes and proprieties, and without hating and rejecting them because they have the opposite tendencies. (Quoted in [54], pp. 127–128)

Once science is associated with the shunning of emotion, and women are portrayed as especially emotional beings, the conclusion is clear: women do not make good scientists.

How can the inference that emotion disqualifies women from science be resisted? One way is to accept that emotion is antagonistic to science, and to deny that women's behaviour is affected disproportionately by emotion. The thesis that women are more emotional than men indeed deserves critical scrutiny: although some empirical evidence suggests that females are more accomplished than males at expressing emotions and recognizing them in others, the further claim that women's behaviour is more strongly affected by emotion than that of men is implausible ([5], [19], [22]). There is, however, a more radical way to undermine the inference that emotion precludes women from making good scientists: by challenging the presupposition that emotion is inimical to science. If it is established that emotion plays a positive role in the exercise of scientific rationality, it follows that there is no tension between being partly influenced by emotion and practising good science. To the contrary, a capacity to experience and express emotion is revealed as an asset to scientists, both women and men.

In sum, the noncognitivist view of emotion has had a profound influence on popular images of science, on the academic disciplines devoted to the study of science, and on the politics of science. The effect in all these domains has been identical: to portray science as devoid of emotion and to portray emotion as a threat to science. It is time to reassess the adequacy of this view.

## 4. The Cognitivist Approach to Emotion

The noncognitivist view of emotion rests on two principal theses, as we saw in the previous section: the empirical claim that emotion and cognition operate independently of one another and are produced by distinct neurophysiological structures, and the conceptual claim that emotion is opposed to and conflicts with cognition. Research since around 1980 has undermined both these theses. The empirical claim has been challenged by the finding that cognitive and emotive functions are distributed over a large number of mutually complementary and interdependent cortical and subcortical systems and processes. The conceptual claim has been weakened by the realization that emotion is cognitive in several respects, and thus cannot be regarded in a straightforward manner as standing in opposition to reason. This research yields a new picture of emotion and of the relation between emotion and cognition. Let us now review these developments.

Evidence from research in neuroscience and from clinical experiments assembled by Antonio R. Damasio ([10]; [11], pp. 35–81), Joseph LeDoux [38], and others suggests that a subject's cognitive abilities are impaired by damage to certain brain areas responsible for the processing of emotion. One such area is the ventromedial region of the frontal lobe. Subjects in whom this area is damaged become emotionally flat: they show emotion infrequently and within a limited repertoire. More unexpectedly, they also show impaired reasoning ability.

The impairment manifests itself more strongly in practical reasoning, which consists of planning and decision making in practical and concrete contexts, than in theoretical reasoning, which consists of drawing inferences from premises in theoretical and abstract contexts. Damasio reports that subjects with frontal lobe damage generally perform well in tests of theoretical reasoning. Indeed, some demonstrate heightened analytical powers in such contexts. By contrast, these subjects exhibit deficiencies in practical reasoning: they often delay a decision inordinately or fail to make a decision in practical tests. In many cases, they exhibit a kind of hyperrationality, tackling trivial dilemmas with disproportionate analytical resources, such as elaborate cost-benefit analyses that have little practical bearing on the outcome. Where they eventually reach a decision, this is often flawed and contrary to their best interests.

One of Damasio's subjects is Elliott, a businessman whose prefrontal cortex was damaged by a brain tumor. Tests indicated that Elliott's intelligence, memory, and attention remained unaffected by the lesion. He performed well in theoretical tasks. By contrast, Elliott's ability to solve practical problems was impaired. For example, he proved incapable of following rational strategies in gambling games, consistently incurring large losses. Damasio interprets his findings as evidence that a capacity to experience emotion, though not essential in theoretical reasoning, is indispensable for effective reasoning in practical contexts.

Emotion appears to contribute in two ways to a capacity for practical reasoning. First, emotion sets the goals, values, and preferences that guide the subject's reasoning and behaviour. In other words, it makes the subject care about the outcome of a given situation and provides the motivation to seek an outcome that satisfies certain requirements. Second, emotion enables the subject to identify the salient features of a situation. Emotion thereby enables the subject to take decisions by focusing on a few decisive elements, instead of having to behave as a rational calculator who carries out a cost-benefit analysis of all available options. Without a functioning emotion system, the subject lacks both the stable goals that would move him or her to make determinate decisions, and the ability to analyze a given situation with a view to attaining these goals. This explains why subjects who lack a functioning emotion system are liable to postpone decisions, performing aimless and interminable analyses of theoretically possible options.

Emotion plays an especially important role in decision making in situations where the subject's goals and information are not complete, consistent, and reviewable. These situations occur when the subject's values, priorities or criteria underdetermine the decision or conflict with one another, or the information on which the decision must be based is incomplete, contradictory, ambiguous or too abundant to review in full. In such circumstances, it is not possible to make a decision by systematically taking all relevant factors into account. Instead, subjects in such situations make decisions mostly by selectively focusing on some factors to the neglect of others. As emotion has the role of attributing salience to some features of situations, a capacity to experience emotion is essential for making decisions in such circumstances.

These findings give a new picture of the relation between emotion, cognition, and rationality. In particular, they blur the opposition between these categories posited by the noncognitivist view of emotion. If emotion plays a crucial role in guiding the analysis of situations and decision making in practical contexts, then it is difficult to draw a sharp distinction between emotion on the one hand and cognition and rationality on the other hand. Emotion is now seen as a function on which the subject can draw to carry out cognitive and rational tasks, alongside functions that are conventionally labelled as cognitive and rational.

The implications of the cognitivist view of emotion for the relation between emotion and cognition must be spelled out with care. This is underlined by an exchange between Robert B. Zajonc and Richard S. Lazarus, who defend the noncognitivist and the cognitivist approaches to emotion, respectively. Zajonc ([61], [62]) argues that affect and cognition constitute separate and partially independent systems. He reports empirical evidence that certain pathways lead directly from the perceptual system to responses regarded as emotional. Conscious appraisal and affect are often uncorrelated and disjoint, in his view, and affective reactions can be triggered without a prior cognitive process. Zajonc concludes that it is inappropriate to regard emotional responses as cognitive. In reply, Lazarus ([36], [37]) emphasizes the primacy of cognition and maintains that emotional responses are supported by brain systems devoted to cognition.

Irrespective of which side can claim better empirical support, the exchange between Zajonc and Lazarus does not address the crucial conceptual difference between cognitivist and noncognitivist views of emotion. To advocate the cognitivist view of emotion is not primarily to assert that emotion is produced by brain systems categorized as cognitive. Rather, it is to assert that emotion is capable of informing the subject about the world. The cognitivists' main claim is—or ought to be—not that the brain structures responsible for cognition also contribute to producing emotion, but that emotion constitutes a cognitive system or has cognitive functions of its own. Thus, Zajonc's claim that emotional responses can be generated without the participation of brain structures labelled as cognitive does not license the conclusion that emotion has no cognitive aspects: it can also be taken as indicating that the category of cognitive functions should be broadened to include emotion. On the other hand, Lazarus's reply that emotion is intertwined with cognitive brain processes fails to establish the conclusion that emotion has cognitive aspects: such a conclusion can be reached only by examining the roles of emotion.

## 5. Emotion in Scientists' Decisions

The cognitivist approach to emotion has implications for all disciplines concerned with cognition, rationality, and decision making. If emotion is now to be seen as a function that has cognitive aspects and that plays an essential role in guiding rational reasoning and decision making, then an account of these phenomena in any domain that fails to take account of emotion will not be complete.

Researchers in several disciplines have begun to consider the significance of the cognitivist approach to emotion for their field of study. For instance, after reviewing the cognitive aspects of emotion for clinical researchers and medical ethicists, Louis C. Charland [8] argues that a person cannot fully grasp the consequences of a clinical procedure without experiencing emotion: he concludes that emotional factors should be given more weight in determining whether a person is competent to consent to participate in a clinical trial. Similarly, Jon Elster [16] introduces economists to the cognitivist approach to emotion, exploring the ways in which the economic behaviour of agents is influenced by their emotions and discussing how taking account of emotion could improve economic modelling.

Since cognition, rationality, and decision making are also central concerns of philosophy of science, this discipline too must inquire what implications the cognitivist approach to emotion has for its theorizing. The cognitivist approach to emotion provides strong reasons for thinking that scientific practice depends crucially on emotion. To appreciate this, it is sufficient to consider some features of the decisions made by scientists.

Scientists make many purely theoretical decisions, such as determining whether a certain conclusion follows logically from given premises or whether a particular mathematical calculation has been performed correctly. Scientists' most interesting and important decisions, however, involve practical reasoning. The most obvious examples are decisions how to explain a finding, how to establish or undermine a claim, how to construct a theory, how to solve a problem, how to design and perform an experiment, and how to develop a research programme. The outcomes of such decisions consist of choices of course of action, plans, and strategies. Many further decisions of scientists, despite superficially appearing theoretical, have practical implications. Deciding whether to trust an empirical finding and whether to accept a theory may appear to involve merely judging whether to lend assent to a certain proposition: in reality, however, these are also practical decisions, determining one's attitudes, commitments, and actions in scientific work.

Furthermore, scientists are often required to make such decisions in complex and confused circumstances, in which the relevant goals and information are not complete, consistent, and reviewable. The goals and priorities of scientists frequently conflict with one another: scientists must often trade accuracy for simplicity in theories, for example, or precision for feasibility in experiments. Moreover, scientists must base many decisions on inadequate information. Empirical data are often incomplete, inconsistent, ambiguous, not fully reliable or too abundant to review in full. Some scientific theories and models are likewise logically incomplete, inconsistent, lend themselves to different interpretations or have implications that cannot be surveyed [43].

According to the cognitivist view of emotion, decision making in practical contexts depends crucially on emotion, especially where goals and information are not complete, consistent, and reviewable. On these grounds, we should expect emotion to play an important role in determining the principal decisions made by scientists.

This hypothesis is supported by evidence about scientists' decision making. Scientists often make important decisions not by systematically taking all relevant factors into account, but by selectively focusing on a specific aspect of the situation. A single consideration is often sufficient to convince a scientist that a certain course of action is justified, for example. In other words, scientists appear to have a mechanism that tells them that a certain aspect of a situation is crucial for a certain decision, and that other aspects—which are a priori equally relevant—can be neglected. Such a mechanism would be lacking in a rational calculator, who would insist on reviewing all relevant factors, but is provided by a capacity to experience emotion. The emotion system attributes salience to some aspects of a situation, leading the subject to neglect other aspects. This is how scientists are able to make decisions even in circumstances in which their goals and information are not complete, consistent, and reviewable, and in which therefore it is impossible to weigh all relevant factors.

Notwithstanding their preferred self-image as objective discoverers of facts, scientists sometimes show awareness that they make decisions partly with the aid of emotion. Some scientists acknowledge having based decisions on hunches, gut feeling or intuition ([51]; [7]; [27]; [57], pp. 25–74). Decisions described as based on these factors seem to have the following characteristics. First, a particular consideration determines the decision in the scientist's mind in a complete and global manner. Second, the force of this consideration becomes apparent to the scientist suddenly or in a flash. Third, the grounds that the scientist is able to offer for the decision do not meet the normal standards of explicit and rigorous justification. One would expect to see these characteristics in decisions based partly on emotion.

Examples are provided by scientists' decisions whether to trust empirical findings and whether to accept theories. In these decisions, scientists use emotion to circumvent the methodological problem of underdetermination. The standard view of science calls for theories to be accepted and rejected on the basis of empirical evidence. However, this view suffers from various well-known problems. First, empirical findings may be incorrect. Second, empirical findings may be susceptible of differing interpretations. Third, distinct empirical findings may conflict with one another. Fourth, empirical findings constitute a test of a theory only under particular circumstances and on a limited part of its predictive range: a complete empirical test of a theory is never available. Fifth, the credibility and significance of empirical findings sometimes depend on the theory under test. Sixth, a theory's agreement or disagreement with empirical findings may be the consequence in part of auxiliary or background hypotheses. Because of these problems, a set of empirical findings and a theory fall short of logically determining a scientist's decisions. A decision whether to trust an empirical finding or whether to accept a theory must be based on something more. It is plausible to identify this extra factor as an emotional response to the empirical finding or the theory.

An examination of scientists' decisions suggests that their emotional responses to empirical findings and to theories include surprise, pleasure, certainty, understanding, puzzlement, doubt, and distaste. These constitute a specific class of cognitive emotions, prominent in scientific practice. A scientist may experience the emotion of pleasure at perceiving an empirical finding or a theory, for example, and the emotion of certainty that it is correct. An empirical finding or theory may evoke the emotion of understanding or of puzzlement in the scientist. Alternatively, the scientist may experience doubt or distaste in considering an empirical finding or theory.

A scientist experiences such emotions not normally on the basis of a systematic review of all aspects of an empirical finding or theory. Instead, these emotions are usually evoked by a small number of individual aspects, to which the scientist's emotion system attributes salience. The aspects of empirical findings that are likely to evoke emotional responses include their pattern and structure, their apparent reasonableness, their degree of consistency with relevant theories, and the method by which the findings were made. The aspects of theories that are likely to evoke emotional responses include individual empirical attainments, such as predictive success in particular experiments. Further aspects are more loosely related to empirical performance, such as forms of simplicity, symmetries, consistency with models, explanatory power, and metaphysical and aesthetic properties.

Faced with the need to decide whether to trust an empirical finding or whether to accept a theory, a rational calculator would be incapable of making a decision. Such an agent would need to review all aspects of the situation, or else to be given sufficient reason to regard some aspects as irrelevant. By contrast, a real scientist is able to experience an emotion in response to one or more aspects of the empirical finding or theory. This emotional response leads the scientist to decide whether to trust the empirical finding or whether to accept the theory. Only by employing emotion are scientists able to make such decisions.

The role of scientists' emotional responses to empirical findings and theories is evident, for instance, in the development of quantum theory. Submicroscopic physics between 1925 and 1927 offers a good example of a situation in which scientists' goals and information are not complete and consistent. Physicists' goals in this domain repeatedly conflicted with one another: the priorities of empirical adequacy, consistency, simplicity, objectivity, intelligibility, and mathematical tractability often pulled in different directions. For example, empirical adequacy and simplicity seemed to require an abstract approach that limited itself to relating the magnitudes of observable quantities, like that which yielded matrix mechanics, whereas intelligibility seemed to require an approach that provided visualizations of quantum phenomena, like that of wave mechanics. Empirical data about submicroscopic phenomena, furthermore, were inconsistent and ambiguous: they gave no univocal message about the nature and properties of the quantum world. Some experiments suggested that the submicroscopic world had a particle structure, for example, while others suggested that it had a wave structure. A rational calculator would have ground to a halt under these circumstances. Physicists such as Niels Bohr, Werner Heisenberg, and Erwin Schrödinger were compelled to base their decisions to a large extent on their emotional responses to empirical findings and theories. It is no exaggeration to say that the path taken by quantum physics in the period 1925–1927 was determined substantially by the emotions of physicists ([3]; [4], pp. 30–39).

## 6. The Warrant of Emotion

The evidence reviewed above establishes, I contend, that scientists rely partly on emotion in making decisions of certain kinds, and furthermore that it would not be possible for scientists to make these decisions without relying partly on emotion. However, the evidence so far does not establish whether and to what extent reliance on emotion in scientific practice is warranted. This issue must now be addressed.

Use of a certain procedure to reach a conclusion is warranted if the fact that a conclusion was reached by this procedure constitutes, under certain conditions, grounds for believing the conclusion to be justified. For example, use of induction is warranted if the fact that an inference was reached by induction constitutes, under certain conditions, grounds for believing the inference to be justified.

Does the fact that a scientist makes a decision by relying partly on emotion constitute, under certain conditions, grounds for believing the decision to be justified? This question is a final test of the noncognitivist view of emotion. If the answer is negative, the noncognitivist view survives: emotion can then be said to be cognitive in the sense that it is an ineliminable element of cognition, perhaps, but not in the sense that it is a cognitive resource. If the answer is positive, the noncognitivist view is definitively rebutted: on the noncognitivist view, any decision made partly on the basis of emotion should be mistrusted for that reason.

To be able to answer the question in the affirmative, we must uncover some feature of emotion that raises the chance that decisions made by relying partly on emotion are justified. Let us return to the example of scientists' decisions whether to trust empirical findings and whether to accept theories. To establish that reliance on emotion in making decisions of this kind is warranted, we must show that scientists' emotional responses to empirical findings and theories are reliable detectors of desirable cognitive properties of empirical findings and theories, such as truth, validity, empirical adequacy, trustworthiness, and the like. If, on the contrary, scientists' emotional responses to empirical findings and theories are not correlated with any desirable cognitive properties, then reliance on emotion in these decisions is not warranted.

The question whether scientists' emotional responses are reliable detectors of desirable cognitive properties is partly empirical, and should be addressed in part on the basis of empirical research in psychology and history of science. Philosophical work can also shed light on the question, however, by hypothesizing mechanisms that would ensure that emotional responses detect desirable cognitive properties.

Such mechanisms establish under what conditions scientists' emotional responses would be trustworthy. The case is complete once it is ascertained that some such mechanism actually governs scientists' emotional responses.

One mechanism has been hypothesized by Paul Thagard ([58], pp. 165–221; [59]). Thagard's starting point is the general principle that coherence is the ultimate criterion of justification of claims and inferences. Any justification of a claim or inference consists at root in showing that its acceptance maximizes the coherence of some conceptual system. Thagard notes further that coherence in various domains evokes positive emotional responses, such as satisfaction or a sense of aesthetic pleasure. He links these two insights by hypothesizing that scientists' emotional responses are sensitive to conceptual coherence. Scientists can thus rely on their emotional responses to detect coherence and incoherence in empirical findings and theories. Emotion, moreover, represents an augmentation of a scientist's cognitive faculties, as it can detect incoherence of which the scientist is not consciously aware. On this analysis, scientists are warranted to trust their emotional responses to empirical findings and theories, as these are tuned to the central cognitive desideratum of coherence. The fact that a scientist's decision to trust an empirical finding or accept a theory is made partly on the basis of emotion constitutes grounds for believing that the decision is justified, as a positive emotional response is evidence that the finding or theory exhibits coherence.

Thagard's approach assumes that the emotion system is permanently tuned to coherence and that it therefore responds identically to the same stimuli in all circumstances and at all times in the history of science. I feel that it thereby underestimates the diversity of scientists' emotional responses, especially in the historical dimension. For this reason, I incline to a mechanism that allows scientists' emotional responses to change in time.

Exposure to situations of various kinds can induce a particular pattern of emotional responses in persons. This process may be wholly unconscious or partly conscious. For example, if a person discovers that situations showing certain features constitute a threat, that person will tend to develop an emotional response of fear to subsequent situations showing the same features. Similarly, a person may learn to overcome feelings of jealousy in situations of a certain kind, having ascertained that jealous feelings in situations of this kind in the past were unwarranted [26].

I conjecture that a similar phenomenon occurs in science. Scientists are taught to pursue and value empirical success in its various forms: they value empirical findings that are correct and theories that are empirically adequate. When an empirical finding is confirmed correct or a theory is confirmed empirically adequate, scientists attach positive affect to the notable features of those constructs. This means that scientists are primed to experience a positive emotional response to future empirical findings or future theories that show the same features. For example, if a body of experimental data that shows a certain pattern or structure is confirmed correct, scientists attach positive affect to those features: they thereby become disposed to experience a positive emotional response to subsequent bodies of data that show the same pattern or structure. Similarly, if a theory that exhibits a certain symmetry is confirmed empirically adequate, scientists attach positive affect to that feature and become disposed to experience a positive emotional reaction to further theories that exhibit the same symmetry. Scientists' emotional responses are thus biased in favour of empirical findings and theories that resemble those confirmed correct in the past.

There is abundant evidence that scientists' aesthetic responses to theories, which are a species of emotional response, are shaped by this inductive mechanism, which in the aesthetic domain I have named the "aesthetic induction" [42].

This mechanism not only accounts for the development of emotional responses in scientists, but also provides a justification of them. If empirical findings and theories that were confirmed correct in the past showed certain features, then it is rational to prefer further findings and theories that show the same features, on the chance that these features are in some way conducive to or correlated with correctness. For example, if previous experimental data in some domain that have been confirmed correct showed a certain pattern or structure, it is rational to lend credence to further data that show the same pattern or structure. If previous theories in a domain that were confirmed empirically adequate showed a certain symmetry, it is rational to require new theories to share that symmetry. There is prima facie evidence that such features reflect the structure of the world. Emotional responses that lead scientists to value features of empirical findings and theories that were confirmed correct in the past thus reinforce rational behaviour.

Even rational expectations may be disappointed, of course: the best new empirical findings and theories may show different features from those confirmed correct in the past. The inductive mechanism means that scientists initially experience a negative emotional response to such empirical findings or theories. However, the mechanism ensures also that their emotional responses, after some delay, retune to the new features.

On my analysis, in sum, scientists are warranted to trust their emotional responses to empirical findings and theories, because these are tuned to the features found in empirical findings and theories that were confirmed correct in the past. The fact that a scientist's decision to trust an empirical finding or accept a theory is made partly on the basis of emotion constitutes grounds for believing the decision to be justified, since a positive emotional response is evidence that the finding or theory shows such features.

## Acknowledgements

This paper is a revised and expanded version of my invited lecture at the 12th International Congress of Logic, Methodology and Philosophy of Science, Oviedo, August 2003, in Section D3, "Philosophical Questions Raised by the History and Sociology of Science". I am indebted to the Section Programme Committee, consisting of James R. Brown, Catherine Chevalley, and Oswaldo Pessoa, and the General Programme Committee, chaired by Petr Hájek, for the invitation. I thank the audience for helpful comments. I am grateful to Luis M. Valdés for his careful reading of the revised text and to Daniela Bailer-Jones for valuable suggestions and encouragement.

## REFERENCES

1. L. M. Antony and C. Witt, eds. *A Mind of One's Own: Feminist Essays on Reason and Objectivity*. Boulder, Col.: Westview Press, 1993.
2. B. Barber. *Science and the Social Order*. Glencoe, Ill.: Free Press, 1952.
3. M. Beller. The Conceptual and the Anecdotal History of Quantum Mechanics, *Foundations of Physics*, 26:545–557, 1996.
4. M. Beller. *Quantum Dialogue: The Making of a Revolution*. Chicago, Ill.: University of Chicago Press, 1999.
5. L. R. Brody and J. A. Hall. Gender and Emotion. In: M. Lewis and J. M. Haviland, eds., *Handbook of Emotions*. New York: Guilford Press; 447–460, 1993.
6. J. R. Brown. *The Rational and the Social*. London: Routledge, 1989.
7. M. Bunge. *Intuition and Science*. Englewood Cliffs, N.J.: Prentice-Hall, 1962.
8. L. C. Charland. Is Mr. Spock Mentally Competent? Competence to Consent and Emotion, *Philosophy, Psychiatry, and Psychology*, 5:67–81, 1998.
9. I. B. Cohen, ed. *Puritanism and the Rise of Modern Science: The Merton Thesis*. New Brunswick, N.J.: Rutgers University Press, 1990.
10. A. R. Damasio. *Descartes' Error: Emotion, Reason, and the Human Brain*. New York: G. P. Putnam, 1994.
11. A. R. Damasio. *The Feeling of What Happens: Body and Emotion in the Making of Consciousness*. New York: Harcourt Brace, 1999.
12. L. Darden. Recent Work in Computational Scientific Discovery. In: M. Shafto and P. Langley, eds., *Proceedings of the Nineteenth Annual Conference of the Cognitive Science Society*. Mahway, N.J.: Lawrence Erlbaum; 161–166, 1997.
13. L. Daston. Fear and Loathing of the Imagination in Science, *Daedalus*, 127(1):73–85, 1998.
14. L. Daston. The Moralized Objectivities of Science. In: W. Carl and L. Daston, eds., *Wahrheit und Geschichte: Ein Kolloquium zu Ehren des 60. Geburtstages von Lorenz Krüger*. Göttingen: Vandenhoeck und Ruprecht; 78–100, 1999.
15. P. Ekman. *Emotions Revealed: Recognizing Faces and Feelings to Improve Communication and Emotional Life*. New York: Times Books, 2003.
16. J. Elster. Emotions and Economic Theory, *Journal of Economic Literature*, 36:47–74, 1998.
17. E. Fee. Women's Nature and Scientific Objectivity. In: M. Lowe and R. Hubbard, eds., *Woman's Nature: Rationalizations of Inequality*. New York: Pergamon Press; 9–27, 1983.
18. G. J. Feist and M. E. Gorman. The Psychology of Science: Review and Integration of a Nascent Discipline, *Review of General Psychology*, 2:3–47, 1998.
19. L. Feldman Barrett, L. Robin, P. R. Pietromonaco and K. M. Eyssell. Are Women the "More Emotional" Sex? Evidence from Emotional Experiences in Social Context, *Cognition and Emotion*, 12:555–578, 1998.
20. L. S. Feuer. *The Scientific Intellectual: The Psychological and Sociological Origins of Modern Science*. New York: Basic Books, 1963.

21. L. S. Feuer. Science and the Ethic of Protestant Ascetism: A Reply to Professor Robert K. Merton. In: R. A. Jones and H. Kuklick, eds., *Research in Sociology of Knowledge, Sciences and Art: A Research Annual.* Volume 2. Greenwich, Conn.: JAI Press; 1–23, 1979.

22. A. Fischer, ed. *Emotion and Gender: Social Psychological Perspectives.* Cambridge: Cambridge University Press, 2000.

23. N. H. Frijda. *The Emotions.* Cambridge: Cambridge University Press, 1986.

24. B. Gholson, W. R. Shadish, Jr., R. A. Neimeyer and A. C. Houts, eds. *Psychology of Science: Contributions to Metascience.* Cambridge: Cambridge University Press, 1989.

25. P. E. Griffiths. *What Emotions Really Are: The Problem of Psychological Categories.* Chicago, Ill.: University of Chicago Press, 1997.

26. J. J. Gross. The Emerging Field of Emotion Regulation: An Integrative Review, *Review of General Psychology*, 2:271–299, 1998.

27. H. E. Gruber. On the Relation between "Aha Experiences" and the Construction of Ideas, *History of Science*, 19:41–59, 1981.

28. A. L. Hinton, ed. *Biocultural Approaches to the Emotions.* Cambridge: Cambridge University Press, 1999.

29. A. O. Hirschman. *The Passions and the Interests: Political Arguments for Capitalism before Its Triumph.* Princeton, N.J.: Princeton University Press, 1977.

30. P. Hoffman. *The Man Who Loved Only Numbers: The Story of Paul Erdös and the Search for Mathematical Truth.* New York: Hyperion, 1998.

31. A. C. Houts. Contributions of the Psychology of Science to Metascience: A Call for Explorers. In: [24], 47–88, 1989.

32. S. James. *Passion and Action: The Emotions in Seventeenth-Century Philosophy.* Oxford: Clarendon Press, 1997.

33. H. Kragh. *Dirac: A Scientific Biography.* Cambridge: Cambridge University Press, 1990.

34. T. S. Kuhn. Reflections on My Critics. In: I. Lakatos and A. Musgrave, eds., *Criticism and the Growth of Knowledge.* Cambridge: Cambridge University Press; 231–278, 1970.

35. L. Laudan. *Progress and Its Problems: Towards a Theory of Scientific Growth.* Berkeley: University of California Press, 1977.

36. R. S. Lazarus. On the Primacy of Cognition, *American Psychologist*, 39:124–129, 1984.

37. R. S. Lazarus. *Emotion and Adaptation.* New York: Oxford University Press, 1991.

38. J. LeDoux. *The Emotional Brain: The Mysterious Underpinnings of Emotional Life.* New York: Simon and Schuster, 1996.

39. A. Lefford. The Influence of Emotional Subject Matter on Logical Reasoning, *Journal of General Psychology*, 34:127–151, 1946.

40. G. Lloyd. *The Man of Reason: "Male" and "Female" in Western Philosophy.* 2nd ed. Minneapolis: University of Minnesota Press, 1993.

41. M. J. Mahoney. *Scientist as Subject: The Psychological Imperative.* Cambridge, Mass.: Ballinger, 1976.

42. J. W. McAllister. *Beauty and Revolution in Science*. Ithaca, N.Y.: Cornell University Press, 1996.

43. J. Meheus, ed. *Inconsistency in Science*. Dordrecht: Kluwer, 2002.

44. R. K. Merton. The Normative Structure of Science. Reprinted in [47], 267–278, 1942.

45. R. K. Merton. Priorities in Scientific Discovery. Reprinted in [47], 286–324, 1957.

46. R. K. Merton. *Science, Technology, and Society in Seventeenth-Century England*. New York: Harper, 1970.

47. R. K. Merton. *The Sociology of Science: Theoretical and Empirical Investigations*. Edited by N. W. Storer. Chicago, Ill.: University of Chicago Press, 1973.

48. I. I. Mitroff. *The Subjective Side of Science: A Philosophical Inquiry into the Psychology of the Apollo Moon Scientists*. Amsterdam: Elsevier, 1974.

49. T. Nickles, ed. *Scientific Discovery, Logic, and Rationality*. Dordrecht: Reidel, 1980.

50. A. Ortony, G. L. Clore and A. Collins. *The Cognitive Structure of Emotions*. Cambridge: Cambridge University Press, 1988.

51. W. Platt and R. A. Baker. The Relation of the Scientific "Hunch" to Research, *Journal of Chemical Education*, 8:1969–2002, 1931.

52. M. Polanyi. *Personal Knowledge: Towards a Post-Critical Philosophy*. Chicago, Ill.: University of Chicago Press, 1958.

53. R. N. Proctor. *Value-Free Science? Purity and Power in Modern Knowledge*. Cambridge, Mass.: Harvard University Press, 1991.

54. M. W. Rossiter. *Women Scientists in America: Struggles and Strategies to 1940*. Baltimore, Md.: Johns Hopkins University Press, 1982.

55. I. Scheffler. In Praise of the Cognitive Emotions. Reprinted in I. Scheffler, *Science and Subjectivity*. 2nd ed. Indianapolis, Ind.: Hackett, 1982; 139–157, 1977.

56. S. Shapin and S. Schaffer. *Leviathan and the Air-Pump: Hobbes, Boyle, and the Experimental Life*. Princeton, N.J.: Princeton University Press, 1985.

57. D. K. Simonton. *Origins of Genius: Darwinian Perspectives on Creativity*. New York: Oxford University Press, 1999.

58. P. Thagard. *Coherence in Thought and Action*. Cambridge, Mass.: MIT Press, 2000.

59. P. Thagard. The Passionate Scientist: Emotion in Scientific Cognition. In: P. Carruthers, S. Stich, and M. Siegal, eds., *The Cognitive Basis of Science*. Cambridge: Cambridge University Press; 235–250, 2002.

60. M. Weber. *The Protestant Ethic and the Spirit of Capitalism*. London: Unwin, 1930.

61. R. B. Zajonc. Feeling and Thinking: Preferences Need No Inferences, *American Psychologist*, 31:151–175, 1980.

62. R. B. Zajonc. On the Primacy of Affect, *American Psychologist*, 39:117–123, 1984.

# Vulnerability – A Principle of the Ethics of Science & Technology

Ren-Zong Qiu

*Peking University Health Science Center*
*rzq@chinaphs.org*

**Abstract.** Protecting the vulnerable is a feature of the Confucian ideal society labelled as Great Harmony. The concept of vulnerability may come from the nature of medicine as a special kind of human activity. Observing the vulnerability principle is one of the axioms necessary to attain the goal of the medical encounter. The argument for the principle of vulnerability can be constructed on naturalist, contractualist, solidary, teleological or Confucian approaches. Everybody becomes vulnerable when facing an epidemic of HIV/AIDS and SARS. Human beings and their future generations are also vulnerable in the face of the advances of science and technology in the sense that all of their parts including genes and brain/psyche can be manipulated. In other words, an increasing number of human beings can be human subjects in experimentation. The same applies to a variety of existing animal and plant species as well as the ecological environment: all of them became increasingly vulnerable. Vulnerability entails our special obligation to protect them and ourselves. An appropriate strategy is therefore needed for the application of a technology which will lead to a broad and profound change of our nature as well as our body: guilty until proven innocent.

## 1. Foreword

Protecting the vulnerable is a feature of the Confucian ideal society labelled as "Great Harmony" (*Da Tong*). Confucius said that "in a society called Great Harmony all old widowers, old widows, orphans, childless elderly and disabled are well care for. " (*Book of Rites*) In *Analects of Confucius* he told his disciples that "my aspiration is that the elderly be cared, friends be trusted and children be loved." (*Four Books and Five Classics*, vol. 1, p. 21) When one of his disciples asked him about filial piety, he answered that "a filial person only worries about the illness of his parents." (*Four Books and Five Classics*, vol. 1, p. 5). For Confucianism, the duty to the vulnerable is implied in its fundamental concept *ren* – love, care, respect and doing good to people. Does the Confucian idea regarding duty to the vulnerable have any implication for our modern society with sophisticated science and technology? E. Pellegrino and D. Thomasma raised in their book *Helping and Healing* the question, "Should a consistent ethic of life target only the vulnerability of human life, or should it include animal life as well, or also environmental ethics?" (Pellegrino and Thomasma 1997, p. 58) Moreover, the documents provided by the Conference on Future Food & Bioethics, held in Copenhagen on 23-25 October 2002, suggested that "Uses of genetic engineering must take into consideration man's right to self-determination and dignity, and to the integrity and vulnerability of man, animals and nature" (Danish Documents 2002: Government Statement on Ethics and Genetic Engineering) and,

If genetic engineering is to be accepted. it will have to be developed and used: *with respect for the vulnerability of life, provided that*

a this vulnerability is not just considered as a fact, but also as an appeal for care and consideration,

b impoverishment and impairment of nature is avoided (Danish Documents 2002: Proposed Ethical Guidelines for Genetic Engineering).

The issue raised is: Can the application of the principle of vulnerability to future generations, non-human animals and the eco-system be justified? This paper attempts to provide an answer.

## 2. Concept of Vulnerability and Its Normative Implications

The concept of vulnerability seems to be derived from medicine: that is to say, from the meaning of illness. Being ill not only limits the patient's freedom to act and make free choices; it forces her/him to trust in others. The resultant vulnerability adds to the patient's plight. Those with power to heal also have the power to harm or to exploit the patient's vulnerability. So, it can be said that disease creates the power imbalance between patient and physician, and makes the former vulnerable. Besides, the power imbalance is contributed to by an imbalance in knowledge, education, capacity, resources etc.

Who is vulnerable sometimes depends on the context. All members of a family that is marginalized in a given society are vulnerable in contrast with other members of the society who are in an advantageous status. However, in this family some members, say, women, may be more vulnerable because of gender prejudice.

A principle of vulnerability requires that having recognized the power imbalance and patient's vulnerability, health professionals ought to help and care for her/him, instead of exploiting the patient's disadvantaged status for their own interests. It is one of the axioms by which the goal of the medical encounter—a right and good action for a particular patient—can be attained. (Pellegrino and Thomasma, 1997, p. 54) Otherwise, the abuse of this power imbalance and patient's vulnerability would be fatal to medicine, to the physician-patient relationship, and also to human society at large.

## 3. Extending Vulnerability to Other Human Contexts

The first attempt to extend the concept of vulnerability to other human contexts was made in human research. CIOMS/WHO's International Ethical Guidelines for Biomedical Research Involving Human Subjects include a special guideline for vulnerable persons/people.

*Guideline 13: Research involving vulnerable persons/people*
Special justification is required for inviting vulnerable individuals to serve as re-

search subjects and, if they are selected, the means of protecting their rights and welfare must be strictly applied.

Who are vulnerable persons/people? In the Commentary on Guideline 13 vulnerable persons/people are defined as:

> Vulnerable persons/people are those who are relatively (or absolutely) incapable of protecting their own interests. More formally, they may have insufficient power, intelligence, education, resources, strength, or other attributes required to protect their own interests. (CIOMS 2002, pp. 64-66)

Vulnerability entails special safeguards. The central problem presented by plans involving vulnerable persons/people as research subjects is that such plans may entail an inequitable distribution of the burdens and benefits of research participation. So, special justification and safeguards are required to protec their rights, interests and welfare. However, in a sense, all human subjects are vulnerable in the research context as are all patients in the clinical context. Human subjects are in a disadvantaged status of power, knowledge, education, capacity, imbalance of resources as patients, let alone the fact that many human subjects are patients per se.

Can the concept of vulnerability be extended to all human relationships? Obviously, there exists a greater power imbalance between all other human relationships than between clinical and research ones. As Pellegrino and Thomasma argued, the principle of vulnerability can be stated in this way: in human relationships in general, if there exist inequalities of power, knowledge, or material means, the obligation is upon the stronger party to respect and protect the vulnerability of the other and not exploit the less advantaged party. (Pellegrino and Thomasma, 1997, p. 55 )

## 4. Arguments for Vulnerability as an Ethical Principle

As an ethical principle vulnerability entails the obligation of the stronger to respect and protect the vulnerable. So, the argumentation for vulnerability as an ethical principle amounts to the argumentation for obligation to the vulnerable.

There are several arguments for vulnerability that will be examined below: the naturalistic argument, contractualist argument, solidary argument, teleological argument, and Confucian argument.

### Naturalistic argument

. The proponents of naturalist argument claimed that the obligation to the vulnerable is grounded on the vulnerability *per se* of people themselves. "It is the vulnerability of the beneficiary rather than any voluntary commitment *per se* on the part of the benefactor which generates these special responsibilities." (Goodin 1985, pp. xi, 42-108)

> However, the vulnerability of the persons/people does not necessarily entail this obligation or responsibility. Some people may see in vulnerability a road to redemption, thinking that it would be counterproductive

to help them; some may see them as losers in a natural or social lottery, we have no obligation to them, if we do something to help them, it is superogatory, but not compulsory.

### Contractualist argument.

Contractualists argued that the basis for special responsibilities to protect the vulnerable from harm come from self-assumed duties and obligations often through implied or explicit contracts. For example, the obligation for a health professional to his or her own patient over other needy persons/people in society. (Veatch 1986)

> However, not all human relationships in which one party is vulnerable can be reduced to the contractual relationship, e.g. the relationship between mother and child is not a contractual one in which each party protects her/his own interest without harming the other. A typical mother is an altruist, even a non-human mother.

### Solidary argument.

It has been argued that the common capacity to suffer confers prima facie rights not to cause suffering; and the community has the obligation not only to refrain from causing suffering but also to ameliorate suffering, and prevent its occurrence among its members. Solidarity in the community will arise from individuals' perceptions that the community has contracted with them in order to address their suffering and try to prevent it. (Loewy 1985) This argument is plausible, but only applicable to members of a human community.

### Teleological argument.

An ethical theory is teleological if it says that one always ought to do the act that is better. Once we know what is good, or more exactly what is better than what, we shall know the right way to live and the right way to act. The notion of good can include many ethical values, e.g. utility, welfare, equality, fairness, and justice. All these goods can be put together to determine overall good. "Two alternatives are equally good if they are equally good for each person; and if one alternative is at least as good as another for everyone and definitely better for someone, it is better." (Bloome 1992, pp. 41-44 ; also see 1995, pp. 1-20). Therefore, the alternative of respecting and caring for the vulnerable is better than not respecting and caring for them, so the right way to act is the former.

### Confucian argument.

For Confucianism helping the vulnerable is part of person-making. We are born as human beings only biologically, but not morally. To be a moral human being we have to learn – a dynamic process of person-making. This person-making process is fundamentally an integrative one in which the self is transformed into the profoundly relational person. The feature of this relational person is the ability to show deference to others, including respect, help, care, doing good etc. *Ren* (kind-heartedness, benevolence, humanness) in the Confucian sense is to take the other into one's own sphere of concern, interest, value and viewpoint, and in the process, the other becomes an integral part of one's self. Care for the vulnerable is ethically required by *ren*.

Human beings are potentially vulnerable. Factors making human beings vulnerable include:

– Natural factors, such as HIV, SARS, earthquakes, floods, droughts etc.

– Social, economic and political factors, such as slavery, tyranny, war, drug trafficking, women and children trafficking, racism, class and gender discrimination, exploitation, oppression, poverty etc.

– Scientific/technological factors:

We are in a new era in which science/technology and society are integrated which each other more closely than before. The impact of the application of science/technology on human beings, other species, the eco-environment, society and the risks it causes are greater than ever before. Sometimes, the risks are so extraordinary and irreversible that they are called "mega-risks". In view of this, contemporary science and technology deserve a new name. In this paper, it is referred to as integrated science and technology (IST) which indicates the integration of science with technology. IST have made and will continue to make human beings and their stakeholders vulnerable in such situations as:

> Deterioration of the external environment (global warming, chemical and nuclear pollution);
>
> Possible threats to biosafety and biodiversity & to existing species (GMO);
>
> New pandemics (cross-species infection caused by xenotransplantation or modified viruses);
>
> Control of body, psyche, behavior by manipulating brain and genes (electronic/chemical stimulus of brain, brain chips, genetic engineering, human reproductive cloning, baby design, massive human research);
>
> Possible threats to future generations (germ-line gene therapy and enhancement).

## 5. Further Extending the Principle of Vulnerability

In what follows I will address the following questions:

Q1: Can the concept of vulnerability be extended to those entities beyond human beings of the present generation?

Q2: Which kinds of arguments can or cannot support the extending of vulnerability principle to those entities? (But I will not make an overall assessment of these arguments.)

## 6. Vulnerability of future generations

The environment and its resources available to each generation are inherited from the previous generation, and thus cannot be chosen. The following generation has to depend upon the previous generation to a varying degree. The next generation is relatively powerless and vulnerable. It may be refuted that the next generation would possess more advanced scientific knowledge and technological skills than the previous one, and so they will not be so vulnerable. However, the fact that the next generation will possess more advanced science and technology does not refute the vulnerability of this next generation regarding the environment and its resources. It is noticeable that the current possible negative impact of IST on the environment is greater than the impact of science and technology in previous generations; although the possible positive impact may also be greater than before. It is uncertain to what degree the possessing of more advanced science and technology could compensate for the possibly irreversible and destructive deterioration of the environment. As Broome argued, although the effects of our action on future generations are highly unpredictable and uncertain, however as far as something like human-induced global warming is concerned, "we can say that the effects will certainly be long-lived, almost certainly large, probably bad, and possibly disastrous." (Broome 1992, p. 12)

### Arguments for duty to future generations.

Contractualism does not entail the obligation to future generations. For Rawls, the rules which guarantee justice are produced between people in an "original position" with a veil of ignorance". He also agreed with Hume in that justice only arises in particular circumstances where people are roughly equal in power but limited in their mutual generosity, and face conditions of moderate scarcity. These exclude the justice between different generations and the obligations to future generations. Gauthier suggested that people give up the veil of ignorance, and the principles of justice are to be derived from bargaining between people who are fully aware of their own situation, but in the bargaining process one person is not allowed to make another person worse off than she/he would have been in the first person's absence. Then, one generation is not allowed to use up any of the earth's exhaustible resources, unless it compensates later generations in some way. Otherwise, it makes later generations worse off than they would have been in its absence. However, the problem is that there is not much interaction between generations, how is it possible to bargain between different generations? (Gauthier 1985, pp. 204-205, 303-305)

### Right argument.

People in the future have the right to unpolluted air, so we have the obligation not to pollute the air with smoke. Moreover, for Nozick and other philosophers right or justice is absolute, not negotiable and must be satisfied (Nozick 1974, pp. 28-33). Therefore, the appeal to rights is supposed to give us a reason to control pollution. However, this appeal will be undermined by Parfit's nonidentity problem (Parfit 1984, pp. 364-366). It is difficult to argue in favour of our obligations to future people on the basis of their rights. The reason is that if we don't control contaminating emissions, some people must be suffering. If we did control them, there would be no such people and they would have no right to request us to control such contamination. Unfortunately, some philosophers argued from this viewpoint that we have no obligations to do good to future generations (Schwartz 1978). One way to get rid of the nonidentity problem has to be recognized in this way: "the owners of rights are not necessarily individual people: nations have rights" (Bloome

1992, pp. 34-35). or the future generations' problem is a problem of humankind-the future of humankind (Qiu 1998).

### Teleological argument.

Broome argued that the future generations issue can be better handled with the teleological argument in which the priority would be placed on good. And good is agent-relative, but we can commit to generation-relative good. "Equality between generations can be taken as a sort of good to be included in the objective to be maximized" (Broome 1992, p. 43). How can people be motivated to act so as to reduce our emissions of carbon dioxide in such a way that it benefits future generations but brings loss to us? Because people recognize that this action is good. This good may or may not be good for us, but it is so for future generations, for humankind.

### Confucian argument.

For Confucianism the approach to actualizing *ren* is from near to far: from family to strangers, from this generation the next generation. In this process, a person is being integrated with humankind as a whole. A person with *ren* will do something, such as the "previous generation plants the trees, future generations rest in their shadow", which will benefit future generations even though the present generation may not rest in their shadow.

## 7. Vulnerability of Non-Human Animals

It is obvious that non-human animals are vulnerable and become more vulnerable with the advances of science and technology, such as massive animal experimentation, and recently fussed over xenotransplantation. Betrand Russell said:

> There is no impersonal reason for regarding the interests of human beings as more important than those of animals. We can destroy animals more easily than they can destroy us; that is the only solid base of our claim to superiority. We value art and science and literature. But whales might value spouting, and donkeys might maintain that a good bray is more exquisite than the music of Bach. We cannot prove them wrong except by the exercise of arbitrary power (Russell 1932).

There are by no means few arguments for human obligations to non-human animals, such as utilitarian, deontological and Confucian arguments. Utilitarians, such as Bentham argued that "Nature has placed mankind under the governance of two sovereign masters, pain and pleasure. It is for them alone to point out what we ought to do, as well as to determine what we shall do" The question to ask about animals "...is not Can they *reason*? Nor Can they *talk*? But, Can they *suffer*?" (cited from Wacks). A similar argument was made in (Singer 1975). The above entails humans having obligations to reduce animals' pain and increase their pleasure. Each of them is a sort of good to be included in the objective to be maximized. There exist a scientific problem: which animals can suffer? As a version of utilitarianism, the teleological argument is applicable if an act should be concerned with the good for animals. If so, the problem of which animals can suffer can be avoided.

### Deontological argument.

Non-human animals, like human beings, are "subjects-of-a-life". They have an inherent, not merely instrumental, value or worth. This entitles them to the absolute right to live their lives with respect and autonomy. "All those beings (and only those beings) which have inherent value have rights" (Regan 1984, p. 397). It will be from value that we will derive duty. (Rolston 1988, p. 2) Others have argued that non-human animals have interests and needs. In particular, they have a clear interest in avoiding pain and an untimely death. (Wacks 1996, p. 44)

### Confucian argument.

For Confucianism, the approach to actualizing *ren* is from near to far: including from human beings to non-human animals. Everybody has a heart which cannot bear another's suffering. When Mencius met King Xuan of Qi Kingdom, the King told him that during the sacrifice ritual he had seen the ox suffering so much, and thought it innocent, that he replaced it with a sheep. Mencius answered that the sheep was also suffering and was also innocent; however, there is *ren* in your heart. Mencius said: "Doing no harm is the art of *ren*. A person with *ren* has a heart which cannot bear to see an animal's death when he sees it alive, and to eat its meat when he hears its voice." (*Mencius*)

For Confucianism, human beings in nature are like an eldest son in a family who has obligations to protect and care for his younger brothers/sisters. In the process of fulfilling responsibilities to animals, the person with *ren* is being integrated with nature.

## 8. Vulnerability of the Eco-System

The eco-system has already been made highly fragile and vulnerable, and IST is currently making it and will make it more vulnerable in the future. Industrial and rapidly developing societies around the world are faced by massive pollution legacies from their emphasis on chemical and nuclear industrial activities The environmental problems posed by GMO are likely to be substantially more intractable than those posed by these earlier instances of pollution because genetic wastes multiply, migrate and mutate. A genetically engineered organism once free in the environment is impossible to recover. The possible negative impacts are irreversible.

Obligations to the eco-system. The eco-system is our Mother-Nature, different from human the mother-child relationship in which after delivery the child is separated from her/his mother. However, humankind grows up within its Mother-Nature, never separate from her. The deterioration of the eco-system threats the present generation as well as future generations, human beings as well as non-human animals and other species. For utilitarians, protecting the eco-system can be taken as a sort of good to be included in the objective to be maximized. An act which reduces the vulnerability of the eco-system is better than another act which makes it more vulnerable. For Confucianism, by analogy with the human mother-child relationship, a filial child has obligations to her/his mother, and so we, human beings, have obligations to Mother-Nature. *Ren* should be extended to the eco-system. In the process of fulfilling responsibilities to eco-systems the person with *ren* is being further integrated with nature. Then, the person reaches an ideal spiritual sphere – *Tian Ren He Yi* (Unity of human being with nature)

So, our conclusion is: extending of the principle of vulnerability to future generations, non-human animals and eco-system is ethically justifiable.

## 9. Policy Implications of the Extending of the Vulnerability Principle

In Northwest China, deforestation is very serious because the mountains have been over-reclaimed. Now, there is a project to recover the forests from grain fields. In these project, farmers are required to change their job to planting trees in the mountains instead of reclaiming grain fields with governmental grain and financial support. Some farmers are better-off, some worse-off, the deforestation is being slowly but gradually reversed. According to the principle of vulnerability, this project can be ethically justified.

The principle of vulnerability requires a policy which should be taken in decision-making in a project with the application of IST that may have wide-ranging and long-lived consequences for human beings and nature: guilty until proven innocent. This means that it is the responsibility of researchers, sponsors and companies to provide the evidence to confirm that the project will not make human beings, future generations, non-animals and the eco-system more vulnerable. In contrast with this, for the conventional technologies the policy should be "innocent until proven guilty".

In spite of *prima facie* evidence to confirm that the project will not make human beings, future generations, non-animals and the eco-system more vulnerable, once the project has been implemented, strict surveillance and close monitoring should be enforced. The project should be responsive to constant checking and ready to be changed, including the possible moratorium or even the complete end of the project in the case of a seriously adverse event.

## REFERENCES

1. *The Analects of Confucius*, in Zhu Xi (ed.): 1985, *Four Books and Five Classics*, volume 1, Beijing: Chinese Bookstore.
2. Bereano, P.: 1996, Some Environmental and Ethical Considerations of Genetically Engineered Plants and Foods, in Becker, G. (ed.): *Changing Nature's Course: The Ethical Challenge of Biotechnology*, Hong Kong: Hong Kong University Press, pp. 27-36.
3. *Book of Rites*, in Zhu Xi (ed.): 1985, *Four Books and Five Classics*, volume 2, Beijing: Chinese Bookstore, p. 120.
4. Broome, J.: 1992, *Counting the Cost of Global Warming, Cambridge*: The White Horse Press.
5. Broome, J.: 1995, *Weighing Goods,* Oxford: Blackwell. CIOMS: 2002, International Ethical Guidelines for Biomedical Research Involving Human Subjects, Geneva.
6. Danish Documents: Government Statement on Ethics and Genetic Engineering

& Proposed Ethical Guidelines for Genetic Engineering, Conference on Future Food and Bioethics, 23-25 October 2002, Copenhagen, Denmark.

7. Gauthier, D.: 1985, *Morals and Arguments*, Oxford University Press.
8. Goodin, R.E.: 1985, Protecting the Vulnerable, Chicago: University of Chicago Press.
9. Loewy, E.: 1991, *Suffering and the Beneficent Community*, Albany: State University of New York Press.
10. *Mencius*, in Zhu Xi (ed.): 1985, *Four Books and Five Classics*, volume 1, Beijing: Chinese Bookstore, p. 5.
11. Nozick, R.: 1974, *Anarchy, State and Utopia*, Basic Books.
12. Parfit, D.: 1984, *Reasons and Persons/people*, Oxford University Press.
13. Pellegrino, E. and Thomasma D.: 1997, *Helping and Healing*, Washington, DC: Georgetown University Press.
14. Qiu, R-Z: 1994, Conceptual Issues in Ethics of Science and Technology, in Prawitz, D., Skyms, B. and Westerstahl, D. *et al.* (eds.): *Logic, Methodology and Philosophy of Science IX*, Elsevier Science B.V., pp. 537-551.
15. Qiu, R-Z: 1998, Germ-Line Engineering as the Eugenics of the Future, in Agius, E. and Busuttil, S. (eds.): *Germ-Line Intervention and our Responsibilities to Future Generations*, Kluwer, pp. 105-116.
16. Regan, T.: 1984, *The Case for Animal Rights*, London: Routledge.
17. Rolston, H.: 1988, *Environmental Ethics: Duties to and Values in the Natural World*, Philadelphia: Temple University Press.
18. Russell, B.: 1932, If animals could talk, in Ruja, H. (ed.): 1975, Mortals and Others: Bertrand Russell's American Essays 1931-1935, London: Allen & Unwin, vol.1, pp. 120-121.
19. Schwartz, T.: 1978, Obligations to posterity, in Sikora R. and Barry, B. (eds.): Obligations to Future Generations, Temple University Press, pp. 3-13.
20. Singer. P.: 1975, *Animal Liberation*, New York: Avon.
21. Veatch, R.: 1986, *The Foundation of Justice,* New York: The Oxford University Press.
22. Wacks, R.: 1996, Sacrificed for Science: Are Animal Experiment Morally Defensible? in Becker, G. (ed.): *Changing Nature's Course: The Ethical Challenge of Biotechnology*, Hong Kong: Hong Kong University Press, pp. 27-36.
23. Zhu Xi (ed.): 1985, *Four Books and Five Classics*, Beijing: Chinese Bookstore.

SYMPOSIUM 1:

# SCIENTIFIC (EVIDENCE-BASED) MEDICINE, 19th–20th CENTURIES

# Evidence-Based Medicine and the Classification of Chronic Lymphocytic Leukemias

Claude Debru

*Département de philosophie, Ecole normale supérieure*
*45 rue d'Ulm, F - 75005 Paris*
*claude.debru@ens.fr*

The establishment and revision of the classification and subclassification of leukemias is a good example of evidence-based medicine because it involves the study of large groups of patients by cooperative groups of physicians working together to compare their data and interpretations. This collaboration is not entirely new in the field of leukemia classification. At the end of the nineteenth century and at the beginning of the twentieth, hematologists did compare their clinical data and sometimes disagreed on their diagnostics, which were based on the existing classifications (see [2], chapter III). Leukemia classification is a natural consequence of the very definition of leukemias as disorders of the differentiation and proliferation processes of blood cells, a definition which was already reached by Rudolf Virchow in 1849. Indeed, the different types of mature blood cells, which are recognized by their different morphologies and are endowed with different functions, like myelocytes, lymphocytes, or red cells, derive by a differentiation process from a common stem cell, which is the hematopoietic stem cell located in the bone marrow. Since there are different types of differentiating and mature blood cells, there are different types of pathologies which are occurring in the differentiation process, different classes of leukemias which have been distinguished during the history of hematology according to the technical means which were available at different periods, be it cytomorphological, cytochemical, cytogenetical, or immunological. As such, the pathology does not or not necessarily mirror normal cell types described in their classification. During the course of the disease, the progression phenomenon may occur, which means that new cell types are affected by the leukemic process. For instance, lymphoid cells may be affected by the leukemic process during the course of a primarily myeloid leukemia. Moreover, leukemic cells may show characters of both myeloid and lymphoid lineages. This is the case in the biphenotypic leukemias which were identified in 1988 by Estela Matutes in Catovsky's group. The existence of biphenotypic leukemias was predicted by thoughtful hematologists like Marcel Bessis to whom I wish to pay tribute. Clinical characters also play a role in the classification of leukemias. The distinction between acute and chronic leukemias, the recognition of acute leukemias as a special group was made by Wilhelm Ebstein at the end of the nineteenth century.

The chronic lymphocytic leukemias (CLL) are a quite interesting group to study from an epistemological viewpoint. As chronic diseases they are incurable, which means that it is impossible to eradicate them from the patient. However, they can be treated in a useful way. CLLs are a good subject for evidence-based medicine, because the results of therapeutics in terms of survival and life expectancy can be evaluated by doing accurate statistics. There are several types of CLLs, according to the kinds of lymphocytes which are affected, be it B lymphocytes, T lymphocytes, or prolymphocytes. 95 % of CLLs are B-CLLs. It was recently found that T lymphocytes function was also affected in B-CLL. William Dameshek used to characterize CLL as an accumulation of lymphocytes endowed with an abnormally long life duration and an altered immunological function. Thus, CLL is different from the other leukemias, which are only proliferative diseases. The reason why lymphocytes accumulate in CLL is that they live much longer than normal lymphocytes, due to the fact that they do not undergo apoptosis. Indeed, in CLL, the anti-apoptotic Bcl-2 protein is overexpressed or upregulated. Moreover, the cell cycle is arrested in the G0/G1 phase. Thus, CLL seems to be more an accumulative than a proliferative disorder. Eugene Cronkite discovered that CLL is a group of different diseases with different survival rates after diagnosis. Hence a classification of CLL in stages, which means predictions regarding survival rates at the time of diagnosis, as functions of certain symptoms. The classification of CLL in stages has been founded by Kanti Rai and modified by Jacques-Louis Binet in 1977. Since that time, the Rai-Binet classification has been further modified and improved. The Rai-Binet clinical staging system was based on a statistical correlation between survival time and anatomical features like the number of territories, ganglia, spleen etc. which are affected. Other features have been more recently included : the immunophenotype of surface CD antigens, or cytogenetical features, like chromosomal translocations, deletions etc. However, the Rai-Binet staging system keeps its predictive value. Rai pointed out recently that the earlier staging systems were useful in helping therapeutic decisions and providing an opportunity for testing new treatments "through the exclusive enrollment of patients with similar prospects for survival. However, these systems fail to predict accurately the course of the disease in individual patients and do not take into account new discoveries about the molecular pathology of this disease" ([3], p.1797). During the last ten years, unexpected findings about the molecular pathology of CLL, the definition of its subgroups due to immunophenotyping and genetics much improved the prediction of individual cases and the testing of new treatments.

B-CLL for instance is no longer considered as a single homogeneous disease. Genetic data have shown that it is composed of different groups within which there seems to be a very strong correlation between biological, genetical features and survival. For instance, patients with mutations on specific immunoglobulin genes have a much better survival (28 years) than patients without these mutations (6 to 8 years). Patients without this mutation may have other phenotypic or cytogenetic abnormalities, like the ZAP-70 protein normally present in T-cells only, which are associated with a poorer clinical outcome. These patients constitute a new subgroup. Other markers, like the CD38 membrane protein, do also show a strong correlation with the clinical outcome - thus having a predictive value of their own. CD38 often correlates with immunoglobulin mutations but sometimes does not correlate, and has indeed a clinical value of its own in correlating with the outcome. The pathogenetic mechanisms involving these proteins remain hypothetical, but could be used in the future to provide more effective therapeutic choices.

The issue of evidence-based medicine is therapeutic success. For a chronic disease like CLL, therapeutic success means remissions and possibly longer survival for the patient. Longer survival may be assessed more accurately thanks to the new, more predictive classifications. Classification remains an essential tool for medicine, in spite of unclassifiable cases. Many different kinds of characters may be used to categorize B-CLL in subgroups which are different according to the characters used. Immunoglobulin genes and CD38 proteins do not correlate entirely. Different chromosomal aberrations may be used to create subgroups with different survival rates. Mutations of the p53 gene which induces apoptosis and suppresses tumours may also be used. How do these different classifications, staging systems or more recent molecular studies, help in evaluating therapeutics? This most interesting question clearly belongs to evidence-based medicine. Staging systems were primarily aimed at helping the physician to decide whether the patient should be treated or not, by stratifying patients into different prognostic groups. For instance, it turned out unexpectedly that it was better not to treat the earlier stage of the disease, Binet stage A or Rai stage 0, first because treatment does not improve the patient's prospect, and second because treated patients may be affected by other malignant phenomena after having received chemotherapy. The efficacy of treatments is assessed by large-scale national or international study groups or networks following-up large numbers of patients.

Presently available treatments include conventional drugs like prednisone, alkylating agents, purine analogs, and more recently chimeric or humanised monoclonal antibodies. Polychemotherapy seems to be increasingly used. The purine analogue fludarabine appears to be the most active single agent in the treatment of CLL. As a primary treatment, it is able to induce up to 60 % of complete remissions, in which no detectable pathological cell is found (apparently, complete remissions do not last forever). Fludarabine induces longer progression-free survival than chlorambucyl, although overall survival time is not different (according to one study) from survival in patients treated otherwise. In these comparisons, a significant factor is introduced when you consider previously treated or previously untreated patients. Most of these treatments are empirically based, their action is not entirely understood. However, more rationally designed treatments are presently available. These are the monoclonal antibodies, which are produced by immunological reactions in mice in order to bind specific antigens. They are "humanised" by genetical engineering, so that they are composed of a specific variable region corresponding to the desired reaction, and another region belonging to human antibodies. The use of these new drugs in combination with more conventional drugs increases the percentage of positive responses without improving the median response duration. One day or the other, the disease starts again.

There are several national and international study groups working on CLL. I would like to borrow my conclusion from one of these hematologists, Federico Caligaris-Cappio, who wrote recently: "CLL is a paradigmatic example of chronic indolent B-cell malignancies that have a prolonged and substantially indolent course but eventually always relapse... The classical approach - first do no harm : watch and wait until you are forced to act - has been gradually replaced by a less pessimistic and more audacious attitude because of at least three major advances. First, the clinical results of treatment with purine nucleoside analogues and monoclonal antibodies have demonstrated that it is possible to obtain complete remissions... Second, CLL is a heterogeneous disease, with some patients having a long survival and never requiring treatment and others running an aggressive course and demanding intensive therapy. Until recently, it was not possible at diagnosis to assign patients to either group. The observation that two subsets of patients may be recognised on the basis of the presence or absence of somatic mutations of immunoglobulin genes has changed the rules of the game. The presence of somatic mutations has been correlated with CLL clinical course and response to therapy... These data, together with those provided by cytogenetical studies, indicate that dissecting the clinical heterogeneity of CLL is another feasible goal. Such a distinction would allow an individually tailored and presumably more successful treatment. The third point is that the biological basis of CLL is becoming increasingly accessible, suggesting that the identification of new targets may not be a too far-stretched goal" ([1]). Classification is more and more refined thanks to the techniques of cellular and molecular biology, and treatment is progressing also thanks to biological engineering technology. The goal of classification is to reach the individual level so that individually tailored treatments may be used. There are good reasons for hope and plenty of room for evidence-based medicine, as illustrated by the work being done at both the clinical and fundamental levels.

## REFERENCES

1. Caligaris-Cappio, F., A. Cignetti, L. Granziero, and P. Ghia 2002. Chronic lymphocytic leukaemia: a model for investigating potential new targets for the therapy of indolent lymphomas. *Best practice and research clinical haematology* 15(3):565–566.
2. Debru, C. 1998. *Philosophie de l'inconnu. Le vivant et la recherche.* Paris : Presses Universitaires de France.
3. Rai, K. R. and N. Chiorazzi 2003. Determining the clinical course and outcome in chronic lymphocytic leukemoa. *The New England Journal of Medicine* 348(18).

# Evidence-Based Medicine: Its History and Philosophy

Anne Fagot-Largeault

*Collège de France, Chaire de philosophie des sciences biologiques et médicales*
*11 pl Marcelin Berthelot, 75005 Paris*
*anne.fagot-largeault@college-de-france.fr*

## Introduction

The common intuition behind "evidence-based medicine" (EBM) is that medical practice should be scientifically oriented. We do not want medicine to go by opinion, hearsay, preconceived ideas, old recipes applied uncritically. We want our family doctors to be reasonably knowledgeable, and apt to rationally justify what they are doing. Claude Bernard had this strong assertion: "Medicine is a science, not an art. Physicians should only strive to become scientists. Only in case of ignorance, and while waiting for better science, will they concede to *provisionally* go by rule of thumb" (*Principles of Experimental Medicine*, 1947, posthumous, Chap. 4). Science requires that its hypotheses be tested, and its conclusions based on hard data. "Experimental truths are objective" (Bernard, 1865).

## 1. Evidence-based medicine (EBM) : a "new paradigm" for clinical medicine?

The emergence of EBM was described by its promoters as a "paradigm shift" (The EBM working group, 1992). New paradigm, or resurgence of an old one? In fact, throughout the 19th and the 20th Centuries, there have been recurrent attempts at making medicine more scientific. Considering only the recent years, the proponents of EBM usually pay homage to the British physician Archibald L. Cochrane for his pioneer work. Cochrane himself wrote that "the decisive step towards a truly scientific approach of clinical medicine may be variously dated" (1972, Chap. 4). He tends to associate it with the first controlled randomized clinical trials conducted in England around 1950, under the supervision of Sir A. Bradford Hill; those trials established the effectiveness of streptomycin as a treatment of tuberculosis. Cochrane comments that Hill's statistical methodology might very well have potentially "revolutionized" not only the sciences of health, but all human and social sciences (a prediction anticipated in the 19th Century by Bertillon, Quételet, and others).

The EBM movement proper issued from a (mostly) Canadian initiative. From 1992 on, the "EBM working group", led by David L. Sackett, Gordon H. Guyatt, and others, published a series of papers in the *Journal of the American Medical Association (JAMA)*. The manifesto of the group emphasized the importance of radically changing the way physicians are educated: "Evidence-based medicine de-emphasizes intuition, unsystematic clinical experience, and pathophysiologic rationale as sufficient grounds for clinical decision making and stresses the examination of evidence from clinical research". What would be the new way of teaching medicine? First, medical students should not learn and repeat a course; they should be trained in searching through the literature and find what the real "facts" are. Second, they should not believe everything they read, but they should learn how to evaluate the scientific quality of published articles. Third, they should be instructed in never making decisions alone, in consulting with their peers, and with the patients. In brief: "Evidence-based medicine is based on a strong ethical and clinical ideal - that it allows the best evaluated methods of health care to be identified and enables patients and doctors to make better informed decisions" (Kerridge *et al.*, 1998).

EBM medicine currently teaches: strategies for efficient literature searching, rules for evaluating the reliability of claims found in the medical literature, guidance for rational decision making aimed at helping practitioners and patients determine what they want done. There are books (see: Greenhalgh, and Coll., *Clinical Evidence*), journals (see: *Evidence-Based Medicine*, published by the *British Medical Journal*, together with the *American College of Physicians* and the *American Society of Internal Medicine*; and versions or adaptations of this journal in other languages), and websites (see the Cochrane Collaboration <www.hiru.mcmaster.ca/COCHRANE/>; see also <www.cche.net> and/or the James Lind Library <www.jameslindlibrary.org>).

"EBM medicine is actually ... not a revolution", Bruno Housset explains (*Médecine légale & Société*, 2001, 4 (3): 87-90), but it is a decisive step towards strenghtening medical behaviours with more scientific rigour: the obligation to look for validated results will protect us from therapeutic laziness (e.g. persevering in old therapeutic habits long after it has been established that they are ineffective); and getting familiar with the methodology of clinical trials will limit the tendency towards blind risk taking (e.g. we don't know under what conditions strategy x is appropriate, but we prescribe it anyway).

## 2. The "numerical method" and clinical medicine in the early 19th Century

In his paper on the emergence of clinical statistics, Peter Armitage says that the "philosophical origins" of EBM lie in the movement launched by statistically oriented French physicians who, at the beginning of the 19th Century, promoted and practised the "numerical method". (The expression: "the philosophy of medical science" was used by Bartlett (1844), one of Louis' disciples.)

Philippe Pinel, mostly remembered for having contributed to free the mentally ill from their chains, actually devised a system for gathering, preparing and tabulating medical data, based on a "uniform method of description", in order to make some generalisation possible: "all medical knowledge is to be extracted from individual histories of illnesses; let us add that thoses histories, purified from all theoretical preconceptions, must be described with care, day after day from the first day (day of invasion), following the order and succession of symptoms up to the full termination..." (Pinel, 1818).

Pierre C.A. Louis is remembered for having demonstrated, through a small comparative trial (leeches against no treatment), later confirmed by another series of experiments made by Jackson in the US, that blood letting is ineffective as a treatment of pneumonia. (At the time blood letting was the standard treatment of pneumonia: Louis' result sterilized the leeches market.) Louis is the author of many careful clinical studies (on phthisis, on typhoid fever, etc); he initiated a new method for teaching medicine (the bedside teaching rounds); his students created numerous research groups and launched epidemiological studies, on the continent (the 'Société d'observation médicale': d'Espine, Maunoir, Bizot, 1832), in England (the work of E. Seaton, W. Farr, W. Budd, W. Guyand), or in the United States, around Boston (Holmes, Shattuck, Jarvis, Bowdich, and the famous Clinical Conference of the Harvard Medical School, 1857) and in New York (E. Bartlett, S. Smith, E. Harris, etc).

The methodology of "numerical" medicine was worked out by mathematicians. Laplace, when enumerating the benefits of using the probability calculus in "conjectural sciences", had outlined the principles of comparative trials: "in order to identify the best of several treatments currently in use to cure a disease, it suffices to try each treatment on an equal number of patients, and render all circumstances perfectly similar: the superiority of the most advantageous treatment will become more and more manifest as the number of patients grows, and the calculus will both give the probability that it is a better treatment, and measure by how much it is better" (Laplace, 1921). Another mathematician, Jules Gavarret, studied medicine and wrote a textbook (1840) to help medics analyse their data and estimate the error rates. Even Cl. Bernard, although he was not fond of statistics, recognized the value of comparative methods: "scientifically oriented physicians have always been convinced of the necessity of comparative experiments ... When a physician orders a tentative treatment and the patient is cured, he tends to believe that the treatment cured the disease. Oftentimes physicians pride themselves on having cured all their patients with some remedy they used. But the preliminary question they should be asked is, whether they also tried to do nothing, that is, to not give the remedy to some other patients; otherwise, how could they possibly know which of the remedy or mother nature cured the patients?" (1865, III, 3).

## Conclusion

Evidence-based medicine has its limitations and shortcomings, which have been amply discussed in the literature. William Guy, a student of Louis, pointed out the dangers of generalizing from statistical knowledge, and applying the generalizations to the treatment of new cases, a recurrent problem in clinical medicine: "Does the numerical method admit of application to individual cases? It must be conceded by the most strenuous advocate of this method, that such application is limited" (Guy, 1839). A contemporary physician observes that one person can hardly take care of patients, keep up with the medical literature, be an expert at evaluating the quality of scientific publications, decide that more research should be conducted on this or that point, and finally write the protocol of the required trial: "I am so busy applying this new methodology ... that I no longer have time to see patients", he says, with "Socratic irony" (Grahame-Smith, 1995). Doing good medical science is obviously a collective task. A nostalgic objection (often heard) is that the more scientific it tends to be, the less humane medicine will become: one should therefore go back to the 'soft' style of ancient Hippocratic medicine. The conviction of the author of this paper is that medicine can and should become both more scientific and more humane, and that the two are compatible.

## REFERENCES

1.  Armitage Peter, 'Trials and errors. The emergence of clinical statistics', *Journal of the Royal Statistical Society A*, 1983, 146 Part 4: 321-334.
2.  Bartlett Elisha, *An Essay on the Philosophy of Medical Science*, Philadelphia: Lea & Blanchard, 1844.
3.  Bernard Claude, *Introduction à l'étude de la médecine expérimentale*, Paris, 1865.
4.  Bertillon Louis-Adolphe, *Conclusions statistiques contre les détracteurs de la vaccine, précédées d'un essai sur la méthode statistique appliquée à l'étude de l'homme*, Paris: Masson, 1857.
5.  Cochrane Archibald L., *Effectiveness and Efficiency*, Abingdon, UK: Burgess & Son, 1972; transl. *L'inflation médicale. Réflexions sur l'efficacité de la médecine*, Paris: Galilée, 1977.
6.  Coll., *Clinical Evidence*, London: BMJ Publishing Group, 1999; tr. fr. *Décider pour traiter (Décider d'un traitement selon un niveau de preuve explicite. Ce que l'on sait, ce que l'on ignore, ce qui reste incertain)*, Meudon: RanD, 2001.
7.  Gavarret Jules, *Principes généraux de statistique médicale*, Paris: Beché Jeune & Labé, 1840.
8.  Grahame-Smith David, 'Evidence based medicine: Socratic dissent', *British Medical Journal*, 1995, 310: 1126-1127.
9.  Greenhalgh Trisha, *How to Read a Paper. The Basics of Evidence-Based Medicine*, London: BMJ Publishing Group, 1997; tr. fr. Drs D. Broclain & J. Dowbovetzky, *Savoir lire un article médical pour décider. La médecine fondée sur les niveaux de preuve au quotidien*, Meudon: RanD, 2000.
10. Guy W.A., 'On the value of the numerical method as applied to science, but especially to physiology and medicine', *J Statist Soc*, 1839, 2: 25-47.
11. Hill A.Bradford, 'Memories of the British streptomycin trial in tuberculosis: the first randomized clinical trial,' *Controlled Clinical Trials*, 1990, 11: 77-79.
12. Kerridge I., Lowe M., Henry D., 'Ethics and evidence based medicine', *British Medical Journal*, 11 Apr 1998, 316: 1151-1153.
13. Laplace Pierre-Simon, *Essai philosophique sur les probabilités*, Paris: Gauthier-Villars, 1921, 2 vols.
14. Louis P.Ch.A., *Recherches sur les effets de la saignée dans quelques maladies inflammatoires*, Paris: Baillière, 1835. Engl. tr. 'Researches in the effects of blood letting in some inflammatory diseases', *American Journal of the Medical Sciences*, 1836, 18: 102.
15. Louis Pierre Charles Alexandre, *Recherches anatomo-pathologiques sur la phtisie*, Paris, 1825. Engl. tr. *Researches on phthisis*, 2nd ed., tr. by W.H. Walsche, London: Sydenham Society, 1843.

16. Medical Research Council, Streptomycin in Tuberculosis Trials Committee, 'Streptomycin treatment of pulmonary tuberculosis', *Brit Med J*, 1948, 2: 769-782.

17. Pinel Philippe, 'Principes généraux sur la méthode d'étudier et d'observer en médecine', in: *Nosographie philosophique, ou la méthode de l'analyse appliquée à la médecine*, Paris: Brosson, 1818, sixième édition, xxxiij - cxx.

18. The Evidence-Based Medicine Working Group, 'Evidence-based medicine. A new approach to teaching the practice of medicine', *JAMA*, Nov 4, 1992, 268 (17): 2420-2425.

# Conceptions of the Normal and the Pathological

Elodie Giroux

*University of Paris 1-Sorbonne/IHPST Paris-France*
*e.giroux@tiscali.fr*

This paper is a critical analysis of a quotation from Professor Dawber, the person who managed, at its beginning, the Framingham study, the first major prospective study in cardiovascular epidemiology. Before I quote him, I propose rapidly a few historical explanations about this study. After the Second World War, chronic diseases like cancers and cardiovascular diseases have become the predominant diseases but the usual methods of research in physiopathology and in laboratory tests have not led to satisfying and conclusive outcomes to tackle these diseases. New epidemiological approaches designed to learn how and why those who developed heart diseases differed from those who escaped it had been set up: the prospective population survey, a prototype of which is the Framingham study, a survey held since 1948. And observations in this prospective study rapidly led to the recognition that elevated blood pressure and elevated serum cholesterol level, in apparently well population, were important predictors of cardiovascular disease. These factors have come to be known as "risk factors"[1].

Here is what Thomas R. Dawber wrote in 1980 in his book [2] about this study:

**"Better knowledge of the natural history of the atherosclerotic process has led to a different concept of normality: that the normal person is one who not only has no disease but also is unlikely to develop it. At the extreme of this normality is the ideal individual who will never develop disease. The importance of this changing definition is best illustrated by the concept of risk factors as they pertain to the development of atherosclerotic disease."**

According to Thomas Dawber, the notion of "risk factor", which, for him, had been coined by the study, had profoundly changed our conception of normality: it has led to a shift from an average sense to an optimal or ideal meaning of the normal. In my view, this new emphasis on the inter-individual variability called into question the traditional western medical model of health and disease, normal and pathological, in which these two states are seen as two mutually exclusive and distinctive states.[2] Indeed, we can say that this traditional model rests on two components: the identification of health with the absence of disease and the identification of health with biological normality. Then, the notion of "risk factor" would call into question the relevance of the concept of 'natural normality' as a basis

---

[1]A risk factor is defined as any variable which is firmly statistically associated with diseases occurrence.

[2]This traditional model assumes the possibility of drawing a "natural" demarcation between the normal and the pathological. We will call this model a 'dichotomic model'.

for definitions of health and disease. I think that the relevance of this concept used
in order to define health and disease is here at stake, and therefore, the idea that
there are some 'natural ends' for medical intervention. Thus, I would like to ask the
following question: do these new ways of building knowledge of the pathological
lead towards "a different concept of normality", as Dawber said, (which means here
to move the boundaries of normality), or do they simply highlight the ambiguity
and insufficiency of the notion of normality in medicine?

## 1. The problem of the demarcation between normal and pathological

In order to answer this question, let us begin by wondering why T. Dawber
thinks that the notion of "risk factor" has led to "a different concept of normality"
and let us see the different problems raised by this notion of "risk factor". First of
all, with risk factors, we have to face the difficult task of drawing a demarcation
line between the normal and the pathological on a continuous scale. Indeed, for
most risk factors there is a continuous link between the level of the variable and
the increasing risk of disease. We encounter differences of degree and no limits
can be set within which there would be no risk of developing a disease and outside
of which a disease would be much more probable. As with many 'quantitative'
diseases, (also, by the way, essential hypertension), which are biological variables
that are distributed continuously in the population, we cannot observe any clear-cut
separation between two populations, one healthy and the other, ill. But physicians
have to make a diagnostic and to take a decision which is inevitably of a binary
nature. At which level of risk should they decide whether or not the level of blood
pressure (or of cholesterol) is 'hyper' or 'hypo'?

## 2. Normality in the statistical sense

According to the traditional medical thinking, majority decides on normality.
The frequency or usualness is in itself regarded as the mark of normality and of
health; the pathological is a deviation from a statistical norm. Statistical means
are used in order to define "normal intervals". But on which naturalistic and
empirical foundations can we ground and justify this identification of health with
a statistical conception of normality? Does not this identification already rest on
a value judgment?[3]

The philosopher Christopher Boorse holds, in my view, the strongest defence
of a naturalistic account of disease and health. He supports the factuality of the
normal-pathological distinction and a value-free concept of medical normality in
giving a bio-statistical theory of normality. He defends this statistical sense of
norm in medicine as the only one able to support a naturalistic conception of
health and disease. He grounds the current medical use of statistical normality in
a value-free analysis of biological function and in a biological notion of "species
design." Medical normality is, for him, "functioning according to design". Disease
is not only a deviation from statistical norms but a reduction in "normal function".

---

[3]The question of the sufficiency or insufficiency of a statistical definition of normality in medicine
and of the possibility of keeping a value-free and descriptive concept of normality gave rise to controversies between "normativists" and "naturalists". This latter school of thought is principally
represented by Christopher Boorse and his Bio-Statistical theory of disease (BST).

And "normal function" in human species can be statistically defined: "normality of function is a statistical concept based on what is typical of a species (or a subclass)"[1]. Thus his definition rests on statistical normality which is biologically justified by the concept of "species design". And then, in relating the definitions of normality and pathology to a notion of species, C. Boorse gives a natural foundation to the traditional biomedical association between frequency, normality and health.

But this "Bio-Statistical Theory" of disease is put into question when we pay more attention to the variations within the so-called "normal range of variations" and to their prognostic meaning. Indeed, in weighting the statistical force of the link between two variables, the prospective epidemiologic studies highlight the fact that some values contained in the 'normal interval' are already associated with an increased risk of morbidity and mortality. Sir George Pickering, the acknowledged father of modern hypertensiology, considered that precisely these values give to essential hypertension a pathological meaning. Only this correlation between an exposure variable and life expectancy or rates of morbidity allows us to distinguish arbitrarily two populations, healthy and ill [4].[4] Pickering noticed that this correlation comes from insurance companies which had led the way in quantifying the association between this new disease and premature death. Thus, using the statistical notion of norm in order to define disease does not make any sense in this case since it precisely disregards the essential hypertension. In Pickering's view, we shall see that this correlation does not allow us to determine an absolute norm which could define health. Talking about a 'quantitative' condition implies that we should renounce any naturalistic normal-pathological distinction. Normal and pathological are conceived as states that continually pass over into each other and can no longer be considered as different and qualitative states. So a 'continuous model' of health and disease is here required.

## 3. Normality as the "biological optimal value"?

But as physicians must decide in clinical practice, which person needs treatment or not, the value that is used as a criterion is the value which corresponds to the longest probability of living for the individual without disease. We speak about " a biologically optimal value ". To define in such a way the normal amounts to recognizing the normative or evaluative character of norms in medicine. But does it undermine all defence of a naturalistic account of medical normality?

This approach leads to an identification of 'high survival value' with normality as if adoption of the Darwinian concept of natural selection were a solution to the demarcation problem. Health is here regarded in terms of fitness and, as a result, is relative to environment. According to James Lennox, because "the determination of 'appropriate cholesterol levels' is not merely a matter of determining what is statistically normal for the species but a matter of determining what level increases the probability of cardiovascular malfunctioning" [3] we should admit, against C. Boorse, that these concepts of health and medical normality are *evaluative* concepts. To live or to die, a success or a failure in functioning, are not indifferent

---

[4]Talking about his fathers view, Thomas G. Pickering wrote: "The distribution curve of blood pressure in a population shows no clearly defined dividing line separating the two, and the populations can only be separated by including variables other than blood pressure." [5, p. 17]

effects. Nevertheless, in Lennox's view, Darwinian concepts give us the possibility to defend a concept of normality as an *"objective* value". For him, life and health are empirical facts though they are values. These concepts are both biologically grounded and evaluative. The value that we grant to health and disease is the very value we grant to life, to a life that is not compromised. J. Lennox defines health as this condition that "refers to that state of affairs in which the biological activities of a specific kind of living thing are operating within the ranges which contribute to continued, uncompromised living. "[3] So J. Lennox keeps the idea that normality defines health but normality in medicine is recognized as an evaluative concept. Health is then defined as an "un-compromised living".

But it seems to me that this conception of normality on which is also founded Thomas Dawber's assertion quoted in the beginning of this paper meets many difficulties when we come to define what is " an un-compromised life " or " a person unlikely to develop disease ", as T. Dawber said. In addition to the fact that it leads to assert that death in itself is abnormal and to identify health with a perfect health,it is difficult to define empirically this optimal value. In fact, this optimal value is determined by the therapeutic means available at a given time. In order to define this threshold, Georges Pickering was rather talking about an "optimal treatment threshold". Then the pathological threshold is the 'treatment threshold', i.e. an optimal value which is determined with the best benefit/risk ratio. This value is not only an "optimal biological value" but is technologically defined by the therapeutic means a society disposes. Moreover, the determination of this optimal threshold in this context of diseases involving several factors is relative to each individual. Disease is then a relation of variables that can only be defined for a given individual in a given context. Thus, to speak of an absolute norm for a given variable does not make any sense. Georges Pickering would have preferred that we renounce to speak in terms of 'hyper', 'hypo' and 'normo'- tension. For him, the concept of "normality" has no meaning here: "normotension and hypertension are not merely meaningless concepts, they are wrong."[4, p. 11]

## 4. Conclusion

Thus the more one recognizes how biological and behavioral phenomena often express themselves along a continuum, the more difficult it becomes to speak of species typicality or statistical normality without explicit appeal to a particular normative viewpoint (be it individual, collective, economic, subjective, etc.). Instead of speaking of 'a different concept of medical normality' with the importance taken by risk factor and the screening of predisposition in the individuals apparently 'healthy', it seems better to insist on the ambiguity of this notion of normality in medicine. Medical use of this notion oscillates between two principal meanings without clear-cut distinction: the normal is seen as the usual and/or as the ideal. So, in order to answer my question, I would say that this new approach of the etiology of diseases does not entail a shift in the boundaries of normality but it rather leads us to highlight the limits of a naturalistic account of disease only grounded in a concept of biological normality. Above all, it calls into question the concept of biological normality as a basis for definitions of health and disease. Then, should "normality" still be regarded as the end of medical intervention?

## REFERENCES

1. Boorse, C. 1987. Concepts of health. In *Health Care Ethics: an introduction*, ed. Van De Veer, D. and T. Regan, 359–393. Temple University Press Philadelphia.
2. Dawber, T. R. 1980. *The framingham study, the epidemiology of atherosclerotic disease*, Harvard University Press, Cambridge, MA
3. Lennox, J. 1995. Health as an objective value. *Journal of Medicine and Philosophy* 20(5):499–511.
4. Pickering, G. 1995. Definitions, natural histories, and consequences. In *Hypertension: pathophysiology, diagnosis, and management*, ed. Laragh, J. H. and B. M. Brenner.
5. Pickering, T. G. 1995. Modern definitions and clinical expressions of hypertension. In *Hypertension: pathophysiology, diagnosis, and management*, ed. Laragh, J. H. and B. M. Brenner.

# Measuring Qualities: How to Quantify Health, Pain, Well-Being, Etc.

Alain Leplège

*Department of Philosophy and Social Sciences, University of Picardie,*
*Campus, Chemin du Thil, 80025 Amiens CEDEX, France and*
*Institute of History and Philosophy of Sciences and Techniques*
*(UMR 8590 CNRS/U.Paris1), 13 rue du Four, 75006 Paris, France*
*Alain.Leplege@wanadoo.fr*

## 1. Preamble

Measurement of subjective concepts such as pain and more notably well-being and quality of life has become common practice in the health care sector from clinical practice to public health. All these measurements aim at taking a holistic patients' perspective into account in scientific medicine, especially in 'evidence based medicine'. Many perspectives are opened and many questions are being raised by these practices (Fagot-Largeault, 1991, Faden and Leplège, 1992).

First of all, this paper does not address the issue of whether qualities and quantities are distinct instances, or the nature and the implications of this distinction. As a matter of fact, the opposition between qualities and quantities which originates in Aristotle and took its classical form in the Kantian opposition between intensive and extensive magnitude, has long been seen as fundamental to an understanding of measurement. However, it now seems that this distinction, at least in the natural sciences, is, at best, peripheral to the contemporary understanding of scientific measurement. For instance, it didn't seem to have played any part in the works of Hölder (1901), Russell (1903), Campbell (1920), Stevens (1951), Rasch (1959) and Krantz, Luce, Suppes and Tversky (1971) who aimed at understanding measurement in the XXth century. Some historians such as Joel Michell (Michell, 1999) have analysed the reasons of this evolution.

I shall not discuss either –for lack of space- the question whether one can (Andrich, 1989) or cannot measure in the social sciences in the same way as one measures in the natural sciences.

For the purpose of this paper, I shall take for granted –although it can be obviously disputed- that the measurement of qualities such as pain, well-being and quality of life, as evidenced by the extensive literature published in the health care field, is indeed possible. In addition, I shall focus mainly on the practicalities of the measurement of well-being and quality of life in Clinical Research and Public Health.

I shall briefly describe the qualities that are intended to be measured, the methodologies that are applied to this purpose and the research issues that are addressed by using these measurements

Finally I shall expose some questions that are pending as a result of the actual surge of measuring instruments and measurements in social sciences in general and in Medicine and Public Health in particular.

## 2. Qualities

The French bureau of standards defines quality as the 'capacity to satisfy the users' expressed or potential needs'. In a similar way, on an international level, (ISO 8402), 'The quality of an entity (a process, an organism)' is defined as 'the set of characteristics related to its capacity to satisfy expressed or implicit needs'.

These two definitions lead to the same directions: 1/ the quality is a set of characteristics of a product or service whose subjective aspect is objectified by measurement and 2/ the measurement of quality is oriented towards the end-user or client and more specifically towards the satisfaction of their needs, which are either explicit (directly expressed by the client) or implicit (potential needs).

There are many need theories, such as Maslow's Need hierarchy (Maslow, 1954) and several list of major needs. Such lists may include for example: autonomy, competence, relatedness, self-actualisation-meaning, physical thriving, pleasure-stimulation, wealth-luxury, security, self-esteem, popularity-influence (Sheldon et al., 2001 ; Nordenfeld, 1992).

In the health care field, these products can be: Comfort (pain reduction), Care (quality of care), Health and nowadays Quality of life. The characteristics to be measured for such 'products' can be very varied depending of the nature of the 'product'. The list of characteristics to be measured operationalizes the quality to be measured

For example, aiming at measuring subjective health status from a functional perspective, the MOS SF-36 developers took the following dimensions into account: General Health, Physical Functioning, Bodily Pain, Vitality, Mental Health, Social Functioning, Role imitation [Physical] and Role limitation [Emotional] (Steward and Ware, 1992 ; Ware and Sherbourne, 1992).

The objective of measuring qualities is to foster the development of explanatory or even causal hypotheses that can be subjected to empirical tests.

## 3. Measurement

Fundamentally, any measurement is a comparison. From an empirical perspective, every measurement is considered to be the expression of a magnitude (a scalar) by a real number in a reference frame (an object to be measured, an agent, an experimental protocol).

The measuring instruments (the agents) are standardised questionnaires with closed end answer-choices constructed so as to quantify abstract concepts that often cannot be observed directly (latent variables).

The development of any instrument involves three main phases: 1/ *A conceptual phase*. Choosing a conceptual framework, clarifying the concept to be measured, identifying attributes or measurable aspects of this concept, predefining a list of specifications. 2/ *A qualitative phase* realising qualitative surveys, analysing the content of transcripts, selecting a group of candidate questions. 3/ *A quantitative phase*. Analysing the results of population surveys in order to reduce the amount of candidate questions , finalise the questionnaire format and optimise its psycho-metric properties.

Needless to say, supervising and implementing this very technical process remains the prerogative of experts: for each one of the required stages, numerous decisions are taken based on the judgements of experts involved in the development and the implementation of the instruments.

In this regard, it should be noted that professional ethics issues such as those involving the analysis of conflict of interest are particularly crucial.

There are several typologies of measuring instruments depending on the mode of administration (the instruments can be administered by an observer, an interviewer, or self-administered) and the concept being measured (Clinical concept such as pain, Health status or 'Quality of life').

Most instruments are developed using psychometric methodologies (Nunnally, 1978). In this regard, the main measurement model is the true score model (Spearman, 1925). It should be noted that this measurement model tends to be challenged by the Rasch model (Rasch 1959) primarily on the ground that this latter family of measurement models pays more attention to the list of requirements formulated by Thurstone (1925): unidimensionality, the existence of a mathematical model, additivity, and invariance.

Amongst the psychometric properties taken into account are the precision or reliability (is the instrument measuring anything at all ?), the validity (is the instrument measuring what it is supposed to ?), other psychometric properties such as the responsiveness and sensitivity to change.

The history of the instruments displays several intertwined evolution patterns (Greenfield and Nelson, 1992): 1/ The concepts to be measured evolved from disease to health and from health to quality of life. 2/ Many of the earliest measurements were designed by clinicians to be used in clinical setting: for example, the *Karnovsky Performance Status* was published in 1948, the *American Rhumatologist Association Arthritis Classification* in 1949, the *Hamilton Depression Scale* in 1967. In the begining of the 1970's, new health status measurements such as the *Sickness Impact Profile*, the *Health Insurance Experiment Adult Health Status Measurements*, etc. were developed. These instruments are appropriate for clinical or therapeutic research and population surveys. One focus of current research in this area is certainly to develop instruments that can be used to monitor individual health and well being. 3/ In term of scope, the instruments focused from specific measurements (prior to 1970) to generic measurements (until the 1990's) and since then back to specific measurements again; the current frontier being the development of individualised measurements which would involve the calibration of item banks (Hays and Morales, 2000). 4/ One can also observe an increased sophistication of the instruments. This increased sophistication which regards the process of instrument development (eg the use of more sophisticated measurement models) is apparent when one considers the instrument size for a given level of measuring precision: from short measurements to long measurements, then to short measurements again.

Although the evaluation of some processes may involve the measurement of qualities, the majority of the researches involving these measurements are outcome researches, be they clinical research, epidemiological research or health service research.

Naturally, these measuring instruments must be used within research protocols in which these measurements of qualities are collected with as much attention paid to the risk of bias as it is to any other variables.

Amongst these studies, one generally distinguishes: 1/ The descriptive or observational studies which may be transversal or longitudinal (repeated measurement). Such studies would aim at the description of the natural history of diseases and the impact of diseases on health status. 2/ Quasi-experimental studies which involve no randomisation but some design features aiming at controlling at least some of the potential biases. One of the simplest designs would involve the measurement of some qualities, for example of the quality of care, before and after the studied intervention. 3/ The experimental studies involve the random allocation of intervention to the subjects. This design makes it possible to control most biases. It is customarily used to evaluate the efficacy of treatments.

The interpretation of these measurements' meaning involves some kind of decision models. Most of the time, these models are deterministic. They involve few variables. They tend to be implicit and fairly rough (eg. more is better). Quite rarely, explicit and sophisticated decision models (eg. the Von Neuman Morgenstern utility theory, the bayesian decision theory or comprehensive models taking into account cognitive aspects such as the time perspective, etc.) are being invoked.

It can be noticed that most users and observers tend to forget that the presentation of these measurements should be accompanied by a comprehensive consideration of the steps that are necessary to act on that information with regard to specified objectives such as health improvement: 1/ The first step should involve the conceptualisation and measurement of the chosen concept, the examination of the validity or meaningfulness of the measurements. 2/ The second step should involve the interpretation of the results on a clinical level in relation to multiples variables and the explanation of observed variations. 3/ Only the third step should involve the use of these measurements in decision making processes.

## 4. Discussion/Conclusion

The validity of the results obtained by the experimental method, which connects natural laws identified through induction to a deductive system represented by a mathematical model, depends on the accuracy of the measurements.

This means that, when it comes to checking hypotheses, progress in scientific medicine, as in any science, depends, at least partly,on the progress of measurements. However, the relative novelty of this area and the tension between the epistemic and practical aim of medical researches explain that many of the epistemological problems related to the measurement of quality in medicine have not yet been thoroughly addressed (Leplege and Hunt, 1996).

## REFERENCES

1.   Andrich D, Distinction between assumptions and requirements in measurement in the social sciences, in JA Keats, R Taft, RA Heath and SH Lovibond (eds)

*Mathematical and Theoretical Systems*, Elsevier Science Publishers BV, 7-15, 1989.

2. Campbell NR, *Physics : The Elements*, London, Cambridge University Press, 1920.

3. Faden R, Leplège A, Assessing quality of life. Moral implications for clinical practice. *Medical Care.* 30 (Sup 1992): MS166-175.

4. Fagot-Largeault A, Réflexions sur la notion de qualité de la vie, Archives de philosophie du droit, 36 : 138-139, 1991.

5. Greenfield S and Nelson EC, Recent developments and future issues in the use of health status assessment measures in clinical setting, *Medical Care*, S23-41, vol 30, N ° 5, May 1992.

6. Hays RD and Morales LS, Item response theory and health outcomes measurements in the 21th century, *Medical Care*, Vol 38, N ° 9, S28-42, 2000.

7. Hölder O, Die Axiome der Quantität und die Lehre vom Mass. *Berichte über die Verhandlungen der Königlich Sächsischen Gesellschaft der Wissenschaften zu Leipzig, Mathematisch-Physische Klasse,* 53: 1-46, 1901.

8. Krantz DH, Luce RD, Suppes P, Tversky A, *Foundations of Measurement, Vol1*, New-York : Academic Press, 1971.

9. Leplège A and Hunt S, The problem of quality of life in medicine, *Journal of the American Medical Association*, July 2, 1 (278) 47-50, 1997.

10. Maslow AH, *Motivation and Personality*, 1954.

11. Michell J, *Measurement in Psychology : A Critical History of the Methodological Concept*, Cambridge University Press, 1999.

12. Lennart Nordenfelt (ed) : *Concept and Measurement of Quality of Life in Health Care*, Kluwer Academic Publishers, Dordrecht, Boston, London, 1994.

13. Nunnally JC, *Psychometric Theory*, Mc Graw Hill, New York, 1978.

14. Rasch G, *Probabilistic models for some intelligence and attainment tests* Danish Institute of Educational Research, 1960 ; University of Chicago Press, 1980 ; MESA Press, 1993.

15. Russell B, *Principles of mathematics.* Cambridge: Cambridge University Press, 1903.

16. Spearman C, "General intelligence" objectively determined and measured *Amer J Psychol*, 15, 201-293, 1904.

17. Sheldon KM, Elliot AJ, Kim Y, Kasser T. What is satisfying about satisfying events? Testing 10 candidate psychological needs. J Pers Soc Psychol. 2001 Feb;80(2):325-39.

18. Stevens SS, On the theory of scales of measurement, *Science*, 103, 677-680, 1946.

19. Stewart A, Ware JE Jr (Ed) *Measuring Functioning and Well-Being - The Medical Outcome Study Approach*, Duke University Press, Durham NC, 1992.

20. Thurstone LL, A method of scaling psychological and educational tests *Journal of Educational Psychology*,(16), 433-451, 1925.

21. Ware JE, Sherbourne CD, The MOS 36-Item Short-Form Health Survey (SF-36) : I Conceptual framework and item selection, *Medical Care*, 30, 473-483, 1992.

# The Polish School of Philosophy of Medicine and The Concept of Cure

Zbigniew Szawarski

*Institute of Philosophy, Warsaw University, Warszawa 64, Poland*
*z.szawarski@uw.edu.pl*

Polish philosophy of medicine began in Wilno where Jędrzej Śniadecki (1768-1838) was professor of chemistry and physiology. In his *Theory of Organic Beings* he defined life as a form of matter and tried to explain its origin and development referring to solar energy and the relation between an organism and its environment. There are three distinct periods in the history of the Polish school of Philosophy of medicine:

(1) The beginnings (1874 - 90). The main (and only) person representative for that stage was Tytus Chałubiński (1820-1889), professor of special pathology in Warsaw. His *Medical writings* comprise two small volumes only: the first one is the famous and celebrated *Method of arriving at medical indications. The plan of treatment and its execution*; the second is a sort of application of his general theory to one particular disease only and is devoted to diagnosis and treatment of malaria.

(2) The proper school (1890-1920). Władysław Biegański (1857-1917), Edmund Biernacki (1866-1911), Zygmunt Kramsztyk (1848-1920), Henryk Nusbaum (1849-1937), Henryk Hoyer (1834-1907). The dominant figure of this period was W. Biegański, whose *Logic of Medicine* [2, 14] and other logical and philosophical publications were very influential not only in Poland but also in Germany and Russia. This was also the period when Zygmunt Kramsztyk established and edited a monthly journal *Krytyka lekarska* [*Medical Critique*] (1897-1908) devoted especially to philosophy of medicine and related issues. It was a unique journal in the world of that time and the main forum of the debate on the social and economic conditions, main problems, and the vocation of the Polish doctors.

(3) Decline. The institutionalisation of the Polish school of philosophy of medicine. (1920 - 1939). The first chair of philosophy and history of medicine in the world was established in Kraków. History and philosophy of medicine entered into curriculum in all Polish medical schools. A new journal *Archiwum Historii i Filozofii Medycyny* [*Archives of the History and Philosophy of Medicine*] was founded by A.Wrzosek (1875-1965) in Poznań in 1924. This was the time of W.Szumowski, T.Bilikiewicz (1901-1980), A.Wrzosek, and others. Some authors have included here also Ludwik Fleck (1896-1961).

## 1. Chałubiński's concept of disease

Every organism - says Chałubiński - is a self-replicating system composed of material molecules. It interacts with its environment and its structure and behaviour can be explained in terms of the "exact and absolute laws of the natural sciences". From that point of view every organism is merely "a true biological machine" whose essential feature is its ability to feel and move. "Health is a state in which the process of life goes on with energy of all its functions and that energy is properly related to its goals in the physiological plan of life" [*Method*, p. 25]. If a normally functioning organism which lives in "normal conditions of life" is exposed to any factors that disturb the "limits of its physiological functions", then the organism becomes ill and the totality of functional disorders of the organism is called a disease [ibid, p. 32].

## 2. Chałubiński's idea of cure

If disease is a loss of function of the organism or a disturbance in its vital functions, then *science based medicine* (this is Chałubiński's original expression) can do two things at least. First, it can discover and define those conditions in which human individuals and communities can avoid harmful consequences of their environment and achieve the optimal level of health. This is the modern idea of prevention and health promotion. Second, it can investigate and define those external and internal conditions in which the patient should be placed to recover from a disease or at least to maintain his vital functions. This is the idea of treatment or cure. Chałubiński is perfectly aware that there are conditions that may be treated and even cured, and that there are conditions that may be treated but not cured. The number of curable diseases is relatively small and they usually belong to the realm of surgery or self-healing conditions. Such are, for example, all kinds of amputations, setting bones, gastric lavage, or surgical excisions or resections. More difficult is controlling the basic physiological functions of the organism but it is still possible if the doctor knows what are the causes of the disturbance and how to use the medical means "to modify pathological moments by aggravating, stimulating, weakening, or slowing down some functions, organs, or organic systems" [ibid, p.73]. It is instructive to see the way he approaches the first case he mentions in his book: it is a valvular heart defect combined with serious circulation problems, hydropneumopericardium with general dropsy, albuminuria, but no lipomatosis of the heart. Although the defect of the mitral valve is a certain pathological state *it is not by itself a disease*, as one can live a relatively healthy life with it. The problem is how to treat the complications. What should be the treatment plan? Where to begin? How to decide what is the most important pathological moment in the state of the patient and what can be done to restore patients health? If we teach the doctor to see the patient as a complex natural system in which all possible pathological states and disorders are closely interconnected, the first thing he should do is to remove all the obstacles to the self-regulative forces of the system. The old adage *Vis medicatrix naturae* refers to the healing powers of nature. The rationally acting doctor should be able to control and stimulate those forces but he cannot do it unless he learns to realize the real meaning of particular "pathological moments" in a condition. That is, because the moments, although pathological, may be harmful, indifferent, and even sometimes helpful. It is not necessary, e.g., to reduce fever in every case, as the fever may be quite useful for the course of a disease. The same happens with a cough, which may have a positive, or negative function depending on the kind of disease and the general state of the organism.

## 3. The main achievements of Chałubiński

Here are the main achievements of the author of *"Method of arriving at medical indications"*.

1. Disease is a natural result of influence of external and internal conditions upon the organism.

2. The causal network of factors responsible for emerging of disease is complex but with the help of scientific method it is always possible to identify the most meaningful nodes in that network. (Even if science cannot explain the cause of a disease at the present moment, it will certainly do it in the future).

3. "The pathological moment" in a disease can be harmful, indifferent, or beneficial.

4. Therefore, it is up to the doctor to identify its character, relevance, and meaning and decide whether and how it is "therapeutically available" (i.e., whether and how it should be treated).

5. As every disease is individual the result of therapy is only probable but never certain.

6. Disease can be cured only if its causes are relatively simple, easy to identify and remove. However, there are some conditions where due to the general and deteriorating state of the organism, and our ignorance in regard to the causal network of the disease, we may offer symptomatic treatment only.

Chałubiński did not know that beautiful old French saying: "Guerir quelquefois, soulager souvent, consoler toujours" ("to cure sometimes, to relieve often, to comfort always") but all his life he practiced his art exactly that way. We use the same word in Polish both for "disease" and "illness". Chałubiński had no doubts that even if we cannot cure the disease, we still can and should treat and perhaps to some extent cure a patient's illness. It seems to me that the idea of "pathological moments" covers both objective causes and subjective symptoms of the disease. Even if we cannot remove the cause of disease (i.e. even if we cannot successfully cure the patient), we can still control some basic functions of the organism and remove the subjective symptoms of disease. Recovering from disease does not need to mean being cured.

## 4. The main achievements of the Polish school of philosophy of medicine

They are: (1) Introduction of scientific method to research and analysis of the complex set of phenomena that happen at the bedside. (2) A rational and scientific approach to disease which was understood as a natural result of interaction between the organism and its environment. (3) The clear distinction between the causal and teleological models of analysis and explanation of pathological events. (4) The

rejection of clinical nihilism and concentration on the patient and his disease: the
cardinal moral obligation of the physician is to treat the patient and to do it in a
rational way.

## REFERENCES

1. Biegański W. Health and disease from a biological point of view [Zdrowie i
   choroba z biologicznego punktu widzenia] *Krytyka lekarska* 1899, 3:7-17.
2. Biegański W. *Thoughts and aphorisms on medical ethics* [Myśli i aforyzmy o
   etyce lekarskiej]. S.Wende: Warszawa 1899.
3. Biegański W. Neovitalism in contemporary biology [Neowitalism we spółczesnej
   biologii]. *Krytyka lekarska* 1904.
4. Biegański W. Neovitalismus in der modernen Biologie. *Ostwalds Annalen der
   Naturphilosophie* 1904.
5. Biegański W. Über die Zweckmässigkeit in den patologischen Erscheinungen,
   *Oswalds Annalen der Naturphilosophie* 1905.
6. Biegański W. *Medizinische Logik, Kritik der ärztlichen Erkentniss*, Würzburg:
   A.Stubers; 1908.
7. Biernacki E. *Die moderne Heilwissenschaft: Wesen und Grenzen des ärztlichen
   Wissens*, B.G.Tauber: Leipzig; 1901.
8. Chałubiński T. Method of arriving at medical indications. The plan of treatment and its execution [Metoda wynajdywania wskazań lekarskich. Plan
   leczenia i jego wykonanie] , in *Medical writings of T.Chałubiński*, vol. 1, Gebethner and Wolff: Warszawa; 1874.
9. Chałubiński T. Malaria [Zimnica. Studyum ze stanowiska praktycznego]. In
   *Medical writings of T.Chałubiński*, vol. 2, Gebethner i Wolff: Warszawa; 1875.
10. Doroszewski J. Philosophy of medicine in Poland at the turn of the 19th and
    20th centuries, *Metamedicine* 1982;3: 75-86.
11. Löwy I. *The Polish School of Philosophy of Medicine. From Tytus Chałubiński
    (1820-1889) to Ludwik Fleck (1896-1961)*. Kluwer Academic Publish-
    ers:Dordrecht;1990.
12. Szumowski W. La philosophie de la médicine – son histoire, son essence, sa
    dénomination et sa définition, *Archives Internationales d'Histoire des Sciences*
    1947;2:1097-1139.
13. Szumowski W. *History of medicine from a philosophical perspective* [Historia
    medycyny filozoficznie ujęta], 1st ed. Gebenthner and Wolff: Warszawa 1935;
    3rd, Sanmedia: Warszawa 1994.
14. Śniadecki J. *Theory of organic beings*, [Teoria jestestw organicznych], vol. 1, W
    drukarni przy Nowolipiu: Warszawa;1804, vol. 2, J.Zawadzki: Wilno; 1811.

Symposium 4:

# Philosophy and Methodology of Empirical Modeling: Causation, Validation and Discovery

# Rational Reconstruction of Wrong Theories

Jean-Gabriel Ganascia

*LIP6 – University Pierre et Marie Curie (Paris VI)*
*8, rue du Capitaine Scott, 75015 Paris France*
*Jean-Gabriel.Ganascia@lip6.fr*

**Abstract.** "Scientific Discovery" is a subfield of Artificial Intelligence aiming at both the rational reconstruction of old scientific discoveries and the automatic generation of new scientific theories. The ultimate goal is to model creative activities with problem solving processes and to simulate them with computers. Among scientific activities that may be simulated, many rely on inexact reasoning. For instance, empirical induction of laws and generation of taxonomies from examples are uncertain: there are numerous empirical laws and numerous taxonomies that can be generated from any set of observations, while none of them is totally assured. Some of the criteria that can be used to discriminate among possible theories are the "cohesion", the "explanatory power", the "Occam razor" etc.

This paper provides a computational translation of these concepts in an algebraic formalism; the latter being commonly used in "data mining" to automate the knowledge discovery process. It illustrates this general framework by explaining how old wrong scientific medical theories can be automatically rebuilt from examples. It also shows the usefulness of this framework in social sciences to model the generation of social representations from preconceptions and news.

## 1. Introduction

Our aim here is to rebuild wrong theories with artificial intelligence techniques. This goal may seem both odd and trivial; indeed, all theories that are not true can be considered as false. Therefore, one could have the impression that it is easy to build wrong theories, since it is only to generate arbitrary theories and to prove that they are not true, which is usually not difficult.

Moreover, philosophers and logicians, fond of truth, will feel it strange to be guided by the study of wrongness, errors and falsity. Nevertheless, we shall demonstrate that studying erroneous and mistaken theories is neither bizarre nor trivial. More precisely, we are not interested in all incorrect theories: we focus our study on the reconstruction of old theories, those that have, at least one time point in the past, been recognized as possibly true. In other words, we are concerned by wrong theories that people had in mind, and which can be characterized as real or "true" wrongness.

Indeed, many theories, recognized today as wrong, such as the theory of "caloric" or the theory of "ether" in ancient physics, had convinced clever people in the past. We might as well imagine that most of our present scientific knowledge might be considered as erroneous in the future. Additionally, "common sense" knowledge is frequently incorrect, even if it seems evident. In a word, many currently accepted conceptions might or will be proved to be false.

The origin of errors is partly due to the lack of information; when almost nobody experiences some facts, theoretical consequences of those facts cannot be perceived. Most of the time, the *state of the art* is responsible, because it renders observations difficult or impossible. For instance, in the $17^{th}$ century, the development of optics allows Galileo to gather some observations in astronomy that were not accessible before.

However, even while it is possible to derive a correct theory from a set of empirical evidences, it may happen that only erroneous theories are accepted as true. We shall try to understand and to explain this strange phenomenon in this paper. For this purpose, we shall provide some examples drawn from medicine and common sense reasoning, even if it is also the case in other scientific disciplines, e.g. in geology or in physics.

In order to simulate the way people thought and erected wrong theories from facts, we shall automatically reconstruct, with the help of computers, this pathway (leading from the data to the formation of erroneous theory), by using artificial intelligence techniques, such as, machine learning and data mining tools.

The first reason why we are interested in such a study is that it is of cognitive significance to note and understand how people actually derived general statements from facts, and not only to consider how they should do it. In the future, we could envisage many developments in cognitive psychology to test the validity of our model. At the present time, we have chosen to deal with pre-scientific knowledge, trying to explain why some misconceptions dominated the world for centuries, even though it was possible to derive more efficient theories than the dominating ones. So, our work is of epistemological interest.

But, we have also in mind the way people – not only the scientists – speculate from facts. This simulation of inexact reasoning could have many applications in social sciences, where it could help to understand the social representations, their evolutions and the way they spread. Finally, it may also enlighten some rhetorical strategies that prefer to provide well-chosen examples, in spite of demonstration, to convince.

This paper is an attempt to model the way misconceptions emerge from facts with machine-learning techniques that simulate induction, i.e. reasoning from facts to general statements. The key concept is the notion of *explanatory power* with which all conflicting theories will be compared: the explanatory power evaluates the number of observations that could be explained by a given theory, so each of the different theories generated by an inductive engine will be ranked with respect to this index.

The first part of the paper will describe the general framework. Then, we shall show the first model based on the use of supervised learning techniques. The two following parts will provide two examples of rational reconstruction of wrong medical theories using our first model. The first example tackles with misconceptions on the causes of scurvy disease, the second with misconceptions on the transmission of leprosy. Then, we shall consider an application to social sciences, more precisely to model the political beliefs in France, at the end of the $19^{th}$ century, a few months before the Dreyfus affair burst.

## 2. General framework: automatic induction of theories from facts

### 2.1. Machine Learning

Machine learning aims at building learning machines (cf. [18,17]) i.e. machines that progressively modify their behavior as they become more experienced. There obviously exist numerous approaches to fulfill this objective. In a way, there are as many machine learning approaches as general learning theories. For instance, one traditionally considers learning by introspection, which is learning by observing its own behavior and its own unconscious knowledge, as Plato referred in the Meno (cf. [20]), and learning by observing the outside world, i.e. by empirical investigation. Machine learning techniques mimic both approaches. The so-called "learning by doing", "explanation-based generalization", "reinforcement learning" etc. reproduce the introspective learning in the sense that the machine observes traces of its own behavior, evaluates its efficiency and modifies its program hoping to improve its behavior in the future. On the other hand, the most widespread machines learning techniques, for instance Top-Down Induction of Decision Trees, Neural Networks, Genetic Algorithms, Association Rules etc. tend to derive general knowledge by observing cases stored in databases, which are, most of the time, drawn from the outside world. Among those techniques, we may distinguish *supervised* and *non-supervised* learning.

In the case of supervised learning, a *learning set*, which is a set of facts, is given to the machine using a representation language $\mathcal{L}$. These described facts are generally referred as *examples* and associated to each example, i.e. to each described fact, there is a label called its *class*. Briefly speaking, the inputs of any supervised learning system are:

1. A description language $\mathcal{L}$
2. A learning set, i.e. a set of examples $\{E_1, E_2, \ldots, E_n\}$ represented in the description language $\mathcal{L}$
3. A label $C_i$, associated to each example $E_i$ [i.e. $\forall i \in [1,n]$, $C_i = \text{class}(E_i)$]

Then, the aim is to automatically build a procedure P able to associate a class label to each description D of the description language $\mathcal{L}$. The procedure P is said to be consistent if and only if it classifies correctly all examples belonging to the learning set, i.e. if the computed label is equal to the given label, for all examples of the learning set:
$\forall i \in [1,n]$, $P(E_i) = \text{class}(E_i)$

The case of non-supervised learning is a little bit different since there is no label associated to the examples of the learning set. Then, being given a distance on $\mathcal{L}$, the goal is to build an arrangement on $\mathcal{L}$, for instance a partition, a hierarchy or a pyramid, that organizes the set of examples with respect to this distance. In other words, being given:

1. A description language $\mathcal{L}$
2. A learning set, i.e. a set of examples $\{E_1, E_2, \ldots, E_n\}$ represented in the description language $\mathcal{L}$
3. A distance d on $\mathcal{L}$

It is to automatically build a structure S that organizes the description space consistently with the distance d

Since we are only interested in the derivation of general knowledge from facts, we shall take into consideration only inductive machine learning techniques, i.e. those supervised or non-supervised techniques that simulate induction. Both supervised and non-supervised learning can be used for our purpose, which is to generate theories from facts. Each has its own advantages and disadvantages. On the one hand, supervised learning procedures are more efficient and easier to program, on the other hand, they require, from the user, to associate a label to each example, which is not always possible as we shall see in the following. In this paper, we shall restrict us to supervised techniques, but, in the future, we plan to extend our model to integrate non-supervised learning techniques.

## 2.2. Sources of induction

Whatever technique we use, a description language is always needed; sometimes, additional background knowledge is also necessary. Therefore, the generated theory depends on all this additional knowledge, which biases the learning procedure. In other words, there is no pure induction because the way facts are given to an inductive machine influences considerably the induced theory.

Moreover, many empirical correlations may be observed, which lead to many different possible theories. Since most of the machine learning programs aim at building efficient and complete (i.e. that recognize all the examples) recognition procedures, they tend to preclude most of the possible correlations, using some general criteria to prune and eliminate them. For instance, in case of TDIDT – Top-Down Induction of Decision Trees – *information entropy* is a very efficient heuristic making the generated decision tree quite small, decreasing the number of leaves. Nevertheless, our goal here is totally different: first we aim at generating all possible theories and then discriminating explanation patterns among those different generated theories, by using a criteria based on the notion of explanatory power. To summarize, being given a set of known facts, we shall build different learning sets, using different representation languages and different background knowledge. Then, for each representation language with additional background knowledge, we shall study the different generated theories by comparing them with the different systems of hypothesis given by people to explain the examples. The general schema presented in figure 1 offers an overview of our global model.

In order to validate our model, we shall show how changing knowledge representation and background knowledge affects the generated theories. More precisely, it means to explain common sense reasoning by taking into account other implicit data, i.e. not only the given facts, but also the description language and all possible sources of associated knowledge. To support this thesis, we shall demonstrate many computer simulations where, by modifying the implicit knowledge, the "explanation power" of the different generated hypothesis will be modified, which means that, with respect to the notion of explanatory power, the respective ranking of each hypothesis generated by our inductive engine will be modified by the introduction of background knowledge, making artificially one more satisfying than the others.

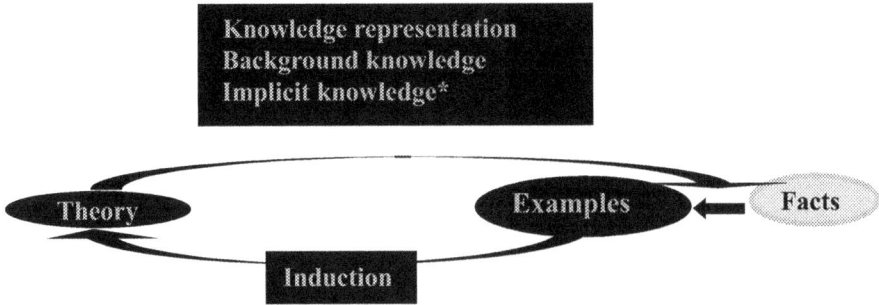

Figure 1. Overview of our general model

As we already said, the key concept here is the notion of explanatory power drawn from [25]: it corresponds to the ratio of the learning set explained by a theory, i.e. to the number of examples belonging to the learning set which are covered by this theory. In other words, our inductive engine generates many conflicting theories that can be compared with respect to their explanatory power, i.e. to the number of examples they cover.

In case of supervised learning, an example E is said to be *covered* or *explained* by a theory T if and only if the label associated to the example, i.e. class(E), is automatically generated by the theory, which means $T(E) = \text{class}(E)$. Then, $E_p(T)$ the explanatory power of the theory T is the number of examples belonging to the learning set that are covered by the theory T: $E_p(T) = \Sigma_{E \in learningset} \delta(T(E) = \text{class}(E))$ where $\delta(\text{true}) = 1$ and $\delta(\text{false}) = 0$.

In case of non-supervised learning, there is no class *a priori* associated with examples, so the preceding definition cannot be in use. However, it is possible to compute the number of examples covered by each generated class. We can then introduce the notion of *cohesion* of a class, which, roughly speaking, corresponds to the sum of average similarities between the examples of a class. It follows that the explanation power of a set of classes is the sum of the cohesions of all classes. Therefore, higher the cohesion of generated classes is, higher is the explanation power.

## 3. Association rules

Our experiments make all use of association rules. These techniques, developed more than 15 years ago [8,10], became very popular with the emergence of data mining. Their goal is to detect frequent and useful patterns in databases. The main difference between the classical supervised learning techniques and inductive engines used in data mining processes is that in the former, the goal is to build an efficient classifier, i.e. a procedure that classifies consistently with the learning set, while, in the latter, it is to extract some remarkable patterns from the data.

As a consequence, an example may be covered by many extracted patterns, in data mining, while it is rarely the case in classical machine learning.

|     | E1 | E2 | E3 | E4 | E5 | E6 | E7 |
|-----|----|----|----|----|----|----|----|
| A   | a1 | a1 | a1 | a2 | a3 | a2 | a2 |
| B   | b1 | b3 | b1 | b2 | b2 | b1 | b1 |
| D   | d1 | d2 | d1 | d2 | d1 | d1 | d2 |
| Cl  | c1 | c1 | c1 | c2 | c2 | c3 | c3 |

Figure 2. A small training set

The basic step in building associated rules is the detection of correlations: if all the examples associated with a descriptor d are also associated with a description d', then it is possible to generate the rule **If** d **then** d'. By taking into account some technical conditions that we shall not describe here, it is also possible to generate "indulgent rules", i.e. rules affected with some degree of plausibility $w$, when *most* of the examples covered by d are also covered by d'. The degree of plausibility $w$ is a number weighting the correlation with respect to the proportion of examples covered by d that are also covered by d'.

Without going into details, the main problem now, either with exact or with inexact rule generation, is to extract the prominent patterns from huge data sets. To do this, it is necessary to enumerate many descriptions d without enumerating all of them, which would be impossible.

An algebraic framework makes the systematization of the enumeration procedure possible. It is based on the notion of Galois connection.

For the sake of clarity, let us consider the table given in figure 2. It corresponds to a small training set which could easily be used as input of classical machine learning programs in order to induce rules or decision trees. From the table, it appears that the whole learning set could be classified using three production rules:

**R1:** **If** $A = a1$ **then** Class $= c1$,
**R2:** **If** $B = b2$ **then** Class $= c2$,
**R3:** **If** $A = a2$ & $B = b1$ **then** Class $= c3$.

The formal framework is based on the use of two lattices, the so-called *relation lattice*, noted $\mathcal{R}$, which contains all the possible descriptions and the *instance lattice*, noted $\mathcal{J}$, which corresponds to the set of parts of the training set. For instance, let us consider our training example: $((A = a2)\&(B = b1))$ and $(D = d1)$ belong to $\mathcal{R}$ whereas {E1 E2 E4} and {E5, E7} belong to $\mathcal{J}$.

Afterwards, two correspondences between these two lattices are introduced, $\gamma :$ $\mathcal{R} \rightarrow \mathcal{J}$ and $\beta : \mathcal{J} \rightarrow \mathcal{R}$:

- The function $\gamma$ associates to each description the set of all the training examples which are covered by this description, so $\gamma(((A = a2)\&(B = b1))) =$ {E6, E7} and $\gamma((B = b2)) = $ {E4, E5}.

- The function $\beta$ generates the most specific description common to all the examples of a subset of the training set. For instance, $\beta(\{E1, E3, E6\}) =$ $((B = b1) \& (D = d1))$ and $\beta(\{E1, E2, E3\}) = ((A = a1) \& (\text{Class} = c1))$.

These two correspondences define what is called a *Galois connection*, whose properties enable an induction mechanism to be built. As an example, we can see that $\beta \circ \gamma((A = a1)) = ((A = a1)\&(Class = c1))$. More generally, $\beta \circ \gamma(d)$ gives the most specific description which is implied by the description d, i.e. the set of all descriptors which are correlated with d in the learning set. This can then be simplified further, giving $\lambda_\mathcal{R}((A = a1)) = \beta \circ \gamma((A = a1)) - (A = a1) = (Class = c1)$. In a first approximation, the operator '—' may be assimilated to the subtraction.

The reader can also verify that it is possible to have $\lambda_\mathcal{R}((B = b2)) = (Class = c2)$ and $\lambda_\mathcal{R}(((B = b1)\&(A = a2))) = (Class = c3)$, which allows the generation of the knowledge base containing rules R1, R2 and R3 since $\lambda_\mathcal{R}(D) = d$ means **If D Then** d.

More precisely, the analytical expression of $\lambda_\mathcal{R}$ takes into account the properties of the lattice structure of $\mathcal{R}$, i.e. the ordering relationship and the existence of a least upper bound [noted $(a \vee b)$] and a greatest lower bound [noted $(a \wedge b)$] for each pair (a, b) of elements:

$$\lambda_\mathcal{R}(D) = \beta(\gamma(D)) - \bigvee_{D' \leq D} \beta(\gamma(D')) = \beta(\gamma(D)) - \bigvee_{D' \leq D} \lambda_\mathcal{R}(D')$$

Using this formal framework, it is possible to generate a minimal and complete set of rules, i.e. all the possible rules, without redundancy. For instance, if the rule "**If D Then** d" is generated, no rule of the form "**If D&E Then** d" will be generated.

## 4. Discovering the cause of scurvy

Our first experiment was an attempt to discover the cause of scurvy and to understand why it took so long to realize that fresh fruits and vegetables could cure the disease.

Let us remember that, many people, more than hundred of thousands, especially in the navy, contracted the disease and perished in the past. There were many possible explanations for this, for instance a "physical explanation" connecting disease to a cold temperature or to humidity, a "physiological explanation" making the lack of food responsible, or even a "psychological explanation". However, until the beginning the $20^{th}$ century, and the discovery of the role of vitamin C, physicians did not agree how to cure the disease, even when empirical evidence and clinical experiments confirmed the relation between the disease and the presence of fresh fruits and vegetables in the alimentary diet (cf. [2]).

We tempted to understand why it was not possible to induce the correct theory. We first consulted the 1880 *Dictionnaire Encyclopédique des Sciences Médicales* [16] which provides relatively precise description of 25 cases of scurvy, and we introduced those descriptions in our inductive engine [5,3]. More precisely, we used a small description language derived from the natural language expressions employed in the medical encyclopedia to describe those 25 cases. This language contained the ten following attributes, *year, location, temperature, humidity, food-quantity, diet-variety, hygiene, type-of-location, fresh-fruit/vegetables, affection-severity*, each of them being affected by one or more values according to its type (see figure 3). For instance, an ordered attribute may be affected by values belonging to an ordered scale.

| **Attribute** | **Type** | **Domain** |
|---|---|---|
| year | integer | N |
| location | string | NA |
| temperature | ordered set | severe-cold < cold < average <hot < very-hot |
| humidity | ordered set | low < high < very-high |
| food-quantity | ordered set | starvation < severe-restriction < ok |
| diet-variety | ordered set | low < average < high |
| hygiene | ordered set | very-bad < bad < average < good < very-good |
| type-of-location | unordered set | land, sea |
| fresh-fruit/vegetable | Boolean | yes, no |
| affection-severity | integer | 0, 1, 2, 3, 4, 5 |

Figure 3. Attributes used in the scurvy experiment

The 25 cases drawn from the medical encyclopedia were all described within this language. This original description may have been automatically completed with respect of the properties of the descriptors, which made the inductive engine able to take into account those properties. For instance, the following figure (see figure 4) provides an original description and its automatic completion.

The attribute "affection-severity" quantified the evolution of the disease, which was of crucial interest since it determined the factors that had influenced the evolution. In our experiment, we restricted our induction engine to generate only rules concluding on this last attribute.

Once those rules have been induced, it was possible to distribute them into small subsets, according to the attributes present in their premises. For instance, the attribute diet-variety being present in the condition of rule R8 (cf. figure 5), it was possible to aggregate it to the "diet-variety" cluster. Each of those clusters corresponded to some explanation schema of the disease, since it was the set of rules concluding to the severity of the disease, which contained a given attribute. For instance, in case of the "diet-variety" set, it corresponded to the theory that explained the evolution of the disease with the "diet-variety". The figure 5 shows the rules generated from the 25 examples of the encyclopedia, classified according the attributes they contain in their premises.

Note that our rules may have contained multiple attributes in their condition, in which case they were distributed in multiple explanatory schemata, i.e. in more than one rule set. Once those explanatory schemata were generated, it was possible to compute their explanation power as it was defined in section 2: to enumerate the number of examples which were covered by each of them.

The results showed that the "best theory", i.e. the theory with the higher explanation power, was the set of rules that contained the attribute "fresh fruits and vegetable" in their premise.

Moreover, it was possible to compare the different explanations given in the encyclopedia with the explanatory schemata generated from the 25 cases given in the same encyclopedia. It appeared that each set of rules corresponded to some

| Original description | Additional description obtained with only in-built general knowledge |
|---|---|
| (affection-severity = 0)<br>(fresh-fruits/vegetables = yes)<br>**Completion** →<br>(diet-variety = low)<br>(food-quantity = ok)<br>(type-of-location = land)<br>(location = Californie)<br>(year = 1604)<br>(hygiene = average) | (affection-severity < 1) (affection -severity < 2)<br>(affection-severity < 3) (affection -severity < 4)<br>(affection-severity < 5)<br><br>(diet-variety < average) (diet-variety < high)<br><br>(food-quantity > restrictions)<br>(food-quantity > severe-restrictions)<br><br>(hygiene > very-bad) (hygiene > bad)<br>(hygiene < very-good) (hygiene < good) |

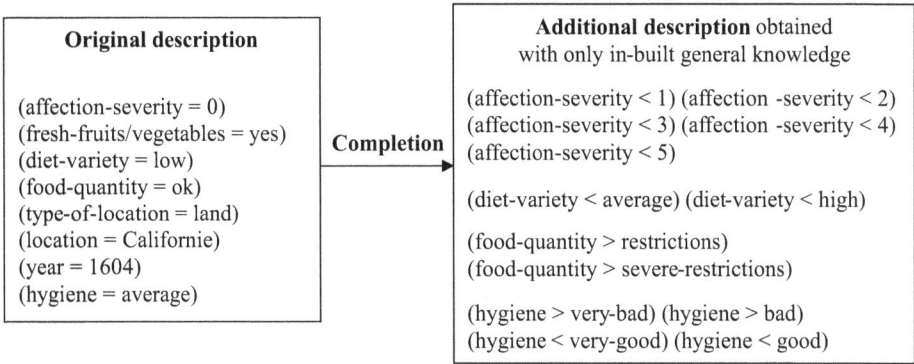

Figure 4. An example and its automatic completion

**Set I: Rules 3,4,8 use in their premises the variety of the diet.**
R3: IF diet-variety $\geq$ high THEN disease-severity $\leq$ 0. [5]
R4: IF diet-variety $\leq$ average THEN disease-severity $\geq$ 3. [4]
R8: IF diet-variety $\geq$ average THEN disease-severity $\leq$ 2. [11]

**Set II: Rules 7, 10 use in their premises the presence (or absence) of fresh fruits and vegetables in the diet.**
R7: IF fresh_fruits/vegetables = no THEN disease-severity $\geq$2. [5]
R10: IF fresh_fruits/vegetables = yes THEN disease-severity $\leq$2. [13]

**Set III: Rule 2 uses in its premises the quantity of food available.**
R2: IF food-quantity $\geq$ ok THEN disease-severity $\leq$0. [4]

**Set IV: Rules 5,6,9,12 use in their premises the level of hygiene.**
R5: IF hygiene $\leq$ bad THEN disease-severity $\geq$ 3. [3]
R6: IF hygiene $\leq$ average THEN disease-severity $\geq$ 2. [4]
R9: IF hygiene $\geq$ average THEN disease-severity $\leq$ 2. [7]
R12: IF hygiene $\geq$ good THEN disease-severity $\leq$ 1. [6]

**Set V: Rules 1, 11 use in their premises the temperature.**
R1: IF location = land, temperature $\geq$ hot THEN disease-severity $\leq$0. [4]
R11: IF temperature $\leq$ severe-cold THEN disease-severity $\geq$1. [5]

Figure 5. Rules generated without background knowledge

explanation given in the encyclopedia [Mahé 1880]. Let us quote here the mention of those explanations:

- Diet variety and fresh fruits and vegetables: *"It was J.F. Bachström (1734) who first expressed the opinion that, "Abstinence of vegetables is the only, the true, the first cause of scurvy.""*

- Food quantity: *"We are lead to conclude that a decrease in quantity of food, or to speak clearly, starvation, can occasionally serve the cause of scurvy, but it cannot produce it by itself."*

- Hygiene: *"If Cook's crews were entirely spared from scurvy, in a relatively large extent considering the times, it is thought that these great results were precisely the happy consequence of the care given to the cleanliness and drying of the ships."*

- Temperature: *"Spring and winter are obviously the seasons of predominance for scurvy."*

The explanation power (see figure 6) ordered those four explanatory schemata in accordance to the preference expressed by the authors of the medical encyclopedia even if the theory considered as the most plausible explanation of the scurvy, i.e. the theory of humidity, did not appear at all in this list. This was because there was no direct correlation between the disease severity and the humidity. But, it appears that the humidity was the most currently accepted hypothesis. Here is the quotation of the encyclopedia that mentions the theory of humidity as the most plausible: *"The influence of a cold and humid atmosphere has been said to be the key factor for the apparition of scurvy. "Air humidity is the main predisposing cause of this disease", according to Lind."* (cf. [Mahé 1880])

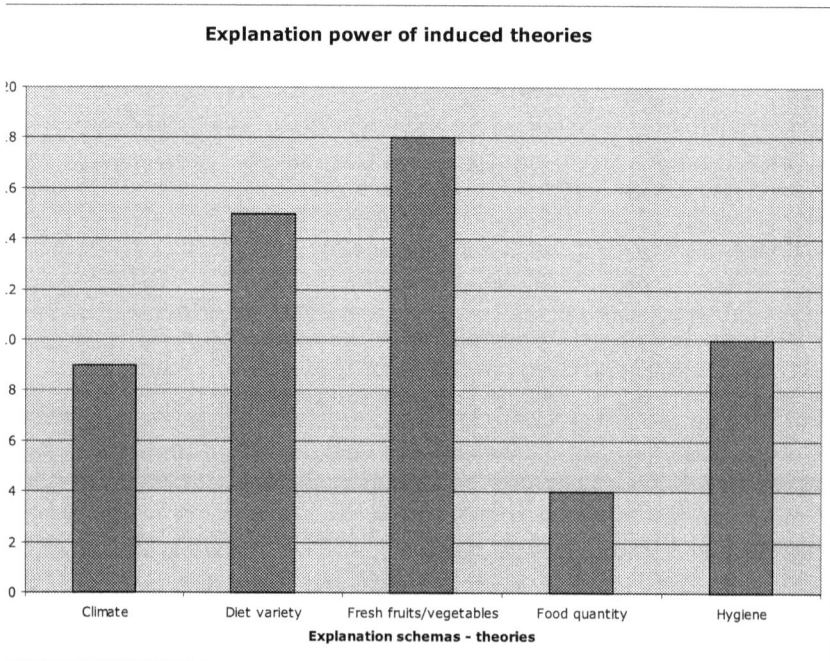

Figure 6. Explanation power of theories induced without background knowledge

In a sense, this first result was a good thing for artificial intelligence: it showed a machine able to induce the correct theory while people, with the same material, were not. However, it did not explain why, in the past, people adopted the humidity theory to explain the apparition and the evolution of scurvy. Because our goal is to model these kinds of wrong reasoning and the way people reason, we considered the result unsatisfactory by itself. Therefore, we tried to understand what biased their inductive ability. Then, we looked for some implicit medical theory that could influence induction. We found as a candidate "the blocked perspiration theory" that was prevalent in medical schools for centuries. This conception was based on the old theory of fluids introduced by Galien (131-201), during the $2^{nd}$ century. According to this hypothesis, without excretions and perspiration, the internal body amasses humors, especially bad humors, which result from fluid corruption and cause diseases. Since humidity and bad hygiene tend to block up pores of skin, it makes perspiration difficult and consequently it leads to accumulation of bad humors. Furthermore, lack of fresh fruits and vegetables thicken internal humors, which render theirs excretions more difficult.

We translated this theory by using two new attributes (cf. figure 7) and a few production rules which were introduced as background knowledge in our induction engine (cf. figure 8)

| Attribute | Type | Domain |
|---|---|---|
| perspiration | ordered set | normal < hard < blocked |
| fluids | ordered set | healthy < corrupted |

Figure 7. New attributes introduced to express blocked perspiration theory

---

IF humidity = high THEN perspiration ≥ hard
IF hygiene ≥ good, humidity ≤ high THEN perspiration ≤ hard
IF humidity ≥ very-high THEN perspiration ≥ blocked
IF perspiration ≤ hard THEN fluids ≤ healthy
IF fresh_fruits/vegetables = yes THEN fluids ≤ healthy
IF fresh_fruits/vegetables <> yes, perspiration ≥ blocked THEN fluids ≥
    corrupted
IF hygiene ≤ average, location = sea THEN humidity ≥ very-high
IF hygiene ≥ good THEN humidity ≤ high

---

Figure 8. Axiom set describing the "blocked perspiration" theory

Then, in addition to the rules generated previously, the inductive engine induced five more rules (cf. figure 9). Taking into account these rules, it appeared that the rules containing the attribute humidity constituted one of the possible explanatory schemata whose explanation power was higher than of other theories (cf. figure 10).

IF humidity ≥ high, fresh_fruits/vegetables = unknown
  THEN disease-severity ≥ 2. [4]
IF humidity ≤ high, hygiene ≥ average THEN disease-severity ≤ 1. [6]
IF perspiration ≤ hard THEN disease-severity ≤ 1. [6]
IF fluids ≥ corrupted THEN disease-severity ≥ 2. [9]
IF fluids ≤ healthy THEN disease-severity ≤ 2. [14]

Figure 9. New rules produced when the domain knowledge is given to the system

As a conclusion, we see here how adding some implicit knowledge during the inductive process may change the results: the theory that appears to be prevailing without background knowledge is dominated by another explanation that seems more satisfying in the sense that it explains more examples than the first.

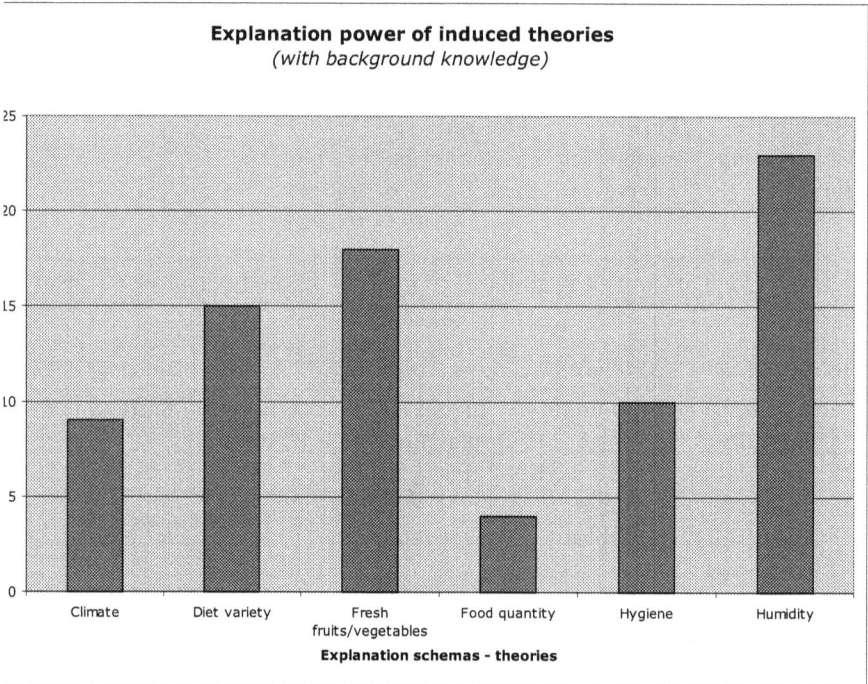

Figure 10. Explanation power of theories induced with background knowledge

This induction bias was caused both by the way the rules were induced, i.e. by the used induction engine, which was based on the notion of association rules, and by the lack of information. More precisely, it was mainly due to the partial description of examples. For instance, the alimentary diet and the presence of fresh fruits and vegetables were not always inserted in cases descriptions.

## 5.  A second medical example: the leprosy

To pursue our investigation, we shall now modify the representation language itself, i.e. the way examples are given to the machine. The effect of such transformation will be illustrated on another medical example: the problem of leprosy (cf. [4,3]).

History of leprosy dates back to ancient China and India. We focus here our study to the $19^{th}$ century medical views on this disease and to the conflict between two theories, the theory of contagion (cf. [6]) which explains the propagation of the disease by a mysterious agent that can pass from one person to another by physical contact, and an hereditary conception (cf. [22]) in which some people are genetically predisposed to contract the disease.

In 1874, a Norwegian physician, Gerhard A Hansen (cf. [12]), discovered the infectious agent, but, for ethical reasons, it was impossible to realize *in vivo* experiments that could validate or invalidate the still existing conflicting theories.

It was only during the second half of the $20^{th}$ century that researchers identified individual immune reactions, which could possibly be inherited. In other words, both theories were justified even if none of them was true. In order to understand both way of reasoning, we tried to apply our inductive engine to a case based on leprosy.

More precisely, the used training set contained 118 cases of leprosy in the Tamtaran Asylum (Punjab) reported by Gulam Mustafa (cf. [19]). The representation language contained 14 attributes (cf. figure 11).

| Attribute | Type | Domain |
|---|---|---|
| name | string | NA |
| sex | unordered set | m, f |
| caste | unordered set | Mussulman Sweeper Jheur Kohle Jat Rajpoot Musician Do_potter Do_teli Bahte |
| Age | Integer | NA |
| disease | Unordered set | mixed do_ anaesthetic tuberc |
| duration | integer | NA |
| father_affected | Boolean | yes no |
| mother_affected | Boolean | yes no |
| father_side | Boolean | yes no |
| mother_side | Boolean | yes no |
| Spouse | hierarchic | no yes (healty, sick) |
| children | unordered set | some_sick all_healthy |
| fish_diet | Ordered set | never rarely sometimes often very_often plenty in_excess |
| initial_location | unordered set | body arm leg hand foot joints face |

Figure 11. Attributes used in the leprosy experiments

Without background knowledge, the induction engine generated two main "indulgent" rules, R1 and R2 plus three minor rules:

R1: IF father_affected = No THEN children = all_healthy
R2: IF father_affected = No & Mother_side = yes &
    disease_type = anaesthetic & age > 35 THEN children = some_sick

Nowadays, those two rules could easily be interpreted as an hereditary reaction of the immune system to the presence of the bacillus. It also appears that the disease could be classified according to the reaction which corresponds to the $20^{th}$ century theory (cf. [21]).

As with the scurvy, we wanted to understand why $19^{th}$ century physician had not discovered this simple hereditary immunity. The first answer was that, for centuries diseases were only considered as positive entities, either animated material being, materiel things or immaterial being, for instance a demon (cf. [11]). Therefore, hereditary immunity, i.e. transmission of a negative entity, was not conceivable.

We have then reconstructed the path from those cases to the hereditary theory without reference to negative entities. It was done by introducing in the background knowledge some rules establishing relation between the symptoms and the affection itself. All these rules had the form: IF leprosy_symptom_X THEN X_affected where X may be replaced by mother, father, father_side, mother_side or spouse. Within this configuration, i.e. with those constraints and this background knowledge, the induction engine gave six rules which could be interpreted as a hereditary theory of the disease transmission:

R2: IF disease_type = do. THEN children = all_healthy
R4: IF disease_type = do & mother_affected = yes THEN children = some_sick
R5: IF disease_type = do & father_affected = yes THEN children = some_sick
R1: IF disease_type = anesth. THEN children = all_healthy
R3: IF disease_type = tuberc. THEN children = all_healthy
R6: IF disease_type = mixed THEN children = all_healthy

The last problem was to simulate the generation of the contagious theory. In order to do that, we introduced a new descriptor called the contagious index which roughly enumerates the number of contacts with people affected by the disease. More precisely, this contagious index was computed according to the following rule: *contagious_index = 2, plus 2 if husband or wife contracted the disease, plus 1 if father contracted, plus 1 if mother contracted, plus 1 if father's family contracted, plus 1 if mother's family contracted.* As a result, we had seven induced "indulgent" rules among which two were prominent, rules R1 and R2 that expressed the role of the contagious index:

R1: IF father_affected = yes THEN children = all_healthy
R2: IF father_affected = yes & contagious_index > 5 THEN children = some_sick

As a conclusion, it appears that by modifying the background knowledge, it was possible to change the way examples were interpreted by the induction engine, and, consequently to change the induced knowledge. One of the causes of this inductive bias was that examples were incompletely specified. The reason of these incomplete specifications was that men noticed only details that seemed relevant. Therefore, observation were not neutral; most of the time, they reflected the implicit scientific theory of that time.

Then, it should be of interest to compare the way examples are given to some implicit theories, and to see if some example sets are more adequate to some particular theory. Our last set of experiments is an attempt to investigate such a comparison.

## 6. Application to social sciences

We shall try now to study common sense reasoning. The goal is both to model the way people reason and to confront different inductions with different example sets. It is to know how preconceived ideas bias the judgements and the interpretation of facts. On the one hand, it is to extend our simulation of wrong reasoning to common sense knowledge. In this respect, it is an application of artificial intelligence techniques to social sciences where it could help to apprehend the way people react to singular cases. In the past, many mathematical and computer science models were used in sociology. However, those models were mainly based on statistical analysis. Our perspective is totally different: it is to model the way individuals reason and how they interpret facts, with respect to implicit theories they have in mind. In other word, it is to model social representations.

On the other hand, this application is an opportunity to compare induction with different data sets and to see how the way data are given influences the induced knowledge.

We focused here on xenophobia in France at the end of the $19^{th}$ century. We have chosen the first decade of September 1893, a few months before the Dreyfus affair burst. For all those ten days, three daily newspapers were fully scanned (cf. [26]), a conservative newspaper, "Le Matin" (cf. [15]), an anti-semitic strong right newspaper, "La Libre Parole" (cf. [14]) and a catholic one, "La Croix", also very conservative (cf. [13]). We gathered all published articles of social dysfunctions, such as political scandals, corruptions, bankrupts, robberies, murders etc. Each of those articles was viewed as a single case, described with a small representation language, similar to those used in the Scurvy and in the Leprosy experiments. This language contains 30 attributes corresponding to the political engagement of protagonists (socialist, radical, or conservative), their religion, their foreign origin, if they are introduced abroad, etc...

Sets of articles from each daily newspaper (here "Le Matin", "La Libre Parole" and "La Croix") were represented in the same way, with the same description language, but they were considered separately, each of them constituting a separate learning set.

Our goal was both to induce rules and theories, with each of those learning sets, but also, to introduce different implicit theories and to compare the adequacy of each learning set, i.e. of each set of examples, to each theory. Four different theories were considered to explain social disorders:

1. The first theory explains the deterioration of the society by an international Jewish and Freemason conspiracy.

2. The second theory mentions the loss of national traditions and qualities.

3. The third refers to incompetence and inability of politicians.

4. The last relies disorders to corruption

Those four theories were drawn from some historical studies (cf. [1,23]). We simplified and translated all of them into a set of production rules (cf. figure 12)

---

**International Jewish and Freemason conspiracy**
IF patriot = No & Introduced_abroad = Yes THEN Traitor = Yes
IF foreign_origin = Yes THEN Introduced_abroad = Yes
IF connection_with_affairs = Yes THEN connection_with_jews = Yes
IF traitor = Yes ∨ Internationalist = Yes ∨ connection_with_jews = Yes ∨
    connection_with_protestant = Yes ∨ freemasonry_involved = Yes ∨
    singular_action = demission ∨ singular_action = suicide THEN conspiracy = Yes

**National traditions and qualities**
IF traitor = Yes ∨ patriot = No ∨ favoritism = Yes ∨ respect_legislation = No ∨
    untouchable = Yes ∨ tendency = opportunist THEN morality = No

**Incompetence and inability**
IF political_scandal = Yes ∨ singular_action = demission ∨ singular_action = suicide
    Internationalist = Yes ∨ personal_problems = Yes ∨ health = bad ∨
    dangerous = Yes THEN incompetence = Yes

**Corruption**
IF connection_with_jews = Yes THEN connection_with_affairs = Yes
IF connection_with_affairs = Yes ∨ favoritism = Yes ∨ tendency = opportunist
    THEN corruption = Yes

---

Figure 12. Translation of each initial theory into production rules

Our aim here was not to study the effect of background knowledge on the explanation power, as it was the case in the two last studies, but to investigate the implicit knowledge concealed behind the examples. This is the reason why we needed different data sets, which correspond here to different sets of articles from different daily newspapers. For each of those data sets, we first induced explanation patterns, as we have done previously, by inserting our examples in the induction engine, without background knowledge. Then, we evaluated the explanation power of all generated explanatory schemata. We wanted to investigate here not those explanation patterns by themselves, but the implicit theory hidden in the back. In other words, newspapers seemed to be read by people with some embedded assumptions. To validate this idea, we introduced successively each of the four initial theories mentioned previously, in our induction engine, as background knowledge. Then, we computed again, for each of those theories, and for each of the data set, the explanation power of each explanation pattern.

For the sake of clarity, let us take an example: figure 13 shows the explanation power of explanation patterns built on four attributes, *tendency, morality, corruption* and *connection_with_Jews* without background theory (blue line) and with theory of corruption as background theory (red line).

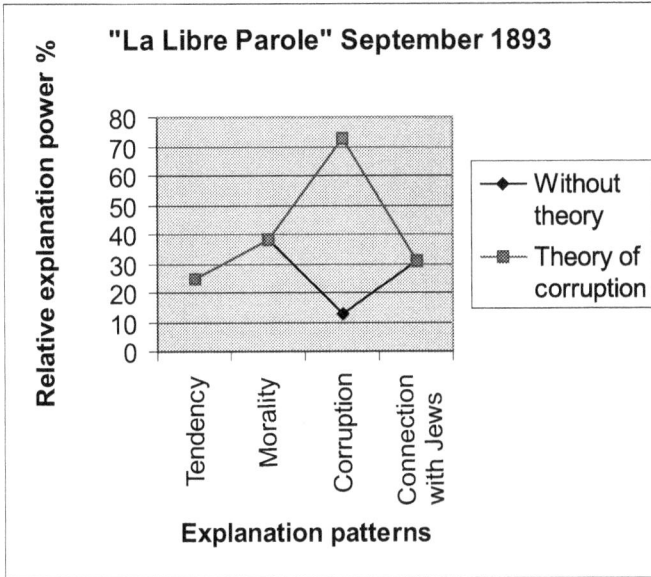

Figure 13. Relative explanation power of four different explanation patterns with and without the theory of corruption

It clearly appears that the presence of the theory of corruption makes higher the explanation power of the attribute corruption, and this renders all the examples more understandable. More technically, with this background theory, the percentage of examples that can be explained by some explanation pattern increases considerably. This remark may be generalized: for each theory, the optimal explanation power is noted, i.e. the highest explanation power, among all explanation powers of all explanation patterns.

The figure 14 summarizes the results that we have obtained. Each curve corresponds to one newspaper. The X-axe is associated with the different initial theories, the Y-axes, with the optimal relative explanation power, i.e. with the percentage of examples of the training set explained by the explanation patterns that has the highest explanation power.

The figure shows that the value of the optimal relative explanation power is in accordance with the tendency of the corresponding newspaper. For instance, the theory of corruption and the theory of conspiracy have a very high relative explanation power for "La Libre Parole", which is an Anti-Semitic extreme right newspaper. On the opposite, the relative explanation power of the theory of corruption is relatively low for "Le Matin" and "La Croix", two traditional and conservative newspapers. It means that the theory of corruption and the theory of conspiracy are implicit for most of the readers of "La Libre Parole", why both theories are not implicit for the remaining two.

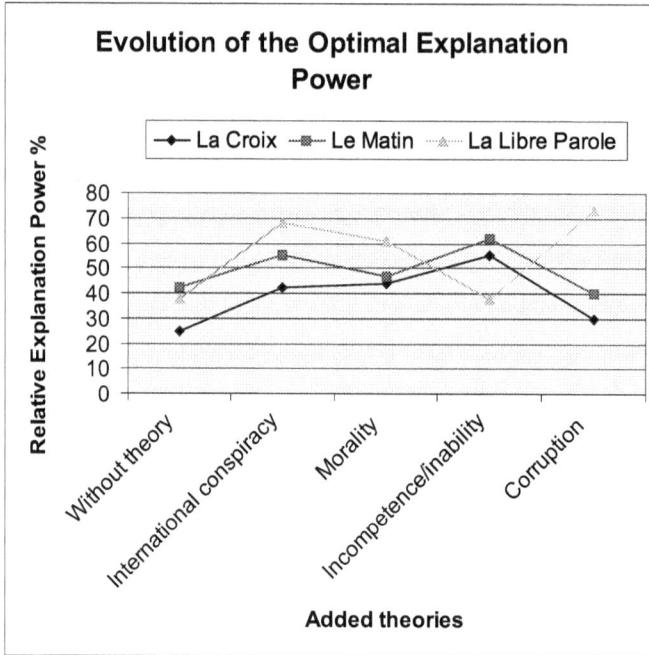

Figure 14. Evolution of relative explanation power with different theories

On the other hand, the theory of incompetence, that has the lower value for "La Libre Parole", seems to explain many examples drawn from "Le Matin" and "La Croix", even if it is less significant for "La Croix". Last point, the theory of morality appears to be more explicative than the theory of conspiracy for "La Croix" while it is the contrary for "Le matin". Since "La Croix" is a catholic newspaper and "Le Matin", just a conservative newspaper, this difference could be easily understandable. For more details concerning this study see [26].

As a conclusion, we observed that, simulating our model on different data sets with different implicit theories, it become apparent that some data sets were more easily understandable with one implicit theory than with the others, which means that data sets predisposes to some interpretations.

Since those implicit theories were directly connected with the political tendency of daily newspapers from which examples were drawn, it validates our model. In other words, it explains how examples induce misrepresentations.

Even if none of the examples is false, the way they are represented, the lack of description and the presence of implicit knowledge may considerably influence the induction. More precisely, examples lead people to construct an implicit theory, by abduction, and this implicit theory will then contribute to facilitate induction and generalization from examples.

## 7. Conclusion

We aimed here at explaining how wrong theories could be generated. We first presented a general framework where induction played a key role. This induction is simulated by an induction engine whose inductions are biased by a representation language, the implicit knowledge and the data sets.

We validated our model on three concrete examples, two in rational reconstruction of old medical theories, one in rational reconstruction of mentalities.

From a technical point of view, each of those three examples deals with some particular question. The first example focuses on the role of background knowledge, which modifies the explanation power of attributes, making artificially one explanation schema more explanatory than the others. The second example shows how changing the description language transforms the results: some theories may appear – as the theory of propagation by contagion for the leprosy – with adequate parameters. Finally, we saw how unseen theories are implicitly present in the way examples are given.

In the future, it would be of interest to pursue the validation of this model in two different directions. On the one hand, we want to evaluate its relevancy in different situations. The one is to systematize and extend our social sciences studies, the ultimate goal being to build some rational reconstruction of mentalities and social representations. The second is to confront this general model of wrong reasoning with psychological experiments.

On the other hand, from a technical perspective, we are currently extending the model with new induction engines based on the notion of *default generalization* prompted by the default logic theory and using non-supervised learning techniques. Then, our model will not be restricted to induce correlations onto some predefined attribute describing, for instance, deterioration of society or the evolution of the disease. It will induce some free associations corresponding to implicit stereotypes. For instance some family names refers to some ethnics groups and then to poverty, robbery and crime, or to richness etc. In other words, it is to automatically induce some kinds of abstracts caricatures from examples, and to confront them with social representations.

## REFERENCES

1. Bredin J. D., *L'affaire*, Julliard, 1983
2. Carpenter K. J., The history of scurvy and vitamin C (Cambridge University Press, 1986).

3. Corruble V., Une approche inductive de la découverte en médecine: les cas du scorbut et de la lèpre (in French), PhD thesis, Université Pierre et Marie Curie - Paris 6 (1996).

4. Corruble V., Ganascia J. G., Discovery of the Causes of Leprosy: a Computational Simulation, in: Proceedings of 13th National Conference on Artificial Intelligence, Portland (OR), USA (1996) 731-736.

5. Corruble V., Ganascia J. G., Induction and the discovery of the causes of scurvy: a computational reconstruction. *Artificial Intelligence Journal*, Special issue on scientific discovery. Elsevier Press, (91)2 (1997) pp. 205-223

6. Drognat-Landré, C. L., 1869. De la contagion, seule cause de la propagation de la lèpre. Paris: G. Baillère

7. Drumont E., *La France Juive*, Paris, V. Palmé, 1886

8. Ganascia J. G., CHARADE : A rule System Learning System. 10th IJCAI, Milan, 1987

9. Ganascia J. G., Deriving the learning bias from rule properties. Hayes J. E., Michie D. & Tyugu E., eds., *Machine intelligence* 12. Towards an automated logic of human thought.- Clarendon Press, 1991

10. Ganascia J. G., TDIS: An algebraic generalization. IJCAI-93 International Joint Conference on Artificial Intelligence, Chambéry, France, 1993

11. Grmek, M. D., 1995. Le concept de maladie. *Histoire de la pensée médicale en Occident*. Ed. M. Grmek. v. 1. Seuil, 1995

12. Hansen, G. Armauer. 1875. On the etiology of leprosy. *British and foreign medico-chirurgical review*, 55.

13. La Croix, daily newspaper from September the $1^{st}$ 1893 to September the $10^{th}$ 1893

14. La Libre Parole, daily newspaper from September the $1^{st}$ 1893 to September the $7^{th}$ 1893

15. Le Matin, daily newspaper from September the $1^{st}$ 1893 to September the $10^{th}$ 1893

16. Mahé J., Le scorbut (in French), in: *Dictionnaire Encyclopédique des Sciences Médicales*, Série 3, Tome 8 (Masson, Paris, 1880) 35-257.

17. Michalski, R., Carbonell, J. G., & Mitchell, T., eds., *Machine Learning: An Artificial Intelligence Approach*, Los Altos, CA: Morgan Kaufmann, 1986.

18. Mitchell T., *Machine Learning*, McGraw Hill, 1997

19. Phineas, S. A., 1889. Analysis of 118 cases of leprosy in the Tarntaran Asylum (Punjab) reported by Gulam Mustafa, Assistant Surgeon. *Transactions of the Epidemiological Society of London*. v. 9 (1889-1890)

20. Plato, Meno

21. Ridley D. S. and Jopling W. H., 1966. Classification of Leprosy According to Immunity, A Five-Group System. *International Journal of Leprosy*. v. 54, n. 3. pp. 255-273.

22. Royal College of Physicians, 1867. Report on leprosy. London.

23. Taguieff P.-A., *La couleur et la sang, Doctrines racistes à la française*, Paris, Mille et une nuits, 2002.

24. Thagard P., *Computational Philosophy of Science* (MIT Press, Cambridge, 1988).

25. Thagard P. and Nowak G., The Conceptual Structure of the Geological Revolution, in: J. Shrager and P. Langley, eds., *Computational Models of Scientific Discovery and Theory Formation* (Morgan Kaufmann, 1990) 27-72.
26. Velcin J., Reconstruction rationnelle des mentalités collectives: deux études sur la xénophobie, DEA report, Internal Report University Paris VI, Paris, 2002.

# Structural Equation Modeling, Causal Inference and Statistical Adequacy

Aris Spanos*

*Department of Economics, Virginia Tech, Blacksburg, VA 24061, USA,*
*aris@vt.edu*

**Abstract.** The primary aim of the paper is to use the last century of experience in econometric modeling as a backdrop, in order to draw valuable lessons concerning the reliability of empirical modeling that can be of value to Graphical Causal Modeling. It is argued that the fundamental weakness of empirical modeling in economics has been the unreliability of inference arising from the fact that estimated models are invariably (statistically) misspecified. In an attempt to address this issue, the author has proposed a particular approach to empirical modeling, the Probabilistic Reduction approach, which is summarized and illustrated using an empirical example. Some of the lessons learned, including the role of statistical adequacy, are then used to shed light on certain aspects of Graphical Causal modeling.

## 1. Econometrics: a <u>very brief</u> history

### 1.1. Early pioneers, 1895-1935

The roots of modern econometrics, defined as probability-based modeling and inference in economics, can be traced back to the late 19th and early 20th century. The primary focus of empirical research during the early period was on two primary areas (see Morgan, 1990, Hendry and Morgan, 1995):

(a) **business cycles:** the ups and downs of economic activity, and

(b) **demand functions:** intention scheduless to buy particular quantities of a commodity.

The business cycle research began as *descriptive statistics* using economic time series, but culminated in the form of *chain-causal macroeconometric models* of Tinbergen (1937,1939). The research on demand curves began as *structural modeling* using regression, and reached a certain level of maturity with the work of Schultz (1938) in the form of *choice of regression.*

This early empirical research in economics was firmly rooted in the 'curve fitting' (*descriptive statistics*) tradition associated with Gauss, Edgeworth and Karl Pearson:

---

*I would like to thank Deborah G. Mayo for encouraging me toward the graphical causal modeling, as well as for many valuable suggestions. Thanks are also due to my fellow symposiasts on "Philosophy and Methodology of Empirical Modeling: Causation, Validation, and Discovery", Clark Glymour, Peter Spirtes and Jim Woodward, for valuable discussions and suggestions.

observed data are used to 'depict' a theory-model in statistical
terms.

During the early period (up until the 1930s) the focus was almost exclusively on
*estimation* with only informal references to testing.

## 1.2. The Cowles Commission, circa 1940s
### 1.2.1. From 'curve fitting' to statistical inference proper

The foundations and the overarching framework for quantitative research in
economics were to change during the 1940s. The Cowles Commission (see Koopmans, 1939,1950), Hood and Koopmans, 1953) introduced formal *statistical inference* procedures associated with Fisher (maximum likelihood estimation) and
Neyman-Pearson (hypothesis testing) into econometrics in their attempt to formalize a blueprint of econometrics revolving around the simultaneous equations
model proposed by Haavelmo (1943,1944).

The transition from 'curve fitting' to 'statistical inference' proper during the
1940s and early 1950s was rather subtle and led to a number of tensions concerning the proper role of theory vs. data in empirical modeling which linger on even
today. These tensions were first articulated in two exchanges between Keynes and
Tinbergen (1939, 1940) and Koopmans and Vining (1947,1949), but were never
fully resolved in the econometric literature. Due to the influence of the Cowles
Commission, the early emphasis on **regression** shifted toward **Structural Equations Models** (SEMs) by the 1950s.

**Causal chain models.** The only dissenting view against Haavelmo's simultaneous equations, giving rise to structural models with feedback, was Wold (1952,1954,
1956,1960). He argued passionately against *simultaneous equations models* (SEMs
with feedback) and proposed the *causal chain* models that gave rise to **recursive
structural models** as a more appropriate alternative. He did not convince econometricians of the merits of his case, and after Basmann (1965) demonstrated that
any causal chain model can be recast as an observationally equivalent simultaneous
equations model, he lost the argument.

## 1.3. The textbook econometric tradition, *circa* 1960s

The textbook econometric tradition took shape in the early 1960s (see Johnston,1963, Goldberger, 1964) and amounted to an incongruous hybrid of the 'curve
fitting' and statistical inference traditions, without resolving the inherited tension
between theory vs. data in empirical modeling. The primary focus of the econometric literature during the period 1960-1980 were the problems of **identification**
and **estimation** in SEMs with dozens of estimators being proposed; see Dhrymes
(1994). The proliferation of estimators slowed down when it was demonstrated
that they were all numerical approximations to Maximum Likelihood Estimators;
see Hendry (1976).

### 1.3.1. Simultaneous Equations models in econometrics *circa* 1980

A Structural Equation Model in econometrics is usually specified in the form of a
deterministic theory model with added structural errors. In econometrics *Structural
Models* are often conflated with *Simultaneous Equations* models; ignoring single
equation structural models.

**Example.** Consider the two-equation structural model:

$$
\begin{aligned}
y_{1t} &= \gamma_{12}y_{2t}+ \delta_{11}x_{1t}+ \delta_{12}x_{2t}+ \delta_{13}x_{3t}+ \varepsilon_{1t} \\
y_{2t} &= \gamma_{21}y_{1t}+ \delta_{21}x_{1t}+ \delta_{24}x_{4t}+ \delta_{25}x_{5t}+ \varepsilon_{2t}
\end{aligned}
$$

$y_{1t}$ - aggregate money, $y_{2t}$ - interest rate, $x_{1t}$- price level, $x_{1t}$- price level, $x_{2t}$- aggregate income, $x_{3t}$- net exports/imports, $x_{4t}$- government budget deficit, $x_{5t}$- Fed-controlled interest rate.$( \ y_{1t} \ , y_{2t} \ )$ are the **endogenous** variables (their behavior is described by the system of equations), and $( \ x_{1t}, \quad x_{2t}, \quad x_{3t}, \quad x_{4t}, \quad x_{5t} \ )$ are the **exogenous** variables (their behavior determined outside this system).

**Structural form:** $\mathbf{\Gamma}^{\top}\mathbf{y}_t = \mathbf{\Delta}^{\top}\mathbf{x}_t + \varepsilon_t, \ \varepsilon_t \backsim N(\mathbf{0}, \mathbf{\Omega}), (\mathbf{\Gamma}, \mathbf{\Delta}, \mathbf{\Omega}) \in \Phi,$

$$
\mathbf{y}_t : m \times 1, \mathbf{x}_t : k \times 1, \ \mathbf{\Gamma}^{\top} := \begin{pmatrix} 1 & -\gamma_{12} \\ -\gamma_{21} & 1 \end{pmatrix}, \mathbf{\Delta}^{\top} := \begin{pmatrix} \delta_{11} & \delta_{12} & \delta_{13} & 0 & 0 \\ \delta_{21} & 0 & 0 & \delta_{24} & \delta_{25} \end{pmatrix}.
\tag{1}
$$

**Reduced form.** Corresponding to each Structural form there is a *Reduced form*, which expresses the 'endogenous' variables as functions only of the 'exogenous' variables

$$
\begin{aligned}
y_{1t} &= \beta_{11}x_{1t}+ \beta_{12}x_{2t}+ \beta_{13}x_{3t}+ \beta_{14}x_{4t}+ \beta_{15}x_{5t}+ u_{1t} \\
y_{2t} &= \beta_{21}x_{1t}+ \beta_{22}x_{2t}+ \beta_{23}x_{3t}+ \beta_{24}x_{4t}+ \beta_{25}x_{5t}+ u_{2t}
\end{aligned}
$$

**Reduced form:** $\mathbf{y}_t = \mathbf{B}^{\top}\mathbf{x}_t + \mathbf{u}_t, \quad \mathbf{u}_t \backsim N(\mathbf{0}, \mathbf{\Sigma}), \ (\mathbf{B}, \mathbf{\Sigma}) \in \Theta.$

**Structural vs. Reduced form:** $\mathbf{y}_t = \left(\mathbf{\Gamma}^{\top}\right)^{-1} \mathbf{\Delta}^{\top}\mathbf{x}_t + \left(\mathbf{\Gamma}^{\top}\right)^{-1} \varepsilon_t$

(i) $\mathbf{B}\mathbf{\Gamma} = \mathbf{\Delta}$     (ii) $\mathbf{\Omega} = \left(\mathbf{\Gamma}^{\top}\mathbf{\Sigma}\mathbf{\Gamma}\right)$

**Example.** Consider these 'implicit restrictions' for the example given above.

$$
\begin{array}{lllll}
\beta_{11}=\beta_{21}\gamma_{12} + \delta_{11} & \beta_{12}=\beta_{22}\gamma_{12} + \delta_{12} & \beta_{13}=\beta_{23}\gamma_{12} + \delta_{13} & \beta_{22}=\beta_{12}\gamma_{21} & \beta_{14}=\beta_{24}\gamma_{12} \\
\beta_{21}=\beta_{11}\gamma_{21} + \delta_{21} & \beta_{24}=\beta_{14}\gamma_{21} + \delta_{24} & \beta_{25}=\beta_{15}\gamma_{21} + \delta_{25} & \beta_{23}=\beta_{13}\gamma_{21} & \beta_{15}=\beta_{25}\gamma_{12}
\end{array}
\tag{2}
$$

These 'restrictions' are defined via *implicit mappings* between the reduced ($\boldsymbol{\theta} :=(\mathbf{B}, \mathbf{\Sigma})$) and structural form parameters $\boldsymbol{\alpha}$ (the unknown parameters in $(\mathbf{\Gamma}, \mathbf{\Delta}, \mathbf{\Omega})$).

**Identification**

Assuming that $\mathbf{\Sigma} > \mathbf{0}$, the reduced form parameters $\boldsymbol{\theta} :=(\mathbf{B}, \mathbf{\Sigma})$ are uniquely determined by the data $(\mathbf{X}, \mathbf{y})$ via the likelihood function to yield the Maximum Likelihood Estimators (MLE): $\widehat{\mathbf{B}}= (\mathbf{X}^{\top}\mathbf{X})^{-1}\mathbf{X}^{\top}\mathbf{y}, \ \widehat{\mathbf{\Sigma}}=\frac{1}{T}(\mathbf{Y} - \mathbf{X}\widehat{\mathbf{B}})^{\top}(\mathbf{Y} - \mathbf{X}\widehat{\mathbf{B}}).$

*Identification* of the structural parameters $\boldsymbol{\alpha}$ boils down to: "can we 'solve' *uniquely* for $\boldsymbol{\alpha} = \mathbf{H}(\boldsymbol{\theta})$ the implicit system of equations?

(i) $\mathbf{B}\mathbf{\Gamma}(\boldsymbol{\alpha}) = \mathbf{\Delta}(\boldsymbol{\alpha}),$   (ii) $\mathbf{\Omega}(\boldsymbol{\alpha}) = \left(\mathbf{\Gamma}^{\top}(\boldsymbol{\alpha})\mathbf{\Sigma}\mathbf{\Gamma}(\boldsymbol{\alpha})\right)$

**Observational equivalence.** For every identifiable structural model $\mathbf{\Gamma}^\top \mathbf{y}_t = \mathbf{\Delta}^\top \mathbf{x}_t + \boldsymbol{\varepsilon}_t$, there exist an infinite number of *observationally equivalent* reduced form models since, for any *non-singular* $m \times m$ matrix $\mathbf{D}$, the structural model
$\mathbf{D}\mathbf{\Gamma}^\top \mathbf{y}_t = \mathbf{D}\mathbf{\Delta}^\top \mathbf{x}_t + \mathbf{D}\boldsymbol{\varepsilon}_t$, has an identical reduced form since:

$$\mathbf{y}_t = \left(\mathbf{\Gamma}^\top \mathbf{D}\right)^{-1} \mathbf{D}\mathbf{\Delta}^\top \mathbf{x}_t + \left(\mathbf{\Gamma}^\top \mathbf{D}\right)^{-1} \mathbf{D}\boldsymbol{\varepsilon}_t = \left(\mathbf{\Gamma}^\top\right)^{-1} \mathbf{\Delta}^\top \mathbf{x}_t + \left(\mathbf{\Gamma}^\top\right)^{-1} \boldsymbol{\varepsilon}_t.$$

**Example.** The two-equation structural model:

$$
\begin{aligned}
y_{1t} &= \gamma_{12} y_{2t} + \delta_{11} x_{1t} + \delta_{12} x_{2t} + \delta_{13} x_{3t} + \varepsilon_{1t} \\
y_{2t} &= \gamma_{21} y_{1t} + \delta_{21} x_{1t} + \delta_{24} x_{4t} + \delta_{25} x_{5t} + \varepsilon_{2t}
\end{aligned}
\tag{3}
$$

has 11 unknown *structural parameters:* $\boldsymbol{\alpha} := (\gamma_{12}, \gamma_{21}, \delta_{11}, \delta_{12}, \delta_{13}, \delta_{21}, \delta_{24}, \delta_{25}, \omega_{11}, \omega_{12}, \omega_{22})$. The corresponding reduced form:

$$
\begin{aligned}
y_{1t} &= \beta_{11} x_{1t} + \beta_{12} x_{2t} + \beta_{13} x_{3t} + \beta_{14} x_{4t} + \beta_{15} x_{5t} + u_{1t} \\
y_{2t} &= \beta_{21} x_{1t} + \beta_{22} x_{2t} + \beta_{23} x_{3t} + \beta_{24} x_{4t} + \beta_{25} x_{5t} + u_{2t}
\end{aligned}
\tag{4}
$$

has 13 *statistical parameters:* $\boldsymbol{\theta} := (\beta_{11}, \beta_{12}, \beta_{13}, \beta_{14}, \beta_{15}, \beta_{21}, \beta_{22}, \beta_{23}, \beta_{24}, \beta_{25}, \sigma_{11}, \sigma_{12}, \sigma_{22})$.

Using the order and rank conditions for identification (see Dhrymes, 1994), it can be shown that both equations are overidentified; they have one overidentifying restriction each.

### Estimation

**Maximum Likelihood Estimation.**

Assuming that the statistical assumptions underlying the reduced form are valid for data $(\mathbf{X}, \mathbf{y})$, one can use the Likelihood Method to estimate the structural parameters $\boldsymbol{\alpha}$ by maximizing the log-likelihood function:

$$\ln L(\boldsymbol{\alpha}) = \text{const} - \frac{T}{2} \ln(\det \mathbf{\Omega}) + T \ln(|\det \mathbf{\Gamma}|) - \frac{1}{2}\text{tr}\left[\mathbf{\Omega}^{-1}(\mathbf{Y}\mathbf{\Gamma} - \mathbf{X}\mathbf{\Delta})^\top (\mathbf{Y}\mathbf{\Gamma} - \mathbf{X}\mathbf{\Delta})\right]$$

**Modified Least-Squares**

The estimation of the structural parameters $\boldsymbol{\alpha}$ can also be seen as constrained least-squares based on the log-likelihood of the reduced form:

$$\ln L(\boldsymbol{\theta}) \propto -\frac{T}{2} \ln(\det \mathbf{\Sigma}) - \frac{1}{2}\text{tr}\left[\mathbf{\Sigma}^{-1}(\mathbf{Y} - \mathbf{X}\mathbf{B})^\top (\mathbf{Y} - \mathbf{X}\mathbf{B})\right]$$

subject to the 'constraints': (i) $\mathbf{B}\mathbf{\Gamma}(\boldsymbol{\alpha}) = \mathbf{\Delta}(\boldsymbol{\alpha})$, (ii) $\mathbf{\Omega}(\boldsymbol{\alpha}) = \left(\mathbf{\Gamma}^\top(\boldsymbol{\alpha})\mathbf{\Sigma}\mathbf{\Gamma}(\boldsymbol{\alpha})\right)$.

The easiest of these methods is the **Two Stage Least-Squares (2SLS)**. Let us illustrate the method by estimating the structural model (3). Ordinary Least-Squares (OLS) applied to the above system will give rise to **inconsistent estimators** because the error terms are correlated with all the Right Hand Side (R.H.S.) variables. In particular, Haavelmo (1943) showed that:

$$Cov(y_{2t} \cdot \varepsilon_{1t} \mid x_{1t}, x_{2t}, x_{3t}) \neq 0, \, Cov(y_{1t} \cdot \varepsilon_{2t} \mid x_{1t}, x_{4t}, x_{5t}) \neq 0.$$

Solution: Estimate the structural parameters
$\boldsymbol{\alpha} := (\gamma_{12}, \gamma_{21}, \delta_{11}, \delta_{12}, \delta_{13}, \delta_{21}, \delta_{24}, \delta_{25}, \omega_{11}, \omega_{12}, \omega_{22})$ in two stages.

**Stage 1.** Estimate the reduced form by least-squares:

$$y_{1t} = \widehat{\beta}_{11} x_{1t} + \widehat{\beta}_{12} x_{2t} + \widehat{\beta}_{13} x_{3t} + \widehat{\beta}_{14} x_{4t} + \widehat{\beta}_{15} x_{5t} + \widehat{u}_{1t} = \widehat{y}_{1t} + \widehat{u}_{1t}$$
$$y_{2t} = \widehat{\beta}_{21} x_{1t} + \widehat{\beta}_{22} x_{2t} + \widehat{\beta}_{23} x_{3t} + \widehat{\beta}_{24} x_{4t} + \widehat{\beta}_{25} x_{5t} + \widehat{u}_{2t} = \widehat{y}_{2t} + \widehat{u}_{2t}$$

where $(\widehat{y}_{1t}, \widehat{y}_{2t})$ denotes the *fitted values*.

**Stage 2.** Replacing the R.H.S. $(y_{1t}, y_{2t})$ in (3) with $(\widehat{y}_{1t} + \widehat{u}_{1t}, \widehat{y}_{2t} + \widehat{u}_{2t})$ :

$$y_{1t} = \gamma_{12} \widehat{y}_{2t} + \delta_{11} x_{1t} + \delta_{12} x_{2t} + \delta_{13} x_{3t} + \varepsilon_{1t} + \gamma_{12} \widehat{u}_{2t}$$
$$y_{2t} = \gamma_{21} \widehat{y}_{1t} + \delta_{21} x_{1t} + \delta_{24} x_{4t} + \delta_{25} x_{5t} + \varepsilon_{2t} + \gamma_{21} \widehat{u}_{1t}$$

Estimating this transformed system by OLS gives rise to consistent estimators for the structural parameters $\boldsymbol{\alpha}$. **How does it work?**

The substitution achieves two things (Spanos, 1986):

(i) **It redefines the error terms**, transforming the structural errors into reduced form errors, from $(\varepsilon_{1t}, \varepsilon_{2t})$ to $(\varepsilon_{1t} + \gamma_{12}\widehat{u}_{2t}, \varepsilon_{2t} + \gamma_{21}\widehat{u}_{1t})$ to render them *orthogonal* to the R.H.S. variables $\mathbf{x}_t$. Given that $\boldsymbol{\Gamma}^\top \mathbf{u}_t = \boldsymbol{\varepsilon}_t$, $u_{1t} = \varepsilon_{1t} + \gamma_{12} u_{2t}$, $u_{2t} = \varepsilon_{2t} + \gamma_{21} u_{1t}$.

(ii) Moreover, the substitution **imposes the 'identifying' restrictions** (2):

$$y_{1t} = \left[\gamma_{12}\widehat{\beta}_{21} + \delta_{11}\right] x_{1t} + \left[\gamma_{12}\widehat{\beta}_{22} + \delta_{12}\right] x_{2t} +$$
$$+ \left[\gamma_{12}\widehat{\beta}_{23} + \delta_{13}\right] x_{3t} + \gamma_{12}\widehat{\beta}_{24} x_{4t} + \gamma_{12}\widehat{\beta}_{25} x_{5t} + u_{1t}$$
$$y_{2t} = \left[\gamma_{21}\widehat{\beta}_{11} + \delta_{21}\right] x_{1t} + \gamma_{21}\widehat{\beta}_{12} x_{2t} + \gamma_{21}\widehat{\beta}_{13} x_{3t} +$$
$$+ \left[\gamma_{21}\widehat{\beta}_{14} + \delta_{24}\right] x_{4t} + \left[\gamma_{21}\widehat{\beta}_{15} + \delta_{25}\right] x_{5t} + u_{2t}$$

as can be seen by comparing this with the reduced form (4).

**Testing the overidentifying restrictions**

Most of the above constraints are needed to identify the structural parameters and these are not testable. Any restrictions, however, over and above the ones needed for identification, the so-called *overidentifying* restrictions, are testable!

As shown above, the structural parameters $\boldsymbol{\alpha}$ constitute a *reparameterization/restriction* of the statistical parameters $\boldsymbol{\theta} := (\mathbf{B}, \boldsymbol{\Sigma})$. When the structural model is **just identified**, the mapping: $\boldsymbol{\alpha} = \mathbf{H}(\boldsymbol{\theta})$, $\boldsymbol{\theta} \in \boldsymbol{\Theta}$, is bijective (one-to-one and onto) - defining a *reparameterization* with $\boldsymbol{\theta} = \mathbf{H}^{-1}(\boldsymbol{\alpha})$, $\boldsymbol{\alpha} \in \boldsymbol{\Phi}$. When the structural model is **overidentified** the mapping is surjective (many to one) - defining a *reparameterization/restriction*, because the *pre-image* of the mapping $\mathbf{H}(.)$ imposes restrictions on the statistical parameters: $\boldsymbol{\theta}^* = \mathbf{H}^-(\boldsymbol{\alpha})$, $\boldsymbol{\alpha} \in \boldsymbol{\Phi} \Rightarrow \boldsymbol{\theta}^* \in \boldsymbol{\Theta}_1 \subset \boldsymbol{\Theta}$. The test of overidentifying restrictions is based on the hypotheses: $H_0 : \boldsymbol{\theta} = \boldsymbol{\theta}^*$ vs. $H_1 : \boldsymbol{\theta} \neq \boldsymbol{\theta}^*$. The idea is that structural models with *more such restrictions* are less data-specific and thus more *informative!*

### 1.3.2. Problems with Structural Equation Modeling (SEM) in Econometrics

**The early promise of SEM that never materialized.**

In the early 1960s econometricians were promising politicians and business executives a reliable new tool that would help them tremendously with their policy decisions. All they had to do was to invest in macro-econometric models build on SEM, and econometricians will deliver empirical models that can be used for forecasting and policy simulation purposes with reliability and accuracy never seen before! By the late 1970s, after investing of a few billion dollars, politicians and business executives in most developed countries ended up with 'monstrosities' running up to 2400 equations whose forecast accuracy could not even compare with a univariate ARIMA model and their policy simulation reliability was highly questionable! **What went wrong?** The jury is still out, more than 20 year later!

### A personal diagnosis

The diagnosis that follows is based on Spanos (1986, 1989,1990). Theories were 'inflicted' upon the data, ignoring two fundamental problems that render the resulting inference unreliable and often misleading:

(a) the sizeable gap between theory concepts and observed data, and

(b) the validity of the premises for inference - the probabilistic assumptions underlying the statistical models utilized.

Largely as a result of these weaknesses, empirical modeling in econometrics is currently viewed with suspicion and often (justifiably so) considered unreliable. The key to unraveling the methodological issues arising from these fundamental problems is the distinction between a theory or a *structural model* and a *statistical model*.

A **structural model** provides an idealized description of certain aspects of the phenomenon of interest in the form of a 'nearly isolated' mathematical system. As such a structural model demarcates the segment of reality to be modeled and suggests the facets of the phenomenon to be measured, the end result being a particular data set $(\mathbf{z}_1, \mathbf{z}_2, ..., \mathbf{z}_n)$.

A **statistical model** is an internally consistent set of probabilistic assumptions defining a stochastic process $\{\mathbf{Z}_t, \ t \in \mathbb{T}\}$ such that the data $(\mathbf{z}_1, \mathbf{z}_2, ..., \mathbf{z}_n)$ could be realistically viewed as a 'truly typical realization' of.

The statistical analysis of the structural model requires the modeler to embed it into a statistical model from which it derives its statistical content. That 'embedding' can often be a source of confusion between *structural* and the corresponding *statistical concepts, issues and problems.*

**Important lessons** *I learned* after studying 100 years of empirical modeling in econometrics:

> Lesson 1. Underlying the inference associated with any 'structural model' there exists a 'statistical model' (implicitly or explicitly), which provides the link between theory and reality (phenomenon of interest) as 'captured' by the data.

> Lesson 2. A necessary condition for the link between theory and reality to be 'trusty' is that the probabilistic assumptions defining the statistical model are valid vis-a-vis the data in question. Whenever the statistical model is misspecified, the link is tenuous at best.

**The misconstrued 'reduced form'.**

The statistical model underlying a structural model is the 'reduced form'. Because it is viewed as *derivable* from the structural model, its **statistical adequacy** is often neglected, with dire consequences for the reliability of inference. The reduced form: $\mathbf{y}_t = \mathbf{B}^\top \mathbf{x}_t + \mathbf{u}_t$, $\mathbf{u}_t \backsim N(\mathbf{0}, \boldsymbol{\Sigma})$, is nothing but a Multivariate Linear Regression (MLR), whose complete set of probabilistic assumptions, is given in table 1.

### Table 1 - The Multivariate Linear Regression (MLR) Model

$$\mathbf{y}_t = \beta_0 + \mathbf{B}_1^\top \mathbf{x}_{1t} + \mathbf{u}_t, \; t \in \mathbb{T},$$

| | | |
|---|---|---|
| (1) | **Normality:** | $\mathbf{u}_t \backsim N(.,)$ is Normal |
| (2) | **Zero mean:** | $E(\mathbf{u}_t \mid \mathbf{X}_{1t} = \mathbf{x}_{1t}) = \mathbf{0}$, |
| (3) | **Homoskedasticity:** | $Cov(\mathbf{u}_t \mid \mathbf{X}_{1t} = \mathbf{x}_{1t}) = \boldsymbol{\Sigma}$, free of $\mathbf{x}_{1t}$, |
| (4) | **Independence:** | $\{(\mathbf{u}_t \mid \mathbf{X}_{1t} = \mathbf{x}_{1t}), \; t \in \mathbb{T}\}$ - independent process, |

NOTATION: $\mathbf{B}^\top = \left( \beta_0^\top, \mathbf{B}_1^\top \right)$, $\mathbf{x}_t := (1, \mathbf{x}_{1t})$.

**Statistical perspective.** The structural model: $\boldsymbol{\Gamma}^\top \mathbf{y}_t = \boldsymbol{\Delta}^\top \mathbf{x}_t + \boldsymbol{\varepsilon}_t$, $t \in \mathbb{T}$, *derives its statistical meaning* from the Multivariate Linear Regression (MLR) model, as specified in table 1, and the former confers theoretical meaningfulness upon the latter. Whenever the LRM is misspecified, any inference concerning the structural parameters $\boldsymbol{\alpha}$ is likely to be unreliable; see Spanos (1990). In practice, the statistical model should be estimated first, and its statistical adequacy established, before it's connected to the structural model via reparameterization/restriction. If any of the assumptions [1]-[5] comprising the reduced form model is invalid, inferences concerning the structural model are likely to be unreliable. Hence, before any form of structural inference is implemented, the assumptions (1)-(4) (see table 1) should be thoroughly probed for possible departures using effective misspecification tests. For an extensive discussion of *misspecification testing* for the MLR model see Spanos (1986), ch. 24.

What if the modeler ignores the potential misspecifications of the reduced form?

**Identifiability.** The structural model is identified with respect to an 'imagined' reduced form that has no veridical connection to the phenomenon of interest!

**Estimation and Testing.** The *notional* error probabilities are, in general, very different from the *actual* error probabilities, rendering any inference based on the estimated model unreliable - Mayo (1996).

> Lesson 3. *Observational equivalence* among structural models is meaningful only if the common 'reduced form' (statistical model) is *statistically adequate*. Moreover, unless the structural model is build upon a statistically adequate reduced form, structural inference is likely to be *unreliable*.
>
> Lesson 4. The reliability of structural inference depends on four pre-conditions:
>
> (i) the accordance between the conditions envisaged by the structural model, vis-a-vis the phenomenon of interest, and the probabilistic assumptions comprising the statistical model, vis-a-vis the data,

(ii) the 'similitude' between the variables envisaged by the theory and those measured by the data,

(iii) the statistical adequacy of the estimated statistical model,and

(iv) the appropriateness of the reparameterization/restriction.

Lesson 4 had a profound effect on the author's thinking about empirical modeling in econometrics and as a result a new modeling framework, the **Probabilistic Reduction**, was proposed in Spanos (1986). The new approach was designed to take account as well as address these issues.

## 2. The Probabilistic Reduction (PR) Approach

A complete discussion of the Probabilistic Reduction (PR) approach is given in Spanos (1986,1995,2001). What follows is a summary of the main features of the approach.

A. The PR approach views a statistical model as a set of internally consistent probabilistic assumption, specifying a stochastic generating mechanism. Its primary objective is to model is to capture the **systematic features** of the observable phenomenon of interest by modeling the systematic (recurring) information in the observed data. Intuitively, **statistical information** is any *recurring chance regularity pattern* which can be modeled via probabilistic concepts such as those in table 2, placed into three broad categories.

| (D) **Distribution** | (M) **Dependence** | (H) **Heterogeneity** |

There is a direct connection between these concepts and chance regularity patterns in observed data; see chapters 5-6 in Spanos (1999).

B. The PR procedure from the data to a reliable structural model has four interrelated facets.

1. **Specification** refers to the choice of a statistical model based on the information provided by the theoretical model and the probabilistic structure of the observed data in question.

2. **MisSpecification** (M-S) refers to informal graphical checks and the formal M-S testing of the assumptions underlying the statistical model.

3. **Respecification** refers to the choice of an alternative statistical model when the original choice is found to be inappropriate for the data in question. This process from specification to misspecification testing and respecification will continue until a **statistically adequate model** is found.

4. **Identification** constitutes the last stage of empirical modeling at which the theoretical model is related to the statistically adequate estimated statistical model.

**Challenge.** How can one arrive at a model for the stochastic process underlying the data and *infer with severity* (Mayo (1996)) that potential violations of its assumptions have been well-probed?

Model Specification: Reduction by partitioning the space of all possible models

The traditional analysis of the MLR model has already, implicitly, *reduced the space of models* that could be considered. It reflects a particular way of reducing the set of all possible models of which the data $\mathbf{Z} := \{(\mathbf{x}_t, \mathbf{y}_t), \; t = 1, 2, ..., n\}$ can be considered to be a 'truly typical' realization; this provides the primary motivation for the PR approach. The set of all possible statistical models, $\mathcal{P}$, can be delineated by imposing one or another set of probabilistic assumptions on the **Haavelmo distribution**: the *joint distribution of all the observable random variables involved.*

The reduction from $\mathcal{P}$ to the particular model amounts to imposing probabilistic assumptions which 'reduce' the space of models by partitioning.

This should be contrasted with the traditional way of statistical model specification which proceeds with the introduction of some arbitrary *ad hoc modifications* of the Linear Regression model, i.e. introduced some arbitrary non-linearity and/or heteroskedasticity, as well as 'modeling' the error! There is an *infinite* number of ad hoc modifications one can envisage, but no *effective strategy* to choose among the possible alternatives.

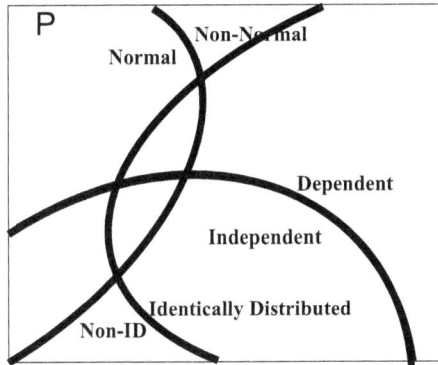

Model specification by partitioning

The reduction assumptions come from a menu of three broad categories: (D) **Distribution**, (M) **Dependence**, (H) **Heterogeneity**.

For example, the MLR model is identified with the *reduction assumptions: (D) Normal, (M) Independent, (H) Identically Distributed (NIID).*

Imposing the NIID assumptions simplifies (or reduces) $D(\mathbf{Z}; \phi)$ into a product of conditional distributions $D(\mathbf{y}_t \mid \mathbf{x}_t; \theta)$ : $D(\mathbf{Z}_1, \mathbf{Z}_2, ..., \mathbf{Z}_n; \phi) \overset{\text{NIID}}{\leadsto} \prod_{t=1}^{n} D(\mathbf{y}_t \mid \mathbf{x}_t; \theta)$; see Spanos (1995) for the details of the reduction that gives rise to the MLR model given in table 2. The MLR model is specified exclusively in terms of the conditional distribution $D(\mathbf{y}_t \mid \mathbf{x}_t; \varphi_2)$ : $(\mathbf{y}_t \mid \mathbf{X}_t = \mathbf{x}_t) \backsim \mathsf{N}\left(\beta_0 + \mathbf{B}_1^\top \mathbf{x}_{1t}, \mathbf{\Sigma}\right)$, $\theta :=$ $(\beta_0, \mathbf{B}_1^\top, \mathbf{\Sigma})$.

By choosing different combinations of reduction assumptions one can generate *numerous statistical models* that would have been impossible to 'devise' otherwise.

## Table 2 - The Multivariate Linear Regression (MLR) Model

$$\mathbf{y}_t = \boldsymbol{\beta}_0 + \mathbf{B}_1^\top \mathbf{x}_{1t} + \mathbf{u}_t, \ t \in \mathbb{T},$$

| [1] | **Normality:** | $D(\mathbf{y}_t \mid \mathbf{x}_{1t}; \boldsymbol{\psi})$ is Normal |
|---|---|---|
| [2] | **Linearity:** | $E(\mathbf{y}_t \mid \mathbf{X}_{1t} = \mathbf{x}_{1t}) = \boldsymbol{\beta}_0 + \mathbf{B}_1^\top \mathbf{x}_{1t}$, linear in $\mathbf{x}_t$, |
| [3] | **Homoskedasticity:** | $Cov(\mathbf{y}_t \mid \mathbf{X}_{1t} = \mathbf{x}_{1t}) = \boldsymbol{\Sigma}$, free of $\mathbf{x}_{1t}$, |
| [4] | **Independence:** | $\{(\mathbf{y}_t \mid \mathbf{X}_{1t} = \mathbf{x}_{1t}), \ t \in \mathbb{T}\}$ - independent process, |
| [5] | **t-homogeneity:** | $(\boldsymbol{\beta}_0, \mathbf{B}_1^\top, \boldsymbol{\Sigma})$ are not functions of $t \in \mathbb{T}$. |

where the unknown parameters are given a *statistical interpretation*:

$$\boldsymbol{\beta}_0 = E(\mathbf{y}_t) + \mathbf{B}_1^\top E(\mathbf{X}_{1t}), \ \mathbf{B}_1 = [Cov(\mathbf{X}_{1t})]^{-1} Cov(\mathbf{X}_{1t}, \mathbf{y}_t),$$
$$\boldsymbol{\Sigma} = Cov(\mathbf{y}_t) - Cov(\mathbf{y}_t, \mathbf{X}_{1t}) [Cov(\mathbf{X}_{1t})]^{-1} Cov(\mathbf{X}_{1t}, \mathbf{y}_t).$$

### Misspecification (M-S) testing

The question posed by M-S testing is in the form of:

$$H_0 : f_0(\mathbf{z}) \in M_0, \text{ against } H_1 : f_0(\mathbf{z}) \in [\mathcal{P} - M_0], \tag{5}$$

where $\mathcal{P}$ denotes the set of all possible statistical models and $M_0$ the assumed one.

**Problem**: how can one probe $[\mathcal{P} - M_0]$ adequately. The key to the effecting probing of the space of alternative models is provided by the conditional expectation orthogonality lemma which suggests misspecification tests based on 'auxiliary regressions'; see Spanos (1999), and Spanos and McGuirk (2001).

## 3. Empirical example of econometric modeling

### 3.1. The Consumption function

Keynes' **Absolute Income Hypothesis** (AIH), in the form of the structural model:
$$C = \alpha + \beta Y^D, \ \alpha > 0, \ 0 < \beta < 1.$$

**Primary question**: is the AIH supported by empirical evidence? Estimating this relationship using annual USA data for the period 1947-1998: $y_t$ - real consumer's expenditure and $x_t$ - personal disposable income, yields:

$$y_t = \underset{(16.930)}{45.279} + \underset{(.007)}{.936}\, x_t + \widehat{u}_t, \quad R^2 = .997, \ s = 49.422, \ DW = .250, \ T = 52.$$

Does the apparently 'excellent' goodness of fit ($R^2 = .997$) and the 'highly significant' coefficients (numbers in square brackets denote p-values):

$$\tau_0(y) = \tfrac{45.279}{16.930} = 2.675[.004], \quad \tau_1(y) = \tfrac{.936}{.007} = 133.71[.000],$$

**provide good evidence for the AIH**? No! The estimated model is **badly misspecified**, as exemplified by the misspecification results in table 3. Hence, any inferences based on it are likely to be unreliable because the *actual error probabilities* are very different from their assumed *nominal* values (see Mayo, 1996).

| Table 3 - MisSpecification (M-S) tests | |
|---|---|
| **Non-Normality:** $D'AP = 2.837[.242]$ | **Heteroskedasticity:** $F(2,47) = 7.344[.002]^*$ |
| **Non-linearity:** $F(1,49) = 154.822[.000]^*$ | **Autocorrelation:** $F(1,48) = 159.274[.000]^*$ |

Lesson 5. The reliability of inference depends on the statistical adequacy of the estimated model, and not on 'goodness of fit' measures such as $R^2$, chi-square measures and Akaike-type information criteria; the latter pre-suppose the statistical adequacy of the model!

**Ad hoc 'fixes'.** What if the Linear Regression (LR) model is 'corrected' for autocorrelation? That is, estimate the Autocorrelation-Corrected (**A-C**) **LR model:**

$$y_t = \underset{(14.367)}{13.574} + \underset{(.026)}{.938}\, x_t + \hat{u}_t \quad u_t = \underset{(.005)}{.931}\, u_{t-1} + \hat{\varepsilon}_t, \quad R^2 = .9994, \; DW = 1.915, \; s = 24.841,$$

This model is also **misspecified** and thus, any inferences based on it are likely to be unreliable; see Spanos (1988).

Lesson 6. In a M-S test one should *not* adopt the alternative when the null is rejected, because the hypothesis has not, usually, passed a severe test.

What can one do next?

The **traditional econometric approach** does *not* provide guidance for:

(a) an exhaustively complete probing strategy for MisSpecification (**M-S**) **testing**, (the above Normality test is misleading because it assumes IID), or

(b) satisfactory answers to the **respecification** question (ad hoc adjustments and fallacious respecifications A-C LR model).

### 3.2. The Probabilistic Reduction (PR) approach
### 3.2.1. Specification

The **Probabilistic Reduction approach** is designed to address the modeling issues raised above by proposing an effective probing strategy to decide on the directions in which the primary statistical model might be potentially misspecified. This strategy relies heavily on graphical techniques, manipulations on paper, and qualitative severity assessments, which raise a number of methodological issues including 'double-uses' of data - see Mayo (1996) and Mayo and Spanos (2003) for more details.

In the case of the LR model, the underlying assumptions are specified in table 4. The relationship between reduction and model assumptions which plays an important role in the PR approach is given in table 5.

**Table 4 - The Linear Regression (LR) Model**

$$Y_t = \beta_0 + \beta_1 x_t + u_t, \ t \in \mathbb{T},$$

| | | |
|---|---|---|
| [1] | **Normality:** | $D(Y_t \mid x_t; \psi)$ is Normal |
| [2] | **Linearity:** | $E(Y_t \mid X_t = x_t) = \beta_0 + \beta_1 x_t$, linear in $x_t$, |
| [3] | **Homoskedasticity:** | $Var(Y_t \mid X_t = x_t) = \sigma^2$, free of $x_t$, |
| [4] | **Independence:** | $\{(Y_t \mid X_t = x_t), \ t \in \mathbb{T}\}$ - independent process. |
| [5] | **t-homogeneity:** | $(\beta_0, \beta_1, \sigma^2)$ are not functions of $t \in \mathbb{T}$. |

**Table 5 - Reduction vs. Model assumptions**

| Reduction: $\{\mathbf{Z}_t, \ t \in \mathbb{T}\}$ | | Model: $\{(y_t \mid \mathbf{X}_t = \mathbf{x}_t), \ t \in \mathbb{T}\}$ |
|---|---|---|
| (Normal) N | $\longrightarrow$ | [1], [2], [3] |
| (Independent) I | $\longrightarrow$ | [5] |
| (Identically Distributed) ID | $\longrightarrow$ | [4] |

It is important to emphasize that assumption [5], which is usually omitted from the traditional specification of the LR model (see table 1), has a crucial role to play in discussions of Invariance in causal modeling (see Woodward (1997)).

As argued above, in the context of the PR approach one can assess the appropriateness of the model assumptions by assessing the validity of the reduction assumptions. Are the reduction assumptions for the LR model appropriate? In the case of the LR model the reduction assumptions are that $\{\mathbf{Z}_t, \ t \in \mathbb{T}\}$ is a Normal, Independent and Identically Distributed vector process.

| (D) Normal | (M) Independent | (H) Identically Distributed |
|---|---|---|

**Fig. 1** exhibits a typical realization of a NIID process. When we compare fig. 1 with the t-plots of the consumption and income data $\{(y_t, x_t), \ t = 1, 2, ..., n\}$ in figures 2-3, we can see significant departures from the NIID assumptions. In particular, it's clear that the two data series exhibit mean-heterogeneity.

In order to get a better picture of departures from the Independence assumption we detrend the original data giving rise to the t-plots in figures 4-5.

The cycles exhibited by both series indicate the presence of positive temporal dependence (see Spanos (1999), ch. 5), suggesting some form of Markov dependence.

| (D) Normal? | (M) Markov | (H) mean-heterogeneous |
|---|---|---|

In addition to detrending the data, if we proceed to dememorize ('subtract' the temporal dependence) the t-plots in figures 6-7 indicate that the data exhibit a trending variance – the variance seems to increase over time.

| (D) Non-symmetric | (M) Markov | (H) { | mean-heterogeneous variance-heterogeneous |
|---|---|---|---|

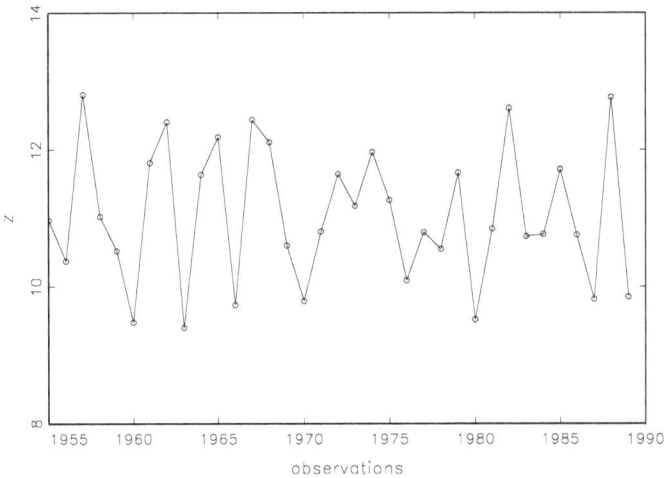

Figure 1. A typical realization of a NIID process

In addition, the scatter plot of the detrended and dememorized data suggests the presence of some degree of asymmetry; joint Normality requires that the scatter plot indicates a symmetric elliptical shape with high concentration of points along the principal axis.

In view of the relationship between reduction and model assumptions given in table 5, Non-Normality leads to drastic respecification because both the regression and skedastic functions need to be re-considered.

### 3.2.2. Misspecification (M-S) testing

In the context of the PR approach ones views the probabilistic assumptions as interrelated in light of the reduction assumptions.

The PR approach provides an <u>exhaustively complete</u> probing strategy for **M-S testing**, by using:

(i) the reduction assumptions to guide the probing in directions of potential departures,

(ii) graphical techniques for *informed probing*,

(iii) a judicious combination of ordered parametric and non-parametric tests to *avoid circular reasoning*, and

(iv) joint M-S tests (testing several assumptions simultaneously) to avoid erroneous diagnoses.

**Regression function**. In view of the chance regularity patterns exhibited by the data in figures 1-8, the test that suggests itself would be based on testing $H_0 : \gamma_2 = 0, \ \gamma_3 = 0, \ \gamma_4 = 0, \ \gamma_5 = 0$, in the context of the auxiliary regression:

$$\widehat{u}_t = \gamma_0 + \gamma_1 x_t \overbrace{+\gamma_2 t}^{\text{trend}} \overbrace{+\gamma_3 x_t^2 +}^{\text{non-linearity}} \overbrace{\gamma_4 x_{t-1} + \gamma_5 x_{t-1}}^{\text{dependence}} +v_t,$$

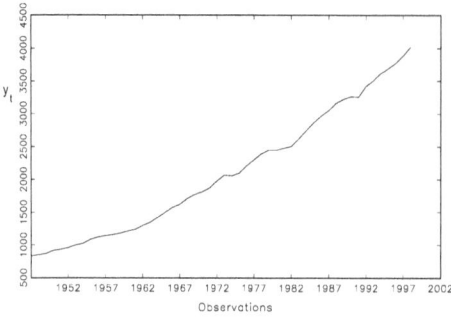

Figure 2. t-plot of real consumers' expenditure

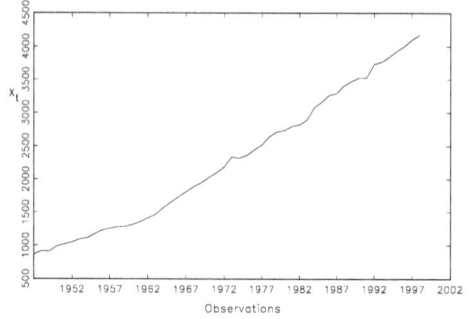

Figure 3. t-plot of real personal disposable income

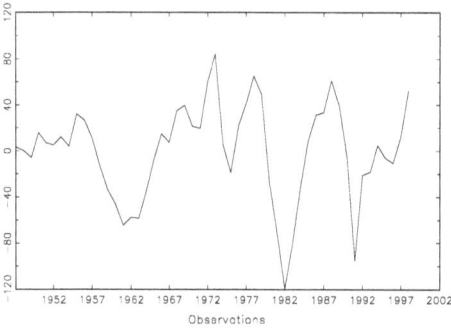

Figure 4. Detrended $y_t$

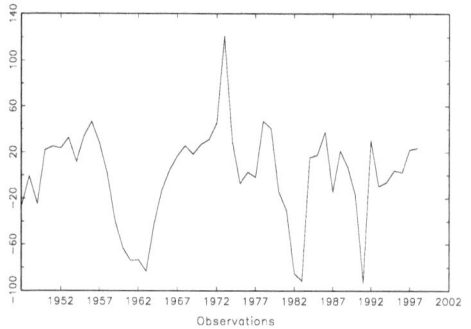

Figure 5. Detrended $x_t$

The F test for the joint significance of the terms $t$, $x_{t-1}$, $y_{t-1}$, $x_t^2$ yields: $F(4, 46) = 50.682[.00000]^*$, where the contribution of each of the terms separately is:

Mean heterogeneity: $H_0 : \gamma_2 = 0$, $F(1, 46) = 7.069[.011]^*$,

Non-linearity: $H_0 : \gamma_3 = 0$, $F(1, 46) = 21.366[.000]^*$,

Temporal dependence: $H_0 : \gamma_4 = 0$, $\gamma_5 = 0$, $F(2, 46) = 7.348[.002]^*$.

**Skedastic function**. The auxiliary regression that suggests itself is:

$$\widehat{u}_t^2 = \delta_0 \overbrace{+\delta_1 t}^{\text{trend}} \overbrace{+\delta_2 x_t^2 + \delta_3 \widehat{u}_{t-1}^2}^{\text{heteroskedasticity}} +v_t,$$

The F test for the joint significance of the terms $t$, $x_t^2$ and $\widehat{u}_{t-1}^2$ yields: $F(4, 46) = 27.630[.00000]^*$, where the contribution of each of the terms separately is:

Variance heterogeneity: $H_0 : \gamma_2 = 0$, $F(1, 46) = 3.208[.080]$,

Heteroskedasticity: $H_0 : \gamma_3 = 0$, $F(2, 46) = 11.901[.000]^*$, i.e. the departure is *heteroskedasticity and not heterogeneity!*

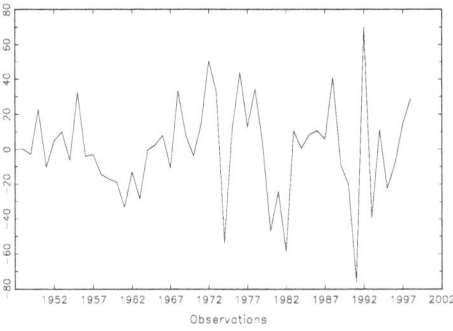

Figure 6. Detrended and dememorized $y_t$

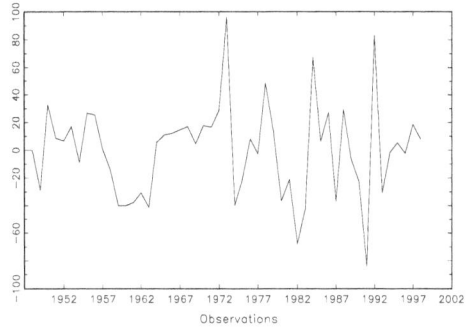

Figure 7. Detrended and dememorized $x_t$

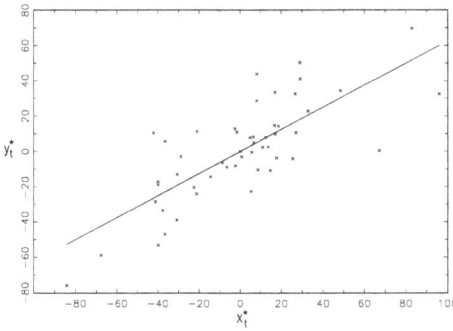

Figure 8. Scatter-plot of detrended and dememorized series

**Caution.** If one were to use the auxiliary regression: $\widehat{u}_t^2 = \delta_0 + \delta_1 t + v_t$, would have *erroneously* concluded that the conditional variance is *heterogenous:* $F(1, 46) = 9.085[.004]^*$.

**Informed probing.** If the modeler were to ignore the **non-Normality** detected above, and went ahead to estimate the Dynamic Linear Regression (DLR) Model with a trend the result will be:

$$y_t = \underset{(18.489)}{14.137} - \underset{(.327)}{.717}\,t + \underset{(.082)}{.673}\,x_t + \underset{(.084)}{.957}\,y_{t-1} - \underset{(.109)}{.610}\,x_{t-1} + \widehat{\varepsilon}_t, \quad R^2 = .9995, \quad s = 21.328, \quad T = 52.$$

(6)

The PR approach suggests that there will be departures from the **Linearity** and **Homoskedasticity** assumptions which are confirmed in table 6.

| Table 6 - Misspecification tests | |
|---|---|
| **Non-Normality:** $D'AP = .347[.841]$ | **Heteroskedasticity:** $F(2, 43) = 7.137[.002]^*$ |
| **Non-linearity:** $F(1, 45) = 14.933[.000]^*$ | **Autocorrelation:** $F(1, 44) = 1.438[.237]$ |

Any inference based on the estimated model (6) is likely to be misleading.

Lesson 7. The probabilistic assumptions comprising a statistical model are interrelated and should not be probed in isolation. Effective probing requires informed searches that involve 'partitioning' the set of all potential alternative models.

### 3.2.3. Respecification

In view of the above misspecification testing results and the relationship between reduction and model assumptions (see table 4), the respecification that suggests itself is to use the logarithm as a variance stabilizing transformation; see Spanos (1986), p. 487-8. Figures 9-10 confirm the variance stabilizing effect of the log transformation because the detrended and dememorized series no long exhibit variance heterogeneity.

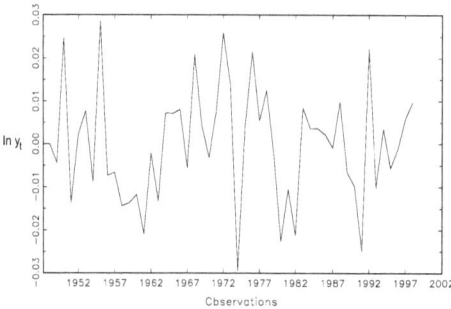

Figure 9. Detrended and dememorized $\ln y_t$

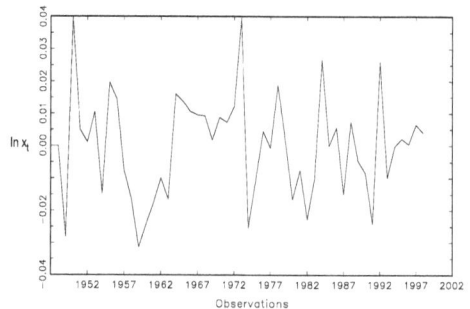

Figure 10. Detrended and dememorized $\ln x_t$

The respecification that suggests itself is to adopt an alternative to the LR model based on the reduction assumptions that $\{\mathbf{Z}_t,\ t \in \mathbb{T}\}$ is a Normal (in logs), Markov and variance homogeneous but mean-heterogeneous vector process.

| (D) Normal (in logs) | (M) Markov | (H)$\Big\{$ | mean-heterogeneous variance-homogeneous |
|---|---|---|---|

Imposing these Reduction assumptions on the Haavelmo distribution $D(\mathbf{Z}; \phi)$ (gives rise to the Dynamic Linear Regression (DLR) model given in table 7.

| Table 7 - The Dynamic Linear Regression model with a trend |
|---|
| $y_t = \beta_0 + \beta_1 x_t + \delta_1 t + a_1 y_{t-1} + \gamma_1 x_{t-1} + \varepsilon_t, \ , \ t \in \mathbb{T},$ |

| | | |
|---|---|---|
| [1] | **Normality:** | $D(y_t \mid x_t, y_{t-1}, x_{t-1}; \psi)$ is Normal |
| [2] | **Linearity:** | $E(y_t \mid X_t = x_t)$ is linear in $x_t, y_{t-1}, x_{t-1}$, |
| [3] | **Homoskedasticity:** | $Var(y_t \mid x_t, y_{t-1}, x_{t-1}) = \sigma^2$, free of $x_t, y_{t-1}$. |
| [4] | **Markov:** | $\{(y_t \mid x_t, y_{t-1}, x_{t-1}), \ t \in \mathbb{T}\}$ is Markov, |
| [5] | **t-homogeneity:** | $(\beta_0, \beta_1, \delta_1, \delta_2, a_1, \gamma_1 \sigma^2)$ are $t$-invariant. |

Estimating this model using the original data yielded:

$$\ln y_t = \underset{(.272)}{.912} + \underset{(.001)}{.005}\,t + \underset{(.069)}{.708} \ln x_t + \underset{(.108)}{.565} \ln y_{t-1} - \underset{(.097)}{.413} \ln x_{t-1} + \widehat{\varepsilon}_t \tag{7}$$
$$R^2 = .9997, \ s = .0084, \ T = 52.$$

The statistical adequacy of this model is assessed by testing assumptions [1]-[5] in table 7; the misspecification results are shown in table 8.

| Table 8 - Misspecification tests | |
|---|---|
| **Non-Normality:** | **Heteroskedasticity:** |
| $D'AP = .872[.646]$ | $F(2, 43) = .317[.730]$ |
| **Non-linearity:** | **Markovness:** |
| $F(1, 45) = 1.142[.291]$ | $F(1, 44) = 1.129[.293]$ |

These misspecification tests indicate no significant departures from the underlying assumptions. Hence, on the basis of a statistically adequate model given in (7) we can conclude that **the AIH is not supported by the data!** This reverses the original inference based on a misspecified model.

> Lesson 8. The respecification of statistical models based on 'fixing' *model assumptions* that appear to be violated, often leads to other misspecified models and/or internally inconsistent probabilistic assumptions. Proper respecification requires the modeler to consider alternative models whose *reduction assumptions* accommodate departures from the original reduction assumptions.

One of the advantages of the Probabilistic Reduction approach is that by choosing different combinations of reduction assumptions one can *generate* numerous *statistical models* that would have been impossible to 'dream up' otherwise - see Spanos (1994).

## 4. Structural and Graphical Causal models

### 4.1. Statistical adequacy and Graphical Causal (GC) Models

The statistical model that underlies the overwhelming majority of GC modeling (see Spirtes et al (2000), Pearl (2000)) is a special case of the MLR model where there are no exogenous variables.

## Table 9 - The Multivariate Normal (MN) Model

$$\mathbf{y}_t = \boldsymbol{\mu} + \mathbf{u}_t, \ t \in \mathbb{T},$$

| | | |
|---|---|---|
| [1] | Normality: | $D(\mathbf{y}_t; \boldsymbol{\psi})$ is Normal |
| [2] | Constant Mean: | $E(\mathbf{y}_t) = \boldsymbol{\mu}$, |
| [3] | Constant Covariance: | $Cov(\mathbf{y}_t) = \boldsymbol{\Sigma}$, |
| [4] | Independence: | $\{\mathbf{y}_t, \ t \in \mathbb{T}\}$ - independent process. |

The discovery algorithms proposed by Spirtes et al (2000) and Pearl (2000) for causal modeling take the estimated variance-covariance matrix:

$$\widehat{\boldsymbol{\Sigma}} = \tfrac{1}{n}(\mathbf{Y} - \mathbf{1}\widehat{\boldsymbol{\mu}})^\top (\mathbf{Y} - \mathbf{1}\widehat{\boldsymbol{\mu}}), \quad \text{where} \quad \widehat{\boldsymbol{\mu}} = \tfrac{1}{n}\sum_{t=1}^{n} \mathbf{y}_t, \tag{8}$$

as their point of departure. Given that the discovery algorithms rely on inference concerning conditional independence, it goes without saying that their reliability depends crucially on the validity of assumptions [1]-[4] (see table 9).

Common appeals to 'robustness' arguments and 'asymptotic results' tend to be based on vague statements about 'small' violations and the resulting 'approximate' reliability of inference. However, 'small' departures from the underlying assumptions can have a substantial impact on the reliability of inference.

The literature on robust statistics based on *influence functions* (see Hampel et al (1986)) has demonstrated how difficult it is to render 'small' violations and 'approximate' reliability of inference more precise in order to be useful in practice. Ultimately, quantification of small departures and its effect on reliability requires one to find a statistically adequate model for the data in hand. Appeals to 'asymptotic results' are often mistakenly based on central limit theorems which will can only offer relief for non-Normality violations; they do not apply to several violations of Independence and/or Heterogeneity. Indeed, the impact of violations of heterogeneity get worse as the sample size increases.

For econometric time series data the assumptions of *Independence* and *Identically Distributed* are invariably false and the use of (8) can give rise to very misleading results.

**Example.** Consider the following annual time series for the USA economy for the period 1947-1999: $C_t$- real consumers' expenditure (in 1987 dollars), $Y_t^D$ - real personal disposable income, $P_t$ - price level , $CC_t$ - consumers' credit outstanding, $r_t$ - short run interest rate (Moody's AAA yield).

| Table A | $C$ | $Y^D$ | $P$ | $CC$ | $r$ |
|---|---|---|---|---|---|
| $C$ | 287.98 | .881[.00] | -.647[.19] | .294[.00] | -8.35[.00] |
| $Y^D$ | 1.12[.00] | 367.11 | .783[.16] | -0.31[.00] | 10.5[.00] |
| $P$ | -0.06[.19] | .052[.16] | 24.59 | .105[.00] | 1.14[.09] |
| $CC$ | 1.29[.00] | -1.07[.00] | 5.43[.00] | 1268.1 | -5.50[.25] |
| $r$ | -.034[.00] | .034[.00] | .054[.09] | -.005[.25] | 1.171 |

The inverse of the estimated covariance matrix on the basis of which inferences concerning conditional independencies are made is given in table A. The covariance matrix was estimated assuming that assumptions [1]-[4] (see table 9) are valid. The number in parentheses denote the p-values of the t-tests for the significance of the coefficients and the diagonal elements are estimates of the conditional variances. It turns out that several of these conditional independence inferences are unreliable because all four *probabilistic assumptions* are *invalid* for the above data.

| Table B | $\ln C$ | $\ln Y^D$ | $\ln P$ | $\ln CC$ | $\ln r$ |
|---|---|---|---|---|---|
| $\ln C$ | .000064 | .575[.00] | -.164[.08] | .086[.00] | -.014[.41] |
| $\ln Y^D$ | 1.02[.00] | .00011 | .181[.15] | -0.045[.28] | .014[.55] |
| $\ln P$ | -0.41[.08] | .255[.15] | .00016 | .042[.40] | .097[.00] |
| $\ln CC$ | 1.95[.00] | -.57[.28] | .376[.40] | .00144 | .055[.52] |
| $\ln r$ | -1.02[.41] | .552[.55] | 2.65[.00] | .167[.52] | .00437 |

After respecification that involved taking into account the non-Normality by taking logs, the non-Independence by allowing for temporal Markov dependence (up to order 2) and non-ID by allowing for polynomial trends (up to order 2), a statistically adequate NM is achieved. The estimated covariance matrix based on the statistically adequate model is shown in table B. If we compare the two estimated covariances we can see that *10* of the original conditional independence inferences were unreliable!

> Lesson 9. Ensuring the statistical adequacy of the underlying Statistical Model reduces the set of observationally equivalent Structural Models considerably and avoids needless confusions and pointless discussions.

## 4.2. Recursive Structural Models

The use of Recursive Structural Equation Models (RSEMs) in econometrics was championed by Wold (1954,1956, 1960), formalizing and extending the work of Tinbergen (1938,1939) on *causal-chain models*. Consider the **recursive SEM**:

$$y_{1t} = d_1^\top x_t + \epsilon_{1t}, \quad y_{2t} = c_{21} y_{1t} + d_2^\top x_t + \epsilon_{2t}, \quad \cdots \quad y_{mt} = \sum_{i=1}^{m-1} c_{mi} y_{it} + d_m^\top x_t + \epsilon_{mt}$$

In matrix form: $\tilde{\Gamma}^\top y_t = \tilde{\Delta}^\top x_t + \tilde{\epsilon}_t$, $\tilde{\epsilon}_t \backsim N(0, \tilde{\Omega})$, where $\tilde{\Gamma}^\top$ and $\tilde{\Omega}$ are *lower triangular* (with $-1$ along the main diagonal) and *diagonal matrix*, respectively. This form of a structural model constitutes a *reparameterization* of the statistical model (reduced form) $y_t = B^\top x_t + u_t$, $u_t \backsim N(0, \Sigma)$, based on sequential conditioning. It can be shown that for *any ordering* of the 'endogenous' variables $y_t := (y_{1t}, y_{2t}, y_{3t}, ..., y_{mt})$, the reduced form distribution $D(y_t \mid x_t; \varphi_2)$ can be sequentially conditioned to give rise to the reduction:

$$D(y_t \mid x_t; \varphi_2) = D(y_{1t} \mid x_t; \varphi_{21}) \prod_{i=2}^{m} D(y_{it} \mid y_{(i-1)t}, \dots, y_{1t}, x_t; \varphi_{2i}).$$

This reduction induces the reparameterization: $\varphi_2 \rightsquigarrow (\varphi_{21}, \varphi_{22}, \dots, \varphi_{2m})$, where the *number* of parameters in $\varphi_2$ remains the same after the reparameterization into $(\varphi_{21}, \varphi_{22}, \dots, \varphi_{2m})$. The error terms $(\epsilon_{1t}, \epsilon_{2t}, \dots, \epsilon_{mt})$ are defined via the orthogonal decomposition: $y_{it} = E(y_{it} \mid \mathcal{D}_i) + \epsilon_{it}$, $i = 1, 3, \dots, m$, where the **systematic component** is:

$$E(y_{it} \mid \mathfrak{D}_i) = \quad \mu_i(y_{(i-1)t}, y_{(i-2)t}, \ldots, y_{1t}, \mathbf{x}_t), \; i = 1, 3, \ldots, m,$$
$$\mathfrak{D}_0 := \{S, \emptyset\}, \quad \mathfrak{D}_i := \{\sigma(y_{(i-1)t}, y_{(i-2)t}, \ldots, y_{1t}), \mathbf{X}_t = \mathbf{x}_t\},$$

and the **error term** takes the form: $\epsilon_{it} = y_{it} - E(y_{it} \mid \mathfrak{D}_i)$, $i = 1, 3, \ldots, m$. By construction:

(i)    $E(\epsilon_{it} \mid \mathfrak{D}_i) = 0,$

(ii)   $E(\epsilon_{it}^2 \mid \mathfrak{D}_i) = Var(y_{it} \mid \mathfrak{D}_i) = \sigma_{ii},$     $\Big\}$ $i = 1, \ldots, m.$

(iii)  $E(\epsilon_{it}\epsilon_{(i-j)t} \mid \mathfrak{D}_i) = E(\epsilon_{it}\epsilon_{(i-j)t} \mid \mathfrak{D}_j) = 0, \; j < i,$

The first two properties follow from the definition of the error term.

(iii) $E(\epsilon_{it}\epsilon_{(i-j)t}|\mathfrak{D}_i) = E([y_{it} - E(y_{it}|\mathfrak{D}_i)]\epsilon_{(i-j)t}|\mathfrak{D}_i) = \epsilon_{(i-j)t}([E(y_{it}|\mathfrak{D}_i) - -E(y_{it}|\mathfrak{D}_i)]) = 0$. Moreover, from the fact that $\mathfrak{D}_j \subset \mathfrak{D}_i$ and thus: $E(\epsilon_{it}\epsilon_{(i-j)t} \mid \mathfrak{D}_j) = E(\epsilon_{it}\epsilon_{(i-j)t} \mid \mathfrak{D}_i) = 0, \; j < i = 1, \ldots, m$; see Doob (1953), pp. 21-2.

The above derivation shows most clearly that *statistical errors terms* (in contrast to structural) are **not autonomous** random components, but they are 'derivable' from the distribution of the observable random variables; the derivation itself determines the probabilistic structure of the statistical error terms.

In this sense, the **recursive structural model with correlated errors** ($\tilde{\Omega}$ is assumed non-diagonal), first proposed by Strotz (see Strotz and Wold, 1960)), is *not* 'identified', because no such model is derivable from $D(\mathbf{y}_t \mid \mathbf{x}_t; \varphi_2)$; see Cartwright (1989) for a different justification.

> Lesson 10. Attaching error terms to structural equations is not as arbitrary as it is often assumed. Such error terms are statistically 'meaningful' to the extent that they can be related to the error terms of the 'underlying' statistical model.

The more general object lesson is the following:

> Lesson 11. Structural concepts and procedures gain statistical 'operational meaning' to the extent that they can be related to statistical concepts and procedures definable in terms of the joint distribution of the observable random variables.

### 4.3. Recursive SEMs and Directed Acyclic Graphs (DAG)

The recursive SEM in the form of: $\tilde{\Gamma}^\top \mathbf{y}_t = \tilde{\Delta}^\top \mathbf{x}_t + \tilde{\varepsilon}_t$, $\tilde{\varepsilon}_t \frown N(\mathbf{0}, \tilde{\Omega})$, should be viewed as a statistical model because it's simply a *reparameterization* of: $\mathbf{y}_t = \mathbf{B}^\top \mathbf{x}_t + \mathbf{u}_t$, $\mathbf{u}_t \frown N(\mathbf{0}, \Sigma)$, without any additional structural restrictions. This is because the reparameterization can be arrived at for *any ordering* of the 'endogenous' variables $(y_{1t}, y_{2t}, y_{3t}, \ldots, y_{mt})$, irrespective of any structural merit. This suggests that the recursive SEM is observationally equivalent to all SEMs which share the same reduced form - assuming the latter is statistically adequate! In order to restrict the equivalence class of structural models, one needs to impose further structural restrictions such as 'dropping' $(y_{it}, \; i = 1, .., m)$ and $(x_{jt}, \; j = 1, \ldots, k)$ variables from specific equations.

The DAGs of Graphical Causal (GC) models belong to this category of structural models. The discussion of SEMs from the PR perspective suggests that the statistical adequacy of the recursive reduced form should be established first and then any conditional independence restrictions associated with dropping ($y_{it}$, $i = 1, .., m$) and ($x_{jt}$, $j = 1, ..., k$) variables from specific equations become properly *testable*.

**Testing overidentifying restrictions.** A major advantage of DAG formulations of structural models is that one can separate the **identifying** from the **overidentifying restrictions** and test the latter before imposing them in order to determine the *empirical adequacy* of the particular DAG model.

It should be noted that structural equation model with feedback no such separation of identifying and overidentifying restrictions is possible; see Dhrymes (1994).

> Lesson 12. DAG models constitute a particular form of structural models which can be viewed as reparameterizations/restrictions of recursive reduced forms (statistical models). Once the statistical adequacy of the recursive reduced form is established, the overidentifying restrictions associated with DAGs become directly testable not only individually but also collectively - and should be tested in order to establish the empirical adequacy of the structural model.

## 5. Summary and Conclusions

The most valuable lesson learned in examining more than a century of empirical modeling in economics is that for reliable structural inference the modeler needs to ensure that the underlying statistical model is not statistically misspecified. Using this lesson the paper has attempted to shed light on certain aspects of Graphical Causal (GC) modeling that can be improved by paying due attention to the statistical adequacy issue. In particular, the algorithms for recovering DAG structures are especially vulnerable to misspecifications of the underlying statistical model of covariance structures. In addition, the distinction between structural and statistical models can be used to shed light on several issues raised by the controversy concerning causal vs. statistical concepts (see Pearl, 2000, pp. 38-40). These include structural vs. statistical **parameters, error terms, exogeneity, correlation, specification error** and **intervention** vs. **conditioning**. The cross-fertilization between GC modeling and econometrics will be very beneficial to both fields.

## REFERENCES

1.  Basmann, R. L. (1965), "A Note on the Testability of 'Explicit Causal Chains' Against the Class of 'Interdependent' Models," *Journal of the American Statistical Association*, **60**, 1080-1093.
2.  Cartwright, N. (1989), *Nature's Capacities and Their Measurement*, Clarendon Press, Oxford.
3.  Dhrymes, P. J. (1994), *Topics in Advanced Econometrics, vol. II: Linear and Nonlinear Simultaneous Equations*, Springer-Verlag, New York.
4.  Doob, J. L. (1953), *Stochastic Processes*, John Wiley & Sons, New York.
5.  Goldberger, A. S. (1964), *Econometric Theory*, John Wiley & Sons, New York.
6.  Haavelmo (1943), "The Statistical Implications of a System of Simultaneous Equations," *Econometrica*, **11**, pp. 1-12.
7.  Haavelmo, T. (1944) "The probability approach to econometrics", *Econometrica*, **12**, suppl., pp. 1-115.

8.  Hampel, F. R., E. M. Ronchetti, P. J. Rousseeuw and W. A. Stahel (1986), *Robust Statistics*, Wiley, New York.

9.  Hendry, D. F. (1976), "The Structure of Simultaneous Equations Estimators," *Journal of Econometrics*, **4**, 51-88.

10. Hendry, D. F. and M. S. Morgan (1995) (editors) *The foundations of econometric analysis*, Cambridge University Press, Cambridge.

11. Hood, W. C. and Koopmans, T. C. (1953), (eds.), *Studies in Econometric Method*, Cowles Commission Monograph, No. 14, John Wiley & Sons, New York.

12. Johnston, J. (1963), *Econometric Methods*, McGraw-Hill Book Co., New York.

13. Keynes, J. M. and Tinbergen (1939-40), "Professor's Tinbergen's Method," *Economic Journal*, **49**, 558-568, "A Reply," by J. Tinbergen, and "Comment," by Keynes, **50**, 141-156.

14. Koopmans, T. C. (1939), *Linear Regression Analysis of Economic Time Series*, Netherlands Economic Institute, Publication No. 20, Haarlem, F. Bohn.

15. Koopmans, T. C. (1950), (ed.), *Statistical Inference in Dynamic Economic Models*, Cowles Commission Monograph, No. 10, John Wiley & Sons, New York.

16. Koopmans, T. C. (1947), "Measurement Without Theory," *Review of Economics and Statistics*, **17**, 161-172.

17. Mayo, D. G. (1996), *Error and the Growth of Experimental Knowledge*, The University of Chicago Press, Chicago.

18. Mayo, D. G. and A. Spanos (2004), "Methodology in Practice: Statistical Misspecification Testing," forthcoming, *Philosophy of Science*.

19. Morgan, M. S. (1990) *The history of econometric ideas*, Cambridge University Press, Cambridge.

20. Pearl, J. (2000), *Causality: Models, Reasoning and Inference*, Cambridge University Press, Cambridge.

21. Schultz,H. (1938), *The Theory and Measurement of Demand*, University of Chicago Press, Chicago.

22. Spanos, A. (1986) *Statistical foundations of econometric modelling*, Cambridge University Press, Cambridge.

23. Spanos, A. (1989) "On re-reading Haavelmo: a retrospective view of econometric modeling", *Econometric Theory*, **5**, 405-429.

24. Spanos, A. (1990) "The Simultaneous Equations Model revisited: statistical adequacy and identification", *Journal of Econometrics*, **44**, 87-108.

25. Spanos, A. (1994) "On modeling heteroskedasticity: the Student's $t$ and elliptical regression models", *Econometric Theory*, **10**, pp. 286-315.

26. Spanos, A. (1995) "On theory testing in Econometrics: modeling with nonexperimental data", *Journal of Econometrics*, **67**, 189-226.

27. Spanos, A. (1999), *Probability Theory and Statistical Inference: econometric modeling with observational data*, Cambridge University Press, Cambridge.

28. Spanos, A. (2000), "Revisiting data mining: 'hunting with or without a license",
*Journal of Economic Methodology*, **7**, July 2000, pp. 231-264.

29. Spanos, A. and A. McGuirk (2001), "The Model Specification Problem from a Probabilistic Reduction Perspective," *Journal of the American Agricultural Association*, **83**, 1168-1176.

30. Spirtes, P., C. Glymor and R. Scheines (2000), *Causation, Prediction and Search*, 2nd edition, The MIT Press, Cambridge.
31. Strotz, R. H. and H. O. A Wold, (1960), "Recursive vs. Nonrecursive Systems: An Attempt at Synthesis," *Econometrica*, **28**, 417-427.
32. Tinbergen, J. (1937), *Econometric Approach to Business Cycle Problems*, Herman, Paris.
33. Tinbergen, J. (1939), *Statistical Testing of Business Cycle Research*, 2 vols., League of Nations, Geneva.
34. Vining, R. and Koopmans, T.C. (1949), "Methodological Issues in Quantitative Economics, " *Review of Economics and Statistics*, **31**, 77-94.
35. Wold, H. (1953), Demand Analysis: A Study in Econometrics, John Wiley & Sons, New York.
36. Wold, H. (1954), "Causality and Econometrics," *Econometrica*, **22**, 62-174.
37. Wold, H. (1956), "Causal Inference from Observational Data: A Review of Ends and Means," *Journal of the Royal Statistical Society*, Series A, **119**, 28-60.
38. Wold, H. (1960), "A Generalization of Causal Chain Models," *Econometrica*, **28**, 443-463.
39. Woodward, J. (1997), "Causal Models, Probabilities, and Invariance," in *Causality in Crisis? Statistical Methods and the Search for Knowledge in the Social Sciences*, eds. V. R. McKim and S. P. Turner, 1997, University of Notre Dame Press, Indiana.

Symposium 5:

# The Unusual Effectiveness of Logic in Computer Science

# Fixed-Point Logics and Computation

Anuj Dawar[*]

*University of Cambridge Computer Laboratory, Cambridge CB3 0FD, UK*
*anuj.dawar@cl.cam.ac.uk*

**Abstract.** Fixed-point logics are logics that incorporate an explicit construction for recursive or inductive definitions. Such logics play an increasingly important role in a number of areas of the computational and cognitive sciences. These logics have been most extensively studied in the context of finite model theory where they have been shown to have a close connection with the study of computational complexity. Here we give an exposition of fixed-point logics and examine the role they play in describing computation, review some results concerning descriptive complexity and explore some avenues for further research.

## 1. Introduction

This paper arose out of a talk given at the symposium on *the Unusual Effectiveness of Logic in Computer Science*. For those who are familiar with theoretical computer science or indeed recent trends in formal logic, there is nothing unusual about the effectiveness of logic as it is now a commonplace. Methods of logic developed originally for the study of metamathematics have found their most important applications in the context of computing. In the process, logic itself has been immensely enriched by new methods, questions and research directions. Nevertheless, the wide sweep of the applications of logic in computer science remains surprising, if not unusual.

In this paper, we will examine one particular class of logics whose study has developed through the interaction of logical and computational concerns. The logics we are concerned with are broadly termed fixed-point logics. They are obtained as extensions of a base logic (such as propositional logic, modal logic or first-order predicate logic) by means of a mechanism for defining recursion or induction. The questions that are our main concern with regard to these logics are ones of definability and expressive power. While fixed-point logics have their origins in the metamathematical study of induction principles their recent development owes much to connections with computational concerns. On the one hand fixed-point predicate logics have been central to descriptive complexity theory, providing model-theoretic analogues to computational complexity classes and, on the other hand fixed-point modal logics have played an important role among logics for the specification and verification of concurrent systems. In this paper we look at a sample of results from the two areas and make connections between them which suggest future lines of investigation.

---

[*]Research supported by EPSRC grant GR/S06721.

Naturally, the work on fixed-point logics is only one small strand in the larger fabric of work on logic in computer science. We begin by a discussion that places it in this larger context. This is followed by an exposition of the background in finite model theory where the study of fixed-point logics developed. Sections 4 and 5 examine work in descriptive complexity theory centred on fixed-point logics. Finally, in Section 6 we look at fixed-point logics in the modal context.

## 2. Logic and Computation

A central motivating concern of mathematical logic as a field of study is to give a formal account of the process of mathematical reasoning. Through this formalisation, mathematical reasoning is itself turned into an object that can be subjected to the same methods of mathematical enquiry as other mathematical objects such as the number line, Euclidean space or permutation groups.

For our purposes, it's useful to view mathematical reasoning so formalised as being constituted of three elements: Structure, Language and Proof. Logic gives an account (indeed, various accounts) of each of these elements and the relationships between them. In one standard account, a mathematical *structure* is understood to be an interpretation of a vocabulary consisting of relation and function symbols over some fixed set or universe of discourse. The *language* in which mathematical statements are formulated is first-order predicate logic which is equipped with a *proof* system consisting of axioms and rules of inference. Language and structure are tied together with a Tarskian definition of truth and Gödel's completeness theorem ensures that the proof system is adequate to deriving truth.

Of course, a large number of variations of this basic scheme have been investigated. It is useful in particular to distinguish the model-theoretic from the proof-theoretic lines of investigation. Proof-theoretic considerations focus on the relationship between language and proof. Questions of interest include how the collection of provable statements changes with the choice of axioms and rules of inference, how the lengths of proofs depend on these choices and the development of proof systems for non-standard logic. In contrast, model-theoretic questions are concerned with the relationship between language and structures. Among the central concerns are issues of definability, dealing with questions such as: what statements are expressible in first-order logic; what structures and relations can be defined in this logic; and how does the expressive power of different logics compare.

### Computation as Logic

One way of understanding the deep relationship between logic and computing is to look at the project of reducing mathematical reasoning to formal symbol manipulation. In this view, by presenting logical inference in terms of symbol manipulation, formal logic reduces reasoning to steps that can be performed mechanically. This naturally raises the question of what operations are permitted as mechanical steps. If the reduction is to have meaning, we need an account of computation that will make sense of the term "symbol manipulation" and indeed, it is out of just such considerations that the Church-Turing account of computability arose.

On the other hand, the development of electronic computing machines which are the most concrete application of the theory of computation has allowed the construction of customised mechanical systems of inference. Such "logic engineering" provides one means by which concerns of computer science feed back into the study of logic.

In the context of viewing computation as a form of logic, it is worth stressing the importance of Gödel's completeness theorem along with Church's proof of the undecidability of first-order logic. Together these statements establish that the validities of first-order logic are r.e.-complete. The fact they are recursively enumerable tells us that computation is an adequate means of obtaining the truths of first-order logic. The fact that the problem is complete tells us that any computational process can be, in principle and in a precise sense, reduced to that of inference in first-order logic. Taken together, the results tie logic and computation in a tight embrace. The elegance of this formulation helps to explain, to some extent, the significance attached to first-order logic in the classical development of mathematical logic.

### Proof Theory in Computation

While the main bulk of this paper is concerned with model-theoretic issues in computation, we begin with a brief look at a proof-theoretic view of computation which has been highly influential and is covered in detail elsewhere. As all programs and data involved in computation are ultimately strings of symbols in a formal system, it is natural to see all computation as consisting of symbol manipulation. Indeed, it can be seen as inference in a suitable formal system.

A paradigmatic example is the view of functional programming which sees programs as constructive proofs of propositions, which themselves represent the types of the programs. The correspondence between proofs and propositions is given by the Curry-Howard isomorphism (see [23]). Computation is now seen as a process which transforms a proof into one in normal form through a series of steps.

### Model Theory in Computation

It is clear that all programs and data involved in computation can be suitably seen as strings of symbols and hence all computation treated at a purely syntactic level, as in the proof-theoretic view. At the same time, it is often useful, from the conceptual point of view, to distinguish certain of these syntactic constructs and regard them as computational *structures* that provide a semantics for other constructs that we regard as syntactic.

| Structure | Language |
| --- | --- |
| Data Structure | Programming Language |
| Database | Query Language |
| Program/State Space | Specification Language |

Table 1
Model theory in computation

Table 1 shows some examples of structure/language pairs whose relationship can be seen as raising model-theoretic questions. Given a program that manipulates data structures, it is natural to think of the latter as structures in the mathematical sense of the word that provide semantics to the program, which is a syntactic object. On the other hand, a program, or the state space it defines, can itself be seen as a semantic structure providing an interpretation for a statement in a specification language. The second entry in the table illustrates an instance where the model-theoretic view has been particularly productive. A relational database can be quite naturally seen as a relational structure in the Tarskian sense and database query languages are indeed based on predicate logic, making this a rich area for applications of model-theoretic methods.

One important respect in which the model-theoretic concerns raised by the above correspondences differ from classical questions is that the structures involved are often required to be finite. A relational database or a data structure is naturally finite (though there are instances where an infinite database is a useful abstraction) and while the state space of a program may often be thought of as infinite, some of the most successful applications of model-checking have concerned finite-state systems. This is in sharp contrast to classical model theory where infinite structures are the main concern. Model-theoretic questions arising in computer science have led to the development of a model theory specific to finite structures. This is where we next turn our attention.

## 3. Finite Model Theory

Finite model theory has developed as a distinct subject of study over the last three decades or so. It is not merely the model theory of finite structures as it has evolved a distinctive set of questions, methods and results that are rather different from those that exercise the minds of model theorists. One distinguishing feature is that first-order logic, which plays an important role in model theory, has a somewhat lesser role in finite model theory. Nevertheless, first-order logic will form the starting point of our discussion and we begin with a definition of its syntax. Though the logic is no doubt familiar to all readers, laying out the definition serves two purposes: it allows us to fix notation and it gives us a classical example (indeed two separate examples) of an inductive definition, which will be a launching point for later discussions.

### First-Order Logic

**Definition 1.** *We are given a vocabulary, which consists of relation symbols $R_1$, $R_2, \ldots$, each of a fixed arity, function symbols $f_1, f_2, \ldots$, again each of a fixed arity, constant symbols $c_1, c_2, \ldots$ and variable symbols $v_1, v_2, \ldots$. The collection of first order formulas is defined by two successive inductions. First, we define the set of valid terms by:*

- *any constant or variable is a term; and*

- *if $f$ is a function symbol of arity $n$, and $t_1, \ldots, t_n$ are terms, then so is $f(t_1, \ldots, t_n)$.*

*Secondly, we define the set of formulas by:*

- *if $t_1$ and $t_2$ are terms, then $t_1 = t_2$ is a formula;*

- *if $R$ is a relation symbol of arity $n$ and $t_1, \ldots, t_n$ are terms, then $R(t_1, \ldots, t_n)$ is a formula;*

- *if $\varphi$ and $\psi$ are formulas then so are $(\varphi \wedge \psi)$, $(\varphi \vee \psi)$ and $\neg \varphi$; and*

- *if $\varphi$ is a formula and $x$ is a variable, then $\exists x \varphi$ and $\forall x \varphi$ are formulas.*

For the rest of this paper, we will assume that the vocabulary consists of a finite number of relation, function and constant symbols. Formulas of first-order logic are interpreted in a *structure*,

$$\mathbb{A} = (A, R_1, \ldots, R_l, f_1, \ldots, f_m, c_1, \ldots, c_n)$$

where $A$ is a set, $c_i \in A$ for all $i$ and $R_i$ and $f_i$ are interpreted as relations and functions over $A$ of the appropriate arity. The equality symbol $=$ is always interpreted as element identity on $A$. Here and elsewhere in the paper, we do not distinguish between a symbol in the vocabulary and its interpretation in a structure in the expectation that this will cause no confusion.

First-order logic has been extremely successful for its intended purpose, namely the formalisation of mathematics. Many natural mathematical theories can be naturally expressed through a set of first-order axioms including, perhaps most importantly of all, *set theory* with its fundamental role in the foundations of mathematics. Moreover, Gödel's completeness theorem tells us that the consequences of a first-order theory can be effectively obtained. This provides another reason why a first-order formalisation of a mathematical theory is seen as an adequate outcome of the reductionist project.

### Finite Structures

If, in the definition of a *structure* given above, we restrict the universe $A$ to be a finite set, a rather different picture emerges. One of the central tools of model theory, the Compactness Theorem, no longer holds and most of its important consequences fail as well. Moreover, we no longer have a completeness theorem. Indeed, Trakhtenbrot has shown [26] that the sentences of first-order logic that are valid on finite structures are not recursively enumerable.

With the vital tools of compactness and completeness no longer available, first-order logic loses much of its significance in finite model theory. Indeed, as this author has said elsewhere [10], when restricted to finite structures, first-order logic is both too strong and too weak.

It is too strong in the sense that the relation of elementary equivalence between structures is trivial on finite structures. A large part of classical model theory can arguably be described as the study of this equivalence relation and the structure of its equivalence classes. Two structures $\mathbb{A}$ and $\mathbb{B}$ are elementarily equivalent if, for every first order sentence $\varphi$,

$$\mathbb{A} \models \varphi \text{ if, and only if, } \mathbb{B} \models \varphi.$$

This is crucial in establishing inexpressibility results. For instance, by proving that all dense linear orders without endpoints are elementarily equivalent, we establish that other properties that might distinguish such orders (such as Dedekind completeness) are not definable.

On finite structures, the elementary equivalence relation is trivial, in that any two elementarily equivalent structures are isomorphic. Indeed, any finite structure is described up to isomorphism by a single sentence. Given a structure $\mathbb{A} = (A, R_1, \ldots, R_m)$ (ignoring function and constant symbols for the moment), where $A$ is a set of $n$ elements, we can construct a sentence

$$\delta_{\mathbb{A}} = \exists x_1 \ldots \exists x_n (\psi \wedge \forall y \bigvee_{1 \leq i \leq n} y = x_i)$$

where, $\psi(x_1, \ldots, x_n)$ is the conjunction of all atomic and negated atomic formulas that hold in $\mathbb{A}$. Now, for any structure $\mathbb{B}$, $\mathbb{B} \models \delta_{\mathbb{A}}$ if, and only if, $\mathbb{A} \cong \mathbb{B}$.

This means that first order logic can make all the distinctions that are to be made between finite structures. Still, first-order logic is very weak in terms of its expressive power. For any first-order sentence $\varphi$ consider the collection of its finite models:

$$\mathrm{Mod}_{\mathcal{F}}(\varphi) = \{\mathbb{A} \mid \mathbb{A} \text{ finite, and } \mathbb{A} \models \varphi\}.$$

It turns out that this class is trivially decidable, in the sense that it can be decided by a deterministic Turing machine with logarithmic work space. Moreover, there are computationally very simple classes of finite structures which cannot be expressed using a first-order sentence. Among them are: the class of sets with an even number of elements; and the class of graphs $(V, E)$ that are connected.

Essentially, in the model theory of infinite structures, as we classify structures by elementary equivalence, we are looking at the expressive power of theories, i.e. possibly infinite sets of sentences. Two structures that are elementarily equivalent cannot be distinguished by any first order theory. In contrast, any isomorphism closed class $S$ of finite structures is defined by the set of negations of the sentences $\delta_{\mathbb{A}}$, as above, for finite structures $\mathbb{A}$ not in $S$. Certainly, this theory may have infinite models, but the collection of its finite models is exactly $S$. In contrast, the expressive power of single sentences is weak. The interesting model-theoretic questions on finite structures have to do with the expressive power of restricted theories where the set of sentences is *regular* in some sense that needs to be made precise. That is, we are interested in *finitely generated* theories.

An alternative way of looking at the same questions is to say that we need to look at definability by single sentences in logics that are more expressive than first-order logic. These single sentences generate first-order theories of a particular regular form.

## 4. Inductive Definitions

In computer science as well as in logic, many interesting objects and relations are naturally defined *inductively*. The definition of terms and formulas of first-order logic in Definition 1 is a typical example. Indeed, the definition of the syntax and semantics of almost any language used in logic and computer science is given inductively as are the definitions of data structures such as trees and lists.

Another classical example is the inductive definition of arithmetic functions. Suppose that we are given the structure $(N, s, 0)$. That is to say, we have a constant symbol for 0, and a symbol for the successor function. There is a natural inductive definition of the addition function in this structure, namely:

$$\begin{aligned} x + 0 &= x \\ x + s(y) &= s(x+y). \end{aligned} \tag{1}$$

Though addition is thus inductively defined using very simple formulas, it is not first-order definable. That is to say, there is no first order formula $p(x, y, z)$ with three free variables such that

$$(N, s, 0) \models p[a, b, c] \quad \text{if, and only if,} \quad a + b = c.$$

To formalise the inductive definition of addition, we say that addition is the *least* function that satisfies the equations in (1). Or, equivalently, the equations define a *monotone operator* on the space of partial functions on the natural numbers and the addition function is the least fixed point of this operator.

Similarly, the definition of first-order terms given in Definition 1 can be read as stating that the collection of first-order terms is *the least set* containing all constants, all variables and such that $f(t_1, \ldots, t_a)$ is a term whenever $t_1, \ldots, t_a$ are terms and $f$ is a function symbol of arity $a$. Once again, fixing a universe $S$ that consists of all strings over a suitable infinite alphabet one can view the definition as giving a monotone operator on the power set of $S$ (ordered by inclusion). The set of terms is then the least fixed point of this operator.

While inductive definitions are ubiquitous in the metalanguage of logic, it is in taking them into the object language that we obtain fixed-point logics. This transfer of fixed-point constructors into the object language is the key insight (originally presented by Aho and Ullman [5]) that led to the flowering of the study of fixed-point logics in finite model theory.

### Least Fixed Point Logic

The first of the fixed-point logics we consider is LFP, which is the result of adding to first-order logic an operator for forming least fixed points as used in our two examples above. This logic was defined by Chandra and Harel [8]. Least fixed points are, in principle, defined for all monotone operators. However, monotonicity is not a syntactic property (for instance, it is undecidable whether an operator defined by a first-order formula is monotone). Thus, in order to be able to effectively define the syntax of our logic, we allow ourselves only to take fixed-points of operators defined by *positive* formulas. A formula $\varphi$ is said to be positive in a relation symbol $R$ if $R$ does not occur in $\varphi$ within the scope of a negation sign.

Formally, the logic LFP is obtained by closing first order logic simultaneously under all the formula forming operations of first order logic along with the rule:

if $R$ is a $k$-ary relation variable, $\mathbf{x}$ is a $k$-tuple of first order variables, $\mathbf{t}$ is a $k$-tuple of terms and $\varphi$ is a formula in which $R$ occurs only positively, then

$$[\mathbf{lfp}_{R,\mathbf{x}} \; \varphi](\mathbf{t})$$

is a formula, in which all occurrences of $R$ are bound, and all occurrences of the variables in $\mathbf{x}$ except those occurring in $\mathbf{t}$ are bound.

The intended semantics of this formula formation rule is that, for any structure $\mathbb{A}$, $\mathbb{A} \models [\mathbf{lfp}_{R,\mathbf{x}} \, \varphi](\mathbf{t})$ if, and only if, $\mathbf{t}^{\mathbb{A}}$—the tuple of elements of $\mathbb{A}$ defined by the terms $\mathbf{t}$—is in the least fixed point of the monotone operator defined by $\varphi(R, \mathbf{x})$ on $A^k$.

As an example, the following formula:

$$\forall u \forall v [\mathbf{lfp}_{T,xy} \, (x = y \vee \exists z(E(x, z) \wedge T(z, y)))](u, v)$$

is satisfied in a graph $(V, E)$ if, and only if, the graphs is connected. Indeed, the least relation $T$ that satisfies the equivalence $T(x, y) \equiv x = y \vee \exists z(E(x, z) \wedge T(z, y)))$ is the reflexive and transitive closure of $E$. This is, therefore the least fixed point defined by the operator $\mathbf{lfp}$. The formula can now be read as saying that every pair $u, v$ is in this reflexive-transitive closure.

As we had earlier remarked, the class of connected graphs is not definable by a first-order sentence. Thus, the above example demonstrates that the expressive power of LFP properly extends that of first-order logic.

### Immerman-Vardi Theorem

The importance that is attached to LFP in the realm of finite model theory is explained in part by the Immerman-Vardi theorem. This result was established independently by Immerman [19] and Vardi [27] in 1982 (a similar result is shown by Livchak [22]). It demonstrates a close relationship between inductive definitions and feasible computation.

Consider only finite structures with a distinguished relation $<$ that is interpreted as a linear order of the universe.

**Theorem 2** (Immerman-Vardi). *A class of finite ordered structures is definable by a sentence of* LFP *if, and only if, membership in the class is decidable by a deterministic Turing machine in* polynomial time.

If we do not restrict ourselves to structures which interpret a linear order, it is still the case that every class of structures definable in LFP is decidable in polynomial time, but the converse fails. There are properties that are easily computable (such as the property of a set having an even number of elements) that are not expressible in the logic. It seems that the order is required if the logic is to be powerful enough to simulate a mechanical computation. For a fuller discussion of these issues see [15].

### 5. Iterative Fixed-Point Logics

Following Chandra and Harel's definition of LFP and the results obtained with it, a number of other extensions of first-order logic by means of inductive operators have been considered. Most of these flow from the observation that the least fixed point of an operator can be obtained by means of an iterative process.

Suppose we consider a formula $\varphi(R, \mathbf{x})$ in which $R$ occurs positively and which therefore defines a suitable monotone operator. By a theorem of Knaster and Tarski [25], the least fixed point of the operator is obtained as a limit of the following sequence:

$$R^0 = \emptyset$$
$$R^{m+1} = \{\mathbf{a} \mid (\mathbb{A}, R^m) \models \varphi[\mathbf{a}/\mathbf{x}]\}$$

Because of the monotonicity of $\varphi$, the above sequence is increasing, i.e. if $m < n$ then $R^m \subseteq R^n$. In general, we would have to continue the iteration through a transfinite sequence of stages, taking unions of previous stages at limit ordinals. However, when we are only concerned with finite structures $\mathbb{A}$ the above suffices. Indeed, if $k$ is the arity of the relation symbol $R$ and $\mathbb{A}$ has $n$ elements then the fixed point is reached in at most $n^k$ stages. This is, in fact, the crucial step in the proof that every formula of LFP can be evaluated in polynomial time.

### Inflationary Fixed-Point Logic

The sequence of stages obtained by iterating a formula $\varphi$, starting with the empty relation, is increasing when $\varphi$ is monotone. However, when $\varphi$ is not monotone it is still possible to force an increasing sequence by the simple device of taking, at each stage, the union with the previous one. Thus, suppose $\varphi(R, \mathbf{x})$ is a formula that is not necessarily positive in $R$, and therefore not necessarily monotone. The following iterative process still gives an increasing sequence of stages:

$$R^0 = \emptyset$$
$$R^{m+1} = R^m \cup \{\mathbf{a} \mid (\mathbb{A}, R^m) \models \varphi[\mathbf{a}/\mathbf{x}]\}.$$

Thus, interpreted on a structure $\mathbb{A}$ with $n$ elements, the sequence must once again converge in a number of steps bounded by $n^k$, where $k$ is the arity of the relation symbol $R$. The limit of the sequence, i.e. the least relation $R^m$ such that $R^m = R^{m+1}$ is termed the *inflationary fixed point* of the operator defined by $\varphi$.

This definition allows us to define a logic IFP with a syntax similar to LFP, but with an operator **ifp**, which allows us to form formulas of the form $[\mathbf{ifp}_{R,\mathbf{x}} \, \varphi](\mathbf{t})$. The semantics of the formula is given by the rule that $\mathbb{A} \models [\mathbf{ifp}_{R,\mathbf{x}} \, \varphi](\mathbf{t})$ if, and only if, the tuple of elements interpreting $\mathbf{t}$ is in the inflationary fixed point of the operator defined by $\varphi(R, \mathbf{x})$ on $\mathbb{A}$.

It is easy to see that every formula of LFP is equivalent to a formula of IFP. Indeed, since the inflationary fixed point of a monotone operator coincides with its least fixed point, directly replacing all occurrences of the operator **lfp** with **ifp** yields an equivalent formula. Gurevich and Shelah [17] proved that the converse is true when we restrict ourselves to finite structures. That is, the inflationary fixed-point of an LFP defined operator can be defined in LFP and hence, by induction, any IFP formula can be translated to an equivalent formula of LFP, though the translation is not straightforward as it is in the other direction. Their result crucially uses the restriction to finite structures. In particular it depends on the fact that, on any finite structure, there is a *last* stage to the inductive process and this stage is itself uniformly definable.

Despite the result of Gurevich and Shelah, the relationship between the expressive power of LFP and IFP on infinite structures remained an open question for many years. It was widely believed that IFP is strictly more expressive than LFP. In particular, it is known that there are first-order definable operators in arithmetic whose inflationary fixed point cannot be obtained as the least fixed point of any monotone first-order operator (see [4]). In contrast, on finite structures it is a consequence of the Gurevich-Shelah result that every IFP definable property is definable by a single application of the **lfp** operator to a first-order formula. The question was recently settled by Kreutzer [20], who showed, through an ingenious adaptation of the method of Gurevich and Shelah, that the restriction to finite structures is not necessary. Every formula of IFP is indeed equivalent to an LFP formula over all structures.

### Partial Fixed-Point Logic

Another fixed-point logic that arose in the study of descriptive complexity theory is partial fixed-point logic. Like IFP it is formed around the idea of constructing relations by iteration. However, rather than forcing the sequence of stages defined by an operator to be increasing in order to guarantee convergence, we allow the sequence to diverge. More formally, for any formula $\varphi(R, \mathbf{x})$, where $R$ is a relational variable and $\mathbf{x}$ a tuple of first-order of variables whose length matches the arity of $R$ and any structure $\mathbb{A}$, we can define the iterative sequence of stages

$$R^0 = \emptyset$$
$$R^{m+1} = \{\mathbf{a} \mid (\mathbb{A}, R^m) \models \varphi[\mathbf{a}/\mathbf{x}]\}.$$

As $\varphi$ does not necessarily define a monotone operator, this sequence is not necessarily increasing and may or may not converge to a fixed-point. The *partial fixed point* is defined to be the limit of this sequence if it exists, and $\emptyset$ otherwise.

The logic PFP is obtained by closing first order logic simultaneously under the formula formation rules of first order logic and the rule that allows us to form the formula $[\mathbf{pfp}_{R,\mathbf{x}} \varphi](\mathbf{t})$ from the formula $\varphi$. This is used to denote that $\mathbf{t}$ is a tuple in the partial fixed point of $\varphi(R, \mathbf{x})$.

The significance of PFP in descriptive complexity theory comes from the fact that it bears a relationship to the class PSPACE (of properties decidable by a Turing machine in polynomial space) similar to that between LFP and PTIME. This was shown by Abiteboul and Vianu, who introduced PFP in the context of query languages for relational databases [2]:

**Theorem 3.** *A class of finite ordered structures is definable by a sentence of* PFP *if, and only if, membership in the class is decidable by a deterministic Turing machine in* polynomial space.

On the other hand, when an order is not available, PFP is still unable to express some computationally simple properties such as the evenness of the size of a set. This leads to the question of how the expressive power of PFP and LFP are related in the absence of order. Here, Abiteboul and Vianu [3] proved the following remarkable result (see also [14] for an alternative treatment).

**Theorem 4** (Abiteboul-Vianu). *The expressive power of* LFP *and* PFP *are equivalent on finite structures if, and only if, the complexity classes* PTIME *and* PSPACE *coincide.*

Thus, one of the most important open questions in complexity theory is reduced to a question about the comparison of two different kinds of fixed-point operators.

Indeed, this programme has been carried further and other complexity theoretic questions (including the question of whether P=NP) have also been similarly characterised (see [1,9]). In particular, the class NP is characterised on finite ordered structures by a *nondeterministic* inflationary fixed-point logic NFP where the fixed-point operator is applied to a pair of formulas (see [12] for a formal definition). The question of whether P=NP has been shown to be equivalent to the expressive equivalence of LFP and NFP.

It turns out that while establishing separations between the fixed-point logics on finite structures is hard (being equivalent to notoriously difficult complexity-theoretic questions), the situation is different when we allow infinite structures. It is shown in [12] that NFP is strictly more expressive than LFP. The case of PFP is somewhat trickier as the definition of partial fixed-points seems to rely on the assumption that structures are finite: a sequence of increasing relations can be extended to the transfinite by taking unions of the relations at limit ordinals but it is not clear what one would do with a sequence of stages that are not increasing. Nevertheless, Kreutzer [21] presents an alternative definition of the semantics of PFP that can be extended to the infinite. He shows that PFP under this revised semantics has the same expressive power as under the Abiteboul-Vianu semantics and moreover that it is strictly more expressive than LFP when infinite structures are permitted.

## 6. Modal Fixed-Point Logics

In this section we look at fixed-point logics in a rather different setting from the foregoing. We look at logics that are formed by adding a fixed-point to propositional modal logic. These fixed-point extensions of propositional modal logic have attracted great interest in the context of formal verification of computing systems as they are useful in specifying the behaviour of such a system.

### Modal Logic

For our purposes, a system is modelled as a particular kind of structure. Such structures are variously referred to as *state transition systems* or *Kripke structures*. The structure interprets a vocabulary consisting of a set $A$ of *actions* and a set $P$ of *propositions* over a universe $S$ of *states*. Each $a \in A$ is interpreted by a binary relation $E_a \subseteq S \times S$ and each proposition $p \in P$ by a subset $p \subseteq S$ (note that we do not distinguish notationally between atomic propositions and their interpretations). The basic propositional multi-modal logic, also known as Hennessy-Milner logic (in reference to [18]) is formed by the following formula-formation rules. A formula is any one of the following (where $\varphi$ and $\psi$ are formulas):

- $T$ and $F$

- $p$ $\qquad$ $(p \in P)$

- $\varphi \wedge \psi; \varphi \vee \psi; \neg \varphi$

- $[a]\varphi; \langle a \rangle \varphi$ $\qquad$ $(a \in A)$.

For the semantics, we interpret formulas in a given state of a state transition system. We write $\mathbb{A}, s \models \varphi$ to denote that the formula $\varphi$ holds in state $s$ in system

A. The formula $T$ is true in all states and $F$ is false in all states. $\mathbb{A}, s \models p$ if, and only if, $s \in p^{\mathbb{A}}$. The Boolean connectives have their usual interpretation and $\mathbb{A}, s \models \langle a \rangle \varphi$ if, and only if, for some $t$ with $s \xrightarrow{a} t$ (i.e. $(s, t) \in E_a$), we have $\mathbb{A}, t \models \varphi$. $[a]$ is the dual of $\langle a \rangle$. For background on propositional modal logic, consult [7].

### Modal Fixed-Point Logic

In general, for the specification of system properties and their automatic verification, more expressive languages than Hennessy-Milner logic are considered. These include languages with quantification over paths (such as CTL and CTL*) or other forms of recursion, such as propositional dynamic logic (PDL). The logic we are concerned with here is the modal $\mu$-calculus (or $L_\mu$) which extends Hennessy-Milner logic with an operator for forming least fixed points. In terms of expressive power $L_\mu$ subsumes CTL, CTL* and PDL (see [24]) while remaining relatively tractable from a computational point of view.

Formally, we form the formulas of $L_\mu$ by taking, in addition to the vocabulary of actions and propositions a collection $X_1, X_2, \ldots$ of propositional variables and adding to the rules of Hennessy-Milner logic the formula formation rule that admits $\mu X : \varphi$ as a formula whenever $\varphi$ is a formula containing only positive occurrences of $X$

For the semantics, we say that $\mathbb{A}, s \models \mu X : \varphi$ if, and only if, $s$ is in the smallest set $X$ of states such that the equivalence $X \leftrightarrow \varphi$ holds in $(\mathbb{A}, X)$. Or, in other words, $s$ is in the least fixed point of the monotone operator defined by $\varphi$ on the power set of the set of states of $\mathbb{A}$ (see [6] for a book-length treatment of the $\mu$-calculus).

There is a natural translation of formulas of Hennessy-Milner logic into first-order logic, where each action is treated as a binary relation symbol and each proposition as a unary relation symbol. Indeed, it is not difficult to see that the translation can be carried out so that the resulting formula of first-order logic (with one free variable) uses no more than two variables altogether. The translation extends further to $L_\mu$, whose formulas can be translated into LFP using only two first-order variables.

The correspondence between LFP with a bounded number of variables and $L_\mu$ also runs in the other direction. Suppose $\varphi$ is a formula of LFP with no more than $k$ first-order variables (and no parameters to fixed-point operators, a restriction we need for technical reasons). Suppose furthermore that there is no relation symbol of arity greater than $k$ in the relational vocabulary. We can then associate with every structure $\mathbb{A}$ in this vocabulary a transition system $\hat{\mathbb{A}}^k$ whose states are $k$-tuples of elements of $\mathbb{A}$. There are $k$ actions, corresponding to substitution in each of the $k$ positions in a tuple and there are propositions that correspond to the relations on $\mathbb{A}$. With this translation, it can be shown that there is a formula $\hat{\varphi}$ of $L_\mu$ such that $\mathbb{A} \models \varphi$ if, and only if, $\hat{\mathbb{A}}^k \models \hat{\varphi}$. This translation gives a means of showing that various problems involving $L_\mu$ and LFP are computationally equivalent. For instance, it is a long-standing open question whether there is a polynomial-time algorithm for the model-checking of $L_\mu$ formulas. This is equivalent to the question of whether there is such an algorithm for the model-checking problem for bounded-variable parameter-free LFP.

## Modal Inflationary Fixed Points

As we observed in Section 5, there is a translation that will convert any formula of IFP into an equivalent LFP formula. The translation does not, however, preserve the number of variables. Thus, the equivalence that allows us to lift computational properties of $L_\mu$ to LFP does not extend to IFP. Indeed, the computational properties of IFP seem less tractable. The problem of model-checking for formulas of IFP, even with a fixed number of variables and no parameters turns out to be PSPACE-complete.

Indeed, it turns out that an extension of propositional modal logic with an inflationary fixed-point operator is already far more expressive than $L_\mu$. That is, the equivalence between LFP and IFP fails when the base logic is reduced from predicate logic to a propositional modal logic. The inflationary extension of the Hennessy-Milner logic, termed MIC (for modal inflationary calculus) is defined in [11] and its properties studied. They are in marked contrast to $L_\mu$:

- $L_\mu$ has the finite model property, that is to say every sentence that has a model has a finite model. This fails for MIC.

- The satisfiability problem for formulas of $L_\mu$ is known to be decidable (and indeed, in the complexity class EXPTIME [16]). In contrast, the satisfiability problem for MIC is not only undecidable, it is not even in the arithmetic hierarchy [11].

- The model-checking problem, i.e. the problem of deciding, given a structure and formula pair whether the formula is satisfied in the structure is in NP∩co-NP for $L_\mu$ and conjectured to be in polynomial time. For MIC the problem is PSPACE-complete.

- Consider transition systems consisting of a finite sequence $s_1, \ldots, s_n$ of states with a single action such that $s_{i+1}$ is the only state to which there's an action from $s_i$. These can be viewed as strings over the alphabet $2^P$. It is therefore natural to ask what languages (i.e. sets of strings) are definable in a given logic. It is known that $L_\mu$ defines exactly the regular languages. In MIC one can define more, including all languages decidable in linear time and even including some non-context-free languages.

These results demonstrate that structural differences between least and inflationary fixed points can be studied by restricting the base logic on which these operators are applied. Indeed differences with other fixed-point operators such as partial and nondeterministic fixed points could also be examined in this context (see [13] for the beginnings of such a study). Such a study may reveal structural properties of the various fixed-point operators which would also shed light on differences between them in the context of predicate logic.

## 7. Conclusion

We have examined how model-theoretic questions, concerned with studying the expressive power of languages, appear in the context of computer science. The interaction of these questions with computational concerns has led to a concentration on finite structures which in turn leads to entirely new model-theoretic methods. One significant feature that is a departure from standard model theory is that first order logic no longer occupies a central place. A variety of extensions of first-order logic by means of inductive or iterative operators have been studied, especially with a view to providing logical characterisations of computational complexity. The study of fixed-point logics in the context of complexity has seen a convergence of methods with fixed-point modal logics studied for verification of systems. This convergence has also enabled us to examine the fine structure of fixed-point logics in a modal context with the hope that differences revealed will cast light on questions of descriptive complexity.

## REFERENCES

1. S. Abiteboul, M. Y. Vardi, and V. Vianu. Fixpoint logics, relational machines, and computational complexity. *J. ACM*, 44(1):30–46, 1997.
2. S. Abiteboul and V. Vianu. Datalog extensions for database queries and updates. *Journal of Computer and System Sciences*, 43:62–124, 1991.
3. S. Abiteboul and V. Vianu. Computing with first-order logic. *Journal of Computer and System Sciences*, 50(2):309–335, 1995.
4. P. Aczel. An introduction to inductive definitions. In J. Barwise, editor, *Handbook of Mathematical Logic*, pages 739–782. North Holland, 1977.
5. A.V. Aho and J.D. Ullman. Universality of data retrieval languages. In *6th ACM Symp. on Principles of Programming Languages*, pages 110–117, 1979.
6. A. Arnold and D. Niwinski. *Rudiments of μ-calculus*. North Holland, 2001.
7. P. Blackburn, M. de Rijke, and Y. Venema. *Modal Logic*. Cambridge University Press, 2001.
8. A. Chandra and D. Harel. Structure and complexity of relational queries. *Journal of Computer and System Sciences*, 25:99–128, 1982.
9. A. Dawar. A restricted second order logic for finite structures. *Information and Computation*, 143:154–174, 1998.
10. A. Dawar. Types and indiscernibles in finite models. In J.A. Makowsky and E.V. Ravve, editors, *Logic Colloquium '95*, Lecture Notes in Logic, pages 51–65. Springer-Verlag, 1998.
11. A. Dawar, E. Grädel, and S. Kreutzer. Inflationary fixed points in modal logic. *ACM Transactions on Computational Logic*, 2004. to appear.
12. A. Dawar and Y. Gurevich. Fixed-point logics. *Bulletin of Symbolic Logic*, 8:65–88, 2002.
13. A. Dawar and S. Kreutzer. Generalising automaticity to modal properties of finite structures. In *Proc. of the 22nd Conf. on Foundations of Software Technology and Theoretical Computer Science (FSTTCS)*, volume 2556 of *LNCS*, pages 109–120. Springer, 2002.
14. A. Dawar, S. Lindell, and S. Weinstein. Infinitary logic and inductive definability over finite structures. *Information and Computation*, 119(2):160–175, 1995.
15. H-D. Ebbinghaus and J. Flum. *Finite Model Theory*. Springer, 2 edition, 1999.
16. A. Emerson and C. Jutla. The complexity of tree automata and logics of

programs. In *Proc. 29th IEEE Symp. on Foundations of Computer Science*, pages 328–337, 1988.

17. Y. Gurevich and S. Shelah. Fixed-point extensions of first-order logic. *Annals of Pure and Applied Logic*, 32:265–280, 1986.

18. M. Hennesey and R. Milner. Algebraic laws for nondeterminism and concurrency. *J. Assoc. of Computing Machinery*, 32:137–162, 1985.

19. N. Immerman. Relational queries computable in polynomial time. *Information and Control*, 68:86–104, 1986.

20. S. Kreutzer. Expressive equivalence of least and inflationary fixed-point logic. In *Proc. of the 17th IEEE Symp. on Logic in Computer Science (LICS).*, pages 403–410, 2002.

21. S. Kreutzer. Partial fixed-point logic on infinite structures. In *CSL '02, Proc. of the Annual Conference of the European Association for Computer Science Logic*, 2002.

22. A. Livchak. The relational model for process control. *Automated Documentation and Mathematical Linguistics*, 4:27–29, 1983.

23. J.P. Seldin and J. R. Hindley, editors. *Essays on Combinatory Logic, Lambda Calculus and Formalism*. Academic Press, 1980.

24. C. Stirling. *Modal and Temporal Properties of Processes*. Springer, 2001.

25. A. Tarski. A lattice-theoretic fixpoint theorem and its applications. *Pacific Journal of Mathematics*, 5(2):285–309, 1955.

26. B. A. Trakhtenbrot. Impossibility of an algorithm for the decision problem in finite classes. *Doklady Akademii Nauk SSSR*, 70:569–572, 1950.

27. M. Y. Vardi. The complexity of relational query languages. In *Proc. of the 14th ACM Symp. on the Theory of Computing*, pages 137–146, 1982.

# Index

www.ingramcontent.com/pod-product-compliance
Lightning Source LLC
Chambersburg PA
CBHW052139070326

40690CB00047B/1046